T0189013

Graduate Texts in Physics

Graduate Texts in Physics

Graduate Texts in Physics publishes core learning/teaching material for graduate- and advanced-level undergraduate courses on topics of current and emerging fields within physics, both pure and applied. These textbooks serve students at the MS- or PhD-level and their instructors as comprehensive sources of principles, definitions, derivations, experiments and applications (as relevant) for their mastery and teaching, respectively. International in scope and relevance, the textbooks correspond to course syllabi sufficiently to serve as required reading. Their didactic style, comprehensiveness and coverage of fundamental material also make them suitable as introductions or references for scientists entering, or requiring timely knowledge of, a research field.

More information about this series at http://www.springer.com/series/8431

R Paul Drake

High-Energy-Density Physics

Foundation of Inertial Fusion
and Experimental Astrophysics

Second Edition

 Springer

R Paul Drake
University of Michigan
Ann Arbor, MI, USA

ISSN 1868-4513 ISSN 1868-4521   (electronic)
Graduate Texts in Physics
ISBN 978-3-319-67710-1   (hardcover) ISBN 978-3-319-67711-8   (eBook)
ISBN 978-3-319-88473-8   (softcover)
https://doi.org/10.1007/978-3-319-67711-8

Library of Congress Control Number: 2017958139

1st edition: © Springer-Verlag Berlin Heidelberg 2006
© Springer International Publishing AG 2018, First softcover printing 2019

Printed on acid-free paper

This Springer imprint is published by Springer Nature
The registered company is Springer International Publishing AG
The registered company address is: Gewerbestrasse 11, 6330 Cham, Switzerland

Dedicated to Kent Estabrook

Preface to the Second (2018) Edition

This second edition is an attempt at making improvements, correcting errors, and adding some important material without adding dross. This new addition incorporates several changes and a few additions. The (alas, many) typos discovered by many people in the previous edition are all fixed. Having taught this material for another decade, I have found more effective ways of teaching some of the material. This especially affected Chap. 4 on 1D hydrodynamics, in which the new text allows one to avoid teaching Riemann invariants. In other cases, I have learned new things in the past decade, some of which were unknown to anyone previously. Chapter 3 on equations of state includes a discussion of the suppression of ionization by degeneracy, which was not known to me at the time, and of the continued evolution of material structure at very high pressure, which was known to almost no one. This discussion of boundary conditions in Chap. 5 includes an examination of pressure variations that is not included in standard treatments and that resolves some issues that had bothered me for a long time. Chapter 6 now includes a much more extensive discussion of atomic processes and their rates, a topic I had to abandon for lack of time previously.

I decided to split the material on radiation hydrodynamics and to provide a thorough discussion of radiative heat fronts in what is now Chap. 8. This provided better support for some topics discussed later in the text. I discovered over time that some things I "learned" in the 1980s about high-Z targets were not correct and read with interest new research on double ablation fronts in mid-Z targets, all of which affected the discussion of these topics in Chap. 9. The material on Z-pinches was moved to the new Chap. 10, which includes a discussion of basic aspects of magnetized flows. A decade ago there was almost no research in this area. Thanks both to advances in technology and to improved perspective on the part of funding agencies, this has now become a very active area. The middle portions of the chapter on inertial confinement fusion (Chap. 11) were substantially reworked, in response to experience to date with the National Ignition Facility and to my own finding time to further explore the modeling of that type of system. The chapter on laboratory astrophysics (Chap. 12) is now united by a more systematic and more extensive discussion of scaling across a wider range of experiments. This reflects both the explosion of work on the laboratory study of dynamical processes that are relevant to astrophysics and my own evolving opinion that, although we are 20 years into that enterprise, our community does not yet do well enough on scaling arguments. The final chapter on relativistic systems (Chap. 13) is changed mainly through an added discussion of betatron radiation, which has become important. My sense of that area is that the fundamental things one needs to understand have not changed much, although certainly the detailed understanding and specifically work on particle acceleration have evolved greatly.

I want to comment on the voice in the text. In the classroom, I teach in the second person. My goal is that the students and I will join in an intellectual journey of discovery. My intent has been to write in just this sort of voice. I also greatly admire the books of Chandrasekhar, who writes this way too. Unfortunately, too many decades of writing multiauthor papers for publications make it perilously easy to slip into the kind of second-person discussions one uses there. Please forgive me if you find this discordant.

I would like to include a few words related to homework. Several of those who have taught from this book would like to have had homework solutions, and under pressure from the editors, I agreed to provide solutions for a significant fraction of the problems. I certainly understand how a time-starved professor would welcome anything that can help keep all the balls in the air, including homework solutions. It is worth asking what the point of homework is in a specialized graduate course such as those taught from this book. Getting the answer in the back of the book is surely not relevant as such. Gaining an increased understanding of the material is what seems relevant to me, and surely this must require working some things out for oneself. This was always the point for me in formulating the homework problems provided, whether they involve fleshing out a derivation or thinking through some out-of-the-box question. Consistent with this goal, the solutions provided generally represent an end point of part of the work on the problem, intended to allow one to see whether work on that problem is headed in the right direction. The solutions provided also include very little of the conceptual discussion that should be part of work turned in.

Once again the text benefitted from discussions with or review by a large number of colleagues. This has included Igor Sokolov, Guy Malamud, Bart van der Holst, Pat Hartigan, Justin Wark, Eric Johnsen, and Ryan McBride, and students Jack Hare and Kevin Ma. Again, there are many others. Of course the book benefited strongly from many further students and other mentees with whom I have worked in the past decade, including Mike Grosskopf, Forrest Doss, Channing Huntington, Christine Krauland, Eliseo Gamboa, Carlos DiStefano, Michael MacDonald, Rachel Young, Willow Wan, Mario Manuel, Josh Davis, Alex Rasmus, Laura Elgin, and Patrick Belancourt. In addition to these, Jeff Fein, Matt Trantham, Joseph Levesque, and Robert VanDervort provided specific material.

The work itself would not have been possible without the efforts of several key people in tending to the research and administrative needs of our program even when I did not. Senior Administrative Assistant Jan Beltran and Financial Administrator Kathy Norris played their key roles, as they have for many years. Carolyn Kuranz and Paul Keiter have met the needs of the students, while Sallee Klein and Matt Trantham have contributed in important ways. The sabbatical leave provided by the University of Michigan was also essential. Finally, as most authors report, family support was crucial. I thank my wife Simona for her understanding and accommodation even when this project kept me distracted for long periods.

Ann Arbor, MI, USA   R Paul Drake
July 2017

Preface to the First (2006) Edition

This book has two goals. One goal is to provide a means for those new to high-energy-density physics to gain a broad foundation from one text. The second goal is to provide a useful working reference for those in the field.

This book has at least four possible applications in an academic context. It can be used for training in high-energy-density physics, in support of the growing number of university and laboratory research groups working in this area. It also can be used by schools with an emphasis on ultrafast lasers, to provide some introduction to issues present in all laser–target experiments with high-power lasers, and with thorough coverage of the material in Chap. 11 on relativistic systems. In addition, it could be used by physics, applied physics, or engineering departments to provide in a single course an introduction to the basics of fluid mechanics and radiative transfer, with dramatic applications. Finally, it could be used by astrophysics departments for a similar purpose, with the parallel benefit of training the students in the similarities and differences between laboratory and astrophysical systems.

The notation in this text is deliberately sparse, and when possible, a given symbol has only one meaning. A definition of the symbols used is given in Appendix A. In various cases, additional subscripts are added to distinguish among cases of the same quantity, for example, in the use of ρ_1 and ρ_2 to distinguish the mass density in two different regions. With the goals of minimizing the total number of symbols and of using them uniquely, the text avoids various common usages. An example is the use of μ for the coefficient of viscosity, which is avoided, with the viscosity expressed always as the product or $\rho\nu$, where ν is the kinematic coefficient of viscosity.

Much of the homework throughout this text is only feasible using a computational mathematics program. I prefer *Mathematica*, which has been an essential tool in the preparation of this text, but there are now and will be several such programs available. This departure from traditional norms reflects the emergence of such programs as effective tools. They should be part of the standard toolkit of all future scientists. This dramatically changes the meaning of "simple" solutions to problems. For example, an eighth-order polynomial equation is not necessarily difficult to deal with.

A word on the use of units is in order. The metric system in a broad sense is the common language of science. But the world in general and high-energy-density systems, in particular, are not conveniently analyzed within any single standard subsystem of these units. Each of the SI system, the Gaussian cgs system, and other systems are the most convenient for certain problems, as are a few other specific units such as the electron volt. This is why these systems exist. It is an *essential* tool for a practicing scientist to be able to readily convert between systems of units "on the fly." This is true because the existing literature is presented for the most part in convenient units, which working scientists use because they are convenient. But comparisons of one system to another are very important as checks on one's reasoning, and this often leads to the need to convert units. Thus, I am an adamant opponent of the SI purists who would commit nothing to print that is not in SI units and an adamant advocate for defining one's units in all work one does. When feasible, the equations in this book are written in a unit-independent form. When this is not possible, for example, with the Lorentz force, the units are specified and are usually in the Gaussian cgs system, which is the most convenient for most plasma applications. The units are also specified when practical equations are given. At least this was the my intent. Please let me know where I failed. Finally, the appendix on units in Jackson's *Classical Electrodynamics* is an excellent reference on this subject.

Bibliographic references are sparse in most chapters of this text. Most of the references are published books that address a certain topic in more detail than is feasible here. The journal literature is cited only when there are as yet no relevant books, and such citations often fail to reflect the scope of work in the journal literature. This was deliberate for several reasons. One of my goals has been to write a book that will prove useful for many years. The archival literature changes rapidly, and the present era is one of the very rapid advances in high-energy-density physics and in astrophysics. As a result, any references to the current literature will rapidly become dated. In addition, the era of immediate bibliographic database searches is here to stay, so future readers will readily be able to find up-to-date references in the archival literature of their time.

The second goal has been to present the material here with a common voice, because in my opinion, this is pedagogically most effective. A book that ties itself too closely to published literature can become disjointed. The third goal has been to show that this material is "simple," in the sense that a physicist would use. A rich panoply of phenomena evolves straightforwardly from what are at root a few and simple starting equations. In the spirit of Richard Feynman, one can understand a great deal without needing more than clear thinking (though one must add that a computational mathematics program helps a lot for some nonlinear problems). The greatest departure from this goal has been in Chaps. 3 and 8, where to avoid very protracted discussions, we have been forced to ask the reader to accept some details without much explanation.

Throughout this text, there are a number of figures showing the results of computer simulations, in order to display hydrodynamic and radiation hydrodynamic phenomena. Unless otherwise noted, these simulations were done using the HYADES computer code authored by Jon Larsen and available at this writing from Cascade Sciences Inc. A number of similar tools exist; they prove very useful for calculations to evaluate possible experiments and to identify the most important physical mechanisms in specific physical systems of interest.

Writing acknowledgments is rather daunting, given the many individuals who contribute to a project such as this. To those overlooked, remind me and I will at least buy you dinner. I must thank my family and my current research group for tolerating the time required for such a project. Dmitri Ryutov has been a source of inspiration, instruction, and encouragement, in addition to a vital collaborator, for a number of years, and also reviewed two chapters. Alexander Velikovich reviewed two chapters, made time for several delightful conversations, and significantly broadened my understanding of several issues. Harry Robey provided valuable insight into hydrodynamic instabilities and found an important error. Robert Kauffman and David Montgomery provided specific useful figures. Enam Chowdhury provided useful input and graciously allowed me to use some of his work. Michael DesJarlais, Warren Mori, Mordecai Rosen, Mark Hermann, James Knauer, Riccardo Betti, and Bedros Afeyan found time to comment on or discuss some of the material. Farhat Beg and William Kruer taught from the draft text. Ralph Schneider was a source of enduring encouragement.

The students in the lectures at Michigan in 2003 and 2005 and the 28 attendees of the summer school in 2004, though too numerous to list, helped identify errors and provided opportunities to improve the text. My own current graduate students Amy Reighard, Carolyn Kuranz, Eric Harding, and Tony Visco suffered through working with the draft while providing continuing motivation. Korbie Dannenberg, in addition to having done some of the work reflected in examples herein, kept my group moving forward when I was off writing. Jan Beltran provided a wide range of administrative assistance with the summer school and with the book, all of which I greatly appreciate. Of course, the responsibility for the errors in the text rests solely with me.

Beyond this specific group, I have enjoyed collaborations with a large community of scientists, engineers, and technicians during the past 20-plus years. A few of the key individuals not mentioned above are Dave Arnett, Jim Asay, Hector Baldis, Steve Batha, Bruno Bauer, Serge Bouquet, Jim Carroll, John DeGroot, Kent Estabrook, Adam Frank, Gail Glendinning, Martin Goldman, Tudor Johnston, Jave Kane, Paul Keiter, Alexei Khokhlov, Marcus Knudson, Barbara Lasinski, Sergey Lebedev, Dick McCray, Tom Mehlhorn, Aaron Miles, Steve Obenschain, Ted Perry, Diana Schroen, Wolf Seka, Bob Turner, David Villeneuve, Russell Wallace, Bob Watt, James Weaver, and Ed Williams. There are many others. I also appreciate the positive interactions and encouragement from my editor at Springer, Dr. Chris Caron.

I love to work in coffee shops and was fortunate that my local favorite, Espresso Royale, opened a branch in Plymouth Road near my home early during this project. I did a lot of writing, editing, and deriving at their tables. To Sarah and all the staff who have worked there, thanks for the hospitality.

Finally, this book would not exist without two people. E. Michael Campbell talked me into entering this field when it had troubled times, supported doing the science needed to make inertial fusion succeed, and helped me move on when the time came for that. Bruce A. Remington talked me into jumping into the astrophysical applications of high-energy-density tools when this was a new idea and has continued to be a valuable collaborator since that time. I thank them both.

Ann Arbor, MI, USA   R Paul Drake
December 2005

About the Front Cover Image

The image on the cover shows a portion of the Crab nebula as we see it today. Many of the mechanisms that contribute to its structure also appear in this book. The small fingers seen in the image are often produced by the magnetized Rayleigh Taylor instability. This and other instabilities were active during and after the explosion of the star, as were radiative shocks. The pulsar at the core of the nebula generates the electrons and positrons that drive the jets which emerge along its axis.

Contents

Chapter 1
Introduction to High-Energy-Density Physics

Abstract This chapter begins by introducing and defining the field of high-energy-density physics. It goes on to compare its domain with those of other areas of science and to survey the historical developments that led to the emergence of this field as a discipline. The chapter then identifies and discusses the regions within which various physical effects, such as Fermi degeneracy, become important. Brief introductions to inertial confinement fusion and to laboratory astrophysics at high energy density follow. The chapter concludes with a discussion of the connections between the present book and various other books, and some discussion of variables and notation.

This book concerns itself with the physics of matter within which the density of energy is high. The first formal definition of the field, early in the twenty-first century, was in a report of the National Research Council of the United States, entitled *Frontiers in High-Energy-Density Physics: the X-games of Contemporary Science*. It suggested a definition of *high-energy-density* systems as those having a pressure above one million atmospheres. The units of this pressure can be designated by 1 Mbar, 10^{11} Pa, 10^{11} J/m^3, 10^{12} dynes/cm^2, or 10^{12} ergs/cm^3. We will tend to express the pressure in Mbars, as this is the most common of these units found in the relevant literature. This definition reflected several observations. For example, one learned in school that solids and liquids are "incompressible," but this is not strictly true. If one applies a pressure exceeding ~ 1 Mbar to ordinary solid matter, it compresses. Another way to make this point is to say that the internal energy density of a hydrogen molecule is ~ 1 Mbar.

Thinking further, one might realize that once the energy that holds a collection of particles together, whether as applied pressure or as binding energy, becomes of the order of the internal energy of the molecules and atoms, their behavior will change. Beyond this point, the system will behave more as ions and electrons than as neutral particles. In the decade since the definition suggested above, it has become clear that solid matter exhibits novel behavior once the medium begins to ionize, creating free electrons. This can occur at a pressure as low as about 0.1 Mbar. This has led some researchers to suggest revising the above definition by lowering the threshold pressure. But in the absence of newly freed electrons, one does not get any novel behavior. In addition, the field as it functions includes the study of plasmas

© Springer International Publishing AG 2018
R P. Drake, *High-Energy-Density Physics*, Graduate Texts in Physics,
https://doi.org/10.1007/978-3-319-67711-8_1

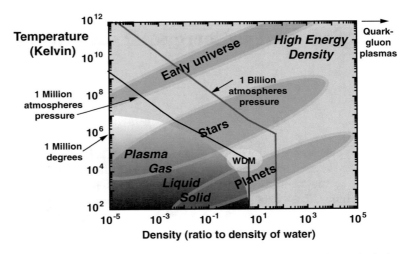

Fig. 1.1 Connection of the high-energy density regime to other physical and astrophysical systems

produced using devices that can create high-energy-density conditions, even when their density of energy is somewhat smaller. So the present author suggests the following revised definition of high-energy-density physics:

> High-energy-density physics is the laboratory study of matter that has a pressure of at least 0.1 Mbar (10 Gpa) and contains free electrons not present in the solid state, and of lower-pressure matter produced using experimental systems that can produce pressures above 0.1 Mbar.

Figure 1.1 shows how the realm of high-energy-density physics connects to other physical and astrophysical systems. At temperatures below about 10,000 K and densities below some multiple of the density of water, one finds the realm of solids, liquids, and gasses one learns of early in school. So-called "ideal plasmas," discussed in Chap. 2, exist at higher temperatures and lower densities. The high-energy-density regime is found to the right of the approximate boundary labeled "1 Million atmospheres pressure." As one crosses the vertical portion of this boundary, increased compression leads to delocalization of increasing numbers of electrons. As one crosses the diagonal portion, thermal ionization has the same effect. After the big bang, the early universe expanded and cooled along a path toward lower density at very high temperature. Stars are hottest and densest at their cores, becoming progressively less hot and less dense toward their surfaces. Planets have a similar structure at lower temperature. There is a small bubble labeled "WDM" near the kink in the 1 Mbar curve. This stands for *warm dense matter*. In this region, the models of condensed matter that work well to the left of and below this bubble fail, as ionization begins, while the ideal-plasma models that work to the left and above the bubble fail, as particle correlations become important. Some authors extend the WDM region to the right of and below the bubble. This did not seem well-justified in 2005, but seems more so now. The succeeding decade has seen the discovery of unanticipated complexity throughout this region, as discussed in Chap. 3. Finally, quark-gluon plasmas are a form of high-energy-density matter produced by colliding relativistic heavy ions. They have a temperature of order 10^{12} K, and a density far to the right of those shown here.

We will discuss some further aspects of the transition into the high-energy-density regime in Sect. 1.2. Matter at high energy density is ionized, with the exception of a small region at low temperature discussed in Chap. 3. The author believes that the proper definition of a plasma is "an ionized medium", and so high-energy-density matter is plasma. In contrast, traditional plasma theory developed in the mid-twentieth century, concerned only with space plasmas and with laboratory devices at very low pressure, defined a plasma as "an ionized medium for which the theory is easy". (These were not the words used, but that is the upshot.) Plasmas for which the theory is relatively easy, whose details we will discuss further in Chap. 2, are known as "ideal plasmas". We will see that ideal-plasma theory is not valid for a large fraction of the high-energy-density regime. Thus, another way to characterize high-energy-density systems is as plasma that is too dense for traditional plasma theories, or that has other features not included in such theories. One of the key aspects of these differences, having several components, is that the matter itself often does not behave as an ideal gas. We explore these components and the fundamental description of high-energy-density matter in Chap. 3. In brief, Chaps. 4 through 5 then discuss how such matter moves and Chaps. 6 through 8 discuss how radiation affects it.

Chapter 9 discusses how to create high-energy-density conditions. One might like to take this up sooner, but in fact the concepts developed in Chaps. 3–8 are essential to presenting a comprehensive and comprehensible discussion. For example, one might launch a shock wave that converts ordinary matter into high-energy-density matter. Such shock waves have velocities above 10 km/s. At constant pressure, shock velocities increase as density decreases, so that shock waves above 100 km/s (>360,000 km per hour) are common in high-energy-density physics. Alternatively, one might produce an intense beam of photons, electrons, or ions that can penetrate the matter and directly heat it, and particle beams themselves may reach this regime.

High-energy-density physics encompasses more than the regime of dense plasma, in the sense just described. It also includes conditions in which pressures >0.1 Mbar result from very high temperature at very low density. For example, air at a density of 1 mg/cm^3, of the order of atmospheric density, reaches a pressure of ∼0.1 Mbar at a temperature above 1 keV. (Throughout this text we express temperature in the energy unit of an electron Volt, so that the Boltzmann constant is 1.6×10^{-12} ergs/eV or 1.6×10^{-19} J/eV.) A temperature of 1 keV is roughly 10 million Kelvin, so that temperatures of millions of Kelvin or more are common in high-energy-density physics. As the density decreases further, conditions in which the pressure remains above one Mbar soon become relativistic, and thus also outside the realm of traditional plasma theory. Overall, what high-energy-density systems have in common with traditional plasmas and with condensed-matter systems is that collective effects are an essential aspect of the behavior. The difference from traditional plasma physics is that the particles are more correlated, and/or relativistic, and/or essentially radiative. The difference from traditional condensed-matter physics is that ionization and Coulomb interactions are essential. Various chapters that follow are focused on one or more aspects of these differences.

The present text was the first book to be written as a textbook in high-energy-density physics. (We place it in the context of prior work later in this chapter.) This reflects the fact that high-energy-density physics is in some sense a new field. One can see that the regimes just discussed offer some challenges beyond established areas of physics, but one might wonder both in what sense this is new and why. The material discussed here, as in condensed-matter physics and other areas, is entirely built on the foundations of classical and modern physics as established from the mid-nineteenth to the mid-twentieth century. In addition, much of the material discussed herein is discussed in more depth in one of a dozen or so more-advanced books. The fundamental sense in which this is a new area is that there are new tools and that new tools beget new areas of science. It is now practical for scientists in an academic or laboratory setting to perform experiments to study the fundamental behavior of high-energy-density systems over a significant range of parameters. This creates a need for the treatment of this material as an integrated subject, moving from fundamentals to their applications, for the presentation of the material in a common voice suitable for graduate courses and as a first working reference, and for a discussion that spans the range of conditions now (or soon to be) available for study. Hence the emergence of high-energy-density physics as a distinct field and hence this text.

1.1 Some Historical Remarks

Let us consider the new tools and some key people that brought this about, with the goal of giving the flow of key developments as opposed to a thorough historical review. The development of spectroscopy and modern astronomy in the later nineteenth century led Eddington, Schwarzschild, Chandrasekhar, and others to seek to understand the structure of stars through the first half of the twentieth century. Key outcomes of this work included the theory of the radiative transfer of energy, discussed in Chap. 6, and understanding on the production of energy in stars by nuclear fusion. This last was a key step leading toward research into inertial confinement fusion, discussed in Chap. 11. The development of particle accelerators in the 1930s began the effort to focus large numbers of particles to small areas. The advent of nuclear weapons in the 1940s produced high-energy-density conditions, but not in a way that permitted systematic study. The development of the light-gas gun in the 1950s and 1960s eventually led to the ability to study matter at the low-pressure edge of the high-energy-density regime. These were all key technical developments, but it was the invention of the laser that most directly led to the emergence of this field.

Work in the 1950s led to the invention of the laser by Townes, Gould, and others late in that decade. It then became sensible to ask whether lasers might be used to produce controlled thermonuclear fusion. This would be accomplished by creating momentary collections of burning fusion fuel, held together by only their inertia. In their book Atzeni and Meyer-Ter-Vehn (2004) review the foundations for this specific development, some of which were in classified research programs, and credit a number of key contributors not mentioned here. The effort to address this question in the USSR was led by Basov. In 1972, Nuckolls, Thiessen, and Wood published the key paper in *Nature* arguing that this approach to fusion might be feasible. Programs to pursue what became known as *inertial confinement fusion* were begun in the U.S., the Soviet Union, Europe, and Japan. A key figure during the development of the necessary lasers was John Emmett, who led the program at Lawrence Livermore that first produced lasers delivering >1 kJ in 1 ns. By the end of the 1970s, there were lasers in several countries that could deliver a number of kJ to volumes of a cubic mm or less in pulses of order 1 ns in duration. One Mbar is 100 J/mm^3, so these systems produced high-energy-density conditions.

In the 1950s the ill-fated Z pinch had been developed, in pursuit of controlled thermonuclear fusion for power production. The Z pinch was intended at first to gradually compress and heat matter through the attraction of parallel channels of current, but this failed notoriously as a path to controlled thermonuclear fusion in the laboratories. But by the late 1970s, devices that could drive currents above 1 MA for short periods, known as pulsed-power devices, were developed. These were motivated in part by their potential application to inertial confinement fusion. In the U.S. this was done at Sandia National Laboratories. The initial intent was that these devices could create inertial fusion using particle beams, but in the end they contributed to the revolution in Z-pinches described below. Other lasers were also developed as high-energy sources, including CO_2 lasers at the Los Alamos Laboratory and Iodine lasers in the Soviet Union. These lasers did not work out for fusion, but in some sense they encouraged the development of KrF lasers, pursued further in the U.S. at the Naval Research Laboratory and at this writing a potentially of use in producing electricity powered by inertial fusion.

These tools could create high-energy-density systems, but not in a way that permitted systematic study. During the 1970s a few research projects and programs began to do systematic fundamental science at high energy density, notably in Europe and at the Naval Research Laboratory. This is perhaps too harsh, as one can find a sequence of refereed journal papers tracing progress in the science from all of the participants. But all such efforts were hampered by a lack of experimental technique and diagnostic hardware. They were also hampered by a tendency to focus on the goal of fusion to the exclusion of its fundamental underpinnings. (As an extreme example, the head of the project to build the Nova laser, completed in the mid-1980s, once told the author that the only diagnostic needed by Nova for the success of inertial fusion was a single neutron detector. This proved to be far from the truth. It took a decade to develop diagnostics and perform experiments addressing a range of issues, and to make substantial improvements to the facility itself, before Nova could achieve the compression of DT fusion fuel to 100 times liquid density—a remarkable accomplishment.)

I date the emergence of high-energy-density physics as a field from 1979, for two reasons. First, this was approximately the first year one could do a range of *physics* experiments at high energy density using lasers. The reason is that multi-beam, high-energy laser facilities now existed, and one could use these beams for several independent purposes. Some beams could strike a target to produce a desired system, other beams might be used to drive some process or event in that system, and still others might be used as diagnostics, often by producing X-rays whose transmission or scattering could be measured. (Low-energy lasers had been in use as probe beams for some time, and this continued.) These sophisticated experiments required that the beams be independently timed and controlled, which is easy to say but imposes considerable additional cost and complexity. Second, this was the year that the first user program for such a laser facility came into existence. This was the National Laser User Facility, which provided facility access and financial support for investigator-driven research at the Omega laser. This peer-reviewed program, and a later version of Omega, still exist at this writing. While other facilities have contributed much science, Omega is widely recognized as the most successful and productive research facility for high-energy-density physics.

The early 1980s saw the realization and demonstration of affordable instrumentation that could obtain data on a sub-ns timescale, including snapshots and time histories as images or as spectra. This was in part driven by an increased focus on studying the elements of the physics that were required for inertial fusion. While many researchers around the world contributed to this developing focus, the one individual who had the biggest impact in the biggest program was a young group leader named Michael Campbell, at the Lawrence Livermore National Laboratory.

The 1980s also saw the invention of *chirped pulse amplification* by Gerard Morou, described in Chap. 13. This made it practical to drive the energy flux of lasers above 10^{18} W/cm^2, and to begin to produce relativistic effects. Such systems have short pulses, typically below 1 ps, and so are known as *ultrafast lasers*. They have contributed the tools that enable exploration of the low-density, relativistic regime of high-energy-density physics. We will discuss some basic aspects of this regime in Chap. 13.

The available experimental tools for high-energy-density physics expanded again in the 1990s with the development of the wire-array Z pinch. A modification known as the fast Z pinch had been under exploration for fusion since the late 1970s. A fast Z pinch avoids the magnetohydrodynamic instabilities that disabled the Z-pinches of the 1950s, by using the pinch to briefly accelerate material inward, after which the stagnation of the imploding material converts kinetic energy to internal energy. There is hope that this might provide an alternative approach to fusion. Whether or not this works out, such pinches are large and efficient radiation sources. When they distribute the current across hundreds of metallic wires, they can produce energies of MJs in volumes of cubic centimeters. This development, also discussed further in Chap. 10, provided yet another environment for the pursuit of high-energy-density physics, and there was a veritable explosion in such activity using pinches at around the turn of the century.

Meanwhile, particle accelerators continued to develop, driven primarily by the needs of particle physics. By the 1990s, these accelerators could also produce sufficient densities of relativistic particles to be the high-energy-density regime. Both ion beams and electron beams can produce high-energy-density conditions. As one example, the Stanford Linear Accelerator has been used to produce bunches of order 10^{10} electrons at an energy of 50 GeV, in a 5 ps pulse. These bunches form a 3 µm spot, and so have an energy flux of 10^{20} W/cm^2. They deliver 150 J per pulse to a target, and they arrive at a rate that can exceed 100 Hz. These electrons are themselves a high-energy-density medium and can be caused to interact with materials of choice. The more-recent emergence of free-electron lasers that produce coherent, intense beams of X-rays has provided yet another way to produce and study matter at high energy density.

All of the above developments produced an environment within which it became possible to pursue questions in high-energy-density physics for their own sake. Researchers around the world can now address the properties of matter, the development of dynamic structure and of instabilities, the properties and transport of radiation, the effect of radiation on the dynamic behavior, and relativistic phenomena in this regime. These fundamentals are what we take up in the next eight chapters and Chap. 13. Researchers can then use this knowledge to invent novel approaches to inertial fusion (Chap. 11), to learn things relevant to astrophysics (Chap. 12), and to develop technologies ranging from improved lithographic systems to novel medical therapies. Before turning to these tasks, the following provides some further overview of the regimes of high-energy-density physics and of its applications to fusion and to astrophysics.

1.2 Regimes of High-Energy-Density Physics

Figure 1.2 shows some important physical regimes and boundaries for high-energy-density physics. This figure merits an extensive discussion, which will point the way to much of our work throughout this text. The horizontal axis shows ion density in cm^{-3}. The vertical axis shows temperature in eV. The axes are logarithmic, so that this figure spans 11 orders of magnitude in density and more than 10 orders of magnitude in temperature. It shows a number of boundaries and curves. These boundaries and curves nearly all assume that the matter and radiation are approximately in equilibrium. We will work our way through these and see what they tell us. One can see the dark black curve labeled "1 Mbar" from Fig. 1.1. Focus first on the diagonal line crossing this curve and labeled $T_e = \epsilon_{Fermi}$, which shows the boundary where the electron temperature, T_e, equals the Fermi energy, ϵ_{Fermi}. Below this boundary the electrons are *Fermi-degenerate*, and above it they are not. We discuss in Chap. 3 how sharp the transition is. When the electrons are Fermi-degenerate, the pressure is above and often far above the ideal-gas value one would find in an ideal plasma.

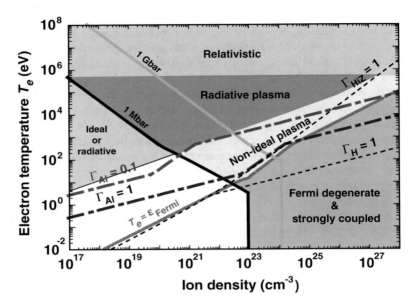

Fig. 1.2 Regimes of high-energy-density physics. The curves are primarily based on the simple models of Chap. 3, as applied to Al. Two dashed lines correspond to H or to a generic, very heavy, "Hi-Z" medium. Other aspects are discussed in the text

Another fundamental aspect of dense plasmas is the magnitude of the energies associated with Coulomb forces. When these are strong, the particle locations become correlated and ideal-plasma theory fails. If one expresses the average electric potential experienced by a particle as ϕ, then the average energy of interaction is $|e\phi|$, where the electronic charge is e. The key comparison is between this energy and the thermal energy, $k_B T_e$, and the *strong coupling parameter* is defined as $\Gamma = |e\phi|/(k_B T_e)$. The plot shows three curves where $\Gamma = 1$. The straight, thin-dashed line labeled $\Gamma_{HiZ} = 1$ would correspond to an imaginary very heavy atom whose free electrons result strictly from ionization, and which ionizes according to the simplest equilibrium model. The curve that splits off from this, labeled $\Gamma_H = 1$, changes its slope at the place where H becomes ionized. The curve $\Gamma_{Al} = 1$ has different behavior at low density and temperature, reflecting the fact that Al in the solid-state has about three, free, conduction electrons. Above this curve is another one, labeled $\Gamma_{Al} = 0.1$, which shows that the transition to conditions where the particles are weakly correlated is fairly gradual. Thus, there may be a region where the electrons are not Fermi-degenerate but the plasma is not an ideal plasma. The gray region in the lower-right corner of the plot is where the matter is both Fermi-degenerate and strongly coupled.

As the plasma heats to hundreds of eV (or sometimes less, especially at lower densities), the transport of energy by radiation becomes important or even dominant. The structure and dynamics of the plasma are affected in turn. In this regime the plasma is a radiative plasma, shown on the plot above the region of non-ideal plasma just discussed. The radiative energy transfer alters the plasma dynamics, and so the system is within the realm of radiation hydrodynamics, discussed in Chap. 7. At high-enough temperatures, the radiation pressure will directly affect the dynamics. At pressures below 1 Mbar, radiative effects may still be dominant, for which cases one will find the relevant models within books and papers on high-energy-density physics but not within those on the physics of ideal plasmas. The plot shows this region, labeled "Ideal or radiative", in light gray. Within this region, plasmas of sufficient size must be described using radiation-hydrodynamic theory. In the upper part of this region, plasmas of very large size (too large to be laboratory plasmas) will have radiation pressures above 1 Mbar. Finally, when the electron temperature reaches hundreds of keV the electrons become relativistic. This region is accessed using ultrafast lasers and particle accelerators, and is discussed in Chap. 13.

However, Fig. 1.2 also provides an incomplete picture of high-energy-density physics, because it assumes equilibrium. What is missing is dynamical processes. Phenomena such as shock waves, radiation waves, material ablation, radiative cooling, and hydrodynamic instabilities are not included. Much of Chaps. 4, 5, and 7 and 8 are concerned with these dynamical processes. In addition, and independent of the information provided by Fig. 1.2, the dynamics of interest may involve magnetized flows, discussed in Chap. 10, Dynamical processes are also essential to the production of high-energy-density conditions (Chap. 9), to the achievement of inertial fusion (Chap. 11), and to experiments relevant to astrophysical phenomena (Chap. 12). In what follows we provide a summary introduction to inertial fusion and to experimental astrophysics before we proceed to our detailed task.

1.3   An Introduction to Inertial Confinement Fusion

We mentioned in Sect. 1.1 that inertial confinement fusion or ICF is the application that has driven much of the development of high-energy-density technology and science. Chapter 11 discusses ICF, beginning with fundamentals. ICF can produce a net energy gain because light elements release energy when they are combined to form heavier elements. This requires a high temperature, to overcome the Coulomb repulsion of the nuclei, but the energy released can be used to sustain the high temperature. This is very much like ordinary combustion, and so fusion fuel is said to ignite and to burn under the proper conditions. The applications of ICF will expand as the energy gain of ICF systems increases. Here energy gain is the ratio of the electricity used to produce an ICF event to the energy of the neutrons (or X-rays) it produces. At modest gains, even of order 1, ICF will produce large amounts of neutrons and/or X-rays that can be used for further areas of research. At larger gains (of order 10), the neutrons from ICF events might be used to breed fuel for electric power plants powered by nuclear fission. At large enough gains (of order 100), ICF events might be used directly as the energy source in electric power plants.

Fig. 1.3 Inertial fusion
targets. (**a**) A hohlraum target
containing a fusion capsule.
(**b**) Image of a thin-walled
hohlraums showing where the
laser beams strike. Credit:
Lawrence Livermore National
Laboratory

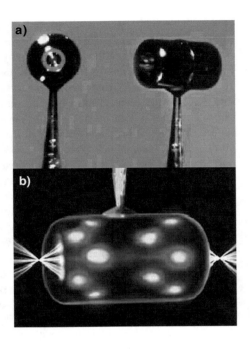

An ICF power plant would operate on a rhythmic cycle that has been compared
to the cycle of an internal combustion engine. (The author is unsure whom to
credit for this analogy.) Let us work our way through this cycle, discussing its
elements. The cycle of an ICF power plant would begin with injection of an ICF
target into a reactor chamber, in analogy to the injection of fuel into the cylinder of
an engine. The target is a structure designed to produce energy by fusion when
energy is delivered to it in a specific way. The target will include a *capsule* of
fusion fuel (probably DT fuel—deuterium and tritium—and very likely spherical
in shape) covered by a material known as an *ablator*. The target may also include
other structures necessary for the operation of a specific fusion design. Figure 1.3a
shows an example. Here the spherical object is a fusion capsule while the cylindrical
structure surrounding it is an object known as a *hohlraum* (see Chap. 9 and the
discussion below). In Fig. 1.3a the target is mounted on a glass stalk; in a power
plant it would be dropped into position, very likely while spinning to help maintain
its orientation.

The phase in internal combustion that follows injection is compression of the
fuel, generally by the motion of a piston. This provides energy to the fuel as
it compresses the fuel. This phase exists in ICF, though it is somewhat more
involved. The first element is to deliver energy to the target with the required spatial
distribution and uniformity. The device that provides the energy is called a *driver*
in the jargon of ICF. It might be a laser beam or a particle beam or conceivably an
intense photon source. The target might be driven by direct irradiation (a condition
known as *direct drive*) or indirectly (known as *indirect drive*). Indirect drive might

be accomplished, for example, through the conversion of the energy from the driver into some other form of energy such as thermal X-ray photons. The hohlraum of Fig. 1.3a converts laser energy into X-rays, producing an X-ray environment with a temperature of order 2 million degrees (200 eV). Figure 1.3b, obtained with a thin hohlraum that allows one to see where the laser beams strike the interior walls, shows an example of such irradiation.

This delivery of energy to the ablator causes it to ablate away (hence the name). This, however, is not a passive process. The delivery of energy to the ablator produces temperatures of millions of degrees and pressures of order 100 million atmospheres. High-velocity, ablated material is propelled away from the hot, high-pressure material at the surface of the ablator. In reaction, the remaining material and the fuel are accelerated inward. This process is identical to rocket propulsion, so that an ICF capsule is sometimes described as "spherical rocket". The material moving inward is accelerated to hundreds of km/s. When the fuel converges at the center, it must stop. This stagnation event converts the kinetic energy of motion to internal energy, enabling high compression even of Fermi degenerate fuel. This completes the compression phase of the cycle. The fuel is compressed somewhat more than hydrocarbon fuels are, reaching a final density of 1000 to several thousand times the density of liquid DT.

The next phase of the cycle in both internal combustion and ICF is ignition. One can have spontaneous ignition, as in diesel engines, or spark ignition, as in gasoline engines. Both these approaches are possible in principle for ICF. On the one hand, one can design the target so that the fuel at the center of the imploding capsule ignites when the fuel stagnates. This is known as ignition from a central hot spot. On the other hand, one can compress the fuel and then use an external energy source to ignite it. Variations on this approach are known as *fast ignition* and *shock ignition*. At this writing, it is not clear which of these will prove at first most productive, or in the long run most practical. The ignition and subsequent burning of the fusion fuel creates a large quantity of energy, completing this phase of the cycle.

The next phase of the cycle is energy extraction. In internal combustion, this occurs as the expanding hot gas does work on the piston. It is more complicated in ICF, as neither neutrons nor X-rays, nor even high-energy particles, are able to push effectively on solid matter. They penetrate rather than push. Instead one must extract their energy in some other way. When ICF uses DT fuel, most of the energy emerges as neutrons, and the only known way to extract their energy is to use them to heat a large volume of matter. This hot matter can in turn heat water to drive a steam cycle, in which energy is extracted by driving large turbines with steam. This may or may not prove economical for electric power production. Steam cycles are not particularly efficient, so one may hope that in the long run one can use fusion fuel that produces only charged-particle output. This is more demanding but offers the potential of directly extracting the energy, with high efficiency. In the absence of new physics, the compact fusion plants that drive many spaceships in science–fiction would only be possible using these advanced fuels.

Returning our focus to high-energy-density physics, many elements of the ICF process depend upon such physics for their success. These include the production of the energy that drives the target, the delivery of energy to the ablator, the implosion process itself, the final properties of the fuel, and any attempt at fast ignition. ICF depends upon the properties of high-energy-density matter (Chap. 3), on the production of shock waves and related effects (Chap. 4), on limitation of hydrodynamic instabilities (Chap. 5), on the transport of radiation within the target (Chap. 6), and on the impact of radiation on material motions (Chap. 7). Furthermore, the basic approaches to ICF reflect the various options for producing high-energy-density conditions (Chap. 9). This makes it fairly clear why the science of high-energy-density physics grew out of ICF and its facilities, and why knowledge of this physics is essential if one is to deeply understand ICF.

1.4 An Introduction to Experimental Astrophysics

We made the point above that new sciences arise from new tools. A second and more specific example of this is the emergence of high-energy-density experimental astrophysics. One might say that the human brain as a tool gave rise to astronomy. Adam Frank has observed that the spectrometer can be argued to have given rise to astrophysics. In the same sense, the ability to do high-energy-density physics in the laboratory has given rise to this branch of experimental astrophysics. Remington et al. (2006) provide a review of work in this area through 2004. Some summary remarks follow here. Various examples are given throughout the text. In Chap. 12 we will take up the specific problem of doing experiments that are sensibly scaled from the astrophysical system to the laboratory. This field has blossomed in the early twenty-first century. In the chapter, we address four specific types of experiment, to show how to construct the scaling arguments that are necessary to understand the relation between the laboratory and astrophysical systems.

The potential for contributions to astrophysics from high-energy-density physics is clear from Fig. 1.1. That figure leads one to focus on the possibility of measuring the equation of state, which for example might relate the pressure, density, and temperature of materials of astrophysical interest. At this writing, for example, it is not yet clear just when and how dense hydrogen enters its metallic state. This is particularly important for gas giant planets such as Jupiter. Figure 1.4 shows a pie-shaped slice to illustrate a segment of the spherical cross section of this planet. The equation of state of hydrogen determines whether Jupiter must have a rock core. The nature of the transition to the metallic state constrains how the dynamo in Jupiter produces its magnetic fields. This connection has been recognized for some time but equation of state experiments are difficult, as they require very high precision. By the 1980s one could begin to attempt measurements using the tools capable of such work at that time. These were principally devices called gas guns and rail guns that can accelerate slabs of material to high velocity, producing high pressure when they collide with other slabs of material. Until recently, such guns could not

Fig. 1.4 Schematic interior of Jupiter

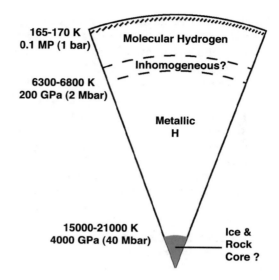

165-170 K
0.1 MP (1 bar)

Molecular Hydrogen

Inhomogeneous?

6300-6800 K
200 GPa (2 Mbar)

Metallic
H

15000-21000 K
4000 GPa (40 Mbar)

Ice &
Rock
Core ?

produce pressures as large as 1 Mbar. The emergence of pulsed-power devices that could drive 10 MA currents, and of lasers capable of creating very large pressures led at around the turn of the century to work at pressures of several Mbar.

The equation of state is one example of a property of matter in its equilibrium state that has implications for astrophysics. We discuss the connections in Chap. 3 and the most common approach to measurements, which uses flyer plates, in Chap. 4. Another example is the X-ray opacity, discussed further in Chap. 6. Much of the energy transported within stars and other hot, dense objects is carried by radiation. The absorption of this radiation turns out to control some of the properties of these objects. Yet this radiation absorption is often dominated by the electronic transitions in various ions, especially those up to and including ions of iron. The measurement of the opacity of materials of astrophysical interest, and the comparison of these measurements with newly available computer codes that could calculate the opacities accurately, began in the early 1990s. Measurements of equation of state and opacity carry out the research suggested naturally by Fig. 1.1, determining important equilibrium properties of materials. More recently, researchers have devised ways to use fusion capsules to measure some cross-sections for nuclear reactions, some of which are important for the evolution of the elements in the Universe. Aside from a few reactions relevant to fusion energy, though, this book will not pursue this topic further.

An explosion of work in laboratory astrophysics followed the realization, independently by Hideaki Takabe in Japan and Bruce Remington in the U.S., that one could also use high-energy-density tools to explore the large-scale dynamics that matters for astrophysics. To that point, our Earth-bound knowledge of astrophysical dynamics depended entirely on computer simulations that could be tested against one another but not against any benchmark data. Some of the resulting work is discussed further in Chap. 12, with a focus on how one establishes the scaling between systems. Here we provide a partial overview.

There are a sequence of steps required to have meaningful scaling between a laboratory system and an astrophysical one. First, the same type of equation must be sufficient in both systems. For example, fluid equations must suffice. Second, the same terms in the equations must be important in for both systems. For example, if the growth of fluid structures at small scales is important, then the Reynolds number, which measures the weakness of viscous dissipation and diffusion, must be large in both systems. Third, the key parameters in the two systems must have the same value of a scaling parameter that we will call a *Ryutov number*, after Dmitri Ryutov, who initiated the work that defined this approach to scaling. Fourth, the detailed behavior of the specific dynamical process of interest must be in the same regime, measured by one or more dimensionless parameters that must be thoughtfully determined. We will call this the specific scaling.

It is most straightforward to produce a well-scaled experiment in systems that are purely hydrodynamic, so that viscous dissipation, heat conduction, and radiation are negligible. This may seem like too much simplification to be worthwhile, but that is not the case. It turns out that many astrophysical systems behave mostly or entirely as hydrodynamic systems, in certain portions of their structure and evolution. These include supernovae, some supernova remnants, blast waves, and some astrophysical jets. In addition, the complex three-dimensional instabilities within these systems and interactions among them are beyond the capabilities of turn-of-the-century simulations to reproduce. This creates a role for experiments. Early work, around the turn of the century, focused on the unstable phenomena that occur during supernova explosions. Figure 1.5 shows an example. As a result of the passage of a blast wave like that in supernovae, the structure at a first interface affected the evolution of a second interface. Other early experiments in this area also explored the simulation of processes in supernova remnants, the dynamic behavior of hydrodynamic astrophysical jets, and the crushing of clouds by shock waves.

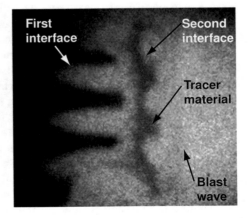

Fig. 1.5 Structure produced when a blast wave crosses two interfaces. The second interface is visible only where a diagnostic tracer is present behind it. There is little contrast across the blast wave itself, making it difficult to see

Despite the comparative simplicity of hydrodynamics, we begin Chap. 12 with a thorough discussion of scaling between systems for such cases.

Once the transport of energy by radiation becomes important in the evolution of the matter, one has entered the regime of radiation hydrodynamics (Chaps. 7 and 8). Important astrophysical phenomena in this regime include radiation waves, radiative shock waves, and radiative jets, not to mention the interiors of stars. Here, however, one nearly always loses the ability to do a precisely scaled experiment. Instead, one can hope to scale the essential dimensionless parameters so that the dynamics of interest to astrophysics are present in the laboratory. One example is the radiative jet, in which there are three key parameters: (i) the internal Mach number (the ratio of the jet velocity to the sound speed of the material in the jet), (ii) the ratio of density in the jet to the ambient density around it, and (iii) the ratio of the distance along the jet required for significant cooling to occur to the radius of the jet. An experiment can in principle scale these three numbers. Such an experiment and an astrophysical jet would be expected to show qualitatively similar behavior, even if some detailed processes, such as the nature of the radiative cooling, were different. Similarly, for radiative shock waves there are three key parameters. The first two are the transmission of radiation by the regions ahead of and behind the density jump produced by the shock wave. The third parameter is the ratio of the radiation energy flux produced in the shocked matter just behind the shock to the material energy flux coming to the shock (in a frame of reference in which the shock is at rest). These parameters can also in principle be scaled from an experiment to an astrophysical case.

Magnetic fields play a significant role in the dynamics of many astrophysical systems, notably including astrophysical jets and accretion flows. We also construct, in Chap. 12, the requirements for scaling of experiments producing MHD flows. We apply these to the case of magnetized astrophysical jets and jet launching. It has proven possible to create, in the laboratory, episodic jets that develop structures similar to those seen emerging from young stellar objects and other systems. The analysis of scaling shows that these systems can be well-scaled to the astrophysical dynamics of interest.

We also provide one example from the scaling of the Weibel instability, relevant to the formation of collisionless shocks in plasmas that are initially unmagnetized. This is an example of a case in which one is attempting to produce and study a process that is important in astrophysics as a fundamental mechanism, with less emphasis on modeling any particular astrophysical case. Even so, one must develop dimensionless parameters that enable one to understand how to produce a laboratory experiment that is in the correct physical regime.

We do not discuss in Chap. 12 experiments driven by ionizing radiation. These often are driven by radiation from Z-pinches. This is a method used to study opacities, but also can examine other systems of interest. One of these is *photoionized plasmas*. Most Earth-bound plasmas are produced by electron-impact ionization, including those in fluorescent light bulbs, in magnetized laboratory plasmas, and in laser-irradiated materials. This is also true of many astrophysical environments, such as the solar corona. However, there are a number of astrophysical environments in which the dominant source of plasma is photoionization. (This is also true of portions of the ionosphere above the Earth.) Indeed, since we depend primarily upon spectroscopy for the study of astrophysical objects, the data we get from very energetic environments such as neutron stars and the space near black holes come from photoionized plasmas. Without well-grounded knowledge of the properties of such plasmas, we cannot hope to interpret the data to see, for example, whether Einstein's theories of gravity accurately describe black holes. Beginning in the mid 1990s, experiments began to produce photoionized plasmas in well scaled environments and to measure their properties.

Finally, astrophysical environments often produce relativistic effects. These are many. A few examples include the production of electron–positron pairs in intense radiation environments, the alteration of atomic structure in very strong magnetic fields, the propagation of relativistic jets through magnetic fields, and the evolution of gamma-ray bursts, which seem likely to involve relativistic radiation hydro-dynamics. At this writing, the use of ultrafast lasers to explore such phenomena has begun. We discuss in Chap. 13 the production of positrons using such lasers, which a step toward being able to study electron-positron plasmas. Such plasmas are important near pulsars and in other astrophysical systems.

1.5   Background Needed for This Book

This book is heavy on conceptual explanations, and so may provide some benefit to those who want to know the qualitative description of some topics. However, just as in other basic graduate books in physics, its fundamental intent is to develop the combined conceptual and mathematical framework that enables us to understand the systems we study. Since our primary concern is with the dynamical behavior of continuous systems, our mathematical language is that of partial differential vector calculus. The mathematical expertise typically learned in a course on mathematical methods for physics and engineering is essential, if one is to follow the mathematical conversation. No background training is needed in fluid dynamics or plasma physics as such; these are covered here as needed. Background in two other areas is fairly important. The first of these is statistical and thermal physics. A familiarity with this subject at the advanced undergraduate level is sufficient for nearly everything discussed herein. The author is partial to the iconic text by Reif (1965). Only the most-advanced sections of Chap. 3 would require some graduate-level background to be understood. The second of these is electricity and magnetism. Here the knowledge from an advanced undergraduate course is probably sufficient, although graduate coursework would help, especially in the area of wave behavior. The book by Jackson (1999) has become the standard reference, and is cited at times throughout this text.

1.6 Some Connections to Prior Work

As indicated above, high-energy-density physics is a new field in some sense, having evolved out of some historical precursors. It might have emerged sooner in the absence of classification of fusion research. However, it might not have, because the ability to do systematic experiments remains relatively recent. Even so, there are a number of historical books in related fields and notably in astrophysics that provide a deeper discussion of some of the material covered here. This section provides an overview of some of this work, oriented to the order of the following chapters. There are two areas for which few or no more detailed books are available. These are experimental astrophysics (Chap. 12), for which no books exist, and relativistic high-energy-density physics (Chap. 13), for which there are as of 2016 at least three (Gibbon 2006; Seryi 2015; Gamaly 2011). These are both relatively new areas of research at this writing. Some review papers exist, no doubt soon to be superseded by others. One will have to wait for more book-length treatments.

The properties of matter at high energy density (Chap. 3) could be addressed more thoroughly by a detailed study of the statistical mechanics of matter in the high-energy-density regime. Such a work does not exist but various references approach aspects of this problem. Many books in statistical mechanics address various fundamentals; one good example is Volume 5 of the series by Landau and Lifshitz (1987). Even so, the specific issues that arise in ionized media and otherwise at high energy density are not addressed there. The book on equations of state by Eliezer et al. (1986) considers some of these issues, as does the book on plasma spectroscopy and atomic physics by Salzman (1998). They are also dealt with, on a fairly *ad hoc* basis, in the books to be highlighted later by Zel'dovich and Razier (1966) and Mihalas and Weibel-Mihalas (1984). Griem (1997) addresses several specific issues in his book on plasma spectroscopy, which, alas, is not so easy to read. Quantum mechanical aspects of the matter also become important at high density and low temperature, as we also discuss in Chap. 3, but the author is not aware at present of a book focused on these issues.

The behavior of shock waves, rarefactions, and self-similar systems (Chap. 4) is addressed to various degrees in a number of texts, notably including those by Sedov (1959) and by Whitham (1974). Principal among such works, for problems of interest to high-energy-density physics, is the book by Zel'dovich and Razier (1966). This book quite rightly has been venerated for many years. If there is a single text that might be said to be a precursor to the field of high-energy-density physics, it is this. It has also long been the best general introduction to high-energy-density physics, despite being too detailed in many respects and incomplete in others. The author hopes that the present text might take over this function. Even so, Zel'dovich and Razier (1966) will remain the next book of choice for details of one-dimensional hydrodynamics and approaches to the key physics based on similarity solutions.

In the area of hydrodynamic instabilities (Chap. 5), the work that stands clearly at the forefront in the analysis of hydrodynamic instabilities is that of Chandrasekhar (1961). It is thorough, it is fundamental, and it works a number of problems of central interest to high-energy-density systems. Unfortunately, having been written in the 1950s this text does not discuss some aspects of compressible hydrodynamic instabilities that are also of substantial interest. For these, to go beyond what little is done in the present text one must head to the archival literature. In the specific area of incompressible hydrodynamic turbulence, the outstanding introductory text is by Tennekes and Lumley (1972) and the thorough and definitive tome is by Hinze (1959).

For issues relevant to high-energy-density physics, radiative transfer (Chap. 6) and radiation hydrodynamics (Chaps. 7 and 8) have never been treated in independent books, so we will discuss them together. The field of radiative transfer has many applications, as this kind of process is essential to the behavior of planetary atmospheres, stellar interiors, energetic astrophysical events, and high-energy-density laboratory systems. Chandrasekhar for example also contributed a book on radiative transfer. However, the book that is without question most relevant to radiative transfer and to radiation hydrodynamics in high-energy-density physics is that by Mihalas and Weibel-Mihalas (1984). The material on these areas in Zel'dovich and Razier (1966) is also relevant and is very insightful. Unfortunately, Mihalas and Weibel-Mihalas (1984) is well known to be difficult to read. This may be in part because of the excellent connections it makes with the literature that existed when it was written. The author hopes that the present text, with its more pedagogical focus and simplified presentation, will provide a foundation that enables a better appreciation of their thorough discussions. Because of its emphasis on the essential physics, the present text has steered clear of the issue that occupies most of the time of most of the people working in radiative transfer and radiation hydrodynamics, which is the discovery and implementation of computer algorithms that can produce practical approximate solutions of radiative transfer problems, either in isolation or in the context of radiation hydrodynamics. There is some discussion of these issues in Mihalas and Weibel-Mihalas (1984). In addition, Castor (2004) has published the definitive text in this area in 2004.

We now turn to experiments and applications. The production of high-energy-density systems touches on a number of areas of research. Some of these have many more-detailed books while others have none. There are many books, for example, on lasers, no one of which seems uniquely relevant. Much less is available on what happens when the laser strikes the target, with the principal reference being the text on laser–plasma interactions by Kruer (1988). The book on Z pinches by Lieberman et al. (1999) has some material that is relevant to high energy density, and much that is not.

It is no surprise that more is available in the area of inertial fusion, as this has been the driving application. Two books in this area deserve specific mention. Atzeni and Meyer-Ter-Vehn (2004) published in 2004 an extensive book whose focus is inertial fusion. This book takes up most of the topics covered here, and a number of topics not covered here, all in the context of how they impact inertial fusion as an application and a goal. It is heavy on formulae, including parametric fits to many complex relationships, and comparatively light on discussion and explanations. As a result, it will be a very useful tool for experts. An earlier book by Lindl (1995), published soon after the declassification of fusion using hohlraums, has more of an engineering orientation. It addresses primarily the specific issues involved in producing fusion by this method, about which it includes much useful detail. It also includes a number of useful formulae based on studies by computer simulation. In a number of cases, we compare results from a fundamental and simplified analysis with those given in one of these two books.

1.7 Variables and Notation

One difficulty with work that involves contributions from various disciplines is conflict of variable definitions. Certain symbols find very common use to mean different things. For example, γ is used as the ratio of specific heatsthroughout fluid dynamics but is also used extensively as the symbol for a growth rate in instability theory and as the Lorentz factor in relativity. The goal here has been to use a notation that was consistent throughout the text while keeping close to standard usage where feasible. For this reason, the book makes extensive use of subscripts. Appendix A includes a list of variables used in the text, and of the symbols used for a number of common constants. This list is intended to include all variables that are used in more than one section and many of the dimensional variables. It may not include some variables that are used solely within the context of a single derivation or a single section of the text, especially when these are nondimensional. As discussed at more length in the preface, the intent in the equations has been to either ensure that an equation has evident and consistent units, so that any system of units can be used, or to give the units used explicitly. In nearly all cases, we will write equations involving a temperature as equations in energy units, representing the contribution of the temperature as $k_B T$, where k_B is the Boltzmann constant and T is the temperature.

In writing variables and equations, this book uses italic text for scalar quantities and boldface text for vector quantities. Tensors of the second rank are written either as underlined boldface symbols, such as $\underline{\boldsymbol{P}}$, or as dyadic notation, such as $\boldsymbol{uw}$. (The element in the ith column and jth row of a tensor written in dyadic notation are ($\boldsymbol{uw}_{ij} = u_i w_j$). In writing differential equations the derivatives are written as fractions rather than using a more compact notation.

Finally, geometric unit vectors are represented using a hat, as for example in $\hat{x}, \hat{y}, \hat{z},$ and $\hat{r}$.

References

Atzeni S, Meyer-Ter-Vehn J (2004) The physics of inertial fusion. Oxford, New York
Castor J (2004) Radiation hydrodynamics. Cambridge University Press, Cambridge
Chandrasekhar S (1961) Hydrodynamic and hydromagnetic stability. Dover, New York

Eliezer S, Ghatak A, Hora H (1986) An introduction to equations of state: theory and applications. Cambridge University Press, Cambridge

Gamaly E (2011) Femtosecond laser-matter interaction: theory, experiments and applications. Pan Stanford, Singapore

Gibbon P (2006) Short pulse laser interactions with matter. Imperial College Press, London

Griem H (1997) Principles of plasma spectroscopy. Cambridge University Press, Cambridge

Hinze JO (1959) Turbulence. McGraw-Hill, New York

Jackson JD (1999) Classical electrodynamics. Wiley, New York

Kruer WL (1988) The physics of laser plasma interactions. Addison-Wesley Publishing Company, Inc., Redwood City, CA

Landau LD, Lifshitz EM (1987) Statistical physics, course in theoretical physics, vol 5, 2nd edn. Pergamon Press, Oxford

Lieberman MA, De Groot JS, Toor A, Spielman RB (1999) Physics of high-density Z-pinch plasmas. Springer, Berlin/Heidelberg/New York

Lindl JD (1995) Development of the indirect-drive approach to inertial confinement fusion and the target physics basis for ignition and gain. Phys Plasmas 2(11):3933–4024

Mihalas D, Weibel-Mihalas B (1984) Foundations of radiation hydrodynamics, vol 1, Dover (1999) edn. Oxford University Press, Oxford

Reif F (1965) Fundamentals of statistical and thermal physics, 56946th edn. Waveland Pr Inc, Long Grove

Remington BA, Drake RP, Ryutov DD (2006) Experimental astrophysics with high power lasers and z pinches. Rev Mod Phys 78:755–807

Salzman D (1998) Atomic physics in hot plasmas. Oxford University Press, New York

Sedov LI (1959) Similarity and dimensional methods in mechanics, vol 1. Academic, New York

Seryi A (2015) Unifying physics of accelerators, lasers and plasma. CRC Press, Boca Raton, FL

Tennekes H, Lumley JL (1972) A first course in turbulence. MIT Press, Cambridge, MA

Whitham GB (1974) Linear and nonlinear waves. Wiley, New York

Zel'dovich YB, Razier YP (1966) Physics of shock waves and high-temperature hydrodynamic phenomena, vol 1, 2002nd edn. Dover, New York

Chapter 2
Descriptions of Fluids and Plasmas

Abstract This chapter begins by introducing the simplest equations that are useful in describing high-energy-density systems (the Euler equations). It proceeds toward more complex models, introducing a range of forces and energy transport mechanisms, and ultimately treating the energy content of the electrons, ions, and radiation independently. Along the way it discusses linearization and the dimensionless analysis of differential equations. It then introduces multi-fluid models and kinetic models. The chapter includes a brief discussion of simulations. It leaves the treatment of magnetized plasmas and flows to Chap. 10 and of most aspects of radiation to Chaps. 6 through 8.

Not long ago, one could say that 99% of the known universe is plasma. The recent discovery of dark matter and dark energy imply that this may no longer be true, but it will remain the case that 99% of the readily observed universe is plasma. The interstellar medium, stars, and more exotic compact objects are all composed of or surrounded by ionized matter. Without understanding something about plasmas, one cannot hope to understand the universe.

It is equally true that knowledge of plasmas is essential to high-energy-density physics. To reach pressures above a megabar at densities of a few times solid density or smaller requires temperatures large enough to ionize the matter. Thus, in most high-energy-density systems the matter is in the plasma state. In various contexts and regimes, this plasma may behave as a simple fluid, as an ideal plasma, or as a plasma beyond the scope of traditional plasma theories. In addition, plasma behavior is essential to the use of lasers to produce high-energy-density systems (Chap. 9). When the behavior of high-energy-density systems departs from that of simple hydrodynamic fluids, either plasma effects or radiation effects are typically responsible. In addition, the models of use in describing plasmas are supersets of those used to describe simpler fluids. We discuss various approaches to describing plasmas and fluids here. Radiation and radiation hydrodynamics are described in Chaps. 6 through 8.

If the reader has studied plasma physics, then parts of this chapter will be review. If the reader has studied electrodynamics or fluid dynamics, but not plasma physics as such, then other parts of this chapter will be review. In addition, our focus here is not at all to span fluid dynamics and plasma physics in a few pages, but rather to

R P. Drake, *High-Energy-Density Physics*, Graduate Texts in Physics,
https://doi.org/10.1007/978-3-319-67711-8_2

introduce the models we will need in the pages that follow. Moreover, we assume here that all motions are non-relativistic. Relativistic motions are considered in Chaps. 6 and 13.

2.1 The Euler Equations for a Polytropic Gas

In books on plasma physics, it is common to begin with collections of individual particles, to determine how to describe their behavior statistically using the Boltzmann equation, and then to average their behavior in ways that produce simpler models of plasma dynamics. This is not always possible in fluid dynamics or in high-energy-density physics, as some forces and energy transport mechanisms are not readily found by averaging over particle distributions. Here we take the reverse path, beginning with the very simple averaged equations that are useful in many high-energy-density contexts, and working our way toward more-complex descriptions that are more powerful but also less-often necessary. In this spirit, we begin with the Euler equations for a polytropic gas:

$$\frac{\partial \rho}{\partial t} + \nabla \cdot \rho \boldsymbol{u} = 0, \tag{2.1}$$

$$\rho \left(\frac{\partial \boldsymbol{u}}{\partial t} + \boldsymbol{u} \cdot \nabla \boldsymbol{u} \right) = -\nabla p, \text{ and} \tag{2.2}$$

$$\frac{\partial p}{\partial t} + \boldsymbol{u} \cdot \nabla p = -\gamma p \nabla \cdot \boldsymbol{u}, \tag{2.3}$$

where $\boldsymbol{u}$, ρ, and p are the velocity, density, and pressure, respectively. Here (2.1) is the continuity equation, describing conservation of mass, (2.2) is the equation of motion, derived from the conservation of momentum, and (2.3) is the energy equation, derived from the conservation of energy. Equation (2.3) assumes that the fluid is polytropic so $p \propto \rho^\gamma$ where γ is the *adiabatic index* (the ratio of specific heats). In their volume on fluid mechanics, Landau and Lifshitz (1987) report that "polytropic processes" is a historic term for processes in which pressure is proportional to some inverse power of volume. This is why a fluid or gas with $p \propto \rho^\gamma$ is described as a polytropic gas, and why γ is often called the *polytropic index* or *polytrope*. For a fully ionized nonrelativistic gas (at a high enough temperature and a low enough density) γ is equal to 5/3; for a gas where radiation pressure is dominant, γ is equal to 4/3; for a diatomic molecular gas, γ is equal to 7/5. We discuss more general versions of these equations in Sect. 2.3 and more complex models in Chap. 3 for circumstances when γ is not constant. In some of the later discussion, especially of shock waves, it will prove useful to have the energy equation in conservative form, discussed next. In this form the energy equation is

$$\frac{\partial}{\partial t}\left(\rho\epsilon + \frac{\rho u^2}{2}\right) = -\nabla \cdot \left[\rho u \left(\epsilon + \frac{u^2}{2}\right) + pu\right],\tag{2.4}$$

where the specific internal energy is ϵ. Other useful forms of the energy equation are discussed later in this section.

Although it is not evident at first glance, all three of these equations are at root continuity equations, in which the total amount of something is changed only by sources (sinks are negative sources). They all can be written in "conservative form", in which the change in the density, ρ_Q, of some quantity Q is determined by the flux of that quantity, $\boldsymbol{\Gamma}_Q$, having units of Q per unit area per unit time, and the net volumetric sources of that quantity, S_Q. The conservative form is then

$$\frac{\partial}{\partial t}\rho_Q + \nabla \cdot \boldsymbol{\Gamma}_Q = S_Q.\tag{2.5}$$

If one integrates such an equation over some volume with a surface, σ, and applies Gauss' law, one obtains

$$\frac{\partial}{\partial t}Q + \oint_\sigma \boldsymbol{\Gamma}_Q \cdot d\boldsymbol{A} = \text{Net source in volume},\tag{2.6}$$

in which the second term is the net flow of Q into or out of the volume and $\oint_\sigma$ represents the integral over the closed surface σ.

Equation (2.1) is often referred to as the continuity equation. It constrains the dynamics of a fixed amount of matter. One can recognize the second term on the left as the divergence of a flux. As a result, if one integrates this equation over a finite volume, the change of mass within the volume will equal the flow of mass into or out of the volume. If there were mass sources or mass sinks, these would appear on the right-hand side of (2.1). This equation, though simple, is a key factor in the complex behavior of hydrodynamic systems, because in many cases the variation of both ρ and u is important. This makes (2.1) an essentially nonlinear partial differential equation, not readily solved by any analytic technique.

Equation (2.2) is the momentum equation, or more accurately is an equation derived from the calculation of the rate of change of momentum density. This specific equation applies when electric and magnetic fields, viscous momentum transfer, and radiative forcing are all negligible. The only remaining momentum source is the pressure gradient, which causes compression or decompression of the plasma. Its effect is represented by the term on the right-hand side. The second term on the left arises from the convection of momentum density.

An important simple application of (2.1) and (2.2), for which an analytic solution is straightforward, is the description of small-amplitude, acoustic waves. We consider this here as an example of the technique called "linearization", because linearization will be important in numerous contexts later in the book. Linearization is possible when the variation of every variable in a problem can be described as a small deviation from a constant average value (which might be zero). The essence of linearization is the realization that terms that are linear in the small deviations are vastly larger than terms that are nonlinear in these quantities. Here we use the subscript o for the average values and the subscript 1 for the small deviations, and we also assume that p can be described as a function of ρ. Then with $\boldsymbol{u}_o = 0$, (2.1) becomes

$$\frac{\partial \rho_1}{\partial t} + \rho_o \nabla \cdot \boldsymbol{u}_1 + \boldsymbol{u}_1 \cdot \nabla \rho_1 + \rho_1 \nabla \cdot \boldsymbol{u}_1 = 0 \tag{2.7}$$

and (2.2) becomes

$$\rho_o \left(\frac{\partial \boldsymbol{u}_1}{\partial t} + \boldsymbol{u}_1 \cdot \nabla \boldsymbol{u}_1 \right) + \rho_1 \left(\frac{\partial \boldsymbol{u}_1}{\partial t} + \boldsymbol{u}_1 \cdot \nabla \boldsymbol{u}_1 \right) = -\frac{\partial p}{\partial \rho} \nabla \rho_1. \tag{2.8}$$

Then the nonlinear terms, with products involving ρ_1 and/or $\boldsymbol{u}_1$, can be discarded as small. To show this formally one should rework these equations so that every small quantity is expressed as a ratio that is actually small. This is left as a homework problem, as is the derivation from these two equations of the acoustic wave equation

$$\frac{\partial^2}{\partial t^2} \rho_1 - \frac{\partial p}{\partial \rho} \nabla^2 \rho_1 = 0. \tag{2.9}$$

Here the square of the sound speed, c_s, is $c_s^2 = \partial p / \partial \rho$, which equals $\gamma p / \rho$ for a polytropic gas. The definition of the sound speed squared is often expressed as the partial derivative of pressure with respect to density at constant entropy. In more detail, this partial derivative is taken according to the properties of the system under study. If the fluctuations are adiabatic, then it is taken at constant entropy. If rapid heat transport keeps the temperature constant, then it is taken at constant temperature, and so on.

It is also useful preparation for later analysis to discuss the solutions of (2.9). The standard method of finding the normal modes of a system described by this equation involves a decomposition of the spatio-temporal structure into plane waves, so that the parameters, such as $\rho_1(\boldsymbol{x}, t)$, are expressed as a sum or integral over plane waves whose amplitudes can be expressed as $\tilde{\rho}_1(\omega, \boldsymbol{k}) \times \exp[i(\boldsymbol{k} \cdot \boldsymbol{x} - \omega t)]$, in which the angular frequency of the oscillation is ω and its wavevector is $\boldsymbol{k}$ (whose magnitude k is related to the wavelength λ by $k = 2\pi / \lambda$). Throughout this book $\boldsymbol{x}$ and t are used as variables for position and time, with boldface indicating vector quantities. A formal solution also must include a consideration of how real physical quantities are to be related to the complex mathematics. (For a discussion of this last point see for example Chap. 6 of the electrodynamics text by Jackson (1999)). One can find the dispersion relation for acoustic waves either by considering a single plane wave, replacing ρ_1 by $\tilde{\rho}_1$, or by taking the Fourier transform of (2.9). One obtains

$$\omega^2 - c_s^2 k^2 = 0, \tag{2.10}$$

from which one sees that c_s is both the phase velocity (ω/k) and the group velocity $(\partial\omega/\partial k)$ of these waves. Note that the fluctuating amplitude $\tilde{\rho}_1$ cancels out of the dispersion relation. This quantity is thus not constrained by (2.1) and (2.2), until it becomes large enough that linearization becomes invalid.

An additional aspect of the behavior of acoustic waves of larger amplitude is worth mentioning. This is *acoustic wave steepening*, encountered in various fluid contexts but of unclear importance at high energy density. The crest of a large acoustic wave is at a higher density than the trough, and the sound speed is correspondingly larger. This has the effect that any given wave crest tends to overtake the trough ahead of it. This in turn causes a steepening of the "front" of the wave. In a system determined by the Euler equations, ordinary acoustic waves steepen until they become a series of smooth increases in the plasma pressure and density connected by abrupt decreases. In any actual plasma, finite viscosity will limit the steepening of the wave fronts.

Equation (2.3) explicitly describes the variation of the plasma pressure but in fact is the simplified equation obtained by calculating the rate of change of the energy density in the plasma. For any polytropic medium, the total internal energy density is proportional to the pressure and is given by $p/(\gamma - 1)$. In (2.3) the left-hand side describes the temporal and convective variation of the plasma pressure. In an incompressible fluid, for which $\nabla \cdot \boldsymbol{u}$ is zero, this is the entire story of energy conservation. In a compressible fluid, in contrast, the work done during compression or decompression, during which $\nabla \cdot \boldsymbol{u}$ is nonzero, is part of the flow of energy. An important additional point here is that (2.2) and (2.3) describe the pressure as a scalar quantity. In general the pressure is a tensor, $\boldsymbol{P}$, and what we write as ∇p would in general be the vector given by $\nabla \cdot \boldsymbol{P}$. In most circumstances in high-energy-density physics, the concept of an isotropic scalar pressure applies very well. We allow for the tensorial nature of pressure when including viscous effects in Sect. 2.3. Other cases where the pressure is not scalar include work with solids, which can sustain shear stresses and other asymmetric internal forces, and with well-magnetized plasmas, in which case the behavior along the magnetic field is fundamentally different than behavior perpendicular to it.

Equation (2.3) also has an important relation to the entropy of the plasma. To be specific, we consider a polytropic gas, although the conclusion is more general than this. The specific entropy (the entropy per unit mass) of a polytropic gas, s, can be expressed (Sedov 1959, p. 261) as

$$s = c_V \ln\left(\frac{p}{\rho^\gamma}\right) - c_V \ln\left(\frac{p_o}{\rho_o^\gamma}\right) + s_o, \tag{2.11}$$

in which s_o is the value of s in a reference state, for which $p = p_o$ and $\rho = \rho_o$, and c_V is the specific heat at constant volume. If one evaluates the total derivative of s, $Ds/Dt = \partial s/\partial t + \boldsymbol{u} \cdot \nabla s$, one finds

$$\frac{1}{c_V}\frac{Ds}{Dt} = \frac{1}{p}\frac{\partial p}{\partial t} + \frac{1}{p}\boldsymbol{u} \cdot \nabla p + \gamma\nabla \cdot \boldsymbol{u}. \tag{2.12}$$

Equation (2.3) then implies that $Ds/Dt = 0$, so one concludes that specific entropy is conserved across regions of time and space where (2.3) is continuously valid. Note that entropy is not conserved across transitions with a discontinuous change in the fluid parameters, such as shock waves. In addition, heat transport or other dissipative processes lead to a change of entropy. In simulations of hydrodynamic systems that include dissipative processes, one can evaluate their importance by examining the change in entropy of the fluid elements. The ways in which these equations apply to shock waves are discussed in Chap. 4. (Note that the total derivative, discussed at more length in texts on fluid dynamics, is the rate of change of some quantity within some specific parcel of fluid. It includes the inherent time dependence of the quantity and also the rate of change resulting from fluid motion in the presence of a spatial derivative.)

The energy equation can be expressed in other useful ways, which will matter for later chapters. If one expands (2.4) and collects all the terms involving derivatives of density, one finds that these terms sum to zero by the continuity equation. By taking the dot product of u with (2.2), one obtains an equation equivalent to conservation of mechanical energy in the plasma. (A complete equation for mechanical energy requires taking the dot product of u with the conservative form of the momentum equation.) Subtracting this from the energy equation, and substituting for $\nabla \cdot u$ from the continuity equation, one obtains what is sometimes called the gas-energy equation,

$$\left(\frac{\partial}{\partial t} + u \cdot \nabla\right)\epsilon - \frac{p}{\rho^2}\left(\frac{\partial}{\partial t} + u \cdot \nabla\right)\rho = \frac{D\epsilon}{Dt} - \frac{p}{\rho^2}\frac{D\rho}{Dt} = 0. \tag{2.13}$$

In the presence of energy sources or heat transport, the right-hand side of this equation would not be zero. For an ideal gas, with $\epsilon = p/[\rho(\gamma - 1)]$, this equation reduces to a particularly useful form:

$$\frac{Dp}{Dt} - c_s^2\frac{D\rho}{Dt} = 0. \tag{2.14}$$

In this last equation, one has multiplied the equation by $(\gamma - 1)$ to simplify it. This factor must be accounted for when sources of heat make the right hand side non-zero.

2.2 The Maxwell Equations

Many simplified fluid equations have some electrodynamic component. To understand these, we will need the Maxwell equations of electrodynamics. They are written here for reference and to allow some discussion of Gaussian cgs and other units. We write these equations assuming that the media are not inherently magnetized or electrically polarized, so that we can account explicitly for all charges and currents, and can take the polarization and magnetization fields to be zero. This is common in plasma physics but not in other areas where polarization electric field and material magnetization are useful concepts. We have the Poisson equation,

$$\nabla \cdot \boldsymbol{E} = 4\pi k_1 \rho_c, \tag{2.15}$$

the absence of magnetic monopoles,

$$\nabla \cdot \boldsymbol{B} = 0, \tag{2.16}$$

Faraday's law

$$\nabla \times \boldsymbol{E} = -k_3 \frac{\partial \boldsymbol{B}}{\partial t}, \tag{2.17}$$

and Maxwell's generalization of Ampere's Law

$$\nabla \times \boldsymbol{B} = \frac{k_2}{k_1 k_3} \frac{\partial \boldsymbol{E}}{\partial t} + 4\pi \frac{k_2}{k_3} \boldsymbol{J}. \tag{2.18}$$

Here ρ_c is the charge density and $\boldsymbol{J}$ is the current density. Here we have followed Jackson (1999) in expressing these equations in unit-independent form. In most applications in this text, the equations will be written in Gaussian cgs units, which turn out to produce convenient expressions for plasma phenomena. In such equations, B is in Gauss and other quantities are in cgs units. The constants are $k_1 = 1, k_2 = 1/c^2$, and $k_3 = 1/c$. In these cases one tends not to be interested in the electric quantities—few researchers actually use statvolts/cm as a unit of electric field but researchers working with cgs equations often need to express the electronic charge as 4.8×10^{-10} statcoul rather than as 1.6×10^{-19} C. The cgs unit of energy is the erg, which may be of use or may be converted to eV or keV ($1\,\text{eV} = 1.6 \times 10^{-12}\,\text{ergs} = 1.6 \times 10^{-19}\,\text{J}$).

Researchers who need to calculate magnetic fields typically work with the SI mks version of Ampere law:

$$\frac{1}{\mu_o} \nabla \times \boldsymbol{B} = \epsilon_o \frac{\partial \boldsymbol{E}}{\partial t} + \boldsymbol{J}, \tag{2.19}$$

in which $\mu_o = 4\pi \times 10^{-7}$ H/m, $\epsilon_o = 10^7/(4\pi c^2)$ F/m (Farads/m) with c in m/s, B is in Tesla, E is in V/m, and J is in A/m^2 (Amps/m^2). More generally, for SI mks units $k_1 = 1/(4\pi\epsilon_o)$ m/F $= 10^{-7}c^2$, $k_2 = \mu_o/(4\pi) = 10^{-7}$ H/m, and $k_3 = 1$.

One also needs the Lorentz force, which acts on any charge q with velocity $\boldsymbol{v}$ and is

$$\boldsymbol{F}_L = q \left(\boldsymbol{E} + \frac{\boldsymbol{v} \times \boldsymbol{B}}{c} \right)_{(\text{cgs})} = q\left(\boldsymbol{E} + \boldsymbol{v} \times \boldsymbol{B}\right)_{(\text{SI})}, \tag{2.20}$$

in which the units are designated by the text in parentheses.

We will also find it convenient to work with the vector potential, A, so that

$$B = \nabla \times A \tag{2.21}$$

and

$$E = -\nabla \Phi - \frac{1}{c} \frac{\partial A}{\partial t}, \tag{2.22}$$

in which the scalar potential is Φ, using the Coulomb gauge, so $\nabla \cdot A = 0$.

We will also need at times to work with the energy density and energy flux of electromagnetic field. In cgs units the energy densities W_E and W_B of the electric and magnetic fields, respectively, are

$$W_E = \frac{E^2}{8\pi} \tag{2.23}$$

and

$$W_B = \frac{B^2}{8\pi}. \tag{2.24}$$

Note that when one averages over many cycles of a fluctuating field to obtain an averaged energy density, the results equal these quantities are divided by 2. This is also the case for the Poynting flux, the energy flux carried by electromagnetic fluctuations,

$$S = \frac{v_g}{4\pi} E \times B, \tag{2.25}$$

in which the group velocity of the wave is v_g.

2.3   More General and Complete Single-Fluid Equations

Figure 2.1 shows an image of the Cygnus loop. This object, six times the size of the moon when viewed from the Earth, is the result of a supernova that occurred about 15,000 years ago. It features very thin, crenellated layers of matter. Spectroscopic imaging of the emission from different elements shows where they are produced. Analysis of these emissions has found that the various features are produced by shock waves and has revealed some of their properties. The Cygnus loop is of note here because it cannot be described using only the Euler equations. In this case, the process that is missing and that matters most is radiative heat transport. There are many cases in which one or more processes, beyond the interplay of pressure and

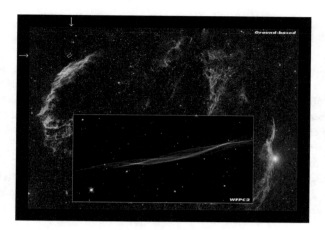

Fig. 2.1 The Cygnus loop supernova remnant. The background image is an optical image from a ground-based telescope. The inset, from the WFPC2 instrument on the Hubble Space Telescope, shows the very thin layer of emission (by hydrogen at 656.3 nm) produced by the upward-moving shock wave in the small box aligned with the arrows. The ground-based image of the Cygnus loop (shown in the background) measures 3° × 2° and was taken by CalTech with the Oschin Schmidt Telescope and scanned as part of the Digitized Sky Survey. Hubble Image credit: European Space Agency. http://origins.jpl.nasa.gov/library/story/101100-a_old.html

momentum, are essential to the behavior of a system of interest. Even so, one very often can ignore the fact that real plasmas include some combination of ions and electrons. A great deal of the behavior of plasmas, especially including high-energy-density ones, can be described by treating the plasma as a single fluid that can be charged, carry currents, and interact with radiation. In this section we discuss such single-fluid equations and a few specific limits of interest.

2.3.1   General Single-Fluid Equations

We discussed, with reference to (2.4), the general structure of transport equations. This structure still applies here. The more complex element is that other sources and fluxes of any given quantity are considered. The problems of concern in this book do not involve interior mass sources, so the continuity equation,

$$\frac{\partial \rho}{\partial t} + \nabla \cdot (\rho \boldsymbol{u}) = 0, \tag{2.26}$$

remains unchanged. The general transport equation for momentum, in the nonrelativistic limit, is

$$\rho \left(\frac{\partial \boldsymbol{u}}{\partial t} + \boldsymbol{u} \cdot \nabla \boldsymbol{u} \right) = -\nabla p + \boldsymbol{F}_{\mathrm{EM}} + \nabla \cdot \underline{\boldsymbol{\sigma}}_\nu + \boldsymbol{F}_{\mathrm{other}}. \tag{2.27}$$

Here (2.26) was used to simplify the left-hand side of (2.27); the more complex expression, involving the time dependence and divergence of the momentum density, is often more useful for computer simulations. The viscous stress tensor, $\underline{\boldsymbol{\sigma}}_\nu$, is further discussed below. The term designated $\boldsymbol{F}_{\mathrm{other}}$ represents the sum of all other forces, such as gravity. The radiation momentum density is not included, because it is insignificant except in relativistic systems.

The total force density due to the interaction of charges with the electromagnetic fields is designated as $\boldsymbol{F}_{\mathrm{EM}}$. The exact form of this term depends upon the regime. In the unlikely event that the fluid is charged, it may include the Lorentz force,

$$\rho_c \left(\boldsymbol{E} + \frac{\boldsymbol{u} \times \boldsymbol{B}}{c} \right),$$

in which ρ_c is the density of charge, and as always throughout this book the electric field is $\boldsymbol{E}$, the magnetic field is $\boldsymbol{B}$, and the speed of light is c. This form of the Lorentz force (with the c) is expressed in Gaussian cgs units. In the magnetohydrodynamic limit, the current density is $\boldsymbol{J}$, but the flow of current does not require motion of the single-fluid plasma, and the associated force is $\boldsymbol{J} \times \boldsymbol{B}/c$. The fluid flow velocity does not enter because the electrons carry nearly all the current but nearly none of the momentum. Similarly, the presence of a significant charge density does not require a significant accumulation of mass.

Under most conditions of interest throughout this book, we will represent $\boldsymbol{F}_{\mathrm{EM}}$ as $-\nabla p_R$, and will identify p_R as the radiation pressure. We discuss the definition and evaluation of p_R in Chap. 6. Here we note that p_R represents a sum over all modes present of their pressure, averaged over a time long compared with one cycle and short compared with hydrodynamic timescales. Using p_R is an effective way to capture the effect of broadband, incoherent thermal radiation on the plasma. In our typical regime where $\boldsymbol{F}_{\mathrm{EM}}$ includes $-\nabla p_R$ and possibly other terms, the total-energy equation is

$$\frac{\partial}{\partial t} \left(\rho \epsilon + \frac{\rho u^2}{2} + E_R \right) + \nabla \cdot \left[\boldsymbol{u} \left(\rho \epsilon + \rho \frac{u^2}{2} + E_R \right) + (p + p_R) \boldsymbol{u} \right] \tag{2.28}$$
$$= -\nabla \cdot \boldsymbol{H} - \boldsymbol{J} \cdot \boldsymbol{E} + (\boldsymbol{F}_{\mathrm{EM}} - \nabla p_R + \boldsymbol{F}_{\mathrm{other}}) \cdot \boldsymbol{u}.$$

The energy density of the radiation field, E_R, is usually ignorable and is also discussed further shortly. These equations can be derived either by taking moments of particle distribution functions or by reasoning about the behavior of small elements of fluid. Versions of these equations, including or excluding various specific source terms, can be found in any plasma physics or fluid dynamics text.

The divergence of the energy flux, $\boldsymbol{H}$, which enters (2.28), is

$$\nabla \cdot \boldsymbol{H} = \nabla \cdot \left[\boldsymbol{F}_R + \boldsymbol{Q} - \underline{\boldsymbol{\sigma}}_\nu \cdot \boldsymbol{u} \right], \tag{2.29}$$

in which the radiative energy flux is $\boldsymbol{F}_R$ and the energy flux from thermal heat conduction is $\boldsymbol{Q}$. The penultimate term in (2.28), $-\boldsymbol{J} \cdot \boldsymbol{E}$, describes the volumetric (Joule) heating by the currents driven by electromagnetic waves. This term is typically negligible except in any plasma penetrated by intense laser light.

We can obtain a gas-energy equation from (2.28) by subtracting from it the equation obtained by taking the dot product of $\boldsymbol{u}$ and (2.27), which is equivalent to an equation for conservation of mechanical energy density. This can then be cast in the form of (2.14), for a polytropic equation of state such that $(p + p_R) = (\widetilde{\gamma} - 1)(\epsilon + E_R/\rho)$, to obtain

$$\frac{D}{Dt}(p + p_R) - \frac{\widetilde{\gamma}(p + p_R)}{\rho} \frac{D\rho}{Dt} = (\widetilde{\gamma} - 1) \left[-\nabla \cdot (\boldsymbol{F}_R + \boldsymbol{Q}) + \underline{\boldsymbol{\sigma}}_\nu \cdot \cdot \nabla \boldsymbol{u} - \boldsymbol{J} \cdot \boldsymbol{E} \right], \tag{2.30}$$

in which $\underline{\boldsymbol{\sigma}}_\nu \cdot \cdot \nabla \boldsymbol{u}$, the viscous energy dissipation, can be expressed in component form as $\sigma_{ij} \partial u_j / \partial x_i$. When the energy density of the radiation is negligible, $\widetilde{\gamma}$ is just γ as introduced above. In this limit, and when viscosity and Joule heating are negligible, a form of the gas-energy equation that will be useful to us later is

$$\frac{Dp}{Dt} - \frac{\gamma p}{\rho} \frac{D\rho}{Dt} = -(\gamma - 1)\nabla \cdot (\boldsymbol{F}_R + \boldsymbol{Q}), \tag{2.31}$$

Complications that arise when γ is not constant are discussed in Chap. 3, and $\widetilde{\gamma}$, when radiation is significant, is discussed in Chap. 7.

We now turn to a discussion of the relative magnitudes of the various terms in (2.27) and (2.28). It helps develop understanding and intuition to discuss these equations while considering a dimensional analysis in which we identify a characteristic velocity of the system, U, and a characteristic dimension, L, which together give a timescale L/U. To make scaling arguments one replaces u by U, ∇ by $1/L$, and $(\partial/\partial t)$ by U/L. There is a sound reason for this. In any profile shaped as an exponential or linear function, the derivative is equivalent to division by whatever scale length is present in the profile. (In any power-law profile other than linear, the scale length is the distance variable itself, such as x or r, to within numerical factors.) These scale lengths are not the wavelength and frequency of local fluctuations but rather are the global scales that define the overall system evolution. Given this identification, one can say that the characteristic global, convective rate of change of momentum density and energy density are $\rho U^2/L$ and $\rho U^3/L$, respectively. If one divides any given equation by the relevant one of these, then one obtains a dimensionless equation from which one can assess the relative contributions of the various terms.

The use of a scalar pressure, p, is a simplification that is usually justified. The exceptions are systems involving solid-state or strongly magnetized matter. In this case ∇p must be replaced by the divergence of the pressure tensor. The material experiences forces, for example, when the force per unit area in one direction has a gradient in an orthogonal direction. The term involving pressure in the energy equation represents the work of compression or expansion, often referred to as pdV work. The fact that this work enters the energy equation in this way implies that the specific enthalpy, $h = \epsilon + p/\rho$, is often a useful variable in describing how hydrodynamic systems behave. If one takes the energy flux, H, to be zero and assumes the medium to be a polytropic gas, one can recover (2.3) from (2.28). To evaluate the dimensionless scaling of the pressure term, one takes $\nabla p \sim p/L$ and divides by $\rho U^2/L$ to find its normalized amplitude, which is $(p/\rho)/U^2$. This is proportional to the inverse of the internal Mach number, U/c_s, squared. Thus, pressure gradients are of decreasing importance as the internal Mach number increases.

The flow of heat is described in (2.29) as the divergence of a heat flux, Q. The heat flux is very important in the heating of plasma by laser light, and in some of the phenomena observed in plasmas produced from gases at a low enough (less than atmospheric) pressure. It is not important in the behavior of plasmas at near solid density or (for reasons discussed in Chap. 12) in the behavior of most astrophysical plasmas. In many cases, the heat flux can be related to the gradient in fluid temperature, T, using an equation of state to relate T to p or ϵ:

$$Q = -\kappa_{th}\nabla T, \tag{2.32}$$

in which the coefficient of heat conduction is κ_{th}. In a stationary fluid in which only the temperature variation is important and $\epsilon \propto T$, this yields a diffusion equation, $\partial T/\partial t \propto -\nabla^2 T$, so the heat transport from such a description is essentially diffusive. For scaling arguments, it is useful to identify and calculate the kinematic coefficient of thermal diffusivity, χ, which has purely the dimensions of a diffusion coefficient (step size squared/collision time). The relation between χ and κ_{th}, developed in Landau and Lifshitz (1987), is

$$\kappa_{th} = \chi\rho c_p = \chi n k_B \gamma/(\gamma - 1), \tag{2.33}$$

in which c_p is the specific heat at constant pressure and n is the density of particles in the fluid, each of which is part of a distribution with the common temperature T. The second equality gives the result for a polytropic gas whose pressure is described by Boyle's law. The Boltzmann constant is k_B, which can be combined with T to give $k_B T$ in energy units. In practical units, one has

$$\begin{aligned}
\chi(\text{cm}^2\,\text{s}^{-1}) &= 2 \times 10^{21} \frac{[T(\text{eV})]^{5/2}}{\ln \Lambda Z(Z+1)n_i(\text{cm}^{-3})} \\
&= 3.3 \times 10^{-3} \frac{A[T(\text{eV})]^{5/2}}{\ln \Lambda Z(Z+1)\rho(\text{g cm}^{-3})},
\end{aligned} \tag{2.34}$$

in which A and Z are the average atomic mass and ionic charge of the plasma ions, $\ln \Lambda$ is the Coulomb logarithm discussed in Sect. 2.4, and the particle and mass densities of the ions are n_i and ρ, respectively. These specific formulae are based on the analysis of processes dominated by Coulomb collisions in the book chapter by Braginskii (1965). Precise values in sufficiently dense plasmas might be different.

To evaluate the dimensionless scaling of the heat transport term, one finds $\nabla \cdot Q \sim \chi\rho(k_B T)/(Am_p L^2)$, divides by $\rho U^3/L$, and notes that $k_B T/(Am_p) \sim U^2$. The normalized amplitude of the heat transport term is thus the inverse of the *Peclet* number, $Pe = UL/\chi$. When Pe is large, heat transport can be neglected.

These equations include several terms describing the effects of radiation. Their derivation and more general forms are discussed in Mihalas and Weibel-Mihalas (1984). Here we define these terms and consider when they matter. In general, F_R is the radiative energy flux. F_R is in fact equal to S, the Poynting flux, given by $(E \times B)c/(4\pi)$ (when we explicitly account for all charges and the group velocity of the light is c), when S is evaluated for all the radiation present. However, in practical applications one uses the traditional form of S only when there are few waves in the problem, as for example in laser–plasma interactions. When there is broadband or line radiation from emission and absorption by dense plasma, one works instead with expressions for F_R that formally represent the integral of the Poynting flux, averaged over appropriate time and spatial scales. For example, the radiative flux emitted by a blackbody at a temperature T is σT^4. Fluids cool by emitting radiation. They emit blackbody radiation when they are sufficiently opaque. Otherwise, their cooling is often dominated by emission from atomic lines. We discuss this further in Chap. 6. The radiative energy flux is often significant in high-energy-density experiments. Note that $-\nabla \cdot F_R$ is the net rate of absorption, per unit volume, of radiative energy by the fluid. In Chap. 9 we will consider cases in which the absorption of laser light or the absorption of X-rays are important.

The terms involving E_R and p_R are important much less often, and it is easy to show why. When the radiation and the fluid are in equilibrium with a temperature T, then one has

$$p_R = E_R/3 = 4\sigma T^4/(3c). \tag{2.35}$$

The ratio of radiation pressure to plasma pressure is of order

$$\frac{4m_p \sigma T^4}{3c\rho k_B T} = 0.05T^3/\rho, \tag{2.36}$$

in which the proton mass is m_p and on the right-hand side T is in keV and ρ is in g/cm^3. Plastics often have densities of ~ 1 g/cm^3, as does water, so one can see from (2.36) that temperatures above 1 keV are required for radiation pressure and energy to be important in the fluid dynamics. Radiation pressure is dominant over material pressure in the shocked material in supernovae (at somewhat lower density). The readers of this book may well be producing and studying radiation-dominated plasmas using facilities of the twenty-first century.

Let us more formally explore the dimensionless scaling of the radiative terms. The normalized radiation pressure term, for radiation in equilibrium with the fluid, is $p_R/\rho U^2$, which is approximately the same ratio as in (2.36) (with $k_B T/m_p \sim U^2$). The energy flux term in (2.29) is larger than the enthalpy ($p_R + \epsilon$) term by roughly U/c, and has a normalized value of $\sigma T^4/(\rho U^3) \sim m_p \sigma T^4/(\rho k_B TU) \sim 1/Bo$, in which Bo is known as the Boltzmann number (see Mihalas and Weibel-Mihalas 1984) and is small when the energy flux due to radiation affects the dynamics significantly. Note that $1/Bo$ is c/U times larger than the ratio in (2.36), which reflects the fact that radiative energy fluxes become significant at temperatures much lower than those required for radiative pressures to be significant. We give specific examples of this in Chaps. 6 through 8.

In some systems, the relative importance of radiation may need to be evaluated by other measures. One can construct a radiation Peclet number when one can identify a (kinematic) radiative thermal conductivity, $\chi_r \sim \ell c$, where the mean free path ℓ might be due to bremsstrahlung interactions, to Compton scattering, or to atomic emission and absorption. Alternatively, one can compare the radiative cooling time, defined as the ratio of energy content to blackbody energy flux, to the hydrodynamic time, L/U. There is a further discussion of these points in Ryutov et al. (1999).

In general, fluids also possess internal friction. The collisions of the particles in the fluid resist its motion, a process known as viscosity. These effects are generally small for plasmas, but we will see when we discuss turbulence in Chap. 5 that they can have important consequences for the structures that develop. In general, collisional viscous effects create forces in a given direction due to gradients in velocity in orthogonal directions. This may be easiest to see by imagining a simple shear layer, in which the velocity is entirely in a direction we label as z, but there is a gradient in velocity in the orthogonal direction, x. When collisions move particles in the x direction, they cause a net transport of momentum. This creates a force. The elements of the viscous stress tensor are

$$\sigma_{vij} = \rho \nu \left(\frac{\partial u_i}{\partial x_j} + \frac{\partial u_j}{\partial x_i} - \frac{2}{3} \delta_{ij} \frac{\partial u_k}{\partial x_k} \right) + \zeta \delta_{ij} \frac{\partial u_k}{\partial x_k}, \qquad (2.37)$$

in which δ_{ij} is the Kronecker delta, the kinematic viscosity is ν, and the second coefficient of viscosity (often ignorable) is ζ, also known as the bulk viscosity. In the usual case that particulate viscosity dominates, the kinematic viscosity is approximately the mean free path squared divided by the collision time; the quantity $(\rho \nu)$ is the *dynamic viscosity*.

In vector notation, the viscous stress tensor is

$$\underline{\sigma}_v = \rho \nu \left(\nabla u + (\nabla u)^T - \frac{2}{3} (\nabla \cdot u) \underline{I} \right) + \zeta (\nabla \cdot u) \underline{I}, \qquad (2.38)$$

in which $\underline{I}$ is the identity tensor and the superscript T designates the transpose. Most theories of turbulence are developed for incompressible fluids, which have $\nabla \cdot u = 0$. In other cases, such as the damping of acoustic waves, the compressible terms are essential. The gradient of the viscous stress is experienced by the fluid as a force density.

The presence of the viscous stress also contributes to the energy content of the fluid. Energy is transported as the stressed fluid moves. The contribution of viscosity to the increase in total energy, $\nabla \cdot (\underline{\sigma}_\nu \cdot \boldsymbol{u})$, includes both contributions to the rate of increase of mechanical energy, which is $\boldsymbol{u} \cdot \nabla \cdot \underline{\sigma}_\nu$ and to the rate of increase of internal energy, $\underline{\sigma}_\nu \cdot \cdot \nabla \boldsymbol{u}$. Interested readers can find derivations of these results in graduate textbooks on fluid dynamics.

In practical units, the kinematic viscosity for unmagnetized plasmas (see Braginskii 1965) is given by

$$
\begin{aligned}
\nu(\text{cm}^2\,\text{s}^{-1}) &= \frac{1.9 \times 10^{19}}{\ln \Lambda \sqrt{A} Z n_i(\text{cm}^{-3})} \left[\frac{[T_i(\text{eV})]^{5/2}}{Z^3} + \frac{0.013[T_e(\text{eV})]^{5/2}}{\sqrt{A}} \right] \\
&= \frac{3.1 \times 10^{-5}}{\ln \Lambda Z \rho(\text{g cm}^{-3})} \left[\frac{\sqrt{A}[T_i(\text{eV})]^{5/2}}{Z^3} + 0.013[T_e(\text{eV})]^{5/2} \right].
\end{aligned}
\tag{2.39}
$$

Here the definitions are those used for (2.34). The viscosity is dominated by the ions for Z below about 6. In plasmas of a high enough temperature, the photon viscosity can be important (in this case one adds the kinematic viscosities). The kinematic photon viscosity (Jeans 1926; Thomas 1930) is

$$
\nu_{\text{rad}}(\text{cm}^2\,\text{s}^{-1}) \approx \frac{\bar{l} c \sigma T^4}{\rho c^3} = 3 \times 10^{-9} \frac{A[T(\text{eV})]^4}{Z[\rho(\text{g cm}^{-3})]^2}.
\tag{2.40}
$$

Here $\bar{\ell}$ is the photon mean-free path and other quantities have their standard definitions.

To evaluate the dimensionless scaling of the viscous effects in the momentum equation, one takes $\nabla \cdot \underline{\sigma}_\nu \sim \rho \nu U/L^2$ and divides by $\rho U^2/L$ to find the normalized amplitude, which is $1/Re$, where Re is the Reynolds number, $Re = UL/\nu$. (The viscous effects in the energy equation have this same scaling, as the extra factor of U from the viscous term in the energy equation divides out when one normalizes.) The Reynolds number is perhaps the most well-known dimensionless parameter, because it has proven very useful in characterizing qualitative regimes of turbulent behavior. In practical units, one has

$$
Re = 5.4 \times 10^{-20} n_e UL \frac{AZ^3 \ln \Lambda}{\sqrt{A} T_i^{5/2} + 0.013 Z^3 T_e^{5/2}},
\tag{2.41}
$$

with temperatures in eV and other quantities in cgs units. When Re is large, viscous effects can be ignored in (2.27)–(2.29). However, turbulence phenomena may inherently involve viscous dissipation on some smaller scale, a topic discussed further in Chap. 5.

The momentum and energy equations also include the electromagnetic force F_{EM} and the plasma heating that results. The divergence of the radiative flux, $-\nabla \cdot F_R$, which was already discussed, is the heating related to the $J \times B$ force. The electrostatic force, $\rho_q E$, produces volumetric heating that can be expressed as $J \cdot E$. In the simple case that the current is resistive and given by $\eta J = E$, where η is the resistivity, this power dissipation is also resistive and of magnitude ηJ^2. The heating need not be resistive in the general case, however. For problems with an electromagnetic component, one finds J, E, and B, in addition to the fluid quantities, by solving the Maxwell equations (see Sect. 2.3) in addition to the single-fluid equations, often in the simplified form of the magnetohydrodynamic equations discussed in Chap. 10. The dimensionless parameter that in most circumstances relates to the scaling of these forces is the magnetic Reynolds number, also discussed there.

2.3.2 Single Fluid, Three Temperature Models

This class of models introduces an additional element of complexity that often is essential for modeling systems at high energy density. In principle one can identify a distinct "temperature" for the electrons, the ions, and the radiation. "Temperature" is in quotes here because this concept is routinely abused in practice in comparison to its pure definition in thermodynamics or statistical mechanics. The meaning of temperature in routine practice is "the value of the temperature of an equilibrium thermodynamic system that would have the same mean energy as that of the actual system being described." The actual system, which might be an energy distribution of electrons or photons, typically is not in equilibrium and very often has an energy spectrum that departs significantly from the equilibrium energy spectrum. Identifying three temperatures in a plasma is a particularly paradoxical action, because the thermodynamic definition of temperature only strictly applies when they are all equal. Nonetheless, the "three-temperature" description of a single-fluid plasma is particularly useful, especially for computer simulations.

It is accurate to employ the single-fluid Euler equations with a single temperature when two conditions apply: radiation must be either negligible or dominant, and if radiation is negligible then the collisional coupling of the electron and ion temperatures must be strong. Under most circumstances in high-energy-density systems, the electrons are very strongly coupled to the ions by collisions, having the same temperature and a local density that is $n_e = Z n_i$, where n_i is the ion density and Z is the average charge. For any given density, as the temperature of a system increases the coupling of the electrons and the ions decreases. One of the first places one sees this is at a shock front, because shock waves directly heat the ions. The ion energy is shared with the electrons by collisions, and the electrons are in turn the primary source of radiation. Figure 2.2 shows how the ion temperature can deviate from the electron temperature at a shock front, after which collisions equalize the two over some distance.

Fig. 2.2 At shock fronts, the ion, electron, and radiation temperatures may differ significantly. The figure shows results from a multi-temperature, single-fluid simulation which a shock wave is driven at ~260 km/s through xenon gas of density 0.006 g/cm³. The ion temperature is shown in gray, the electron temperature is a solid black curve, and the radiation temperature is the dashed curve. The shock front, where the ion temperature increases, would be much steeper in the corresponding physical system

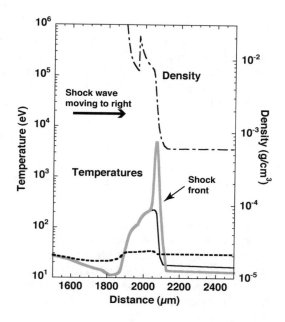

In a three-temperature, single-fluid description, (2.1) and (2.2) are unchanged, although the pressure must now be determined by adding the contributions from the three species—electrons, ions, and radiation—independently. One then replaces (2.3) with one equation for each species. For each species, the temporal and convective rate of change of temperature (or perhaps energy density) are equal to the terms involving the sources of energy from the spatial flow of heat within the species, from exchanges of energy with other species, and from any external sources.

2.4 Multi-Fluid Models

One encounters some phenomena that have timescales shorter than those required for the validity of any single-fluid model. The next level of higher complexity is found in multi-fluid models. In high-energy-density physics, these models are needed to describe phenomena in which the independent, rapid motions of the electrons are important. Principal among these is the interaction of laser light with the plasma it penetrates. The resulting processes, discussed in Chap. 9, can reflect a large majority of the laser light. Multi-fluid models enable one to derive many of the basic phenomena occurring during laser-plasma interactions. Here we will explore these equations. We will discuss multi-fluid models, the electron plasma oscillations that occur in most plasmas, and the scaling of collisional coupling.

As density decreases or temperature increases, the collisional coupling of the electrons and ions becomes smaller. Eventually the electrons and ions begin to act independently, and some phenomena appear in which there are important differences in their densities. Most of these phenomena can be successfully described using continuity and momentum equations like the following for each of the species in the plasma:

$$\frac{\partial n_j}{\partial t} + \nabla \cdot (n_j \boldsymbol{u}_j) = 0 \text{ and} \tag{2.42}$$

$$m_j n_j \frac{\partial \boldsymbol{u}_j}{\partial t} + m_j n_j \boldsymbol{u}_j \cdot \nabla \boldsymbol{u}_j = n_j q_j \left(\boldsymbol{E} + \frac{\boldsymbol{u}_j}{c} \times \boldsymbol{B} \right) - \nabla p_j + \sum_l \boldsymbol{R}_{jl}, \tag{2.43}$$

in which the subscript j (often e for electrons and i for ions) designates the species, $\boldsymbol{E}$ and $\boldsymbol{B}$ are the electric and magnetic fields, respectively, and $\boldsymbol{R}_{jl}$ is the rate of momentum density exchange between species j and l, discussed further shortly. Equation (2.42) is obvious, but there are some new features in the right-hand side of (2.43) by comparison to (2.27). The electromagnetic effects now appear in the form of the complete Lorentz force density, $nq(\boldsymbol{E} + \boldsymbol{u} \times \boldsymbol{B}/c)$ in Gaussian cgs units. As a result, and unlike the case of the MHD theory, the two-fluid theory describes phenomena in which there is a dynamic or static electric field. Examples include plasma oscillations, discussed below, and the Debye sheath that forms at bodies immersed in plasmas.

The final term on the right-hand side describes the momentum exchange with other species. The sum is over the other species in the plasma, designated by l. In an ideal plasma, which may include low-density plasmas produced by lasers or other devices, one can write this term as a drag term:

$$\boldsymbol{R}_{jl} = m_j n_j (\boldsymbol{u}_j - \boldsymbol{u}_l) \nu_{jl}, \tag{2.44}$$

in which the rate of momentum exchange through interaction of species j and the other species l is ν_{jl}. We discuss this rate further below. In more general cases, important in denser plasmas, which may carry current or have temperature gradients and may be magnetized, one has for the electrons specifically

$$\boldsymbol{R}_{el} = n_e e \underline{\boldsymbol{\alpha}} \cdot \boldsymbol{J} - n_e \underline{\boldsymbol{\beta}} \cdot \nabla T_e, \tag{2.45}$$

for current density $\boldsymbol{J}$ and electron temperature T_e. The tensor coefficients $\underline{\boldsymbol{\alpha}}$ and $\underline{\boldsymbol{\beta}}$ are discussed further in Chap. 10.

Successful analysis using these equations depends upon having a qualitative sense of the differences between electron and ions. (I have yet to see a student in a qualifying exam who did not know the approximate ratio of electron to ion mass, but I regret to report that I did encounter one student who seemed to have no sense of the implications of this. He did not pass.) Two very important points are that the electrons nearly always move much faster than the ions, but the momentum of an

ion is huge compared to that of an electron. A related point is that the radius of the
ion orbits in a magnetic field is much larger than that of the electron orbits.

2.4.1   Electron–Plasma Waves

A simple implication of (2.42) and (2.43) is the presence of electron plasma waves
in plasmas with weak enough collisions. Suppose that the final term in (2.43) can be
ignored and that we are looking for very fast, fluctuating phenomena so that the ion
density can be assumed to be fixed and unvarying. Also suppose there is no ambient
E or B. Then by linearizing these two equations, taking the partial derivatives in
time of the first and the divergence of the second, then simplifying, we find

$$\frac{\partial^2 n_{e1}}{\partial t^2} = \frac{n_{eo}e}{m_e}\nabla \cdot E_1 + \frac{1}{m_e}\nabla^2 p_{e1}. \tag{2.46}$$

Here again, the subscript 1 designates the small-amplitude modulations while o
designates the averaged, zero-order quantities.

This particular equation helps one see the physics of the wave we are finding. It
is a purely longitudinal wave like an acoustic wave, in which the fluctuating electric
field and compression by the electron pressure both cause the electron density
to vary. The first term on the right-hand side can be evaluated from the Poisson
equation (Sect. 2.2), which gives in this case

$$\nabla \cdot E_1 = 4\pi(Zen_{io} - en_{eo} - en_{e1}), \tag{2.47}$$

in which the first two terms in parentheses cancel because the plasma is quasi-
neutral. Then assuming the electrons behave as a polytropic gas with index γ_e, we
obtain a wave equation

$$\left(\frac{\partial^2}{\partial t^2} + \omega_{pe}^2 - \frac{\gamma_e p_{eo}}{n_{eo}m_e}\nabla^2\right)n_{e1} = 0, \tag{2.48}$$

in which we have introduced the electron plasma frequency,

$$\omega_{pe} = \sqrt{4\pi e^2 n_{eo}/m_e} = 5.64 \times 10^4 \sqrt{n_{eo}} \text{ rad/s}, \tag{2.49}$$

with n_{eo} in cm^{-3} and using Gaussian cgs units. Equation (2.48) describes waves
known as electron–plasma waves. (For historical reasons, they are also known as
"Langmuir waves" and as "Bohm-Gross waves.") By comparison with the derivation
of acoustic waves in Sect. 2.1, one can see that the pressure term in (2.48) will
introduce terms involving the wavenumber into the dispersion relation. For high-
frequency, plane waves which involve adiabatic compression of the electrons only
along k, there is only one degree of freedom and $\gamma_e = 3$. This result can also be

confirmed using kinetic theory (Sect. 2.5). Then with $p_{eo} = n_{eo}k_B T_e$, where T_e is the electron temperature, one finds from 2.4.17 a dispersion relation

$$\omega^2 - \omega_{pe}^2 - 3\frac{k_B T_e}{m_e}k^2 = 0. \tag{2.50}$$

Equation (2.50) is generally known as the Bohm–Gross dispersion relation. In the limit that T_e or k are small, one obtains the so-called cold-plasma oscillations, with $\omega = \omega_{pe}$. This emphasizes that plasma with weak collisions tend to sustain oscillations at $\omega \sim \omega_{pe}$. The discussion here is introductory. A lot more can be said about electron plasma waves and their interactions with other waves. We will take up some of these effects in Chap. 9. The reader who needs to work with these waves seriously should consult plasma-physics books on their damping and laser–plasma-interactions books on their interactions.

2.4.2 Ion Acoustic Waves

A slightly more complex, but also quite important, application of (2.42) and (2.43) is that of waves featuring ion-density modulations. These are sound waves in the traditional sense. Because they must be distinguished from the electron plasma waves, which are also longitudinal waves driven by pressure relaxations, these waves in plasma are most often referred to as ion-acoustic waves. The ions move much more slowly than the electrons, thanks to the enormous difference in mass. As a result, the electrons remain in steady state relative to the ions, so that the linearized version of (2.43) for the electrons implies

$$n_{eo}e\mathbf{E_1} = -\nabla p_{e1}. \tag{2.51}$$

The physical situation is that the electrons seek to escape the ion-density maxima, establishing an electric field that holds them in place. A simple calculation can show that the difference between the densities of ion and electron charge required to do this is negligible.

Manipulating (2.42) and (2.43) just as one does to produce (2.46) and using (2.51) to substitute for $\mathbf{E}_1$, one obtains

$$\frac{\partial^2}{\partial t^2}\frac{n_{i1}}{n_{io}} = \frac{1}{n_{io}M}\nabla^2(p_{e1} + p_{i1}) = \left(\frac{\partial p_i}{\partial \rho_i} + \frac{Zm_e}{M}\frac{\partial p_e}{\partial \rho_e}\right)\nabla^2\frac{n_{i1}}{n_{io}}. \tag{2.52}$$

Note that this equation is substantially similar to (2.9), with fluctuations in total pressure driving fluctuations in ion density. The final term here shows the division of the pressure derivative into electron and ion parts, indicated by subscripts, with ionic charge Z, ion mass M, and electron mass m_e. The partial derivatives are rather tricky but can be expressed in terms of a polytropic index γ such that $\partial p/\partial \rho =$

$\gamma p/\rho$ for either species. Definitive results for the evaluation of γ can come from the kinetic theory described in the next section. Under most conditions, the fluid-dynamics behavior is that the plane-wave modulations of the ions involve motion in one dimension, so $\gamma_i = 3$ (see Chap. 3), while the electrons have plenty of time for heat conduction to maintain a constant temperature, so that $\gamma_e = 1$. This gives an expression for the sound speed that we will use in Chap. 9,

$$c_s = \sqrt{\frac{Zk_BT_e + 3k_BT_i}{M}}. \tag{2.53}$$

2.4.3 Collisions in Plasmas

We now return to the final term in (2.43) and discuss collisional momentum exchange between species. Note that this term gives a rate of change of momentum that is measured with respect to the momentum of the designated species. As a result, the coefficient ν_{jl} is not symmetric in the exchange of j and l. This is trivial to visualize, if one imagines for example that one throws bowling balls into a room full of bouncing ping–pong balls. The effect of the ping–pong balls is to make tiny, and perhaps negligible changes in the momentum of the bowling balls relative to their initial momentum. In contrast, the bowling balls make enormous changes in the momentum of those ping–pong balls they interact with. Mathematically, with b for bowling balls and p for ping–pong balls, one can see that $\nu_{bp} \ll \nu_{pb}$. Similarly, for electrons and ions in (2.43), $\nu_{ie} \ll \nu_{ei}$. In fact, the final term in (2.43) is nearly always negligible in the ion equation, but often important in the electron equation. Let us consider further ν_{ei}, which enters the electron equation.

First recall some of the fundamental relations involving collisional interactions. If particles of type a and density n_a, having a single, fixed relative velocity, $v_{ab} = |\boldsymbol{v}_a - \boldsymbol{v}_b|$, are interacting with particle of type b and density n_b, and the interaction cross section at this velocity is σ_{ab}, then the mean free path for this interaction is

$$\lambda_{mfp} = 1/(n_b\sigma_{ab}), \tag{2.54}$$

the interaction time is $1/(n_b\sigma_{ab}v_{ab})$, and the interaction rate is

$$\nu_{ab} = n_b\sigma_{ab}v_{ab}. \tag{2.55}$$

In many cases, including the one of interest here, the interaction cross section depends upon v_{ab} and v_{ab} is not fixed. In this case, describing the distribution in velocity of the two species by distribution functions $f_a(\boldsymbol{v}_a)$ and $f_b(\boldsymbol{v}_b)$, normalized to unity so that e.g. $\int f_z(\boldsymbol{v}_a)d\boldsymbol{v}_a = 1$ with the integral over all velocities, one has in general

$$v_{ab} = n_b \int \int f_b(\boldsymbol{v}_b) f_a(\boldsymbol{v}_a) \times \sigma_{ab} \left(|\boldsymbol{v}_a - \boldsymbol{v}_b| \right) \times |\boldsymbol{v}_a - \boldsymbol{v}_b| \mathrm{d}\boldsymbol{v}_a \mathrm{d}\boldsymbol{v}_b. \qquad (2.56)$$

Next consider some of the specific properties of Coulomb collisions in plasmas. They involve the interaction of particles in the presence of an inverse-square-law force. This is Rutherford scattering, with results that are typically derived in either classical mechanics or plasma-physics texts. The force between two isolated particles extends to infinity, but the presence of other particles creates a shielding effect, so that the collision only has an effect until the particle separation reaches a distance comparable to this shielding distance. The net result is that the cross section for momentum transfer is

$$\sigma_{ab} = 4\pi \ln \Lambda \left(\frac{q_a q_b}{m^* v_{ab}^2} \right)^2, \qquad (2.57)$$

in which m* is the reduced mass $m_a m_b / (m_a + m_b)$ and $\ln \Lambda$ is the Coulomb logarithm, which accounts for the effects of shielding. We refer the reader to any plasma-physics text for a partial discussion of $\ln \Lambda$, and to the book by Shkarofsky et al. (1966) for a complete one. Such discussions are lengthy, as a number of factors must be considered. In addition, in high-energy-density systems the shielding distance often becomes so small that $\ln \Lambda$ approaches its limiting small value of order 1. In high-energy-density research, it is generally sufficient to take

$$\ln \Lambda = \mathrm{Max} \left[1, \{ 24 - \ln \left(\sqrt{n_e}/T_e \right) \} \right], \qquad (2.58)$$

with n_e in cm^{-3} and T_e in eV.

The most important point about (2.57) is that the cross section is proportional to $1/v_{ab}^4$, so that the contribution to the overall rate at each velocity is proportional to $1/v_{ab}^3$. Thus, pairs of particles having low relative velocities dominate the effects of Coulomb collisions, and high-velocity particles contribute little. One consequence of this is that all Coulomb processes, from momentum exchange to ionization or excitation by electron impact, become much weaker as the plasma temperature increases.

The net result of the integral in (2.56), giving the change of electron momentum by interaction with ions, when evaluated for Maxwellian distributions of particles, is

$$v_{ei} = \frac{1}{3(2\pi)^{3/2}} \frac{Z \omega_{pe}^4}{n_e v_e^3} \ln \Lambda = 3 \times 10^{-6} \ln \Lambda \frac{n_e Z}{T_e^{3/2}} \ (1/\mathrm{s}), \qquad (2.59)$$

in which $v_e = \sqrt{k_B T_e / m_e}$ and on the right n_e is in cm^{-3} and T_e is in eV. This traditional way of writing v_{ei} may tend to obscure the fact that it is a rate for a binary collision process between electrons and ions. The rate of momentum change for the electrons involves the product of v_{ei} and n_e, but fundamentally this rate must be proportional to $n_e n_i$ times a rate coefficient. From this perspective it would be

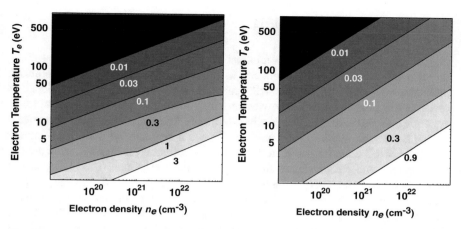

Fig. 2.3 Curves of constant collisionality, ν_{ei}/ω_{pe}, as labeled, for low-Z (left) and high-Z (right) plasmas

physically more transparent to replace $n_e Z$ by $n_i Z^2$ in (2.59). This matters in plasmas with multiple ion species, in which it is the average of Z^2, not Z, that determines the average collision rate. A typical way to account for this across plasma physics is to define an effective charge as $Z_{eff} = \langle Z^2 \rangle / \langle Z \rangle$, where $\langle \rangle$ denotes the average over all ion species, and to replace Z in (2.59) by Z_{eff}.

Figure 2.3 shows contours of constant ν_{ei}/ω_{pe} as a function of n_e and T_e. Two different evaluations of Z are used. Part (a) shows results for $Z = 3.5$, typical of low-Z materials such as plastic. Part (b) shows results for $Z = 0.63\sqrt{T_e}$, typical of higher-Z materials as is discussed in Chap. 3. Wherever this quantity exceeds 0.1, any electron plasma wave will damp within ten cycles, i.e., on femtosecond timescales. (The drag term at the end of (2.43) introduces a term proportional to $\partial n/\partial t$ into the wave equation, which in turn introduces an imaginary term into the dispersion relation so that the implied frequency is no longer purely real, which leads to damping.) One can see that under most high-energy-density conditions this damping is very strong.

One is at times interested in the collisional mean free path. One may need for example to assess whether heat transport might matter or to compare the size of a computational zone to this distance. This can be estimated as

$$\lambda_{mfp} = v_e / \nu_{ei} = (1/\nu_{ei})\sqrt{k_B T_e/m_e}. \tag{2.60}$$

2.5   The Kinetic Description

Figure 2.4 shows the supernova remnant SNR 1006. At the edges of this image one can see the shock wave produced as the disturbance caused by the remnant propagates outward into the interstellar medium. As a first level of description, one can treat this object and its shock wave as a hydrodynamic structure using the Euler equations, as we discuss further in Chaps. 4 and 5. And there are weaker shocks and other phenomena within the solar system that can be modeled with fair accuracy using the MHD equations. But in fact these are collisionless shocks. Structures in the magnetic and electric fields are essential to their existence. The electrons and ions interact with them very differently. In addition, there is a third group of particles that often are analyzed as a separate species—the energetic ions accelerated by such shocks, some of which eventually become cosmic rays. One could obtain a better description of this system by using a multi-fluid model, but such a model will still not manage to accurately calculate the shock structure, which generates strongly non-Maxwellian particle distributions. In this case, one needs to use some sort of kinetic model. We very briefly introduce this approach here.

All of the above equations are strictly correct only if the velocity distributions of all the particles are Maxwellian. This means that the number of particles of species s within an interval dv around v is given by

$$f_s(v) = \left(\frac{m_s}{2\pi k_B T_s} \right)^{3/2} \exp\left(\frac{-m_s v^2}{2 k_B T_s} \right), \tag{2.61}$$

in which the temperature and mass of species s are T_s and m_s, respectively. Any velocity distribution must be properly normalized. In this case the normalization is

Fig. 2.4 SNR 1006. X-ray image from the ROSAT satellite, the remnant of the widely observed supernova from AD 1006. The image convolves emission in several energy bands. The brighter emission at the upper left and lower right is attributed to cosmic ray acceleration. Image courtesy of University of Leicester, X-ray Astronomy Group http://wave.xray.mpe.mpg.de/rosat/calendar/1997/jul

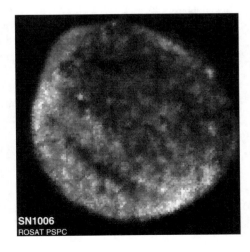

$$\int f_s(\boldsymbol{v})\mathrm{d}\boldsymbol{v} = 1. \qquad (2.62)$$

The reader should note that it is common for $f_s(v)$ to be normalized to 1, as shown here, or to the particle density, n_s. In the literature the specific normalization is often not defined; one can even find papers that switch normalizations in the course of their work. If the particle distributions are not Maxwellian, but the interactions of interest are determined by the average energy that they carry, then the fluid and other equations above are accurate, though perhaps with some changes in the value of some coefficients and certainly with a nonthermodynamic definition of "temperature", as discussed in Sect. 2.3.2.

In addition, there are circumstances in which it is an energetic "tail" on the distribution function that produces the phenomena of interest. At low energies the distribution usually has a Maxwellian shape. Whenever there are waves or instabilities that affect the particles, one frequently sees a surplus of particles at energies above the thermal energy. This is the *tail*. Laser–plasma instabilities, discussed in Chap. 9, typically produce such exponential tails (though not for reasons that are well understood). In space systems and astrophysics, one typically encounters power-law tails. An important example of this is the distribution of cosmic rays, whose flux falls as $1/v^3$, implying a distribution function scaling as $1/v^4$. In other cases of interest, such as the transport of heat into a target (Sect. 9.1.5), the structure of the tail is more complicated.

When distributions of energetic particles, or any other deviation from Maxwellian distributions, are important to the dynamics of interest, then to investigate their effects one must work with the Boltzmann equation:

$$\frac{\partial f_s}{\partial t} + \boldsymbol{v} \cdot \nabla f_s + \frac{\boldsymbol{F}}{m_s} \cdot \nabla_v f_s = \left(\frac{\delta f_s}{\delta t}\right)_C, \qquad (2.63)$$

in which $\boldsymbol{F}$ is the sum of all forces acting on each particle and ∇_v is the gradient operator in velocity space, sometimes written as $\nabla_v = \partial/\partial\boldsymbol{v}$. The symbol $\boldsymbol{F}$ is typically the Lorentz force, $q(\boldsymbol{E}+(\boldsymbol{v}/c)\times\boldsymbol{B})$, but also would include any other forces that are present. Equation (2.63) is fundamentally a continuity equation relating the local rate of change of f_s in time, the flow of f_s within the six-dimensional phase space of $\boldsymbol{x}$ and $\boldsymbol{v}$, and the source of f_s on the right-hand side. Particles suddenly appear in an element of velocity space as a result of collisions, that the right-hand side of (2.63) is the rate of change of f_s due to collisions. This term is discussed in detail in the book by Shkarofsky et al. (1966). When the right-hand side is zero, (10.3) is known as the Vlasov equation. It effectively describes many phenomena in collisionless plasmas.

2.6 Approaches to Computer Simulation

Because of its emphasis on conceptual descriptions, this book includes relatively little material on computer simulations. Yet it does include a number of examples produced from computer simulations, and many readers will proceed to work extensively with them. The purpose of the present section is to provide some initial context regarding what computer simulations are and some alternative approaches to them. We will also see that these are limiting cases and that many other possibilities exist.

Those computer simulations of interest here seek to represent a physical system using a much smaller number of computational elements. When one includes the material particles and the photons (or electromagnetic properties, depending on the context), the simulations of interest in high-energy-density physics typically represent the behavior of systems having a number of elements within a few orders of magnitude of Avogadro's number. In contrast, at this writing (2016) typical nodes have 20 cores, and up to 1 TB per node of memory. A computing cluster could have any number of nodes; a couple hundred nodes is not uncommon. Even so, the implication is that the description of the physical system must be an approximate treatment of the behavior of aggregates of particles.

Fundamentally one desires that the simulation follow the evolution of the physical system in time and in space. The fundamental description of the system is always based on a set of differential equations like those we have discussed above. But the simulation must necessarily take one step backward in the calculus, and work with a *discretized* set of equations. This always involves dividing the system into components. In many cases these components are physical cells, in which case the distance between cells establishes the spatial increment used in defining derivatives. For example, if this distance in one direction is δx, then the derivative of pressure across the boundary from cell i to cell j is evaluated as $\partial p/\partial x = (p(j) - p(i))/\delta x$. We discuss some aspects of this below. Other aspects, such as whether to evaluate quantities in the center of a cell or at cell boundaries (and why), we leave to deeper discussions.

One also must establish the temporal increment used to determine how the value of the variables changes in time. This is known as a *timestep*. It must be small enough to give reasonably accurate dynamics yet large enough that the simulation will finish in a reasonable time. One constraint on the timestep is known as the *Courant condition*. The fastest wave of interest must cross no more than one cell in one timestep, otherwise the simulation will artificially retard the propagation of this wave. In cases of interest here, this wave is usually a sound wave. If it has speed c_s, then the Courant condition for the timestep is $\delta t < \min(\delta x/c_s)$.

A calculation (or portion of one) in which variables are advanced in time based on increments found from the differential equation using a timestep is described as an *explicit scheme*. Very fast waves, such as the light waves that make up the radiation, are often treated by an *implicit scheme*. In this case one realizes that the response of the radiation is very fast, reaching a steady state so quickly that the dynamic behavior does not affect the hydrodynamics. As a result, one can find new values of the hydrodynamic variables explicitly and then can solve for the new steady state of the radiation. This is only one example of the use of an implicit scheme. There are many other more complicated examples, but these are not common in high-energy-density physics.

The simulator also faces the problem of figuring out how to divide up the system of interest and deciding what equations to use to describe its evolution. The most natural choice might be to divide the space of interest into small regions. This is known as the *Eulerian* approach. For example, if the system is contained within a volume of a cubic mm, one might divide this volume into 10^9 cubic cells each 1 μm on a side. One could describe the initial condition of the system by giving the variables a value for each cell.

On the other hand, we might choose to divide the matter into cells, so that each cell permanently followed the evolution of a given quantity of material. This is known as the *Lagrangian* approach. This would mean that the cells could move as the material moved. The equations solved in this case would not be precisely those discussed above. Instead one would recast these equations using the Lagrangian mass variable, often written as m, and defined by $dm = \rho dx$.

Simple Eulerian codes have the strength that they can handle arbitrary motions of the material. Their main weakness is that they are inherently diffusive. Once material enters a cell it is treated as though it is spread evenly across the cell. Then in the next timestep some of this matter can move by another cell. The result of enhanced mass diffusion is that materials interpenetrate one another far faster than they would physically. The result of momentum diffusion is that the numerical viscosity is large, typically corresponding to a Reynolds number of about 10^3. One way to minimize these effects is to locate many cells at regions where materials meet or where gradients are steep. This can be done by using a moving *grid* (the grid is the set of lines demarking the cell boundaries). A more advanced approach is to use an adaptive grid, which changes the distribution of the cells during the simulation. Examples of this at this writing include the FLASH hydrodynamic code developed at the University of Chicago and the BATS-R-US MHD code and CRASH radiation hydrodynamic code developed at the University of Michigan. A different advanced approach is to explicitly track any material interfaces in a problem and to treat them distinctly so that the materials do not interpenetrate. This is done for example by the FRONTIER code developed at SUNY Stony Brook.

Simple Lagrangian codes have the strength that they follow the motions of the actual material, allowing an accurate description of complex systems involving a number of components with different properties. Such systems are common in high-energy-density experiments. Such codes allow *no* diffusion, which is usually a strength but can at times be a weakness. They are outstanding tools for one-dimensional modeling of experiments. They have a major weakness in two or three dimensions, however, because they cannot follow swirling (vortical) motions. When the material tries to form a vortex, it tries to send matter from one zone through another zone. This could result in overlapping zones but usually causes one corner of a cell to overtake another corner, so that the new cell is no longer rectangular but instead looks like a twisted rectangle (called a "bow tie" because it looks like one). Examples of Lagrangian codes include HYADES, used to produce a number of figures in this book, and HELIOS.

Simulators have been inventing improvements on these techniques for decades; the description above is necessarily sparse. One can for example make a code that incorporates both Lagrangian and Eulerian elements. Examples of such codes are the RAGE code developed at Los Alamos National Laboratory and the CALE code developed at the Lawrence Livermore National Laboratory.

In addition, beyond these core techniques the treatment of various specific physical mechanisms can be a major challenge. For example, accurately treating the absorption of laser light requires implementing additional physical models along the lines of those described in Chap. 9. As another example, the treatment of thermal radiation or line radiation is a major issue in all these codes. We discuss some aspects of describing radiation in Chap. 6. The book by Castor (2004) addresses at length the difficult problem of treating radiation and hydrodynamics computationally.

A very different alternative, not useful for hydrodynamic systems but useful for the relativistic behavior of small number of particles is the *particle in cell* or PIC approach. In a PIC code one solves for the electric and magnetic fields using a discretized system of differential equations in some way, but one approximates the particles by using sample particles taken to represent the behavior of all the particles within a sphere of one electrostatic shielding distance (the Debye length: see Sect. 3.2) in radius. One explicitly follows the motion of all these sample particles within the environment of the electric and magnetic fields. A variation on PIC, usually described as a hybrid code, is a code that treats the electrons and perhaps some of the ions as a fluid, treating the remaining ions as PIC particles.

Homework Problems

2.1 One approach to deriving the Euler equations is to identify the density, flux, and sources of mass, momentum, and energy and then to use (2.5). Do this for a polytropic gas and then simplify the results to obtain (2.1) through (2.3).

2.2 Linearize the Euler equations to derive (2.7) and (2.8). Find appropriate divisors to make the physical variables in these equations nondimensional. Then derive the equivalent of (2.9) and an equation for a normalized velocity variable. Comment on the result. (Hint: this is a wave problem not a global-scaling problem, so what you are looking for is not U, L, and etc. as used in that part of the chapter.)

2.3 Take the actual, mathematical Fourier transform of (2.9) to find (2.10). Comment on the connection of the result to the substitution used in the text.

2.4 Substitute, for the density in (2.9), the actual, mathematical Fourier transform of the spectral density $\tilde{\rho}(k, \omega)$. Show how the result is related to (2.10).

2.5 The Euler equations apply to an ideal gas with $\epsilon = p/[\rho(\gamma - 1)]$, so they should imply (2.14). Demonstrate this by deriving (2.14) from (2.1), (2.2), and (2.4).

2.6 Begin to explore the behavior of longitudinal waves in a charged fluid. Specifically, derive (2.46) from the equations for number and momentum for an electron fluid.

2.7 Collisions do affect electron plasma waves. To see how, derive a replacement for (2.48), keeping an appropriate version of the drag term at the end of (2.43). Comment on the results.

References

Braginskii SI (1965) Transport processes in a plasma. Rev Plasma Phys 1:205

Castor J (2004) Radiation hydrodynamics. Cambridge University Press, Cambridge

Jackson JD (1999) Classical electrodynamics. Wiley, New York

Jeans JH (1926) The radiation from a pulsating star and from a star in process of fission. Mon Not R Astron Soc 86:86–93

Landau LD, Lifshitz EM (1987) Fluid mechanics, course in theoretical physics, vol 6, 2nd edn. Pergamon Press, Oxford

Mihalas D, Weibel-Mihalas B (1984) Foundations of radiation hydrodynamics, vol 1, Dover (1999) edn. Oxford University Press, Oxford

Ryutov DD, Drake RP, Kane J, Liang E, Remington BA, Wood-Vasey M (1999) Similarity criteria for the laboratory simulation of supernova hydrodynamics. Astrophys J 518(2):821

Sedov LI (1959) Similarity and dimensional methods in mechanics, vol 1. Academic, New York

Shkarofsky IP, Johnston TW, Bachynski MP (1966) The particle kinetics of the plasmas. Addison-Wesley, Reading, MA

Thomas LH (1930) The radiation field of a fluid in motion. Quart J Math 1:239–251

Chapter 3
Properties of High-Energy-Density Plasmas

Abstract This chapter is concerned with the properties of high-energy-density matter, and with how it differs from ideal plasmas and solids. It introduces the concept of equations of state that relate various thermodynamic variables. After reviewing some simple equations of state and some aspects of ideal plasmas, the chapter proceeds to build up an understanding of high-energy-density matter. It discusses the behavior of the electrons, the degree of ionization, continuum lowering, and Coulomb interactions. This enables the generation of a very simple model for the equations of state of these systems. Some additional features are revealed by a more complex model based on statistical mechanics. The chapter then discusses generalized polytropic indices, the degenerate and strongly coupled regime, tabular equations of state, and equations of state in some physical contexts.

The discussion of energy in Sect. 2.1 was entirely based on the notion of a polytropic gas. The speed of sound waves, which we found by examining fluctuations in density and velocity, was found to depend upon the derivative of pressure with density, and the solution was dependent upon having this derivative be constant. This was our first encounter with what is known as the closure problem. The physical densities of interest involve progressively higher powers of velocity: mass density (ρ), momentum density ($\rho \mathbf{u}$), energy density ($\propto \rho u^2$), energy flux density ($\propto \rho u^3$), and so on if needed. But the equations for any given density always involve the divergence of a flux, as we discussed with reference to (2.5), and this flux corresponds to the density involving the next higher power of velocity. Thus, for example, the equation for mass density (2.1) involves the mass density flux, which is also the momentum density. The result is that there is, in general, no way to obtain a closed set of equations simply by including new conservation equations for additional quantities. This is the closure problem.

The only way to ultimately obtain a closed set of equations is to find a way to relate all variables in some set of equations to other variables already defined. We did this twice in Chap. 2. In finding the dispersion relation for acoustic waves, we assumed the pressure (a momentum density flux) to be a knowable function of density only. In relating the Euler equations to the more general equation for energy density, we assumed $\rho \epsilon = p/(\gamma - 1)$ and also assumed the heat flux (and other source terms) to be zero. As another example, sometimes the energy equation is

© Springer International Publishing AG 2018
R P. Drake, *High-Energy-Density Physics*, Graduate Texts in Physics,
https://doi.org/10.1007/978-3-319-67711-8_3

expressed as an equation for temperature, pressure is written as a function of density and temperature, and the heat flux is written as $-\kappa_{th}\nabla T$, which also produces a closed system of three equations. The relations necessary to obtain closed sets of equations, in order to describe behavior of some specific fluid medium, are among its equations of state.

Readers with a plasma-physics background may find this discussion a bit convoluted, since in plasma physics one typically derives all the fluid equations by taking moments of the distribution of the particles in velocity, v, as is done in graduate courses in plasma physics. The continuity equation is the moment taken with v^0, the momentum equation is the moment with v, the energy equation is the moment with v^2, the heat transport equation is the moment with v^3, and one can keep going. The closure problem then arises because every moment equation contains terms involving the next higher moment. But this point of view does not readily allow for some of the more complicated aspects of high-energy-density fluids, such as complex relations between $\rho\epsilon$ and p, cases where the pressure is tensorial, and viscosity.

To successfully solve the continuity, momentum, and energy equations, even numerically, one must typically understand the relation among internal energy, pressure, density, temperature, and ionization. This chapter discusses the equations of state (EOS) that specify these relations. It begins by discussing simple equations of state, which are often useful in limited regimes and for estimates generally. It then considers the conditions for ideal-plasma theory to be valid, because the equations of state are simpler where it is valid.

What follows is an examination of the various issues that lead high-energy-density plasmas to be more complex than low-density plasmas composed of hydrogen. First is a more extensive discussion of electrons, aimed at understanding their transition from ideal-gas behavior to degenerate behavior. Section 3.4 then takes up issues that also involve the ions. The first of these is the degree of ionization, because high-energy-density plasmas are always somewhat ionized but only occasionally fully ionized. The next is how the ions behave when the Debye length is less than the size of an atom. The third is how strong Coulomb interactions manifest themselves in thermodynamic behavior. By the end of this section we will understand the fundamental elements of the equations of state of high-energy-density plasma. This will enable us, in Sect. 3.5, to develop two models for the EOS of high-energy-density matter, and to see how pressure, internal energy, and ionization vary with density and temperature.

Following this, we consider more specifically the high-pressure, low- temperature conditions where the matter is both strongly coupled and Fermi-degenerate, about which we learned in the decade preceding 2015 that our prior understanding was substantially incorrect. We will at that point be in a good position to consider one approach to generalizing the polytropic indices that will prove useful later in the book. Section 3.6 does this. At this point, we will have completely addressed the problem of what high-energy-density matter is, at the level of conceptual

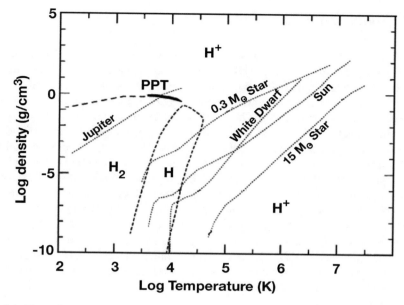

Fig. 3.1 Phase diagram of hydrogen. The dark curve segment shows a theoretical plasma phase transition. Dotted curves show the theoretical path of various astrophysical objects. Adapted from Saumon et al. (1995)

discussions and simple models. Section 3.9 then discusses briefly approaches to equations of state in support of computer simulations while Sect. 3.10 discusses the relation of EOS measurements in the laboratory to astrophysical questions. Specific experimental methods for measuring some aspects of the EOS are discussed in Sect. 4.2 after we explore shock waves.

Before turning to the details, consider this example of the relevance of EOS to astrophysics. Figure 3.1 shows a theoretical phase diagram for hydrogen, and also shows where various interesting objects lie in this diagram. The objects include Jupiter, a typical brown dwarf, and a typical dwarf star. The phase diagram is a model, and the location of the curves depends on the model. These might be wrong, but the range of pressures is correct. The phase diagram of hydrogen includes a region of molecular hydrogen, of atomic hydrogen, and of so-called *metallic hydrogen* in which the electrons are free to move and to conduct electricity. Metallic hydrogen carries the currents that sustain Jupiter's large magnetic field. These regions have boundaries, which might on the one hand be gradual transitions and might on the other hand be abrupt phase transitions. In this particular model, the molecular-to-metallic transition is a phase transition. Evidently a thorough understanding of the EOS will be essential to thoroughly understand astrophysical objects.

3.1 Simple Equations of State

Figure 3.2 shows an image of a Type Ia supernova explosion. This explosion is brighter than the entire galaxy that surrounds it, which is not uncommon. Current understanding is that a Type Ia supernova occurs when a white dwarf star, accumulating mass from its environment, reaches a total of just over 1.4 solar masses. This is enough for gravitational forces to overcome the pressure of the degenerate electrons (Sect. 3.4), which initiates the gravitational collapse of the star. However, the star does not fully collapse. Instead, the energy released as collapse begins heats the C and O that make up the white dwarf, which initiates the violent fusion burning that blows the star apart. The properties of the star as its explosion begins are very relevant to this chapter. Its outer layers are accurately described using the polytropic equation of state for an ideal gas (Sect. 3.1.1). To describe its core, one must use the Fermi-degenerate equation of state (Sect. 3.1.3). And the region heated by the fusion burning requires an equation of state for a radiation-dominated plasma (Sect. 3.1.2). These simple models introduce the relevant regimes and concepts; detailed treatment of white dwarf stars and Type Ia supernovae requires more-sophisticated models.

3.1.1 Polytropic Gases

The polytropic equation of state (EOS), often also described as an ideal-gas EOS, is a useful approximation under many circumstances. At a high enough temperature, any material will behave like an ideal gas. In practice, once the temperature is far enough above the value required to fully ionize any material, its behavior is well described by a polytropic EOS. As we shall see, even radiation-dominated plasmas can be described that way. Moreover, for conceptual and analytic calculations we often use a polytropic description even when it is not precisely accurate. A polytropic gas having n degrees of freedom has certain simple and interconnected properties. The pressure is

Fig. 3.2 A Type Ia
supernova produced the
bright spot of emission near
the edge of this galaxy.
Credit: Jha et al., Harvard
Center for Astrophysics

$$p = \rho RT = Nk_BT = \frac{\rho(1+Z)k_BT}{Am_p},\qquad(3.1)$$

where for the moment N is the total number density of particles, and k_B and T are the Boltzmann constant and the temperature, respectively. The final expression for pressure often applies to a high-energy-density plasma, having an average level of ionization Z, an average atomic mass of the ions in the fluid A, and where the proton mass is m_p. This equation also implies that the gas constant is $R = \rho(1 + Z)k_B/(Am_p)$. In the discussion that follows in the present subsection, we assume Z and thus R to actually be constant. *This is often a poor assumption in high-energy-density physics*, as will become clear in later sections. We will see below that (3.1) also fails to apply when Coulomb interactions become too important or if the radiation pressure becomes too high. Equation (3.1) can be recognized as essentially Boyle's law.

The internal energy, for a system of particles having n degrees of freedom, is

$$\rho\epsilon = \frac{n}{2}\rho RT = \frac{n}{2}Nk_BT = \frac{n}{2}\frac{\rho(1+Z)k_BT}{Am_p},\qquad(3.2)$$

reflecting the basic result from statistical physics that the mean energy of a particle in equilibrium is one-half k_BT per degree of freedom. From this, the specific heat at constant volume is

$$c_V = \left(\frac{\partial\epsilon}{\partial T}\right)_\rho = \frac{n}{2}R = \frac{n}{2}\frac{(1+Z)k_B}{Am_p},\qquad(3.3)$$

if Z (R) is independent of T. Evaluating the specific heat at constant pressure, c_p, is a more complex result from thermodynamics, discussed later in this chapter in Sect. 3.6. The result still depends only on (3.1) and (3.2), and is

$$c_p = \left(\frac{\partial\epsilon}{\partial T}\right)_p = \left(\frac{n}{2}+1\right)R,\qquad(3.4)$$

where similarly R must be independent of T. Thermodynamic arguments also imply a result for the sound speed, specifically

$$c_s^2 = \left(\frac{\partial p}{\partial\rho}\right)_s = \frac{c_p}{c_V}\left(\frac{\partial p}{\partial\rho}\right)_T = \frac{c_p}{c_V}\frac{p}{\rho} = \frac{\gamma p}{\rho},\qquad(3.5)$$

which defines

$$\gamma = \frac{c_p}{c_V} = 1 + \frac{2}{n}.\qquad(3.6)$$

This is an important result. Note that for $n = 3$ one finds the familiar consequence that $\gamma = 5/3$. But as the degrees of freedom become larger, γ decreases toward one. The importance of (3.5) was seen in the discussion of sound waves surrounding (2.6)–(2.9), and we note that here the partial derivative is taken at constant entropy, designated here by the subscript s.

Equation (3.5) also implies that

$$p \propto \rho^{\gamma} \tag{3.7}$$

for isentropic (i.e., adiabatic) changes over a range of pressures for which γ is constant.

Equations (3.1), (3.2), and (3.6), imply that $\rho\epsilon = p/(\gamma - 1)$. In addition, we can obtain the same, self-consistent result by evaluating the internal energy as the integral of the pdV work required to assemble an element of fluid from infinity to some volume V. Note that the conserved mass, $M = \rho V$ so $dV = -Md\rho/\rho^2$. The work is

$$\rho\epsilon V = -\int_{\infty}^{V} p dV' = \int_{0}^{\rho} \frac{Mp}{\rho'^2} d\rho', \text{ so} \tag{3.8}$$

$$\epsilon = \int_{0}^{\rho} \frac{p}{\rho'^2} d\rho' = \frac{p}{\rho(\gamma - 1)}. \tag{3.9}$$

Calculations using polytropic models can become tricky in the important case of an isothermal system. From the perspective of (3.1), an isothermal system would have $p \propto \rho$ and thus $\gamma = 1$. Then (3.9) would imply that the internal energy is infinite. In contrast, for a system whose particles have only kinetic degrees of freedom (3.2) would imply that $\gamma = 5/3$. The key here is that (3.9) describes the adiabatic assembly of the system and such a process is not isothermal. To change compression while maintaining constant temperature requires heat transport, and indeed isothermal systems are those having very rapid heat transport. In a typical case of an isothermal system, one would describe small variations in ρ, such as those due to acoustic phenomena, using $\gamma = 1$, but would still evaluate the portion of the internal energy due to thermal motions as $(3/2)\rho RT$.

Thus, the basic properties of polytropic gases involve a self-consistent set of relationships any one of which can be described as an equation of state. In the event that R and thus γ are not constant, however, one no longer has such a simple story. This important and realistic case motivates the discussion in Sect. 3.6.

3.1.2   Radiation-Pressure-Dominated Plasma

The properties of blackbody radiation and of systems in which radiation is important or dominant are discussed in Chaps. 6 through 8. The radiation pressure p_R is 1/3 the radiation energy density and may be expressed as

$$p_R = \frac{4}{3}\frac{\sigma}{c}T^4,\tag{3.10}$$

where T is the temperature, c is the speed of light, and σ is the Stefan–Boltzmann constant familiar from blackbody emission. Because this pressure depends upon T to the fourth power, while material pressures depend upon T to the first power, at a high enough temperature the radiation pressure will be completely dominant. This is the case, for example, within matter shocked during supernova explosions and near neutron stars and black holes. The transition temperature can be determined by asking when the radiation pressure equals the material pressure. Assuming (3.1) to be accurate, one finds

$$T = \frac{1}{k_B}\left(\frac{3k_B^4 c\rho(1+Z)}{4\sigma m_p A}\right)^{1/3} = 2.6\left(\frac{\rho(1+Z)}{A}\right)^{1/3}\text{keV},\tag{3.11}$$

in which ρ is in g/cm^3. Here, outside the parenthesis in the middle term, $k_B = 1.6 \times 10^{-9}$ ergs/keV to find the temperature in energy units (keV). Within the parentheses, the units of energy and temperature in k_B must be consistent with those in σ and with the other units used there. For laboratory systems or within stars where ρ is within a few orders of magnitude of 1 g/cm^3, keV temperatures are thus required for radiation to dominate. At typical astrophysical densities much lower temperatures would be required, except that such systems tend to be "optically thin" (see Chap. 6), implying that the radiation pressure is far below the value given by (3.10).

To utilize simple equations in describing radiation-dominated plasmas, one desires to determine $\partial p / \partial \rho$ for this case—that is, to determine how the radiation pressure varies with plasma density. This is often feasible, because in order for the radiation temperature to remain large enough that the system stays radiation-dominated, the mean free path for the radiation must be small on the scale of the physical system of interest. This in turn implies that the material is strongly coupled to the radiation and will have the same temperature. In addition, because the material is strongly coupled to the radiation, changes in the density of the material involve changes in the volume containing a fixed amount of radiation. This prepares us to identify a polytropic index for the radiation-dominated plasma, as follows.

Standard arguments in statistical mechanics lead to an expression for the pressure of the photon gas as

$$p = -\sum_j \bar{\sigma}_j \frac{\partial \epsilon_j}{\partial V},\tag{3.12}$$

in which the sum is over all possible states j, the mean occupancy of each state is $\bar{\sigma}_j = 1/[\exp(\epsilon_j/k_B T) - 1]$, and the energy of each state is ϵ_j. Equation (3.12) makes sense when one recalls that the pressure is the negative of the change in internal energy as volume increases. The energy of a state varies with volume as the wavelength of the light in that state is reduced or increased by the compression or expansion. One can see how by considering the simple example of a cubic box with an edge of length L, in which a given state has an integer number of wavelengths along each side of the box. The wavenumber of each state, k_j, is then proportional to $1/L$, so one has $\epsilon_j \propto hck_j \propto L^{-1} \propto V^{-1/3}$, where h is the Planck constant. Thus

$$-\frac{\partial \epsilon_j}{\partial V} \propto V^{-4/3} \propto \rho^{4/3} \qquad (3.13)$$

and $p \propto \rho^{4/3}$, showing that $\gamma = 4/3$ for a radiation-dominated plasma. The Euler equations can be applied to such a system using $\gamma = 4/3$. One can obtain the same result more simply by recognizing that the photons have 6 degrees of freedom, thanks to their two possible polarizations. Further details can be found in the chapter on radiation hydrodynamics.

3.1.3 Fermi-Degenerate EOS

In ordinary plasmas it is the thermal pressure, experienced by the particles through Coulomb collisions, that resists compression of the plasma. This is a classical effect, and the properties of the electrons are described by Boltzmann statistics. But when plasma or other matter becomes dense enough, then quantum mechanical effects involving the electrons create pressure and resist compression. The electrons are subject to the Pauli exclusion principle, which prevents more than one of them from occupying the same quantum state. As we will see, this implies that the most energetic electron in cold, high-density matter can be quite energetic indeed. Matter in which nearly all of the electrons are in their lowest-energy states is described as Fermi-degenerate matter. The EOS of Fermi-degenerate matter is of substantial importance in massive planets, white dwarf stars, and inertial fusion implosions or other high-energy-density experiments that compress solid matter. The fact that electrons are fermions has an impact over a broader range of conditions, as we will see in Sect. 3.5. Fundamental derivations of the electron behavior can be found in any book on statistical physics, including for example Reif (1965) and the relevant volume by Landau and Lifshitz (1987).

Figure 3.3 shows the energy distributions of free electrons in dense matter, for several temperatures. In very cold, dense matter the energy distribution is a step function—all the electrons are in the lowest accessible state. As temperature increases, some of these states are depleted and a tail of electrons develops at higher energy. The energy of the state whose occupancy is 50% is known as the Fermi energy. The Fermi energy at absolute zero, ϵ_F, is

$$\epsilon_F = \frac{h^2}{2m_e} \left(\frac{3}{8\pi} n_e \right)^{2/3} = 7.9 \, n_{23}^{2/3} \text{ eV}, \qquad (3.14)$$

Fig. 3.3 Electron energy distributions in dense matter. The distribution function, normalized to be 1 at zero energy, is shown against energy, normalized to the Fermi energy, for $k_B T = 0.01$, 1, and 10 ϵ_F. The gray curve shows a Maxwellian distribution for $k_B T = 10\epsilon_F$

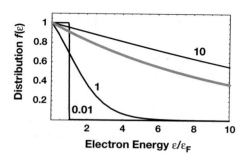

in which m_e is the electron mass, n_e is the number density of electrons, and n_{23} is the electron density in units of 10^{23} cm^{-3}. This value (10^{23} cm^{-3}) is of order both the density of electrons in low-Z plasmas with a mass density near 1 g/cm^3 and the density of conduction electrons in a typical metal. In any material there may also be bound electrons, attached to specific atoms. These electrons do not contribute to the electron density n_e in (3.14). If we displayed the bound electrons on the scale of Fig. 3.3, they would appear as spikes at negative electron energy. We discuss the degree of ionization (and hence the relative numbers of free and bound electrons) in Sect. 3.4. Equation (3.14) has a number of consequences for physical systems of interest here. It implies that the electrons are not Fermi degenerate in plasmas with densities well below solid density, heated to temperatures of tens to hundreds of eV. In contrast, compressed plasmas at densities of more than 100 times solid density, produced in inertial fusion implosions, have a Fermi energy of hundreds of eV. Such plasmas are often cool enough that the EOS of the electrons is the Fermi-degenerate EOS. The *degeneracy temperature*, T_d, above which the electrons can be approximated as a classical gas, is found by setting $k_B T_d = \epsilon_F$.

Despite its obvious differences from an ordinary gas, the equation of state of Fermi-degenerate matter is quite similar to that of an ideal polytropic gas with $\gamma = 5/3$. Equation (3.8) applies in both cases, so $p = (2/3)\rho\epsilon$. In addition, while the electron pressure in an ideal gas is $p = n_e k_B T$, the electron pressure in Fermi-degenerate matter is $p_F = (2/5)n_e\epsilon_F$. Evaluating this one finds

$$p_F = \frac{2}{5}n_e\epsilon_F = \frac{h^2}{20m_e}\left(\frac{3}{\pi}\right)^{2/3}n_e^{5/3}, \tag{3.15}$$

or in practical units

$$p_F = 0.50n_{23}^{5/3} = 9.9\left(\frac{\rho}{A/Z}\right)^{5/3} \text{ Mbar}, \tag{3.16}$$

in which $A/Z \sim 2$ and the units of density are cgs. The transition from (3.16) to (3.1) occurs approximately when $T = T_d$, although one can see in Fig. 3.3 that the electron distribution still departs significantly from a Maxwellian even at $T = 10T_d$.

3.2   Regimes of Validity of Traditional Plasma Theory

High-energy-density systems are nearly always plasmas, in the sense that they are ionized and that electromagnetic interactions at a distance can play a role in their dynamics, at least in principle. Unfortunately, the theory of plasmas, as covered in traditional texts such as Krall and Trivelpiece (1986), has a range of validity that only partly overlaps the regimes of high-energy-density physics. Even so, plasma concepts have tremendous utility when they are valid. This motivates a discussion of these issues.

Traditional plasma theory faces the challenge of describing a system composed of mobile charged particles and capable of dramatic electrodynamic effects. The particles quickly scurry over to surround any exposed charge, yet also can carry currents that produce magnetic fields which can store immense energy. The eruptions on the surface of the sun are an example of the potential consequences. The shielding of exposed charges is one of the fundamental aspects of plasmas. Yet even as the charges try to cluster about one another, their thermal motions limit the clustering. The competition between these gives rise to a characteristic shielding distance, known as the *Debye length*. The Debye length is defined in Gaussian cgs units by

$$\lambda_D^{-2} = 4\pi e^2 \left(\frac{n_e}{k_B T_e} + \sum_\alpha \frac{n_\alpha Z_\alpha^2}{k_B T_\alpha} \right), \tag{3.17}$$

in which the sum is over all ion species, the subscript e designates electrons while α designates an ion species, n is a number density, T is a temperature, Z is a number of unit charges, k_B is the Boltzmann constant, and e is the electronic charge (4.8×10^{-10} statcoul here). On the one hand, when one considers fast enough timescales, the ions cannot move and the *electron Debye length*,

$$\lambda_{De} = \sqrt{\frac{k_B T_e}{4\pi n_e e^2}}, \tag{3.18}$$

(in the same units) becomes relevant. This is the only Debye length defined in the *NRL Plasma Formulary*, among other references. In addition, traditional plasma texts often assume all plasmas to be pure hydrogen, replacing the 4 with an 8 in (3.18). On the other hand, there are cases in dense plasmas when ion–ion shielding determines the behavior, as for example when the electrons cluster poorly because they are Fermi degenerate (Sect. 3.1.3). Then the *ion Debye length*,

$$\lambda_{Di}^{-2} = 4\pi e^2 \sum_\alpha \frac{n_\alpha Z_\alpha^2}{k_B T_\alpha}, \tag{3.19}$$

(in the same units) comes into play.

High-energy-density plasmas, like most plasmas, are quasi-neutral, so that

$$n_e = \sum_\alpha n_\alpha Z_\alpha. \tag{3.20}$$

In addition, in such plasmas collision rates are large (Sect. 2.4) so the temperatures of the particle species are usually equal and designated by T. When this is the case, one can use the standard definition of the effective charge, Z_{eff}, as

$$Z_{\text{eff}} = \frac{\sum_\alpha n_\alpha Z_\alpha^2}{\sum_\alpha n_\alpha Z_\alpha} = \frac{\sum_\alpha n_\alpha Z_\alpha^2}{n_e}, \text{ to write} \tag{3.21}$$

$$\lambda_D = \sqrt{\frac{k_B T}{4\pi n_e (1 + Z_{\text{eff}}) e^2}}, \tag{3.22}$$

again in Gaussian cgs units. This is a form we will use in later discussions. For calculations involving binary collisions, Z_{eff} is the appropriate average charge, while for calculations involving particle counting, $Z = n_e/n_i$ is the appropriate average charge.

The Debye length arises quite naturally in the most-sophisticated developments of plasma theory. It also can be found from a simple calculation that can be used to highlight the limitations of traditional plasma theory for us. We consider a two-species plasma, in which the ions have charge Z, and we also suppose that the particles are distributed by classical statistics with a common temperature T. This implies that the density of particles with charge q, at a location with a potential ϕ relative to the potential at some reference location is proportional to $\exp[-q\phi/(k_B T)]$. Then the charge density ρ_c in the vicinity of an ion at $x = 0$ is

$$\rho_c = Ze\delta(0) - n_e e \exp\left[\frac{e\phi}{k_B T}\right] + n_i e Z \exp\left[\frac{-eZ\phi}{k_B T}\right]. \tag{3.23}$$

If we assume that $|q\phi| \ll k_B T$ and that the plasma is quasi-neutral, then this becomes

$$\rho_c = Ze\delta(0) - \frac{e^2\phi}{k_B T}\left(n_e + n_i Z^2\right) = Ze\delta(0) - \frac{\phi}{4\pi\lambda_D^2}. \tag{3.24}$$

At this point we can write the Poisson equation in spherical coordinates, assuming that the charges are distributed with spherical symmetry, as

$$\frac{1}{r^2}\frac{d}{dr}\left(r^2\frac{d\phi}{dr}\right) = -4\pi Ze\delta(0) + \frac{\phi}{\lambda_D^2}, \tag{3.25}$$

which (in cgs units) has the solution

$$\phi = \frac{Ze}{r}e^{-r/\lambda_D}. \tag{3.26}$$

Equation (3.26) displays the standard result that the potential of any given charge falls away exponentially faster in a plasma than it would in vacuum. But what is relevant to our interests is two aspects of this derivation. First, (3.23) only makes sense in the end if there are numerous particles within a sphere whose radius is the Debye length. Second, the key assumption in this derivation is that $|q\phi| \ll k_BT$, which must be violated if the particles are cold enough. These turn out to be related, and we will explore them in turn.

The number of particles in a Debye sphere, in a quasi-neutral plasma, is $n_e(1+1/Z)(4\pi/3)\lambda_D^3$. The inverse of this, sometimes defined without the numerical coefficients, is a fundamental expansion parameter for traditional plasma theory (see Krall and Trivelpiece 1986). A plasma is known as an *ideal plasma* when the number of particles in a Debye sphere can be taken to approach infinity. In this case collective effects, involving all the particles, remain, while effects relating to particle correlations vanish. Figure 3.4 shows the number of particles in a Debye sphere in the high-energy-density regime. There are not many. The number varies from tens of particles in the upper-left corner of the regime shown to less than 0.01 particles in the lower-right corner. In the lower-right corner the electrons are Fermi degenerate (Sect. 3.1.3), which reduces even further their ability to shield the ions. The ion density in a typical solid is also shown. It is evident from this figure that high-energy-density plasmas are rarely ideal plasmas.

Now consider the assumption that $|q\phi| \ll k_BT$. We can take a typical value of ϕ to be the electrostatic interaction of two particles at their average spacing. We find

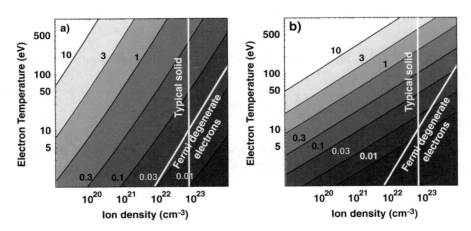

Fig. 3.4 Contours of the number of particles in a Debye sphere. The contours show 0.01, 0.03, 0.1, 0.3, 1, 3, and 10 particles, and increase to the upper left. (**a**) A high-Z plasma with $Z = 0.63\sqrt{T_{eV}}$ (see Sect. 3.4). (**b**) A low-Z plasma with $Z = 4$

the average spacing by giving each particle a spherical volume of radius r_{av}, so that
the average spacing is $2r_{av}$. Thus we take $4\pi r_{av}^3/3 = 1/[n_e(1 + 1/Z)]$. Then we find

$$\phi = \frac{k_1 Z e}{2 r_{av}} = \frac{k_1 Z e}{2} \left(\frac{3}{4\pi n_e(1 + 1/Z)} \right)^{1/3}, \tag{3.27}$$

so that the assumption becomes

$$\frac{|q\phi|}{k_B T} = \frac{k_1 Z e^2}{2 r_{av} k_B T} = \frac{k_1 4\pi n_e(1 + Z)e^2}{2 k_B T} \frac{Z/(4\pi)}{r_{av} n_e(1 + Z)} = \frac{\lambda_D}{6 r_{av}} g = \frac{g^{2/3}}{6} \ll 1, \tag{3.28}$$

where g is the inverse of the number of particles in a Debye sphere, $1/g = n_e(1 + 1/Z)(4\pi/3)\lambda_D^3$. Thus, the two requirements of the Debye-shielding analysis
are intimately connected. It is no surprise that this assumption (3.28) is violated
over about half the parameter space shown in Fig. 3.4. The ratio $|q\phi|/k_B T$ is often
known as the *strong coupling parameter*, Γ, first introduced in Chap. 1. Salzman
(1998) discusses this parameter, which he calls the *plasma coupling constant*, at
more length. Like the Debye length, Γ comes in different flavors depending upon
whether one evaluates ion–ion coupling, ion–electron coupling, or electron–electron
coupling. To be precise, one must evaluate Z and r_{av} for a specific, chosen set of
particles. The most common type of Γ found in the literature is that for ion–ion
coupling. Across much of the parameter space of Fig. 3.4, the ions are strongly
coupled but the ions and electrons are not. We specifically discuss the regime where
the ions are strongly coupled and the electrons are Fermi degenerate in Sect. 3.7.
The pressure and energy of the plasma will depart from their ideal-gas values across
a large fraction of this parameter space. In order to enable us to understand the
actual behavior of matter at high energy density, Sect. 3.3 considers the behavior
of electrons across this regime and Sect. 3.4 considers ionization and of energies
associated with Coulomb forces. This will enable us, in Sect. 3.5 and beyond, to see
the combined effect of all these elements.

3.3 Electrons at High Energy Density

We begin by returning to the behavior of the electrons. We know that at high-
enough temperature they behave as an ideal gas, and that they are Fermi-degenerate
at low-enough temperature and high-enough density, with the boundary between
these regimes running directly across the middle of the parameters of interest for
high-energy-density physics. What we have not yet determined is how abrupt the
transition is. In the limit that it is very abrupt, one could model the electron effects
discontinuously, just switching models at the boundary. In contrast, if the transition
is very gradual, then one might have to implement a more-complicated treatment
of the electrons to even come close to the correct behavior. So our task here is to
determine how abrupt the transition from ideal-gas to Fermi-degenerate behavior
actually is.

The detailed properties of the partially degenerate matter at a temperature near the degeneracy temperature involve some straightforward numerical integrals. The ion density range of interest to high-energy-density physics spans 10^{19} to 10^{24} cm^{-3}, but reaches $\sim 10^{26}$ cm^{-3} in compressed inertial fusion capsules. All of the electrons participate in Fermi-degenerate behavior, so this corresponds to a range of electron densities from 10^{19} to 10^{26} cm^{-3}, where the upper limit might correspond either to high-Z matter at an ion density of 10^{24} cm^{-3} or to low-Z matter compressed for inertial fusion. The electron temperatures of interest span 1–1000 eV. Let us examine the behavior of the electrons over this range of conditions.

The electron density is given by the integral over all momenta, χ_e, of the probability that an electron will have a specific momentum. With the electron energy given by $\mathcal{E}_e = \chi_e^2/(2m_e)$, this is

$$n_e = \frac{8\pi}{h^3} \int_0^\infty \frac{\chi_e^2 d\chi_e}{\exp[\frac{(-\mu+\mathcal{E}_e)}{k_B T_e}] + 1}, \tag{3.29}$$

in which μ is the chemical potential, which has energy units. Within this integral, the factor equal to $\chi_e^2 d\chi_e$ gives the scaling of the density of states while the remaining factor gives the probability that a certain state is occupied by an electron. Equation (3.29) can be put in the useful form

$$\Theta = \frac{T_e}{T_d} = T_e\left[\left(\frac{8\pi}{3n_e}\right)^{2/3}\frac{2m_e k_B}{h^2}\right] = \left[\frac{3}{2}F_{1/2}\left(\frac{\mu}{k_B T_e}\right)\right]^{-2/3}, \tag{3.30}$$

which defines the ratio of electron temperature to degeneracy temperature as Θ. We also define in general $F_n(\phi) = \int_0^\infty x^n [\exp(x-\phi)+1]^{-1} dx$. This will have further application below. Our parameter range of interest corresponds to $\Theta = 10^{-3}$ to 10^4.

The chemical potential is the internal energy required to add a particle to the system at constant entropy and constant volume. For a Fermi-degenerate system the chemical potential is positive; a new particle goes in at the Fermi energy even at zero entropy, so one must invest energy to put a new particle into the system. For a Boltzmann system μ is negative: a new particle can be added at zero energy but to keep entropy constant the internal energy of the system must decrease. The limiting behavior of $\mu/(k_B T_e)$ is of some interest. In the degenerate regime, $\mu = \epsilon_F$ so

$$\frac{\mu}{k_B T_e} = \frac{\epsilon_F}{k_B T_e} = \frac{1}{\Theta}. \tag{3.31}$$

In the Boltzmann limit, designating the Boltzmann chemical potential as μ_c, one has

$$e^{\mu_c/(k_B T_e)} = \frac{n_e h^3}{2(2\pi m_e k_B T_e)^{3/2}}, \tag{3.32}$$

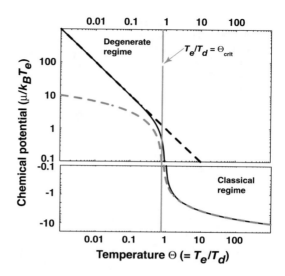

Fig. 3.5 The chemical potential is shown vs $\Theta = T_e/T_d$. In the classical (Boltzmann) regime, μ is negative

so

$$e^{\mu_c/(k_B T_e)} = \frac{4}{3\sqrt{\pi}\Theta^{3/2}}, \tag{3.33}$$

so μ_c is zero when $\Theta = \Theta_{\text{crit}} = 0.827$. Atzeni and Meyer-Ter-Vehn (2004) give a fit due to Ichimaru that spans both limits. This is

$$\frac{\mu}{k_B T_e} = -\frac{3}{2}\ln(\Theta) + \ln\left(\frac{4}{3\sqrt{\pi}}\right) + \frac{0.25954\Theta^{-1.858} + 0.072\Theta^{-1.858/2}}{1 + 0.25054\Theta^{-0.858}}. \tag{3.34}$$

One can vary $\mu/(k_B T_e)$ and calculate the integral (3.30). Figure 3.5 compares the result of this calculation with the values implied by (3.31) and (3.33). The solid curve shows the actual value, with the gray, dashed curve showing the Boltzmann limit and the black, dashed curve showing the Fermi limit. The result is rather dramatic. The electron chemical potential has the Boltzmann value for $\Theta > \Theta_{\text{crit}}$, where it abruptly transitions to the degenerate value. *This observation is the most important result of this section.* If the temperature is further than a factor of two from the T_d, then the behavior of the electrons is solidly in the corresponding limit. Only very near the transition need one consider using a more complex model.

One can evaluate the electron pressure by averaging the energy of each state over the probability that the state is occupied. The general integral for the internal energy density, $n_e \epsilon_e$, where ϵ_e is the specific internal energy per electron, and the pressure, p_e, is

$$n_e \epsilon_e = \frac{3}{2} p_e = \frac{8\pi}{h^3} \int_0^\infty \frac{\mathcal{E}_e \chi_e^2 d\chi_e}{\exp[\frac{(-\mu + \mathcal{E}_e)}{k_B T_e}] + 1}, \tag{3.35}$$

Fig. 3.6 Normalized electron
pressure *versus* chemical
potential. This asymptotes to
$(2/5)\mu/(k_BT_e)$ at large
$\mu/(k_BT_e)$ and approaches 1 as
$\mu/(k_BT_e)$ approaches 0 (and
is 1 in the Boltzmann regime)

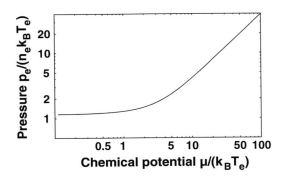

which can be written as

$$n_e\epsilon_e = \frac{3}{2}p_e = \frac{3}{2}n_ek_BT_e\Theta^{3/2}F_{3/2}\left(\frac{\mu}{k_BT_e}\right) = n_ek_BT_e\frac{F_{3/2}(\frac{\mu}{k_BT_e})}{F_{1/2}(\frac{\mu}{k_BT_e})}. \tag{3.36}$$

Figure 3.6 shows how the normalized pressure, $p/(n_ek_BT_e)$, increases with $\mu/(k_BT_e)$, for $\Theta < \Theta_{\mathrm{crit}}$. The electron contribution to the pressure and internal energy is classical for $\Theta > \Theta_{\mathrm{crit}}$. Despite the difference in the pressure, the electrons behave as a gas with $\gamma = 5/3$ throughout. Under strongly Fermi-degenerate conditions, the electron pressure and energy completely dominate those of the ions. However, because of the energy associated with ionization, the electrons do not necessarily dominate the internal energy of the plasma throughout our regime of interest. We explore this further in Sect. 3.4.

For various applications, including inertial fusion, it is worthwhile to understand the heat capacity and entropy of electrons. For this purpose it helps to understand $F_n(\phi)$ more thoroughly. One can show that $F_n'(\phi) = n\phi'F_{n-1}(\phi)$. In addition, if T_e is near zero, then $F_n(\phi) = \phi^{n+1}/(n+1)$. It is also useful to know that if ϕ is zero, then $F_{3/2} = 1.153$ while $F_{1/2} = 0.678$. In the Boltzmann limit (3.32) implies that

$$F_n\left(\frac{\mu}{k_BT_e}\right) = \frac{n_eh^3}{2(2\pi m_ek_BT_e)^{3/2}}\Gamma(1+n) = \frac{4}{3\sqrt{\pi}\Theta^{3/2}}\Gamma(1+n), \tag{3.37}$$

so in the Boltzmann limit $F_{3/2} = \Theta^{-3/2}$ while $F_{1/2} = (2/3)\Theta^{-3/2}$.

Turning to the heat capacity, one finds

$$C_V = \frac{\partial}{\partial T_e}(n_e\epsilon_e)\bigg|_{n_e} = \frac{3}{2}n_ek_B\left[\frac{5}{2}F_{3/2}\Theta^{3/2} + T_e\frac{\partial}{\partial T_e}\left(\frac{\mu}{k_BT_e}\right)\right], \tag{3.38}$$

where C_V has units of energy per unit volume per unit temperature and the argument of both $F_{3/2}$ and $F_{1/2}$ is $\mu/(k_BT_e)$. In the Boltzmann limit this becomes $C_V = (3/2)n_ek_B$. In the degenerate limit and for small temperatures, one can expand the integrals to find $C_V = (3/2)n_ek_B[\pi^2k_BT_e/(3\epsilon_F)]$. This is the electronic contribution, which is dominant for strongly degenerate matter, As any book on statistical physics will discuss the ionic contribution at small temperature.

The entropy per unit volume of the electrons, S/V, may be found from

$$\frac{S}{V} = \left(-\frac{1}{V}\frac{\partial(pV)}{\partial T_e} \right)_{\mu,V} = \frac{2}{3}\frac{\partial}{\partial T_e}(n_e\epsilon_e)\Big|_{\mu,V}, \tag{3.39}$$

in which $-pV$ is one of the thermodynamic potentials discussed by Landau and Lifshitz (1987) in their volume on statistical physics. Note that holding μ and V constant is not identical to holding n_e constant, as was done to find C_V. This implies

$$\frac{S}{V} = \frac{5}{2}n_e k_B\left[\frac{2}{3}\frac{F_{3/2}(\frac{\mu}{k_B T_e})}{F_{1/2}(\frac{\mu}{k_B T_e})} - \frac{2}{5}\frac{\mu}{k_B T_e} \right], \tag{3.40}$$

which for the Boltzmann limit is

$$\begin{aligned}
\frac{S}{V} &= n_e k_B\left(\frac{5}{2} + \ln\left[\frac{2(2\pi m_e k_B T_e)^{3/2}}{n_e h^3} \right] \right) \\
&= n_e k_B\left[\frac{5}{2} + \ln\left(\frac{3\sqrt{\pi}\Theta^{3/2}}{4} \right) \right]
\end{aligned} \tag{3.41}$$

or in the degenerate limit where $T_e \ll \epsilon_F$ is

$$\frac{S}{V} = \frac{3}{2}n_e k_B\left(\frac{\pi^2}{3}\Theta \right). \tag{3.42}$$

The entropy approaches zero as the temperature approaches absolute zero, as it should.

In the context of inertial fusion, one cares about the relation of pressure and entropy, because the shock waves produced during compression increase the entropy (see Chap. 4). Since the pressure is proportional to $F_{3/2}$, while density is proportional to $F_{1/2}$, (3.40) can be rearranged to obtain

$$p = \frac{2}{5}\frac{S}{V}T_e + \frac{2}{5}n_e\mu. \tag{3.43}$$

As T_e and S approach zero, this reduces to (3.15). One sees that the pressure is not sensitive to the value of the entropy until the entropy reaches a threshold value given by setting the two terms of this equation equal. This is evident in Fig. 3.6, where we see that the pressure begins to depart from $2n_e\mu/5$ when $\mu \sim 5k_B T_e$ or $\Theta \sim 0.2$ so $T_e \sim 0.2T_d$.

The quantity p/p_F is known in inertial fusion as the *degeneracy parameter*. It has important practical consequences as the fusion gain decreases for increasing p/p_F. In general $p/p_F = 1$ for degenerate matter and increases with Θ, equaling $(5/2)\Theta$

in the Boltzmann regime. The practical importance of this quantity makes it useful to have approximate estimates of p/p_F. Atzeni and Meyer-Ter-Vehn (2004) give the following fit for p/p_F:

$$\frac{p}{p_F} = \frac{5}{2}\Theta + \frac{0.27232\Theta^{-1.044} + 0.145\Theta^{0.022}}{1 + 0.27232\Theta^{-1.044}}. \tag{3.44}$$

The present section has provided a variety of useful models and limiting cases for the behavior of the electrons. If we pull back to our overall mission, what matters is this: The behavior of the electrons changes quite abruptly from ideal-gas behavior to Fermi-degenerate behavior as T crosses T_d. Only if the temperature is within a factor of about 2 from this boundary need one consider their behavior in more detail. Taking the point of view that our goal here is to see the overall behavior, in our further discussions we will treat the transition in electron behavior to be abrupt.

3.4 Ionizing Plasmas

Mid-Z and high-Z ions in high-energy-density plasmas are rarely *fully stripped*, meaning that all their electrons have been removed. Only as temperatures approach and exceed 1 keV, or as compressions exceed ten times solid density will one encounter completely stripped ions of any except very-low-Z species. When it becomes routine to work far above solid density at temperatures of many keV, the materials may become fully stripped, although the increased role of radiation will provide ample new complications. We discuss some of these in Chap. 7. For the moment, it is clear that we must understand the behavior of partially ionized plasmas, which we will describe as *ionizing plasmas*, if we are to succeed in understanding high-energy-density phenomena.

One needs to estimate the degree of ionization for a variety of reasons. The most important is that their thermodynamic properties also depend upon ionization, as we discuss in the next two sections. The internal energy of fully stripped ions also includes a major contribution from ionization. While the behavior of actual materials is complicated and difficult to calculate accurately, there are some simple models that can capture aspects of their behavior. These we discuss here.

The electron density is Zn_i, but the value of the average charge Z depends upon the temperature. To know Z precisely, one must evaluate the *ionization balance* to determine the relative populations, N_i, of the various ionization states. Then one has Z as a sum over ionization states,

$$Z = \frac{1}{N} \sum_i Z_i N_i, \tag{3.45}$$

in which the state populations can be either a number or a density, and N is either the total number of ions or the ion density n_i, respectively.

We will designate the various ionization states of a given species by their net charge Z_i. The electrons in any given ion may reside in the ground state or in an excited state. These of course are designated precisely by the necessary quantum numbers, such as the principal quantum number, n, the quantum number for orbital angular momentum, ℓ, and the spin quantum number, s. In the present discussion, we will occasionally have reason to specify the principal quantum number. We will often, however, ignore excited states and implicitly treat all ions as ground state ions. In most cases this is reasonable. The minimum excited state energy, with $n = 2$, has an energy above the ground state that is 3/4 of the ionization energy, E_i. On the one hand, if the ion is in an environment where E_i is well above T_e, as is common, then the excited state population is smaller than the ground state population by a factor smaller than $\exp[-3E_i/(4k_B T_e)]$, which is fairly small. On the other hand, if E_i is small relative to T_e, then it is more likely that the electrons striking the ion will deliver its outer electron into one of the indefinite number of free states as opposed to one of the few and definite excited states.

The exact ionization energy required to remove the outermost electron from a given ionization state does depend on the number and arrangement of the remaining electrons, but we will ignore this here and adopt a *hydrogenic atom* analysis. In such a treatment, all atoms and all ions are treated as hydrogenic systems, having one electron and a nucleus with the appropriate net charge. This approach is more accurate as the net charge on the atom increases (so that the inner electrons are more tightly bound). This approach allows comparatively tractable computational models to work with a wide range of atoms and ionization states, giving qualitatively correct answers. In our work here we will primarily use the ionization energy associated with a hydrogenic atom model, which is energy $E_i = Z^2 E_H$, where $E_H = 13.6\,\text{eV}$ and Z is the net charge on the atom after ionization (and thus is consistent with our use of "Z" elsewhere).

The simple view of atomic structure we will use here is distinct from the computational "average atom model" (discussed in Salzman 1998). The computational model provides a physically consistent approach to the definition of an "average atom", including both bound states and free electrons, that characterizes each element.

The density of ions will play an important role in our discussions of ionization, as this scales the electron density. A factor-of-two estimate of the typical ion density, for a solid, can be made by taking $\rho = Z_n/4\,\text{g/cm}^{-3}$ and $A = 2Z_n$. Then

$$n_i = \frac{\rho}{Am_p} \sim \frac{Z_n}{8Z_n m_p} = 7.5 \times 10^{22}\,\text{g/cm}^3. \tag{3.46}$$

This density is indicated in several of the plots in the following.

3.4.1 Ionization Balance from the Saha Equation for Boltzmann Electrons

Determining the exact degree of ionization is a difficult problem involving sophisticated calculations, but we can arrive at a reasonable approximation on very simple grounds. We can expect that the ionization energy of the ions in a plasma will have some typical relation to the electron temperature. If we approximate the ion as a hydrogenic ion of charge Z (after ionization), then the ionization energy $E_i = Z^2 E_H$, where $E_H/k_B = 13.6\,\text{eV}$. Thus, we expect $Z^2 E_H/(k_B T_e) \sim C^2$, where C is a constant, so $Z = C\sqrt{k_B T_e/E_H}$, which is $Z = 0.27 C\sqrt{T_{eV}}$. The problem is to find C. On the one hand, if we recall that Coulomb processes often are effective at energies of about $3 k_B T_e$, as is the case for heat transport (see Chap. 9), then we would say $C \sim \sqrt{3}$, which is not far from the better estimates discussed next.

More sophisticated estimates of the ionization involve balancing ionization and recombination or assuming that the distribution of ions is in equilibrium. These turn out to be equivalent at high enough densities, but not at low densities. Griem (1997) and Salzman (1998) discuss in detail the dynamics that are involved, in their books. Here, and at more length in Chap. 6, we discuss the basic phenomena that are important for high-energy-density systems. In low-density plasmas, the archetype of which is the solar corona, collisional ionization is balanced by radiative recombination, establishing a steady state known as *coronal equilibrium*. An additional process, dielectronic recombination, is of increasing importance as the density increases, particularly in the range of densities found in magnetic fusion devices. But at most densities found in high-energy-density systems, the relevant balance is between collisional ionization and collisional (three-body) recombination. In equilibrium, collisional ionization and collisional recombination are equal by the principle of detailed balance.

At high enough density and temperature the distribution of ions, and the distribution of electrons within energy levels, approaches the equilibrium distribution given by the Saha equation, derived in statistical mechanics. For an estimate of the ionization balance we will ignore the distribution of electrons among the excited states, and will focus only upon the distribution of ions among the ionization levels. We work here with the Saha equation to estimate Z. The Saha equation gives the ratio of the population of ions in state j, N_j, to those in state k, N_k, as

$$\frac{N_j}{N_k} n_e = \frac{g_j}{4 g_k a_o^3} \left(\frac{k_B T_e}{\pi E_H} \right)^{3/2} e^{\frac{-E_{jk}}{k_B T_e}}, \tag{3.47}$$

in which in cgs units $a_o = \hbar^2/(m_e e^2) = 5.29 \times 10^{-9}\,\text{cm}$ is the Bohr radius, E_{jk} is the energy required to go from state k to state j. The symbols g_j and g_k are the degeneracies of the ions. These are the number of distinct states of the ion having the same energy for states j and k, respectively. For a hydrogenic ion with an electron having principal quantum number n, this is $2n^2$. Thus, for our assumption of hydrogenic ions in the ground state, discussed above, $g_j = g_k = 2$.

(To help interpret various references, it may also help to know that $E_H = e^2/(2a_o)$ so $E_H a_o^2 = \hbar^2/(2m_e)$, ignoring a very small center-of-mass correction.) For simple calculations, the only practical choice is to assume that the ions are hydrogenic, so that the ionization energy from state k to state $k + 1 = j$, in an isolated ion, is $E_{(k+1)k} = Z_{k+1}^2 E_H$. We will discuss below the consequences of the fact that the ions are not isolated. At a high enough temperature, this has a small effect on the average ionization.

We can determine a characteristic charge, not far from the actual average charge, from this equation as follows. There will be some value, Z_{bal}, not necessarily an integer, for which the ratio $N_j/N_k = 1$ for two imaginary ionization states having charge $Z_{bal} + 1/2$ and $Z_{bal} - 1/2$. Then Z_{bal} should be close to, but may not equal, the average charge Z. Recalling that $n_e = Z n_i$, we can solve (3.47) for Z_{bal} to find

$$ Z_{bal} = \sqrt{\frac{k_B T_e}{E_H}} \sqrt{\ln \left[\frac{1}{n_e} \frac{g_j}{4 g_k a_o^3} \left(\frac{k_B T_e}{\pi E_H} \right)^{3/2} \right]} - \frac{1}{2}, \tag{3.48} $$

which is

$$ Z_{bal} = 0.63 \sqrt{T_{eV} \left[1 + 0.19 \ln \left(\frac{(T_{eV}/100)^{3/2}}{n_{21}} \right) \right]} - \frac{1}{2}, \tag{3.49} $$

with T_{eV} in eV and n_{21} being the electron density in units of 10^{21} cm^{-3}, and (3.49) assuming $g_j = g_k$. One might approximate this as $Z_{bal} = 0.63\sqrt{T_{eV}}$, for $Z_{bal} \leq Z_n$, where Z_n is the nuclear charge.

The first estimate is to assume $Z = Z_{bal}$, in which case one can either approximate n_{21} or solve (3.49), which becomes an implicit equation for Z, through the electron density (with $n_e = Z_{bal} n_i$). In terms of the initial formulation of this problem above, the coefficient in (3.49) corresponds to $C \sim 2.3$, which is not far from our initial guess of $\sqrt{3}$. Figure 3.7 shows how Z_{bal} varies as ion density and temperature vary, solving implicitly for Z_{bal}. If the result were strictly $0.63\sqrt{T_{eV}}$, the contours would be vertical. The curve crossing the plot shows where the solution for Z_{bal} does equal $0.63\sqrt{T_{eV}}$. One can see that using $0.63\sqrt{T_{eV}}$ is accurate to about 50% over most of the parameter space shown, with a greater error at ion densities above 10^{23} cm^{-3}. One would expect the ions to exist primarily in the one or two states for which ionization and recombination nearly balance, so the value of Z from (3.49) ought to be close to the actual average ion charge in the plasma.

One can demonstrate that $Z \sim Z_{bal}$, when the ionization energies are as assumed above, as follows. One can use the definition of Z_{bal} to rewrite (3.46), for arbitrary j and k, as

$$ \frac{N_j}{N_k} = \exp \left[-\frac{(E_{jk} - Z_{bal}^2 E_H)}{k_B T_e} \right]. \tag{3.50} $$

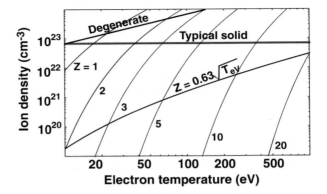

Fig. 3.7 Ionization from the Saha equation. Curves of constant Z_{bal} are shown. The electrons are Fermi degenerate in the region above the line labeled "Degenerate". The lower curve shows where Z_{bal} equals the approximate value $0.63\sqrt{T_{eV}}$

Fig. 3.8 Normalized relative populations of ionization states, for $T_e = 1$ keV and $Z_n = 30$

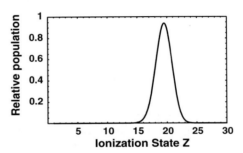

Note that this corresponds to a distribution of ions peaked around $Z_k \sim Z_{bal}$, since $N_j < N_k$ for $E_{jk}/E_H > Z_{bal}^2$, and $N_j > N_k$ for $E_{jk}/E_H < Z_{bal}^2$. Figure 3.8 shows the ratio N_j/N_1 for $T_e = 1$ keV and $Z_{bal} = 20$. Note that to obtain this one must apply (3.50) repeatedly, obtaining

$$N_j/N_1 = \prod_{k=1}^{j-1} N_{k+1}/N_k. \qquad (3.51)$$

This gives a sum in the exponent that can be evaluated, as follows:

$$N_j/N_1 = \prod_{m=2}^{j} \exp\left[-\frac{m^2 - Z_{bal}^2}{k_B T_e/E_H}\right] = \exp\left[-\frac{(j-1)\left(6 + 5j + 2j^2 - 6Z_{bal}^2\right)}{6k_B T_e/E_H}\right]. \qquad (3.52)$$

Figure 3.8 shows a plot of this distribution, which turns out to be very strongly peaked, with nearly all of the ions having a charge within a few unit charges of Z_{bal}. As it should, the peak of the distribution corresponds almost exactly to Z_{bal} as given by (3.49). One could formally evaluate the average charge using (3.52). For the ratio of ionization state populations, and a nuclear charge Z_n, one has

$$Z = \sum_{j=1}^{Z_n} j \frac{N_j}{N_1} \Bigg/ \sum_{j=1}^{Z_n} \frac{N_j}{N_1}. \tag{3.53}$$

One can show that Z determined by this method is quite close to Z_{bal}.

It is worthwhile to emphasize that the fundamental basis for our estimate of Z here is the Saha equation. However, the Saha equation is not an inviolate law of the universe, even for equilibrium systems. It is a consequence of statistical mechanics when the only important energies are the ionization and excitation energies and when the electrons and ions both follow Boltzmann statistics. As plasmas become denser or colder, energies associated with the interaction of the particles become important. Some aspects of this are discussed in the next section and further below. To some extent, these could be accounted for within the framework of the Saha equation. However, once quantum effects become essential to the behavior of the particles, whether through Fermi degeneracy or through ion–ion correlations, their partition functions change significantly and the Saha equation is no longer the relevant statement of equilibrium. This will be true well before (3.48) and (3.49) find Z_{bal} to decrease to zero and then become imaginary at high enough density or low enough temperature. The curve in the upper left corner of Fig. 3.7 shows where the electrons become Fermi degenerate based on the discussion of Sect. 3.1.3 (and assuming $Z = 0.63 \sqrt{T_{eV}}$, although the curve placement on such a log–log plot is not very sensitive to the specific assumption about Z). It remains worth noting, though, that Z decreases as density increases, even solely as a consequence of Boltzmann statistics. This in turn reflects the presence of n_e in (3.47), which arises from the degeneracy of the free electrons themselves.

Following through on the question of when the electrons dominate the internal energy of high-energy-density plasmas, we can compare the total energy of ionization, which is part of the internal energy of the plasma, with the internal energy of the electrons. The ionization energy is the sum of $Z_i^2 E_H$ over the ionization states up to Z. Here we will use the integer part of Z_{bal} as Z for this energy. The electron energy per ion is $(3/2)Z k_B T_e$, where we will use $Z = Z_{bal}$. Figure 3.9 shows the comparison of these two energies. The ionization energy forms a stairstep in such

Fig. 3.9 The increase of internal energy and ionization energy (the stairstep) and electron kinetic energy (the line) in eV with increasing T_e

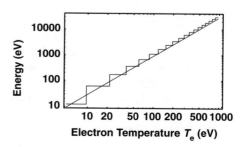

a model, though in reality the fact that several ionization states are present would smooth this out. The important conclusion is that, so long as the ion can keep on ionizing and the electrons are not Fermi degenerate, the ionization energy is the larger contribution to the internal energy. Only once the ions become fully stripped will the electron energy come to dominate. This is a major difference in comparison to low-density laboratory or space plasmas, in which the internal energy can usually be ignored.

3.4.2   The Ion Sphere Regime and Coulomb Effects

Equation (3.49) becomes inaccurate in compressed, denser matter with a high nuclear charge and low temperature. One reason is that the electrons become Fermi degenerate. Another reason is that the ions in high-energy-density plasmas do not exist in isolation. Even though plasmas are charge-neutral on a volume-averaged basis, in detail the particles arrange themselves so that a particle with any given charge is closer on average to particles of the opposite charge. As a result, one would have to invest energy to pull the plasma apart, so that the particles were far enough away from one another that their interactions were negligible. That is to say, the potential energy of the plasma is negative relative to vacuum. The introduction of new particles or charges to the plasma, as occurs during ionization, lowers the potential even further. This effect was long known as *continuum lowering*, but more recently is typically labeled *ionization potential depression*. It has consequences for the ions or atoms in the plasma—the vacuum energy levels having energies between the plasma potential and vacuum no longer exist. Figure 3.10 shows an energy level diagram to illustrate this point. With regard to ionization, the consequence is that the energy required to ionize is reduced relative to its value in vacuum.

The amount by which the ionization potential is lowered can be evaluated by determining the change in electrostatic potential energy produced by the ionization of an atom or ion. There are two basic approaches to this calculation, corresponding to two regimes of validity. For low-density plasmas, in which the Debye length exceeds the spacing of the ions, one can calculate the changes to the shielding potentials and the corresponding electrostatic energy introduced by ionization. Equivalent treatments of this regime can be found in Zel'dovich and Razier (1966), Griem (1997), and Krall and Trivelpiece (1986). We will discuss only the case most relevant to high-energy-density plasmas, in which the spacing of the ions is more than a Debye length. This has the consequence that the shielding occurs in the vicinity of each ion individually. This will still be true if the electrons are Fermi degenerate, but the electron density will be more uniform in space than it would be otherwise. The fact that the shielding is local around each ion gives rise to the ion-sphere model and variations on it. Figure 3.11 shows that the boundary between the Debye shielding regime and the ion-sphere regime lies at lower densities than those of primary interest in high-energy-density physics.

Fig. 3.10 A lowered continuum can eliminate some excited states and reduce the ionization energy

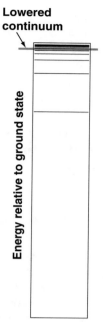

Fig. 3.11 Boundary between ion-sphere and long-range Debye-shielding regimes of ionization potential depression. The ion density is shown for reference, inferred from $Z_i = 0.63\sqrt{T_{eV}}$ and n_e

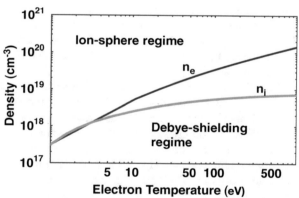

In the ion-sphere model, each ion is assumed to influence only a region within a radius R_o given by

$$\frac{4\pi}{3}R_o^3 n_i = 1,\tag{3.54}$$

in which n_i is the particle density of the ions. Beyond this distance, the positive and negative charge densities, as seen by the ion, are equal, so these make no contribution to the electrostatic potential energy. Recalling that the typical ion density is 7.5×10^{22} cm^{-3}, one can see that $R_o \sim 10^{-8}$ cm ~ 1 Å for solids, as one would expect since atoms are about 1 Å in size. Within R_o, the charge due to the free electrons must balance that of the ion, Z_i, and for consistency (with the viewpoint of other ions) the average free electron density must equal that throughout the entire plasma, so

$$Z_i = \frac{4\pi}{3} R_o^3 n_e. \tag{3.55}$$

The ion-sphere viewpoint provides the context for modern Thomas-Fermi models, for estimates of continuum lowering, and for an approximate EOS for high-energy-density plasmas, which are our next three topics.

3.4.3 The Thomas–Fermi Model and QEOS

The *Thomas–Fermi model* provides a way to account for the impact of ion-sphere effects on electron behavior. Based on a few very simple relations, the Thomas–Fermi model accurately includes the effects of ionization, excitation, Fermi-degeneracy, Coulomb interactions, self-consistent electron density structure, and to some extent ion–ion coupling. In various versions it may also include some quantum-mechanical effects such as those of shell structure. There is a nice summary of the Thomas–Fermi model in Salzman (1998) and more detail in Eliezer et al. (1986). The model itself requires nontrivial numerical solution, but we reproduce below some fits due to Salzman that provide a useful way to connect temperature and ionization.

The Thomas–Fermi model is a self-consistent combination of the ion–sphere model and the treatment of the electrons as fermions. The key to its power is that it demands that the electrostatic interaction of the electrons as fermions and the nucleus be self-consistent within this context. It naturally accommodates, in a classical context, the increase in electrostatic energy associated with increasing density or temperature. This allows one to ignore ionization and excitation as separate processes. They are accounted for, on average, by the expansion of the heated electrons or the lowering of the continuum as conditions change. The model can be formulated, in a simple form, as follows. It assumes spherical symmetry.

The electric potential, $\Phi(r)$, is given by the Poisson equation,

$$\nabla^2 \Phi(r) = 4\pi e n_e(r) - 4\pi Z_n e \delta(r), \tag{3.56}$$

with the boundary condition that $\partial \Phi / \partial r = 0$ at the boundary of the ion sphere, $r = R_o$, which follows from the net charge neutrality of each ion sphere. The electron density is given by the generalization of (3.29) to include a varying potential energy:

$$n_e(r) = \frac{8\pi}{h^3} \int_0^\infty \frac{\chi_e^2 d\chi_e}{\exp[(-\mu - e\Phi(r) + \mathcal{E}_e)/(k_B T_e)] + 1}$$

$$= \frac{4\pi(2m_e k_B T_e)^{3/2}}{h^3} \int_0^\infty \frac{\sqrt{x}dx}{\exp[x - (\mu + e\Phi(r))/(k_B T_e)] + 1}. \tag{3.57}$$

Thus, the nature of electrons as fermions is accounted for. The net neutrality of each ion sphere sets a constraint on the density,

$$Z_n = 4\pi \int_0^{R_o} n_e(r)r^2 dr, \tag{3.58}$$

which determines the chemical potential. These three equations are all that must be solved to describe the system. Once computational mathematics programs evolve beyond root finding to profile finding, this model may become simple to implement.

Now, supposing we have solved the above equations, we consider how the results may be used. The potential is only defined in the above to within an arbitrary constant, although the choice of this constant will affect the value of μ. It is conventional to choose $\Phi = 0$ at the boundary of the sphere. As a result, the potential throughout the sphere becomes increasingly positive as the density increases. This is how this model captures the effects of ion-electron interactions. The charge state is calculated as

$$Z = \frac{4\pi}{3} R_o^3 n_e(R_o), \tag{3.59}$$

which amounts to assuming that the free electrons flow freely between ions and thus establish the density at the ion-sphere boundary.

Some other thermodynamic quantities are as follows. The electron pressure is

$$p_e(r) = \frac{8\pi(2m_e)^{3/2}(k_B T_e)^{5/2}}{3h^3} \int_0^\infty \frac{x^{3/2}dx}{\exp[x - (\mu + e\Phi(r))/(k_B T_e)] + 1}, \tag{3.60}$$

and the electron kinetic energy in each ion sphere is given by

$$K_e = \frac{4\pi(2m_e)^{3/2}(k_B T_e)^{5/2}}{h^3}$$

$$\times \int_0^{R_o} \int_0^\infty \frac{x^{3/2}dx}{\exp[x - (\mu + e\Phi(r))/(k_B T_e)] + 1} dr. \tag{3.61}$$

This is not, however, the entire energy, because the Coulomb energy of attraction remains to be accounted for. This can be calculated directly, with the interaction energy of the electrons and the nucleus, per atom, being U_{en} while the energy per atom of the interactions among the electrons is U_{ee}. One has

$$U_{en} = -4\pi Z_n^2 e^2 \int_0^{R_o} n(r) r \mathrm{d}r \tag{3.62}$$

and

$$U_{ee} = \frac{e^2}{2} \int_0^{R_o} \int_0^{R_o} \frac{n(r)n(r')}{|r-r'|} \mathrm{d}^3 r \mathrm{d}^3 r'. \tag{3.63}$$

With these definitions, the total specific internal energy is

$$\epsilon = (K_e + U_{en} + U_{ee})/(Am_p). \tag{3.64}$$

The sum $U_{en} + U_{ee}$ is the electrostatic energy per atom.

In the limit that $n(r)$ is constant (and thus equal to $Z_n n_i$, where n_i is the ion density), then one finds $U_{en} = -(3/2)Z_n^2 e^2/R_o$ and $U_{ee} = (3/5)Z_n^2 e^2/R_o$. As More et al. (1988) discuss, the assumption of constant density for all electrons is most applicable for either fully stripped ions at high temperature or strongly Fermi-degenerate electrons at high density. Under other conditions, the actual total energy for all electrons will reflect the actual profile of electron density. However, these results are also relevant to the free electrons, assumed in the model to have constant density. Their total electrostatic energy per atom, under this assumption, is given by $U_{net} = U_{en} + U_{ee}$, evaluated for the constant (free) electron density with Z_n equal to the net charge Z. Here again, the actual energy will be different to the extent that the free-electron density is not in fact uniform.

To avoid confusion in connecting this section with others, we should note that the zero of the energy scale here for the electrostatic energies is *not* consistent with the conventions used in other discussions in this book. In most of those other discussions, the implicit point of view is that a state of zero energy and pressure is a neutral gas nominally at zero temperature (but without quantum effects). At higher temperatures, positive energy is invested to ionize the gas. At high densities, the Coulomb interactions of the ionized gas provide some binding energy and reduce the energy input that would otherwise be required. In contrast, the state of zero energy in conventional Thomas–Fermi models has all the particles dispersed to infinity. In the ion-sphere applications of Thomas–Fermi models, it is further assumed that the ions are shielded from one another and that the potential of the ion-sphere boundary is equivalent to the potential at infinity. To convert the Thomas–Fermi result to the standard scale, one would have to add the ionization energy necessary to ionize the atom in vacuum, up to its ionization state Z, to the internal energy per atom. (See Sect. 3.5 below.) This matters to account for the internal energy properly.

Salzman (1998) provides a more extensive discussion of Thomas-Fermi models, and of many issues involving atomic physics in plasmas. He includes a fit to the ionization produced by such models, attributed to a laboratory report by R.M. More. My students have found this quite useful, and so it is included here for convenience. The fits are provided for T in eV and ρ in g/cm^3. Using the notation above when feasible, one has an average charge

$$Z = f(x)Z_n, \tag{3.65}$$

in which

$$f(x) = x/(1 + x + \sqrt{1 + 2x}), \tag{3.66}$$

in which for $T = 0$ one has $x = \alpha \mathcal{R}^\beta$, where $\alpha = 14.3139, \beta = 0.6624$, and $\mathcal{R} = \rho/(Z_n A)$.

For $T > 0$, has $x = \alpha \mathcal{Q}^\beta$ but the evaluation of $\mathcal{Q}$ is rather involved. Take $T_0 = T/Z_n^{4/3}$. Then

$$A = a_1 T_0^{a_2} + a_3 T_0^{a_4}, \tag{3.67}$$

where $a_1 = 3.323 \times 10^{-3}, a_2 = 0.971832, a_3 = 9.26148 \times 10^{-5}$, and $a_4 = 3.10165$. Then with $T_F = T_0/(1 + T_0)$ one has

$$B = -\exp[b_1 + b_2 T_F + b_3 T_F^7], \tag{3.68}$$

where $b_1 = -1.7630, b_2 = 1.43175$, and $b_3 = 0.315463$. One also defines

$$C = c_1 T_F + c_2, \tag{3.69}$$

where $c_1 = -0.366667$, and $c_2 = 0.983333$. Further defining $Q_1 = A\mathcal{R}^B$, one has at last $\mathcal{Q}$ as

$$\mathcal{Q} = (\mathcal{R}^C + Q_1^C)^{1/C}. \tag{3.70}$$

One can set up these relations in a computational mathematics program so that they quickly can be evaluated to find $Z(T)$ and thus $T(Z)$ for conditions of interest, for example in seeing what can be inferred from experimental data.

Finally, there are a class of computer models known as *QEOS* models, where QEOS stands for quotidian equation of state, where quotidian means "everyday" or "routinely usable". The Thomas–Fermi model is often incorporated into these (see for example the description in More et al. (1988)). Such models are likely to include additional terms or equations intended to account for the solid, liquid, and gaseous states and for the transitions between them. They can be a useful way to bridge the wide range of parameters that simulations must deal with.

3.4.4 Ionization Potential Depression

Here we consider the amount of continuum lowering in the ion-sphere regime. There have been several approaches. A simple estimate would be that an electron is free if it has enough energy to reach the boundary of the ion sphere, and that the difference between the potential energy there and at infinity (for an isolated ion of initial charge Z_i) is

$$\Delta E = (Z_i + 1)e^2/R_o = (Z_i + 1)(2E_H a_o)/R_o. \qquad (3.71)$$

The generalization of (3.71), to include the behavior at lower densities where the Debye length, λ_D, exceeds R_o, is

$$\Delta E \approx (Z_i + 1)E_H \min\left(\frac{2a_o}{\lambda_D}, \frac{2a_o}{R_o}\right). \qquad (3.72)$$

There are a number of models that have attempted to do better for the ion sphere regime, and at this writing (2014) it has only recently become possible to do well-resolved measurements of continuum lowering in this regime. Some measurements seem to support one or the other of the available models. But all the models are simplifications of various types. One can expect a more definitive understanding to arise over the next few years.

A brief survey of the historic models follows here. (A reader not interested in these details could skip to the discussion of the net ionization with reference to the next figure.) To compare these models we generalize (3.71) as follows:

$$\Delta E = CZ^* e^2/R_o = CZ^*(2E_H a_o)/R_o. \qquad (3.73)$$

Griem (1997) makes an approximate calculation of the shift in the energy levels of the ion by determining from the Poisson equation the electrostatic potential surrounding the ion, assuming a constant electron density, and by using the first-order perturbation theory of hydrogenic ions from basic quantum mechanics. One finds the principal quantum number of the highest remaining bound state to be

$$n_c = \sqrt{(Z_i + 1)R_o/a_o}. \qquad (3.74)$$

Zel'dovich and Razier (1966) find the same result from the semiclassical argument that the highest quantum number will be the one for which the semimajor axis of the orbit equals R_o. The corresponding reduction in ionization energy is

$$\Delta E \approx (Z_i + 1)E_H a_o/R_o = (Z_i + 1)e^2/(2R_o). \qquad (3.75)$$

Thus, such models find $C = 1/2$ and $Z^* = Z_i + 1$ in (3.73). One can expect that this might under-estimate the lowering, because most of the electrons having principal quantum number n_c will have larger orbital angular momentum and have orbits that attempt to extend beyond R_o. But on the other hand, the actual interaction with the other bound electrons might reduce the lowering.

An important historic approach is that of Ecker and Kroll (1963). Their calculation is a statistical mechanical one not structurally unlike the calculation we discuss in Sect. 3.5. One difference is that they treat the electrons (and ions) as a classical gas, so their results ought not to apply when the electrons are degenerate (as they are for much of the ion-sphere regime of interest). Another difference is that they seek to calculate the effects of Coulomb forces by a calculation of the microfields in the plasma, using traditional techniques of plasma physics. A third difference is that their calculation is not based on the ion-sphere model. Even so, they are forced to approximate the calculation and in so doing they choose, as a characteristic distance, the average interparticle distance among all particles,

$$R_a = \left(\frac{3}{4\pi n_i(1 + Z_i)} \right)^{1/3} = \frac{1}{(1 + Z_i)^{1/3}} R_o. \tag{3.76}$$

The also find that it is the charge of the ion resulting from the ionization process $(1 + Z_i)$ that matters, and ultimately they find

$$\Delta E = C_{EK}(1 + Z_i)e^2/R_a = C_{EK}(1 + Z_i)^{4/3}e^2/R_o, \tag{3.77}$$

where their $Z^* = (1 + Z_i)^{4/3}$. Their C_{EK} is of order 1 but is a function that depends on Z_i and Z_n in some way that is not very clear.

Other approaches are based on the semi-classical Thomas-Fermi model discussed in the previous section. The historic paper of this type is by Stewart and Pyatt (1966). Their calculation differs in some ways from the now-standard approach to Thomas-Fermi, ion-sphere calculations described in the previous section, but what is important is that they apparently do not consider the effect of electron-electron interactions. They find, in the ion-sphere limit,

$$\Delta E = (3/2)(Z_i + 1)e^2/R_o, \tag{3.78}$$

and thus conclude that $Z^* = (Z_i + 1)$ and $C = 3/2$, so the lowering is more than three times larger than in our simple result above. They confirm that, within their assumptions, the results are correct by numerical integration. Note that the ΔE found in (3.78) equals exactly $-U_{en}/Z^*$, the average electrostatic energy per electron for Z^* electrons of constant density interacting with a central ion of the same charge in an ion sphere (from the previous section). This raises issues discussed more clearly with regard to the next model.

More et al. (1988) report results of a more modern numerical Thomas-Fermi calculation, consistent with the description in the previous section. For a nuclear point charge of value $Z_n e$, the net electrostatic energy per atom is

$$U_{\text{net}} = -(9/10)Z_n^2 e^2/R_o, \tag{3.79}$$

where coefficient of $(9/10)$ is the combination of coefficients of $3/2$ for the electron-ion interactions and $-3/5$ for the electron-electron interactions. The negative of this is the amount of energy that would be required to remove the electrons to infinity from an initial state in which they have a uniform density within the ion sphere. Thus U_{net} is the difference between the energy required to fully strip the atom in vacuum vs in the ion-sphere environment, and so represents the total continuum lowering for all the electrons. But as the electrons ionize successively this cannot be equal for each electron. If one imagines that the bound electrons are localized in orbits at small radii (which is necessary to get the ionization energy right in a semiclassical model), then it is a sensible approximation to treat the net charge of the ion, Q, as though it were a point charge, and to assume the density of the free electrons to be constant. In this case the continuum lowering for the next electron will correspond to the difference in U_{net} as Z_n increases by one. Mathematically, for ionization out of state Z_i, this implies

$$\Delta E = \frac{9}{10} \frac{[(Z_i + 1)^2 - Z_i^2]e^2}{R_o} = 1.8 \frac{(Z_i + 0.5)e^2}{R_o}, \tag{3.80}$$

which at large Z_i is even larger than the Stewart-Pyatt value. Thus in this case we find $C = 1.8$ and $Z^* = (Z_i + 0.5)$.

All these estimates for continuum lowering imply that there can be conditions where ΔE is larger than the vacuum ionization energy, producing some ionization even at zero temperature. This effect is known by the somewhat misleading name of *pressure ionization*. (The name is misleading since only density enters. However, in dense, Fermi-degenerate matter the pressure can be substantial even at zero temperature.) To assess this, one can write the ratio of ΔE to the ionization energy in vacuum ($\sim (Z_i + 1)^2 E_H$). Specifically, using (3.80) for ΔE one finds

$$\frac{\Delta E}{(Z_i + 1)^2 E_H} = 3.6 \frac{(Z_i + 0.5)a_o}{(Z_i + 1)^2 R_o} = 3.04 \frac{(Z_i + 0.5)}{(Z_i + 1)^2} n_{24}^{1/3}, \tag{3.81}$$

in which the approximation uses the ion density in units of $10^{24}\,\mathrm{cm}^{-3}$. In our hydrogenic model a state will be ionized when the left-hand term here equals 1 and we will designate this ionization Z_{TF} for later reference. This gives

$$Z_{TF} = 1.52 \left[n_{24}^{1/3} + \sqrt{n_{24}^{2/3} - 0.66 n_{24}^{1/3}} \right] - 1, \tag{3.82}$$

In this model the ionization goes to zero at an ion density of $2.8 \times 10^{23}\,\mathrm{cm}^{-3}$. Figure 3.12 shows the ionization as calculated from this model.

One can see in this figure that the amount of ionization remains fairly small for compression to only a few times solid density. In the regime where the ions behave like hydrogen, as a crude model one could take Z to be the maximum of the values implied by (3.81) and Z_{bal} from the Saha model (3.49). Near the transition between the two models, this will underestimate Z, because plasma effects will reduce the ionization energy of the next couple of ionization states significantly. In addition, once the electrons become degenerate the thermal ionization is further reduced. We discuss the behavior found from a better, statistical-mechanical model below, in Sect. 3.5.2. By the time T_e increases much above $10\,\mathrm{eV}$, most materials of interest in present-day experiments will be in the Saha regime.

However, Fig. 3.12 can be quite misleading, because most ordinary materials, at temperatures of order an eV, do not behave like a simple hydrogenic model would

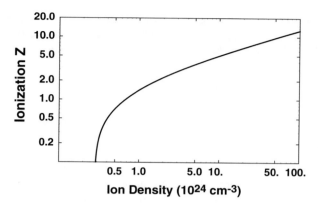

Fig. 3.12 The ionization from (3.82)

predict. One might say, for example, that conductors have an effective ionization state corresponding to the number of free electrons per atom that exist in the conduction band. In the case of solid-density aluminum ($Z_n = 13$), for example, this is about 3 electrons per atom. One can reasonably describe this as "pressure ionization", recalling again that a better term would be "density ionization". To some degree, the appearance of ionization where the hydrogenic model would not find it could be due to subtle quantum effects. But more important is the classical impact of the existence of multiple electrons around the nucleus. In a classical context, one would say that the inner electrons act to shield the outermost electrons from the nuclear charge. This effect is accounted for in the Thomas–Fermi model, which does find approximately the correct number of free electrons for aluminum. (The difference from Fig. 3.12 presumably reflects the evaluation of self-consistent electron density profiles in the full model.) While all materials behave in similar ways in a global sense, the exact density or temperature where certain transitions occur varies greatly. Quantum effects in the ions may be very important, especially in the regime known as *warm dense matter*, corresponding to densities of order solid density and temperatures below a few eV. We discuss this regime of Fermi degenerate, strongly coupled matter in Sect. 3.7.

3.4.5 Coulomb Contributions to the Equation of State

The Coulomb interactions discussed in the previous section also contribute to the EOS. The corresponding contribution to the internal energy, for ions of charge Z, is

$$\rho\epsilon_{Coul} = n_i U_{net} = -(9/10)n_i \langle Z^2 \rangle e^2 / R_o, \tag{3.83}$$

where $\langle Z^2 \rangle$ should be an average of j^2 over the distribution of states of charge j, but will be quite close to Z^2 for a narrowly peaked state distribution of average ionization Z. The pressure of N_a such atoms in a volume V, such that $n_i = N_a/V$, is

$$p_{\text{Coul}} = -\frac{\partial}{\partial V}(N_a U_{\text{net}}) = -(3/10)n_i Z^2 e^2/R_o, \tag{3.84}$$

3.4.6 The Ions

For most conditions of interest, the ions can be treated as an ideal, classical gas. When this is the case, we can take the ion contribution to the pressure, p_i, and specific kinetic energy, ϵ_{ik}, to be

$$p_i = \frac{\rho k_B T_i}{A m_p} \tag{3.85}$$

and

$$\epsilon_{ik} = \frac{3}{2}\frac{k_B T_i}{A m_p}, \tag{3.86}$$

knowing that the Coulomb binding energy associated with close packing of ions will be included in the accounting for the electrons. Here the ion temperature is T_i. In addition, in the context of our convention that the initial material state is a low (or zero) but positive energy state, the ions also contribute energies of ionization and excitation. We ignore excitation here, for reasons discussed near the start of Sect. 3.4, and once again use a hydrogenic model, describing the internal energy of the ions as

$$\epsilon_{ii} = \frac{R_i}{A m_p} = \frac{E_H}{A m_p}\sum_{k=0}^{Z} k^2 = \frac{E_H}{6 A m_p}Z(1 + Z)(1 + 2Z), \tag{3.87}$$

where the maximum allowed value of Z is Z_n and we define R_i as the internal energy per ion and we have evaluated it for our hydrogenic model of the ions. One could do better by including the continuum lowering in the sum making up R_i, and even better by using actual ionization energies for the species in question.

3.5 Approximate Equations of State for High-Energy-Density Plasmas

In this section we work with the above results to find an approximate EOS for high-energy-density matter at two levels of sophistication. We begin with results for the regime of classical electrons, which spans most of our region of interest, applying as well the simple expression for Saha ionization $Z \sim 0.63\sqrt{T_{eV}}$. We then develop a more sophisticated model, working with the Helmholtz Free Energy. The reader can find an excellent discussion of the departure from an ideal plasma in Krall and Trivelpiece (1986), while Griem (1997) provides a connection with the more-recent literature.

There are two limiting cases where the simple models used here break down. One of them, discussed in Chap. 11, is connected with inertial fusion. At the temperatures of cryogenic fusion fuel, one must consider the nature of the fuel as quantum particles. In particular, one needs to examine the behavior of deuterons as bosons and of tritons as fermions. We consider this issue in Chap. 11. The other case is that of Fermi-degenerate, strongly coupled matter. New effects appear when matter is pushed into this regime, which was not known at the time of writing of the first edition of this book. We discuss this regime in Sect. 3.7.

3.5.1 The Simplest EOS Model

Here we combine the results of the previous two sections to assemble an EOS for the regime across which the Saha model is valid, but throughout which the matter may be ionizing and Coulomb effects may not be negligible. This spans most of our parameter space of interest. One could replace the thermal components of pressure and internal energy with results for Fermi-degenerate matter, when $T_e < \epsilon_F$, to have a crude model that also spanned the Fermi-degenerate regime. For simplicity, we have chosen here not to do this. A more complex model, able to deal with the degenerate regime more accurately, is presented in the following subsection.

We can represent the electrons as a classical gas so that the electron thermal pressure is $n_i Z k_B T$ and the electron internal thermal energy is $3 n_i Z k_B T / 2$. We can anticipate that these values will become inaccurate when the electrons become Fermi degenerate, and we know that pressure ionization may increase Z at temperatures below some value when the ion density exceeds 10^{24} cm^{-3}. However, the full impact of ionization is more complex, as we discuss in Sect. 3.5.2. In most cases, the electrons dominate the pressure and the kinetic energy. Based on the discussion above, we can represent the pressure as

$$p_{tot} = p_{th} + p_{\mathrm{Coul}} = n_i(1 + Z)k_B T - \frac{3}{10}\frac{n_i Z^2 e^2}{R_o} \tag{3.88}$$

and the internal energy as

$$\rho\epsilon_{tot} = \rho\epsilon_k + \rho\epsilon_{Coul} + \rho\epsilon_{ik} =$$

$$\frac{3}{2}n_i(1+Z)k_BT - \frac{9}{10}\frac{n_iZ^2e^2}{R_o} + n_i\frac{E_H}{6}Z(1+Z)(1+2Z). \tag{3.89}$$

Here we have expressed the results in terms of the ion density $n_i = \rho/(Am_p)$. In this simplest model, we give formulae for these results for two cases: an ionizing regime in which $Z = 0.63\sqrt{T_{eV}}$ and a regime with fully stripped ions, for which $Z = Z_n$. With n_i in cm^{-3}, we have:

For the ionizing regime

$$p = 1.6 \times 10^{-12}n_iT_{eV}\left(1 + 0.63\sqrt{T_{eV}} - 2.76 \times 10^{-8}n_i^{1/3}\right) \tag{3.90}$$

and

$$\rho\epsilon = 1.6 \times 10^{-12}n_i\left[1.44\sqrt{T_{eV}} + 4.22T_{eV}\right.$$

$$\left. +2.09T_{eV}^{3/2} - 8.29 \times 10^{-8}n_i^{1/3}T_{eV}\right], \tag{3.91}$$

and for the fully stripped regime

$$p = 1.6 \times 10^{-12}n_i\left(T_{eV}(1+Z_n) - 6.96 \times 10^{-8}n_i^{1/3}Z_n^2\right) \tag{3.92}$$

and

$$\rho\epsilon = 1.6 \times 10^{-12}n_i\left(1.5T_{eV}(1+Z_n)\right.$$

$$\left. +2.28Z_n + (6.85 - 2.09 \times 10^{-7}n_i^{1/3})Z_n^2 + 4.57Z_n^3\right). \tag{3.93}$$

We can evaluate the specific heat at constant volume, c_V, for these cases as well. For the ionizing plasma case of (3.91), we find

$$c_V = \frac{1.6 \times 10^{-12}}{Am_p}\left(4.2 + \frac{0.715}{\sqrt{T_{eV}}} + 3.1\sqrt{T_{eV}} - 8.3 \times 10^{-8}n_i^{1/3}\right), \tag{3.94}$$

while for the fully stripped case of (3.93), we have

$$c_V = \frac{2.4 \times 10^{-12}(1+Z_n)}{Am_p}. \tag{3.95}$$

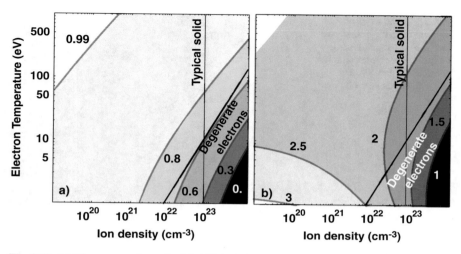

Fig. 3.13 (a) The pressure for an ionizing high-Z element is shown, normalized to the ideal-gas pressure. The contours increase from the lower right, and are at 0, 0.1, 0.3, 0.5, and 0.99. (b) The internal energy density for an ionizing high-Z element is shown, normalized to the ideal-gas value. The contours increase from right to left, and are at 3, 5, and 6

We will first explore the implications of the first two equations for an element of high enough Z to keep ionizing. Then we will consider carbon as an example of an element that can be fully stripped. Considering a high-Z element, it is informative to compare the pressure and energy from (3.90) and (3.91) with their ideal-gas equivalents, which are $p = n_i(1 + Z)k_B T_e$ and $\rho\epsilon = (3/2)n_i(1 + Z)k_B T_e$, respectively. Figure 3.13a, b shows the ratio of the more-complete estimates in (3.90) and (3.91) to these ideal-gas values. One can see that the model for the pressure fails badly in the Fermi-degenerate region, which is no surprise. In actuality, pressure and internal energy both increase in that region as one moves down and to the right. Otherwise, the pressure across the space of Fig. 3.13a is typically between 50% and 100% of the ideal-gas value. Thus, the ideal-gas value is not too bad but it may overestimate the pressure. In contrast, Fig. 3.13b shows that the internal energy is typically twice or more the ideal gas value, as indeed one would expect from Fig. 3.9.

With an increased internal energy and a decreased pressure, the value of γ evaluated from $\rho\epsilon = p/(\gamma - 1)$ must decrease. Figure 3.14 shows the values of γ obtained from (3.90) and (3.91). Here again, the quantitative value (~ 1.25) should not be taken too seriously but the qualitative point, that γ should be reduced substantially compared to the ideal-gas value of 5/3, should be real. Standard EOS evaluations for xenon ($A = 130, Z_n = 54$), for example, give $\gamma \sim 1.2$ to 1.3. We will see in Chap. 4 that this implies increased compression by shocks.

Now consider carbon, an element with six electrons that can become fully stripped at modest temperatures. Using our estimate that $Z = 0.63\sqrt{T_e}$, Carbon will ionize fully at $T_e = 91$ eV. At higher temperatures, the internal energy still

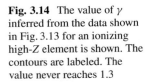

Fig. 3.14 The value of γ inferred from the data shown in Fig. 3.13 for an ionizing high-Z element is shown. The contours are labeled. The value never reaches 1.3

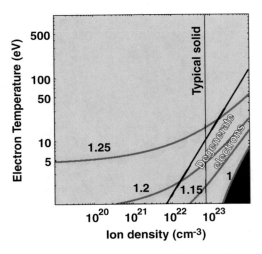

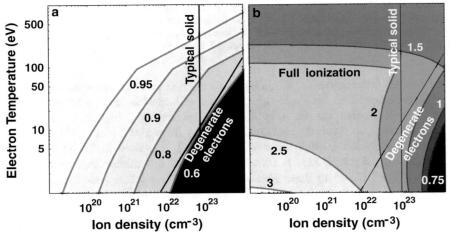

Fig. 3.15 (**a**) The pressure for carbon is shown, normalized to the ideal-gas pressure. The contours increase from the lower right, as labeled. (**b**) The internal energy density for carbon is shown, normalized to the ideal-gas value. The contours are labeled

includes the energy of ionization, but this contribution does not increase any further. To estimate the properties of carbon, we use (3.90) and (3.91) until $T_e = 91$ eV, then (3.92) and (3.93) at higher temperatures. This produces Figs. 3.15 and 3.16. The pressure shows a structure similar to that of the ionizing case. In contrast, the internal energy shows more structure, as Fig. 3.15b shows. At temperatures where carbon is not fully stripped, the ionization energy is a dominant factor and the internal energy substantially exceeds the ideal-gas value. As ion density increases, though, this effect becomes smaller. Then, once the temperature has increased

Fig. 3.16 The value of γ inferred from the data shown in Fig. 3.15 for carbon is shown. The contours are labeled

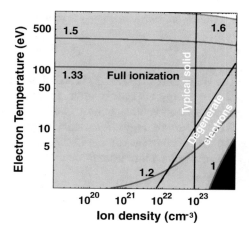

enough to fully strip the material, the internal energy decreases toward the ideal-gas value.

This behavior leaves its footprint on the inferred γ, shown in Fig. 3.16. At temperatures below 91 eV, one sees behavior very like that of Fig. 3.14. The inferred value of γ is generally near 1.25. Then, once the element becomes fully stripped, γ begins to increase, although in this model it does not reach the ideal-gas limit of 5/3 even at a temperature of 1 keV. In optically thick media, at true LTE this will begin to become artificial, because the coupling to the radiation will begin to reduce γ toward 4/3 by the time T_e reaches 1 keV.

3.5.2 An EOS Model Based on the Helmholtz Free Energy

The assumptions of the prior section are that each ion species and the electrons behave as a classical gas and that there is some energy of Coulomb interaction that lowers the internal energy. This applies across much of the high-energy-density regime. A more broadly applicable set of altered assumptions are that the electrons behave as fermions within each ion sphere and that the Coulomb effects also act to lower the ionization energies. We can analyze such a system using very standard statistical mechanics. We begin with a very brief summary of the statistical mechanical context. Our point of view will be that there are N_a atoms within a volume V. The statistical analysis is based on the assumption that we know the probability that the entire system will be found in some state having total energy E_T. For particles obeying Boltzmann statistics, this is proportional to $e^{-E_T/(k_B T)}$, with the constant of proportionality established so that the sum (or integral) over all possible states, S, yields the correct total number of particles. The discussion in this section in part follows that of More et al. (1988) and also (unpublished) work by Igor Sokolov.

The *Helmholtz free energy* is then given by

$$F = -k_B T \ln \mathcal{S}. \tag{3.96}$$

The Helmholtz Free Energy is useful in determining quantities of interest. The units of free energy are energy units, though it may be expressed per particle, per unit mass, per unit volume, or as a total for a system of particles. We did the latter above. From this free energy one can find the pressure, the internal energy density $(\rho\epsilon)$, heat capacity at constant volume C_V, the entropy density (ρs), and the electron chemical potential, μ_e on the assumption that $T_e = T_i = T$, as follows:

$$p = -\left(\frac{\partial F}{\partial V}\right)_{N_a, T}, \tag{3.97}$$

$$(\rho\epsilon) = F - TS = -\frac{T^2}{V}\left[\frac{\partial}{\partial T}\left(\frac{F}{T}\right)\right]_{N_a, V}, \tag{3.98}$$

$$C_V = \left(\frac{\partial(\rho\epsilon)}{\partial T}\right)_{N_a, V}, \tag{3.99}$$

$$(\rho s) = -\frac{1}{V}\left(\frac{\partial F}{\partial T}\right)_{N_a, V}, \tag{3.100}$$

and

$$\mu_e = \left(\frac{\partial F}{\partial N_e}\right)_{N_a, V}. \tag{3.101}$$

Equation (3.98) first relates the internal-energy density to F and to entropy density S, then gives a simpler combined expression. (One property of F is that $S = \partial F/\partial T$ at constant N_a and V.)

For a system having independent species labeled $A, B, \ldots$, each of which has a total number of particles $N_A, N_B, \ldots$, one has

$$S = \frac{\mathcal{P}_A^{N_A} \mathcal{P}_B^{N_B}}{N_A! \ N_B!} \cdots, \tag{3.102}$$

in which the partition function of species k is given by $\mathcal{P}_k$. The partition functions are in turn sums over the probability that given states are occupied, but here we will use results for the values of such sums and will not derive them. Using Stirling's formula, $N! = (N/e)^N$, where here e is the base of the natural logarithm, one finds

$$F = -k_B T \left[N_A \ln\left(\frac{e\mathcal{P}_A}{N_A}\right) + N_B \ln\left(\frac{e\mathcal{P}_B}{N_B}\right) + \ldots\right], \tag{3.103}$$

where one assumes all species are equilibrated to the same temperature T.

The specific case of a single atomic species that is ionizing, which is our focus here, deserves some further discussion. One could write down a partition function for all possible states of the atom including all degrees of ionization. The corresponding value of N in these equations would be the total number of atoms, N_a. Thus $N_a!$ would appear in the denominator. However, such a sum would include many very unlikely states, such as states near complete ionization having total energies very far above $k_B T$. Since we will in fact be interested only in equilibrium and perhaps near-equilibrium states, we divide the states of the atom into distinct ionization states, each of which has a population N_j and (in principle) a range of energy levels accessible to it. And we will find the equilibrium populations by seeking a minimum of F. We then, for the ions, have a sum like that of (3.103), in which there is a term for each ionization state with its corresponding population.

How to handle the electrons is a somewhat more subtle question. The standard derivation of the Saha equations assumes that the electrons freely sample the entire volume, and implicitly takes the electron density to be constant throughout. In an ion-sphere environment, however, this is not really a sensible assumption. In the Thomas–Fermi model, for example, the chemical potential is determined within an individual ion sphere. Even in the classical limit, the electron partition function in an ion-sphere environment varies with the ionization state. We will approach this as follows. We will assume that the distribution of the electrons corresponds quite closely to having local quasi-neutrality, so that $N_{ej} = jN_j$ electrons associated with atoms having ionization j and the corresponding electron partition function. This leads us to write

$$ F = \sum_{j=0}^{Z_n} \left[-k_B T N_j \ln \left(\frac{e \mathcal{P}_j}{N_j} \right) + j N_j F_e(j) \right], \qquad (3.104) $$

where $F_e(j)$ is the free energy of the electrons in an atom of ionization state j, which we will later evaluate for the two limiting cases.

It is standard practice to recognize that the partition function of each species can often be factored into terms representing distinct physical mechanisms, such as the energies associated with translation and rotation. In the present context, the question of how to treat the contribution to the free energy of the electrostatic energies that produce continuum lowering is also significant. In a low-density plasma, this represents averages over many atoms, and thus reflects the average state of ionization of the system. In the ion-sphere limit, however, each ionic species has its own electrostatic energy and its own electron chemical potential (here μ_e), reflecting the specific properties of that ion species. In effect, the statistical analysis here must be viewed as an average over a large number of distinct atoms or over an ensemble of possible states of a given atom. The implication is that we must write the partition function of the species having ionization j as

$$ \mathcal{P}_j = \mathcal{P}_{\text{trans}} \mathcal{P}_{\text{ionize}} \mathcal{P}_{\text{Coul}}, \qquad (3.105) $$

where

$$\mathcal{P}_{\text{trans}} = g_j \left(\frac{2\pi M k_B T}{h^2} \right)^{3/2} V \tag{3.106}$$

$$\mathcal{P}_{\text{ionize}} = e^{-E_j/(k_B T)} \tag{3.107}$$

$$\mathcal{P}_{\text{Coul}} = e^{-U_{\text{net}}(j)/(k_B T)} = \exp \left[\frac{9}{10} \frac{j^2 e^2}{R_o k_B T} \right], \tag{3.108}$$

in which the ion mass is M, the Planck constant is h, the Boltzmann constant is k_B, and the ion-sphere radius is R_o. Note that the electronic charge is (italic) e, while e as in (3.104) is the base of the natural logarithm. We expand ln(e) to 1 in the following to minimize confusion. Here as in Sect. 3.4.1 we consider only ground-state electrons so that $g_j = 2$. The total ionization energy of state j is the sum over the ionization energies, in vacuum, of all lower ionization states,

$$E_j = \sum_{k=1}^{j} E_{k(k-1)}, \tag{3.109}$$

with $E_{k(k-1)}$ being the energy to ionize from state $(k-1)$ to k, as it was above. Here also $E_0 = 0$.

Motivated by the results shown in Sect. 3.1.3, we will consider the electron free energy F_{ej} in only its two limits, adding to the subscript when needed to discriminate between them. For the degenerate limit we have

$$F_{ed}(j) = \frac{3}{5} \epsilon_F(j) = \frac{3}{5} \frac{h^2}{2m_e} \left(\frac{3jN_a}{8\pi V} \right)^{2/3}, \tag{3.110}$$

which depends on the electron density of an ion sphere having j free electrons. For the classical limit we have

$$F_{ec}(j) = -k_B T \ln \left(\frac{2V(2\pi m_e k_B T)^{3/2}}{jN_a h^3} \right), \tag{3.111}$$

in which the electron mass is m_e. Note that neither of these electron free energies depends on N_j. The partition function of the bound electrons is accounted for by g_j in $\mathcal{P}_{\text{trans}}$, and there is no further electron contribution for the neutral state so $F_{ed(0)} = F_{ec(0)} = 0$. The Fermi energy ϵ_F is defined above in Sect. 3.1.3. For the degenerate case, (3.110) gives the correct value for the chemical potential, as is seen below. But it is significant that the Fermi energy in an ion sphere depends on j, and must be evaluated using an electron density of $n_e = jN_a/V$, as shown explicitly in (3.110).

Regarding the classical case, note that the electron partition function is defined by integrating over all the states corresponding to a single ion sphere, of volume

V/N_a, and then recognizing that the total volume corresponding to the N_j such ion spheres is VN_j/N_a. Then when the partition function corresponding to this volume is divided by $N_{ej} = jN_j$ one obtains the final result in (3.111). In instead one were to demand that the electron density in this term be constant independent of ionization state, which is done implicitly in the standard derivation of the Saha equation, then the analysis below would obtain Eq. (3.47) to describe the ionization states.

With all the above, we have for the Helmholtz free energy

$$F = \sum_{j=0}^{Z_n} N_j \left(-k_B T - k_B T \ln \left[g_j \frac{V}{N_j} \left(\frac{2\pi M k_B T}{h^2} \right)^{3/2} \right] \right.$$

$$\left. + E_j + U_{\text{net}}(j) + j F_e(j) \right). \tag{3.112}$$

It is worth noting that one would obtain the same result by defining $E_{(k+1)k}$ to include the contributions from continuum lowering from (3.80). Equation (3.112) differs from the results shown in More et al. (1988) in two ways, both reflecting the considerations about how to include the electron effects. They have the Coulomb energy depending on $\langle j \rangle^2$, with $\langle \ \rangle$ denoting an average over ionization states, while here we have energy proportional to $\langle j^2 \rangle$ through the average of U_{net}. This is, in most cases, a small difference. Similarly, they define F_{ej} in terms of an average charge, while the average shown here is a more complex one.

3.5.2.1 Ionization from the Helmholtz Free Energy

The equilibrium ionization corresponds to a minimum in the free energy. We thus find the variation in F with respect to a change in the populations of two adjacent ionization states, N_j and N_{j+1} and set this equal to zero. Because we associated the electrons with each ionization state above, there is no separate contribution from them. An ionization event reduces N_j and increases N_{j+1} by one. We have

$$\delta F = \delta N_j \frac{\partial F}{\partial N_j} + \delta N_{j+1} \frac{\partial F}{\partial N_{j+1}} = 0. \tag{3.113}$$

Since $\delta N_j = -\delta N_{j+1}$ this implies

$$\frac{E_{j+1} - E_j}{k_B T} + \frac{U_{\text{net}}(j+1) - U_{\text{net}}(j)}{k_B T} + \frac{(j+1)F_{e(j+1)} - jF_{e(j)}}{k_B T} = \tag{3.114}$$

$$\ln \left[g_{j+1} \frac{V}{N_{j+1}} \left(\frac{2\pi M k_B T}{h^2} \right)^{3/2} \right] - \ln \left[g_j \frac{V}{N_j} \left(\frac{2\pi M k_B T}{h^2} \right)^{3/2} \right],$$

from which

$$\frac{N_{j+1}}{N_j} = \frac{g_{j+1}}{g_j} \exp\left[-\frac{E_{j+1} - E_j}{k_B T}\right] \times \tag{3.115}$$

$$\exp\left[-\frac{U_{\mathrm{net}}(j+1) - U_{\mathrm{net}}(j)}{k_B T} - \frac{(j+1)F_{e(j+1)} - jF_{e(j)}}{k_B T}\right].$$

We apply this relation recursively to obtain N_j/N_0, which will prove most useful below. Since $E_o = U_{\mathrm{net}}(0) = 0$ this gives

$$\frac{N_j}{N_0} = \frac{g_j}{g_0} \exp\left[-\frac{E_j}{k_B T} - \frac{U_{\mathrm{net}}(j)}{k_B T} - \frac{jF_{e(j)}}{k_B T}\right]. \tag{3.116}$$

We can proceed to evaluate the average ionization. Just as we discussed in Sect. 3.4.1, we can now find

$$Z = \langle j \rangle = \left(\sum_{j=0}^{Z_n} j \frac{N_j}{N_0}\right) \Big/ \left(\sum_{j=0}^{Z_n} \frac{N_j}{N_0}\right), \tag{3.117}$$

where N_0/N_0 is of course 1. We can note that the denominator here is equal to the ratio N_a/N_0, which will also prove useful below. A complication for high densities is that continuum lowering may remove one or more of the lower ionization states. For such cases one must modify this calculation accordingly, setting N_j to zero for the quenched ionization states and using a reference state other than state zero in (3.116). For this purpose it is helpful to realize that one can multiply the numerator and denominator of (3.117) by any quantity whatsoever without changing its validity.

To this point these equations are completely general, within the validity of their assumptions. Thus, for example, one could use known values of the ionization energies and a general treatment of F_e to determine the thermodynamic quantities needed for a simulation and to accurately capture the behavior across the transition from classical to degenerate behavior. This is the approach taken by some EOS models for computations (for the regime of ionized matter), such as the PROPACEOS model (developed by PRISM Scientific). Or one could develop more general models in the same spirit, to include for example electronic excitation.

If instead we assume hydrogenic ions and so take $E_{(j+1)j} = (j + 1)^2 E_H$, then we obtain the distributions shown in Figs. 3.17 and 3.18. Here in evaluating (3.124) we have applied the form of F_e appropriate to whether or not any given ionization state is degenerate. The trends seen in the Saha regime we explored in Sect. 3.4.1 remain and are labeled in Fig. 3.17: over much of the space the ionization increases gradually with temperature by thermal ionization. In the degenerate regime, ionization is suppressed until the density becomes so large that pressure ionization sets in. At high enough temperatures, the carbon becomes fully ionized. In Fig. 3.18, one can see similar trends, although the large thermal ionization at low density and high temperature makes them less evident.

Fig. 3.17 Ionization level shown against electron temperature and ion density for carbon with $Z_n = 6$

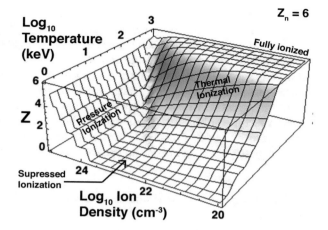

Fig. 3.18 Ionization level shown against electron temperature and ion density for ionizing high-Z matter having $Z_n = 80$

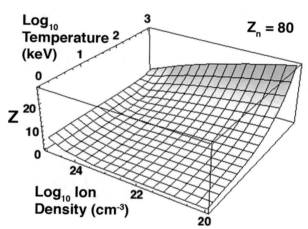

Degeneracy acts to strongly reduce ionization, as we can see by considering the purely degenerate case, when all ionization states of the atom are degenerate. (In the non-degenerate case, despite the modest effects of continuum lowering, we obtain results very similar to those of Sect. 3.1.3). In the degenerate limit (3.115) can be written

$$\frac{(j+1)F_{e(j+1)} - jF_{e(j)}}{k_B T} = \frac{\epsilon_F(j)}{k_B T} \frac{3}{5} \frac{(j+1)^{5/3} - j^{5/3}}{j^{2/3}}, \tag{3.118}$$

which has the correct limit for large j, when the change in electron free energy should equal the chemical potential, $\mu_e = \epsilon_F$. This gives, with $g_{j+1} = g_j$,

$$\frac{N_{j+1}}{N_j} = \exp\left[-\frac{E_{(j+1)j}}{k_B T} + \frac{9}{10}\frac{(2j+1)e^2}{R_o k_B T} - \frac{\epsilon_F(j)}{k_B T}\frac{3}{5}\frac{(j+1)^{5/3} - j^{5/3}}{j^{2/3}}\right]. \tag{3.119}$$

We can view these population balance equations as having three factors—an ionization factor, a Coulomb factor, and an electron factor. In order to produce a population distribution that is peaked above whatever minimum is allowed by continuum lowering, the other factors must exceed the factor involving the ionization energy, so that the net argument of the exponent is positive for some ionization states. As a practical matter, the Coulomb factor cannot do this for states beyond those already quenched by continuum lowering. In the classical limit the electron factor can be negative enough for the average Z to be peaked well above zero, just as we saw in Sect. 3.4.1. In contrast, in (3.119) the electron factor has the wrong sign to produce a distribution peaked above the minimum.

3.5.2.2   Thermodynamic Properties from the Helmholtz Free Energy

We can now write expressions for the thermodynamic parameters of interest, working with (3.97) through (3.101). We have the pressure,

$$p = n_i k_B T - \frac{3e^2 n_i \langle Z^2 \rangle}{10 R_o} - n_i \sum_{j=0}^{Z_n} \left(\frac{j N_j}{N_a} V \frac{\partial F_e(j)}{\partial V} \right), \tag{3.120}$$

the internal energy

$$(\rho \epsilon) = \frac{3}{2} n_i k_B T - \frac{9 e^2 n_i \langle Z^2 \rangle}{10 R_o} + n_i E_H \langle \frac{E_i}{E_H} \rangle \tag{3.121}$$

$$+ n_i \sum_{j=0}^{Z_n} \left[\frac{j N_j}{N_a} \left(F_e(j) - T \frac{\partial F_e(j)}{\partial T} \right) \right],$$

the heat capacity at constant volume

$$C_V = \frac{3}{2} n_i k_B - \frac{9 e^2 n_i}{10 R_o} \left(\frac{\partial \langle Z^2 \rangle}{\partial T} \right)_\rho + n_i E_H \left(\frac{\partial}{\partial T} \langle \frac{E_i}{E_H} \rangle \right)_\rho \tag{3.122}$$

$$+ n_i \sum_{j=0}^{Z_n} \left[\frac{j N_j}{N_a} \left(T \frac{\partial^2 F_e(j)}{\partial T^2} \right)_\rho \right],$$

and so on, where for any function D

$$\langle D \rangle = \frac{1}{N_a} \sum_{j=0}^{Z_n} (N_j D) . \tag{3.123}$$

Here to be consistent with usage elsewhere in the text we write $N_a/V = n_i$, noting that here this is the total density of particles having a nucleus, including neutral atoms.

We have chosen in the above to write the sums in terms of $F_e(j)$, and not yet to evaluate F_e, for the following reason. As the ionization increases, the electron density in the ion sphere increases. As a result, an ion sphere can transition from the classical limit to the degenerate limit in consequence of becoming more ionized. An accurate EOS model would evaluate $F_e(j)$ using the methods discussed in Sect. 3.1.3. To obtain the approximate numerical results given below, we apply a sharp transition from F_{ec} to F_{ed} as the Fermi temperature rises above T with increasing density. For the limiting cases is worth noting that

$$\frac{\partial F_{ed}(j)}{\partial T} = 0 \tag{3.124}$$

and that

$$\left(F_{ec}(j) - T\frac{\partial F_{ec}(j)}{\partial T}\right) = \frac{3}{2}k_B T, \tag{3.125}$$

which gives the intuitive result for the internal energy of the electrons. One also finds

$$T\frac{\partial^2 F_{ec}(j)}{\partial T^2} = \frac{3}{2}k_B. \tag{3.126}$$

One can evaluate the above quantities numerically. For (hydrogenic) carbon, Figs. 3.19 and 3.20 show plots for the present model like those shown in Figs. 3.15 and 3.16 for the simpler model above. The ideal-gas pressure and internal energy are based on the calculated ionization. The density scale is shifted here, to show more of the behavior at higher densities where degeneracy may matter. In the Saha regime, at relatively low density and high temperature, the two sets of figures are similar, showing a moderate pressure decrease that is due to Coulomb effects, and an internal energy that decreases from near twice the ideal gas value (because of energy in ionization) as temperature increases. There also remains a substantial region of reduced γ, although the values are larger than they were in the simpler model. The figures show more structure, some of which reflects the explicit calculation of the ionization. One sees this effect most strongly as temperature increases at the lowest density. The curve in all these figures showing the boundary for electron degeneracy is drawn on the assumption that $Z = 0.63\sqrt{T_{eV}}$, which ignores the decrease in ionization as the electrons become degenerate, so the actual transition to degenerate behavior occurs at a few times higher density, as the contours in the figures suggest.

The degenerate regime shows additional effects reflecting the additional physics included in the present model. Both pressure and internal energy increase strongly above ideal-gas values at low temperature and high density. This reflects the Fermi pressure and Fermi energy of the electrons freed by continuum lowering. At higher

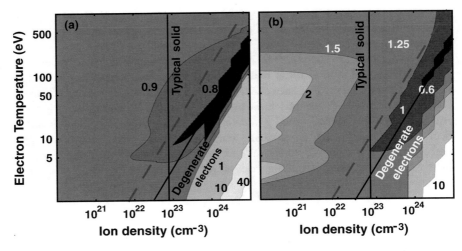

Fig. 3.19 (a) The pressure for carbon is shown, normalized to the ideal-gas pressure. The contours increase from the lower right, as labeled. (b) The internal energy density for carbon is shown, normalized to the ideal-gas value. The contours are labeled. The abrupt changes at some boundaries reflect either the threshold for pressure ionization or the finite grid of the calculation. The dashed, gray, diagonal lines show the location where $k_B T_e = \epsilon_F$, evaluated using $Z = 0.63\sqrt{T_{eV}}$. The solid diagonal lines, at about three times higher density, show the approximate location where Fermi degeneracy becomes important in this calculation, where the internal energy begins to increase

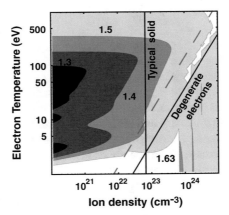

Fig. 3.20 The value of γ inferred from the data shown in Fig. 3.15 for carbon is shown. The contours are labeled. The dashed, gray, diagonal line shows the location where $k_B T_e = \epsilon_F$, evaluated using $Z = 0.63\sqrt{T_{eV}}$. The solid diagonal line, at about three times higher density, shows the approximate location where Fermi degeneracy becomes important in this calculation, where the internal energy begins to increase

temperatures, there is a modest reduction of pressure and internal energy by a combination of degeneracy and Coulomb effects. The Coulomb aspect was already present in the Saha model.

3.6 Generalized Polytropic Indices

Both ionizing and radiating plasmas, unfortunately, have pressures and internal energies that change in complex ways until after the plasma is fully ionized or completely radiation dominated. As a result, the assumption of constant polytropic index is a poor one for such systems. In this case, the question is whether there is any fairly simple way to treat the behavior of the system that might still allow simple models to be developed. Fortunately, one is able to do so, and several approaches are worked out in the literature. The best choice depends on the application. Here we identify three generalized polytropic indices for specific contexts of interest.

The derivation of shock behavior in Chap. 4 depended on the explicit expression $\rho\epsilon = p/(\gamma - 1)$. Accordingly, we define γ as the shock index via this relation. This is what we used in Sects. 3.5.1 and 3.5.2 to plot γ based on our EOS models. This index gives correct results for the change in properties across a shock front. For sound-wave applications, the derivation in Chap. 2 makes it clear that the relevant index (under isentropic conditions) is

$$\gamma_s = \left(\frac{\partial \ln p}{\partial \ln \rho}\right)_s. \tag{3.127}$$

However, our EOS does let us take this derivative directly so there will be some work to do to figure out how to evaluate this. For heat-transport applications, we need to find a thermodynamically correct generalization of (2.31)), which will define a heat-transport index, γ_h. For an energy density flux $-\nabla \cdot \mathbf{H}$, we will find an equation of the form of (2.31),

$$\frac{Dp}{Dt} - c_s^2 \frac{D\rho}{Dt} = -(\gamma_h - 1)\nabla \cdot \mathbf{H}. \tag{3.128}$$

We seek to know how this equation relates to our typical, known EOS equations. To apply the thermodynamic analysis, note that $\nabla \cdot \mathbf{H} = \rho dq/dt$, where an increment of specific heat input is dq. If γ_s as defined by (3.127) is constant, then $\gamma = \gamma_s = \gamma_h$.

Finding useful expressions for these quantities, and in particular a useful equation for heat transport, takes one into the realm of thermodynamic functions. It is easy to get lost in the forest where one seemingly can take the partial derivative of anything with respect to everything. Our job here is not to visit all the trees in this forest, but rather to develop specific equations that we will use later. Remarkably, aside from some patience, all the fundamental information we need to do this is a pair of equations from the first and second laws of thermodynamics,

$$d\epsilon - \frac{p}{\rho^2}d\rho = dq = Tds, \tag{3.129}$$

where $d\epsilon$, dq, and ds are the specific internal energy, heat input, and entropy, respectively, and two mathematical relations, specifically

$$\left(\frac{\partial a}{\partial b}\right)_c = 1 \Big/ \left(\frac{\partial b}{\partial a}\right)_c, \tag{3.130}$$

and

$$\left(\frac{\partial a}{\partial b}\right)_c \left(\frac{\partial b}{\partial c}\right)_a \left(\frac{\partial c}{\partial a}\right)_b = -1. \tag{3.131}$$

As is usual in thermodynamic calculations, at any given moment we express the thermodynamic functions in terms of two independent variables chosen from the three quantities ρ, p, and T. We proceed at first by expressing ϵ as $\epsilon(p, \rho)$; so from (3.129) we find

$$Tds = dq = \left(\frac{\partial \epsilon}{\partial p}\right)_\rho dp + \left[\left(\frac{\partial \epsilon}{\partial \rho}\right)_p - \frac{p}{\rho^2}\right]d\rho. \tag{3.132}$$

We also have, as ds is an exact differential,

$$dq = Tds = T\left(\frac{\partial s}{\partial p}\right)_\rho dp + T\left(\frac{\partial s}{\partial \rho}\right)_p d\rho. \tag{3.133}$$

The specific heats involve the use of T and ρ or T and p as the thermodynamic variables. Equation (3.132) implies that the specific heat at constant volume is

$$c_V = \left(\frac{dq}{dT}\right)_\rho = \left(\frac{\partial \epsilon}{\partial T}\right)_\rho, \tag{3.134}$$

while the specific heat at constant pressure is found by writing ϵ as $\epsilon(T, \rho)$ in (3.129) and then differentiating, to obtain

$$c_p = \left(\frac{dq}{dT}\right)_p = \left(\frac{\partial \epsilon}{\partial T}\right)_\rho + \left[\left(\frac{\partial \epsilon}{\partial \rho}\right)_T - \frac{p}{\rho^2}\right]\left(\frac{\partial \rho}{\partial T}\right)_p. \tag{3.135}$$

Note that we can evaluate the coefficients in (3.132), (3.134), and (3.135) from any EOS like those above that relates p, ϵ, ρ, and T. Multiplying (3.135) by $(\partial T/\partial \rho)_p$, and using the definition of c_V, one finds

$$\left(\frac{\partial \epsilon}{\partial \rho}\right)_p = \left(\frac{\partial \epsilon}{\partial T}\right)_\rho \left(\frac{\partial T}{\partial \rho}\right)_p + \left(\frac{\partial \epsilon}{\partial \rho}\right)_T = c_p \left(\frac{\partial T}{\partial \rho}\right)_p + \frac{p}{\rho^2}, \tag{3.136}$$

while from the chain rule

$$\left(\frac{\partial \epsilon}{\partial p}\right)_\rho = \left(\frac{\partial \epsilon}{\partial T}\right)_\rho \left(\frac{\partial T}{\partial p}\right)_\rho = c_V \left(\frac{\partial T}{\partial p}\right)_\rho. \tag{3.137}$$

This then gives for the heat input per (3.132)

$$dq = c_V \left(\frac{\partial T}{\partial p}\right)_\rho dp + c_p \left(\frac{\partial T}{\partial \rho}\right)_p d\rho = T ds. \tag{3.138}$$

To simplify this further note that

$$\frac{c_p}{c_V} = -\left(\frac{\partial p}{\partial \rho}\right)_s \left(\frac{\partial \rho}{\partial T}\right)_p \left(\frac{\partial T}{\partial p}\right)_\rho = \left(\frac{\partial p}{\partial \rho}\right)_s \left(\frac{\partial \rho}{\partial p}\right)_T, \tag{3.139}$$

obtained by substituting from (3.136) and (3.137) into (3.132) and taking $(\partial p / \partial \rho)_s$. The isentropic sound speed and thus γ_s can be found from (3.139) and (3.135), by

$$\left(\frac{\partial p}{\partial \rho}\right)_s = \frac{c_p}{c_V}\left(\frac{\partial p}{\partial \rho}\right)_T = \left(\frac{\partial p}{\partial \rho}\right)_T - \frac{1}{c_V}\left[\left(\frac{\partial \epsilon}{\partial \rho}\right)_T - \frac{p}{\rho^2}\right]\left(\frac{\partial p}{\partial T}\right)_\rho = \gamma_s \frac{p}{\rho}. \tag{3.140}$$

This expression for the sound speed is readily evaluated from expressions for p and ϵ.

We pursue the heat-transport coefficient as follows. We substitute for c_p in (3.138), which becomes

$$dq = c_V \left(\frac{\partial T}{\partial p}\right)_\rho \left[dp + \left(\frac{\partial p}{\partial \rho}\right)_s \left(\frac{\partial \rho}{\partial p}\right)_T \left(\frac{\partial p}{\partial T}\right)_\rho \left(\frac{\partial T}{\partial \rho}\right)_p d\rho\right], \text{ or} \tag{3.141}$$

or

$$dq = c_V \left(\frac{\partial T}{\partial p}\right)_\rho \left[dp - \left(\frac{\partial p}{\partial \rho}\right)_s d\rho\right], \tag{3.142}$$

using (3.131). This is the form we were seeking. The quantity in square brackets has the form of (2.31), as desired. It also shows that the sound speed in (3.128) is the isentropic sound speed. One can convert to an expression for the total heat input by multiplying (3.142) by ρ. Thus one finds

$$(\gamma_h - 1)^{-1} = \rho c_V \left(\frac{\partial T}{\partial p}\right)_\rho. \tag{3.143}$$

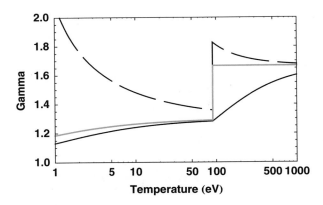

Fig. 3.21 Values of the shock γ (solid black), sound-speed γ_s (dashed), and heat flux γ_h (gray), for carbon at $1\,\mathrm{g\,cm^{-3}}$

So long as $(\partial p/\partial \rho)_T = p/\rho$ and $(\partial \epsilon/\partial \rho)_T = 0$, one has $\gamma_s = \gamma_h$. This is the case for simple ideal gasses but not in the regimes where more complex effects are important or in the radiating plasmas considered in Chap. 7.

We can then apply our simplest model—(3.140) and (3.143)—to find γ_s and γ_h. This produces rather messy expressions, but they are readily evaluated by computer. Figure 3.21 show the results for carbon at $1\,\mathrm{g\,cm^{-3}}$, which has an ionizing regime followed by a fully ionizing regime. The shock index, γ, increases slowly in the ionizing regime and then more quickly once the carbon is fully ionized. It will eventually approach 5/3. The heat-flow index, γ_h, is close to γ in the ionizing regime but jumps abruptly to 5/3 when the carbon becomes fully ionized. This is sensible—beyond that point energy is not being absorbed by further ionization. These trends in γ and γ_h remain present at lower density. The sound-speed index, γ_s, starts much higher than γ and decreases to approach it with increasing temperature. Then upon full ionization the index jumps abruptly to near 5/3 (and thus the sound speed jumps too). At lower density, γ_s remains close to γ in the ionizing regime but still jumps upon full ionization. At higher density the behavior becomes more complex, but the model becomes poor at low temperature when the electrons in fact become degenerate. The major conclusion here is that the sound speed and heat-flow rate can vary substantially across high-energy-density systems, especially as they transition to a fully ionized state.

3.7   The Degenerate, Strongly Coupled Regime

At the time of the writing of the first edition of this book, the community believed that material physics would become simple as density increased. The electrons would be strongly degenerate and tend toward constant density, and the ions would

Fig. 3.22 Charge clustering
is a key feature of
Fermi-degenerate, strongly
coupled matter

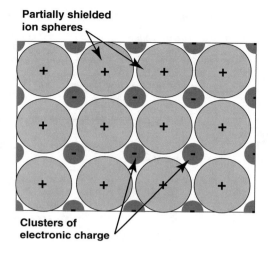

be happily isolated in their ion spheres, not unlike the monads of Leibniz. The
model described above, based on the Helmholtz Free Energy, would work well.
But we have learned since that this model also fails at high enough density. We
came to know this because experiments began to observe behavior showing that
late-twentieth-century models were *qualitatively* incorrect. They observed changes
in material structure that were entirely unexpected, and that the transition to a
liquid state occurred at very different temperatures than had been predicted. We
came to understand what is happening qualitatively by means of advanced computer
simulations that only become possible during this same period.

Figure 3.22 illustrates the fundamental reason why unanticipated behavior arose.
When one fills some volume with spheres, there remain gaps between them. The
models above assumed that the electron-density at and beyond the ion-sphere
boundaries was constant, but in fact what happens is that the electrons cluster in
the gaps. Once the ions are incompletely shielded, they are affected by other ions as
well as by the clusters of electronic charge. This enables the formation of chemical
bonds, not unlike ionic bonds. The difference relative to ordinary chemistry is that
these bonds have transition energies in the keV range as opposed to the eV range.
These interactions are now sometimes described as "kilovolt chemistry."

One might describe Fermi-degenerate, strongly coupled matter at high energy
density as a "quasi-solid", because it can have various crystal structures and
can transition between them, despite the presence of additional freed electrons.
Table 3.1 shows some of the standard structural designations. Modeling these
structures is challenging, as one must account for the quantum mechanical effects
including the interactions of several (or more) ions. The theoretical calculations
employ two methods. One of these is Density Functional Theory, a quantum-
mechanical approach that represents the electron density using functions. The other
is Molecular Dynamics, which models the interactions of multiple molecules (or
ions) from first principles, using potentials to describe the forces between them. In

Table 3.1 Table of geometric material structures

Acronym	Description	Acronym	Description
sc	Simple cubic	bcc	Body-centered cubic
sh	Simple hexagonal	fcc	Face-centered cubic
hcp	Hexagonal close packed	BC8	Tetragonally bonded structure

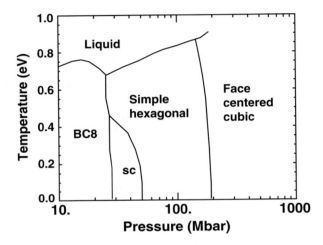

Fig. 3.23 Phase diagram of Fermi-degenerate, strongly coupled carbon, from calculations using Density Functional Theory

the regime of interest, the potentials must be quantum-mechanical and may come from calculations using Density Functional Theory. Only in the twenty-first century have computers become capable enough to enable these calculations for high-energy-density conditions. The interested reader will have to seek further discussion of these methods elsewhere, as this is a topic rather afar from the primary focus of the present book. Simple models that capture the essential physics have not yet emerged, and by analogy with chemistry may not.

Figure 3.23 is based on the results of calculations by Martinez-Canales et al. (2012), using Density Functional Theory, for carbon. One sees that the structure, at low temperature, is predicted to change three times between 10 and 200 Mbar pressure. Experiments with carbon, at this writing, have reached pressures of 50 Mbar (Smith et al. 2014). There remains much to be learned about the behavior of quasi-solids at high energy density.

3.8 The EOS Landscape

At this point it may be useful to summarize what we have learned about the equation of state in high-energy-density systems. Figure 3.24 provides this summary. The specific lines in the figure are drawn for an ionizing plasma, assuming $A = 2Z_n$, but the relative orientation of the various elements in this log–log space is not sensitive to these assumptions. At the upper left is the ideal-plasma regime. Examples are hot enough coronal plasmas, as for example in the laser-heated zone in front of a dense target, or the plasma generated in Z pinches during their implosion (see Chap. 9). At the right pressure ionization becomes important, as occurs when solids are sufficiently compressed. Throughout the lower right region the electrons are Fermi degenerate, which determines the pressure needed to compress solid-density matter including the fuel for inertial fusion.

In between these limits is the realm of many experiments in the early twenty-first century. Here the matter is partly ionized but probably is not fully stripped, the ions live in the privacy of their own ion spheres but represent much of the internal energy in the system, and the electrons are a Fermi gas whose pressure is reduced by Coulomb interactions. The internal energy is roughly twice that of an ideal gas, because of the energy invested in ionization. As a result, γ is in the vicinity of 4/3. Across much of this regime, a model based on the Saha equation should work well. But the Saha model fails at high density and temperature because of the combined effects of degeneracy and continuum lowering, as we have seen.

This completes our discussion of specific models of equations of state. In the following chapters, we will typically take $\gamma = 4/3$ or $5/3$ for our examples. We will not need to distinguish among the different polytropic indices until we work with radiation hydrodynamics in Chap. 7. But it should be clear from the above that γ

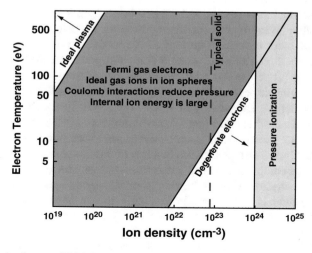

Fig. 3.24 The landscape of EOS for high-energy-density plasmas

can be substantially less than 5/3, that these dense plasmas are *not* ideal-gases, and that it is not so easy to know just what the equation of state is. The next section discusses tabular equations of state.

3.9   Tabular Equations of State

The chapter thus far has made it evident that equations of state in the dense-plasma regime are complicated. The appeal of using a polytropic index, at the expense of detailed accuracy, is quite clear. Indeed, this will be our approach throughout much of the text. But if one is to try to simulate these systems with computers, then one would hope to be more accurate. It is evidently a great challenge to accurately simulate the behavior of materials at high energy density. One has Coulomb energy corrections, degenerate electrons, pressure ionization, and ionization potential depression, among other effects. To be fully accurate one would need to include several effects that we mentioned but did not incorporate, such as the impact of bound electrons. One would also need to handle the transitions between regimes more accurately. But the actual problem is worse than this, because high-energy-density matter nearly always evolves out of and is adjacent to matter that is not at high energy density, but rather is in a solid or liquid state. So realistic computations must also be able to account for these states of matter and for their transition to hotter and perhaps denser conditions. A particularly difficult example at this writing is that of the behavior of the wires in Z-pinch plasmas (see Chap. 9). These begin as solids, ablate and (perhaps) explode, creating the material that the Z pinch accelerates inward. Modeling these dynamics is a severe challenge.

One approach to addressing these issues for simulations is to use a tabular EOS. The idea behind a tabular EOS is that one can work with experimental data, molecular dynamics simulations, and the best possible models. From them one can construct a table giving two of the thermodynamic variables (ρ, p, ϵ, and T) as a function of the other two. As is true of all the models we have discussed, this is necessarily done in equilibrium. Then a computer code can interpolate from the tables to find the properties it needs with adequate accuracy.

One challenging aspect of constructing such a table is the need for thermodynamic consistency. The table will show how some thermodynamic quantities vary when others are held constant. These variations must be thermodynamically consistent. As one does work on the material or adds heat to it, the changes of state that result must be consistent with the first law of thermodynamics. If this were not the case, then the computer code using the table would mysteriously create or absorb energy in an unphysical way. Achieving thermodynamic consistency in practice, while merging models that cover adjacent regimes, can be very difficult. One can check for thermodynamic consistency by applying the first law of thermodynamics to the table. One way to do this is to evaluate the local deviation from the first law of thermodynamics. Landau and Lifshitz (1987) show in Vol. 5 that one can write the first law of thermodynamics as $d(\rho\epsilon)/dV + p - T(dp/dT) = 0$. One can evaluate

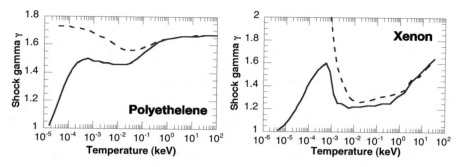

Fig. 3.25 For polyethylene (C_1H_1) on the left and xenon ($A = 131, Z = 54$) on the right, these figures show the inferred γ from the SESAME table. The lower curve is at 0.1 g/cm³ density, while the upper curve is at solid density. Credit: Carolyn Kuranz

this quantity throughout a candidate EOS table and display the results as curves or a contour plot.

The most widely used EOS tables are the SESAME tables, available from the Los Alamos National Laboratories. These tabulate specific pressure (pressure per unit density) and specific energy as functions of density and temperature, over several orders of magnitude in density and in temperature. Figure 3.25 shows two examples based on these tables. In each case, we have used the equation of state to plot γ. The range of temperatures in the table is shown. The densities shown are solid density (dashed) and 0.1 g/cm³, which are relevant to laboratory work in high-energy-density physics. One sees first that the behavior at low temperatures is quite different. This reflects the presumed development of a gaseous state (and perhaps even clusters) at low densities, with many degrees of freedom, which forces γ close to 1. In contrast, the solid becomes more ordered as temperature decreases. From traditional thermodynamics, one would expect γ to approach 3 at low temperatures if the solid forms a lattice with tightly bound planes. In the tables, γ sometimes exceeds 3 at low temperatures.

At the highest temperatures, the materials seem to approach $\gamma = 5/3$, which would correspond to a fully stripped, ideal gas. We comment more on this below. At intermediate temperatures, between a few eV and 100 eV for polyethylene and a few eV and 1000 eV for xenon, γ is reduced. This is as expected from the previous discussion in this chapter. Indeed, the result for xenon is not far above the value we inferred for an ionizing, high-Z material. The value of γ for polyethylene, on the other hand, is not so far below 5/3. One might be skeptical as to whether this decrease is in fact large enough.

These figures also provide one example of the limitations of these tables. If the high-temperature states (above 1 keV) were truly in equilibrium, as is assumed, then the presence of the radiation field would be driving γ to 4/3. So these tables ignore the radiation field. The problem is that they have to make some specific assumptions, though in this case they do not assume true equilibrium. Real systems do vary greatly with regard to the coupling of the radiation field and the matter. There is no way that one table can account for this. Any given computer code may or may not handle it well.

There are other problems with the use of EOS tables in particular, and equilibrium models in general, in simulations of real systems. Real systems are almost never in equilibrium. They are often in steady state, or nearly in steady state, but not in equilibrium. A good example is a plasma that expands from a hot surface but is not actively heated. The expanding plasma cools, and after a time its properties slowly evolve. Even so, on the scale of tens of ns that often applies, the ions and electrons may not recombine and the plasma certainly will not reach its equilibrium state. The EOS table, on the other hand, presumes the plasma is instantaneously in equilibrium. Thus, if it reaches a condensation temperature, the table will make it condense, no matter how unrealistic this may be. This, and theoretical equilibrium phase changes in general, can be a source of abrupt density changes in simulations that are completely unreal. There are times when an ideal-gas model with fixed γ provides a much more realistic approach to simulating a time-varying system. The main point is that one must pay attention, think about what one sees, and not assume that the code reveals truth.

In addition, you may have noticed that some of the equations above would produce regimes where the pressure from a given model became negative. This happens with the models used for the EOS tables as well. In some cases, this is sensible. For example, the only realistic way to incorporate tension in a material, in the context of a hydrodynamic model, is by adding negative terms to the pressure. If the material is tightly enough bound and cold enough, it may be sensible in this sense to treat the pressure as negative. However, the existence of negative pressure regions in EOS tables can create serious problems when simulating real, nonequilibrium systems. In the example of the previous paragraph, for example, the plasma expanding from a surface may have a temperature and density that would correspond to a condensed state with tension in equilibrium, yet in actual fact may be more accurately treated as an ideal-gas. In some contexts, it is sensible to modify the EOS tables to destroy the tension regimes and maintain positive pressure. When the EOS table works well, it will do a better job of reproducing the dynamics than any simpler model can. But it cannot be counted on to always work well. It is very often sensible to compare simulations using EOS tables for various similar materials and also using a fixed γ to help determine which aspects of the observed dynamics are due to the specifics of the EOS table.

Finally, tables do not typically exist for novel materials, such as low-density foams. These materials are not microscopically uniform. They are unlikely to behave like a uniform, low-density material. There is some discussion of foam behavior in Zel'dovich and Razier (1966), but it applies only to foams that are compressed very gently by comparison to the behavior of typical high-energy-density systems. Indeed, in high-energy-density experiments to date with foams, the uniform-density models fail to accurately predict phenomena such as shock-wave propagation. Whether in the end new tables or some other approach proves the best for working with them remains to be seen.

3.10 Equations of State in the Laboratory and in Astrophysics

A moment's thought will show that equation of state (EOS) properties are quite important in astrophysics. In gravitationally bound objects, such as planets, white-dwarf stars, or neutron stars, the interior pressure is determined primarily by gravity. However, to know the density, and hence the volume of the material in any given pressure range one must know the equation of state. Direct astronomical measurements can determine the mass, and sometimes the size, of such objects, and may be able to learn about the surface composition from spectroscopy. But there is usually neither direct nor indirect information relating to the interior. (An exception is the Sun, for which seismology is possible and productive, producing data that greatly constrain the EOS.)

Assuming that one knows the EOS, one can construct a model of a planet in which the known mass of the planet is distributed in radius as gravitational pressure and the EOS dictate, based on assumptions about what the composition of the planet is. Uncertainties in the EOS make this more difficult. In the case of Jupiter, for example, it is an interesting question whether an entire planet of its size and mass might be composed of hydrogen or whether there must be an ice and rock core. This certainly has implications for theories of planet formation. With sufficient knowledge of the hydrogen EOS, one will be able to answer this question. At the turn of the twenty-first century, such knowledge was insufficient.

In addition, the EOS affects one's ability to understand magnetic fields, as we discussed briefly with reference to Fig. 1.4. Planetary magnetic fields are produced by interior currents, known as dynamos. The theory of planetary dynamos unfortunately requires complex three-dimensional calculations. Nonetheless, the possibilities for magnetic field generation are constrained by the locations where the planetary interior is conducting, and this is constrained by the EOS. Here again Jupiter provides an interesting way to frame the puzzle. Jupiter has an extremely strong magnetic field, producing very-large-scale effects within the solar system. At the surface of Jupiter, hydrogen is an insulator. The nature of the hydrogen EOS will determine how close to the surface of Jupiter currents can flow and what volume of the planet can participate in the dynamo. This will constrain the possibilities for the production of Jupiter's magnetic field.

3.10.1 The Astrophysical Context for EOS

To illustrate the importance of EOS, consider Jupiter in more detail. Figure 1.4 showed a schematic of its interior based on one specific model (for more discussion, see Saumon and Guillot 2004). Jupiter has an outer envelope of dielectric molecular H_2, believed to transition to metallic atomic hydrogen at a radius of $0.75R_J$ and pressure of $p \sim 2\,\mathrm{Mbar}$, and ending in an ice–rock core, in this model, when the

Fig. 3.26
Temperature–pressure
profiles in Jupiter and brown
dwarf Gl 229B, for various
ages, from models in
Hubbard et al. (1997)

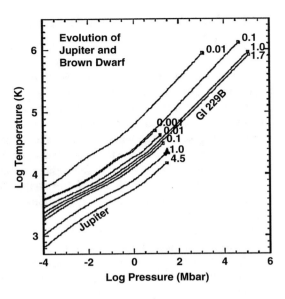

pressure reaches $\sim$40 Mbar. There is uncertainty about whether such an ice–rock core actually is present, and if so how massive it is. In part this reflects uncertainty in the EOS of H. The mass of Jupiter is $M_J \sim 10^{-3} M_S$, (where M_S is the mass of the sun) and its intrinsic radius is $R_J = 7.2 \times 10^4$ km. Some model calculations for the interior of Jupiter are shown as temperature–pressure (T–p) profiles as a function of age in Fig. 3.26. Profiles for the brown dwarf Gl229B are also shown in this figure. Under these conditions, molecular hydrogen (H_2) dissociates to atomic hydrogen and ionizes deeper in the mantle, changing from a dielectric to a conductor. The pressure and temperature in the mantle of Jupiter near the surface are in the range of 1–3 Mbar at temperatures of a fraction of an eV. Deeper in the interior, the pressure and temperature increase, rising to $\sim$80 Mbar at a couple of eV at the center. (The corresponding numbers for the brown dwarf Gl 229 are similar in the mantle, but it has four orders of magnitude higher pressures in the core, $p_{\rm core} \sim 10^5$ Mbar.)

One of the key questions about the interior of Jupiter is whether there is a sharp boundary between the molecular hydrogen mantle and the monatomic hydrogen core, caused by a first-order plasma phase transition. This has significance for the exact internal structure, as the discontinuities caused by such a phase transition tend to inhibit convective heat transport, modifying the thermal profile of the planetary interior. This also affects the degree and rate of gravitational energy-release due to sedimentation of He and heavier elements. Jupiter and Saturn's atmospheres are observed to contain less helium than is believed to have been present at their formation. This is thought to be due to a H–He phase separation. The presence of a helium-poor outer region, and helium-rich inner region is important, both because it has implications for the amount of heavier elements contained deeper in the interior of the planet, and also because of the gravitational energy released as heat during

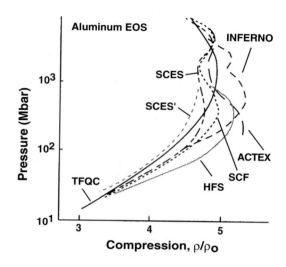

Fig. 3.27 Various theoretical models of the shock Hugoniot of Al, as described in the text. Note the considerable uncertainty, which only experiments can resolve. Adapted from Avrorin et al. (1987)

helium sedimentation. Helium sedimentation is required to explain Saturn's intrinsic heat flux, and may also be significant in Jupiter. The important point in the present context is that all of the detailed issues of hydrogen behavior are quite uncertain at present. The nature of the transition from molecular to monatomic hydrogen, the existence of a metallic phase, the possibility of a H–He phase separation, and other factors are not known.

The EOS of elements heavier than H and He, relevant to Earth-like planets, is even more complex at ultrahigh pressures. To illustrate this, we show in Fig. 3.27 a plot of a number of different theoretical models for the behavior of Al at very-high pressures and compressions, $p > 10$ Mbar, $\rho/\rho_o > 3$. These models calculate the shock *Hugoniot*, which is the locus of the points in pressure and density that can be reached from a single initial condition by means of shock waves of varying strength. The various models (see Avrorin et al. 1987; Hicks et al. 2009) exhibit significant differences. The simplest and most widely used of the models is the statistical Thomas–Fermi model with quantum corrections (TFQC), shown by the solid curve. This model does not include atomic shell structure, but rather treats the electron states as a continuum. The self-consistent field (SCF), Hartree–Fock–Slater (HFS), and INFERNO models treat the electron shells quantum mechanically, but differ in their handling of close-packed levels corresponding to energy bands. The semiclassical equation of state (SCES) model treats both the discrete electron shells and the energy bands semiclassically. The ACTEX model is an ionization equilibrium plasma model which uses effective electron–ion potentials fitted to experimental spectroscopic data. These models typically include the nuclear component using the ideal-gas approximation. An exception is a Monte Carlo treatment of the thermal motion of the nuclei implemented in one of the versions of the semiclassical equation of state model (SCES).

The oscillations in the theoretical pressure versus compression curves shown in Fig. 3.27 result from the pressure ionization of the K- and L-shell electrons of Al. At pressures of 100–500 Mbar, ionization of the L-shell electrons occurs as the high compression forces neighboring atoms sufficiently close together to disrupt the $n = 2$ electron orbital. When the shock places the material in a state where these electrons are becoming free, more of the energy flowing through the shock must go into internal energy. This leads to a larger density increase, exactly as we discuss in Sect. 4.1. Hence, at the onset of pressure ionization of a new shell in a model, the postshock density increases more rapidly with postshock pressure, behavior known as a "softer" EOS. This pressure-ionization effect on the EOS is qualitatively similar to that due to molecular dissociation of N_2 and D_2, which has been experimentally observed at lower pressures (see Nellis 2006). Once ionization from the shell is complete, the effect is a "hardening" of the EOS, as the fraction of the energy flowing through the shock that is converted to internal energy decreases. This is why, above ~ 1 Gbar, some of the $p - \rho$ curves turn back toward lower compression. A similar softening–hardening oscillation is predicted at pressures of 3–5 Gbar due to ionization of the K-shell electrons, though the magnitude of the effect is smaller due to the lower number of K electrons. How real such oscillations in the Hugoniot are is unclear at this writing. If the actual process of liberating new electrons develops more gradually than it does in the model, this may smooth out the response and avoid the oscillation.

3.10.2 Connecting EOS from the Laboratory to Astrophysics

The EOS describes the equilibrium properties of any large aggregation of atoms of a given type. Even microscopic quantities of matter typically include enormous numbers of atoms. As a result, measurements using aggregations of matter that are very small on a human scale can provide results which apply directly to aggregations of matter on a planetary or stellar scale. In this sense, it is straightforward to make a laboratory measurement that applies directly to astrophysics.

Unfortunately, however, laboratory measurements can only achieve a limited range of pressures and densities by comparison with those existing in astrophysical systems. It would be desirable to be able to scale the equation of state in pressure and density, so that laboratory measurements could be applied to a wider range of astrophysical conditions. This is possible but unnecessary in the case of simple equations of state, such as an ideal-gas or a radiation-dominated system. In more-complex cases, however, the dynamics of the material is specific to the material conditions. The chemical structure of a material is not easily scaled to other conditions, and processes such as dissociation and ionization occur only at specific energies. Thus, laboratory measurements can only address astrophysical issues in EOS at pressures they can actually achieve.

Given the technologies of the early twenty-first century, it seems likely that the pressures employed for EOS studies during this period, using planar targets, will be in the range of 1 to less than 100 Mbar. It may prove feasible, using implosions, to access pressures of a Gbar or even more. These are suitable for addressing issues in planetary equations of state. One can expect this to be the primary focus of such studies.

Homework Problems

3.1 Inertial fusion designs typically involve the compression of DT fuel to about 1000 times the liquid density of $0.25\,\mathrm{g\,cm^{-3}}$. Assuming that this compression is isentropic and that the fuel remains at absolute zero, determine the energy per gram required to compress this fuel. Compare this to the energy per gram required to isentropically compress the fuel to this same density, assuming the fuel is an ideal gas whose final temperature is to be the ignition temperature of 5 keV.

3.2 Generalize the derivation of the Debye length in Sect. 3.2 to a plasma with an arbitrary number of ion species, each of which may have a distinct temperature.

3.3 Examine the behavior of the integrals for Fermions. Argue conceptually that the contribution of the denominator in (3.29) at large $\mu/(k_B T_e)$ is a step function. Evaluate this integral numerically to determine how rapidly it becomes a step function as $\mu/(k_B T_e)$ increases.

3.4 Examine the limiting behavior of the internal energy of Fermi degenerate electrons. Show, in the limit as $T_e \rightarrow 0$, that $n_e \epsilon_e = (3/5) n_e \epsilon_F$.

3.5 What is the relation of heat capacity and entropy? Derive 3.38 and 3.40 and discuss their differences.

3.6 Make plots comparing Z_{bal} from (3.49) with the estimate $20\sqrt{T_e}$ as a function of T_e, for ion densities of 10^{19}, 10^{21}, and $10^{23}\,\mathrm{cm^{-3}}$. Discuss the results.

3.7 Carry out the evaluation of the average charge, Z, in (3.53) and compare the result to Z_{bal}, for $T_e = 1\,\mathrm{keV}$, $Z_n = 30$, and $n_i = 10^{21}\,\mathrm{cm^{-3}}$.

3.8 Plot the ratio of ΔE to the ionization energy versus ion density for the various models described in Sect. 3.4.4. Discuss the results.

3.9 The value of R_i used in Sect. 3.4.6 ignores the internal energy in excited states (as well as the energy lost by radiation during ionization, which would properly have to be treated by more general equations). Again assuming hydrogenic ions, estimate what fraction of the internal energy is present in excited states, and how this varies with Z.

3.10 Complete the derivation of the polytropic index for heat conduction. Derive (3.143) from relations (3.130)–(3.134).

References

Atzeni S, Meyer-Ter-Vehn J (2004) The physics of inertial fusion. Oxford, New York

Avrorin E, Vodolaga BK, Voloshin NP, Kovalenko GV, Kuropantenko VF, Simonenko VA, Chernovolyuk BT (1987) Experimental study of the influence of electron shell structure on shock adiabats of condensed materials. Sov Phys JETP 66:347–354

Ecker G, Kroll W (1963) Lowering of the ionization energy for a plasma in thermodynamic equilibrium. Phys Fluids 6(1):62–69

Eliezer S, Ghatak, Hora H (1986) An introduction to equations of state: theory and applications. Cambridge University Press, Cambridge

Griem H (1997) Principles of plasma spectroscopy. Cambridge University Press, Cambridge

Hicks DG, Boehly TR, Celliers PM, Eggert JH, Moon SJ, Meyerhofer DD, Collins GW (2009) Laser-driven single shock compression of fluid deuterium from 45 to 220 GPa. Phys Rev B 79(1), times Cited: 2

Hubbard WB, Guillot T, Lunine JI, Burrows A, Saumon D, Marley MS, Freedman RS (1997) Liquid metallic hydrogen and the structure of brown dwarfs and giant planets. Phys Plasmas 4:2011–2015

Krall NA, Trivelpiece AW (1986) Principles of plasma physics. San Francisco Press, San Francisco

Landau LD, Lifshitz EM (1987) Statistical physics, course in theoretical physics, vol 5, 2nd edn. Pergamon Press, Oxford

Martinez-Canales M, Pickard CJ, Needs RJ (2012) Thermodynamically stable phases of carbon at multiterapascal pressures. Phys Rev Lett 108(4):045704

More RM, Warren KH, Young DA, Zimmerman GB (1988) A new quotidian equation of state (qeos) for hot dense matter. Phys Fluids 31(10):3059–3078

Nellis WJ (2006) Dynamic compression of materials: metallization of fluid hydrogen at high pressures. Rep Prog Phys 69(5):1479–1580

Reif F (1965) Fundamentals of statistical and thermal physics, 56946th edn. Waveland Pr, Long Grove, IL

Salzman D (1998) Atomic physics in hot plasmas. Oxford University Press, New York

Saumon D, Guillot T (2004) Shock compression of deuterium and the interiors of Jupiter and Saturn. Astrophys J 609(2, Pt 1):1170–1180

Saumon D, Chabrier G, Van Horn HM (1995) An equation of state for low-mass stars and giant planets. Astrophys J Suppl 99:713–741

Smith RF, Eggert JH, Jeanloz R, Duffy TS, Braun DG, Patterson JR, Rudd RE, Biener J, Lazicki AE, Hamza AV, Wang J, Braun T, Benedict LX, Celliers PM, Collins GW (2014) Ramp compression of diamond to five terapascals. Nature 511(7509):330–333

Stewart JC, Pyatt KD (1966) Lowering of ionization potentials in plasmas. Astrophys J 144(3):1203–1211

Zel'dovich YB, Razier YP (1966) Physics of shock waves and high-temperature hydrodynamic phenomena, vol 1, 2002nd edn. Dover, New York

Chapter 4
Shocks and Rarefactions

Abstract This chapter discusses the fundamental elements of one-dimensional, compressible fluid dynamics. These are essential to the behavior of high-energy-density matter. It begins by developing the theory of shock waves in fluid media, developing results for shock waves of arbitrary strength, entropy generation by shock waves, oblique shock waves, shock waves at interfaces, and the use of flyer plates to drive shock waves for measurements of equations of state. The chapter then introduces self-similar dynamics, which turns out to describe the expansions of matter known as rarefactions, and also to describe the blast waves produced by a brief deposition of energy. It then discusses the interaction of shock waves and rarefactions with each other and with interfaces where the density changes.

The word "shock" is used very widely in common experience. One is shocked by an unexpected event; a wounded victim goes into shock; and one shocks a material by suddenly cooling it. A "shock wave" is a sudden transition in the properties of a fluid medium, involving a difference in flow velocity across a narrow (ideally, abrupt) transition. In high-energy-density physics, nearly any experiment involves at least one shock wave. Such shock waves may be produced by applying pressure to a surface or by creating a collision between two materials. In astrophysics, nearly every sudden event produces a shock wave. Yet in common experience one encounters very few shock waves. We hear thunder after lightning, which is a long-term consequence of the shock wave produced by the lightning channel, but as we shall see below one would hope never to directly experience this shock wave. Most of us hear sonic booms infrequently, but they are the only shock wave of human origin we typically encounter.

We have more direct experience with rarefactions or "rarefaction waves," in which a fluid begins to move, expanding and becoming less dense, with the edge of the moving region propagating into an initial body of fluid. Household drafts may be due to rarefactions, which can occur in a house, for example, when a gust of wind drops the pressure at an open door, by the Bernoulli effect. Rarefactions also have real practical uses, notably in refrigeration where they are used to produce expansion cooling. It is also true that nearly every high-energy-density experiment involves at least one rarefaction wave.

R P. Drake, *High-Energy-Density Physics*, Graduate Texts in Physics,
https://doi.org/10.1007/978-3-319-67711-8_4

Moreover, most high-energy-density experiments involve at least one interface, where the density (and perhaps the equation of state) changes. Whenever a shock wave or a rarefaction wave reaches an interface, there are transmitted and reflected waves in response. In each of these two directions, these waves might be either shock waves or rarefaction waves, so that there are four possible responses. Which of these four occurs depends on the details. One can find systematic discussions of this in books on shock physics. As we proceed to consider various cases, we will encounter specific examples. It should be clear that shock waves, rarefaction waves, and their interactions merit a serious examination, which we undertake in this chapter.

4.1   Shock Waves

Figure 4.1 shows an image of the supernova remnant known as Tycho. The remnant shown in the image has sharp edges, where spectral measurements show that the temperature reaches 20 million degrees. This is one of many examples of an astrophysical shock. The magnetic field in such a shock is not dynamically important, except that it localizes the particles as discussed in Chap. 10. For this reason, a laboratory experiment can hope to produce dynamics similar to those of this shock. (There are caveats—it is possible that the dynamics of the shock in the remnant causes the magnetic field to grow, and it is also possible that cosmic ray acceleration at the shock has an effect on the shock itself. Both are active areas of research at this writing.) In contrast, the weaker shock waves produced by the sun are very much affected by the magnetic field. We discuss some elements of magnetized shocks in Chap. 10, leaving many of the details for books on space plasma physics.

Even the simpler, unmagnetized shock waves are at first glance mysterious. Why would a fluid decide to abruptly change its properties? When we make music louder, the energy flux of the sound waves carrying energy to our ears increases. Why does

Fig. 4.1 An image of the
Tycho supernova remnant.
Credit: National Aeronautics
and Space Administration,
Chadra X-ray Center,
Smithsonian Center for
Astrophysics

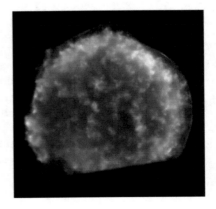

the energy flux of sound waves not just increase as necessary to transport as much energy as is needed? The fundamental answer is that sound waves move at the sound speed, and that the largest pressure modulation they can transport is of order the initial pressure of the fluid. This pressure is of order ρc_s^2, so the largest energy flux the sound waves could imaginably carry is of order ρc_s^3. But one can readily force a fluid to carry more energy than this, either by rapidly moving its boundary or by releasing energy within it. For example, the pressure in a singly ionized plasma at a temperature of 1 eV and a density of 1 g/cm^3 is of order a million atmospheres (1 Mbar). In high-energy-density experiments, much larger pressures are easy to obtain. The plasma cannot respond to such pressures by radiating sound waves. Instead, a shock wave forms. We discuss its basic properties in this section.

4.1.1 Jump Conditions

The shock wave does three things. First, it carries energy forward at the shock velocity, which is supersonic. Second, it heats and accelerates the medium as it passes, so that the fluid behind the shock carries kinetic energy and internal energy (their energy densities are equal in strong shocks, in the frame of reference in which the upstream fluid is at rest, known as the *laboratory frame*). Third, the shock wave heats the fluid behind the shock so that the motion of the shock wave relative to the heated fluid is subsonic. As a result, changes in the source of the energy are communicated to the shock front at the (new, higher) sound speed. For the original fluid, though, the arrival of the disturbance comes as a shock. Figure 4.2 shows a schematic diagram of a shock wave, in the frame of reference in which the shock is at rest, known as the *shock frame*. We will work consistently from a viewpoint in which the shock moves from left to right so that in the shock frame the fluid flows from right to left.

Establishing a discontinuity does not in any way relieve the system from the conservation of mass, momentum, and energy, however. To explore this, we begin with the Euler equations in conservative form (with a scalar pressure and explicit internal energy terms):

Fig. 4.2 Diagram of an isolated, steady shock, in a reference frame that moves with the shock. Here u_1 and u_2 are < 0 and we define the shock velocity as $u_s = -u_1$ in this reference frame

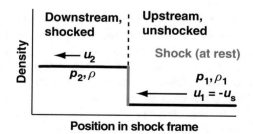

Continuity

$$\frac{\partial \rho}{\partial t} = -\nabla \cdot (\rho \boldsymbol{u}), \tag{4.1}$$

Momentum

$$\frac{\partial}{\partial t}(\rho \boldsymbol{u}) = -\nabla \cdot (\rho \boldsymbol{uu}) - \nabla p, \text{ and} \tag{4.2}$$

Energy

$$\frac{\partial}{\partial t}\left(\frac{\rho u^2}{2} + \rho \epsilon\right) = -\nabla \cdot \left[\rho \boldsymbol{u}\left(\epsilon + \frac{u^2}{2}\right) + p\boldsymbol{u}\right]. \tag{4.3}$$

Now consider a planar disturbance and integrate any one of these equations across a small region that may include an abrupt change in parameters. In the notation of Sect. 2.1 for a general equation in conservative form, we will have

$$\int_{x_1}^{x_2} \frac{\partial}{\partial t}\rho_Q dx' = -\int_{x_1}^{x_2} \frac{\partial}{\partial x}\Gamma_Q(x')dx' = \Gamma_Q(x_2) - \Gamma_Q(x_1). \tag{4.4}$$

The integral on the left approaches zero as $x_2 - x_1$ becomes infinitesimal, but the fluxes Γ_Q need not. Instead, in the limit that $x_2 - x_1 \to 0$, one has $\Gamma_Q(x_2) = \Gamma_Q(x_1)$. This analysis evidently applies to a fixed location within one's coordinate system, and this is part of the importance of the shock frame, in which the shock location remains fixed at some x (typically $x = 0$). Applying (4.4) to (4.1)–(4.3), one finds the *jump conditions* for a shock wave, which are

$$\rho_1 u_1 = \rho_2 u_2, \tag{4.5}$$

$$\rho_1 u_1^2 + p_1 = \rho_2 u_2^2 + p_2, \text{ and} \tag{4.6}$$

$$\left[\rho_1 u_1\left(\epsilon_1 + \frac{u_1^2}{2}\right) + p_1 u_1\right] = \left[\rho_2 u_2\left(\epsilon_2 + \frac{u_2^2}{2}\right) + p_2 u_2\right], \tag{4.7}$$

in the event that the motion is one-dimensional. More generally, for shock jump conditions we need to integrate in the direction of propagation of the shock wave. The two vector quantities, $\boldsymbol{u}$ and ∇p, may or may not have components transverse to this direction. If all we care about is a single shock wave interacting with a planar interface in a planar system, then we go into the shock frame by choosing a reference frame that is moving in the transverse direction, so as to eliminate the transverse components of $\boldsymbol{u}$. This, however, is not always feasible in practice. We discuss shocks with finite transverse fluid velocity, known as *oblique shocks*, in Sect. 4.1.5.

In a fluid described by the Euler equations, the shock jump must be infinitesimal in width. When one introduces additional phenomena that are always present at some level in actual fluids, such as viscosity, then the shock transition becomes gradual. However, so long as viscosity or other effects are only important near the shock front, then in steady state the jump conditions apply equally well to locations that are far enough from the shock front. Shocks that involve radiation are discussed in Chap. 7. Equations (4.5)–(4.7) can be manipulated to find relations that are convenient in a given context. We consider some of these relations in the next section.

Before proceeding, though, note that there is a seemingly paradoxical aspect to our description of shocks so far. On the one hand, we described a shock wave as something that heats and compresses the medium that flows into it. On the other hand, (4.5)–(4.7) are symmetric in the exchange of the indices. From their point of view the matter flowing into a discontinuity could be heated and compressed, or alternatively could be cooled and made less dense (rarefied). An abrupt transition in which matter was cooled and rarefied would be described as a *rarefaction shock*. From the point of view of the conservation equations, a rarefaction shock could exist. However, it is forbidden by the Second Law of Thermodynamics, as we will see in Sect. 4.1.4.

4.1.2 The Shock Hugoniot and Equations of State

A shock wave can place a material in a new state, whose properties depend, for example, on the amount of internal energy the material requires at a given pressure and density. By varying the initial density, pressure, and velocity of the material, one can access a continuous sequence of final states. One of the primary methods used to determine the equation of state involves measurements using shock waves. These measurements determine points along the *Rankine–Hugoniot relation*, which is traditionally identified as the function $p(p_1, 1/\rho_1, 1/\rho_2)$. The inverse of the density is the *specific volume*, often written as $V = 1/\rho$. The use of the postshock (downstream) density is an arbitrary choice. One can use any of the postshock parameters, and indeed one sees Rankine–Hugoniot curves plotted in various ways. (Figures 3.27 and 3.25 show examples.) Let us consider how measurements can determine the Rankine–Hugoniot relation. This relation is also often called the shock Hugoniot, even though Rankine's work (in 1870) came 17 years before Hugoniot's (in 1887).

One often can manage to measure the shock velocity and the postshock fluid velocity. The shock velocity can be determined, for example, by measuring when the shock emerges from varying thicknesses of shocked material. This can often be done using emission from the surface, which is strongly heated by the shock wave. Measurements of the postshock fluid velocity use targets in which the shock wave crosses an interface whose motion can be measured, for example, using the Doppler shift of light reflected from it or measuring its shadow with X-ray radiography.

Fig. 4.3 A useful practical
application of (4.9). Here the
independent variable is the
postshock fluid velocity in the
lab frame, u_p, in km/s, the
ordinate is the pressure
difference across the shock in
Mbars, and the curve is for
$\rho_1 u_1 = 30\,\mathrm{g\,cm}^{-3}\,\mathrm{km\,s}^{-1}$

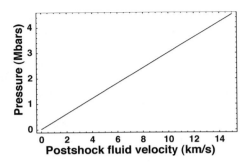

Such measurements are typically done in the inertial frame of the laboratory
where the upstream fluid is at rest. In this case, the postshock fluid velocity, u_p, that
one measures is the difference between the incoming and outgoing velocities in the
shock frame, shown in Fig. 4.2. Thus $u_p = u_1 - u_2$ or $u_2 = u_1 - u_p$. Then from (4.5)
and (4.6) we can find

$$\frac{\rho_2}{\rho_1} = 1 + \frac{u_p}{u_1 - u_p} \quad \text{and} \tag{4.8}$$

$$p_2 - p_1 = \rho_1 u_1 (u_1 - u_2) = \rho_1 u_1 u_p. \tag{4.9}$$

It is a very neat trick to determine the thermodynamic state of the fluid from
two measurements (of u_1 and u_p), but this is the power of shock Hugoniot
measurements. However, determining u_p is often not easy. We discussed some
of the experimental approaches to shock Hugoniot measurements in Sect. 4.2.2.
Figure 4.3 illustrates (4.9), showing how measurements of u_1 and u_p determine
$p_2 - p_1$. Researchers doing shock Hugoniot measurements with flyer plates often
work with this equation using graphs like that of Fig. 4.3, which allow one to think
directly in terms of the measured quantities.

4.1.3 Useful Shock Relations

The jump conditions are sometimes useful as they stand, but there are also useful
alternative solutions of these equations. There is an upstream fluid, within which the
distance from any fluid element to the shock decreases with time, and which we
will designate by the subscript 1. There is also a downstream fluid, within which the
distance from any fluid element to the shock increases with time, and which we will
designate by the subscript 2. The most useful equations relate specific properties
of the upstream fluid to those of the downstream fluid and the *upstream Mach
number*, M_u. This Mach number is defined as the ratio of the rate at which the
upstream material and the shock approach one another to the sound speed in the
upstream material. The rate at which the upstream material and the shock approach

one another is often called the *shock velocity*, which is also the velocity of an isolated shock when the upstream fluid is at rest in the inertial frame of reference of the laboratory. We will designate this as u_s, which will equal $|u_1|$ in the simple case we now discuss. However, when one analyzes complex systems with several shocks, one must think carefully to properly identify the upstream Mach number for each one.

To obtain useful solutions of (4.5)–(4.7), we work in the *shock frame*. In the (one-dimensional) shock frame, one has three equations and four variables in the upstream state, so one needs an equation of state to relate ϵ to p. It is most useful to assume the fluid to be a polytropic gas so that $\rho\epsilon = p/(\gamma - 1)$. Thus, in terms of the various polytropic indices discussed in Chap. 3, we are working with the shock gamma. This one term in the energy equation is the source of all the factors involving γ that appear in the following. However, for high-energy-density physics we should note that the polytropic index sometimes differs greatly across a shock. We may, for example, start with a cold, highly-ordered crystal for which γ approaches 3 and shock it into an ionizing, plasma state for which $\gamma \sim 4/3$. Equations (4.5)–(4.7) still apply in such a case, with γ evaluated appropriately on each side of the interface. In what follows, we provide both the traditional results obtained when γ is taken to be unchanging (and is not subscripted) and also results in which γ is subscripted and applies separately to the fluid on the two sides of the interface.

Thus, the velocity u_1 is the shock velocity, and the upstream Mach number is $M_u = -u_1/c_{s1} = u_s/c_{s1}$, which for a polytropic gas is $u_s\sqrt{\rho_1/(\gamma_1 p_1)}$. Solving these equations for the ratio of the pressures, one can show that

$$\frac{p_2}{p_1} = \frac{\rho_2(\gamma + 1) - \rho_1(\gamma - 1)}{\rho_1(\gamma + 1) - \rho_2(\gamma - 1)}$$

or (4.10)

$$\frac{p_2}{p_1} = \left[\frac{\rho_2(\gamma_1 + 1) - \rho_1(\gamma_1 - 1)}{\rho_1(\gamma_2 + 1) - \rho_2(\gamma_2 - 1)}\right]\frac{(\gamma_2 - 1)}{(\gamma_1 - 1)}.$$

Figure 4.4a shows this pressure ratio as a function of the density ratio ρ_2/ρ_1. Alternatively, if one does the algebra by hand, for constant γ, it is easier to relate the pressures to the specific volumes (the inverse of density) as

$$\frac{p_2}{p_1} = \frac{V_1(\gamma + 1) - V_2(\gamma - 1)}{V_2(\gamma + 1) - V_1(\gamma - 1)}.$$

(4.11)

Note that the postshock pressure implied by these two equations diverges when the denominator becomes zero at a specific density ratio. One can rearrange (4.10) to find the density ratio in terms of the pressures, which is

$$\frac{\rho_2}{\rho_1} = \frac{p_2(\gamma + 1) + p_1(\gamma - 1)}{p_1(\gamma + 1) + p_2(\gamma - 1)}$$

or (4.12)

$$\frac{\rho_2}{\rho_1} = \left[\frac{p_2(\gamma_2 + 1) + p_1(\gamma_2 - 1)}{p_1(\gamma_1 + 1) + p_2(\gamma_1 - 1)}\right] \frac{(\gamma_1 - 1)}{(\gamma_2 - 1)}.$$

This makes it very clear that as p_2 becomes $\gg p_1$, the density approaches a fixed density ratio given by

$$\frac{\rho_2}{\rho_1} = \frac{(\gamma_2 + 1)}{(\gamma_2 - 1)}. \tag{4.13}$$

This density ratio is the physical limit that can be produced by a single shock in a polytropic gas, and only the postshock value of γ enters. Shocks encountered in high-energy-density physics often have density ratios near this limit, and thus are *strong shocks* as defined shortly. Low atomic number materials, subject to strong enough shocks, may behave this way with $\gamma \sim 5/3$. Thus, the density ratio ρ_2/ρ_1 may be 4 to 1. One may encounter differences if the internal energy of the shocked material is a significant fraction of the thermal energy density, as for example in materials that are ionizing as discussed in Chap. 3. In this case, the jump conditions [specifically (4.6) and (4.7)] imply that u_2 must be smaller than it would be otherwise—the increased internal energy comes from the kinetic and thermal energies. As a result, (4.5) implies that ρ_2 must be larger than it would be otherwise. (In some cases, radiation lost during ionization can have a similar effect.) There is still a limiting density ratio, but it is affected by the properties of the material. Xenon in particular is a gas that both absorbs a lot of ionization energy and radiates strongly under typical experimental conditions. Accordingly, a strong shock in xenon produces a larger density jump than is produced in a lower-Z gas such as nitrogen. The way that this enters the mathematics is that the value of γ is smaller in such a material, just as we saw in Chap. 3. In terms of an effective polytropic index, xenon typically would have $\gamma \sim 1.2$ to 1.3 at densities above atmospheric density. At lower densities, weakly ionized xenon can store a great deal of energy in excited states. If one accounts for this by adjusting γ, then γ can be driven down below approximately 1.1.

Returning to (4.12), the traditional approach is to find useful expressions involving M_u, for example, by substituting for p_2 from (4.9). We then can obtain for the density ratio

$$\frac{\rho_2}{\rho_1} = \frac{M_u^2(\gamma + 1)}{M_u^2(\gamma - 1) + 2}. \tag{4.14}$$

This ratio exhibits the behavior we expect, tending to the value given by (4.13) as M_u becomes large. In general, the limit in which M_u is large and only terms in the largest power of M_u need to be kept is referred to as the *strong shock limit*. In a similar way, we find the pressure ratio

$$\frac{p_2}{p_1} = \frac{2\gamma M_u^2 - (\gamma - 1)}{(\gamma + 1)}, \tag{4.15}$$

which increases indefinitely as M_u increases.

Unfortunately, in real high-energy-density systems M_u is often very poorly known. This is because the upstream temperature might be room temperature, at which the system is typically prepared, a significantly smaller temperature through cooling in vacuum, or a significantly higher temperature because of small levels of heating by radiation or electrons in advance of the shock. The uncertainty in M_u can easily be a factor of several. However, the stronger the shock the less this matters. One can show this by working with (4.5), (4.6), and (4.10), dividing the pressures by $\rho_1 u_s^2$. Then one can define

$$S = \sqrt{1 + \left(\frac{\gamma_2 p_1}{\rho_1 u_s^2}\right)^2 + \frac{2p_1(\gamma_1 - \gamma_2^2)}{\rho_1 u_s^2(\gamma_1 - 1)}}. \tag{4.16}$$

Note that S approaches 1 as shock velocity increases. (For $\gamma_1 = \gamma_2$, $S = 1 - \gamma_2 p_1/(\rho_1 u_s^2)$.) This allows one to write the density ratio as

$$\frac{\rho_2}{\rho_1} = \left[\frac{(\gamma_2 + S) + \frac{\gamma_2 p_1}{\rho_1 u_s^2}}{(\gamma_2 - 1) + 2\frac{\gamma_1 p_1}{\rho_1 u_s^2}\frac{(\gamma_2 - 1)}{(\gamma_1 - 1)}}\right]. \tag{4.17}$$

The corresponding density ratio is shown for various values of γ_1 and γ_2 in Fig. 4.4a. While the eventual density ratio reached in a strong shock is not affected by γ_1, the ratio of $\rho_1 u_s^2/p_1$ required to approach this value is affected. The downstream pressure can similarly be written as

$$p_2 = \frac{2}{(\gamma + 1)}\rho_1 u_s^2 \left[1 - \frac{(\gamma - 1)p_1}{2\rho_1 u_s^2}\right]$$

or

$$p_2 = \frac{\rho_1 u_s^2}{(\gamma_2 + S) + \frac{\gamma_2 p_1}{\rho_1 u_s^2}}\left[1 + S\left(1 + \frac{p_1}{\rho_1 u_s^2}\right) + \frac{p_1}{\rho_1 u_s^2}\left(\frac{2(\gamma_1 - \gamma_2)}{(\gamma_1 - 1)} + \frac{\gamma_2 p_1}{\rho_1 u_s^2}\right)\right]. \tag{4.18}$$

The first form is useful for quick estimates assuming a single value for γ. Even more useful is the realization that $2/(\gamma+1)$ is of order 1 so for strong shocks $p_2 \sim \rho_1 u_s^2$, which is easy to remember and to evaluate. Figure 4.4b shows the dependence of $p_2/(\rho_1 u_s^2)$ on $\rho_1 u_s^2/p_1$.

One can proceed to obtain a similar expression for the temperature, taking $p_2 = (Z_2 + 1)k_B T_2 \rho_2/(Am_p)$, where the electrons are assumed to fully equilibrate with the ions, to be non-degenerate, and where we can ignore Coulomb modifications to the pressure. One finds

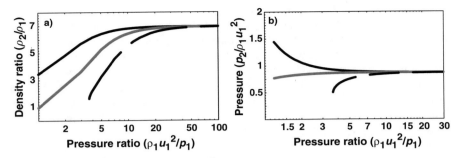

Fig. 4.4 (a) The ratio of postshock to preshock density depends as shown on the ratio of $\rho_1 u_1^2$ to preshock pressure, from (4.17). Here $\gamma_2 = 4/3$. The values of γ_1 are 3 (black curve), 4/3, (gray curve), and 1.1 (dashed curve). (b) The ratio of postshock pressure to $\rho_1 u_1^2$ depends as shown on the ratio of $\rho_1 u_1^2$ to preshock pressure, from (4.18). This ratio asymptotes to 2/(γ+1). Here $\gamma_2 = 5/3$. The values of γ_1 are 3 (black curve), 4/3, (gray curve), and 1.1 (dashed curve)

$$k_B T_2 = \frac{A m_p}{(1 + Z_2)} u_s^2 \frac{2(\gamma_2 - 1)}{\left((\gamma_2 + S) + \frac{\gamma_2 p_1}{\rho_1 u_s^2}\right)^2} \left(1 + \frac{2\gamma_1 p_1}{(\gamma_1 - 1)\rho_1 u_s^2}\right)$$

$$\times \left[\frac{(1 + S)}{2} + \frac{p_1}{\rho_1 u_s^2}\left(\frac{S}{2} + \frac{(\gamma_1 - \gamma_2)}{(\gamma_1 - 1)}\right) + \frac{\gamma_2}{2}\left(\frac{p_1}{\rho_1 u_s^2}\right)^2\right]. \quad (4.19)$$

Here Z_2 is the average ionization of the postshock state. In the strong shock limit we find

$$k_B T_2 = \frac{A m_p}{(1 + Z_2)} u_s^2 \frac{2(\gamma_2 - 1)}{(\gamma_2 + 1)^2}. \quad (4.20)$$

In typical cases, the ion-ion collision length, which sets the distance over which the shock transition occurs, is much shorter than the ion-electron equilibration distance. In addition, the distance required for equilibration of the electron and ion temperatures increases for materials that ionize further as the electrons are heated. The immediate postshock temperature of the ions can be found by setting $Z_2 = 0$ and $\gamma_2 = 5/3$ in this equation. A given measurement or simulation may or may not have sufficiently fine resolution to detect these ions before they equilibrate with the electrons. In astrophysical or other shocks with weak collisionality, because of low density or high temperature, the equilibration distance can become very large and the difference of electron and ion temperatures may be readily observed. For shocks in atomic neutral gasses that are not heated enough to ionize, Z_2 does equal zero, which is why one may encounter $k_B T_2 = (3/16)A m_p u_s^2$ in various places as a standard expression. However, in a highly ionized plasma with strong collisional coupling of electrons and ions, it is evident that this standard expression can greatly overestimate the temperature.

Strong shocks have some additional properties that are worthwhile to know. As always, the velocity ratio is the inverse of the density ratio, in this limit being

$$\frac{u_2}{u_1} = \frac{(\gamma - 1)}{(\gamma + 1)}. \tag{4.21}$$

From this one can find the postshock particle velocity, u_p, in the lab frame (in which the upstream fluid is at rest; see the discussion near (4.8)). This is

$$u_p = u_s + u_2 = \frac{2}{(\gamma + 1)} u_s, \tag{4.22}$$

which is $(3/4)u_s$ for $\gamma = 5/3$ and $(6/7) u_s$ for $\gamma = 4/3$. So the postshock fluid velocity is approximately 80% of the shock velocity in typical materials. It is easy to show that the postshock fluid velocity becomes closer to the shock velocity as the density jump increases. If we examine the postshock pressure p_2 in the strong-shock limit, we can substitute for ρ_1 and u_s to find

$$p_2 = (\gamma - 1)\frac{\rho_2 u_p^2}{2}, \tag{4.23}$$

so for strong shocks in the laboratory frame of reference the kinetic and internal energy densities are equal in a polytropic gas [because $\rho \epsilon = p/(\gamma - 1)$]. (But note from Eqs. (4.3) and (4.7) that the non-kinetic part of the total energy *flux*, which is proportional to $\rho \epsilon + p$, is γ times the kinetic part.) Using (4.23), we can look again at some of the velocities in the strong shock limit. The sound speed in the shocked fluid is $\sqrt{\gamma p_2/\rho_2} = \sqrt{\gamma(\gamma - 1)/2} \times u_p$, which for $\gamma = 5/3$ is $\sqrt{5/9}u_p$. Thus, in the laboratory frame (and for $\gamma < 2$) the flow is supersonic, as the sound speed is less than u_p. However, the source of pressure that sustains the shock must move at u_p, and the distance between it and the shock increases at speed u_2, which can easily be shown to be $(\gamma - 1)/2 \times u_p$. This is $(1/3)u_p$ for $\gamma = 5/3$. Thus the separation of the pressure source from the shock is subsonic. This last statement is completely equivalent to saying that, in the shock frame, the downstream fluid moves subsonically.

We can also look at some typical parameters. A high-energy-density experiment may produce a shock wave in a plastic material having $\rho_1 = 1$ g/cm^3, $A = 6.5$, and $Z = 3.5$ using a pressure of 50 Mbar (i.e., 5×10^{13} dynes/cm^2). The plastic behaves like a polytropic gas with $\gamma \sim 4/3$, so the shocked density is $\rho_2 \sim 7$ g/cm^3. The shock velocity, from (4.18), is approximately 80 km/s (8×10^6 cm/s), so the shock will traverse a 100 μm thick layer in about 1.2 ns. The postshock temperature, from (4.20), is approximately 25 eV. For comparison, consider the shock in the interstellar medium produced by a supernova remnant such as Tycho. The velocity is ≥ 1000 km/s (faster in younger remnants) and $\rho_1 \sim 1$ amu/cm$^3 \sim 10^{-24}$ g/cm^3. Thus $p_2 \sim 10^{-8}$ dynes/cm^2. This seems small, but the postshock temperature is ≥ 1 keV, so the resulting plasma is quite hot. However, even for the velocities approximately

ten times higher that are present in very young supernova remnants, the temperature
is not near relativistic values. Thus, except for the cosmic rays produced at the shock,
nonrelativistic theories and experiments can address the behavior of such systems.

4.1.4 Entropy Changes Across Shocks

While mass, momentum, and energy are conserved across shock transitions, entropy
is not. This should not come as a surprise to anyone who has studied statistical or
thermal physics, as one is increasing the temperature of the fluid in a nonadiabatic
transition. Here we determine and discuss the change in entropy across the shock
transition.

Using (2.11) for the specific entropy and (4.12), one can find the difference in
entropy across the shock wave to be

$$s_2 - s_1 = c_V \ln \left[\frac{p_2}{p_1} \left(\frac{\rho_1}{\rho_2} \right)^\gamma \right]. \qquad (4.24)$$

Thus, in an adiabatic transition that keeps p/ρ^γ constant, the entropy does not
increase. Shocks increase entropy because they are irreversible, non-adiabatic
transitions. For single strong shocks, the argument of the logarithm is dominated
by p_2, as the density ratio varies only over a limited range. Taking the strong shock
limit, we find

$$s_2 - s_1 = c_V \ln \left[\frac{p_2}{p_1} \left(\frac{\gamma - 1}{\gamma + 1} \right)^\gamma \right] \sim c_V \left[\ln \left(\frac{p_2}{p_1} \right) - 2.2 \right], \qquad (4.25)$$

in which the final equality is obtained by setting $\gamma = 5/3$. Noting once again that
in a fully ionized plasma both electrons and ions carry heat, one has

$$c_V = \frac{(Z + 1)k_B}{A m_p (\gamma - 1)} = \frac{(Z + 1)}{A(\gamma - 1)} 9.57 \times 10^7 \text{ J/(keV g)}, \qquad (4.26)$$

expressed to accommodate temperature in keV. As we will see, the entropy in an
inertial fusion capsule needs to be kept below about 4×10^8 J/(keV g). Since the ratio
of pressures is > 1000 in this case, compression for inertial fusion cannot occur by
means of a single shock.

It may seem strange to the reader that entropy is generated by shock waves, when
we have shown that shock waves are consistent with the fundamental equations for
mass, momentum, and energy in the fluid, because these same equations produced
(2.3), which we showed in Chap. 2 to express the conservation of entropy. In
other words, if the conservation of entropy is consistent with and derived from the
equations that include the presence of shocks, why is entropy not conserved across
shocks? The solution to this puzzle is that shocks are in fact dissipative structures

whose details cannot be described by (4.1)–(4.3). From the point of view of these three equations, the shock transition can be taken to be a thin layer of zero thickness, because it does not alter the mass, momentum, or energy fluxes in the flow. However, the shock does convert kinetic energy to heat, and thus is an entropy source. If we were to describe the action of the shock in detail, we would need to add terms to the momentum and energy equations to account for the changes in the fluid. When we then used these equations to obtain an equation for the entropy, we would be left with an entropy source term, and this source term would not have a limit of zero as the thickness of the shock transition approached zero.

Now suppose that instead of using a single strong shock to achieve a desired value of p_2/p_1, which we write here as p_{final}/p_{init}, we use n shocks. Each of these will produce pressure ratio $R_p = (p_{final}/p_{init})^{1/n}$ and a density ratio R_ρ. The density ratio produced by each shock may or may not approach $(\gamma + 1)/(\gamma - 1)$, but the final density ratio ρ_{final}/ρ_{init} will be much larger than $(\gamma + 1)/(\gamma - 1)$ for $n > 1$. The temperature will be correspondingly smaller, as the pressure is the same yet the density is higher. Thus, the entropy increase will be smaller for multiple shocks than for one shock. Specifically,

$$s_2 - s_1 = nc_V\ln\left[R_p\left(\frac{1}{R_\rho}\right)^\gamma\right] = c_V\ln\left[\frac{p_{final}}{p_{init}}\left(\frac{1}{R_\rho}\right)^{n\gamma}\right]. \tag{4.27}$$

Figure 4.5 shows the resulting entropy increase, using (4.12) to compute R_ρ, for a value of $p_{final}/p_{init} = 1000$. One can see that only a few shocks are needed to greatly reduce the total increase of entropy. The limit of a very large number of shocks with progressively smaller individual pressure jumps, as $n \rightarrow \infty$, is an adiabatic compression, which produces no entropy increase.

The use of multiple shocks to apply the available pressure is thus an important design tool. If one desires to achieve low entropy or high densities, as one does, for example, in ICF, then one should use several shocks. If one desires high temperature, as one does to produce fast ejecta, then one should use a single shock.

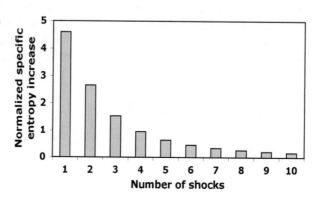

Fig. 4.5 The total increase in normalized specific entropy, $(s_2 - s_1)/c_v$ is shown *vs.* the number of shocks, for an overall pressure increase of a factor of 1000

If one desires to more carefully tailor the density, temperature, and entropy of a final state, then one can use multiple shocks chosen for that purpose.

4.1.5 Oblique Shocks

We now return to the issue of *oblique shocks*, which we must analyze when there is a reason to choose a shock frame with an upstream velocity component transverse to the shock that is nonzero. This issue arose in the context of (4.5)–(4.7), which are valid only for zero transverse velocity. Note that an oblique shock may develop in more than one way. A shock can be launched as an oblique shock, by something as simple as a tilted piston. Also, a shock can become oblique by interacting with an interface. The geometric definitions we need for this problem are shown in Fig. 4.6.

To deal with nonzero transverse velocity, we define the shock normal, a unit vector n, which is normal to the shock front and in the direction of the normal flow into the shock front (thus, n points from right to left in our standard orientation). Then the component of u in the normal direction is $n(u \cdot n)$ while the transverse component of u is $u_\perp = (n \times u) \times n$. Using the subscript n for the normal direction, (4.1)–(4.3) give us the following relations:

$$\rho_1 u_{n1} = \rho_2 u_{n2}, \tag{4.28}$$

$$\rho_1 u_{n1}^2 + p_1 = \rho_2 u_{n2}^2 + p_2, \tag{4.29}$$

$$u_{\perp 1} = u_{\perp 2}, \text{ and} \tag{4.30}$$

$$\rho_1 \epsilon_1 + p_1 + \rho_1 \frac{u_{n1}^2}{2} = \rho_2 \epsilon_2 + p_2 + \rho_2 \frac{u_{n2}^2}{2}. \tag{4.31}$$

By comparison with the previous equations, one can see that we have gained one important new piece of information: the transverse velocity is conserved across the shock. Otherwise we have regained the previous equations (or an equivalent one in (4.31)) with u replaced by u_n. It is not surprising that the change in u_n is responsible for the changes in pressure and internal energy.

Fig. 4.6 Definitions for oblique shocks

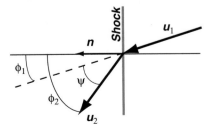

A clear consequence of these equations is that the fluid velocity is bent away from the normal as it crosses the shock. The normal velocity decreases but the transverse velocity does not, and this is the consequence. We can develop the mathematics for this situation as follows. Suppose that $\boldsymbol{u}_1 \cdot \boldsymbol{n} = u \cos \phi_1$, thus defining the angle of the incoming flow, ϕ_1. Then the angle of the outgoing flow can be found from

$$\tan\phi_2 = \frac{u_\perp}{u_{n2}} = \frac{u_\perp}{u_{n1}} \frac{\rho_2}{\rho_1} = \tan\phi_1 \frac{\rho_2}{\rho_1}, \tag{4.32}$$

which for a polytropic gas becomes

$$\tan\phi_2 = \tan\phi_1 \frac{M_{nu}^2(\gamma + 1)}{M_{nu}^2(\gamma - 1) + 2}, \tag{4.33}$$

in which the upstream Mach number, calculated using u_{n1}, is M_{nu}.

One may in some cases need to know how much the flow is (or can be) deflected by the shock, which is relevant, for example, to the supersonic movement of bodies through fluids. For this purpose we seek ψ defined by $\boldsymbol{u}_1 \cdot \boldsymbol{u}_2 = u_1 u_2 \cos \psi$. By expressing the velocities in terms of their normal and tangential components, then dividing by $u_1 u_2$, one can show

$$\cos\psi = \cos\phi_2 \left[\cos\phi_1 + \sin\phi_1 \tan\phi_2\right], \tag{4.34}$$

from which via (4.32) we have

$$\cos\psi = \frac{\left[\cos\phi_1 + \sin\phi_1 \tan\phi_1 (\rho_2/\rho_1)\right]}{\sqrt{1 + \tan^2\phi_1 (\rho_2/\rho_1)^2}}. \tag{4.35}$$

An interesting implication of (4.35) is that there is a maximum possible angle of deflection that can be produced by a shock. Figure 4.7 shows the dependence of ψ on ϕ_1 for several specific density ratios. One can see that the maximum angle of deflection depends on the density ratio across the shock. If a supersonic object is too

Fig. 4.7 Flow deflection vs. incident angle for a density ratio from bottom to top of 2, 4, 7, and 14. The thick gray line shows the maximum deflection, producing flow parallel to the shock front

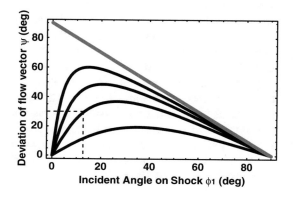

Fig. 4.8 Schematic of object
and resulting shocks

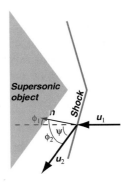

blunt, so that it attempts to deflect the incoming material by more than this angle, then a bow shock forms in front of the object, heating the material so that its flow around the object is subsonic. Figure 4.8 illustrates the case where a deflecting shock can be established, enabling supersonic flow around the object. These two figures are most useful for strong shocks with well-known density ratios. For weaker shocks in polytropic gases one could substitute from (4.33), noting that the upstream Mach number also depends on angle of incidence. This can be put into a standard form known as the *shock polar*, discussed in Landau and Lifshitz (1987) and many other texts on fluid dynamics. We will have an interest in the small-angle limit of (4.35) when we consider shock stability in Chap. 5. Assuming that ϕ_1 is small enough that $-(\phi_1^2/2)(\rho_2/\rho_1)^2$ remains small, this is $\psi = \phi_1(\rho_2/\rho_1 - 1)$.

4.1.6 Shocks and Interfaces

Understanding the basic properties of an isolated shock is an important fundamental building block. There are cases, such as the edge of a supernova remnant or the initial response of a target to laser ablation, in which the dynamics is essentially single-shock dynamics. There are many more cases, however, in which the dynamics is produced by the interaction of shock waves, interfaces, and other phenomena. This occurs, for example, when the shock wave in a supernova remnant encounters a molecular cloud or other dense obstacle, and also in experiments that use shock reverberation to compress and heat a material. The first of these that we discuss is the interaction of a shock wave (or in general, an incoming fluid with specific properties) with an interface at which the density increases. In general, this leads to a "reflected" shock in the initial fluid and a "transmitted" shock in the denser fluid. This is an example of the one possible response when a shock reaches an interface—two shocks are produced. Figure 4.9 shows a schematic of such an interaction. We discuss it further here. Other cases are discussed in Sect. 4.6.

The introduction of the second material and the interaction greatly complicates our bookkeeping. As is shown in Fig. 4.9, we designate the unshocked first fluid as

Fig. 4.9 Schematic of steady shock incident on interface where density increases. (a) Before shock reaches interface. (b) After shock reaches interface

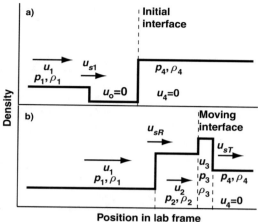

state o, the shocked fluid as state 1, the state of the reflected shock as state 2, the state of the transmitted shock as state 3, and the unshocked second fluid as state 4. We have little interest in the properties of state o, except in whatever way they influence state 1. In addition, the material in state 4 would expand to the left unless this material is a solid (the typical case) or $p_o = p_4$. Our goal is to calculate the properties of the reflected and transmitted shocks, given the properties of the fluids and the initial shock. We can take advantage of two conditions in this calculation. These are $p_2 = p_3$ and $u_2 = u_3$. That is, the pressure and velocity are both continuous across the density interface. If this were not the case, one would produce additional waves (or voids) at the interface in the postshock state, which is not consistent with our definition of the shocks as discontinuous, localized transitions in fluid properties. To develop an analytic treatment of this behavior, we will designate the polytropic index of the first fluid by γ_1 and that of the second fluid by γ_4.

Our approach here will be fairly general, which will give us results that can be applied to a number of specific cases. Some simpler cases, such as the reflection of a shock from a rigid wall or the behavior of an ideal flyer plate, are left as homework problems. In the case of an incoming shock as shown in Fig. 4.9a, one can find ρ_1, u_1, and p_1 from (4.5), (4.14), and (4.15) and the properties of state o. Alternatively, Fig. 4.9b also may describe the interaction between an incoming block of solid material (a *flyer plate*) that strikes a second block of solid material. In this case, the properties of region 1 will not be determined by shock relations. As an extension of this analysis, one can use Fig. 4.9b to approximately describe the interaction of two colliding fluids with more general properties, if it makes sense after some initial transient to ignore the behavior of their leading edges. This may, for example, occur in experiments that produce a flyer plate that is in the plasma state. The net effect is that we start with the properties of state 1, in which the ratio of p_1 to $\rho_1 u_1^2$ will depend on the way in which the interaction develops. In any event we have the following equations, assuming that $\gamma_1 = \gamma_2$ and $\gamma_3 = \gamma_4$:

$$\frac{\rho_2}{\rho_1} = \frac{M_{12}^2(\gamma_1 + 1)}{M_{12}^2(\gamma_1 - 1) + 2}, \tag{4.36}$$

$$\frac{p_2}{p_1} = \frac{2\gamma_1 M_{12}^2 - (\gamma_1 - 1)}{(\gamma_1 + 1)}, \tag{4.37}$$

$$\frac{\rho_3}{\rho_4} = \frac{M_{34}^2(\gamma_4 + 1)}{M_{34}^2(\gamma_4 - 1) + 2}, \tag{4.38}$$

$$\frac{p_3}{p_4} = \frac{2\gamma_4 M_{34}^2 - (\gamma_4 - 1)}{(\gamma_4 + 1)}, \tag{4.39}$$

$$u_{sT} - u_3 = -u_3' = u_{sT}\rho_4/\rho_3, \tag{4.40}$$

$$u_1 - u_{sR} = u_{s12}, \tag{4.41}$$

$$u_2 - u_{sR} = u_2' = u_{s12}\rho_1/\rho_2, \tag{4.42}$$

$$p_2 = p_3, \tag{4.43}$$

$$u_2 = u_3, \tag{4.44}$$

in which $M_{12} = u_{s12}/c_{s1}$ where $c_{s1} = \sqrt{\gamma_1 p_1/\rho_1}$, $M_{34} = u_{sT}/c_{s4}$ where $c_{s4} = \sqrt{\gamma_4 p_4/\rho_4}$, the upstream velocity in the reflected shock frame is u_{s12}, the postshock fluid velocity in the reflected shock frame is u_2', the postshock fluid velocity in the transmitted shock frame is u_3', and the reflected and transmitted shock velocities in the lab frame are u_{sR} and u_{sT}, respectively.

The known quantities in these nine equations are $\rho_1, \rho_4, p_1, p_4, u_1, u_4, \gamma_1, \gamma_4$. The unknowns are $p_2, \rho_3, p_2, p_3, u_2, u_3, u_{s12}, u_{sR}$, and u_{sT}, so the set of equations is closed, even if complex.

One can, with some work, solve (4.36)–(4.44). One approach is to solve for the various quantities in terms of u_2, the postshock fluid velocity in both materials, and then to obtain an equation for u_2 itself. This is productive since most aspects of the shock in one material do not depend on the properties of other material. This produces the following eight (4.45)–(4.52) in addition to (4.44):

$$0 = 16\gamma_1 p_1 + \left[\sqrt{\frac{16c_{s1}^2 + (\gamma_1 + 1)^2(u_1 - u_2)^2}{(\gamma_1 + 1)^2(u_1 - u_2)^2}} - 1 \right] \left[4(\gamma_1 + 1)(p_1 - p_4) \right.$$

$$\left. - (\gamma_1 + 1)(\gamma_4 + 1)\rho_4 u_2^2 \left(1 + \sqrt{1 - \frac{16c_{s4}^2}{(\gamma_4 + 1)^2 u_2^2}} \right) \right], \tag{4.45}$$

$$\rho_2 = \frac{\rho_1 \left[4c_{s1}^2 + (u_1 - u_2) \left((\gamma_1 + 1)(u_1 - u_2) + \sqrt{16c_{s1}^2 + (\gamma_1 + 1)^2(u_1 - u_2)^2} \right) \right]}{2 \left[(\gamma_1 - 1)(u_1 - u_2)^2 + 2c_{s1}^2 \right]},$$

(4.46)

$$p_2 = \frac{p1}{\gamma_1 + 1} \left[(-\gamma_1 + 1) + \frac{2\gamma_1 \rho_2^2 (u_1 - u_2)^2}{c_{s1}^2 (\rho_1 - \rho_2)^2} \right],$$

(4.47)

$$\rho_3 = \frac{(\gamma_4 + 1)\rho_4 \left((\gamma_4 + 1)u_2 + \sqrt{16c_{s4}^2 + (\gamma_4 + 1)^2 u_2^2} \right)^2}{\left[32c_{s4}^2 + (\gamma_4 - 1) \left((\gamma_4 + 1)u_2 + \sqrt{16c_{s4}^2 + (\gamma_4 + 1)^2 u_2^2} \right)^2 \right]},$$

(4.48)

$$p_3 = \frac{p_4}{(\gamma_4 + 1)} \left[(-\gamma_4 + 1) + \frac{\gamma_4}{8c_{s4}^2} \left((\gamma_4 + 1)u_2 + \sqrt{16c_{s4}^2 + (\gamma_4 + 1)^2 u_2^2} \right)^2 \right],$$

(4.49)

$$u_{s12} = \frac{\rho_2(u_1 - u_2)}{\rho_2 - \rho_1},$$

(4.50)

$$u_{sT} = \frac{1}{4} \left((\gamma_4 + 1)u_2 + \sqrt{16c_{s4}^2 + (\gamma_4 + 1)^2 u_2^2} \right),$$

(4.51)

and

$$u_{sR} = \frac{\rho_2 u_2 - \rho_1 u_1}{\rho_2 - \rho_1}.$$

(4.52)

The first of these equations (4.45) is an implicit equation for u_2 in terms of known quantities. It can be converted to a polynomial equation in u_2, whose order depends on the assumptions. The second (4.46) determines ρ_2 based on u_2 and known quantities. The remaining equations determine the other unknowns based on ρ_2, u_2, and known quantities. A convenient and meaningful normalization is to divide velocities by u_1, densities by ρ_1, and pressures by $\rho_1 u_1^2$, since these quantities determine the dynamics. These equations may appear complicated, but this is mainly because of the numerous parameters that must be specified to define the system. We will show results for two examples.

The first example is that of reflected and transmitted shocks, produced when a first shock is incident on an interface where the density is greater than $(\gamma_1 + 1)\rho_1/(\gamma_4 + 1) \sim \rho_1$. Thus, ρ_1, u_1, and p_1 are produced by a shock, so that $p_1 = (\gamma - 1)\rho_1 u_1^2/2$. Figure 4.10 shows how the normalized interface velocity, u_2/u_1, varies with the density ratio ρ_4/ρ_1. One sees that the normalized velocity is quite close to $1/\sqrt{\rho_4/\rho_1}$. This is sensible. Once the density beyond the interface is a few times ρ_1, the incoming plasma is nearly stopped and the pressure on the interface is approximately constant and approximately equals $\rho_1 u_1^2$. For fixed pressure, the transmitted shock velocity scales inversely with the square root of the density. So this is what one sees.

Fig. 4.10 The dependence of
the normalized interface
velocity on the density ratio
when a shock encounters an
interface. The two solid
curves are for $\gamma = 4/3$
(lower) and $\gamma = 5/3$ (upper).
The gray curve shows
$1/\sqrt{\rho_4/\rho_1}$

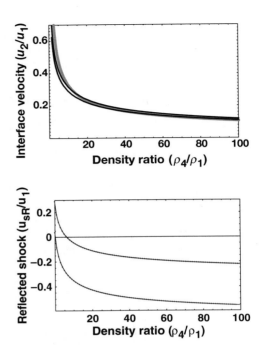

Fig. 4.11 The velocity of the
reflected shock decreases as
the density ratio increases.
The lower curve is for
$\gamma = 5/3$ and the upper curve
is for $\gamma = 4/3$

The transmitted shock velocity is $(\gamma_4 + 1)/2$ times u_2, which is shown in the figure. This is a simple relation because the upstream fluid for this shock is at rest in the lab frame. The reflected shock is more interesting. Its velocity is shown in Fig. 4.11. The velocity and direction of the reflected shock shows a strong dependence on γ. If the material being impacted is sufficiently compressible and low enough in density, then the momentum of the incoming fluid is sufficient to push the interface forward faster than the reflected shock retreats from it. In this case the reflected shock continues to move forward in the lab frame. On the other hand, if the material being struck is sufficiently incompressible or sufficiently dense, then the reflected shock will recoil from the interface in the lab frame. The limiting reflected shock velocities are $-u_1/3$ for $\gamma = 4/3$ and $-2u_1/3$ for $\gamma = 5/3$. This can be a useful limit, for example, to estimate how thick a wall one may need to contain a shocked material in an experiment. One can use shock reflections in measuring EOS. One measures the time it takes for the reflected shock to return to the downstream surface, known as the time for *shock reverberation*. This time is sensitive primarily to the compression produced by the initial shock.

In addition, this case is relevant to interaction of the forward shock in a supernova remnant with a molecular cloud. Figure 4.12 shows an image of such a collision. The spherical object is the remnant, and the figure shows both X-rays and radio emission produced by its interaction with the cloud. Leaving aside the clumpiness of the cloud, we can take γ to be 5/3 for both the interstellar medium and the cloud. Typical densities for the interstellar medium and the cloud are 2.5 and $10^4\,\mathrm{cm}^{-3}$,

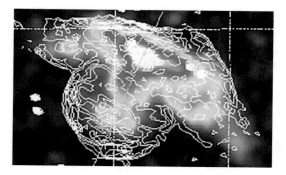

Fig. 4.12 Interaction of a supernova remnant (the spherical object to the lower left) and a molecular cloud (the elongated object above). The figure shows a grayscale image of the X-ray emission, overlaid with a contour image of the radio emission. Credit: D. Burrows of Penn State and T. Landecker of the Dominion Radio Astronomy Observatory

respectively, so that the forward shock in the remnant has a density of $10\,cm^{-3}$. We will estimate its velocity as 1000 km/s; it could be somewhat larger. Given these numbers, we would find the shock transmitted into the cloud to move at about 40 km/s while the reflected shock moves at 330 km/s back into the remnant. Given the comparable scale of these two objects in the image, it is clear that the entire remnant might be affected before the transmitted shock has traversed much of the cloud.

4.2   Flyer Plates and Shock Hugoniot Measurements

Here we consider the flyer-plate problem and then the application of flyer plates to shock Hugoniot measurements, both of which are further examples of shock-wave behavior at interfaces. Flyer plates have long been used to measure the shock Hugoniot. Historical experiments typically used either gas guns or rail guns to launch the flyer plates, and were limited to pressures below 1 Mbar prior to roughly the turn of the century. Hydrogen was a very active area of study in such experiments, because of its application to the interiors of gas giant planets (see Nellis 2006). Some experiments were done using nuclear weapons to launch flyer plates at a higher velocities, producing higher pressures. These experiments were in the high-energy-density regime, but were certainly not "laboratory" experiments in the usual sense. With the advent of modern, high-energy-density laboratory systems, it became possible to launch flyer plates at high enough velocity to access pressures well above 1 Mbar.

There is a key equation that is used for such measurements. It is derived starting from (4.5) and (4.6), using subscripts u for upstream and d for downstream to find

$$p_d = p_u(\rho_u u_u)(u_u - u_d), \tag{4.53}$$

and then recognizing that for a material stationary in the laboratory frame $u_u = -u_s$ and $u_d = u_p - u_s$, where u_p is the post shock fluid velocity in the laboratory frame, also known as the "particle velocity", which is equivalently the velocity of an imaginary piston driving the shock, so that

$$p_d = p_u + \rho_u u_s u_p. \tag{4.54}$$

One can often measure or know u_s, u_p, ρ_u, and p_u, enabling one to infer p_d from (4.54). One can also infer ρ_d from u_s, u_p, and mass conservation.

4.2.1 Flyer Plates

The ideal flyer plate is a cold, planar material moving at a high velocity. An ideal flyer plate, by cleanly striking the surface of a target material, can create a very uniform and well-characterized shock in the target material. Very often the flyer plate does not directly impact the sample to be studied, but instead impacts a fixed layer of the same material, through which the resulting shock propagates to reach the sample. In this case using the notation of Fig. 4.9, $\rho_4/\rho_1 = 1$. Ideally $p_1 \sim 0$ and $p_4 \sim 0$. Figure 4.13 shows the resulting dependence of u_2/u_1 on the initial pressure in the flyer plate. As the pressure becomes small, u_2 approaches $0.5\, u_1$. This limited result is straightforward to obtain from the original equations, which is left as an exercise. In addition, so long as p_1 is small, the interaction of the incident flyer plate with a material at any density produces reflected and transmitted shocks. (We discuss the behavior when p_1 is not small when treating rarefactions at interfaces, below.) In this case one can simplify (4.45) to find that

$$\frac{u_2}{u_1} = \frac{1}{1 + \sqrt{\dfrac{\rho_4(\gamma_4+1)}{\rho_1(\gamma_1+1)}}}. \tag{4.55}$$

Fig. 4.13 The dependence of u_2/u_1 on the normalized initial pressure of a flyer plate

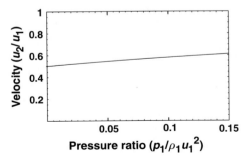

We now designate the material of the flyer plate and target as material A and label the downstream state in the flyer as state 1, with the upstream state being state o, anticipating the application of Fig. 4.13 to the interaction of the shock wave from the target material with the sample. For a flyer plate of initial velocity u_f, we apply (4.54) to the target of the flyer plate. Knowing that $u_{A1} = u_f/2$, we have

$$p_{A1} = p_{Ao} + \rho_{Ao}u_{s1}u_f/2. \tag{4.56}$$

Known variables include ρ_{Ao} and p_{Ao}, which is ideally zero. We can measure u_f optically and can measure u_{s1} from the shock transit time through the target of the flyer plate. As a result, we can know p_{A1} (and $\rho_A 1$ from mass conservation). Because aluminum is a very practical material for flyer plates, the shock Hugoniot of aluminum is now very well known at pressures below several Mbars.

4.2.2   Impedance Matching

Once one has a material with a known shock Hugoniot (and a known relation of pressure and expansion velocity), one can use this to advantage in determining the shock Hugoniot of other materials. This technique is commonly known as *impedance matching*, although it actually depends on measuring the difference in the degree to which two materials impede the shock. The analysis used in impedance matching is as follows, with reference to Fig. 4.14. With reference to Fig. 4.13, our notation includes designating the material in regions 1 and 2 as material A and the material in regions 3 and 4 as material B. One begins by producing a known and steady shock, with a known postshock fluid velocity, u_{p1} in the laboratory frame. It is essential that the shock be steady to very high ($\sim$1%) accuracy, otherwise one finds large errors in the inferred pressure and compression. One then allows this shock to enter a layer of the sample to be measured, of material B, of initial density ρ_{B4} and initial pressure p_{B4} (often negligible). This produces a transmitted shock through the sample and either a reflected shock or a rarefaction (discussed below) in

Fig. 4.14 How impedance matching works. The initial shock in material A is located by the dashed line. The curves for p_{A2} and p_{B3} shows how the pressure varies with velocity. Their intersection gives the values of pressure and of $\Delta u = u_{p3} - u_{p1}$ for the EOS measurement

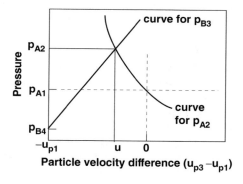

the first material. One measures the shock velocity in the second material, u_{sB}, also in the laboratory frame. The post-shock velocity of material B is u_{B3} in the shock frame and u_{p3} in the laboratory frame. Applying (4.56) to this system to find the post-shock pressure in material B one has

$$p_{B3} = p_{B4} + \rho_{B4} u_{sB} u_{p3}. \tag{4.57}$$

One knows the initial density of this material, ρ_{B4}, so one knows that its postshock pressure, p_{B4}, lies along the line given by (4.57), in which u_{p3} is the (not yet known) postshock fluid velocity of material B in the laboratory frame. Note that p_{B3} is an increasing function of u_{p3}. We can plot this equation against an independent variable $u_{p3} - u_{p1}$, as shown in Fig. 4.14, to allow further analysis.

Consider the reaction of material A when the initial shock wave reaches material B, with reference to the same figure. As discussed in Sect. 4.1.6, one also knows that $u_{p2} = u_{p3}$, so that one can also plot the response of material A against the velocity difference $(u_{p3} - u_{p1})$ which is also $(u_{p2} - u_{p1})$. Given sufficient knowledge of the equation of state of material A, one can plot a curve giving p_{A2} as a function of this variable. One finds that p_{A2} is a decreasing function of u_{p3}, because a decrease of u_{p3} corresponds to an increase in the velocity at which the reflected shock separates from the interface, and thus to a higher reflected-shock pressure. In the event that the reflected wave is a rarefaction rather than a shock, it remains the case that p_{A2} is a decreasing function of u_{p3}, as one can verify from Sect. 4.4 below. The net result is that the curves for p_{A2} and p_{B3} cross at only one point (actually an area whose size is determined by the uncertainties). Because $p_{A2} = p_{B3}$, this point determines both p_{B3} and u_{p3}, giving us the shock Hugoniot at one point. Hugoniot results are often plotted in a space of pressure *versus* density. The postshock density is related to u_{p3} by $\rho_{B3}/\rho_{B4} = (1 - u_{p3}/u_{sB})^{-1}$.

One way to apply the initial pressure to material A, so that one can know its value with high accuracy, is to make material A be Al and to apply the pressure by using an Al flyer plate to strike it. This has been accomplished using magnetically launched flyer plates from pulsed power machines, discussed in Sect. 10.10.2. Figure 4.15 shows a drawing of the experimental system used (Knudson et al. 2001) to determine the Hugoniot of D_2 by this method, at pressures above 1 Mbar. The Al flyer plate impacted a "drive plate", also of Al, producing an interface velocity of half the flyer-plate velocity and a shock pressure (p_{A1}) known from the Al EOS. The flyer plate was thick enough to sustain this pressure throughout the experiment. When the shock in the Al reached the D_2, it drove a transmitted shock through the D_2 (and a rarefaction into the Al). The diagnostics measured the emergence of the transmitted shock after it had propagated through each of two thicknesses of D_2, thus giving u_{sB}, which with the density of D_2 defines the slope of the curve for p_{B2} in Fig. 4.14. From this the authors obtained the shock Hugoniot, to a sufficient accuracy to exclude some models.

An alternative way to apply the pressure to material A is to use a laser or other radiation source to do so. In this case the initial shock velocity in material A is not as well known, so one must also measure it. Figure 4.16 shows a measurement of the

Fig. 4.15 Flyer-plate driven impedance matching experiment. From Knudson et al. (2001)

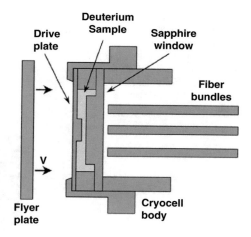

shock Hugoniot of copper by this method. A steady shock wave was driven through a stepped aluminum plate using a pressure source (in this case from laser ablation; see Chap. 9). A sample of Cu abuts part of the surface of the thinner step. The optical emission produced upon the emergence of the shock from each region is detected using an optical streak camera. Such data are shown in Fig. 4.16b. What matters is the time difference between the signal in the middle (from the thin Al step) and the signals through the two thicker layers. An experimental complication is that the edges of the samples affect the shock propagation (slowing it down from the edges inward). This limits how thick one can make the samples, which limits the accuracy of the measurement.

The emergence of the shock is detected from both the thin and the thick aluminum steps. This determines the shock velocity in the Al, from which the known Al EOS implies the postshock fluid velocity and the pressure p_{A1} in the Al. The time of emergence of the shock from the Cu sample then determines the shock velocity in the sample, u_{sB}. Then one applies the analysis illustrated in Fig. 4.14 to find the pressure and postshock fluid velocity in the Cu, shown in part Fig. 4.16c.

4.3   An Introduction to Self-similar Hydrodynamics

The majority of students to whom I have taught this material have been unfamiliar with self-similar behavior until we discussed rarefactions and blast waves. Yet in fact self-similarity appears in many diverse physical systems. A dynamical system is *self-similar* if it has a normalized structure whose shape is fixed in time, being a function of a dimensionless variable that in general depends on both space and time. A very simple example is the planar, isothermal rarefaction we will derive below. The density profile will turn out to be $\rho = \rho_o e^{-\xi}$, where the density of the source material is ρ_o and the dimensionless variable is $\xi = x/c_o t$, with the

Fig. 4.16 Laser-driven
impedance-matching
experiment. (**a**) Sketch.
(**b**) Spatial profile of emission
with time increasing to right.
(**c**) Inferred EOS of Cu. Parts
(**b**) and (**c**) are from Benuzzi
et al.

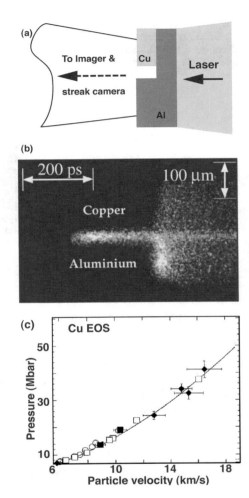

sound speed being c_o. Within the assumptions of the calculation, the density profile
is *always* exponential in terms of ξ. Another example that may be familiar to the
reader is simple diffusion with a constant diffusion coefficient. Simple solutions
of this type of problem often show a Gaussian shape, again independent of time.
Self-similar models are often very useful in the approximate description of how
a system evolves. For example, good experiment design is very often based on
simple physical reasoning, and self-similar models are an important tool for the
experiment designer. They also provide a useful conceptual framework to discuss
(and to estimate) how a system will evolve.

Before we take up the mathematics for fluids, we briefly discuss the broader
context. Dynamic physical systems often evolve into a state that is (nearly) self-
similar for some extended period of time. This behavior is technically described as
intermediate asymptotic behavior. There are typically some initial conditions and

very early evolution that is not self-similar, and there is often some region in space where the influence of the initial conditions remains. Examples include the initial core of a lightning channel or the leading edges of self-similar expansions. But in fact the behavior can be well-described by a self-similar solution for a long period of time and large regions of space. Eventually the system evolves to the point that the assumptions of the solution are violated, beyond which the further evolution is not self-similar. We will see examples of this below. The book by Barenblatt (1996) on intermediate asymptotics provides a thorough and outstanding treatment of this general topic.

One key to self-similarity is that physical systems are often meaningfully characterized by a very small number of properties, even fewer of which are truly independent of one another. In the case of the isothermal rarefaction just mentioned, these are the density and the sound speed. In simple diffusion problems, they are the diffusion coefficient and the total amount of whatever is diffusing (mass, energy, etc.). There is a method for determining which physical properties of a system are independent of one another, often taught in fluid mechanics courses, that involves application of the Buckingham Pi Theorem. This topic is also discussed at length by Barenblatt. Here we will not invest time in this topic, but note that if the reader seeks to know whether some new problem admits self-similar solutions, that is the place to go.

Self-similar solutions are often considered in connection with *similarity transformations*, in which a scaling of some variables leads to identical equations using scaled variables. The general problem of similarity transformations is discussed briefly in Chap. 10, at some length in Zel'dovich and Razier (1966), and often also in books on fluid dynamics.

4.3.1   Self-similarity in Hydrodynamic Flows

We now turn to the mathematics we will need for the remainder of this chapter. Here we are concerned with finding self-similar solutions, in which there is a single variable ξ that describes the shape of the solution for all time and all space. In hydrodynamics, self-similar solutions have shapes in space that are independent of time, with a spatial scale, $R(t)$, that is a function of time. Thus, the generally useful similarity variable, which traces out the shape of the fluid parameters, turns out to be $\xi = r/R$.

We consider the possibility of self-similar motions in systems whose evolution is symmetric (planar, cylindrical, or spherical). These motions are described by the corresponding versions of (2.1), (2.2), and (2.14), which are

$$\frac{\partial \rho}{\partial t} + u \frac{\partial \rho}{\partial r} + \rho \left(\frac{\partial u}{\partial r} + \frac{su}{r} \right) = 0, \qquad (4.58)$$

$$\frac{\partial u}{\partial t} + u\frac{\partial u}{\partial r} + \frac{1}{\rho}\frac{\partial p}{\partial r} = 0, \text{ and} \tag{4.59}$$

$$\frac{\partial p}{\partial t} + u\frac{\partial p}{\partial r} - c_s^2\left(\frac{\partial \rho}{\partial t} + u\frac{\partial \rho}{\partial r}\right) = 0, \tag{4.60}$$

in which $s = 0$, 1, and 2, respectively, for planar, cylindrical, and spherical symmetry. Observe that so long as the EOS that relates sound speed to density and pressure is a function of these quantities and numeric parameters (such as γ), these equations contain only variables. We will use a polytropic equation of state, as is already assumed in (4.60). The parameters having numerical values with physical dimensions (such as a density of 1 g/cm^3) enter through the boundary conditions and initial conditions that are necessary to solve these equations for some specific case. Note that the third equation is trivially zero if the sound speed is independent of space and time, but not otherwise.

The key to finding a self-similar solution to these equations is to transform them to ordinary differential equations involving dimensionless functions of ξ that represent the shape of the fluid variables. We will work with density, velocity, and pressure. Alternatively, we could use the sound speed instead of the pressure. To reduce (4.58)–(4.60) to self-similar form, we must express each of u, ρ, and p as the product of a dimensionless function ξ, here r/R, and other necessary parameters. This turns out to require that we specify a normalizing amplitude for either density or pressure. Here we work with density, which is the most common case. We take

$$u = \dot{R}U(\xi), \quad \rho = \rho_o(r,t)\Omega(\xi), \quad \text{and} \quad p = \rho_o(r,t)\dot{R}^2P(\xi), \tag{4.61}$$

in which the overdot represents a time derivative and the initial density ρ_o is in general a function of both space and time. We convert to a coordinate system, using ξ and t', in which $\xi = r/R$ and $t' = t$. For some general function $g(\xi, t')$, we have by the chain rule

$$\frac{\partial g(\xi, t')}{\partial t} = \frac{\partial g}{\partial \xi}\frac{\partial \xi}{\partial t} + \frac{\partial g}{\partial t'}\frac{\partial t'}{\partial t} \tag{4.62}$$

for time derivatives and a similar expression for the spatial derivatives. Of course, $\partial t'/\partial r = 0$ and $\partial t'/\partial t = 1$. However, the tricky point is that we are seeking a solution in which all the time dependence is included in the dependence on ξ so that $\partial g/\partial t' = 0$. This in the end imposes a restriction on the number of constraints imposed by any other boundary or initial conditions. Recognizing that

$$\frac{\partial h(\xi)}{\partial t} = -\xi\frac{\dot{R}}{R}h'(\xi) \quad \text{and} \quad \frac{\partial h(\xi)}{\partial r} = \frac{1}{R}h'(\xi), \tag{4.63}$$

in which the $'$ designates the derivative with respect to ξ, we can obtain after some algebra the following equations, in which ρ_o' is the spatial derivative of ρ_o,

$$\frac{\dot{\rho}_o}{\rho_o}\frac{R}{\dot{R}}\Omega(\xi) + \frac{\rho_o'R}{\rho_o}U(\xi)\Omega(\xi) + [U(\xi) - \xi]\Omega'(\xi)$$

$$+ \Omega(\xi)U'(\xi) + \frac{sU(\xi)\Omega(\xi)}{\xi} = 0, \tag{4.64}$$

$$\frac{\rho_o'R}{\rho_o}P(\xi) + \frac{R\ddot{R}}{\dot{R}^2}U(\xi)\Omega(\xi) + [U(\xi) - \xi]U'(\xi)\Omega(\xi) + P'(\xi) = 0, \text{ and} \tag{4.65}$$

$$\frac{\dot{\rho}_o}{\rho_o}\frac{R}{\dot{R}}(1 - \gamma)P(\xi) + \frac{\rho_o'R}{\rho_o}U(\xi)(1 - \gamma)P(\xi) + 2\frac{R\ddot{R}}{\dot{R}^2}P(\xi)$$

$$+ [U(\xi) - \xi]\left(P'(\xi) - \gamma P(\xi)\frac{\Omega'(\xi)}{\Omega(\xi)}\right) = 0. \tag{4.66}$$

In (4.64)–(4.66), we have obtained three ordinary differential equations, with explicit dependences on time and space in some terms. These dependences must also cancel out if the evolution is to be self-similar. Consider first (4.65). The dependence on time in the second term cancels out if R is a power law in time or an exponential with an argument that is linear in time. Here we will emphasize power-law solutions, and so we will take

$$R = R_o t^\alpha, \tag{4.67}$$

in which R_o is the position at time $t = 1$. (In some practical applications, it can be useful to specify a starting time, and so to replace t with t/t_o in this equation.) This is sufficient to remove all the time dependences from (4.64) to (4.66) except those involving the density. If the density has a time dependence, one can see that self-similar behavior can follow only if the dependence of ρ_o and R is the same type of function. They can both be power-law dependences, for example, and any difference in the exponent will just produce a constant factor in the equations. If the time dependence of ρ_o does not have the same functional form as R, then the system will not exhibit a self-similar evolution. In other words, its shape will change with time.

In order for the spatial dependence to drop out of (4.64), the quantity $\rho_o'R/\rho_o$ must have a constant value. This will occur if ρ_o is a power law function of position, so that

$$\rho_o(r) = \hat{\rho}r^\delta = \hat{\rho}\xi^\delta R^\delta \text{ and} \tag{4.68}$$

$$\frac{\rho_o'(r)R}{\rho_o} = \frac{\rho_o'(\xi)}{\rho_o R}R = \frac{\delta}{\xi}. \tag{4.69}$$

Thus, any power-law dependences of ρ_o on r and t are consistent with a self-similar evolution involving a scale R that is a power of t. However, the time dependence of R is part of the time dependence of ρ_o, so if ρ_o is $\propto t^\beta$ overall, one must have

$\hat{\rho} \propto t^{(\beta - \alpha\delta)}$. At times, self-similar solutions can be found in two adjacent regions, for example in which ρ_o may have two different dependences on space and time. What is required is that one be able to specify boundary conditions that connect the two regions. This is the case, for example, in the treatment of the structure of young supernova remnants by Chevalier (1982).

To obtain (4.64)–(4.66) in a self-similar form, we must specify an initial density (or pressure) and are restricted to certain types of functional dependences in space and time, as just discussed. This amounts to having specified a single parameter with physical dimensions. If this is all that is specified, then one has a self-similar problem. An example is the propagation of a shock wave through a fluid whose density decreases as a power law in space and has a specified initial profile. This occurs, for example, when a shock wave emerges from a star. This specific case is treated by Zel'dovich and Razier (1966), when they discuss this type of self-similar problem. In this case, one must solve the equations numerically to determine the value of the parameter α. Curiously, the "impulsive loading" problem (the planar blast wave problem), discussed thoroughly there, is also in this category, even though it seems to have two parameters. The open boundary creates special problems, because beyond it the acceleration is large enough to cause a divergence of the energy integral in the self-similar solution. The solution to this problem is that there is always a small initial quantity of mass to which the self-similar solution does not apply, and that this mass contains only finite energy.

If the specification of the system includes a second parameter with physical dimensions, such as a total energy, then one has self-similar behavior of the type discussed by Sedov (1959). The definition of ξ must be related to the known properties of the system, since these establish the relation between r and t in the actual physical system. For example, more energy will lead to faster motions, corresponding to larger values of $\dot{R}$, so if energy is specified then this must be included in the definition of R and hence ξ. One can define R to within a constant by creating a dimensionless combination of the known physical quantities, r, and t. This must be constant for $r = R$, and so can be solved to find the dependence of R on time (i.e., α) and on the specified physical quantities. The constant, unknown coefficient can then be specified during the solution of the problem. We carry out this exercise for the spherical blast wave, below.

If the specification of the system includes a third parameter with physical dimensions, such as the location of an interface in a blast-wave problem, then the evolution is not self-similar. The solution of (4.64)–(4.66) permits only one additional parameter to be defined, whether as a boundary condition, an initial condition, or an integral property of the system. Once this constraint is imposed on the solution, there remain no further undefined variables. Adding an additional constraint causes the self-similar problem to be overconstrained. The system may still evolve to have a fixed shape in space and time, but it will have distinct reference scales for r and for t. We discuss an example of this in Chap. 10.

This makes it fairly easy to tell whether there is a self-similar solution for a hydrodynamic flow. If the system is described by one or two dimensional parameters and these are simple power laws (or perhaps exponentials) in space and time, then a self-similar solution will exist. If there are more than this, then there will be no self-similar solution. Later in the book, and especially in Chap. 7, we will encounter self-similar solutions for other sets of equations.

4.4 Rarefaction Waves

A rarefaction is a decrease in density and pressure caused by expansion of a material. Rarefactions are common to many laboratory and astrophysical systems. Releases of energy, as when a shock wave emerges from a dense material layer or an exploding star, produce expansions. The flow of material, whether emerging from a channel in an experiment or emerging from a star to form a planetary nebula, produces an expansion. The cessation of pressure when a radiation source, whether a laser or a z-pinch or a star, becomes less powerful is followed by an expansion toward the source. Thus, expansions have broad relevance.

A rarefaction wave occurs when the onset of the expansion propagates through the material from one edge. Thus, for example, when the laser pulse that is creating pressure and plasma on the surface of an object ceases, the dense plasma expands outward. The corresponding decrease in pressure propagates into the object at the sound speed, and is accompanied by an outward flow of material and a corresponding density decrease. As another example, when a shock wave emerges from an object into a region of lower density, the high pressure produced by the shock wave causes material to accelerate forward from the object. As we will see, the flow of material outward into the rarefaction begins at a point that propagates inward into the material at the sound speed. In shock physics this behavior is known as the *release* of the shocked material. In high-energy-density physics or astrophysics it is often described as *shock breakout*.

In this section, we will first consider an idealized problem—the isothermal rarefaction—that is a useful model in many cases. This problem will also serve as our first application of self-similar solutions.

4.4.1 The Planar Isothermal Rarefaction

There are cases in which a plasma expands from a planar surface at constant temperature. This requires a continuous supply of heat, to counteract expansion cooling, and so can happen only when the Peclet number (Chap. 2) is small or when the heat transport is very fast compared to the timescale of the expansion. A very common example is the expansion of the low-density, laser-heated plasma from the irradiated surface of a laser target. But there are also other cases when the plasma does not cool too quickly and the very simple isothermal model can be used as a good first estimate. Figure 4.17 shows a sketch of the initial condition for this

Fig. 4.17 Sketch of initial
condition for isothermal
rarefaction

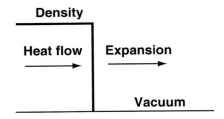

expansion. Our physical system at $t = 0$ has warm, dense matter of uniform density ρ_o to the left of a boundary at which the density drops abruptly to zero, and heat flows into or through our system as necessary to keep the temperature constant, so the system is characterized by a constant sound speed, c_o. Thus, we can expect self-similar evolution. One can recognize that the only sensible normalization of the velocity is to take $\dot{R} = c_o$, so that one also has $R = c_o t$. Since $c_o^2 = p/\rho$ is constant everywhere, Eq. (4.61) implies that $P(\xi) = \Omega(\xi)$. In this case all terms in Eq. (4.66) are zero and Eqs. (4.64) and (4.65) become

$$[U(\xi) - \xi]\,\Omega'(\xi) + \Omega(\xi)U'(\xi) = 0, \qquad (4.70)$$

$$[U(\xi) - \xi]\Omega(\xi)U'(\xi) + \Omega'(\xi) = 0, \qquad (4.71)$$

where for this one dimensional problem $s = 0$. These are very easily solved. Substituting for $\Omega(\xi)U'(\xi)$ from the first equation, the second equation implies that $U(\xi) = 1 + \xi$, and then the first equation implies that $\Omega(\xi) = \Omega_o e^{-\xi}$. Examination of the problem shows that u must be positive or zero everywhere so the solution applies for $\xi \geq -1$. We recognize that the point where $\xi = -1$ moves into the dense material as $r = -c_o t$, and so propagates as a sonic disturbance. This is very sensible behavior, as there is nothing in this problem to drive a shock and so the effect of the boundary should propagate into the material at the sound speed. At the point where $\xi = -1$ we must have $\rho = \rho_o$ so $\Omega_o = \rho_o e^{-1}$. In physical units the solution is thus

$$u = c_o(1 + \xi) = c_o + \frac{r}{t} \text{ and} \qquad (4.72)$$

$$\rho = \rho_o e^{-(1+\xi)} = \rho_o e^{-[1+r/(c_o t)]}, \text{ where} \qquad (4.73)$$

$$r \geq -c_o t. \qquad (4.74)$$

This solution has several features worth mentioning. First, it has a linear velocity profile and an exponential density profile. Linear velocity profiles are common to many free expansions, which is not surprising as the distance any unforced parcel of fluid will travel is its speed times the time. The exponential density profile reflects the specifics of this case; we will see others soon. Second, profiles are often characterized by a scale length, L, typically defined as $(d \ln \rho/dx)^{-1}$, which is the distance over which an exponential profile decreases by a factor of e or a linear profile decreases to 0. A potentially independent definition of L is the distance over which the velocity changes by c_s (thus $L = (dM/dx)^{-1}$). In this case by either definition one finds $L = c_s t$. Thus, the scale length is the distance an acoustic wave would travel in time t. It is also the distance over which the initial material has begun to flow outward. Third, heat must flow outward to sustain an isothermal rarefaction. In the absence of strong heat transport, expansion cooling will reduce the temperature and sound speed, as we will see next. Fourth, the density stays constant at the original interface as the expansion proceeds. A wave of this type is known as a *centered* wave. The value of the density there is $\rho_o e^{-1}$. When an isolated dense block of material expands, ρ_o is the initial density of the material. In other expansions, such as those produced by laser heating, the electron density profile may tend to be exponential below the density at which the laser heating is strongest, but to have a different shape at higher density. In this case one would replace $\rho_o e^{-1}$ in (4.73) by the density below which the profile is exponential.

One can also find isothermal models of cylindrical or spherical self-similar expansions in the specialized literature. However, these do not produce simple solutions. They also have rather limited applicability, as diverging expansions cool much more strongly than planar ones, so that the isothermal assumption is more readily violated. So instead of pursuing them here, we turn to adiabatic expansions, in which there is no heating and no heat transport.

4.4.2 The Planar Adiabatic Rarefaction

Now instead of assuming that the expansion is isothermal, we assume that it is adiabatic with p changing in proportion to ρ^γ and with γ being constant. This is often a sensible assumption when matter that is cool enough to be strongly collisional emerges from the surface of some dense material. A good example is a shock wave emerging from a material at some density near solid density.

We first consider the release of material into vacuum, so that the geometry is just that of Fig. 4.17 above. The problem is characterized again by constant density ρ_o and sound speed c_o in the initial material, and so once again we know that the disturbance will propagate into this material at speed c_o. Also again it is clear we should take $\dot{R} = c_o$, so $R = c_o t$ and we will use $\xi = r/R$. Since $c_o^2 = \gamma p_o / \rho_o$, Eq. (4.61) and the adiabatic equation of state imply that $P(\xi) = [\Omega(\xi)]^\gamma / \gamma$. In this case all terms in Eq. (4.66) are zero and Eqs. (4.64) and (4.65) become

$$[U(\xi) - \xi]\,\Omega'(\xi) + \Omega(\xi)U'(\xi) = 0 \text{ and,} \qquad (4.75)$$

$$[U(\xi) - \xi]\Omega(\xi)U'(\xi) + [\Omega(\xi)]^{(\gamma-1)}\Omega'(\xi) = 0, \qquad (4.76)$$

where for this one dimensional problem $s = 0$. Solving these equations is a bit more involved than was the case for the isothermal rarefaction. One can use Eq. (4.75) to eliminate $\Omega(\xi)U'(\xi)$ from Eq. (4.76) to find

$$U(\xi) = \xi + [\Omega(\xi)]^{(\gamma-1)/2}. \tag{4.77}$$

Then substituting for $U(\xi)$ in Eq. (4.75) enables one to integrate for $\Omega(\xi)$, using the boundary condition that $\Omega = 1$ at $\xi = -1$, finding

$$\Omega(\xi) = \left[\frac{2}{\gamma+1} - \frac{\gamma-1}{\gamma+1}\xi\right]^{2/(\gamma-1)} \quad \text{and so,} \tag{4.78}$$

$$U(\xi) = \frac{2}{\gamma+1}(1+\xi). \tag{4.79}$$

This solution also satisfies the expectation that $U = 0$ at $\xi = -1$. The corresponding solutions for the physical profiles are

$$\frac{\rho}{\rho_o} = \left(\frac{2}{\gamma+1} - \frac{\gamma-1}{\gamma+1}\frac{r}{c_o t}\right)^{2/(\gamma-1)}, \tag{4.80}$$

$$\frac{p}{p_o} = \left(\frac{2}{\gamma+1} - \frac{\gamma-1}{\gamma+1}\frac{r}{c_o t}\right)^{2\gamma/(\gamma-1)}, \quad \text{and} \tag{4.81}$$

$$u = \frac{2}{\gamma+1}c_o\left(1 + \frac{r}{c_o t}\right). \tag{4.82}$$

These profiles have the property that the density and pressure vanish at $r = 2c_o t/(\gamma-1)$, where $u = 2c_o/(\gamma-1)$ is the terminal velocity. The terminal velocity is 3 to 6 times c_o for γ from 5/3 to 4/3. The solutions for the profiles thus apply over the region

$$-c_o t \le r \le 2c_o t/(\gamma-1). \tag{4.83}$$

Also note that this rarefaction is also centered; the density and pressure at $r = 0$ are constant.

The fact that the terminal velocity is $2c_o/(\gamma-1)$ is very useful. It is, for example, the limiting speed when one material expands against a second material of much lower density. An example of this phenomenon is found in the design of experiments. It is common to create a shock wave or a blast wave that produces a rarefaction when it reaches a second material layer of very low density. As we discuss further below, the rarefaction drives a shock wave into the second material. Intuitively, one would expect the driven shock wave to become faster when the density of the second material is reduced. In practice, this often is not the case. If the density of the second material is low enough, then the leading edge of the rarefaction is moving close to the maximum possible speed. In this regime, changing the density of the low-density material has little impact on the speed of the shock driven into it.

On the other hand, any high-energy-density fluid is a plasma, and this single-fluid description does not accurately describe the leading edge of a freely expanding plasma. The behavior of the leading edge of a freely expanding plasma is much closer to the behavior of an isothermal rarefaction, because as the density decreases the electrons are able to transport heat throughout the expanding plasma. The dynamical behavior is that the fast, light electrons try to rush out ahead of the ions, establishing an electric field that accelerates the ions. Thus, the electrons progressively give their energy to the electric field, which gives it to the ions. The electrons get more energy from heating by the ions and from the new electrons reaching low enough density to transport heat readily. The ions are accelerated by their own pressure and by the electric field, and also lose some of their internal energy to electron heating. From the perspective of two-fluid theory as discussed in Chap. 2, the electric field is $eE = k_B T_e d[\ln n_e]/dx$. The limitation on the acceleration of the leading edge is a kinetic one. The electric field keeps nearly all of the electrons in the hot plasma. They reflect at some density and return to the denser region. (This behavior can be visualized as the electrons attempting to climb a long potential hill—most of them roll back down.) Only the most energetic electrons reach the leading edge of the expansion, where interactions with the surrounding gas can cool them and can help shield the potential of the plasma. An accurate analysis of this problem does not exist; it would require an entertaining foray into collisional kinetic theory. But one would roughly anticipate that the leading edge will form where $n_e \sim n_{gas}$, the gas density. Using a typical value of $n_{gas} \sim 10^{10}$ cm^{-3} and a maximum density $n_e \sim 10^{23}$ cm^{-3}, one finds that the *plasma potential* in energy units is roughly $30 k_B T_e$.

We next take up the case of a limited adiabatic expansion. Suppose that instead of the material releasing into vacuum, it instead expands against a piston being withdrawn at some speed $2c_o/(\gamma + 1) < U_p < 2c_o/(\gamma - 1)$. This is supersonic, as indeed is the entire rarefaction for $r > 0$. As a result, information about the piston cannot move back up the rarefaction back toward $r = 0$, but instead continually moves outward in the laboratory frame. So the rarefaction proceeds as though it were in vacuum until u reaches U_p. From Eq. (4.82), this occurs where

$$r = \left(\frac{\gamma + 1}{2} U_p - c_o \right) t. \qquad (4.84)$$

This must be where the expansion ends; the speed of the expanding material cannot exceed that of the piston. But what is remarkable is that this location itself does not keep up with the position of the piston, which is $U_p t$. What happens is that the matter expands until it reaches a final state in which its speed equals that of the piston, and then accumulates in that final state, with constant speed, pressure, and density. Thus the expansion occurs over the range

Fig. 4.18 Density (gray), pressure (black), and velocity (dashed) profiles in a planar adiabatic rarefaction, normalized to the ρ_o, p_o, and U_p, respectively. Here $\gamma = 5/3$, $c_o t = 1$, and $U = 1.5 c_o$

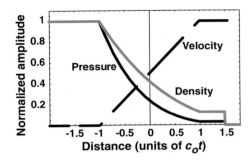

$$-c_o t \leq r \leq \frac{\gamma + 1}{2} U_p t - c_o t \leq \frac{2}{\gamma - 1} c_o t. \tag{4.85}$$

The resulting profiles are illustrated in Fig. 4.18. One can observe that density and pressure reach their initial values at $r = -c_o t$ and reach their steady values near the piston when $u = U_p$. (With the specific choice of parameters in Fig. 4.18, this occurs at $r = c_o t$. This is *not* a general result.) In addition, the density, pressure, and velocity are constant at $r = 0$. In (4.85), the limits on r are imposed by the trailing edge on the left and the leading edge on the right, with the upper limit reached when U_p reaches the maximum value the fluid can accommodate, $2 c_o/(\gamma - 1)$. At this point the leading edge equals the path of the piston. If the piston withdraws faster than that, it will pull away from the expanding fluid. This case is equivalent to a free adiabatic rarefaction.

Now instead of a piston, suppose that there is a zero-pressure, lower-density material to the right of the initial interface, which has density ρ_r and polytropic index γ_r. A strong shock will be driven into this material, at some velocity u_s, with postshock velocity u_r, producing a pressure $p_r = (\gamma_r - 1)\rho_r u_r^2/2$. The motion of the interface then acts like the piston, so the rarefaction in the denser material will proceed as described above, for a value of $u_r = U$ that is self-consistent, so that p_r is equal to the pressure at the piston in the description above.

Adiabatic rarefactions often are produced when shock waves reach an interface where the density drops. This occurs several times during the explosion of a star, with the added complication that the shock wave is a blast wave (below). It happens frequently in experiments as well. We will apply the above equations to this case when we consider the behavior of a shock at a density drop in Sect. 4.6.1.

4.4.3   Riemann Invariants

Another approach to the adiabatic expansion involves the use of *Riemann invariants*. The discussion here of Riemann invariants is included primarily for historical interest. They enable a time-honored approach to some problems, but the author has not found them to be of very common use in the context of high-energy-density research. They also are very time-consuming to teach.

It will turn out that the Riemann invariants are two quantities that (for isentropic flows) do not change along specific trajectories, known as *characteristics*. These invariants also permit the calculation of adiabatic rarefactions, as we will see. Adiabatic processes, which may involve mechanical work but not the flow of heat, are also isentropic. Thus, they are fully described by the Euler equations.

A general disturbance in the properties of a moving fluid can affect the rest of the fluid in two ways. It can generate sound waves, which move at the sound speed relative to the fluid flow, or it can generate local changes in properties (such as composition) that flow with the fluid at its velocity u. Of course, a general disturbance produces sound waves that move in all possible directions. It requires very special conditions to produce sound waves moving in a restricted range of directions. The trajectories of such sonic or fluid disturbances are known as characteristics. The position vector of a characteristic, x, changes for fluid disturbances, as

$$\frac{dx}{dt} = u \tag{4.86}$$

and for sonic disturbances as

$$\frac{dx}{dt} = u + c_s \hat{k}, \tag{4.87}$$

in which $\hat{k}$ is a unit vector defining a direction of propagation and in which the fluid velocity u, the sound speed c_s, and $\hat{k}$ depend, in general, on x and t. The trajectory defined by (4.86) is known as the C_o characteristic. The trajectories defined by (4.87) when $\hat{k}$ is aligned with or opposed to the x-axis are known as C_+ and C_-, respectively. As we will see, it helps one visualize and understand planar rarefactions to plot the evolution of the fluid with position along the abscissa and time along the ordinate. Then a surface moving at constant velocity is a straight line.

We now develop the equations of motion into a form relevant to propagation along characteristics. First recall that in general the derivative of some function $f(x, t)$ along a specific trajectory defined by $dx/dt = w$ is

$$\left(\frac{df}{dt}\right)_w = \frac{\partial f}{\partial t} + w \cdot \nabla f. \tag{4.88}$$

Finding equations of motion that connect with characteristics turns out to be easiest working with pressure rather than density so we take

$$d\rho = \left(\frac{\partial \rho}{\partial p}\right)_S dp = \frac{dp}{c_s^2}, \tag{4.89}$$

where the derivative is taken at constant entropy. Then (2.1) and (2.2) become

$$\frac{1}{\rho c_s}\frac{\partial p}{\partial t} + \frac{1}{\rho c_s}\boldsymbol{u}\cdot\nabla p = -c_s\nabla\cdot\boldsymbol{u} \quad \text{and} \tag{4.90}$$

$$\frac{\partial \boldsymbol{u}}{\partial t} + \boldsymbol{u}\cdot\nabla\boldsymbol{u} = -\frac{\nabla p}{\rho}. \tag{4.91}$$

To seek potential behavior along trajectories, we multiply (4.91) by a unit vector $\hat{\boldsymbol{k}}$ (which could be a direction of sonic propagation), add the equations, and seek an equation in the form of (4.88) to obtain

$$\hat{\boldsymbol{k}}\cdot\left[\frac{\partial \boldsymbol{u}}{\partial t} + \left(\boldsymbol{u} + c_s\hat{\boldsymbol{k}}\right)\cdot\nabla\boldsymbol{u}\right] + \frac{1}{\rho c_s}\left[\frac{\partial p}{\partial t} + \left(\boldsymbol{u} + c_s\hat{\boldsymbol{k}}\right)\cdot\nabla p\right]$$
$$= -c_s\left[-\nabla\cdot\boldsymbol{u} + \hat{\boldsymbol{k}}\cdot\left(\hat{\boldsymbol{k}}\cdot\nabla\right)\boldsymbol{u}\right]. \tag{4.92}$$

Here the quantity in the leftmost square brackets is the vector generalization of 4.88, with $\nabla\boldsymbol{u}$ being the tensor with element (i,j) equal to $(\partial/\partial x_i)u_j$, written in *dyadic notation*. We can recognize that the two square brackets on the left-hand side contain derivatives of the functions $\boldsymbol{u}$ and p along the trajectory given by (4.87), if $\hat{\boldsymbol{k}}$ is a direction of sonic propagation. In other words, we can write (4.92) as

$$\hat{\boldsymbol{k}}\cdot\left(\frac{d\boldsymbol{u}}{dt}\right)_{\boldsymbol{u}+c_s\hat{\boldsymbol{k}}} + \frac{1}{\rho c_s}\left(\frac{dp}{dt}\right)_{\boldsymbol{u}+c_s\hat{\boldsymbol{k}}} = -c_s\left[-\nabla\cdot\boldsymbol{u} + \hat{\boldsymbol{k}}\cdot\left(\hat{\boldsymbol{k}}\cdot\nabla\right)\boldsymbol{u}\right]. \tag{4.93}$$

One might have hoped to find here general three-dimensional invariants of the flow with very broad applicability, which would have required that the right-hand side be identically zero. However, there are special cases for which one does find invariants. They include planar, one-dimensional flow, which is all we will consider from this point forward. In this case we have

$$du + \frac{dp}{\rho c_s} = 0 \text{ along the } C_+ \text{ trajectory} \frac{dx}{dt} = u + c_s, \tag{4.94}$$

and by taking the difference of Eqs. (4.90) and (4.91) we also find

$$du - \frac{dp}{\rho c_s} = 0 \text{ along the } C_- \text{ trajectory} \frac{dx}{dt} = u - c_s \tag{4.95}$$

These two equations, upon integration, yield the Riemann invariants J_+ and J_-, usually written as

$$J_+ = u + \int\frac{dp}{\rho c_s} \quad \text{and} \quad J_- = u - \int\frac{dp}{\rho c_s}. \tag{4.96}$$

Here the integral is the indefinite integral. In effect, it is evaluated only at a specific point of interest. By integrating either (4.94) or (4.95) from an arbitrary starting point to two distinct but arbitrary final points and then subtracting the two results, one can show that the value of J_+ and J_- thus defined must be constant. The meaning of these equations and of the fact that these are invariants is that they remain constant along the trajectories, and can thus be used to help find the properties of the flow. Note that neither ρ nor c_s can be removed from the integral, as both in general are functions of the pressure. The important application of (4.96) is to polytropic gases. In this case one finds

$$J_+ = u + \frac{2c_s}{\gamma - 1} \quad \text{and} \quad J_- = u - \frac{2c_s}{\gamma - 1}. \tag{4.97}$$

To obtain useful information from the Riemann invariants, there are certain properties one must understand. First of all, recalling that we are discussing only planar isentropic flows, once one has specified an initial state of the fluid, only u and one other quantity are needed to specify completely any other state of the fluid. The second quantity can be density, pressure, sound speed, or any combination of these, such as the $\int dp/(\rho c_s)$ in J_+ or J_-. Because of this, it is also true that J_+ and J_- also completely specify the state of the fluid. Thus, if you can follow C_+ and C_- characteristics to specify J_+ and J_- at their intersection, then you can infer the properties of the fluid there.

We generally plot the characteristics in a space of x along the horizontal axis and t along the vertical axis, as in Fig. 4.19. Consider (4.94) and (4.95) for the characteristics. When both u and c_s are constant, the characteristics are straight lines. Along C_+, for example, we know that J_+ is constant, yet both J_+ and J_- depend on similar variables. A change in slope of C_+ requires that either u or c_s changes. Since J_+ is constant, this means that J_- must change in order for the slope to change. This has a very useful implication: *If one Riemann invariant is constant over some region, then the characteristics for the other Riemann invariant as they cross this region are straight lines.* We will see how these two properties can be used in describing the planar adiabatic rarefaction.

Next suppose that either C_+ or C_- starts where $u = 0$, and, for a polytropic gas, $c_s = c_o$. Then ask, what is the largest speed that the fluid can flow? This will occur where the internal energy becomes zero. This is no surprise—the limiting speed of a fluid must be the speed it reaches when it has no internal energy so all its energy is kinetic energy. For a polytropic gas, one finds the maximum speed to be $2c_o/(\gamma - 1)$, which is $3c_o$ for $\gamma = 5/3$. This, as discussed in the previous section, is a useful estimate.

4.4.4 Planar Adiabatic Rarefactions via Reimann Invariants

We are now ready to revisit the planar adiabatic rarefaction. We imagine that there is an initial, semi-infinite, uniform fluid to the left of the origin, bounded on the right

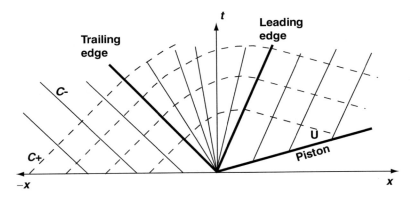

Fig. 4.19 Trajectories of the characteristics in a planar adiabatic rarefaction. C_- characteristics are thin solid lines; C_+ characteristics are dashed lines

by a piston. At time $t = 0$, the piston instantaneously accelerates to a velocity U. (We could consider gradual accelerations of the piston, but this has little importance for our long-term applications of this conceptual model.) We desire to find the profiles of the fluid parameters that result in time and in space.

Figure 4.19 shows such a rarefaction, with position along the abscissa and time along the ordinate. In the figure, the thick line to the right shows the velocity U at which a piston is withdrawing from its initial position at $x = 0$. (The fluid is uniform to the left of the origin at $t = 0$.) The line labeled "Piston" is the boundary of the fluid; it is not a characteristic. Now consider the region to the left of the origin, which has uniform properties at $t = 0$. To the left of the thick line labeled "trailing edge," the C_+ and C_- characteristics that intersect at any point originate from initial points whose properties are identical. Thus, both J_+ and J_- are the same as they were at $t = 0$. In this region, the properties of the fluid are unchanged.

Indeed, changes in the fluid properties can occur only through the arrival of other values of J_-, from points where $x \geq 0$. The earliest this can occur is along the C_- characteristic from the origin that is the line labeled "trailing edge." This is the tail of the rarefaction wave. One often sees such a feature propagate through a region that has first been shocked or otherwise heated, when the edge of the region is allowed to expand. This wave propagates at the sound speed of the initial medium. The C_+ characteristics that reach the edge of the rarefaction wave continue across the system toward the right, everywhere the fluid goes. As a result, because J_+ is constant along $x \leq 0$ at $t = 0$, J_+ is constant everywhere. This implies as well that all the C_- characteristics are straight lines. For the case of a polytropic gas, one has $J_+ = 2c_o/(\gamma - 1)$ so at any location one has

$$J_+ = u + \frac{2c_s}{\gamma - 1} = \frac{2c_o}{\gamma - 1}, \text{ from which} \qquad (4.98)$$

$$c_s = c_o - \frac{\gamma - 1}{2} u. \tag{4.99}$$

This tells us how the sound speed (and thus temperature) varies through the rarefaction. As u increases the medium cools—this is the anticipated expansion cooling. Next consider the fluid properties at the piston. Since u is fixed there, as U, determining either J_+ or J_- determines the state of the fluid. The C_+ characteristics propagate from the initial state to the piston, so J_+ is known. Knowing J_+ and U, one can then find J_- for the C_- characteristics leaving the piston. For the case of a polytropic gas, one has

$$J_- = U - \frac{2c_s}{\gamma - 1} = 2U - \frac{2c_o}{\gamma - 1}, \tag{4.100}$$

using (4.99). Because J_+ is constant, the C_- characteristics emerging from the piston are straight lines with

$$\frac{dx}{dt} = U - \left(c_o - \frac{\gamma - 1}{2} U \right) = \frac{\gamma + 1}{2} U - c_o. \tag{4.101}$$

Along these characteristics, J_+ and J_- are both constant, so the state of the fluid is constant. This portion of the fluid moves with the piston at velocity U and has sound speed

$$c_s = c_o - \frac{\gamma - 1}{2} U. \tag{4.102}$$

This region with constant fluid properties is bounded on the left by the thick line labeled "leading edge," which is the front end of the region of expansion. What remains is to describe the region between the leading edge and the trailing edge. The C_- characteristics form a fan of straight lines emanating from the origin. This is sometimes known as a *rarefaction fan*. The rarefaction wave is *centered*, which refers to the fact that one set of characteristics emerges from a common point. Equation (4.99) gives c_s as a function of u. As a result, the equations for the C_- characteristics are

$$\frac{dx}{dt} = u - c_s = \frac{\gamma + 1}{2} u - c_o = \text{const.} \tag{4.103}$$

We can integrate this equation and rearrange it to find an equation for u:

$$u = \frac{2}{\gamma + 1} \left(c_o + \frac{x}{t} \right). \tag{4.104}$$

For the polytropic gas, we now have a complete description of the fluid because

$$\frac{\rho}{\rho_o} = \left(\frac{c_s}{c_o}\right)^{2/(\gamma-1)} \text{ and } \frac{p}{p_o} = \left(\frac{c_s}{c_o}\right)^{2\gamma/(\gamma-1)}, \text{ which become} \qquad (4.105)$$

$$\frac{\rho}{\rho_o} = \left(1 - \frac{(\gamma-1)u}{2c_o}\right)^{2/(\gamma-1)} \text{ and } \frac{p}{p_o} = \left(1 - \frac{(\gamma-1)u}{2c_o}\right)^{2\gamma/(\gamma-1)}, \text{ or} \qquad (4.106)$$

$$\frac{\rho}{\rho_o} = \left(\frac{2}{\gamma+1} - \frac{\gamma-1}{\gamma+1}\frac{x}{c_o t}\right)^{2/(\gamma-1)} \text{ and} \qquad (4.107)$$

$$\frac{p}{p_o} = \left(\frac{2}{\gamma+1} - \frac{\gamma-1}{\gamma+1}\frac{x}{c_o t}\right)^{2\gamma/(\gamma-1)}, \text{ for} \qquad (4.108)$$

$$-c_o t \le x \le \frac{\gamma+1}{2}Ut - c_o t \le \frac{2}{\gamma-1}c_o t. \qquad (4.109)$$

In this way we recover the same results for the profile that we found in Sect. 4.4.2. It is likewise the same that the velocity of the point where $u = U$ in the rarefaction moves more slowly than the piston itself, so that there is a region of constant properties attached to the piston, as shown in Fig. 4.18.

4.5 Blast Waves

A uniform shock wave can come to an end in two ways. The previous section discussed one of them: the shock wave can reach a lower-density medium and a rarefaction wave can propagate backward into the shocked material. This section discusses the other one: the source of pressure can end, allowing a rarefaction wave to propagate forward and overtake the shock. This forms the structure illustrated in Fig. 4.20. In the figure, the first two curves show the nearly steady shock produced by the pressure of laser ablation (Chap. 8). The curves are not completely flat because, in the simulation and perhaps in a real system, the pressure produced by the laser does evolve with time. The later three curves, all after the end of the laser pulse, show the rarefaction wave developing on the left edge of the structure and soon overtaking the shock to form a *blast wave*. (A word on semantics is in order here. This definition of a blast wave as the structure formed when a rarefaction overtakes a shock is fairly common in the astrophysical literature, and is the one we will use. In the shock physics literature, the term *blast wave* is more often restricted to such structures produced by spherical expansions from a point explosion, while other cases would be described as "waves produced by impulsive loading," or perhaps as "planar blast waves.")

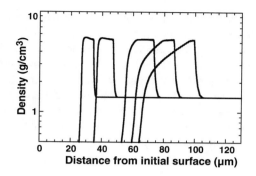

Fig. 4.20 Simulated evolution of a blast wave from a shock wave. Here laser irradiation from the left drives a nearly steady shock wave for 1 ns, after which the rarefaction from the front surface overtakes the shock wave to form a blast wave. The curves show the profile at 0.6, 0.8, 1.2, 1.4, and 1.6 ns after the start of the laser pulse

Blast waves are very common because releases of energy are of limited duration. On a small scale, processes such as solar flares release energy, causing blast waves to form in the solar wind. In some systems, hydrogen can accumulate on the surface of a neutron star, leading to occasional nuclear explosions known as "astrophysical flashes." Such flashes will drive blast waves through the surrounding material. Stellar explosions at first drive shocks, not blast waves. But eventually, when the interior pressure is much reduced and the accumulated interstellar material exceeds the mass of the star, supernova remnants develop a blast-wave structure which they retain for much of their evolution. One can think of many other cases, such as the interaction of jets with clouds, in which blast waves are produced. The most common blast wave in the Earth environment is produced by lightning, which briefly deposits energy within the lightning channel.

Planar blast waves are often useful in high-energy-density experiments. They can drive Rayleigh–Taylor instabilities at interfaces, as is discussed in Chap. 5. In addition, they can be used as timescale converters. One may, for example, have a laser that can provide power most effectively for 1 ns (or a *Z* pinch that can do so for 10 ns), but need to deliver energy over a longer timescale to some object. By forming a blast wave and then letting it propagate, one creates a store of energy. With time and distance, the blast wave carries more mass at lower velocity. If it is then allowed to release that energy, for example, by encountering a lower-density medium, it can drive further hydrodynamics for a much longer period than the duration of the initial energy source.

4.5.1   Energy Conservation in Blast Waves

Because most of the material in a blast wave is near the shock, it is useful and informative to see what one can infer from energy and momentum conservation. Consider again Fig. 4.20. We will discuss the spherical case. Spherical blast waves are often known as Sedov–Taylor (or Taylor–Sedov) blast waves as Sedov and Taylor, along with von Neumann, were the first to discuss their behavior. There is an initial shock transition, followed by a rarefaction. The shock is nearly always a strong shock; the source of energy in the problem is the energy behind the shock. The total mass in the blast wave is the amount of mass that has been swept up by the shock. If the mass density ρ of the external medium is constant and the shock radius is R, then the total mass within a spherical blast wave is $M = 4\pi\rho R^3/3$. In cases where very little energy has been lost by radiation or heat conduction, the energy within the blast wave is approximately

$$E = M\left(\frac{2}{\gamma+1}\right)^2 \dot{R}^2, \tag{4.110}$$

in which we assume the material is a polytropic gas to obtain the fluid velocity in terms of the shock velocity $\dot{R}$. The important result does not depend on this assumption. Equation (4.110) remains valid despite the conversion of thermal energy to kinetic energy in the rarefaction, but does assume that all the matter has been accelerated by the shock wave at its present velocity. In other words, it assumes that the change in shock velocity is slow. Conserving energy, we obtain

$$\frac{\mathrm{d}M}{\mathrm{d}t}\dot{R}^2 = -M2\dot{R}\frac{\mathrm{d}^2 R}{\mathrm{d}t^2}, \tag{4.111}$$

in which $\mathrm{d}M/\mathrm{d}t = 4\pi\rho R^2\dot{R}$, so

$$3\dot{R}^2 = -2R\frac{\mathrm{d}^2 R}{\mathrm{d}t^2}. \tag{4.112}$$

If one seeks a power-law solution of the form $R = R_o t^\alpha$, in which R_o and α are constants, one finds $\alpha = 2/5$. The Sedov–Taylor blast wave thus has a radius that increases in proportion to $t^{2/5}$.

If an energy-conserving blast wave is able to propagate far enough, which happens, for example, with blast waves from lightning, then the shock wave becomes a weak shock wave and eventually a disturbance that propagates at the local sound speed. This case is discussed in more detail in Zel'dovich and Razier (1966). An estimate of the radius at which this will occur can be found by setting the explosion energy E_x per unit volume equal to the thermal energy. For a spherical explosion, this occurs when $r \sim (3E_x/(8p_o))^{1/3}$, where p_o is the initial pressure. Thus, for a typical supernova, which deposits about 10^{51} ergs in exploding stellar material, and a typical interstellar pressure of 1.6 picodynes (1 cm^{-3} and 1 eV), one finds a radius of 6×10^{20} cm, or about 600 light years. For a cylindrical explosion, where E_x is now the energy per unit length, one finds $r \sim \sqrt{E_x/(9p_o)}$. For lightning, which has $E_x \sim 10^5$ J/m $= 10^{10}$ ergs/cm, one finds $r \sim 33$ cm. One can see that one could not experience the shock wave produced by lightning at a safe distance from the lightning itself.

The estimate of $\alpha = 2/3$, obtained in a homework problem, on the assumption that all the mass is near the shock, is only an upper limit because the mass is in fact distributed to minus infinity. (The lower limit is $\alpha = 1/2$, and for $\gamma = 5/3$, $\alpha = 0.611$.)

To examine momentum conservation in a blast wave, it is tempting to take the same approach and to set the derivative of $M\dot{R}$ equal to 0. However, this is not generally valid. While the conversion of thermal energy to kinetic energy in the rarefaction does not affect the overall energy balance, the production of momentum in the rarefaction does affect the momentum balance. While outward momentum is added to the newly shocked matter, inward momentum is generated by the rarefaction. In consequence, momentum conservation is more difficult to calculate and does not generate much insight into the global evolution of an energy-conserving blast wave.

However, momentum conservation is sometimes important because radiative energy losses are not always negligible. There are circumstances in which a blast wave enters a strongly radiating phase, so that it no longer conserves energy. All supernova remnants eventually enter this phase, when they become slow enough that the postshock material cools rapidly by radiation (see Chap. 7). The thermal energy produced by the shock is radiated away, so the energy remaining in the system steadily decreases. At the same time, there is little thermal energy to drive a rarefaction so the shocked material tends to become a dense shell moving with the shock. The pressure of this shell must equal the ram pressure of the incoming material, so as it loses energy by radiation it becomes quite dense and cool. At the same time, its velocity approaches the shock velocity in the frame of the unshocked matter. (In the shock frame, the very dense material that has been shocked and then cooled can move only very slowly away from the shock.) Such a system is often described as a momentum-conserving snowplow.

The time evolution of a momentum-conserving snowplow can be found by setting the derivative of $M\dot{R}$ equal to 0. The approach to the solution is identical to that used for the conservation of energy. For the spherical case, one finds $\alpha = 1/4$. Thus, spherical momentum-conserving snowplows expand with $R \propto t^{1/4}$.

4.5.2 The Sedov–Taylor Spherical Blast Wave

We now turn specifically to the problem of finding the profiles of the fluid variables in a spherical blast wave. This is often known as the *point explosion problem*, as self-similar solutions require one to assume that the energy originated at an initial point (or line). Solutions found under this assumption will apply only when the blast wave is far enough from the source that this assumption becomes accurate. This problem can be solved analytically, as is discussed by Sedov (1959). The first numerical solution was reported by G.I. Taylor, and von Neumann also contributed an early solution. Here we show how to develop a set of ordinary differential equations in terms of an appropriate similarity variable. These equations can typically be integrated quite quickly using a computational mathematics program.

The point explosion problem has only two constraints having physical dimensions. There are the explosion energy, E_x, and the initial density of the surrounding medium (assumed constant), ρ_o. The independent variables are space r and time t. This allows us to carry out the procedure described in Sect. 4.3. The ratio of E_x to ρ_o has units of length to the fifth power divided by time squared, so we can obtain a dimensionless parameter from these quantities. For a self-similar motion, we then have

$$\left(\frac{E_x t^2}{\rho_o r^5}\right) = \text{const}, \qquad (4.113)$$

in the sense that the motion of any feature or point must keep this parameter constant. For this problem it is convenient to consider the position of the shock, $R(t)$, for which

$$R = \frac{1}{Q}\left(\frac{E_x}{\rho_o}\right)^{1/5} t^{2/5}, \qquad (4.114)$$

where Q is a constant to be determined later. Then we can use $\xi = r/R$ as the similarity variable and we see that $\alpha = 2/5$. Note that we have determined the scaling of the radius with time much more easily and exactly than we did in the energy argument of Sect. 4.5.1. This is rather amazing, since we needed only a little simple reasoning to do so. Note also that the shock velocity is $\dot{R} = (2/5)R/t$. We apply the general transformation relations of Sect. 4.3, writing

$$u = \frac{2}{5}\frac{R}{t}U(\xi), \quad \rho = \rho_o\Omega(\xi), \quad \text{and} \quad p = \left(\frac{2}{5}\frac{R}{t}\right)^2 \rho_o P(\xi). \qquad (4.115)$$

Here U, Ω, and P are dimensionless functions providing the shape of each of the fluid variables. We then have for any function $h(\xi)$,

$$\frac{\partial h(\xi)}{\partial t} = h'(\xi)\frac{\partial \xi}{\partial t} = -\left(\frac{2}{5}\frac{\xi}{t}\right)h'(\xi), \quad \text{and} \qquad (4.116)$$

$$\frac{\partial h(\xi)}{\partial r} = h'(\xi)\frac{\partial \xi}{\partial r} = \left(\frac{\xi}{r}\right)h'(\xi). \qquad (4.117)$$

Here the prime designates the derivative of the function with respect to its argument. Using (4.115)–(4.117), Eqs. (4.64)–(4.66) become

$$[U(\xi) - \xi]\xi\Omega'(\xi) + [\xi U'(\xi) + sU(\xi)]\Omega(\xi) = 0, \qquad (4.118)$$

$$-\frac{3}{2}\Omega(\xi)U(\xi) + [U(\xi) - \xi]\Omega(\xi)U'(\xi) + P'(\xi) = 0, \quad \text{and} \qquad (4.119)$$

$$-3\Omega(\xi)P(\xi) + [U(\xi) - \xi][\Omega(\xi)P'(\xi) - \gamma P(\xi)\Omega'(\xi)] = 0. \qquad (4.120)$$

Fig. 4.21 Dimensionless profiles for the Sedov–Taylor blast wave. U (dashed), Ω (black), and P (gray) are shown as a function of r/R. The amplitudes are normalized to unity at the shock radius $r = R$

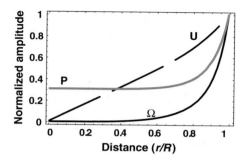

Here $s = 2$ for this spherical case. These are the ordinary differential equations we set out to obtain. Note that γ enters explicitly as a parameter in these equations. Thus, the solution is not independent of the equation of state. The numerical coefficients arise from the scaling of the dimensionless parameter, and thus are specific to this problem.

The boundary conditions required to integrate the equations are obtained at the shock front, where

$$U(1) = \frac{2}{\gamma + 1}, \quad \Omega(1) = \frac{\gamma + 1}{\gamma - 1}, \quad \text{and} \quad P(1) = \frac{2}{\gamma + 1}. \tag{4.121}$$

By numerically integrating (4.118)–(4.121), one finds the profiles of the three dimensionless functions. These are shown in Fig. 4.21 for the spherical case ($s = 2$) and for $\gamma = 5/3$. This value of γ is reasonable for nearly all astrophysical systems and for some laboratory experiments. For explosions in air, it would be better to take $\gamma = 1.4$ and for many laboratory experiments γ could be as low as $4/3$ or even less, as was discussed in Chap. 3. One sees that nearly all the mass is concentrated near the shock (even more than it first appears, when one realizes that the total mass per unit radius is proportional to $r^2 \Omega$).

The above equations are sufficient to give us the profile shapes, but not to determine the quantity Q, which sets the absolute value of R at a given time. Here is where we must make use of the second parameter having physical dimensions, E_x. To find Q, one can evaluate the total energy in the self-similar profile, which must equal the explosion energy E_x. The integral for E_x is

$$E_x = \int_0^R \left(\frac{p}{\gamma - 1} + \frac{\rho u^2}{2} \right) 4\pi r^2 dr, \tag{4.122}$$

from which one can show that

$$Q^5 = \frac{16\pi}{25} \int_0^1 \left(\frac{P(\xi)}{\gamma - 1} + \frac{1}{2} \Omega(\xi) U^2(\xi) \right) \xi^2 d\xi. \tag{4.123}$$

Evaluating this integral for the profiles shown in Fig. 4.21, which assumes $\gamma = 5/3$, one finds $Q = 0.868$. This in turn lets us evaluate from (4.114) the radius for actual cases. For a supernova remnant formed by the release of 10^{51} ergs into a medium with a density of 10^{-23} g/cm^3, one finds $R = 1.2t^{2/5}$, with R in light years and t in years. Obviously this does not apply within 1 year, and in fact hundreds to thousands of years are required to sweep up enough mass that a point-explosion model is appropriate. As another example, a laboratory blast wave experiment might release 100 J into a gas at a density of 10 mg/cm^3. In this case $R = 11.1t^{2/5}$, with R in mm and t in μs.

4.6 Phenomena at Interfaces

All of the discussion above, with the exception of the discussion of reflected shocks, relates only to the behavior of an isolated hydrodynamic phenomenon in an unbounded medium. This is a necessary start, but the features of interest in most physical systems arise from the interaction of hydrodynamic phenomena with structure in the medium or with each other. Understanding these effects is also needed to design clever experiments.

4.6.1 Shocks at Interfaces and Their Consequences

In Sect. 4.1.6, we discussed the generation of reflected and transmitted shocks when a shock wave approaches an interface where the density increases. We also discussed the flyer-plate case, described by the same mathematics, in which a cold, moving material collides with another material. In general, the material approaching (or creating) the interface has a velocity u_1 and a pressure p_1, and the pressure p_1 can range from 0 to $(\gamma - 1)\rho_1 u_1^2/2$, which is the limit obtained when the pressure was produced by a strong shock (and is $\rho_1 u_1^2/3$ for $\gamma = 5/3$). Here we want to consider the more general case in which a shock wave (or in general a moving slab of material) approaches an interface beyond which there is material of arbitrary density, specified as ρ_4 in the notation of Sect. 4.1.5, which we will use in this discussion.

This more general case is of interest in the laboratory and in astrophysical systems. In the laboratory, a radiation source can accelerate material in addition to shocking it, as is discussed in Chap. 9. This can be useful in an experiment if the goal is to make more energy available for the later evolution of the system. Such an object—a shocked and accelerated slab—might be described as a *plasma flyer plate*. In astrophysics, enduring radiation sources can shock and accelerate the objects they irradiate. This is the case, for example, in the star-forming region that includes the Eagle nebula, where the bright young stars have shocked and are now accelerating the nearby molecular clouds. When such an accelerated object

Fig. 4.22 A rarefaction may occur when a shock wave reaches an interface where the density drops, as illustrated here

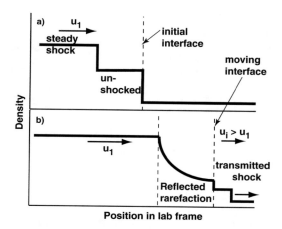

encounters a stationary one, which might be a clump at higher density or a cloud at lower density, then on a global scale the kinds of interactions discussed here will ensue.

If ρ_4 is small enough, we expect to see an adiabatic rarefaction when the shock reaches the interface. Figure 4.22 illustrates the situation in the lab frame. Let us apply the theory of Sect. 4.4.2 to this case. The new feature here is that it is best to do the mathematics in the *downstream frame*, in which the fluid downstream of the shock is at rest. We want to examine the system at the precise moment when the shock has reached an interface where the density decreases. In this frame, the expansion of the shocked matter follows identically (4.79)–(4.83). All the new aspects of this problem then have to do with correctly specifying the properties of the upstream medium.

In the downstream frame, at the moment the shock reaches the interface, the new upstream material (of density ρ_4) is moving toward the shocked material at a velocity u_1. The rarefaction then pushes shocked material forward at a velocity U in the downstream frame, which is $U + u_1$ in the lab frame. Thus, $U + u_1$ is the postshock fluid velocity of the upstream material in the lab frame. This implies that the transmitted shock velocity, in the lab frame, is $(\gamma_3 + 1)(U + u_1)/2$. We described the head of a centered rarefaction wave as moving backward in the material, which it does do in a Lagrangian sense. In the lab frame, however, the head of the rarefaction wave moves forward at u_{rw}, which is

$$u_{rw} = u_1 \left(1 - \frac{\sqrt{\gamma_1(\gamma_1 - 1)}}{\sqrt{2}} \right). \tag{4.124}$$

Note that u_1 can also be written $2u_{s1}/(\gamma + 1)$. The pressure in the shocked upstream medium, p_3, based on (4.18), is

Fig. 4.23 Profiles of the
normalized interface velocity
in the downstream frame
when a shock encounters a
density drop. The black curve
shows $\gamma_1 = 5/3$ and the gray
curve shows $\gamma_1 = 4/3$, with
$\gamma_4 = 5/3$ in both cases

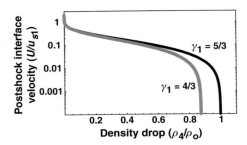

$$p_3 = \frac{\gamma_3 + 1}{2} \rho_4 (U + u_1)^2 . \tag{4.125}$$

Matching this pressure is the value at the interface, p_i. This can be found
from (4.106), evaluated at the piston, which with current variable definitions is

$$p_i = p_1 \left(1 - \frac{\gamma_1 - 1}{2} \frac{U}{c_{s1}} \right)^{2\gamma/(\gamma-1)} , \tag{4.126}$$

in which c_{s1} is the sound speed in the initial shocked matter. One can find U by
setting $p_i = p_3$. For any specific choice of γ, the resulting equation can be converted
to a polynomial equation for U. Figure 4.23 shows, for $\gamma_3 = 5/3$ and for two
values of γ_1, how the resulting value of U/u_{s1} depends on the density ratio ρ_4/ρ_o.
Remember that this is in the downstream frame; in the lab frame, U is increased by
$2u_{s1}/(\gamma_1 + 1)$.

One can see in Fig. 4.23 that there is a limiting value of ρ_4/ρ_o beyond which
U disappears on the plot. In terms of the mathematics just described, U becomes
negative, but this is not the physical solution. Instead, at this point the response of
the system is to produce a reflected shock rather than a rarefaction. This can occur
even if $\rho_4 < \rho_1$. The gray curve in Fig. 4.23 illustrates this case. To see how this
occurs, consider that a reflected shock will form once the pressure in the shocked,
low-density material exceeds p_1. Consider also that the postshock fluid velocity in
the low-density material decreases as ρ_4 increases, and has a limiting value of u_1
at the transition from a rarefaction to a reflected shock. Thus, assuming the shock in
the low-density material to be a strong shock (so (4.23) applies), we would expect
this transition to occur when $(\gamma_3 - 1)\rho_4 u_1^2 / 2 = p_1$. In the specific case in which
p_1 is produced by a strong shock in a stationary material (which then approaches a
stationary interface), one has $p_1 = (\gamma_1 - 1)\rho_1 u_1^2 / 2$, from which one can obtain the
threshold density for a reflected shock as

$$\rho_4 = \frac{(\gamma_1 - 1)}{(\gamma_3 + 1)} \rho_1 = \frac{(\gamma_1 + 1)}{(\gamma_3 + 1)} \rho_o. \tag{4.127}$$

This is no surprise. It just says, for $\gamma_1 = \gamma_3$, that the transition occurs when ρ_4/ρ_o at the interface exceeds 1, or in other words when the interface changes from a density decrease to an increase. But on the other hand if $\gamma_3 > \gamma_1$, then this transition will occur while there is still a density drop.

Consider also the more general case of a plasma flyer plate in which material that has been shocked or otherwise heated is accelerated to a higher velocity before it impacts a second material, creating an interface. In this case $p_1 < (\gamma_1 - 1)\rho_1 u_1^2/2$. If we express this as $p_1 = f\rho_1 u_1^2$, where as discussed above $0 < f < 1/3$ for $\gamma = 5/3$, then the condition to produce a reflected shock becomes

$$\rho_4 = \frac{2f}{(\gamma_3 + 1)}\rho_1. \tag{4.128}$$

Here we see that the transition from a rarefaction to a reflected shock can occur at an arbitrarily low density, which depends on the properties and thus the history of the incoming matter.

4.6.2   Overtaking Shocks

It is not uncommon to find shock waves produced in succession, whether by a sequence of energy releases at the solar surface or by a sequence of irradiation pulses in an experiment. If a second shock is stronger than a first shock, then it moves more rapidly and will overtake the first one. The discussion so far in this chapter makes the qualitative behavior of such systems fairly obvious. We review it briefly here.

As the stronger second shock overtakes the first shock, there is a moment when they coalesce. In Fig. 4.24, the left set of curves show the two shocks, before the stronger one has overtaken the weaker one. The middle set of curves shows them as they coalesce. At this instant, the total density jump is the product of the density jump produced by each of the two shocks. However, unless the two shocks are fairly weak, this density jump is not consistent with a single shock. For example, if a shock with a density jump of 4-to-1 overtakes a shock with a density jump of 3-to-1, the

Fig. 4.24 Dynamics of overtaking shocks. Here a polyimide material of density 1.4 g/cm^3 is driven at 10 Mbar for 1 ns and then is driven at 100 Mbar. The curves show 10.5, 12, and 13.5 ns. The dashed gray curves are pressure

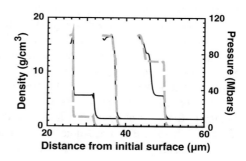

resulting instantaneous density jump is 12-to-1, which few materials can sustain. So what does the system do?

One can see what happens by considering the moment of coalescence as an initial condition, in which the postshock material has a certain density, pressure, and velocity. What we have then is identical to the plasma-flyer-plate problem with a low value of the density beyond the interface created when the flyer plate reaches the object it collides with. The fact that the density of the doubly shocked material, ρ_1, is produced by two shocks guarantees that the unshocked density, ρ_4, will be less than that given by (4.128). As a result, there will be a rarefaction in the shocked material, and a strong shock will be driven forward into the unshocked material. One can see this shock and rarefaction in the rightmost curves in Fig. 4.24. Soon after the rarefaction forms, the head of the rarefaction wave will return to the driven surface where the pressure that drove the shocks was applied. We take up next what happens then.

4.6.3 Reshocks in Rarefactions

The longer-term behavior of shocked layers of material is often very relevant to systems of interest. As shock waves or rarefactions traverse a system, they encounter interfaces or other waves and interact. This can be a complicating factor in any system for which the study of a later interface is of interest. Once more than one wave reaches the interface, its behavior becomes more complex. It is tempting to form the conclusion that every wave always begets a next wave, but this is not correct. In particular, when a rarefaction wave crosses a system, it may or may not produce a subsequent wave. Suppose specifically that a pressure source creates a shock wave in a layer of material, as described in Sect. 4.1. Suppose further that when the shock wave reaches the end of the layer, a rarefaction forms as described in Sect. 4.4. The head of the rarefaction moves back through the shocked material, and eventually reaches the initial surface. What happens then is illustrated in Fig. 4.25.

In these simulations whose results are shown in the figure, a pressure of 30 Mbar drives a shock through C_1H_1 of density 1 g/cm^3. This produces the shock wave that moves up and to the right across the system. Once the shock wave reaches zone 400 (the end of the system), a rarefaction wave returns toward the driven surface. Consider first parts (a) and (b) of the figure. In the simulation producing these results, the driving pressure is always present. One can clearly see the rarefaction structure as a variation in density (shade of gray) between the rarefaction wave and the top of the image. When the rarefaction wave reaches the driven surface, a new shock is launched back into the plasma. This is easy to understand as follows. The rarefaction wave decreases the pressure of the shocked material. Even so, by assumption the driving pressure does not change. Once the pressure in the material begins to decrease at the driven interface, the interface will accelerate in response. This will launch either a sound wave or a shock wave back into the material. One cannot tell which from simple reasoning, but it turns out in this particular case

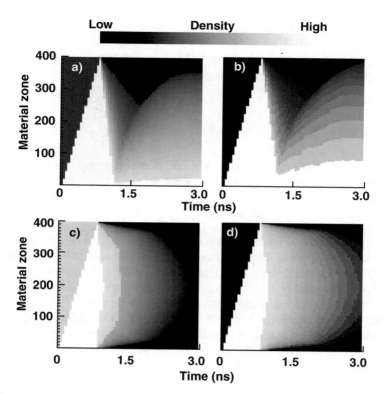

Fig. 4.25 A grayscale display vs. Lagrangian zone and time can be an effective way to see the waves in a hydrodynamic system. In the first row are (**a**) density and (**b**) pressure for a reshock in a rarefaction, created by the continuous application of a 30 Mbar pressure to a CH material at 1 g/cc. In the second row are (**c**) density and (**d**) pressure for a system in which the driving pressure ends before the rarefaction returns to the driving surface, so no reshock is produced. The jagged boundaries are caused by the finite zone size and finite number of time outputs in the simulation

that in response to the first rarefaction a second shock wave is launched. In this specific system, there is no next wave, as the second shock just travels down the density gradient to the end of the expanding plasma where it disappears. If there were further layers of material, then the second shock might produce further shock or rarefaction waves that would traverse the plasma. When this happens, it is known as *reverberation*.

Consider now parts (c) and (d) of Fig. 4.25. In the simulation producing these results, the driving pressure ends after 0.8 ns. One can see the shock and the beginning of the rarefaction wave returning toward the driven surface. However, there is a second rarefaction wave moving upward from the driven surface, due to the expansion of this surface now that it is no longer driven. When rarefaction waves meet, they do not produce further waves. There is no source of pressure to produce a reshock in either rarefaction. Thus, if a rarefaction approaches a surface that was driven but is no longer driven, for example, because the radiation source has turned off, then there will be no further wave. The result in Fig. 4.25 is that the density and pressure of the plasma decrease smoothly in space and time as the plasma expands and cools.

4.6.4   Blast Waves at Interfaces

We discussed above how common blast waves are in astrophysics, because the originating event that produces a shock wave very often is short lived compared to the lifetime of the shock. As a result, the rarefaction from the source overtakes the shock and produces a blast wave. The blast wave then may encounter interfaces where the density changes, in response to which these interfaces will evolve. A very important application of this lies within a Type II supernova, in which the blast wave generated near the core of the star encounters density drops at the boundary between the C–O layer and the He layer, and again at the boundary between the He layer and the H layer. The density also decreases within each layer as radius increases, but this turns out not to be essential to the behavior at the interface.

Figure 4.26 illustrates the behavior. When the blast wave reaches the interface, the density drop at the interface cannot be sustained by a single shock. Just as in the cases of the adiabatic rarefaction and of overtaking shocks, a forward shock is driven into the low-density material. The surprising development is that a shock develops in the high-density material, despite the fact that (4.127) is not satisfied. We can explain what this is and why it is called a *reverse shock* by contrasting this case with the adiabatic rarefaction. In the adiabatic rarefaction, there is an indefinite supply of density and pressure behind the interface. In the adiabatic rarefaction, both the density and pressure increase with distance behind the interface, until they eventually reach their initial values. In contrast, in the blast-wave case, a rarefaction does propagate backwards, causing the plasma behind the interface to accelerate, but the pressure in the blast wave soon drops below the pressure in the shocked low-density material in front of the interface. One then has an expanding and

Fig. 4.26 Development of a reverse shock when a blast wave passes through an interface. This case is produced in a simulation when a laser with an energy flux of 4.2×10^{14} W/cm^2 strikes a 150-μm-thick layer of polyimide, followed by a low-density carbon layer

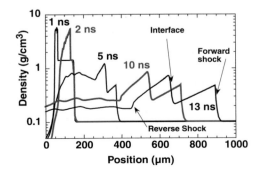

Fig. 4.27 Result of a simulation of the explosion of SN 1987A. The forward and reverse shocks are evident near the outer edges. The dramatic structures are discussed in Chap. 5. Credit: Kifonidis et al. (2003)

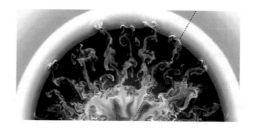

accelerating flow of material that encounters the slower, higher-pressure material near the original interface. A shock develops at this transition. Thus, a *reverse shock* is a shock formed when a freely expanding plasma encounters an obstacle. This is distinct from the reflected shock formed when a shock wave crosses an interface. In the blast-wave case, the blast wave creates its own obstacle through its interaction with the interface.

One can also see in Fig. 4.26 that the density decreases behind the forward shock and behind the interface. This reflects the gradual expansion of the region between the shocks in response to the pressure gradient that develops as the system slows. This has the additional consequence that the forward shock soon develops a blast-wave structure itself. The shape of this structure may differ significantly from the shape of the initial blast wave. It also may not soon be self-similar, as the distance from the interface introduces an additional physical scale into the problem. Even so, if the interface moves so that its position is $\propto t^{\alpha}$, then the profile of the shocked low-density material is a self-similar one.

Systems in which blast waves encounter interfaces have been an important area of activity in the early years of experimental astrophysics. This has been motivated by instabilities in Type II supernovae, which we discuss further in the next chapter, and by the question whether errors in calculations of their nonlinear evolution might explain some discrepancies with data. Figure 4.27 shows results from a calculation of the explosion of a Type II supernova (1987A). The forward and reverse shocks are clearly evident. Figure 4.28 shows data from an experiment in which these two features were also produced. The experiment was a well-scaled reproduction of the supernova explosion for reasons discussed in Chap. 10. This particular experiment was done to confirm that the correct one-dimensional behavior was achieved. It lacked the initial perturbations that would have produced unstable structures.

4.6.5 Rarefactions at Interfaces

Rarefactions never proceed unimpeded forever. Whenever a rarefaction develops in astrophysics, whether at the edge of a supernova when the blast wave emerges from the star, at the edge of a molecular cloud when a shock wave emerges from it, or somewhere else, the rarefaction encounters at minimum the interstellar

Fig. 4.28 Data from experiment sending a blast wave through an interface. Laser irradiation produced a shock in a 150-μm-thick layer of dense plastic (1.41 g/cm³), then ended allowing a blast wave to form. At the time of the image, the interface between the plastic and 50 mg/cm³ foam has moved 650-μm and 2D effects, producing curvature and rollups at the edges, are becoming important

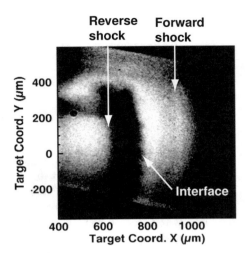

medium. In addition, it may encounter other objects as it propagates. This produces a situation in which a flowing, expanding, cool plasma produces an interface through its interaction with something. After reading the prior section, it will come as no surprise that the interaction produces a forward shock and a reverse shock.

Experiments can produce similar phenomena, by creating rarefactions that encounter a layer of material. In general, this can be a way to produce a high-Mach-number flow and then to let it interact. It also can have the effect of converting a brief source of energy into a lower-pressure source of much longer duration. Among recent applications of this technique have been equation of state studies (see Chap. 3), experiments related to supernova remnants (see below), experiments to produce jets, and experiments to study the long-term interactions of shocks and clumps.

The classic example of a rarefaction that encounters an interface is the young supernova remnant. We can observe the explosion of a star for at most a few years. In contrast, we can observe nearby supernova remnants for centuries if not millennia. Supernova remnants are the observable structures that form through the interaction of the ejecta from a stellar explosion with the surrounding (circumstellar) environment. They are widely believed to produce most of the Cosmic Rays that irradiate the Earth. Despite our ability to observe a number of supernova remnants in considerable detail, the structure and the evolution of supernova remnants pose many challenges to our understanding.

The energy that creates the supernova remnant is the kinetic energy of the exploding star, typically about 10^{51} ergs. An interesting feature is that the "interface" that leads to the structure has neither a decrease nor an increase of the density, but rather has an abrupt decrease in the density gradient. The material emerging from most stellar explosions can be argued to be self-similar (see Zel'dovich and Razier 1966) and to have an inverse-power-law dependence on radius and time. The profile is quite steep, with an exponent of 8 or 9. The stellar ejecta undergo a homologous expansion, with velocity, v, radial distance, r, and time, t, related by $v = r/t$. Expansion cooling reduces the temperature of this material to a low value early on, so that nearly all the energy of the ejecta is kinetic energy.

In contrast, the circumstellar density falls off much more slowly, as $1/r^2$ if it is due to a prior stellar wind or perhaps more slowly if the star has been an inactive white dwarf, as in the case of Type Ia explosions. When the rapidly expanding ejecta from the star interact with the nearly stationary circumstellar matter, forward and reverse shocks develop. This initial velocity of the forward shock is of order 10,000 km/s. This first phase of supernova-remnant evolution is the free-expansion (or "young-remnant") phase. Ignoring clumps and instabilities, the entire structure between the two shocks moves at a velocity that is determined by the properties of the ejecta and the circumstellar material.

The system involves initial densities that are power laws, and the expansion velocity, which is x/t, introduces no additional scales, which suggests that the evolution might be self-similar. In 1982, R.A. Chevalier showed that it can be analyzed as two self-similar regions that are matched across the contact surface. One can find three coupled equations for the self-similar evolution of the density, the velocity, and the sound speed between the reverse shock and the interface, and between the forward shock an the interface, just as we discussed in Sect. 4.3. Here again, modern computational mathematics programs make the integration of these equations straightforward. Figure 4.29 shows the density profile for parameters relevant to SN 1987A, with the ejecta density scaling as r^{-9} and the circumstellar density scaling as r^{-2}.

It may seem strange to treat the supernova remnant as a hydrodynamic object, because the average density of the circumstellar medium may be of order 1 or 10 particles per cubic centimeter. The feature that permits a hydrodynamic treatment is the presence of a magnetic field that is small enough that it does not affect the dynamics yet large enough and structured enough to confine the particles to a very small volume on the scale of the entire supernova remnant. This turns out to be very much the case. The primary uncertainty in the hydrodynamic treatment is the potential effect on the hydrodynamics of a developing population of cosmic rays.

One example of an extremely young supernova remnant is the remnant from SN 1987A (reviewed by Chevalier 1992), shown in Fig. 4.30. At only 150,000

Fig. 4.29 Self-similar profile of density in a young supernova remnant, showing forward and reverse shocks. The supernova ejecta come in from the left

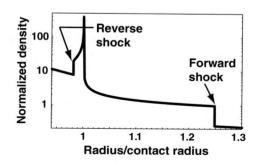

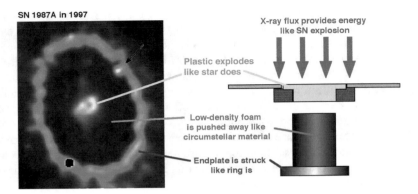

Fig. 4.30 The supernova remnant from SN 1987A and a related experiment. (On left) SN 1987A in 1997. The arrow on the image of SN1987A shows the hot spot where interaction of the shocked matter and the ring had begun. The image of SN 1987A is from the Hubble Space Telescope. It was created with support to the Space Telescope Science Institute, operated by the Association of Universities for Research in Astronomy, Inc., from NASA contract NAS5-26555, and is reproduced with permission from AURA/STScI. (On right) Schematic of the experiment. The thickness of the plastic layer is 200 μm. The diameter of the foam cylinder is 700 μm

light years, this object is far closer and thus far more diagnosable than any other supernova of the modern era. The ring shown, and two larger rings as well, are of unknown origin and provide an added element of excitement. During the years after the explosion, the development of radio and X-ray emission from this object were followed by the advent of visible emission at "hot spots" as the stellar ejecta began to collide with the innermost ring.

Laboratory experiments can help improve our understanding of some of the mechanisms present in supernova remnants, and can help test the computational models we build to interpret their behavior. The design of the first such experiment, by the author and colleagues (Drake et al. 2000b) is also illustrated in Fig. 4.30. The arrows and labels in the figure identify the correspondence between features in the experiment and those in SN 1987A. These experiments were in a planar geometry, intended to simulate a small segment of the overall supernova remnant expansion. The experiment began when an intense X-ray flux, produced by laser heating of a gold hohlraum (see Sect. 9.3), irradiated a 200-μm-thick layer of plastic. The X-rays ablated the plastic, launching a strong shock wave through it, at a pressure of 5×10^{13} dynes/cm^2 (50 Mbar). This was the analog of the initial blast wave produced by the SN explosion. This shock wave compressed, accelerated, and heated the plastic. When the shock broke out of the plastic, the ejecta from its rear expanded, accelerated, cooled, and decompressed across a 150-μm-wide gap. In an actual supernova remnant, spherical expansion provides the decompression (McKee 1974). Here the gap served an analogous function. The ejecta then launched a forward shock into the ambient matter, in this case a foam whose density was less than 1% of the density of the compressed plastic layer. The ejecta stagnated against the

Fig. 4.31 Data from an ejecta-driven shock experiment. The image shows the X-ray transmission through the experimental system. One can clearly see the reverse shock and forward shock. The curves show evaluations of the optical depth of the system as a function of position from an initial surface. The spatial resolution is limited, so the transitions in optical depth are smoothed out somewhat. From Drake et al. (2000a)

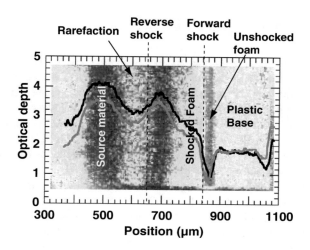

(moving) contact surface with the foam, which launched a reverse shock into the ejecta, just as occurs in a supernova remnant.

Figure 4.31 shows the measured profile, obtained by X-ray radiography. The forward shock and reverse shock are clearly established. This system is a well-scaled model of the basic hydrodynamic structure of a young supernova remnant (see Chap. 12 regarding how to scale such systems). Specifically, with reference to quantities defined in Chap. 2, in the supernova remnant and the experiment, respectively, $Re = 6 \times 10^8$ and 7×10^6, $Pe = 10^7$ and 10^4. Radiative losses are unimportant in both systems. This basic experiment design has subsequently been used to address instabilities in such systems and their interaction with other structures.

The later evolution of the supernova remnant connects with other topics in this book. Eventually, the mass of the accumulated circumstellar matter exceeds the mass of the stellar ejecta. This is generally taken to mark the (gradual) transition to the "Sedov–Taylor" phase. As this phase begins, the reverse shock runs in to the center of the supernova remnant and dissipates, after which the supernova remnant is believed to develop the characteristic structure of the Sedov–Taylor blast wave discussed above. Throughout the development of the supernova remnant, the shocked matter also radiates energy. Radiative losses are in some cases important during the young-supernova-remnant phase, but were not in SN 1987A. They are never important during the Sedov–Taylor phase, but eventually the remnant slows down and cools enough that they become important (they pass through the minimum of the "cooling function" discussed in Sect. 6.3.2). Once enough cooling has occurred, the remnant becomes a momentum-driven snowplow (Sect. 4.5.1) and the shell structure (now much thinner) may become unstable to thin-shell instabilities. The above is the one-dimensional story, but the extent to which three-dimensional effects such as instabilities or interactions with clumps may distort this picture is not entirely known. Not only supernova remnants but also other objects such as molecular clouds are observed to be clumpy in general.

4.6.6 Oblique Shocks at Interfaces

To prepare for the discussion in the next chapter, we also need to consider the behavior that develops when an oblique shock wave arrives at an interface, where the density increases or decreases. One can see from the discussion earlier in this chapter that in general the result will involve a transmitted shock beyond the interface and a reflected wave propagating backwards (relative to the interface). The postshock contact surface will be between these, with the sign of the angles (α and χ in Fig. 4.32) depending on the type of reflected wave. The properties of the interface will determine whether the reflected wave is a shock wave or a rarefaction wave. So long as the EOS is the same on the two sides of the interface, the reflected wave will be a shock when the density increases across the interface and a rarefaction when the density drops. Figure 4.32 is a schematic of the essential geometry, assuming the system to be uniform in the direction out of the page. We will label the shocked or unshocked regions a, b, c, and d, as indicated on the subscripts on the density ρ, and will use the subscript R for the region between the contact surface and the reflected wave. So long as the pressure source driving the shock remains constant, and so long as the edges of the system have no effect, the various waves will each have a constant velocity. As a result, they will radiate in straight lines from the point where the shock and the interface meet, and the fluid velocity will be independent of the distance from this point. The entropy will be constant across each region of this system, changing only at the shocks.

Following Section 109 of the fluid dynamics book by Landau and Lifshitz (1987), we can observe that the sensible way to analyze this system is in a cylindrical coordinate system centered at the point where the shock meets the interface. We can make this coordinate system be stationary by working in a frame in which $\boldsymbol{u} = \boldsymbol{u}_{lab} + \boldsymbol{u}_I$, where $\boldsymbol{u}_I$ is given by $(u_s/\tan\beta)\hat{x} - u_s\hat{y}$, with the x and y directions defined as shown in Fig. 4.32. We define the azimuthal angle ϕ relative to the x-axis, as usual. In this frame of reference, a point on the shock wave or the interface moves radially inward with time, while a point on the transmitted shock or the reflected

Fig. 4.32 Sketch of behavior occurring when a shock approaches an oblique interface. The angles shown are for a reflected rarefaction wave. For convenience, η and β as shown are taken to be positive angles

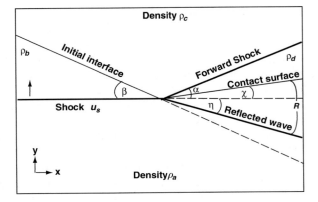

wave moves radially outward. Assuming that u_z is constant everywhere, and based on the assumptions of the previous paragraph, the derivatives in r and z are zero in this coordinate system while u_r and u_ϕ may vary with ϕ. With these assumptions the momentum equation implies

$$\frac{\partial u_r}{\partial \phi} - u_\phi = 0 \text{ and} \tag{4.129}$$

$$u_\phi \frac{\partial u_\phi}{\partial \phi} + u_r u_\phi = -\frac{1}{\rho} \frac{\partial p}{\partial \phi} = -\frac{\partial w}{\partial \phi}, \tag{4.130}$$

where $dw = dp/\rho$ is the differential enthalpy at constant entropy, while the continuity equation implies

$$\left(u_r + \frac{\partial u_\phi}{\partial \phi}\right) + \frac{u_\phi}{\rho} \frac{\partial \rho}{\partial \phi} = 0. \tag{4.131}$$

By combining these equations, we can obtain some insight into the behavior. Equations (4.129) and (4.130) imply that

$$w + \frac{1}{2}\left(u_r^2 + u_\phi^2\right) = \text{constant}, \tag{4.132}$$

while these equations in combination with (4.131) imply that

$$\left(u_r + \frac{\partial u_\phi}{\partial \phi}\right)\left(1 - \frac{u_\phi^2}{c_s^2}\right) = 0 , \tag{4.133}$$

where c_s is the local sound speed. Equation (4.132) connects the changes in velocity with changes in enthalpy as one moves across the system. We will return to it. Equation (4.133) evidently has three solutions. One of these solutions occurs when the argument of the left parentheses is zero. This corresponds to uniform flow.

The flow is uniform in regions a, b, c, and d, with discontinuities at the boundaries between regions. The equations describing these parts of the flow do not depend on the nature of the reflected wave. To simplify the mathematics, we will assume the initial pressure in regions b and c to be negligible and the entire system to have constant γ. Then in regions b and c the velocity equals the velocity of our moving frame of reference,

$$\boldsymbol{u}_b = \boldsymbol{u}_c = \boldsymbol{u}_I = \frac{u_s}{\sin\beta} [\cos\beta\hat{x} - \sin\beta\hat{y}] , \tag{4.134}$$

while in region a we have $\rho_a = \rho_b(\gamma + 1)/(\gamma - 1)$, and we use the results of Sect. 4.1.5 for oblique shocks to find the postshock velocity, which is

$$\boldsymbol{u}_a = \frac{u_s}{\sin\beta}\left[\cos\beta\hat{x} - \left(\frac{\gamma-1}{\gamma+1}\right)\sin\beta\hat{y}\right], \quad \text{and} \tag{4.135}$$

$$p_a = \frac{\rho_b u_s^2}{\gamma+1}. \tag{4.136}$$

In region d we have $\rho_d = \rho_c(\gamma+1)/(\gamma-1)$, and the shock is also oblique in the moving frame. One can show that

$$\boldsymbol{u}_d = \frac{u_s}{\sin\beta}\left[\frac{\gamma\cos\beta + \cos(2\alpha+\beta)}{\gamma+1}\right]\hat{x}$$
$$- \frac{u_s}{\sin\beta}\left[\frac{\gamma\sin\beta - \sin(2\alpha+\beta)}{\gamma+1}\right]\hat{y} \quad \text{and} \tag{4.137}$$

$$p_d = \frac{\rho_c u_s^2}{\gamma+1}\frac{\sin^2(\alpha+\beta)}{\sin^2\beta}. \tag{4.138}$$

The ratio of the component of $\boldsymbol{u}_d$ normal to the surface of the forward shock to the component parallel to this surface gives $\tan(\alpha-\chi)$, as it is the radial flow away from the forward shock that establishes the downstream boundary of region d. This gives

$$\left(\frac{\gamma-1}{\gamma+1}\right)\tan(\alpha+\beta) = \tan(\alpha-\chi). \tag{4.139}$$

All of the above applies whether there is a reflected shock or a rarefaction. We consider these possibilities in turn. In the case of the reflected shock, the flow within the reflected shock is also uniform. The difference with the planar case of Sect. 4.6.1 is that the transmitted and reflected shocks are both oblique. The angles α and χ in Fig. 4.32 are both negative in this case, but (4.137)–(4.139) still apply. For region R we have

$$\boldsymbol{u}_R = \frac{u_s}{(1+\gamma)^2\sin\beta}([\gamma(\gamma+1)\cos\beta + \gamma\cos(\beta-2\eta) + \cos(\beta+2\eta)]\hat{x}$$
$$+ [-\gamma(\gamma-1)\sin\beta + \gamma\sin(\beta-2\eta) + \sin(\beta+2\eta)]\hat{y}), \tag{4.140}$$

$$p_R = \frac{\rho_b u_s^2}{\gamma-1}\frac{[\sin(\beta+\eta) - \gamma\sin(\beta-\eta)]^2}{(\gamma+1)^2\sin^2\beta}, \quad \text{and} \tag{4.141}$$

$$\left(\frac{\gamma-1}{\gamma+1}\right)\frac{\sin(\beta+\eta) - \gamma\sin(\beta-\eta)}{\cos(\beta+\eta) + \gamma\cos(\beta-\eta)} = \tan(\eta+\chi)). \tag{4.142}$$

Given the known parameters, one can solve numerically for α and η by setting $p_d = p_R$ and setting equal the solutions of (4.139) and (4.142) for χ.

In the case of the reflected rarefaction wave, the flow is not uniform in the region between the rarefaction wave and the contact surface. This region, which we can call the rarefaction, is isentropic so $p/c_s^{2\gamma/(\gamma-1)}$ and $\rho/c_s^{2/(\gamma-1)}$ are both constant. In the rarefaction one must replace (4.140) through (4.142) with a description of the rarefaction. Within this region, u_r is clearly positive and the relevant solution of (4.133) has $u_\phi = c_s$. If we designate the properties at the head of the rarefaction wave as w_a, u_{ra}, and c_a, then we can write from (4.132)

$$u_r^2 = u_{ra}^2 + c_a^2 - c_s^2 + 2\left(w_a - w\right). \tag{4.143}$$

Substituting these two results into (4.131), rearranging, and integrating gives

$$\phi + \eta = \int_{\rho_a c_a}^{\rho c_s} \frac{\mathrm{d}(\rho c_s)}{\rho \sqrt{u_{ra}^2 + c_a^2 - c_s^2 + 2(w_a - w)}}. \tag{4.144}$$

Here η is as shown in Fig. 4.32, and would correspond to $-\phi_o$ if we designated the angle of the rarefaction wave as ϕ_o. For a polytropic gas, this can be reduced to

$$\phi + \eta = \sqrt{\frac{\gamma + 1}{\gamma - 1}} \int_{c_s/c_a}^{1} \frac{\mathrm{d}\xi}{\sqrt{Q^2 - \xi^2}}, \quad \text{in which} \tag{4.145}$$

$$Q^2 = 1 + \frac{\gamma - 1}{\gamma + 1} \frac{u_{ra}^2}{c_a^2} > 1 \tag{4.146}$$

Note that Q depends on η, because

$$\frac{u_{ra}^2}{c_a^2} = \frac{(\gamma + 1)^2}{\gamma(\gamma - 1)} \frac{\cos^2(\beta - \eta)}{\sin^2 \beta} > 1 \tag{4.147}$$

We can find the angle η by considering the behavior of the rarefaction wave. In a planar system, the rarefaction wave flows away from the shocked interface into the downstream system at the downstream sound speed. One way to think of this is to realize that sound waves are launched in all directions from the disturbed interface, and that in the planar geometry the leading edge of their phase fronts initiates the rarefaction wave. Taking this same point of view, we can say that sound waves propagate from any point on the interface in the system of Fig. 4.32, beginning at the moment the shock reaches the interface (taken as $t = 0$). The vector f describing their location is then

$$f = \left(u_I + c_a\hat{k}\right)t, \tag{4.148}$$

in which $\hat{k}$ is a unit vector in an arbitrary direction. Defining the angle of this vector as ϕ_s, and expressing both c_a and u_I in terms of u_s, we can find the following

Fig. 4.33 The angle of
rarefaction produced by
oblique interface turns out to
depend only on γ and β.
Solid is $\gamma = 5/3$; gray is
$\gamma = 4/3$

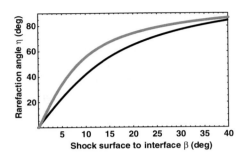

equation for angle of the rarefaction, η:

$$- \tan \eta = \mathrm{Min} \left[\frac{-\frac{\gamma+1}{\gamma-1}\sin\beta + \sqrt{\frac{\gamma}{\gamma+1}}\sin\beta\sin\phi_s}{\cos\beta + \sqrt{\frac{\gamma}{\gamma+1}}\sin\beta\cos\phi_s} \right], \qquad (4.149)$$

where the minimum is found by varying ϕ_s. One typically finds $\phi_s \sim 180\text{–}250°$.
Figure 4.33 shows η vs. β.

Knowing η, (4.99) allows one to evaluate c_s (and hence other quantities) as a
function of the angle ϕ in between the rarefaction wave and the contact surface as

$$c_s = c_o Q \sin\left[\sin^{-1}\left(\frac{1}{Q}\right) - \sqrt{\frac{\gamma-1}{\gamma+1}}\,(\phi + \eta) \right]. \qquad (4.150)$$

At the contact surface, this gives an equation for χ, in terms of the sound speed
c_{end} there:

$$\chi + \eta = \sqrt{\frac{\gamma+1}{\gamma-1}} \left[\sin^{-1}\left(\frac{1}{Q}\right) - \sin^{-1}\left(\frac{c_{\mathrm{end}}/c_a}{Q}\right) \right], \qquad (4.151)$$

where the fact that the pressure in the rarefaction at the contact surface equals p_d
gives

$$\frac{c_{\mathrm{end}}}{c_a} = \left(\frac{p_d}{p_a}\right)^{(\gamma-1)/(2\gamma)} = \left(\frac{\rho_c}{\rho_b}\frac{\sin^2(\beta+\alpha)}{\sin^2\beta}\right)^{(\gamma-1)/(2\gamma)}. \qquad (4.152)$$

Thus, (4.151) gives χ as a function of η (already known as described above) and
α. Equation (4.139) still gives χ as a function of α. This allows one to obtain
numerically a solution for α and χ.

This completes our description of the shock at an oblique interface. When we
consider rippled interfaces in Chap. 5, we will work with the small-angle limits of
the above equations.

Homework Problems

4.1 Add a gravitational force density and gravitational energy input to (4.2) and (4.3) and derive the modified jump conditions.

4.2 Suppose that during the shock transition significant energy is lost by radiation. Write down the modified jump conditions.

4.3 Determine from energy arguments how to generalize (4.20) for a plasma having two ion species.

4.4 For $\gamma_1 = \gamma_2$, derive the equivalent of (4.18) and (4.20). Express the result in physically clear parameters, so the relation among the terms is evident. Check your result by finding it as a limit of (4.19) and by finding (4.20) as a limit from it. Using a computational mathematics program is suggested.

4.5 Find an expression for the entropy production by a shock wave (4.24) as the Mach number approaches 1 from above.

4.6 Derive the jump conditions for oblique shocks, (4.28)–(4.31).

4.7 Derive the relations of the angles for oblique shocks, (4.34) and (4.35).

4.8 Derive (4.42) relating the speeds in different frames of reference. This requires thinking about which frame of reference one is working in, a key element in all such problems.

4.9 Determine the equations and derive the behavior of the simpler case in which a shock is incident on a stationary wall. Let state 0 be the state of the unshocked fluid, state 1 be that of the once-shocked fluid, and state 2 be the state of the reshocked fluid produced when the shock reflects from the wall.

4.10 Derive the differential equations for self-similar motions of fluids, (4.64)–(4.66). Identify the requirements for quantity in the final curved brackets in the third of these equations to vanish.

4.11 Show that the conservation of mass in the planar isothermal rarefaction in fact requires $r \geq -c_s t$ in (4.72) and (4.73).

4.12 Plot the minimum density and pressure in the planar adiabatic rarefaction as a function of U_p. Discuss the meaning of the plots. Reasonable normalizations are recommended.

4.13 Sketch the C_+ and C_- characteristics defined in Sect. 4.4.3 in a fluid flowing uniformly with velocity u.

4.14 Show that the analysis of blast waves that preserves energy conservation produces $\alpha = 1/2$ for cylindrical blast waves and $\alpha = 2/3$ for planar blast waves.

4.15 Find the coefficients α for blast waves treated as cylindrical and planar, momentum-conserving snowplows.

4.16 Use a computational mathematics program to integrate the relevant equations to find and plot the profiles, and to evaluate Q, of (4.123) for a cylindrical blast wave. Apply this to find the behavior of a lightning channel produced by a deposited energy of 10^{10} ergs/cm.

4.17 Assuming that a strong shock reaches an interface beyond which the density (ρ_4) is 0.1 times the density of the shocked material behind the interface (ρ_1), solve for the profiles of the fluid parameters in the rarefaction that results.

4.18 Assuming that $\gamma_1 = \gamma_3$ (or not, if you wish), derive (4.128) from (4.44) to (4.52) by letting p_3 approach p_1 as the definition of the transition to a rarefaction. Hint: This one is not easy. Taking a limit will be necessary and the approach to the solution will matter.

4.19 An entertaining aspect of the problem of reshocks in rarefactions is that it is one case where the traditional model in which shocks are driven by moving pistons does not produce correct qualitative behavior. Consider a rarefaction as it approaches a piston that is moving forward at a constant velocity. What will happen?

4.20 To obtain the behavior of oblique shocks at interfaces, one must evaluate the equations in cylindrical polar coordinates. Beginning with the first two Euler equations, carry out this evaluation.

4.21 Equation (4.133) implies that a property of uniform flow is that $u_r = -\partial u_\phi / \partial \phi$ in any cylindrical polar coordinate system. Landau and Lifshitz use a geometric argument to demonstrate this. Instead, demonstrate this using a vectorial argument. (Hint: Begin by taking dot products of unit vectors along r and ϕ with an arbitrary velocity vector.)

References

Barenblatt GI (1996) Scaling, self-similarity, and intermediate asymptotics: dimensional analysis and intermediate asymptotics. Cambridge University Press, Cambridge

Chevalier RA (1982) Self-similar solutions for the interaction of stellar ejecta with an external medium. Astrophys J 258:790–797

Chevalier RA (1992) Supernova 1987a at five years of age. Nature 355(6362):691–696

Drake RP, Carroll III JJ, Smith TB, Keiter P, Glendinning SG, Hurricane O, Estabrook K, Ryutov DD, Remington BA, (LLNL) RJW, Michael E, McCray R (2000a) Laboratory experiments to simulate supernova remnants. Phys Plasmas 7:2142

Drake RP, Smith T, Carroll III JJ, Yan Y, Glendinning SG, Estabrook K, Ryutov DD, Remington BA, Wallace R, McCray R (2000b) Progress toward the laboratory simulation of young supernova remnants. Astrophys J Suppl 127(2):305–310

Knudson MD, Hanson DL, Bailey JE, Hall CA, Asay JR, Anderson WW (2001) Equation of state measurements in liquid deuterium to 70 GPa. Phys Rev Lett 87(22):225501-1–225501-4

Landau LD, Lifshitz EM (1987) Fluid mechanics, course in theoretical physics, vol 6, 2nd edn. Pergamon Press, Oxford

McKee CF (1974) X-ray emission from an inward-propagating shock in young supernova remnants. Astrophys J 188:335–339

Nellis WJ (2006) Dynamic compression of materials: metallization of fluid hydrogen at high pressures. Rep Prog Phys 69(5):1479–1580

Sedov LI (1959) Similarity and dimensional methods in mechanics, vol 1. Academic, New York

Zel'dovich YB, Razier YP (1966) Physics of shock waves and high-temperature hydrodynamic phenomena, vol 1, 2002nd edn. Dover, New York

Chapter 5
Hydrodynamic Instabilities

Abstract This chapter begins with a discussion of buoyancy as a driving force and proceeds to derive the dispersion relation for the Rayleigh-Taylor instability. It considers Rayleigh-Taylor in several specific contexts, the generalization of Rayleigh Taylor, sometimes called the entropy mode, and the nonlinear behavior of Rayleigh-Taylor-unstable systems. It hen discusses lift as a driving force, proceeds to derive the dispersion relation for the Kelvin-Helmholtz instability, and discusses Kelvin-Helmholtz in several specific contexts. Following that, it discusses the stability of shock waves and then the Richtmyer-Meshkov process, through which deposition of vorticity leads to evolving structure. The chapter concludes with a discussion of hydrodynamic turbulence.

Our discussion of the previous chapter focused on one-dimensional phenomena, in which a physical system was structured as a function of linear or radial distance, but by assumption was not structured in the other two dimensions. Common experience tells us that this will rarely be a good assumption. We see turbulent clouds and whirlwinds in the air and complex eddies in water. We know of three-dimensional turbulent motions within the Earth and within the Sun. For that matter, if we focus our attention, we can see amazing hydrodynamic phenomena every day in the bathroom sink. We know that one can save fuel in cars and in airplanes by careful design that reduces the energy delivered to turbulence in the air. In fact, our experiences and common knowledge would lead us to conclude that hydrodynamic fluids are more often than not unstable and turbulent in some sense.

We might suppose, however, that all such effects are well understood, because they have been studied in depth for more than a century. Surely a series of brilliant humans, armed with modern mathematics and, more recently, with powerful computers, working with phenomena that can very readily be observed in nature or produced in laboratories, will have come to understand this subject thoroughly. As it happens, this is not true. One could nominate hydrodynamic instabilities and turbulence for an award in the category of area of physics in which the least fundamental progress has been made during the last century. There are, of course, positive outcomes of the effort in this area. Much has been learned, much is being

© Springer International Publishing AG 2018
R P. Drake, *High-Energy-Density Physics*, Graduate Texts in Physics,
https://doi.org/10.1007/978-3-319-67711-8_5

learned, and what has been learned has often had real practical importance far beyond any direct impact of the quest for the next quark. But one still must wonder why this has been so difficult.

Much of the answer to this question can be found by contemplating the first Euler equation—the continuity equation. This equation contains the divergence of the product ρu, which makes it a nonlinear equation. It often proves feasible in physics to deal with nonlinear terms in physical equations by assuming that one of the variables is constant or by linearizing both variables. Indeed, these approaches will provide some insight into the fundamental hydrodynamic instabilities as this chapter develops. In some physical systems, the nonlinear terms drive waves that saturate themselves in ways that do not affect the global dynamics of the system. In hydrodynamic systems, the variations in both density and velocity often become large and structured in comparison with the initial values. The resulting dynamics are not tractable. Even computer simulations cannot follow all of the behavior, as the finite size of computer memory and run time imposes significant limits on the resolution with which one can examine the dynamics.

Even so, gaining some understanding of hydrodynamic instabilities is feasible. In particular, one can identify various circumstances that produce unstable behavior; these give us the instabilities with well-known names. One can use linearized theory to evaluate the initial growth rate of the instabilities when their amplitudes are small. We will pursue this for some instabilities that are important to high-energy-density physics. The cases we will examine all involve modulations of a system in only two dimensions. This is where one finds the strong effects that tend to initiate the growth of unstable structures in real fluids. As the modulations grow, they proceed to develop structure in the third dimension, which sometimes sets the stage for secondary instabilities that involve modulations of a system in three dimensions. In contrast to the two-dimensional instabilities, which are few and ubiquitous, the three-dimensional instabilities tend to be specific to a given detailed geometry. We will leave their details to the specialized literature, but we will discuss their consequence, which is a state of the fluid known as hydrodynamic turbulence.

5.1 Introduction to the Rayleigh–Taylor Instability

The *Rayleigh-Taylor instability* causes the interpenetration of fluid regions having different density. Figure 5.1 shows an example. These fluid regions may be two different materials, divided by an interface, or the same material at two different average densities, with a density gradient between the regions. The instability, which we will designate as the RT instability, is often said to occur whenever a less-dense fluid supports a more-dense fluid against gravity. A first generalization of this condition is to say that the RT instability occurs whenever fluid regions that differ in density experience a pressure gradient that opposes the density gradient. A second and broader generalization of this condition is to say that an *entropy mode* occurs whenever the entropy gradient is parallel to the pressure gradient (see Sect. 5.3).

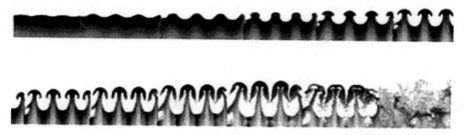

Fig. 5.1 Growth and saturation of the Rayleigh–Taylor instability, observed by acceleration of two fluids in a test facility. Credit: University of Arizona, Jeff Jacobs

5.1.1 Buoyancy as a Driving Force

We will focus here primarily on the condition of opposed pressure and density gradients. This condition is perhaps too general to be immediately clear, but this is necessary to cover most cases of interest in high-energy-density physics. When this condition for instability is satisfied, the system can reduce its potential energy through the interpenetration of the two fluid regions. Our approach to this instability will be to begin with some simple analysis for the sake of improving our intuition about both buoyancy and acceleration. Then we will proceed in the next section with a formal derivation of the unstable behavior.

We begin with bubbles. We know that bubbles of air rise in water, and can define analogies in other fluids. We can understand the upward force on the bubble, as Archimedes did, by thinking about the force required to insert the bubble into the water, thereby lifting the water surface. But this does not give us any understanding of what really goes on in the bubble. So it is natural to ask why, in detail, a bubble experiences a net upward force. We could begin to think about this by considering the fluid momentum equation in a fluid without viscosity,

$$\rho \frac{\partial}{\partial t} u + \rho u \cdot \nabla u = -\nabla p - \nabla \Psi, \tag{5.1}$$

in which Ψ is the potential, equal to $\int \rho g dz$ for gravity. (We ignore surface tension here, although that would not be justified for air bubbles in water.) In steady state, the gradients of pressure and gravitational potential balance one another, and indeed we often determine the pressure at some point by thinking about the weight of the fluid above it. If we think only in the vertical direction, it would seem that any distribution of matter could come to an equilibrium, with as much pressure as is needed to balance gravity. Then there would be no buoyancy (and no Rayleigh–Taylor instability either).

In a structured fluid under gravity, the weight of the matter above various points will vary. If this determined the local pressure, then there would be lateral pressure gradients in the fluid. But the fluid cannot sustain such gradients. In a compressible fluid they would be relaxed through compression and sound waves. In the limit that the fluid becomes incompressible, the pressure is instantaneously constant along surfaces of constant potential energy (i.e., at the same height in a gravitational potential). This implies that the pressure is determined by the average of the weight of the matter above any such surface. In mathematical terms, if z defines an axis parallel to the direction of the potential gradient, then

$$p = \frac{1}{A} \int_A \int_z^{zmax} -\frac{\partial \Psi}{\partial z} dz dA = \frac{-1}{A} \int_A \int_z^{zmax} \rho(z, x, y) g dz dA, \qquad (5.2)$$

in which the area of the fluid over the (x, y) plane is A and $zmax$ is a height beyond which there is no influence on the location of interest, and the rightmost expression is specific to mass in a gravitational field. The corresponding term in (5.1) is

$$-\frac{dp}{dz} = \frac{1}{A} \int_A \frac{\partial \Psi}{\partial z} dA = \frac{1}{A} \int_A \rho(z, x, y) g dA = \overline{\rho} g, \qquad (5.3)$$

which defines the average density as $\overline{\rho}$. For the simple gravitational case, we thus have

$$\rho \frac{\partial}{\partial t} u + \rho u \cdot \nabla u = (\overline{\rho} - \rho) g. \qquad (5.4)$$

Now we can see that the bubble, having $\rho < \overline{\rho}$, experiences an upward force. Also we see that the remaining fluid experiences a downward force. This corresponds to the settling that must occur as the bubble moves higher. For a small bubble in a fluid of density ρ_2, $\overline{\rho} \approx \rho_2$, and by integrating over the bubble volume V one finds the standard result that the upward force is $(\rho_2 - \rho)V$. At this point we understand in detail why the bubble rises. To understand the dynamics of the bubble as an entity, we would also have to consider the consequences of displacing the fluid above the bubble and any other forces such as surface tension. But (5.4) takes us far enough to address RT instabilities, so we stop here for now.

The standard simple example of an RT instability is the evolution of a system in which a denser fluid, such as water, is initially oriented above a less dense fluid, such as oil, in a gravitational field (whose direction defines "above"). A standard demonstration uses a jar filled with oil and water, which is quickly yet smoothly inverted. One can also find toys or desktop knick-knacks that display the resulting dynamics. (For such demonstrations it does matter that the motion which inverts the two fluids does not cause a change of state of the fluid. Thus, one may observe that dark beer is less dense than amber beer, but the attempt by one of my graduate students to invert a glass containing these two unmixed fluids had comic consequences unrelated to RT.) Returning to our reference situation, first note that the pressure increases toward the bottom of this structure, because of the weight of the fluid above it. Thus, the pressure gradient is downward. Second, by assumption the density gradient is upward. Within any ripple at the interface between the two fluids, the less dense fluid will feel an upward force while the more dense fluid will feel a downward force. The result is that small ripples of the interface, always present because of thermal noise, will grow. The comparison with the case of the bubble makes it seem natural to describe the region of less-dense, rising fluid as the *bubble*. This is standard jargon in discussions of RT. The denser material that penetrates into the less-dense fluid is known as the *spike*.

The cases that arise in high-energy-density physics rarely involve gravity as such. Instead, they tend to involve a low-density fluid that is pushing against a higher density fluid, causing the acceleration or deceleration of the higher density fluid. One example is that of hot air beginning to rise against cooler, higher density air. An analogous case is that of a pulsar wind (a very high temperature yet low-density fluid of positrons and electrons) accelerating outward the denser matter surrounding the pulsar. A second analogous case is that of the low-density, laser-heated corona surrounding a laser-fusion capsule pushing the denser capsule material and accelerating it inward. In all these cases the pressure is higher in the low-density fluid, so there is a pressure gradient that opposes the density gradient. One way to see intuitively how the instability works in this case is to realize that the inertia of any fluid will cause it to resist being accelerated. The denser fluid has more inertia and the interpenetration of the two fluids allows some of it to lag behind.

There is a counterintuitive aspect to such systems, though. The potential energy is reduced when the denser fluid "falls" up the pressure gradient. In the context of fluid dynamics, one tends to think that pressure gradients cause the acceleration of material down the pressure gradient. This is true in isolation, but here it is the behavior of each individual fluid parcel relative to the average behavior that matters. If one thinks of the position and velocity of a fluid parcel relative to the position of an accelerating interface, one can see that more energy must be invested to move ahead of the interface rather than to stay with the interface, but less energy must be invested to lag behind. Thus, in the context of a system with a steady, imposed acceleration of an interface, the potential energy is proportional to the distance material has moved ahead of the interface. This defines the analog of "up" in this system. The less-dense fluid will feel a force, relative to the interface position, that is in the direction of the acceleration, while the denser fluid will feel a force in the opposite direction, causing it to lag behind.

A mathematically identical situation develops at a decelerating interface when the densities are reversed. This occurs when denser matter has an initial velocity, so it is moving into a region of less-dense matter. The compression and heating of the less-dense matter, perhaps by a shock, establishes a pressure gradient that acts to decelerate the denser matter, thus opposing the density gradient. Examples occur in the laboratory when a blast wave exits dense matter into less dense matter, as for example, at the interior surface of an inertial-fusion capsule. Astrophysical cases abound, for example, at the head of some astrophysical jets, where the denser material in the jet is decelerated by the less-dense material in front of it. In these cases with a decelerating interface, less energy is again invested to lag "behind" the interface, though now this is accomplished by slowing down less than the interface does. Here the interpenetration of the fluids reduces the potential energy of the matter relative to that generated by the imposed deceleration of the interface.

Another example of a decelerating interface is found in supernovae. Figure 4.28 showed the results of a simulation of one particular supernova explosion (SN1987A). During supernova explosions the blast wave from the explosion crosses the material interfaces in the star, where the density decreases more rapidly. The interface then decelerates, as the blast wave moves outward and the velocity of the interface decreases (see Sect. 4.6.4). In the process, a pressure gradient is established (again this is just part of the blast wave) that points outward, opposing the inward density gradient at the interface. In consequence RT develops at each interface, and in the nonlinear phase spikes of dense material flow outward through the star. The regions between the spikes, where less-dense material moves inward are the bubbles.

We can summarize the above mathematically by noting that the natural frame of reference in which to examine RT growth is that of the interface. This frame is typically accelerating relative to the frame of the laboratory. In the frame of the laboratory, the acceleration of the interface might be written as $\boldsymbol{a} = -\nabla p / \rho$. In the frame of the interface, any modulations see an average "gravitational" acceleration $\boldsymbol{g}$ in the opposite direction. Thus $\boldsymbol{g} = \nabla p_o / \rho_o$, in which the subscript o designates the averaged values. This completes our introductory contemplation. We now proceed to develop a mathematical treatment of the linear phase of the RT instability.

5.1.2 *Fundamentals of the Fluid-Dynamics Description*

In the present section, we take up the fluid-dynamics description of the RT instability, in which we consider the behavior of the entire fluid using the fundamental physical equations. An alternative approach is to use potential-flow theory, requiring that one have, away from any surfaces or interfaces between fluid regions, both $\nabla \cdot \mathbf{u}$ and $\nabla \times \mathbf{u}$ equal to zero everywhere. In this approximation, any vorticity is localized to the interfaces. One ignores the diffusion of vorticity and cannot treat phenomena that involve distributed vorticity, viscosity, density gradients, or compressibility. The potential-flow description, being more limited, can be easier to formulate mathematically and is often used in fluid-dynamics textbooks. However, it is a dead-end approach.

So we proceed here with a full, fluid-dynamics description. We will end up finding solutions for the RT modulations as *surface waves*, which are waves whose influence on the medium decays as one moves away from the surface. We can analyze the dynamics, using the fluid continuity and momentum equations. For momentum, we will use (2.27). We take the radiation pressure, the electromagnetic forces, and the other forces to be negligible, but we will keep the viscous force to explore the effects of viscosity. For comparison with other literature, note that surface tension, treated for example in Chandrasekhar (1961) would be one of the other forces in (2.27). We ignore this force because it has no relevance to high-energy-density systems, which are too hot to allow the molecular interactions that create it.

We define our fluid such that the initial unperturbed interface is in the x–y plane. Our approach will be to linearize the fluid equations, so we take the unperturbed pressure and density to be $p = p(z)$ and $\rho = \rho(z)$, respectively, and we take the first-order perturbations in the same quantities to be $\delta p(x, y, z)$ and $\delta \rho(x, y, z)$. We work in a frame in which the interface is at rest, so the zeroth-order velocity is 0 and we can take the first-order velocity to be $\boldsymbol{u} = (u, v, w)$, with each component a function of (x, y, z). Our initial equations are those of a compressible fluid, and its overall, zeroth-order behavior may involve compression. However, we anticipate that the fluctuations in velocity caused by the instability will be very subsonic. And the analysis of acoustic waves in Chap. 2 implies that very subsonic motions produce very small density fluctuations. Accordingly, we assume the fluctuations to be incompressible. The treatment of the fluctuations as compressible is much more involved than the theory we explore here, and in the end the consequences of compressibility are small, as expected.

The assumption of incompressible fluctuations is expressed as $\nabla \cdot \boldsymbol{u} = 0$, so the continuity equation becomes convective:

$$\frac{\partial \delta \rho}{\partial t} + \boldsymbol{u} \cdot \nabla \rho = 0, \tag{5.5}$$

and with our assumptions the linearized momentum equation becomes

$$\rho \frac{\partial \boldsymbol{u}}{\partial t} = -\nabla \delta p + \nabla \cdot \underline{\boldsymbol{\sigma}}_\nu - g \delta \rho \hat{z}, \tag{5.6}$$

in which $\hat{z}$ is a unit vector in the z direction and the effective gravitational acceleration, in a (noninertial) frame of reference in which the interface is at rest, is g. This can be tricky to apply, as one may be inclined to assume that g is in the direction of the acceleration in the frame of reference of the laboratory. However, in the frame of reference of the interface, g points toward the region of higher pressure, for reasons discussed at the end of the previous subsection. Here $\underline{\boldsymbol{\sigma}}_\nu$ represents the linearized form of the viscosity tensor discussed in Chap. 2, with elements

$$\sigma_{\nu ij} = \rho \nu \left(\frac{\partial u_i}{\partial x_j} + \frac{\partial u_j}{\partial x_i} - \frac{2}{3} \delta_{ij} \frac{\partial u_k}{\partial x_k} \right) + \zeta \delta_{ij} \frac{\partial u_k}{\partial x_k}, \tag{5.7}$$

in which y and z are x_2 and x_3, respectively, one sums over repeated indices, and δ_{ij} is the Kronecker delta function. Here the kinematic viscosity is ν and the second coefficient of viscosity, which is not important here, is ζ. Linearization does not have a major effect here, only requiring that the full density, $\rho + \delta \rho$, be replaced with ρ. The term involving the viscosity tensor simplifies considerably (and the term involving ζ vanishes), because $\partial u_i / \partial x_i = 0$ from incompressibility. Also, it is consistent with our assumptions that the only nonzero derivative of v is $dv/dz(= dv/dx_3)$. With these observations and assumptions, one has for the k component of $\nabla \cdot \underline{\sigma}_v$,

$$
\left(\nabla \cdot \underline{\sigma}_v \right)_k = \rho v \nabla^2 u_k + \frac{\partial(\rho v)}{\partial z} \left(\frac{\partial u_k}{\partial z} + \frac{\partial w}{\partial x_k} \right). \tag{5.8}
$$

There are three very distinct directions in this problem, which are the direction of gravity, the direction of the wavevector of a surface modulation, and the direction perpendicular to these two. We will assume throughout that the mean surface is perpendicular to the pressure gradient. We further assume the surface modulations to be plane waves, expecting to express any actual surface modulation as a sum over all the possible plane waves, which form a complete basis set. Our goal is to find the evolution of an arbitrary plane wave, assuming that it grows in time from a very small initial amplitude. We can now write the components of (5.6), the incompressibility condition, and (5.5) as

$$
\rho \frac{\partial u}{\partial t} = -\frac{\partial}{\partial x} \delta p + \rho v \nabla^2 u + \frac{\partial(\rho v)}{\partial z} \left(\frac{\partial u}{\partial z} + \frac{\partial w}{\partial x} \right), \tag{5.9}
$$

$$
\rho \frac{\partial v}{\partial t} = -\frac{\partial}{\partial y} \delta p + \rho v \nabla^2 v + \frac{\partial(\rho v)}{\partial z} \left(\frac{\partial v}{\partial z} + \frac{\partial w}{\partial y} \right), \tag{5.10}
$$

$$
\rho \frac{\partial w}{\partial t} = -\frac{\partial}{\partial z} \delta p + \rho v \nabla^2 w + \frac{\partial(\rho v)}{\partial z} \left(2\frac{\partial w}{\partial z} \right) - g\delta\rho, \tag{5.11}
$$

$$
\frac{\partial u}{\partial x} + \frac{\partial v}{\partial y} + \frac{\partial w}{\partial z} = 0, \text{ and} \tag{5.12}
$$

$$
\frac{\partial}{\partial t} \delta\rho = -w \frac{\partial \rho}{\partial z}. \tag{5.13}
$$

This is the set of equations that describes the linear phase of the RT instability, including effects of viscosity or density gradients. We look for waves that represent growing modulations of the surface, and thus in general will have amplitudes with an unknown variation in z but proportional to $\exp(ik_x x + ik_y y + nt)$ in x, y, and time t. Here k_x and k_y are the x and y components of the wavevector (which we could have chosen to lie along one of these axes) and n is the exponential growth rate. With these substitutions, we get a new equation set:

$$
\rho u n = -ik_x \delta p + \rho v \left(\frac{\partial^2}{\partial z^2} - k^2 \right) u + \frac{\partial(\rho v)}{\partial z} \left(\frac{\partial u}{\partial z} + ik_x w \right), \tag{5.14}
$$

$$\rho v n = -ik_y \delta p + \rho v \left(\frac{\partial^2}{\partial z^2} - k^2 \right) v + \frac{\partial(\rho v)}{\partial z} \left(\frac{\partial v}{\partial z} + ik_y w \right),$$ (5.15)

$$\rho w n = -\frac{\partial}{\partial z} \delta p + \rho v \left(\frac{\partial^2}{\partial z^2} - k^2 \right) w + \frac{\partial(\rho v)}{\partial z} \left(2\frac{\partial w}{\partial z} \right) - g\delta\rho,$$ (5.16)

$$ik_x u + ik_y v = -\frac{\partial w}{\partial z}, \text{ and}$$ (5.17)

$$n\delta\rho = -w\frac{\partial\rho}{\partial z}.$$ (5.18)

in which $k^2 = k_x^2 + k_y^2$. It may be helpful to note that Eq. (5.18) gives the density change due to the vertical displacement of the background density profile. In atmospheric waves, this can produce a restoring force due to buoyancy that contributes to their dispersion relation.

We can reduce these equations from five to two through the following steps. Multiply the first by $-ik_x$ and the second by $-ik_y$, then add the first two equations and use the fourth equation to simplify them. Also use the fifth equation to eliminate $\delta\rho$ from the third equation. This gives us

$$\rho n \frac{\partial w}{\partial z} = -k^2 \delta p + \rho v \left(\frac{\partial^2}{\partial z^2} - k^2 \right) \frac{\partial w}{\partial z} + \frac{\partial(\rho v)}{\partial z} \left(\frac{\partial^2}{\partial z^2} + k^2 \right) w \text{ and}$$ (5.19)

$$\rho w n = -\frac{\partial}{\partial z} \delta p + \rho v \left(\frac{\partial^2}{\partial z^2} - k^2 \right) w + \frac{\partial(\rho v)}{\partial z} \left(2\frac{\partial w}{\partial z} \right) + w\frac{g}{n}\frac{\partial\rho}{\partial z}.$$ (5.20)

Recalling that $v(z)$ and $\rho(z)$ are given as properties of the unperturbed system, we can see that we need only to eliminate δp to have an equation for w in terms of known parameters. Doing this, we obtain

$$\frac{\partial}{\partial z}\left[-\rho n\frac{\partial w}{\partial z} + \rho v \left(\frac{\partial^2}{\partial z^2} - k^2 \right) \frac{\partial w}{\partial z} + \frac{\partial(\rho v)}{\partial z} \left(\frac{\partial^2}{\partial z^2} + k^2 \right) w \right]$$
$$= k^2 \left[-\rho n w + \rho v \left(\frac{\partial^2}{\partial z^2} - k^2 \right) w + \frac{\partial(\rho v)}{\partial z} \left(2\frac{\partial w}{\partial z} \right) + w\frac{g}{n}\frac{\partial\rho}{\partial z} \right].$$ (5.21)

This equation, along with the previous ones and boundary conditions, provides the tools we need to investigate RT growth rates in the linear regime. The equations apply within the fluid on each side of an interface.

Boundary conditions are essential in finding the dispersion properties of surface waves, including unstable waves. These are of several types. First are geometric boundary conditions. One example is that of an unbounded fluid, in which the vertical velocity must die away at large distances from the interface. Another example is that of a fixed wall, at which some components of velocity (or all of them in viscous fluids) must vanish. As a result, in the present context of incompressible fluctuations, the derivative of the velocity normal to the wall (typically $\partial w/\partial z$) must also vanish there.

The second type of boundary conditions exist because the fluid must remain continuous across the interface between two fluid regions having different properties, not developing gaps or zones of overlap. For non-viscous (inviscid) fluids this requires that the velocity normal to the surface must be continuous, and for small-amplitude waves the resulting, first-order condition is that w be continuous across the interface. There may be discontinuities of u and v in such fluids, as we will see later. In viscous fluids there could be no discontinuities in velocity, so that u and v both would have to be continuous across an interface. This would imply, in our present context, that $\partial w/\partial z = 0$ at the interface as well.

The third type of boundary conditions involves the forces at an interface. In an inviscid fluid, pressure is continuous throughout the fluid (save at shocks). This remains true at interfaces. In viscous fluids, the (tensorial) stress, $\underline{\tau} = p\underline{I} + \underline{\sigma}_v$ is continuous throughout the fluid. This produces two useful boundary conditions . For our surface waves, the continuity of the two tangential components of stress gives us

$$\tau_{xz} = \sigma_{vxz} = \rho v \left(\frac{\partial u}{\partial z} + \frac{\partial w}{\partial x} \right) = \rho v \left(\frac{\partial u}{\partial z} + ik_x w \right) \text{ and}$$

$$\tau_{yz} = \sigma_{vyz} = \rho v \left(\frac{\partial v}{\partial z} + \frac{\partial w}{\partial y} \right) = \rho v \left(\frac{\partial v}{\partial z} + ik_y w \right),$$

$$(5.22)$$

which imply that

$$ik_x \tau_{xz} + ik_y \tau_{yz} = \rho v \left[\frac{\partial}{\partial z} \left(ik_x u + ik_y v \right) - k^2 w \right] = \rho v \left(\frac{\partial^2}{\partial z^2} - k^2 \right) w \qquad (5.23)$$

is continuous across the interface. Specializing to a case with regions designated by the subscript 2 at $z > z_o$ and by the subscript 1 at $z < z_o$, with the interface at z_o, we can subtract (5.23) across the interface to find one boundary condition relating to the forces at the interface. This gives

$$0 = \left[\rho_2 v_2 \left(\frac{\partial^2}{\partial z^2} + k^2 \right) w_2 - \rho_1 v_1 \left(\frac{\partial^2}{\partial z^2} + k^2 \right) w_1 \right]_{z=z_o} \text{ and} \qquad (5.24)$$

Continuity of the normal component of stress gives us another boundary condition. We have

$$\tau_{zz} = p + \sigma_{vzz} = p + 2(\rho v) \frac{\partial w}{\partial z}, \qquad (5.25)$$

and can note that σ_{vzz} is a first-order quantity here. In order to make use of this, it is helpful to explore the precise meaning of δp in the above equations. We designate the initial pressure profile within fluid region i as $p_{oi}(z)$ and the instantaneous pressure profile as $p_i(z)$. We *define* the deviation in the normal component of the stress tensor as $\delta p_i(z) + \sigma_{vzz} = p_i(z) + \sigma_{vzz} - p_{oi}(z)$. For a modulated interface located at $z = \zeta(x, y, t)$, we then have

$$\delta p_i(\zeta) = p_i(\zeta) - p_{oi}(\zeta). \tag{5.26}$$

Taylor expanding, we have

$$p_{oi}(\zeta) = p_{oi}(0) + \left[\frac{\partial p_{oi}}{\partial z}\right]_o \zeta \text{ and} \tag{5.27}$$

$$\delta p_i(\zeta) = \delta p_i(0) + \left[\frac{\partial \delta p_i}{\partial z}\right]_o \zeta = \delta p_i(0) \text{ to first order in } \zeta, \text{ so} \tag{5.28}$$

$$p_i(\zeta) = p_{oi}(0) + \left[\frac{\partial p_{oi}}{\partial z}\right]_o \zeta + \delta p_i(0) \tag{5.29}$$

to first order. We also have $(\partial p_{oi}/\partial z)_o = -\rho_i(0)g$, where g is gravity and $\rho_i(z)$ is the initial density profile and its value at 0 is taken in the limit as the initial, unperturbed interface is approached. Thus

$$p_i(\zeta) = p_{oi}(0) - \rho_i(0)g\zeta + \delta p_i(0). \tag{5.30}$$

In the absence of viscosity, continuity of pressure across the actual location of the interface ($z = \zeta$) implies $p_i(\zeta) = p_j(\zeta)$ so $\delta p_i(0) \neq \delta p_j(0)$, for two fluids i and j. For nonzero σ_{vzz}, τ_{zzi} instead is continuous and we have

$$\tau_{zzi}(\zeta) = p_{oi}(0) - \rho_i(0)g\zeta + \delta p_i(0) + \sigma_{vzz}(0). \tag{5.31}$$

Because $p_{oi}(0) = p_{oj}(0)$, the boundary condition across the interface is

$$\delta p_j - \delta p_i + 2(\rho v)_j \frac{\partial w_j}{\partial z} - 2(\rho v)_i \frac{\partial w_i}{\partial z} - (\rho_{oj} - \rho_{oi})g\zeta = 0, \tag{5.32}$$

where all subscripted quantities are evaluated in the limit that one approaches $z = 0$ from within the fluid region designated by the subscript. Thus, continuity of the normal component of stress at the instantaneous interface location at $z = \zeta$ turns out to imply that the pressure modulations defined as above are not equal on the two sides of the interface. Instead it is the quantity $\delta p_i + \sigma_{zzi} - \rho_i g\zeta$ that is continuous, reflecting how the initial pressure profiles vary with z. We can note that, for inviscid fluids, one can also obtain this same result using the Bernoulli relation.

Again specializing to a case with regions designated by the subscript 2 at $z > z_o$ and by the subscript 1 at $z < z_o$, with the interface at z_o, and recalling that $\partial w/\partial z$ is continuous across the interface, (5.32) becomes the boundary condition:

$$0 = -(\delta p_2 - \delta p_1)_{z=z_o} + \left[2\left(\rho_2 \nu_2 - \rho_1 \nu_1\right)\frac{\partial w_1}{\partial z}\right]_{z=z_o} + w_o \frac{g}{n}(\rho_2 - \rho_1)_{z=z_o},$$

$$(5.33)$$

in which we have used w_o to designate the common value of w at the interface.

We now return to our main line of discussion. Equation (5.33) in particular is essential to the analysis of RT, but is not yet in the useful form of having only one physical variable, specifically w. We need to eliminate δp. To accomplish this we can obtain another equation involving δp, by subtracting (5.19) from itself across the interface to find

$$k^2(\delta p_2 - \delta p_1)$$

$$= \rho_2 \left[-n + \nu_2\left(\frac{\partial^2}{\partial z^2} - k^2\right)\right]\frac{\partial w_2}{\partial z} + \rho_1 \left[n - \nu_1\left(\frac{\partial^2}{\partial z^2} - k^2\right)\right]\frac{\partial w_1}{\partial z}$$

$$+ \frac{\partial(\rho_2 \nu_2)}{\partial z}\left(\frac{\partial^2}{\partial z^2} + k^2\right)w_2 - \frac{\partial(\rho_1 \nu_1)}{\partial z}\left(\frac{\partial^2}{\partial z^2} + k^2\right)w_1. \qquad (5.34)$$

We then combine this with (5.33) to eliminate δp. (Here and henceforth we drop the notation "$z = z_o$", realizing that in such boundary conditions all quantities are evaluated as z approaches the interface from within the fluid designated by the subscript.) After using (5.24) to eliminate two terms, this gives a usable, if complex, boundary condition,

$$w_o k^2 \frac{g}{n}(\rho_2 - \rho_1) + k^2 \left[2\left(\rho_2 \nu_2 - \rho_1 \nu_1\right)\frac{\partial w_1}{\partial z}\right]$$

$$= \rho_2 \left[-n + \nu_2\left(\frac{\partial^2}{\partial z^2} - k^2\right)\right]\frac{\partial w_2}{\partial z} - \rho_1 \left[-n + \nu_1\left(\frac{\partial^2}{\partial z^2} - k^2\right)\right]\frac{\partial w_1}{\partial z}. \quad (5.35)$$

Thus, the boundary conditions we have to work with are (5.24), (5.35), the continuity of w across the boundary, additional constraints imposed by the geometry of the problem, and for viscous flows the continuity of $\partial w/\partial z$ and all components of $\boldsymbol{u}$ across the boundary.

Aside: An Alternative Approach

One can also obtain these and other boundary conditions directly from equations found in the course of the derivation, by integrating them across the interface. Chandrasekhar (1961) uses this powerful method, but it comes

(continued)

at the cost of losing the physical insight of the above discussion. We define and use the method in this brief aside, and apply it below in the derivation of the Kelvin-Helmholtz instability.

Consider the interface to be at $z = 0$. Suppose that $q(z)$ is an arbitrary function that is continuous and differentiable everywhere. Suppose $f(z)$ and $h(z)$ are arbitrary functions that are continuous and differentiable everywhere except at the interface, so we can write, for example, $f(z) = f_1(z)H(-z) + f_2(z)H(z)$, in which $H(z)$ is a Heaviside step function, equal to zero for $z < 0$ and to 1 for $z > 0$, and f_1 and f_2 are continuous, differentiable functions. The derivative of $f(z)$, $g(z) = df/dz$, can be written $g(z) = g_1(z)H(-z) + g_2(z)H(z) + \Delta f \delta(z)$, in which $\delta(z)$ is the Dirac delta function, g_1 and g_2 are continuous, differentiable functions, and $\Delta f = f_2(0) - f_1(0)$. We then take the limit of the integral over a small region about the interface, as the width of the region goes to zero. Evidently this will give zero unless the argument of the integrand includes a delta function. Specifically we find the set of relations shown below.

Here the subscript $z = 0$ indicates that the quantities should be evaluated as the interface is approached. Just as in the case of shock waves, the interface may be treated as a mathematical discontinuity, although in microscopic reality all physical quantities and their derivatives vary continuously across the interface.

Applying the relations of (5.36) and (5.37)–(5.19) and (5.20), realizing that w is continuous across the interface and that all derivatives of w in z are continuous and bounded as one approaches the interface, we again find the same boundary conditions of Eqs. (5.24) and (5.33).

$$\lim_{\epsilon \to 0} \int_{-\epsilon}^{\epsilon} f(z)dz = \lim_{\epsilon \to 0} [\epsilon f_2(\epsilon/2) - \epsilon f_1(\epsilon/2)] = 0,$$

$$\lim_{\epsilon \to 0} \int_{-\epsilon}^{\epsilon} (\partial f(z)/\partial z)dz = f_2(0) - f_1(0),$$

$$\lim_{\epsilon \to 0} \int_{-\epsilon}^{\epsilon} (\partial^2 q(z)/\partial z^2)dz = \lim_{\epsilon \to 0} \left[\frac{\partial q_2}{\partial z} - \frac{\partial q_1}{\partial z} \right] = \left[\frac{\partial q_2}{\partial z} - \frac{\partial q_1}{\partial z} \right]_{z=0},$$

(5.36)

(continued)

$$\lim_{\epsilon \to 0} \int_{-\epsilon}^{\epsilon} q(z) \frac{\partial f(z)}{\partial z} dz = q_s (f_2 - f_1)_{z=0} + \lim_{\epsilon \to 0} \int_{-\epsilon}^{\epsilon} \epsilon \frac{\partial q}{\partial z} \frac{\partial f}{\partial z} dz$$

$$= q_s (f_2 - f_1)_{z=0},$$

$$\lim_{\epsilon \to 0} \int_{-\epsilon}^{\epsilon} f(z) \frac{\partial q(z)}{\partial z} dz = q_s (f_2 - f_1)_{z=0} - \lim_{\epsilon \to 0} \int_{-\epsilon}^{\epsilon} q(z) \frac{\partial f}{\partial z} dz = 0, \text{ and}$$

$$\lim_{\epsilon \to 0} \int_{-\epsilon}^{\epsilon} \frac{\partial f(z)}{\partial z} h(z) dz$$

$$= \lim_{\epsilon \to 0} \int_{-\epsilon}^{\epsilon} (f_1(z)\delta(z)h_1(z)H(-z) - f_2(z)\delta(z)h_2(z)H(z)) \, dz$$

$$= \frac{1}{2} (f_2(0)h_2(0) - f_1(0)h_1(0)).$$

$$(5.37)$$

5.2 Applications of the Linear Theory of the Rayleigh–Taylor Instability

At this point we have the tools we need to address various cases. We now proceed to consider three basic applications of the theory developed in the previous section. We begin with the simple case of an interface separating two fluids of different density. We then discuss the effects of viscosity on the instability. This is important for example in the atmosphere. It seems that it might not matter for high-energy-density systems, which have large Reynolds number (see Chap. 2). However, viscosity can play a role in such systems in altering the growth of small-scale structures. After that, we turn to the impact of density gradients, which are important in many applications.

5.2.1 Rayleigh–Taylor Instability with Two Uniform Fluids

The simplest case is that of two uniform fluids with a boundary at $z = 0$ and with no viscosity. Equation (5.21) then becomes

$$\frac{\partial}{\partial z} \left[-\rho n \frac{\partial w}{\partial z} \right] = k^2 \left[-\rho n w + w \frac{g}{n} \frac{\partial \rho}{\partial z} \right], \qquad (5.38)$$

in which $\partial \rho / \partial z$ also equals zero for the uniform fluids. This then simplifies for uniform density in each region to

$$\frac{\partial^2 w}{\partial z^2} = k^2 w. \tag{5.39}$$

The fluid must be undisturbed at sufficiently large distances, as $z \to \pm\infty$, so the solutions are

$$w_1 = w_o e^{kz} \quad \text{for} \quad z < 0 \text{ and}$$
$$w_2 = w_o e^{-kz} \quad \text{for} \quad z > 0, \tag{5.40}$$

where w_o is the same in both solutions because w must be continuous at the interface (to avoid the creating of voids or the accumulation of matter). Here we have defined fluid 1 as the region below the interface and fluid 2 as the region above it, by using subscripts on w. Our primary differential equation has thus given us the profiles but not the growth rate. To find this, we use the boundary condition (5.35) to find

$$w_o \frac{g}{n} (\rho_2 - \rho_1) = \frac{n}{k^2} (\rho_2 + \rho_1) k w_o, \text{ from which} \tag{5.41}$$

$$n_o = \sqrt{\frac{\rho_2 - \rho_1}{\rho_2 + \rho_1} k g} = \sqrt{A_n k g} \tag{5.42}$$

in which we have labeled the growth rate for this case as n_o and defined the Atwood number, $A_n = (\rho_2 - \rho_1)/(\rho_2 + \rho_1)$, which varies from -1 to 1 and measures how strong the density jump is at an interface. When A_n is negative, meaning that the denser fluid is already "below" the less dense one, then in the simple limit of (5.42) n_o is purely imaginary and the modulations oscillate but do not grow. (If we included finite viscosity such modulations would damp, as is discussed at length in Chandrasekhar (1961).) Equation (5.42) gives the simplest result for the RT growth rate, and for this reason is often referred to as the "classical" RT growth rate. (Thus continuing the flagrant abuse of the term "classical" throughout physics.) This growth rate n_o provides a reference for the growth in more-complicated systems. Adding complications tends to reduce this growth rate below n_o.

Before considering complications, it is worthwhile to point out how RT inherently provides circumstances that may lead to further instabilities. Suppose that the wavevector points in the x direction, so that $k = k_x$. Then in light of the solution given by (5.40), (5.17) implies $i u_1 = w_o$ and $u_2 = -u_1$. The first of these relations implies that u and w are out of phase spatially. The second implies that there is shear flow across the interface. Figure 5.2 illustrates this. Material must flow along the interface to provide the mass that penetrates across the original interface. Correspondingly, the material must flow in opposite directions on the two sides of the interface. This shear flow provides the potential for growth of the Kelvin–Helmholtz instability, discussed later in the chapter.

Fig. 5.2 Shear flow induced
by Rayleigh-Taylor. The
arrows show the location and
direction of the maxima in the
velocity perpendicular to the
interface and along the
interface. The amplitude
shown is nonlinearly large. In
the linear limit, u is
horizontal to first order

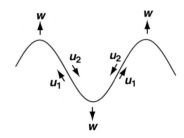

5.2.2   Effects of Viscosity on the Rayleigh–Taylor Instability

As a first example of a complication that reduces the RT growth rate, consider
the effects of viscosity. (We leave to the specialized literature the effects of mass
diffusion due to binary collisions, which complement viscosity and further reduce
the RT growth rate.) As a preliminary exploration, let us assume that the viscosities
on the two sides of the interface are nonzero, and that the densities on the two sides
of the interface are different in magnitude, but that the densities and viscosities are
both otherwise uniform in space. In this case, (5.21) becomes

$$\left[-n + \nu\left(\frac{\partial^2}{\partial z^2} - k^2\right)\right]\frac{\partial^2 w}{\partial z^2} = \left[-nw + \nu\left(\frac{\partial^2}{\partial z^2} - k^2\right)w\right]k^2. \tag{5.43}$$

This equation has a general solution,

$$w(z) = C_1 e^{s_1 z} + C_2 e^{-s_2 z} + C_3 e^{kz} + C_4 e^{-kz} \tag{5.44}$$

in which $s_i = k\sqrt{1 + n/(k^2\nu_i)}$ with i being 1 or 2. To assure that w vanishes at $\pm\infty$,
it is clear that C_1 and C_3 are zero for $z > 0$ and C_2 and C_4 are zero for $z < 0$. Note
that this choice regarding how to satisfy the specific boundary conditions used here
implies that the real parts of k, s_1, and s_2 are all positive. This will be an important
constraint as we develop our solution. We have four boundary conditions. These are
the continuity of $w(z)$, the continuity of $\partial w/\partial z$, (5.24) and (5.35). These with the
definition of s_i give us

$$C_1 - C_2 + C_3 - C_4 = 0, \tag{5.45}$$

$$s_1 C_1 + s_2 C_2 + kC_3 + kC_4 = 0, \tag{5.46}$$

$$n(\rho_2 C_2 - \rho_1 C_1) + \rho_2 \nu_2 k^2 (2C_2 + 2C_4) - \rho_1 \nu_1 k^2 (2C_1 + 2C_3) = 0, \text{ and} \tag{5.47}$$

$$0 = \frac{2A_n g k^2}{n}(C_1 + C_3) - kn\left(C_3(1 - A_n) + C_4(1 + A_n)\right)$$
$$-2k^2(kC_3 + s_1 C_1)(\nu_1 - \nu_2) + 2A_n k^2(kC_3 + s_1 C_1)(\nu_1 + \nu_2). \tag{5.48}$$

Since we have four equations that are linear in the four amplitudes, we can write (5.45)–(5.48) as an equation in which a matrix M multiplies the vector (C_1, C_2, C_3, C_4). Then the determinant of M must be zero, which gives us the following general dispersion relation for this case:

$$0 = n^2 \left[2A_n^2 k - (s_2 + s_1) + A_n(s_2 - s_1) \right]$$

$$+ 2k^2 n \left[(s_2 - s_1) + A_n(2k - (s_2 + s_1)) \right] \left[v_2(1 + A_n) - v_1(1 - A_n) \right]$$

$$+ 2 \left[k^5 - k^4(s_2 + s_1) + k^3 s_1 s_2 \right] \left[v_2(1 + A_n) - v_1(1 - A_n) \right]^2$$

$$+ A_n g k \left[s_2(1 - A_n) + s_1(1 + A_n) - 2k \right]. \tag{5.49}$$

For further discussion here, we will specialize to the case $v_1 = v_2$ and thus $s_1 = s_2$, obtaining

$$0 = 2n^2 \left(A_n^2 k - s \right) + 8nA_n^2 k^2 v(k - s) + 8A_n^2 k^3 v^2 (k - s)^2 - 2A_n g k(k - s). \tag{5.50}$$

This equation is deceptively simple, as s depends on n. We could solve for either variable, but it is most useful to solve for s because s is constrained to have a positive real part. Substituting for n (as a function of s) in (5.50), we obtain a fifth-order polynomial for s,

$$0 = -s^5 v^2 + s^4 k v^2 A_n^2 + 2s^3 k^2 v^2 (1 - 2A_n^2) + s^2 k^3 v^2 (6A_n^2)$$

$$+ s(A_n g k - k^4 v^2 (1 + 4A_n^2)) - A_n g k^2 + A_n^2 k^5 v^2. \tag{5.51}$$

By taking the limit as the viscosity vanishes, one can see from (5.51) that the growth rate goes to $\sqrt{A_n g k}$, as it should. While one can solve (5.51) straightforwardly with a computational mathematics program, it is more useful to first cast it in a nondimensional form. If one compares the terms in the coefficient of s, it is clear that g corresponds to $k^3 v^2$, suggesting that one uses a normalized wavenumber

$$\tilde{k} = \frac{k}{(g/v^2)^{1/3}} = \left[\frac{(k^2 v)}{\sqrt{gk}} \right]^{2/3}, \tag{5.52}$$

so one sees that $\tilde{k}$ depends on the competition between diffusion and growth. Specifically, for a spatial scale of $1/k$, $\tilde{k}$ is the 2/3 power of the ratio of the rate of viscous diffusion to the fundamental RT growth rate $\sqrt{kg}$. Comparing the third term on the right-hand side with the sixth term, one sees that $s^3 v^2$ here corresponds to g, suggesting that one uses a normalized value

$$\tilde{s} = \frac{s}{(g/v^2)^{1/3}}, \text{ implying} \tag{5.53}$$

$$\tilde{n} = \frac{n}{(g^2/v)^{1/3}} = \frac{n}{\sqrt{kg}} \left(\frac{k^2 v}{\sqrt{kg}} \right)^{1/3}, \tag{5.54}$$

which depends on the growth rate per unit fundamental RT growth rate times $\sqrt{\tilde{k}}$. With these normalizations, the zero-viscosity growth rate is $\tilde{n} = \sqrt{A_n\tilde{k}}$, and the dispersion relation becomes

$$0 = -\tilde{s}^5 + \tilde{s}^4 A_n^2 \tilde{k} + 2\tilde{s}^3 \tilde{k}^2 (1 - 2A_n^2)$$
$$+ 6\tilde{s}^2 A_n^2 \tilde{k}^3 + \tilde{s}\tilde{k}\left[A_n - \tilde{k}^3(1 + 4A_n^2)\right] - A_n\tilde{k}^2 + A_n^2\tilde{k}^5. \qquad (5.55)$$

This equation provides a universal relation between the normalized growth rate and the normalized wavenumber, depending only on the value of the Atwood number. Any root of this equation (for $\tilde{s}$), whose real part is positive, corresponds to a physical mode, but this mode is only exponentially growing if $\mathfrak{R}(s) > k$. Otherwise the mode is damped. Any roots with nonzero imaginary parts would correspond to oscillating modes, which might in principle be growing or damped.

Figure 5.3 shows the non-trivial solutions of (5.55) for $\tilde{s}$. One of the roots has a positive real part that always exceeds one. This is the exponentially growing mode. Two of the roots always have negative real parts, and so never correspond to

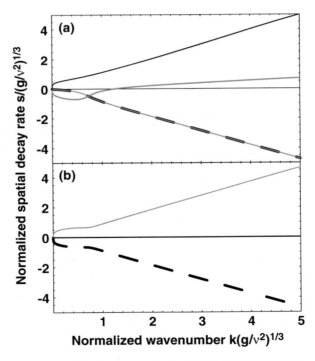

Fig. 5.3 Solutions for the spatial decay rate s for the Rayleigh-Taylor instability with viscosity, for $A_n = 0.5$. The real and imaginary part of each root are shown using the same curve type in both (**a**) and (**b**). (**a**) Real parts. Modes with positive real parts are physical solutions to the problem considered here. (**b**) Imaginary parts

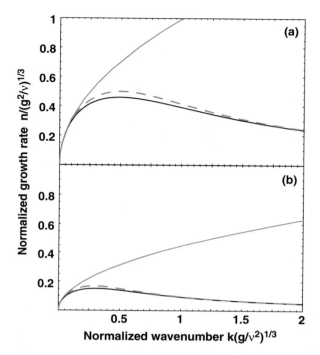

Fig. 5.4 Rayleigh-Taylor instability with viscosity. The black curves show the actual growth rates. The solid gray curves show the zero-viscosity result, $n = \sqrt{A_n k g}$. The dashed gray curves shows the approximation of (5.57). The panels show (**a**) $A_n = 1$ and (**b**) $A_n = 0.2$

solutions of this problem. These two roots also have imaginary parts; the other two are purely real. The final root is negative at small $\tilde{k}$ (small viscosity) but becomes a damped mode as $\tilde{k}$ increases.

Figure 5.4 shows the corresponding values of the normalized growth rate $\tilde{n}$, for the root corresponding to an exponentially growing mode, for two values of A_n. The roots shown in Fig. 5.4 are readily obtained from a computational mathematics program, but are not algebraically simple. However, it turns out that a simpler equation captures much of the behavior with high accuracy except at very small A_n. The physical basis for this is that the growth will approach the zero-viscosity value when viscosity is small and that viscous effects will dominate at high viscosity, so that a solution that joins these two regimes may work well even through the transition between them. To obtain such a solution, one can replace s in (5.50) with the value from an expansion for high viscosity, $s \sim k[1 + n/(2k^2\nu)]$. Making this substitution and solving the resulting equation produces the much simpler dispersion relation

$$n = \sqrt{A_n k g + k^4 \nu^2} - k^2 \nu, \tag{5.56}$$

The figure y-axis is labeled: Normalized growth rate $n/(g^2/\nu)^{1/3}$

The figure x-axis is labeled: Normalized wavenumber $k(g/\nu^2)^{1/3}$

which in our dimensionless units becomes

$$\tilde{n} = \sqrt{A_n \tilde{k} + \tilde{k}^4} - \tilde{k}^2. \tag{5.57}$$

This solution, originally developed by Duff et al. (1962), is shown as the dashed gray curves in Fig. 5.4.

It is no surprise that the wavenumber of the mode with the highest growth rate has a normalized wavenumber that is some fraction of unity. The largest growth occurs at wavenumbers just smaller than those for which viscosity begins to substantially reduce the growth. The wavenumber of maximum growth is approximately that given by (5.56) and (5.57), from which $\tilde{k} = A_n^{1/3}/2$ or

$$k = \frac{1}{2}\left(\frac{A_n g}{\nu^2}\right)^{1/3} \tag{5.58}$$

at the maximum. The magnitude of the growth rate at this wavenumber is

$$n = \frac{(A_n g)^{2/3}}{2\nu^{1/3}}. \tag{5.59}$$

Another observation from (5.56) and (5.57) is that, although the effect of viscosity is to reduce the growth rate, viscosity alone can never reduce it to zero. This makes physical sense, because while the viscosity can resist the flow of fluid and turn some kinetic energy into heat, the system will still seek its minimum-potential-energy state. Turning to real numbers, the viscosity is given in (2.40), and in a very rough estimate has a typical value $\sim 0.01\,\mathrm{cm^2/s}$ in high-energy-density experiments. A characteristic value for g might be $(100\,\mathrm{km/s})/(10\,\mathrm{ns}) = 10^{15}\,\mathrm{cm/s^2}$. With these assumptions, the wavenumber of maximum growth for an RT mode, already well below $\sqrt{A_n k g}$, from (5.58) with $A_n = 0.5$, is of order $10^6\,\mathrm{cm^{-1}}$, so the wavelength is of order $0.1\,\mu\mathrm{m}$. In some experiments this wavelength can be larger, of order $1\,\mu\mathrm{m}$. Wavelengths shorter than this will experience greatly reduced RT growth. This will definitely limit the ability of RT and related mechanisms to produce short-wavelength turbulence. In an astrophysical context, one might have $\nu \sim 10^{20}\,\mathrm{cm^2/s}$ and g might be $(100\,\mathrm{km/s})/(100\,\mathrm{years}) \sim 0.003\,\mathrm{cm^2/s}$. In this case, the maximum growth occurs for a wavenumber of $10^{-15}\,\mathrm{cm^{-1}}$, or a wavelength of order $10^{16}\,\mathrm{cm} \sim 0.01$ light years. Wavelengths much shorter than this would be in the high-viscosity regime and experience reduced growth.

5.2.3   Rayleigh–Taylor with Density Gradients and the Global Mode

Some interfaces are abrupt, and one can design experiments to create abrupt interfaces at least initially. However, the RT instability occurs in many situations that have a gradual interface. Indeed, sometimes there is no "interface" as such but merely an extended density gradient that opposes a pressure gradient. This is the case, for example, in supernovae. So it is worthwhile to explore the effects of a density gradient on this instability. Given that the effects of viscosity are typically small, it is sensible to set $v = 0$ for this calculation. In this case our basic differential equation (5.21) becomes

$$k^2 \left(\rho n - \frac{g}{n} \frac{\partial \rho}{\partial z} \right) w - \frac{\partial}{\partial z} \left(n \rho \frac{\partial w}{\partial z} \right) = 0. \tag{5.60}$$

Before proceeding to specific cases, it is worthwhile to observe that this equation has the solution $w = w_o e^{-kz}$ for an arbitrary density profile, corresponding to a growth rate of $n = \sqrt{kg}$. This mode is known as the global Rayleigh–Taylor mode, and this growth rate is the largest RT growth rate that exists (Bychkov et al. 1990). One might think that this is the end of the story for RT in density profiles. However, this mode does not always exist because it may not satisfy the boundary conditions. On the one hand, whenever a high-pressure region of negligible density is either accelerating or decelerating a fluid layer of some thickness L, the fluid layer will be unstable to the global RT mode for modes with $kL \gg 1$. The maximum amplitude of these modes will be at the free surface where the high pressure is located. This mode can also be viewed as a generalization of the mode we found in Sect. 5.2.2 to an arbitrary density profile and to $A_n = 1$. On the other hand, if there is a nonnegligible density on both sides of the interface, then the boundary conditions do not allow the global mode. We consider next such a case in which the instability develops somewhere on an extended and continuous density profile.

We will assume, as a sensible general case, that the density is exponentially distributed, so that $\rho(z) = \rho_o e^{z/L}$. Thus the density increases with "height", defined as the direction opposite the acceleration g in the frame of reference of the interface. Thus $\rho'(z) = \rho_o/L$. Substituting for $\rho = \rho(z)$, (5.60) has the solution

$$
\begin{aligned}
w = C_1 \exp\left[\left(\sqrt{1 + 4k^2 L^2 - 4gk^2 L/n^2} - 1 \right) \frac{z}{2L} \right] \\
+ C_2 \exp\left[-\left(\sqrt{1 + 4k^2 L^2 - 4gk^2 L/n^2} + 1 \right) \frac{z}{2L} \right],
\end{aligned}
\tag{5.61}
$$

with two constants C_1 and C_2. Here again these constants respond to the boundary conditions. If, for example, the unstable zone is confined between two boundaries, as can happen in the Earth's atmosphere, then one would need the amplitude to be zero at these boundaries (though for a linear theory this would be relevant only to wavelengths of order the distance between boundaries). Such close boundaries are

Fig. 5.5 A Rayleigh-Taylor
mode on a density gradient.
This grayscale representation
shows a vertically exponential
density profile in which there
is a density perturbation due
to a single mode

less common in the systems of interest to us, so we will take $C_1 = 0$ for $z < 0$ and
$C_2 = 0$ for $z > 0$, in order to assure that the perturbation dies out with distance.
Then the remaining constants are of equal magnitude to keep w continuous at the
interface.

It is worth focusing on the fact that the notion of an interface is somewhat
artificial in a continuous density profile. The instability might develop at any
location in the profile. The largest fluctuations in the profile, wherever they may
be, will produce large modulations first. Figure 5.5 illustrates the impact of a
single mode in such an environment. The mode is strongest at some location (about
halfway up the figure) and results in the flow of material both laterally and vertically.
Matter flows into the downward moving spikes and upward moving bubbles. In
reality, instabilities are likely to be seeded throughout the profile, and the entire
unstable region is likely to become very clumpy.

Indeed, the localized modes with $\mathbf{k} \perp \nabla p$ that we consider here are a subset of
the possible modes. In a continuous profile the direction of $\mathbf{k}$ (and thus the "surface"
considered) are not restricted to lie in the plane perpendicular to ∇p, but for a given
magnitude of k this direction will correspond to the direction of largest growth.
In addition, one can find plane-wave solutions to (5.60) in which the growth rate
has both real and complex parts. These correspond to modes that grow while they
propagate. Their growth rate is somewhat smaller than that of the global RT mode.

Returning to a localized RT mode in a continuous density profile, we find the
RT growth rate by applying the boundary condition (5.35), noting that at the chosen
interface $\rho_1 = \rho_2 = \rho_o$, so the term involving g in this equation drops out. Some
simple algebra then gives an equation for the growth rate,

$$0 = n^2(1 + 4k^2L^2) - 4gk^2L, \tag{5.62}$$

with the obvious solution for the growing mode

$$n = \sqrt{gk}\sqrt{\frac{4kL}{1 + 4k^2L^2}}. \tag{5.63}$$

Fig. 5.6 Impact of density gradient on Rayleigh-Taylor growth rate. The black curve shows the result from (5.63). The gray curve shows the approximate relation, $\sqrt{1/(1+kL)}$, discussed in the text

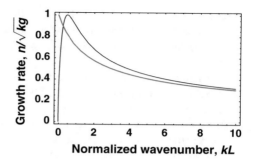

The normalized growth rate, $n/\sqrt{kg}$, is plotted against kL in Fig. 5.6. The growth rate reaches the value for the global mode at $kL = 1/2$, corresponding to a wavelength about ten times the density scale length. (For comparison, Fig. 5.5 shows a mode whose wavelength is about 30% of the density scale length, so $kL \sim 20$.) The normalized growth rate is finite but rapidly decreasing at small kL, becoming proportional to $\sqrt{2kL}$. If one thinks about a steadily increasing wavelength in Fig. 5.5, one can see that more mass has to flow over longer distances as the wavelength increases. On the other hand, at large enough kL the growth rate goes to $\sqrt{g/L}$, losing all dependence on k. This is thought to be the relevant limit for many cases in astrophysics. If we write $g = |\nabla p|/\rho = p/(\rho L_p)$, where L_p is the scale length of the pressure profile, then the growth rate takes a form familiar to astrophysics, becoming $(c_s^2/\gamma)/\sqrt{LL_p}$. (There is also a contribution to growth from the pressure gradient alone in this limit, which we discuss in Sect. 5.3.)

There are cases in which a density gradient exists and may even be exponential, but only over a limited range of densities. A prime example is found in inertial fusion, at the inner surface of the fusion capsule. Unstable wavelengths that are small compared to the density scale length are affected by the density profile, while wavelengths that are long compared to the density scale length tend to respond to the densities at the upper and lower boundaries, behaving as though the interface is abrupt. The gray curve in Fig. 5.6, showing the function $\sqrt{1/(1+kL)}$, is a reasonable compromise to approximate the behavior under such conditions. It is widely used, as is the similar approximation $\sqrt{A_n/(1+A_nkL)}$.

5.3   The Convective Instability or the Entropy Mode

The RT instability is in fact a special case, although it is a very important one. Consider, for example, the behavior of the Earth's atmosphere near the surface. The density gradient is negligible in comparison to the temperature gradient. On a hot day, when the air is hottest and the pressure is greatest near the surface, a parcel of air that rises slightly will expand to equalize its pressure. This in turn reduces the density of the parcel and makes it buoyant. The reverse happens when a parcel of air drops slightly. In short, the air is unstable to convective motions that will have the net effect of bringing cooler air down and hotter air up. Cumulus cloud formation is often a diagnostic of this. This instability is naturally called the *convective instability*.

The general instability of which the convective instability and the RT instability are special cases is the *entropy mode*. The instability occurs when $\nabla s \cdot \nabla p > 0$, where once again s is the specific entropy. This condition is known as the Schwarzschild stability criterion in the Western literature. When $\nabla s \cdot \nabla p < 0$ the fluid supports stable, oscillating waves. The Rayleigh-Taylor instability is all that remains in the fully incompressible limit. The condition $\nabla s \cdot \nabla p > 0$ can be reduced to $\nabla \rho \cdot \nabla p < 0$ by recalling that the specific entropy s can be expressed as $s_o + c_V \ln(p/\rho^\gamma)$ and that in the incompressible limit $\gamma \to \infty$. The more general condition allows for the possibility described above that a fluid parcel may expand or contract adiabatically as it crosses the interface, because of the overall pressure gradient. In that case, interpenetration of the fluids leads to a reduction in potential energy if $\nabla s \cdot \nabla p > 0$ is satisfied. Landau and Lifshitz (1987) anticipate this instability in their section entitled "internal waves in an incompressible fluid." We can develop a linear theory of this mode as follows.

We use the same conventions as in Sect. 5.1.2, with $s(z)$ being the initial entropy profile and δs being the first-order deviation. We also assume that only the first derivatives of p, ρ, and s are nonzero in the initial state. The linearized conservation of entropy can be written as

$$\frac{\partial}{\partial t} \delta s = -\boldsymbol{u} \cdot \nabla s. \tag{5.64}$$

Although we will take the medium to be compressible, we look for fluctuations that involve no compression to simplify the mathematics. There is some chance that these will be the fastest growing modes, as they invest no energy in longitudinal compression. Thus

$$\nabla \cdot \boldsymbol{u} = 0. \tag{5.65}$$

As in the case of Rayleigh–Taylor, we want to consider the motion in the plane of the interface, so the momentum equation in this accelerating frame becomes

$$\frac{\partial}{\partial t} \boldsymbol{u} = -\frac{1}{\rho} \nabla \delta p + \frac{\nabla p}{\rho^2} \delta \rho. \tag{5.66}$$

Here we have explicitly written the force introduced by the accelerating frame in terms of the pressure gradient. Equivalently we could write $\boldsymbol{g} = \nabla p / \rho$.

To solve these equations, we begin by using pressure and entropy as the thermodynamic variables. Thus $\rho = \rho(p, s)$ and

$$\frac{\delta \rho}{\rho} = \frac{1}{\rho} \left(\frac{\partial \rho}{\partial s} \right)_p \delta s + \frac{1}{\rho} \left(\frac{\partial \rho}{\partial p} \right)_s \delta p. \tag{5.67}$$

The second term in this equation is negligible, as $\delta p \ll \rho c_s^2$, which is necessary for a linear theory to be valid. This term also introduces no new dependences in the solution. Thus the momentum equation becomes

$$\frac{\partial}{\partial t} \boldsymbol{u} = -\frac{1}{\rho} \nabla \delta p + \left(\frac{\nabla p}{\rho^2} \frac{\partial \rho}{\partial s} \right)_p \delta s. \tag{5.68}$$

We look for solutions to this equation of the form $\exp[nt + ik_x x]$, allowing for exponential growth and for propagation in some direction perpendicular to z, which we designate as x. The y component of $\boldsymbol{u}$ is not affected by these dynamics, as the right-hand side of (5.68) has no curl. We seek a wave equation for the fluctuating velocity w as in Sect. 5.1.2. We begin by taking the dot product of (5.68) with ∇s, and also use (5.64) to obtain

$$-\frac{\partial^2}{\partial t^2} \delta s = -\frac{1}{\rho} \frac{\partial s}{\partial z} \frac{\partial}{\partial z} \delta p + \left(\frac{1}{\rho^2} \frac{\partial p}{\partial z} \frac{\partial s}{\partial z} \frac{\partial \rho}{\partial s} \right)_p \delta s. \tag{5.69}$$

The time derivative of this equation, with (5.64) and dividing out the common factor, gives

$$n^2 w = -\frac{n}{\rho} \frac{\partial}{\partial z} \delta p - \left(\frac{1}{\rho^2} \frac{\partial p}{\partial z} \frac{\partial s}{\partial z} \frac{\partial \rho}{\partial s} \right)_p w. \tag{5.70}$$

Next we use the x component of (5.68) and (5.65) to eliminate δp, finding

$$\delta p = -\frac{n\rho}{k_x^2} \frac{\partial w}{\partial z}. \tag{5.71}$$

Noting that both w and ρ in this equation have finite derivatives in z, (5.70) becomes

$$n^2 w = \frac{n^2}{k_x^2} \frac{\partial^2 w}{\partial z^2} + \frac{n^2}{\rho k_x^2} \frac{\partial \rho}{\partial z} \frac{\partial w}{\partial z} - \left(\frac{1}{\rho^2} \frac{\partial p}{\partial z} \frac{\partial s}{\partial z} \frac{\partial \rho}{\partial s} \right)_p w, \tag{5.72}$$

which is a wave equation

$$k_x^2 \left(1 + \frac{\omega_s^2}{n^2} \right) w - \frac{1}{L} \frac{\partial w}{\partial z} - \frac{\partial^2 w}{\partial z^2} = 0, \tag{5.73}$$

where for convenience we have defined $1/L = \partial \ln \rho / \partial z$ and

$$\omega_s^2 = \left(\frac{1}{\rho^2} \frac{\partial p}{\partial z} \frac{\partial s}{\partial z} \frac{\partial \rho}{\partial s} \right)_p. \tag{5.74}$$

Equation (5.73), in combination with boundary conditions developed as described above, covers a wide variety of limiting cases. For example, in the limit as $L \to \infty$, one obtains plane-wave solutions having $n^2 = -\omega_s^2 \sin^2 \theta$, where θ is the angle between the z axis and $\boldsymbol{k}$. This is the solution found in the fluid mechanics text of Landau and Lifshitz (1987), which also covers stable gravity waves. We will develop these applications further shortly, but first it is useful to return to the use of p and ρ as the thermodynamic variables, in which case, using the thermodynamic relation

$$\left(\frac{\partial \rho}{\partial s}\right)_p \left(\frac{\partial s}{\partial p}\right)_\rho = -\left(\frac{\partial \rho}{\partial p}\right)_s. \tag{5.75}$$

This allows us to connect several useful forms of ω_s as

$$\omega_s^2 = -\frac{1}{\rho}\frac{\partial p}{\partial z}\left[\frac{1}{\rho c_s^2}\frac{\partial p}{\partial z} - \frac{1}{\rho}\frac{\partial \rho}{\partial z}\right] = -\frac{c_s^2}{\gamma}\left[\frac{1}{\gamma L_p^2} - \frac{1}{L_p L}\right] = -g k_p, \tag{5.76}$$

in which the third term assumes a polytropic gas, $p/L_p = \partial p/\partial z$, and $k_p = |\nabla p|/(\rho c_s^2) - (1/L)$. Note that both L and L_p can be positive or negative. One sees that a pressure gradient is always destabilizing, which is sensible from the discussion at the beginning of this section, and that a density gradient must oppose the pressure gradient to be destabilizing. The frequency ω_s, when real, is called the Brunt–Väisälä buoyancy frequency (see, for example, Tritton 1988).

If we seek a general solution to (5.73), we find

$$w = C_1 \exp\left[\left(\frac{z}{2L}\right)\left(\sqrt{1 + 4k_x^2 L^2(1 - g k_p/n^2)} - 1\right)\right]$$
$$+ C_2 \exp\left[\left(\frac{-z}{2L}\right)\left(\sqrt{1 + 4k_x^2 L^2(1 - g k_p/n^2)} + 1\right)\right], \tag{5.77}$$

in which C_1 and C_2 are constants. For $L \to \infty$ and no pressure gradient, the two terms here are proportional to $\exp[\pm k_x z]$, as they should be. Just as in the above, one needs the boundary condition to find the growth rate. If one finds the growth rate for the simple case that $w \to 0$ at $\pm\infty$, with finite density and pressure gradients near an interface where the Atwood number is A_n, one finds

$$n^2 = g k_x \left(\frac{2 k_x L}{1 + 4 k_x^2 L^2 - A_n^2}\right) \times \left[k_p L - A_n^2 + \sqrt{k_p^2 L^2 + A_n^2(1 + 4k_x^2 L^2 - 2k_p L)}\right]. \tag{5.78}$$

This equation includes most of the cases one may encounter in the laboratory or in astrophysics, with the exception of a density gradient that extends for a finite distance between two layers of constant density.

5.4 Buoyancy-Drag Models of the Nonlinear Rayleigh–Taylor State

Once the amplitude of a single-mode RT instability reaches about 10% of the wavelength of the initial perturbation, nonlinear effects begin to alter the rate of growth. For a purely sinusoidal initial condition, the first development is that harmonics of the imposed wavelength begin to appear as the shape of the perturbation becomes distorted. This development has been studied in experiments and may have some relevance to specific applications. Even so, like the linear phase, the phase when harmonics are important is only transitory. Eventually the instability develops very elongated bubbles and spikes. In addition, the response to the lift induced by the shear flow at the tips of the bubbles and spikes is to broaden the tips until the interface, expressed as a function, becomes double-valued. This evolution can be seen in Fig. 5.1. This phase of the evolution, with elongated bubbles and spikes growing in time and having broad tips, may last for a significant time. The evolution during this phase can be thought of as the buoyancy-driven rising of the bubbles, limited by the drag on their tips. Models that describe this behavior are known as buoyancy-drag models. We discuss an example here. At present, more details can be found only in the literature, for example in Oron et al. (2001) and Dimonte (2000), and in the references these contain.

A buoyancy-drag model describes the velocity of the interface, u_i, with the equation

$$(\rho_1 + \rho_2) \frac{du_i}{dt} = (\rho_2 - \rho_1) g - \frac{C_d}{\lambda} \rho_2 u_i^2, \tag{5.79}$$

for densities $\rho_1 < \rho_2$ and with λ a "wavelength" corresponding to the width of a bubble. This is not quite a simple Newtonian force equation. Here we focus on the evolution of the bubbles. Similar considerations apply to the spikes. The contribution of ρ_2 on the left-hand side represents the fact that as the bubbles rise the denser mass must be displaced sideways. This might not necessarily contribute with a factor of 1 as assumed here. The first term on the right-hand side gives the buoyancy force causing the bubble to rise. Compressible effects might modify this term. The second term on the right-hand side gives the drag force that resists the rise of the bubble. For three-dimensional bubbles, $C_d = 2\pi$. The factor of $1/\lambda$ in this term is not genuinely an inverse wavelength. Physically it represents the ratio of the bubble volume, which contributes to the other two terms, to the bubble area, which produces the drag. However, the drag on the bubble will vary in response to its shape, which varies a great deal across different experiments and computer models. All in all this model is a nice way to capture key physics but it would be foolish to imaging that it captures the absolute truth.

When (5.79) applies, the bubble will accelerate until the two terms on the right-hand side balance. This defines the asymptotic bubble velocity,

$$u_i = \sqrt{\frac{A_n g \lambda}{\pi(1 + A_n)}},$$

(5.80)

in which we have employed the Atwood number. But reaching this saturation velocity is not instantaneous. Estimating the saturation time Δt from (5.79) and (5.80), one finds

$$\Delta t \sim \frac{u_i}{A_n g} = \sqrt{\frac{\lambda}{A_n g \pi(1 + A_n)}}.$$

(5.81)

For $\lambda \sim 2\pi$ μm and $g \sim 10^{14}$ cm/s^2, this gives 1 ns. Saturation might be up to an order of magnitude faster or slower for inertial confinement fusion and decelerating-interface experiments, respectively. So one must pay attention to the question of saturation time in applications.

There are two cases for which one can extend the buoyancy-drag model. Defining the bubble height h as the displacement of the bubble from the mean position of the original interface, one has $u_i = dh/dt$. For isolated single bubbles, which may occur for example in an experiment that seeds a long-wavelength modulation, λ in (5.79) is actually the bubble height, h. One can then convert (5.79) into an equation for $h(t)$ that can be solved numerically.

For broadband spectra in a sufficiently unstable system, one can assume that the bubbles have a characteristic shape, so that the ratio of the inertial and buoyancy terms to the drag term sustains a characteristic value such that $h = \lambda b$, for lateral scale size λ. In addition, it is reasonable to suppose that the bubbles have a characteristic shape (see below). One can express this as a ratio $b = h/\lambda$. Substituting for λ in (5.80) gives an equation one can solve for h, finding

$$h = \left[\frac{1}{2\pi b(1 + A_n)}\right] A_n g t^2 = \alpha_B A_n g t^2,$$

(5.82)

defining a parameter α_B or "alpha bubble". One can observe the growth of unstable structures in experiments or simulations to find a value of α_B. Typical values of α_B are within a factor of 2 of 0.05. There is much physics in the details that can be summarized by a certain value of α_B, but these are not our concern here.

Interfaces that are not prepared with a specific initial mode typically have a broad spectrum of initial modes. Our discussion of RT makes some features of such systems evident. The initial growth of the unstable modes will be most rapid for the short-wavelength modes, whose exponential growth rate is proportional to $\sqrt{A_n k g}$. These modes grow faster and also have a smaller asymptotic velocity from (5.80). Thus, they reach their final velocity first. As time progresses, larger bubbles reach their final asymptotic velocity, overtaking and absorbing the smaller bubbles. This process is known as bubble competition. In what is known as the self-similar regime, the net effect of bubble competition is that the characteristic shape of the bubbles remains constant as they grow in amplitude. Detailed calculations have shown that such a self-similar regime is reached, for an initial broadband spectrum and for constant acceleration. Experiments or other physical systems, however, may not have a broadband initial spectrum, may remain for a very long time in a state in which the initial conditions impact the structure, and may not have truly constant acceleration. As a result, the bubble-competition viewpoint and (5.82) are a useful model but not one that can be assumed to always apply.

5.5 Mode Coupling

Thus far we have considered the RT modes to be independent of one another, even in the nonlinear regime. This is consistent with our treatment of the equations in Sect. 5.1.2. Indeed, it is implied by these equations, because we began by linearizing them. Linearization amounts to assuming that the modes do not affect one another, because it is the nonlinear terms that would permit such effects. From the point of view that the nonlinear terms are small during the initial phases of RT growth, this is perfectly acceptable. However, as the modes grow (or if their initial amplitude is not very small), the existing modes do in fact couple to one another. This produces source terms for modes whose wave vector is the sum or difference of the wave vectors of existing modes. The production (or enhancement) of some modes by this process is known as *mode coupling*. We explore this here.

The general notion that the beating of two waves can produce sum and difference modes is a familiar one from general physics, often discussed in the context of music. In such discussions, it is only sometimes emphasized that it is the nonlinear terms in the underlying equations that create such possibilities. In general, such mode coupling can develop in two ways. On the one hand, the coupling can occur throughout a volume in which waves are present. In this case, the equations of continuity and of momentum are key to describing the interaction of the waves throughout space. Harmonic generation in music, laser scattering, wave coupling in the ionosphere, and the interaction of fluctuations in the solar wind all are examples of such volumetric mode coupling. This type of coupling, however, is of limited importance in RT. On the other hand, the coupling can occur at a surface, where the requirement that the surface move self-consistently and the other boundary conditions may include nonlinear terms and introduce mode coupling. Such a surface is often, but not always, an interface between two regions with distinct properties.

Let us begin by considering the behavior of such a surface, and return to the volumetric behavior later. The velocity of a point on a continuous surface is determined by the combination of its local time variation and motion that propagates to that point from adjacent regions. The requirement for continuity of the surface can be written

$$\frac{\partial x_s}{\partial t} + \mathbf{u}(x_s) \cdot \nabla x_s = \boldsymbol{u}_s, \tag{5.83}$$

in which x_s is the location of a point on the surface, $\boldsymbol{u}(\mathbf{x}_s)$ is the fluid velocity at that point, and $\boldsymbol{u}_s$ is the velocity of the point on the surface at x_s. Note, for example by reference to Fig. 5.2, that the fluid velocity may differ substantially from the velocity of the point on the surface.

The important point about (5.83) is that the second term in this equation is nonlinear. The fluid velocity $\boldsymbol{u}$ includes motion due to all the modes that are present at the surface, as does the location of the surface x_s. This inherently produces coupling of any two modes present at the surface to drive other modes. This is known as *second-order mode coupling*. In more detail, the fact that $\boldsymbol{u}$ is evaluated at x_s and not at an unperturbed, flat, initial interface, creates finite though weaker coupling at all higher orders. We will not discuss this aspect here.

To see more clearly what happens in second-order mode coupling at surfaces, we can explore mathematically a simple case that an experiment might attempt. Suppose we have an interface, separating two uniform fluids of different density and negligible viscosity, initially perturbed by some number of modes of small amplitude. The amplitude of each of these modes is made to be much larger than that of the other modes. All these other modes have finite initial amplitudes, at minimum corresponding to variations in the surface location on the atomic scale, but we take these to be negligible. We then apply and maintain a constant acceleration g to the system, beginning at some time $t = 0$.

We know from the differential equation (5.38) found in Sect. 5.2.1 that all the modes decay exponentially in $\pm z$. We can specify the perturbed velocity as a sum over surface fluctuations involving the possible wavevectors in the $x - y$ plane, $\boldsymbol{k}_m$. As we are considering mode coupling, we cannot use the usual complex notation without thought, but instead must represent the physical variables as real quantities. Taking all this into account, we can write the z-component of the velocity as

$$w = \sum_m w_m(t) e^{-sk_m(z-z_s)} \cosh(\mathrm{i}\boldsymbol{k}_m \cdot \boldsymbol{x} - \mathrm{i}\phi_m), \tag{5.84}$$

in which ϕ_m is the phase of mode m, s is $+1$ for $z > z_s$ and -1 for $z < z_s$, $w_m(t)$ is its time-dependent amplitude, and z_s, also a function of time, is the location of the surface. It would be mathematically simpler but less intuitive to absorb the term involving z_s into the time-dependent function $w_m(t)$. In addition, the present formulation explicitly shows that the behavior of every mode is affected by all the other modes, through z_s.

Since $\nabla \cdot \boldsymbol{u} = 0$ by assumption, the fluctuating velocity along the surface, $\boldsymbol{u}_\perp$ is given by

$$\boldsymbol{u}_\perp = \sum_m \boldsymbol{u}_{m\perp} = \sum_m w_m(t) e^{-sk_m(z-z_s)} (-\mathrm{i}s) \sinh(\mathrm{i}\boldsymbol{k}_m \cdot \boldsymbol{x} - \mathrm{i}\phi_m) \hat{\boldsymbol{k}}_m, \tag{5.85}$$

in which $\hat{k}_m$ is a unit vector in the direction of k_m. Note that $u_{m\perp}$ has this definite direction, but that the mode having a wave vector of $-k_m$ is redundant, as the hyperbolic sine changes sign, compensating for the change in unit-vector direction. One could attempt to sum over only a half space but the bookkeeping would become messy. Instead, we will sum over all directions and realize that the amplitude a measurement would detect is twice that corresponding to any one term in the sum.

We can express the position of the surface as a sum over the same modes

$$z_s = \sum_m z_m = \sum_m z_m(t) \cosh(ik_m \cdot x - i\phi_m), \tag{5.86}$$

in which only some modes have finite initial amplitude at $t = 0$. If we take the average initial position of the interface to be at $z = 0$, then modes with an initial amplitude small enough to be in the linear regime evolve with $z_m(t) \propto \cosh(n_m t)$, where n_m is the linear growth rate, and have $w_m(t) = n_m z_m(0) \sinh(n_m t)$.

Some further discussion of these initial values is worthwhile. Note that they involve functions of $\pm n_m t$. Although in Sect. 5.2.1 we took $n > 0$ to find growing modes, we can observe that the differential equations found there for an interface separating two uniform fluids are unchanged for $n < 0$. Modes with $n < 0$ decay with time and so are not relevant to the behavior after a few growth times. However, they may be important to the initial condition. Examination of our derivation in Sect. 5.2.1 shows that the implicit initial condition in that section is that of a flat interface on which a velocity perturbation has been imposed. Such an initial condition is physically sensible and might be achieved in practice, but is certainly not typical. Much more typical is the case of (5.86), in which the interface is initially structured and the velocity is initially zero.

With these definitions, we can evaluate the z component of (5.83), using an overdot for the partial derivative in time. This gives

$$\sum_m [\dot{z}_m(t) - w_m(t)] \cosh(ik_m \cdot x - i\phi_m) = \sum_\ell u_{\ell\perp}(t, z_s) \cdot \nabla \sum_j z_j$$

$$= \sum_\ell \sum_j w_\ell(t) z_j(t)(\hat{k}_\ell \cdot k_j) s \sinh(ik_\ell \cdot x - i\phi_\ell) \sinh(ik_j \cdot x - i\phi_j), \tag{5.87}$$

which becomes, upon expanding the hyperbolic sines and cosines,

$$\sum_m [\dot{z}_m(t) - w_m(t)] \cos(k_m \cdot x - \phi_m)$$

$$= \frac{s}{2} \sum_\ell \sum_j k_j w_\ell(t) z_j(t)(\hat{k}_\ell \cdot \hat{k}_j)$$

$$\times \big(\cos \big[(k_\ell + k_j) \cdot x - (\phi_\ell + \phi_j) \big]$$

$$- \cos \big[(k_\ell - k_j) \cdot x - (\phi_\ell - \phi_j) \big] \big). \tag{5.88}$$

We want to identify the term in the sum corresponding to any specific mode m. Each possible combination of two modes ℓ and j shows up four times in the sum, in consequence of summing over all directions. To be specific, a sum mode with $k = k_1 + k_2$ shows up twice in each term through various combinations of terms involving $\pm$ each wave vector. The redundant mode with $k = -(k_1 + k_2)$ also appears four times. The result for any one of the redundant modes, summing over only one of the two wave vectors, is to introduce a factor of 2. We get

$$
\dot{z}_m(t) - w_m(t) = s \sum_j k_j w_\ell(t) z_j(t) (\hat{k}_\ell \cdot \hat{k}_j) \Big|_{k_m = k_\ell + k_j; \phi_m = \phi_\ell + \phi_j}
$$
$$
- s \sum_j k_j w_\ell(t) z_j(t) (\hat{k}_\ell \cdot \hat{k}_j) \Big|_{k_m = k_\ell - k_j; \phi_m = \phi_\ell - \phi_j} .
\tag{5.89}
$$

Here the matching condition in wave vector and phase is indicated by the vertical line following each sum. This designates which terms in the sum are selected; these are the terms that contribute to mode m. The other terms in the sum are ignored. (Alternatively, one could devise some more-complicated notation related to a Kronecker delta function.)

When existing modes are creating new modes by beating together, we call them the driving modes and call the beat modes the driven modes. A consequence of the two terms in this equation is that any two driving modes produce driven modes with wave vectors that are the sum or difference of their wave vectors. Any two driving modes will produce one term driving a mode with a larger wave number, said to be upshifted and one term driving a mode with a smaller wavenumber, said to be downshifted.

In addition, the phase ϕ_m of the driven mode is determined by the phases of the driving modes. A specific driven mode may already be present at some amplitude, but how this mode is affected by the driving modes will depend upon the relative phases. In experiments using initial modes to drive others, the phases are chosen. Then the phases of the driving modes determine the phase of the driven mode. In more general circumstances, such as an inertial fusion capsule, the amplitude of the driven mode might be initially increased, initially decreased, or gradually become altered in phase through the influence of the driving modes. Henceforth we will ignore any contributions from the relative phases, assuming the modes to be in phase. This allows us to rewrite (5.89), explicitly specifying k_ℓ in the argument of z_ℓ, as

$$
\dot{z}_m(t) - w_m(t) = s \sum_j z_j(t)
$$
$$
\times \left[(\hat{k}_\ell \cdot \hat{k}_j) w_\ell(t, k_\ell = k_m - k_j) - (\hat{k}_\ell \cdot \hat{k}_j) w_\ell(t, k_\ell = k_j - k_m) \right] .
\tag{5.90}
$$

For our purposes, it will suffice to have a second-order expression for w_m. We can obtain one by realizing that $w_m = \dot{z}_m$ to first order. To second order in the mode amplitudes, this gives us

$$
w_m(t) = \dot{z}_m(t) + s \sum_j k_j z_j(t) \left(\hat{k}_\ell \cdot \hat{k}_j\right) \left[\dot{z}_\ell(t, \mathbf{k}_\ell) + \dot{z}_\ell(t, -\mathbf{k}_\ell)\right] \Bigg|_{\mathbf{k}_\ell = \mathbf{k}_m - \mathbf{k}_j}. \tag{5.91}
$$

For reasons discussed above $z_\ell(\mathbf{k}_\ell) = z_\ell(-\mathbf{k}_\ell)$, but we leave them separate to clarify some of the steps below. We will use this relation in another boundary condition to find an equation for the overall behavior of the modes.

To make further progress, we now must return to the fundamental differential equations. For our special case of constant density, the continuity equation does not produce any contributions to mode coupling. (This is not true if there is a density gradient.) The momentum equation under these assumptions is

$$
\rho \frac{\partial}{\partial t} \mathbf{u} + \rho \mathbf{u} \cdot \nabla \mathbf{u} = -\nabla p - \nabla \Psi, \tag{5.92}
$$

in which Ψ is the gravitational potential, given by $\Psi = \int \rho g \, dz$. Note that Ψ has a discontinuous derivative at the interface, for our assumptions. The initial profile of pressure is determined by the initial gravitational potential, and the gradients of these profiles cancel one another in this equation. This lets us follow only the variations, δp and $\delta \Psi$. We can also expand the convective derivative, using $\mathbf{u} \cdot \nabla \mathbf{u} = \rho \nabla \frac{u^2}{2} - \mathbf{u} \times \nabla \times \mathbf{u}$. This allows us to see, by taking the curl of the resulting equation, that $\nabla \times \mathbf{u}$ must remain zero if it is initially zero. Since $\nabla \times \mathbf{u}$ is zero for our initial conditions, (5.92) becomes

$$
\rho \frac{\partial}{\partial t} \mathbf{u} + \rho \nabla \frac{u^2}{2} = -\nabla \delta p - \nabla \delta \Psi. \tag{5.93}
$$

As an aside, it is worth mentioning that much of the literature takes an alternative approach to this problem of a stationary, structured interface, by exploiting the fact that $\nabla \times \mathbf{u}$, which is known as the *vorticity*, is zero. The vorticity corresponds qualitatively to the degree of swirling present in the motion. The vorticity plays an essential role in the development of hydrodynamic turbulence, as is discussed in Sect. 5.8. We show there that, in the absence of viscosity, vorticity is frozen into the fluid volume. Thus, in this limit the volumetric vorticity is fixed in time (the vorticity on a surface is not fixed). Therefore, if one assumes that the fluid is *inviscid* (which means that the viscosity is zero) and also is incompressible, then the velocity has both zero curl and zero divergence for all time. This in turn implies that the velocity is the gradient of a potential, ϕ_v and that this potential satisfies Laplace's equation, $\nabla^2 \phi_v = 0$. Such models are known as *potential flow* models, since the flow is described by a potential. In a potential flow model, one can write the momentum equation (2.27), under the present assumptions, as the gradient of an equation involving the density and gradients of pressure, velocity potential, and gravitational potential. The resulting equation corresponds to one version of Bernoulli's equation, which can also be used as a starting point. The potential-flow approach enables some simpler approaches to numerical simulation. A drawback is that such a model cannot describe any system containing actual vorticity and thus cannot follow the onset of turbulence.

Returning our attention to (5.93), our assumption of uniform density implies that the first term introduces no mode coupling, so that all the mode coupling enters here through the term involving u^2. In addition, we can identify $\delta\Psi$ as

$$\delta\Psi = -(\rho_2 - \rho_1) g \left[z \left[H(z) - H(z - z_s) \right] + z_s H(z - z_s) \right], \tag{5.94}$$

where again H is the Heavyside step function. As in Sect. 5.1.2, (5.93) has components in the z direction and in the x–y plane, written for example as

$$\rho \frac{\partial}{\partial t} w + \rho \frac{\partial}{\partial z} \frac{u^2}{2} = -\frac{\partial}{\partial z} \delta p - \frac{\partial}{\partial z} \delta\Psi, \text{ and} \tag{5.95}$$

$$\rho \frac{\partial}{\partial t} \boldsymbol{u}_\perp + \rho \nabla_\perp \frac{u^2}{2} = -\nabla_\perp \delta p, \tag{5.96}$$

in which $\nabla_\perp$ is the gradient in the x–y plane. Our first step in Sect. 5.1.2 was to take the versions of these equations corresponding to specific assumptions and to find a single differential equation. The general version of this is left for homework, but the outcome is that no mode coupling remains in the differential equation for w. This is the origin of the statement above that there is no mode coupling in the absence of a density gradient.

To proceed toward a solution for the behavior with mode coupling, we proceed as we did above in finding a boundary condition across the interface. Integrating (5.95) across the interface gives

$$(\delta p_2 - \delta p_1)_{z_s} = (\rho_2 - \rho_1) g z_s, \tag{5.97}$$

while operating on (5.96) with $\nabla_\perp$, using the incompressibility condition, and subtracting across the interface gives

$$-\nabla_\perp^2 (\delta p_2 - \delta p_1)_{z_s} = -\frac{\partial}{\partial t} \left(\rho_2 \frac{\partial w_2}{\partial z} - \rho_1 \frac{\partial w_1}{\partial z} \right) + \rho_2 \nabla_\perp^2 \frac{u_2^2}{2} - \rho_1 \nabla_\perp^2 \frac{u_1^2}{2}. \tag{5.98}$$

Here as before the quantities are evaluated as one approaches the interface from the side designated by the subscript. Together these give

$$-(\rho_2 - \rho_1) g \nabla_\perp^2 z_s = -\frac{\partial}{\partial t} \left(\rho_2 \frac{\partial w_2}{\partial z} - \rho_1 \frac{\partial w_1}{\partial z} \right) + \rho_2 \nabla_\perp^2 \frac{u_2^2}{2} - \rho_1 \nabla_\perp^2 \frac{u_1^2}{2}. \tag{5.99}$$

In the absence of mode coupling, this gives the standard RT growth rate as found in Sect. 5.2.1. Mode coupling appears to second order through (5.91) for w_m and through the final two terms (from the convective derivative). One finds

$$
\frac{(\rho_2 - \rho_1)}{2} \nabla_\perp^2 \frac{u^2}{2}\bigg|_m = \frac{(\rho_2 - \rho_1)}{2} k_m^2 \sum_j \left(1 - \hat{k}_\ell \cdot \hat{k}_j\right) \dot{z}_j
$$

$$
\times \left[\ddot{z}_\ell(t, \mathbf{k}_\ell) + \dot{z}_\ell(t, -\mathbf{k}_\ell)\right]\bigg|_{k_\ell = k_m - k_j} . \tag{5.100}
$$

Now we can use this equation, (5.91), and (5.99) to obtain an equation for the evolution of a mode on the interface having wave vector $\mathbf{k}$, as

$$
\ddot{z}_k - A_n g k_m z_k = -A_n \sum_j k_j \left[\left(\hat{k}_\ell \cdot \hat{k}_j\right) z_j(t) \left[\ddot{z}_\ell(t, \mathbf{k}_\ell) \right. \right. \tag{5.101}
$$

$$
\left. + \ddot{z}_\ell(t, -\mathbf{k}_\ell)\right] + \left(1 + \hat{k}_\ell \cdot \hat{k}_j\right) \frac{\dot{z}_j(t)}{2} \left[\dot{z}_\ell(t, \mathbf{k}_\ell) + \dot{z}_\ell(t, -\mathbf{k}_\ell)\right] \Bigg]_{k_\ell = k - k_j} .
$$

Here again one sees that in the absence of mode coupling, one recovers the usual RT growth rate. The presence of mode coupling can increase or decrease the growth of the mode relative to this, depending on the sign of the right-hand side (and thus on the phases of the modes). The right-hand side must be positive to add to the growth of the mode (with the assumed phase). Note that the sum is over all directions, so that if the term $(1 + \hat{k}_\ell \cdot \hat{k}_j)$ is 0 in one case, it will equal 2 for the opposing mode. The mode-coupling terms will tend to dominate if $k_j z_j z_\ell / z_k \gg 1$. If the driving modes are growing exponentially, the right-hand side will have terms proportional to $n_\ell^2 e^{(n_\ell + n_j)t}$ and $n_\ell n_j e^{(n_\ell + n_j)t}$. In this limit, the driven mode will grow in approximately an exponential way and will have a growth rate larger than that of the driving modes by approximately $k_j z_j z_\ell / z_k$. Thus, mode coupling can rapidly bring the coupled modes up to an amplitude of order kz times that of the driving modes. (Note: The precise result of (5.101) differs from that given in the original paper by Haan (1991). The author asked a colleague to check these results in 2005, and he found that agreement with them. Yet another colleague since remarked that he had done the derivation and agreed with the results in Hahn. As with all theory, any reader who actually needs a complete and accurate result had better derive it for themselves.)

The modes driven as described above then act as driving modes in turn. In this way, two initial modes can over time produce a broad spectrum of modes. These modes will have a sparse spectrum that can be constructed by taking sums and differences of multiples of the initial wave vectors. In applications, mode coupling can play a substantial role in creating more complicated structures at an RT-unstable interface.

5.6   The Kelvin–Helmholtz Instability

The Kelvin–Helmholtz (KH) instability, like Rayleigh-Taylor, is seen frequently in many disparate physical systems. Whenever two fluid regions flow past one another, with a sufficiently narrow transition region at their mutual boundary, fluctuations at the boundary are unstable and will grow. The transition region where the velocity changes quickly in magnitude but remains along the same axis is known as a *shear layer*. Figure 5.7 shows an example of modulations caused by a KH instability. Modulations driven by KH can routinely be seen in clouds, in flowing water, and in the ripples in the sand at the beach. They are also observed at shear layers in the magnetosphere. Throughout astrophysics, there are many systems that produce shear layers, anytime a flow of material from one object or region passes through or around another object or region. In addition, the characteristic mushroom shape that develops at the spike tips in the RT instability is produced by the same sort of lift force that drives KH. (See, for example, the simulation results shown in Fig. 4.27.) One can see that this process is so prevalent that it is worthwhile to understand.

5.6.1   Lift as a Driving Force

Lift is the second primary driving force, after buoyancy, that alters the structure of flows. One often first meets this force in basic physics, when discussing the Bernoulli effect and how airplanes work. By making the fluid flow faster over the upper edge of the wing, the airfoil causes the pressure above the wing to be lower than the pressure below it, and so the wing is pushed upward. A rippled interface in a fluid is like a sequence of airfoils, with each one inverted by comparison to the previous one. Figure 5.8 illustrates the effect by plotting the fluid streamlines near

-- Photograph by Bob Rilling --
-- U. of Illinois Cloud Catalog --

Fig. 5.7   The structures seen along the upper edges of these clouds were produced by the Kelvin-Helmholtz instability

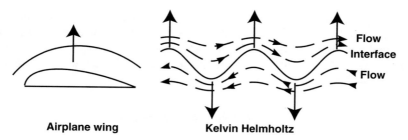

Fig. 5.8 Lift on airplane wings and in fluid flows. On the right, the solid curve is a rippled interface across which the velocity changes (a shear layer). The dashed curves show streamlines in the flow, which close when the surface extends into the fluid and spread out when the surface recedes

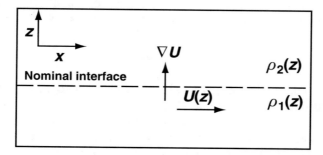

Fig. 5.9 Geometry for Kelvin–Helmholtz instability calculations. The densities and velocities may vary with z. The shear layer and velocity gradient may or may not be localized at the nominal interface

the interface. Where they are tightly spaced the fluid flows quickly and the pressure is reduced, and vice versa. The result is that each region of maximum displacement experiences a force that pulls it further from the mean interface location. Small ripples are caused to grow, until nonlinear effects limit the growth.

5.6.2   Fundamental Equations for Kelvin–Helmholtz Instabilities

The fundamental equations for KH instabilities are similar to those for Rayleigh-Taylor instabilities, but have differences reflecting the presence of a nonzero velocity and velocity gradient in the initial, unperturbed state. As in the case of the Rayleigh-Taylor instability, and for the same reasons, we will develop the theory for incompressible fluctuations. Here again, the unstable behavior is not strongly modified by compressibility. We consider the system sketched in Fig. 5.9. We assume the shear layer to be planar and to lie in the x–y plane, so that the z-direction is perpendicular to it. We further assume the initial flow, designated by

U, to be parallel to the x-axis, and that the zeroth-order gradients of U and of ρ are parallel to the z-axis. As in Sect. 5.1.2, we designate the first-order density and pressure perturbations by $\delta\rho$ and δp, respectively, and the $x, y,$ and z components of the first-order velocity perturbation, $\boldsymbol{u}$, by $u, v,$ and w. With these assumptions the continuity and momentum equations become

$$\frac{\partial \delta\rho}{\partial t} + \boldsymbol{U} \cdot \nabla \delta\rho + \boldsymbol{u} \cdot \nabla \rho = 0, \text{ and} \tag{5.102}$$

$$\rho\frac{\partial \boldsymbol{u}}{\partial t} + \rho \boldsymbol{U} \cdot \nabla \boldsymbol{u} + \rho \boldsymbol{u} \cdot \nabla \boldsymbol{U} = -\nabla \delta p - g\delta\rho\hat{z}, \tag{5.103}$$

where once again $\hat{z}$ is a unit vector in the z direction. In addition, we have the important additional condition expressed in (5.83) above, which is to first order

$$\frac{\partial \delta\boldsymbol{x}_s}{\partial t} + \boldsymbol{U} \cdot \nabla \delta\boldsymbol{x}_s = \boldsymbol{u}_s, \tag{5.104}$$

in which $\delta\boldsymbol{x}_s$ is the location of a point on the interface relative to its initial position and $\boldsymbol{u}_s$ is the velocity of that point. Here x_s and u_s are both first-order quantities. This equation specifies that the interface must move with the fluid self-consistently. We will not consider mode coupling for KH instabilities, but it exists for the same reasons that produce it in RT instabilities. One specific source is the requirement that the interface remain continuous, as represented in its full nonlinear form by (5.83).

In writing (5.102)–(5.104), we have ignored surface tension for the reasons discussed in Sect. 5.1.2. We also have ignored viscosity, for which we have much less excuse. Viscosity can play a role in KH instabilities at short wavelength. However, the mathematics turns out to be particularly intractable. Nonetheless, one aspect of the influence of viscosity can be accounted for using the above equations. Viscous diffusion of momentum causes the transition region in any initially abrupt shear layer to develop a scale length of $\sqrt{\nu t}$. This stabilizes the KH instability for the shortest wavelengths, and the maximum wavelength that is stabilized increases with time. We consider the effect of an extended shear layer below in Sect. 5.6.3.

It is helpful to express (5.102) through (5.104) as equations for the components, and to write out the incompressibility condition, in order to obtain a set of equations we can solve to see the unstable behavior. These are as follows:

$$\rho\frac{\partial u}{\partial t} + \rho U \frac{\partial}{\partial x} u + \rho w \frac{\partial}{\partial z} U = -\frac{\partial}{\partial x}\delta p, \tag{5.105}$$

$$\rho\frac{\partial v}{\partial t} + \rho U \frac{\partial}{\partial x} v = -\frac{\partial}{\partial y}\delta p, \tag{5.106}$$

$$\rho\frac{\partial w}{\partial t} + \rho U \frac{\partial}{\partial x} w = -\frac{\partial}{\partial z}\delta p - g\delta\rho, \tag{5.107}$$

$$\frac{\partial}{\partial t}\delta\rho + U\frac{\partial}{\partial x}\delta\rho = -w\frac{\partial\rho}{\partial z}, \tag{5.108}$$

$$\frac{\partial\delta z_s}{\partial t} + U\frac{\partial}{\partial x}\delta z_s = w_s, \text{ and} \tag{5.109}$$

$$\frac{\partial u}{\partial x} + \frac{\partial v}{\partial y} + \frac{\partial w}{\partial z} = 0. \tag{5.110}$$

Here only the z component of (5.104) is important in a linearized analysis. Note that all three dimensions matter for KH, unlike simple RT. This is because three directions—that of the gradients, that of U, and that of k—all matter independently. Also note that the sign of gravity is such as to produce a downward acceleration. We seek surface waves growing exponentially in time, but possibly also having an oscillatory component, we assume all linearized amplitudes to be proportional to $\exp i(k_x x + k_y y + nt)$. This differs from our assumption in the Rayleigh-Taylor problem. Now a growing instability will be one with negative imaginary n. Our set of equations then becomes

$$i\rho\left(n + k_x U\right) u + \rho w\frac{\partial U}{\partial z} = -ik_x\delta p, \tag{5.111}$$

$$i\rho\left(n + k_x U\right) v = -ik_y\delta p, \tag{5.112}$$

$$i\rho\left(n + k_x U\right) w = -\frac{\partial}{\partial z}\delta p - g\delta\rho, \tag{5.113}$$

$$i\left(n + k_x U\right)\delta\rho = -w\frac{\partial\rho}{\partial z}, \tag{5.114}$$

$$i\left(n + k_x U\right)\delta z_s = w_s, \text{ and} \tag{5.115}$$

$$ik_x u + ik_y v = -\frac{\partial w}{\partial z}. \tag{5.116}$$

One sees that five of these six equations involve the term $(n + k_x U)$. In a system with uniform flow, this type of term introduces a Doppler shift into wave frequencies. Here we have the added complication that U varies with z. We simplify these equations first by obtaining from (5.111), (5.112), and (5.116)

$$-\rho\left(n + k_x U\right)\frac{\partial w}{\partial z} + \rho k_x w\frac{\partial U}{\partial z} = -ik^2\delta p, \tag{5.117}$$

while from (5.113) and (5.114) we find

$$i\rho\left(n + k_x U\right) w = -\frac{\partial}{\partial z}\delta p - ig\frac{w}{\left(n + k_x U\right)}\frac{\partial\rho}{\partial z}. \tag{5.118}$$

Eliminating δp from these equations gives a single differential equation for w in terms of known parameters,

$$-k^2 \rho \left(n + k_x U \right) w + \frac{\partial}{\partial z} \left[\rho \left(n + k_x U \right) \frac{\partial w}{\partial z} \right] - \frac{\partial}{\partial z} \left(\rho k_x w \frac{\partial U}{\partial z} \right) = g \frac{w k^2}{\left(n + k_x U \right)} \frac{\partial \rho}{\partial z}.$$

(5.119)

One can see that this is a second-order equation for w and thus likely to allow solutions that decay away from the interface, or that combine to satisfy specific geometric constraints. Chandrasekhar (1961) points out that it is worthwhile to separate out the role of the density in this equation, obtaining

$$-k^2 \left(n + k_x U \right) w + \frac{\partial}{\partial z} \left(\left(n + k_x U \right) \frac{\partial w}{\partial z} - k_x w \frac{\partial U}{\partial z} \right)$$
$$= \frac{1}{\rho} \frac{\partial \rho}{\partial z} \left[g \frac{w k^2}{\left(n + k_x U \right)} - \left(\left(n + k_x U \right) \frac{\partial w}{\partial z} - k_x w \frac{\partial U}{\partial z} \right) \right].$$

(5.120)

The right-hand side of this equation can be ignored so long as the scale length of the density profile, L, is large compared to the perturbation wavelength of interest and unless the gravitational acceleration is very large ($> k_x^2 U^2 L$).

To develop solutions that involve an interface, we need boundary conditions at the interface. At this point we have incorporated (5.111) through (5.116) except for (5.115), which gives us one boundary condition as the interface position must be the same when approached from either side. This implies that $w/(n + k_x U)$ is continuous at the interface, so

$$\frac{w_2}{\left(n + k_x U_2 \right)} = \frac{w_1}{\left(n + k_x U_1 \right)}.$$

(5.121)

Here and in the following the subscript 1 or 2 indicates the value found by approaching the interface from the side designated by the subscript and the subscript s designates the value of a continuous quantity at the interface. To find another boundary condition, we can proceed as discussed in Sect. 5.1.2. We integrate (5.118) across the interface, then subtract (5.117) from itself across the interface so we can eliminate δp. The resulting boundary condition is

$$g k^2 \left(\frac{w}{n + k_x U} \right)_s \left(\rho_2 - \rho_1 \right) = \rho_2 (n + k_x U_2) \frac{\partial w_2}{\partial z} - \rho_1 (n + k_x U_1) \frac{\partial w_1}{\partial z}$$
$$+ k_x \left(-\rho_2 w_2 \frac{\partial U_2}{\partial z} + \rho_1 w_1 \frac{\partial U_1}{\partial z} \right).$$

(5.122)

We are now prepared to consider specific cases of interest.

5.6.3   Uniform Fluids with a Sharp Boundary

We consider the simplest case first, to determine the most general features of this process. Consider the two fluids to have uniform densities and uniform initial flow velocity, U, supposing that the value of these parameters may change only at an interface. Further assume the gravitational acceleration to be negligible for now. We then find from (5.120) that

$$-k^2 w + \frac{\partial^2 w}{\partial z^2} = 0, \tag{5.123}$$

so that we have solutions that are a linear combination of terms proportional to e^{kz} and e^{-kz}, with coefficients that must be set to match the geometric boundary conditions. We will consider the case with the simplest algebra, in which the boundary condition is that the disturbance become negligible at large distances, so that

$$w = A_2 e^{-kz} \quad \text{for} \quad z > 0 \quad \text{and} \quad w = A_1 e^{kz} \quad \text{for} \quad z < 0, \tag{5.124}$$

where (5.121) implies

$$A_2 = A_1 \frac{n + k_x U_2}{n + k_x U_1}. \tag{5.125}$$

It is convenient to work in a frame of reference corresponding to the average velocity of the two regions, because the velocity difference is what drives the instability and because one often knows the velocity difference in real applications. In this case $U_2 = \Delta U/2$ and $U_1 = -\Delta U/2$. With these results, (5.122) becomes

$$0 = \rho_2 (n + k_x U_2)^2 + \rho_1 (n + k_x U_1)^2, \tag{5.126}$$

which has the solution for n

$$n = -k_x \frac{A_n}{2} \Delta U \pm i k_x \Delta U \frac{\sqrt{\rho_1 \rho_2}}{(\rho_1 + \rho_2)}. \tag{5.127}$$

The real part of n is finite if $A_n \neq 0$, so that in such cases the wave propagates along the surface in this frame of reference. The negative imaginary part of n describes the exponential growth, given our specification of the modulations. For equal densities one finds the standard and very simple result that the exponential growth rate is $k_x \Delta U/2$. (The factor of 2 reflects the definition of ΔU, which varies among references.)

There are some things worth noticing about the result of (5.127). First, this process has no minimum wavenumber. Perturbations at all wavelengths are unstable (until the wavelength approaches the scale of the system, in which case this calculation becomes invalid). Shorter-wavelength perturbations have more-rapid growth rates. On the one hand, if the initial fluctuations present at a sharp interface corresponded to broadband noise, one would expect to see small-scale hair grow first, followed by the evolution of larger scales. On the other hand, one does not typically see this, which probably reflects the fact either that the initial fluctuations are larger at some specific wavelengths or that the shear layer is not indefinitely sharp. Finally, while the component of k along U determines the growth rate, there is no limitation on the y component of k. Fluctuations whose wave vector makes some angle with U grow freely, though more slowly than do fluctuations of the same wavelength for which k is parallel to U.

If we now allow for gravity but change no other assumptions, then (5.123) through (5.125) remain correct, but now (5.122) gives

$$- gk(\rho_2 - \rho_1) = \rho_2 (n + k_x U_2)^2 + \rho_1 (n + k_x U_1)^2 , \tag{5.128}$$

in which we have divided out a factor of $[kw/(n + k_x U)]_s$. The solution for n now becomes

$$n = -k_x \frac{A_n}{2} \Delta U \pm i \frac{\sqrt{k_x^2 \Delta U^2 \rho_1 \rho_2 + gk(\rho_2^2 - \rho_1^2)}}{(\rho_1 + \rho_2)} , \tag{5.129}$$

in which if the argument of the square root is positive then there is an unstable root. Our conventions imply that ρ_2 is from the "upper" region as defined relative to the gravitational acceleration. One sees that instability is always present if the upper density (ρ_2) is higher than the lower density (ρ_1). In this case, the KH and RT instabilities work together to produce larger growth. In contrast, when the lower density exceeds the upper density, this places a condition on the wavenumber for instability,

$$k > \frac{g(\rho_1^2 - \rho_2^2)}{\Delta U^2 \rho_1 \rho_2 \cos^2 \theta} . \tag{5.130}$$

Here θ is defined by $\cos \theta = k_x/k$. Thus, the RT dynamics at a given k opposes the instability growth due to KH, but at large enough k the KH instability dominates and one will see a positive growth rate. When gravity becomes large enough (and $\rho_1 > \rho_2$) the argument of the square root in (5.129) becomes negative. Then any modulations of the interface oscillate but do not grow.

5.6.4 Otherwise Uniform Fluids with a Distributed Shear Layer

The next level of complexity is to assume that the shear layer is not an instantaneous change in velocity, which at the microscopic level is unphysical in any case. Realistic problems with shear can become quite complex. As a first simple problem, we will suppose that the velocity shears but that the density changes abruptly at an interface. This may be relevant, for example, to the KH instability at boundaries between two fluids that are incompressible or that have only slow variations in density. The boundaries created by the RT instability or at bow shocks may be of this type. We mentioned above that the minimum width of the shear layer, in a system that has kinematic viscosity ν and has evolved for time t, is $\sqrt{\nu t}$.

We assume that the right-hand side of (5.120) is small, because the density is constant or slowly varying in the sense required. Further assuming the velocity profile to be given by $U_s(1 + z/L)$, we can observe that the terms involving $\partial U/\partial z$ in this equation cancel out, and that we are left once again with (5.123), solution (5.124), and condition (5.125) on the amplitudes. Here, because there are no boundaries on the flow, L is the distance over which U changes by U_s. Note that this assumption implies working in an inertial frame for which $U = 0$ at $z = -L$ and that $U_1 = U_2 = U_s$ so $w_1 = w_2 = w_s$. The boundary condition of (5.122) then becomes

$$A_n g k^2 + \frac{A_n k_x U_s}{L}(n + k_x U_s) + (n + k_x U_s)^2 k = 0, \qquad (5.131)$$

which can be solved for n to give

$$n = -k_x U_s \left(1 + \frac{A_n}{2kL}\right) \pm i\sqrt{gA_n k - \frac{A_n^2 k_x^2 U_s^2}{4k^2 L^2}}. \qquad (5.132)$$

We see that n has a real part, so these modes oscillate and propagate. For instability, the argument of the square root must be positive. In particular, an interface of this type (continuous linear U, discontinuous ρ) is always stable if the product gA_n is zero or negative. Another way to put this is that modes that perceive the region of velocity shear to be large are stabilized, and here the shear region is indefinitely large. In the next section, we will see that any boundaries, no matter how distant, will have the effect of destabilizing modes whose wavelength is of order the distance between boundaries. In contrast, when $gA_n > 0$ the interface is unstable so long as

$$kL > \frac{A_n \cos^2\theta U_s^2}{4Lg}. \qquad (5.133)$$

This is a curious result, as the shear in this context acts to stabilize long-wavelength modes but not short-wavelength ones. Note that the shear acts to stabilize modes with k along U, but not modes with k perpendicular to U. If k is aligned with U and the interface decelerates over some distance h so $g \sim U_s^2/h$, then modes with wavelength $\lambda > 8\pi L^2/(A_n h)$ are stable.

5.6.5 Uniform Fluids with a Transition Region

The notion of a sharp interface is an approximation, as molecular diffusion always will mix the materials from the two sides of the interface to some extent. This is particularly true in high-energy-density physics, where surface tension does not exist. Unfortunately, when U and ρ both vary, the solutions become much more complex. We will work out one standard case here, following Chandrasekhar (1961), and will leave other and more realistic cases to the specialized literature and to simulations. The geometry of this case is illustrated in Fig. 5.10. One has two layers of fluid, of density ρ_1 and ρ_2, separated by a transition region of width $2L$. The velocity of the layers $\pm U_s$, and with a linear velocity profile $U(z) = U_s z/L$ connecting them through the transition layer. The density of the transition layer is assumed to be constant and equal to $\rho_o = (\rho_1 + \rho_2)/2$. This corresponds to the approximation that the transition layer is fully mixed, presumably through the action of instabilities and turbulence. We will designate the lower and upper regions, using the subscripts 1 and 2, respectively, and the transition region, using the subscript o.

We now can apply our fundamental analysis to this problem. We have three distinct regions with two boundaries, and we will once again assume that the perturbations must vanish at large $|z|$. Under the assumptions stated, (5.120) once again reduces to (5.123), $\partial^2 w/\partial z^2 = k^2 w$, in all three regions. [The right-hand side

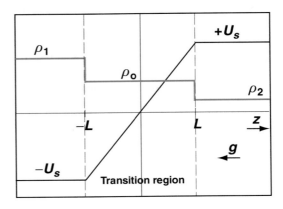

Fig. 5.10 Geometry for
Kelvin-Helmholtz instability
with a transition region
having uniform density

of (5.123) is zero and the other derivatives on the left-hand side cancel one another.] The solutions are $w(z) = Ae^{\pm kz}$, with coefficients chosen so to make w vanish appropriately in the outer regions. This gives

$$w = A_2 e^{-kz} \text{ for } z > L,$$

$$w = A_o e^{-kz} + B_o e^{kz} \text{ for } -L < z < L, \tag{5.134}$$

$$\text{and} \quad w = B_1 e^{kz} \text{ for } z < -L.$$

Thus, we have four unknown amplitudes. Our first boundary condition (5.121) tells us that $w(z)$ must be continuous at both boundaries, because $U(z)$ is continuous by assumption. This gives us two equations,

$$A_o e^{2kL} + B_o - B_1 = 0, \text{ and} \tag{5.135}$$

$$A_o + B_o e^{2kL} - A_2 = 0. \tag{5.136}$$

Our second boundary condition (5.122) applies at each interface, giving two more equations,

$$0 = B_1 e^{-kL} \left[\frac{gk^2(\rho_1 - \rho_o)}{(n - k_x U_s)} - k\rho_1(n - k_x U_s) \right]$$

$$+ \rho_o B_o e^{-kL} \left[nk - \left(k + \frac{1}{L} \right) k_x U_s \right] - \rho_o A_o e^{kL} \left[nk - \left(k - \frac{1}{L} \right) k_x U_s \right], \text{ and} \tag{5.137}$$

$$0 = A_2 e^{-kL} \left[\frac{gk^2(\rho_o - \rho_2)}{(n + k_x U_s)} - k\rho_2(n + k_x U_s) \right]$$

$$- \rho_o B_o e^{kL} \left[nk + \left(k - \frac{1}{L} \right) k_x U_s \right] + \rho_o A_o e^{-kL} \left[nk + \left(k + \frac{1}{L} \right) k_x U_s \right]. \tag{5.138}$$

As before, we can express (5.135) through (5.138) as the product of a matrix and the vector of coefficients (B_1, A_o, B_o, A_2). The determinant of this matrix then gives the dispersion relation, which is fourth order in n.

Certain quantities appear in natural combinations in these equations. It simplifies the resulting expressions to define $\eta = kL$, $v_g = n/(k_x U_s)$, and $\beta = (\rho_1 - \rho_2)/\rho_o$. In addition, one can define a Richardson number, J_r, which measures the ratio of buoyancy to inertia, as

$$J_r = \frac{gk\eta\beta}{2k_x^2 U_s^2}. \tag{5.139}$$

For further discussion of the Richardson number, see Chandrasekhar (1961). With these substitutions, and looking only at the modes with $k = k_x$, the dispersion relation becomes a fourth-order equation for the normalized growth rate v_g:

$$
\begin{aligned}
0 = {}& v_g^4 \left[4\eta^2 e^{4\eta} + \frac{\beta^2 \eta^2}{4} \left(1 - e^{4\eta} \right) \right] + v_g^3 \beta \eta \left(1 - e^{4\eta} \right) \\
& + v_g^2 \left[\left(1 - e^{4\eta} \right) \left(1 - \frac{\beta^2 \eta^2}{2} \right) + 4\eta e^{4\eta} (1 - J_r - 2\eta) \right] \\
& - v_g \left[\beta \eta \left(1 - e^{4\eta} \right) (1 + 2J_r) \right] - 2J_r \left(1 - e^{4\eta} + 2\eta e^{4\eta} \right) \\
& + \left(1 - e^{4\eta} \right) \left[\frac{\beta^2 \eta^2}{4} - 1 - J_r^2 \right] + 4e^{4\eta} (\eta^2 - \eta).
\end{aligned}
\tag{5.140}
$$

The solution of this equation, readily obtained from a computational mathematics program, has only one root with an imaginary part that is at times negative, corresponding to instability by our definition of the modes. It turns out, quite fortuitously, that the growth rate is nearly independent of β. As $|\beta|$ increases, the root develops a finite real part, implying that the growing solution would oscillate once the density difference becomes larger. The growth rates can be accurately obtained from (5.140) assuming β to be small. This equation then becomes quadratic in v_g^2, and the growth rate can be displayed as contours on a plot with $\eta(\sim kL)$ and J_r (buoyancy) as axes. Figure 5.11 shows this.

This plot supports a number of observations. First, if there is no gravity and thus no buoyancy, the KH instability is only unstable up to a maximum kL of about 0.65. This short-wavelength cutoff is the effect of the gradient in velocity. It says that

Fig. 5.11 Kelvin-Helmholtz at an interface with linear velocity shear and a density transition. Working from the interior outward, the contours show a growth rate, in units of $k_x U_s$, of -0.3, -0.1, and 0. Surface modulations outside of the $v_g = 0$ contour are damped

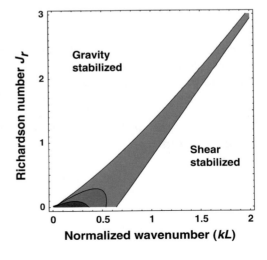

wavelengths shorter than about $10L$ are stabilized. As any given system evolves in time, L, being approximately $\sqrt{\nu t}$, increases, so that growth of the KH instability will be stopped at progressively longer wavelengths.

Second, for finite gravity, our new assumptions have introduced an additional feature that was not present in the absence of velocity shear. In the presence of a sharp interface and gravity, we found (5.130), which says approximately that $kL > J_r$ for instability. This corresponds to the left boundary in Fig. 5.11, and determines the longest wavelength that is unstable. The right boundary is the new feature, introduced by the presence of a velocity gradient that stabilizes the shortest wavelength modes. This is the impact of a limited region of velocity shear, allowing instability for waves to which the change in velocity seems abrupt. One sees that the combination of buoyancy and shear can produce a very narrow range of unstable wavelengths.

Finally, it is of interest to compare the results of these last two calculations. On the one hand, with a velocity gradient that extended over all space and a sharp density change at the interface, we found instability only in the presence of gravity and only when the interface is RT unstable ($A_n > 0$ with our definitions). In effect, the velocity gradient acted to stabilize all the modes we would normally describe as Kelvin–Helmholtz modes. In the presence of shear but only across a transition layer, we find instability whether or not there is gravity and for either direction of the density gradient, but only over a range of wavelengths longer than some multiple of the velocity scale length. In addition, the shear acts to stabilize the RT modes, producing a long-wavelength limit like that we saw in Sect. 5.6.3.

5.7 Shock Stability and the Richtmyer–Meshkov Process

The two instabilities we have now considered develop within some enduring state of a fluid system. Rayleigh–Taylor requires sustained acceleration, to create buoyancy, while Kelvin–Helmholtz depends on sustained shear, to create lift. In both cases, a calculation assuming incompressible fluctuations produces an excellent conceptual model for observed phenomena, even when the actual fluids are compressible. Even so, it was clear in Chap. 4 that real high-energy-density systems nearly always involve some combination of shock waves, rarefactions, and interfaces. This introduces an additional dynamical effect, which is the deposition of *vorticity* in the fluid. (Recall that the vorticity is $\nabla \times \boldsymbol{u}$.) This causes oscillations in shock waves to damp and structure to grow at interfaces. In our description here, we will summarize what has been learned from some complex calculations, and use some very simple models to provide some perspective. A recent paper by Wouchuk and Cobos-Campos (2017) provides an insightful discussion of these effects and a summary of the complex calculations and their results.

5.7.1 Shock Stability

We consider first what may happen to a shock wave that has lateral structure. We will describe this as a rippled shock and assume the rippling (in the z direction) to be proportional to $\cos(kx)$, corresponding to a two-dimensional ripple with no dependence on y. As usual, more-complex structures can be treated as a sum over such plane waves. We will also suppose that the rippling is of a long-enough wavelength that we can think about its effects using our analysis of oblique shocks in Sect. 4.1.5. A rippled shock can be produced by pushing on a fluid with a rippled piston, or by allowing a planar shock to interact with a rippled interface. Here we focus only on the shock; later we will consider the interface. We will assume the ripple to be of initial amplitude a_o, and to be small (so that $a_o k \ll 1$). For our specific calculations, we will also assume the shock wave to be strong, in order to simplify the mathematics.

Figure 5.12 illustrates the deflection of the flow that occurs at a rippled shock. Here a fluid moving in the $-z$ direction approaches a shock whose z location is given (in the shock frame) by $a = a_o \cos(kx)$. This results in a deflection of the flow away from the shock normal. The shock normal is indicated in the figure by arrows attached to the shock. The deflection has three consequences. First, material flows toward the lagging section of the shock; the horizontal arrows in the figure indicate this component of the flow. This causes the shock wave to oscillate. Second, vorticity is deposited and sound waves are driven in the shocked material. This both speeds up the oscillations and leads them to damp. Third, the shock transition is affected by the change in deflection as a function of position. We consider these in turn.

The shock normal vector is shown in the figure and given to first order in $a_o k$ by

$$\boldsymbol{n} = -\hat{x}(a_o k)\sin(kx) - \hat{z}. \tag{5.141}$$

$$\tan\phi_1 \approx \phi_1 = (a_o k)\sin(kx), \tag{5.142}$$

from which the small-angle formula for the deflection ψ of the flow away from the $-z$ direction, $\psi = \phi_1(\rho_2/\rho_1 - 1)$ gives $\psi = 2\phi_1/(\gamma - 1)$ for strong shocks. As is

Fig. 5.12 Sketch showing the horizontal flow produced by a rippled shock

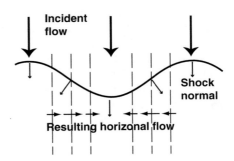

discussed in Sect. 4.1.5, the local transverse component of the flow is unchanged by the shock, while the local normal component of the flow is reduced.

To first order in $a_o k$, the immediate postshock fluid velocity is given by

$$\boldsymbol{u}_2 = \hat{x} u_s(a_o k)\sin(kx) - \hat{z} u_s \left(\frac{\gamma - 1}{\gamma + 1} \right), \tag{5.143}$$

in which the shock speed is u_s and the postshock velocity, in the shock frame, is $\boldsymbol{u}_2$. The lateral (x) component of this, u_{2x}, is proportional to $a_o k \sin(kx)$, illustrated by the horizontal arrows in Fig. 5.12. We can estimate the consequences of the lateral flow by the following simple estimate. We make the oversimplified assumption that the shape of the shock remains sinusoidal and that the amplitude is reduced as material flows laterally. The rate of mass flow per unit length from the leading to the lagging sections, on each side of the minimum of $a(x)$, designated as z_{min}, is approximately the wave amplitude times the transverse mass flux at the mean interface position, or $\rho a_o u_s(a_o k)$. This has units of mass per unit length per unit time. The mass per unit length between z_{min} and $z = 0$, within half of the lagging region of the shock, is $\rho a_o/k$. If this flow of material were the only factor, one would have

$$\frac{d}{dt}a_o \sim -(a_o k)^2 u_s \sim \frac{a_o}{\tau_{fill}}, \tag{5.144}$$

defining a characteristic time for the oscillation to be filled in of

$$\tau_{fill} = \frac{1}{(a_{oo}k)(ku_s t)} \quad \text{so} \quad \frac{u_s \tau_{fill}}{\lambda} = \frac{1}{2\pi(a_o k)}, \tag{5.145}$$

in which the latter expression gives the distance the shock would travel, in wavelengths, before the mass flux could have filled in the trough. This turns out to be an over-estimate, and fails to capture the fact that the shock oscillates as it damps. Both effects are a result of the second factor we mentioned above, discussed next.

The horizontal flow illustrated in Fig. 5.12 produces pressure extrema, as illustrated in Fig. 5.13. These accelerate the flattening of the shock wave and cause it to overshoot, producing oscillations. The resulting pattern of pressure extrema corresponds to a pattern of standing acoustic waves. The figure makes no attempt to show the subsequent fluctuations of the extrema. The variation in lateral flow produced as the shock wave oscillates while moving forward has the effect of continuously depositing vorticity in the fluid. The zones between the pressure extrema each contain vortical flow around their center. The figure shows these zones as symmetric, but they may not be. The net consequence of these dynamics is that the amplitude of the modulations of the shock wave decays rapidly as the shock propagates. Figure 5.14 shows the decay of a shock wave whose initial

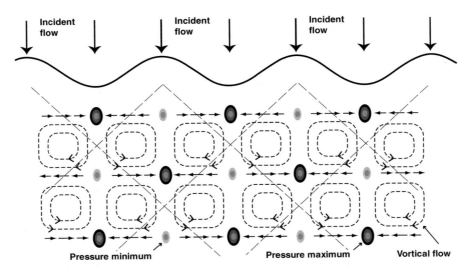

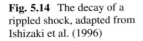

Fig. 5.13 The lateral flow produced by a rippled, oscillating shock creates both vortical flow and pressure perturbations behind the shock wave. Here the upper part of the figure shows the phasing of the shock wave corresponding to the upper row of arrows in the lower part. Figure 1 of Wouchuk and Cobos-Campos (2017) shows a more detailed depiction of the vortical flow, from simulations

Fig. 5.14 The decay of a rippled shock, adapted from Ishizaki et al. (1996)

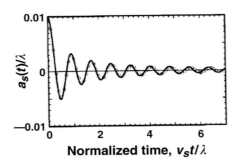

amplitude is $a_o = 0.01\lambda$, from simulations. Note that the decay is rapid, with the amplitude decreasing an order of magnitude after the shock has propagated about six wavelengths.

One often describes this behavior by saying that shock waves *anneal*. They smooth themselves out if allowed to propagate far enough. In some experiments, this can be exploited. If one uses the phase plates discussed in Chap. 9, then structure is initially produced at the scale of the speckles, typically of order $10\,\mu$m. If the shock wave propagates more than about $100\,\mu$m before being used to some experimental purpose, then it will have become substantially smoothed.

The third effect mentioned above is the change in the shock properties due to the ripple. At any given location, the shocked material is diverted toward the unshocked material, altering to some degree the conditions of the adjacent element of material as it is shocked. For ordinary materials and polytropic gases this is a small effect, but for a sufficiently pathological material it can lead the shock to become unstable. Landau and Lifshitz (1987) discuss the necessary conditions.

A qualitative summary is then as follows. Shocks are ordinarily stable. If they become rippled somehow, the ripple damps out as the shock propagates. During that period, the rippled shock wave deposits vorticity in the fluid it traverses. The shock oscillates as it damps (sometimes this is described as *superstable* behavior). The ripple damps and becomes negligible as the shock propagates.

5.7.2   Interaction of Shocks with Rippled Interfaces

In high-energy-density experiments, one is often concerned with the interaction of a shock with an interface and with the structure that may be introduced by that interaction. Here we consider first what structures result from the interaction of a shock with a rippled interface. For modeling, we will assume that the analysis of Sect. 4.6 can be applied point by point along the ripple. However, our conclusions will be more general, as these depend mainly on the relative speed of the various waves in the problem. At the interface of interest, the density may increase or the density may decrease.

Figure 5.15 shows what usually happens when the density increases at a rippled interface. We anticipate that there will be a transmitted shock and a reflected shock, as discussed in Chap. 4. As the shock crosses the interface, the reflected shock moves backward with a faster velocity than the incoming shock. As a result, the phase of the ripple of this wave is the same as that of the interface ripple while the initial ripple amplitude on this shock is larger than the ripple amplitude on the interface; the ratio equals the ratio of reflected-shock velocity to interface velocity. For any ordinary equation of state, the postshock velocity of the interface and the transmitted shock velocity are each smaller than that of the initial shock wave. A first consequence is that the ripple on each of them remains in phase with the ripple on the interface. A

Fig. 5.15 Behavior when a shock reaches a rippled interface where the density increases

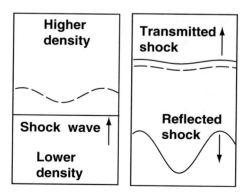

second consequence is that the amplitude of the modulation decreases, in this case in proportion to the ratio of postshock interface velocity to incoming shock velocity. The velocity of the transmitted shock will typically be somewhat larger than that of the interface. As a result, the initial modulations of the transmitted shock will be somewhat larger than those of the interface.

There are some differences in the response when a shock reaches an interface where the density decreases. Figure 5.16 illustrates this case. We will assume that the conditions are such that there will be a transmitted shock and a reflected rarefaction wave; the specialized exceptions of Sect. 4.6.1 are straightforward and we ignore those others that might correspond to a very unusual equation of state. The reflected rarefaction wave again moves faster than the shock wave, though not by much; it moves at the sound speed of the initially shocked matter. As a result, the ripple on the reflected rarefaction wave remains in phase with the ripple on the interface. Velikovich and Phillips (1996) show that such reflected rarefaction waves are weakly unstable. The perturbation amplitude of the "trailing edge" (the one near the interface) grows linearly with time. There are standing but damped sound waves, emitted downstream and propagating in the rarefaction fan toward the interface.

In contrast, the transmitted shock nearly always moves faster than the incident shock. As a result, the ripple of this shock is typically inverted in phase relative to the ripple of the interface. This is equivalent to saying that $\alpha > 0$ in Sect. 4.6.6. However, the postshock behavior of the interface depends on whether its postshock velocity is larger or smaller than the incident shock velocity. The case $\chi > 0$ corresponds to a larger postshock velocity, which will occur for strong shocks if the density ratio is large enough. As the density ratio becomes smaller, and depending on the EOS, eventually one will have $\chi < 0$, and the interface will not be inverted as the shock passes.

These dynamics of the rarefaction wave have interesting consequences for experiments that view such behavior from the side (from above the page here). It is common in such experiments to use a radiographic diagnostic whose X-rays are preferentially absorbed by the material on one side of the interface. If the downstream side is diagnosed in this way, then the signal decreases exponentially as

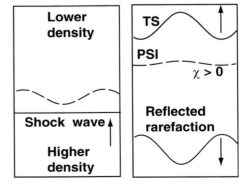

Fig. 5.16 Behavior when a shock reaches a rippled interface where the density decreases. The transmitted shock is labeled TS and the postshock interface is labeled PSI. Note that for nearly all cases of interest, the interface modulations invert as the shock passes

the areal density (mass/area) increases along any given line of sight. In consequence, the diagnostic may be more sensitive to the higher-density material near the head of the rarefaction wave (if the transmission is relatively high) or to the lower-density material near the interface (if the diagnostic X-rays are strongly absorbed). Thus, for large density ratios (so $\chi > 0$) the structure that the diagnostic detects may be in phase or out of phase with the modulations at the interface, or may change phase within the image.

5.7.3 Postshock Evolution of the Interface; Richtmyer–Meshkov Process

When a planar shock encounters a rippled interface, the rippled shocks produced in consequence will then deposit vorticity within the subsequently shocked fluid, by the mechanism discussed above. Imagine inserting a diagnostic tracer layer, transverse to the flow direction, into the fluid of Fig. 5.13. So long as the shock does not happen to be flat when it meets the layer, the post-shock vortical flow will proceed to create structure in the tracer layer. This is essentially what happens to a rippled interface when it is shocked by a shock wave, or to any interface crossed by a shock that is rippled.

At this point we have seen that modulations in shock waves are typically damped, so modulations introduced by an interface on the shock waves will die away in time. We have also seen that the heads of rarefaction waves are stable, so modulations introduced by an interface will not grow further. The remaining issue involved in understanding such systems is the postshock evolution of the interface. It turns out that structure on the interface grows in time after the shock passes. The origin of this is easily seen in Fig. 5.17a. This figure is centered on a trough in the interface modulation. The behavior of the shock waves is clear from our discussion above. As the shock crosses the interface, the postshock velocity slows. This causes a deflection of the flow away from the normal to the interface, toward the trough. As it moves, the transmitted shock wave continues at first to deflect material toward the trough. This corresponds to $\alpha < 0$ in the analysis of Sect. 4.6.6. As the transmitted shock propagates further, it begins to oscillate, depositing vorticity in the fluid (Fig. 5.17b). Similarly, the reflected shock wave initially deflects the flow away from the trough, and also deposits vorticity as it inverts. Thus, one can see that the postshock flow of material on both sides of the interface acts to deepen the valleys and raise the peaks of the initial modulations.

This process is known as the *Richtmyer–Meshkov instability*. We will designate it by the initials RM. Independent of the history and semantics discussed in the gray box, the best intuitive sense of the RM process can be found by thinking of the flow that will develop following the initial conditions created by the shock. For the case of Fig. 5.17, we can use this intuition and the analysis of Sect. 4.6.6 to develop a semiquantitative description of the growth of modulations, as follows.

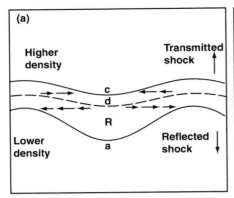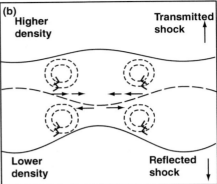

Fig. 5.17 (**a**) Lateral velocities after a shock interacts with an interface where the density increases. This schematic depicts the horizontal velocity components just after shock passage, and does not include any post-shock evolution. (**b**) Schematic of structure after shocks have each inverted and modulation at interface has grown slightly. Additional vortical flows near edges are not shown. In later evolution, the modulation of the interface will continue to grow while the shocks oscillate and damp. The letters show the regions as defined in Sect. 4.6.6

What Is an Instability?

The label "Richtmyer-Meshkov *instability*", is firmly entrenched in the literature, but strikes the author of this book as problematic. We describe systems as *unstable* if small perturbations can lead to large dynamical responses, the classic example being a ball at rest atop a parabolic hill. But in common use we do not describe the motion of the ball as an instability. In any event, RM has little in common with the ball atop the hill. Instead, it is more like analogous to the response of some hockey pucks connected by strings and sitting on ice, once one of more of them has been given an impulse. A common definition of *an instability*, in physics and in circuits, is "a process through which the rate of increase of the magnitude of some physical quantity increases in time." This is typified by the equation $df/dt = \gamma f$, in which γ is the (exponential) growth rate of the quantity f. Instabilities satisfying such a definition inherently involve some feedback mechanism. In the case of Rayleigh–Taylor, for example, this mechanism is the increase in the net buoyant force resulting from increased interpenetration of the two materials. The RM process has no such feedback. A parallel definition of an instability is "a process through which modulations of an initially steady system grow and cause it to reach a state of lower potential energy." RT likewise satisfies this definition while RM does not. It seems to this author that it would be too big a

(continued)

stretch to describe an instability as "a process through which the magnitude of some physical quantity increases in time," which is all that one can say about RM. Many other phenomena, never described as instabilities, would satisfy this definition. The main point is that, unlike every other process described in this book as an instability, one will search in vain for a feedback mechanism that increases the rate of RM growth with time.

There is also a further conceptual difficulty associated with the RM process. The RM instability is often described as the impulsive limit of the RT instability, from the point of view that it corresponds to the limit of RT as the variation of the acceleration in time approaches the delta function corresponding to the shock. This description originates with the original paper of Richtmyer (1960), but he and others have recognized problems with it. The evolution of the structure occurs after the shock passes, and thus is not the limiting case of growth that occurs during acceleration. Correspondingly, one does not do theory of RM by taking a limit of RT theory. In addition, as Velikovich (1996) described, one can produce RM, at least in principle, in a system with two rarefaction waves and no initial acceleration of the interface.

The four regions seen in Fig. 5.17a correspond from bottom to top to regions a, R, d, and c of Fig. 4.33 and Sect. 4.6.6. The small-angle limits of (4.139) and (4.142) and give

$$\chi = \frac{2\alpha - (\gamma - 1)\beta}{\gamma + 1}, \text{ and} \tag{5.146}$$

$$\chi = \frac{2\eta(\gamma + 1) - \beta(\gamma - 1)^2}{(\gamma + 1)^2}, \tag{5.147}$$

respectively. Recall that β is the angle between the initial shock normal and the local interface normal. Setting the pressures in the two postshock regions equal, in the small-angle limit, gives

$$\frac{\rho_c}{\rho_b}\left(1 + \frac{\alpha}{\beta}\right)^2 = \frac{\gamma + 1}{\gamma - 1}\left(\frac{\gamma - 1}{\gamma + 1} + \frac{\eta}{\beta}\right)^2, \tag{5.148}$$

These three equations can be solved to find the ratio of α, η, and χ to β as a function of γ and the density ratio at the interface, ρ_c/ρ_b. Also note that in the geometry of (5.141) and for small β one has $\beta = (a_o k)\sin(kx)$. One ends up, for any given value of γ, with a figure like Fig. 5.18.

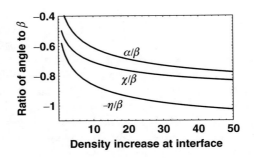

Fig. 5.18 The dependence of the postshock angles of deflection on the density ratio at an oblique interface where the density increases, for strong shocks. The angles of the transmitted shock, the interface, and the reflected shock are α, χ, and η, respectively, with η defined in the opposite direction in Sect. 4.6.6

Fig. 5.19 The lateral flow velocities produced by a shock at an interface where the density increases by a factor of 10. This circumstance produces shear flow at the interface

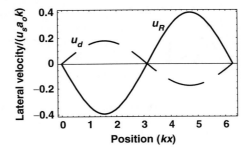

Equations (4.137) and (4.140), after transformation back into the lab frame, give the lateral deviation of the flow as the x-component of the velocity vectors $\boldsymbol{u}_{d\text{lab}}$ and $\boldsymbol{u}_{R\text{lab}}$. In the small-angle limit, these are

$$\boldsymbol{u}_{d\text{lab}} = \frac{-2\beta}{(\gamma+1)}u_s\left[\frac{\alpha}{\beta}\left(1+\frac{\alpha}{\beta}\right)\right]\hat{x} + \frac{2}{(\gamma+1)}u_s\left[1+\frac{\alpha}{\beta}\right]\hat{y} \text{ and} \tag{5.149}$$

$$\boldsymbol{u}_{R\text{lab}} = \frac{2\beta}{(\gamma+1)^2}u_s\left[\frac{\eta}{\beta}\left((\gamma-1)+\frac{\eta}{\beta}(\gamma+1)\right)\right]\hat{x}$$
$$+ \frac{u_s}{(\gamma+1)^2}\left[4\gamma - 2\frac{\eta}{\beta}(\gamma+1)\right]\hat{y}, \tag{5.150}$$

respectively. By substituting the solution to (5.146)–(5.148) into these two equations, one can find the deviations in velocity introduced by the shocks. For any given value of the density ratio ρ_c/ρ_b, one can plot the lateral deviation as a function of distance as Fig. 5.19 shows.

Referring again to Fig. 5.17, one can see that there is a significant difference between the behavior at the interface and the behavior at the shock. The flow at the interface acts to increase (rather than to decrease) the size of the perturbation. We can make an approximate calculation, similar to the one we did for the rippled shock, to estimate how rapidly the ripple amplitude will increase. An important difference is that the lateral velocity is set by the initial amplitude and does not evolve further as the ripple changes, except perhaps due to the effects of the sound waves emanating from the shock waves, which are not accounted for in the present estimate. The lateral velocity produced by the reflected shock tends to be the larger of the two, so we will assume that it is responsible for the flow. Using this velocity to be u_{perp}, taking $\eta/\beta \sim 0.9$ as a typical value, and recalling that $\beta = a_{ps}k$ for small angles, for initial postshock ripple amplitude a_{ps}, one can show

$$u_\perp \approx \frac{3.4\gamma}{(\gamma + 1)^2} k u_s a_{ps} = \frac{1.7\gamma}{(\gamma + 1)} k u_{ps} a_{ps}, \tag{5.151}$$

in which the postshock velocity of the interface in the lab frame is u_{ps}

From the point of view that this inward flow of material from each side produces a corresponding increase in the full amplitude ($2a_o$) of the ripple, we then estimate

$$\frac{d}{dt} a_o = \frac{1.7\gamma}{(\gamma + 1)} k u_{ps} a_{ps}. \tag{5.152}$$

Key qualitative features of this estimate are that the interface ripple grows linearly in time, and that the rate of growth is proportional to the initial normalized amplitude, $a_{ps}k$ and to the post-shock fluid velocity. These features are also present in the widely used formula due to Richtmyer,

$$\frac{d}{dt} a_o = kA^* u_{ps} a_{ps}, \tag{5.153}$$

which also involves the postshock Atwood number at the interface, A^*. We discuss the more general story below, after a look at the other type of interface.

Now we turn to the behavior when a shock reaches a rippled interface where the density decreases. This can be analyzed in a similar way, using the equations for a rarefaction from Sect. 4.6.6. Figure 5.20 illustrates the qualitative behavior, for the typical case of a large enough density decrease that $\chi > 0$. One can see that here again the amplitude of the ripple at the interface will grow with time. The qualitative behavior at the interface is the same in this case as in the previous one. A small ripple grows linearly with time, with the rate of increase of the amplitude being proportional to the initial normalized amplitude and the shock velocity. After this, the transmitted shock continues to deposit vorticity in the newly shocked fluid. In contrast, the rarefaction wave does not deposit additional vorticity in the fluid it affects, but does produce lateral flow that reinforces the growth of the modulations at the interface while producing a standing acoustic structure.

If the density decrease is smaller, so that $\chi < 0$, then the interface at first retains the phase of the initial ripple. The flows then cause the interface to invert before

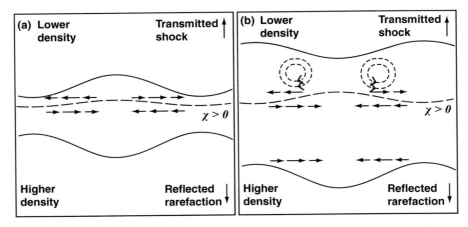

Fig. 5.20 Lateral velocities and vortical flow after a shock interacts with an interface where the density decreases. (**a**) Lateral velocities after a shock interacts with an interface where the density decreases. This schematic depicts the horizontal velocity components just after shock passage, and does not include any post-shock evolution. (**b**) Schematic of structure after shock has inverted and modulation at interface has grown slightly. Additional vortical flows near edges are not shown. The rarefaction does not oscillate. In later evolution, the modulation of the interface will continue to grow while the shock oscillates and damps

the ripples grow larger. For this case, there is a standard theoretical estimate due to Meyer and Blewett, which involves the average of a_{oo} and a_{ps}. It is

$$\frac{\mathrm{d}}{\mathrm{d}t}a_o = kA^* u_{ps}\frac{1}{2}(a_{ps} + a_{oo}), \tag{5.154}$$

in which one should note that $A^* < 0$ for this case so the modulation inverts.

An interesting aspect of RM flows driven by shocks is that the RM structures can overtake the shock waves. When this happens, the increased drag prevents the structures from penetrating the shock front. They have been observed to distort it. The condition for this is straightforward. Using the transmitted shock (of speed u_{TS} as an example), it separates from the interface at $(\gamma-1)u_{TS}/(\gamma+1)$, so the condition for the RM structures to overtake the forward shock is approximately

$$\frac{\gamma-1}{\gamma+1}u_{TS} < \frac{\mathrm{d}}{\mathrm{d}t}a_o = (ka_o)\frac{\gamma+1}{2}u_{TS} \text{ or } ka_o > \frac{2(\gamma-1)}{(\gamma+1)^2} \sim 0.15. \tag{5.155}$$

Such structures are in the weak nonlinear regime, but initial conditions with large enough amplitude and few modes do produce structures that move with the shock wave and distort it.

In the first two decades of the twenty-first century, a more complete understanding of flows of this general type emerged, along with methods for treating them. General, RM-like flows include any flows established so that they contain vortical flows within the fluid near an interface. The may involve rippled shock waves, rippled interfaces, or other special initial conditions. The long-term, asymptotic growth of the amplitude of modulations in such flows, while they stay in the linear regime, is always described by

$$a_o = a_\infty + v_\infty t, \tag{5.156}$$

in which a_∞ and v_∞ are long-term distance and velocity coefficients, respectively. The methods for calculating a_∞ and v_∞, for arbitrary initial circumstances, are now known (Wouchuk and Cobos-Campos 2017), though complex. In general, a_∞ is not a_{ps} and v_∞ is not $(ka_o)u_{ps}$, though these quantities do set the approximate scale. In addition, in at least some cases there is an initial, brief period of slower increase before the system settles into the growth described by (5.156).

Finally, the RM structures do exhibit nonlinear behavior as they grow. Mode coupling exists for the same reason it does for RT and KH—the interface must remain continuous as it evolves. This has the effect of driving up the amplitude of smaller-k structures over time, creating a behavior like that of bubble merger for RT. Mode coupling has been observed. One can also note that the rate of growth of any structure slows once it reaches nonlinear amplitude, which allows smaller-k structures to overtake larger-k structures as time goes on. This behavior is often described as "bubble-merger," but this too is a misnomer since no buoyancy force is at work; calling the lower-density structures "bubbles" is strictly a convention.

5.8 Hydrodynamic Turbulence

We often see phenomena that one might describe as turbulent. This is particularly true when two distinctly observable fluids, such as clouds and air or cream and coffee, mix. But flow in a single fluid also can become in some sense turbulent, as does the airflow behind an airplane wing or a racecar. There are a number of possible definitions of *turbulence*, and one finds that the word has distinctly different meaning in different areas of plasma physics and hydrodynamics. As a result, when reading a wide range of literature one should be somewhat wary of this term. A fairly general definition of turbulence is "the presence of structures having a range of spatial scales that are smaller than the spatial scales of the motions that provide the energy source producing the structures." This may not satisfy the extreme hydrodynamicist, who may insist that a system to be turbulent must have evolved to a state that is independent of its initial conditions. Whether such a state is practically realizable is not so clear. Here we will not trouble ourselves further with definitions. Rather, we will examine the properties of hydrodynamic systems that lead to the presence of structure over a range of small spatial scales. The book by Tennekes and Lumley (1972) provides an excellent introduction to hydrodynamic

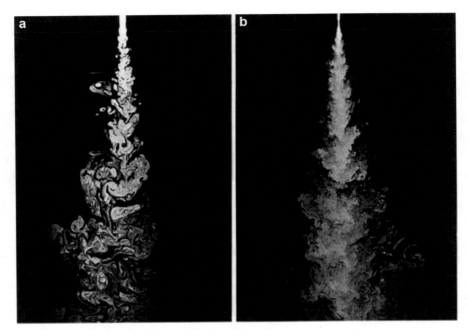

Fig. 5.21 Images of a slice through round turbulent jets in liquids, illuminated by a laser (Dimotakis 2005). (**a**) $Re \approx 2500$. (**b**) $Re \approx 10{,}000$

turbulence. They emphasize that such turbulence is a property of fluid flows and not of the underlying fluid itself.

The basic notion behind descriptions of turbulence is that energy is introduced to a system by some process, such as the RT or KH instability, that this enables processes to occur which produce much smaller-scale fluctuations, and that these fluctuations eventually lose their energy by viscous dissipation. Because vorticity once generated spreads by viscous diffusion and is removed only through viscous heating, swirling patterns of motion generally characterize turbulent systems. Indeed, the presence of varying patterns of vorticity on a range of spatial scales is considered to be an essential property of turbulent hydrodynamic flows.

A simple example of turbulent flows are the jets shown in Fig. 5.21. The Reynolds number Re increases from left to right in the figure. In all similar cases, the jet first produces the KH instability. This sets the stage for further instabilities and for the development of smaller scale structures. At low Re, the flow remains dominated by large-scale structures. As Re increases, the flow develops finer-scale structures and the distribution of these structures becomes more uniform.

As we shall see, turbulence is typically a property of flows with large Re. This may seem paradoxical at first. In Chap. 2 we found that Re is the ratio of the convective momentum transport to viscous momentum transport. We argued that Re is nearly always large in high-energy-density flows, and correspondingly that the Euler equations are typically a good basis for analysis. In detail, however, this argument works only for phenomena whose spatial scale is not too small. One can construct a Reynolds number from any sensible length and velocity scales so that different aspects of a system can have different Reynolds numbers. If we focus on small-enough spatial scales, the corresponding phenomena do experience strong dissipation and cannot be described by the Euler equations. This is what makes it possible for a turbulent flow to dissipate energy.

Before we proceed with some discussion of incompressible hydrodynamic turbulence, it is worthwhile to consider how this might apply to plasma systems. In a sufficiently collisional plasma, the fluid motion is hydrodynamic and collisional damping dominates the dissipation of energy. However, as the plasma becomes less collisional, collective effects begin to occur and the compressive fluctuations in the plasma begin to produce significant electric fields. These electric fields accelerate particles, providing a source of energy dissipation that is distinct from viscous effects. At this writing, the competition between these sources of dissipation is not well understood. This competition could alter the structure of turbulence in plasmas as compared to that in purely hydrodynamic fluids.

Returning to the point of view that turbulent flows are dominated by rotating motions that we can call *vortices*, we can idealize these motions, using the (oversimplified) model of rotating toroids. These donut-shaped structures rotate about the axis of the donut. They can have any aspect ratio, being thin rings, fat rings, or elongated and nearly cylindrical structures. The rotation is essential, though, as this is what makes $\nabla \times \boldsymbol{u}$ nonzero so that there is vorticity. Such vortices can have a range of sizes but the smallest possible vortex is one that is damped by viscosity in of order one rotation. If we use w to represent the rotational velocity and λ to represent the diameter of the vortex, then the rotational timescale is λ/w while the timescale for viscous damping is λ^2/ν, where ν is again the kinematic viscosity. Setting these timescales equal gives $w\lambda/\nu \sim 1$ for the smallest vortex. Thus, the Reynolds number constructed from the characteristic scales of the smallest vortex is of order unity. The reader may recall from Chap. 2 that the typical Reynolds number describing high-energy-density flows is at least several orders of magnitude larger than 1. The consequence is that the smallest vortices are some orders of magnitude smaller than the characteristic scale of the entire system.

A next step in this discussion is to consider the overall rate at which damping must dissipate energy. The largest vortices produced in the system are known as the *eddies*. The eddies typically span the turbulent zone. In simple cases such as KH or RT they are created through the evolution of the large structures generated by these instabilities. We will designate the characteristic speed of the material in the eddy as w_e and the characteristic diameter of the eddy as ℓ. An observed property of turbulence, and an assumption in traditional turbulence theory, is that these large eddies dissipate their energy on a timescale of order 1 circulation time. Their specific energy is of order w_e^2, so the specific energy dissipation rate (a power per unit mass) is of order w_e^3/ℓ. We will designate this turbulent dissipated power by P_t. However, the inherent viscous damping of these structures is small. To see this, one can note that their viscous timescale is ℓ^2/ν, so their viscous damping rate is $w_e^2\nu/\ell^2$. Thus, the ratio of the viscous damping rate to P_t is $v/(w_e\ell) = 1/Re$. Taken together with the previous paragraph, the implication is that dynamical processes must create small-scale structures to dissipate the energy deposited in the eddies by the global processes in the system (such as KH or RT).

The order-of-magnitude size of the smallest structures is one of the *Kolmogorov scales*. These are the length scale, η_k, the time scale, τ_k, and the velocity scale, u_k, that can be constructed from the specific energy dissipation rate and the viscosity. One has

$$\eta_k = (v^3/P_t)^{1/4}, \quad \tau_k = (v/P_t)^{1/2}, \quad \text{and} \quad u_k = (vP_t)^{1/4}. \tag{5.157}$$

To see how small these scales are physically, one can substitute for P_t and obtain results in terms of the Reynolds number corresponding to the eddies, $w_e\ell/v$. This gives

$$\eta_k = \ell/Re^{3/4}, \quad \tau_k = (\ell/w_e)/Re^{1/2}, \quad \text{and} \quad u_k = w_e/Re^{1/4}. \tag{5.158}$$

Thus, for a Reynolds number of 10^5, the size of the smallest vortex will be of order 6000 times smaller than the size of the largest eddy. Note that the Reynolds number corresponding to the Kolmogorov scales satisfies the condition we developed above, having $\eta_k u_k/v = 1$.

We will explore how structures can form on such scales below. To prepare for this, we first will compare them with some other characteristic dimensions and then discuss the dynamics of the fluid in more detail. An eddy may initially form with sharp edges, as during the roll-up produced by Kelvin–Helmholtz, as at the spike tips in Fig. 5.1 and at the shear layer in Fig. 5.7. As the eddy evolves, viscous diffusion smoothes the edge of the shear layer (see Sect. 5.6.4). This diffusion produces a laminar boundary layer within which there is a finite transverse gradient of the velocity. The scale length of this boundary layer is $\sqrt{vt}$. On the timescale of the large eddy, ℓ/w_e, this boundary layer scale length is thus $\ell/\sqrt{Re}$. Comparing this with (5.158), one sees that the Kolmogorov scale length where the dissipation occurs is smaller than the boundary-layer scale length, and that the difference between them increases as Re increases. Externally driven instabilities do not readily occur on the small scales that exist within such a boundary layer, where there is a continuous gradient in flow velocity. KH, for example, is stabilized by this gradient. Thus, the fluctuations within the boundary layer should evolve through local fluid dynamics until they dissipate. The full thickness of the boundary layer, throughout which the velocity gradients may limit the instabilities, is a few times $\ell/\sqrt{Re}$.

To be able to go further in our description, we need to work with the fluid equations. The relevant equations are (2.27) for momentum and the mechanical-energy equation that can be constructed from it by taking the dot product with $\boldsymbol{u}$, keeping the terms involving viscosity but dropping all the terms involving radiation or other forces. One also assumes incompressibility and for simplicity assumes constant ρ, constant v, and that the second coefficient of viscosity is 0. Then one has

$$\frac{\partial \boldsymbol{u}}{\partial t} + \boldsymbol{u} \cdot \nabla \boldsymbol{u} = \frac{-1}{\rho} \nabla p + 2\nu \nabla \cdot \underline{\boldsymbol{s}}, \text{ and} \tag{5.159}$$

$$\frac{1}{2}\frac{\partial u^2}{\partial t} + \boldsymbol{u} \cdot \nabla \left(\frac{u^2}{2}\right) = \frac{-1}{\rho} \boldsymbol{u} \cdot \nabla p + 2\nu \nabla \cdot (\boldsymbol{u} \cdot \underline{\boldsymbol{s}}) - 2\nu(\underline{\boldsymbol{s}} \cdot \cdot \underline{\boldsymbol{s}}). \tag{5.160}$$

in which one has used the incompressibility condition to simplify $\underline{\sigma}_\nu$ from (2.37), and defined the *strain rate tensor* as $\underline{\boldsymbol{s}}$, given by

$$\underline{\boldsymbol{s}} = \frac{1}{2}\left(\nabla \boldsymbol{u} + (\nabla \boldsymbol{u})^T\right). \tag{5.161}$$

With the elements of $\underline{\boldsymbol{s}}$ given as s_{ij}, the expression $\underline{\boldsymbol{s}} \cdot \cdot \underline{\boldsymbol{s}}$ is the sum over both indices of $s_{ij}s_{ij}$.

Since fluid turbulence develops within a fluid flow, it is useful to analyze these equations as the sum of terms describing the mean flow and terms describing the (turbulent) fluctuations. Doing this is known as the *Reynolds decomposition* of the fluid equations. We take $\boldsymbol{u} = \boldsymbol{U} + \boldsymbol{w}, p = P + \delta p$, and $\underline{\boldsymbol{s}} = \underline{\boldsymbol{S}} + \underline{\delta\boldsymbol{s}}$, in which the first, uppercase quantity is the mean value and the second term is the fluctuating term. We substitute these definitions into (5.159) and (5.160) and average over a time long compared with the dissipation time for any specific eddy. The fluctuating terms average to zero individually. However, products of the fluctuating quantities do not, in general, average to 0, but instead have values depending on the degree of correlation of these quantities. Indicating such an average by an overbar, and assuming the overall system to be in steady state for simplicity, we obtain equations for the mean flow and for the turbulent fluctuations. The equations for the momentum and energy of the mean flow are

$$\rho(\boldsymbol{U} \cdot \nabla)\boldsymbol{U} = \frac{-1}{\rho}\nabla P + 2\nu\nabla \cdot \underline{\boldsymbol{S}} - \nabla \cdot (\rho\overline{\boldsymbol{w}\boldsymbol{w}}) \quad \text{and} \tag{5.162}$$

$$\boldsymbol{U} \cdot \nabla \left(\frac{U^2}{2}\right) = \nabla \cdot \left(\frac{-P}{\rho}\boldsymbol{U} + 2\nu\boldsymbol{U} \cdot \underline{\boldsymbol{S}} - \overline{\boldsymbol{w}\boldsymbol{w}} \cdot \boldsymbol{U}\right), \\ -2\nu(\underline{\boldsymbol{S}} \cdot \cdot \underline{\boldsymbol{S}}) + \overline{\boldsymbol{w}\boldsymbol{w}} \cdot \cdot \underline{\boldsymbol{S}} \tag{5.163}$$

respectively, while the equation for the turbulent energy is

$$\boldsymbol{U} \cdot \nabla \left(\frac{\overline{w^2}}{2}\right) = -\nabla \cdot \left(\frac{1}{\rho}\overline{w\delta p} - 2\nu\overline{\boldsymbol{w} \cdot \underline{\delta\boldsymbol{s}}} + \frac{1}{2}\overline{w^2\boldsymbol{w}}\right). \\ -\overline{\boldsymbol{w}\boldsymbol{w}} \cdot \cdot \underline{\boldsymbol{S}} - 2\nu\overline{\underline{\delta\boldsymbol{s}} \cdot \cdot \underline{\delta\boldsymbol{s}}} \tag{5.164}$$

If one pursues the literature of fluid dynamics, one finds various intermediate quantities that are given names and contribute to the jargon. These include, for example, the general *stress tensor*, which includes both pressure and viscous stress terms. We are now prepared to discuss the dynamics of the turbulent flow in more detail.

First consider the mean momentum. Applying the scaling analysis discussed in Sect. 2.3 shows that the viscous term in (5.162) is of order $1/Re$ relative to the convective term and so typically is negligible. The rightmost term in this equation is the divergence of a tensor. This term describes the forcing of the mean flow by the fluctuations. This tensor, having first been developed by Reynolds in 1895, is known as the *Reynolds stress tensor*. It quantifies the effects of the turbulent fluctuations on the mean flow. Unfortunately, the magnitude of this term is not obvious. The turbulent velocity w should be smaller than U and the eddy diameter ℓ smaller than the global scale length of the flow L. In actual turbulence it is common to see $w/U \sim \ell/L \sim$ a few percent. This has the implication that the final term in (5.162) is a few percent of the term on the left-hand side. This is often much larger than the viscous loss term but remains small enough that the flow changes gradually on the scale of L.

The story is similar with regard to the mean energy, as expressed by (5.163). The viscous terms are of order $1/Re$ relative to the energy. The terms involving the correlations of the turbulent velocities are typically much larger, but remain small enough that the energy of the mean flow is only gradually reduced.

The equation for the turbulent energy (5.164) has more to tell us. The first line can be rearranged as the divergence of a vector that contains energy fluxes, pdV work, and energy transport by viscous stresses. These sum to zero over any volume within which the fluctuating mechanical energy is unchanging, as will be the case in steady turbulence. The second line of this equation includes the terms identified with the production of turbulent energy (the first term) and with the dissipation of turbulent energy (the second term). In a crude scaling sense $S \sim U/L \sim w_e/\ell$, so the magnitude of the first term in the second line is w_e^3/ℓ. This equals the specific energy dissipated by turbulence that we obtained above from general arguments.

The final term in (5.164) represents the dissipation by the turbulence, and is the significant new result we obtain from this analysis. Assuming the turbulence to be isotropic, one can show with some algebra that

$$2\nu \overline{\delta s \cdot \cdot \delta s} = 15\nu \overline{(\partial w_1/\partial x_1)^2}, \tag{5.165}$$

in which the subscript 1 designates the first vector component. (For a detailed development of the equations related to turbulence, see Hinze (1959).) Fluctuations at all spatial scales contribute to δs, so that δs is not given by w/ℓ. Instead, the length-scale associated with δs must be smaller than ℓ. Defining this length scale as λ_T, and balancing the production of turbulence with its dissipation, we have

$$w^3/\ell = 15\nu w^2/\lambda_T^2, \text{ so that} \tag{5.166}$$

$$\lambda_T/\ell = \sqrt{15}/Re^{1/2}. \tag{5.167}$$

The length scale λ_T is known as the Taylor microscale, named after G.I. Taylor, who first defined it. However, the curious aspect of the Taylor microscale is that it is not a physical distance that characterizes the turbulence. Rather, it is the maximum size at which the energy from the large eddies can be dissipated by viscosity, if the turbulent fluctuation velocity does not change as the scale decreases. Tennekes and Lumley (1972) would prefer to see λ_T used only in combination with w to give a dissipation rate. Vortices at this scale are able to dissipate all the energy from the eddies. Thus, this is a reasonable estimate of the scale below which the behavior is not influenced by the large-scale dynamics that drives the eddies. However, this dissipation cannot occur within one vortex circulation timescale for eddies with rotation speed w and size scale λ_T. Thus, such vortices fail to satisfy the assumption that any turbulent vortex dissipates its energy in one turnover time. If this presumed property of turbulence, which is experimentally supported, is valid, smaller vortices will continue to form until the Kolmogorov scale is reached.

A notable feature of the Taylor microscale is that it is of the same order as the laminar boundary layer thickness developed during one eddy timescale, discussed above. Any structures that endure for an eddy timescale will develop such boundary layers. This adds a physical basis to the notion that the global unstable dynamics might have limited effects below this scale. To emphasize this correspondence, Dimotakis calls this the Liepmann–Taylor microscale.

Thus, we have a picture in which the fluid dynamics causes the transfer of energy to smaller-scale vortices, although we have yet to discuss the dynamics of this. The result is that the global unstable dynamics produces eddies and also produces vorticity, at least on surfaces. The vorticity spreads by viscous diffusion. The energy created by the global dynamics is transferred to smaller and smaller vortices, which become independent of the global processes once the vortex size drops below the Taylor microscale. The energy is then eventually dissipated when it reaches the Kolmogorov scale. As Re increases, the Taylor microscale and Kolmogorov scale become increasingly separated.

In between the Taylor microscale and the Kolmogorov scale, only the dynamics of the vortices governs the flow of energy to smaller scales. Since the equations do not depend fundamentally on the scale, one might hope to find a fairly simple scaling for the changes in the eddy properties as their diameter decreases. Working from a fairly general set of limiting assumptions, reviewed, for example, in Chap. 8 of Tennekes and Lumley (1972), Kolmogorov showed that the wavenumber spectrum of the kinetic energy, $E(k)$, is proportional to $k^{-5/3}$. The meaning of this statement is that $E(k)dk$ is the kinetic energy of the vortices whose characteristic wavenumber is within $dk/2$ of $k \sim 2\pi/\lambda$, and that this is proportional to $k^{-5/3}$. The corresponding scaling of the fluctuating strain rate is $\delta s \propto k^{2/3}$, so the fluctuating strain rate is largest at the smallest scales. The exponent of 5/3 is not a universal constant; one will see other values in both theory and observations for systems that satisfy assumptions different from those used by Kolmogorov.

Figure 5.22 illustrates the qualitative structure of the resulting spectrum of $E(k)$. The spectrum at the lowest wavenumbers is determined by the processes that create the turbulent energy, in addition to the processes that transfer it to smaller scales. Thus, the structure of this part of the spectrum may vary with conditions. It is shown as flat in the figure. As k increases so that the vortex size λ drops below λ_T,

Fig. 5.22 Structure of the
turbulent spectrum of kinetic
energy

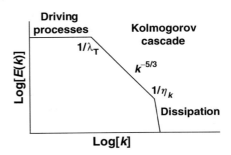

the spectrum becomes a *Kolmogorov spectrum*, with a slope of $-5/3$. Then as the
vortex size approaches the Kolmogorov dissipation scale, the energy is dissipated
and $E(k)$ decreases more rapidly. On the basis of a review of data, Dimotakis
(2000) concludes that dissipative effects begin to alter the spectrum for vortex scales
$\lambda \leq 50\eta_k$. The region of the spectrum that has a power-law shape is known as the
inertial range. This reflects the point of view that the inertial dynamics of the fluid
are responsible for producing this part of the spectrum. Based on this discussion, an
inertial range should appear once Re becomes large enough that $50\eta_k < \lambda_T$. This
requires a value of Re above about 10^4.

An important feature of turbulent systems is the presence or absence of a *mixing
transition*. This is generally observed to occur at some value of Re which depends
on details. Once the mixing transition has occurred, the turbulence causes rapid
mixing of the two interacting fluids and rapid diffusion of each into the other.
This transition is of significant practical importance for systems such as chemical
processors, intended to generate copious interactions between the molecules in two
fluids. This may also be important for high-energy-density systems, as one may in
various contexts desire to encourage or to discourage such mixing. A conjecture
due to Dimotakis is that the mixing transition corresponds to the development of
an inertial range, and that achieving $Re > 10^4$ is a necessary condition for this
development.

We close this section with a brief discussion of the dynamics that produces the
flow of energy from larger to smaller vortices. To see these dynamics, it is useful
to recast the momentum equation, (5.159) in two ways. We keep all the same
assumptions including incompressibility. First, one can manipulate the convective
derivative and the viscous term to highlight the effect of the vorticity $\boldsymbol{\omega} = \nabla \times \boldsymbol{u}$ in
this equation. We obtain

$$\frac{\partial \boldsymbol{u}}{\partial t} = -\nabla \left(\frac{p}{\rho} + \frac{u^2}{2} \right) + \boldsymbol{u} \times \boldsymbol{\omega} - \nu \nabla \times \boldsymbol{\omega}. \tag{5.168}$$

This equation illustrates one important effect of vorticity. The fluid velocity is
redirected in the direction of $\boldsymbol{u} \times \boldsymbol{\omega}$. This is easy to understand if one returns to
our analogy that vortices are spinning donuts. Recall the behavior of topspin shots
in tennis, curve balls in baseball, or slice shots in golf. In all these cases, a spinning

object creates lift by increasing the flow velocity on one side and decreasing it on the other, which creates a pressure difference through the Bernoulli effect. Thus, vortices redirect the flow in a direction perpendicular to the flow and to the vorticity vector.

Second, we can take the curl of (5.159) to develop a dynamic equation for the vorticity itself. This gives

$$\frac{\partial \boldsymbol{\omega}}{\partial t} = \nabla \times (\boldsymbol{u} \times \boldsymbol{\omega}) + \nu \nabla^2 \boldsymbol{\omega}. \tag{5.169}$$

This equation is identical to (10.37), describing the behavior of the magnetic field. The analogy between vorticity and magnetic field is often exploited for both physical explanations and mathematical analysis. Here we note that the vorticity will diffuse if there is no net flow or if its spatial scales are small enough. Otherwise, the first term on the right-hand side causes the vortices to move with the flow. Thus, in the same sense that magnetic field in a plasma is *frozen in*, the vorticity in a fluid is frozen in. In addition, this equation makes it clear that the role of viscosity with regard to momentum is to create diffusion. This transfers momentum (and vorticity) from the structure being damped into the surrounding fluid. The energy involved in vortex motion can be dissipated by its conversion to thermal energy, described by the viscous heating term in the energy equation.

The consequence of (5.168) and (5.169) together is that vortices do not allow a fluid to flow through them undisturbed. They deflect the fluid, and stay with it to deflect it further. However, it remains the case that changes in the flow can affect the local value of $\boldsymbol{\omega}$. This is due to the effects of the gradients in velocity. To see these effects more clearly, it is helpful to recast (5.169) as follows:

$$\frac{\partial \boldsymbol{\omega}}{\partial t} + \boldsymbol{u} \cdot \nabla \omega = \boldsymbol{\omega} \cdot \underline{s} + \nu \nabla^2 \boldsymbol{\omega} . \tag{5.170}$$

Here we see that only if the strain rate is zero (and viscous damping remains negligible) does the local vorticity move with the fluid without changing. To see what kinds of changes may occur, we consider the effects of finite strain rate on the vortices.

First, the off-diagonal elements of $\underline{s}$ act to rotate the vortices. This is simple to understand. We discussed the lift generated by the interaction of $\boldsymbol{\omega}$ and $\boldsymbol{u}$. If $\boldsymbol{u}$ changes along the vortex, then the lift varies along the vortex, which will produce a torque and cause a rotation of it. Since vortices generally involve derivatives of $\boldsymbol{u}$ in all three directions, the distribution of vorticity tends to become isotropic as vortices come to dominate the dynamics. As a result, small-scale turbulence is typically isotropic even when the driving instability at the global scale may not be.

Second, the diagonal elements of $\underline{s}$ produce changes along the direction of the vortices. These act either to stretch or to compress the vortices. These effects are illustrated in Fig. 5.23. As is illustrated, stretching or compressing changes the size of a vortex. This is a simple consequence of the fluid flow. When, for example, fluid

Fig. 5.23 Vortex stretching. Changes in fluid velocity cause vortices to stretch or shrink

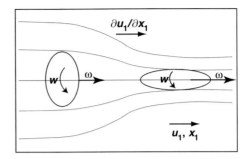

in a pipe speeds up to pass through a narrower section, a cylindrical element of fluid is stretched in length but shrinks in diameter. When this happens to a vortex, the rotation speed also must change to conserve angular momentum. As a result, when the fluid dynamics stretches a vortex, the vorticity increases. Furthermore, note that the increase in vorticity is very rapid. For example, in the case of Fig. 5.23, the vorticity is in the x_1 direction, and the nonzero element of the strain rate tensor that produces the vortex stretching is $s_{11} = \partial u_1 / \partial x_1$. Through (5.170), this produces exponential growth of the vorticity (for constant strain rate). This might potentially produce a turbulent state, as exponential growth often leads to large amplitudes. However, any fluctuations in the strain rate as the flow developed would tend to prevent this outcome.

Even so, vorticity in turbulent systems often increases explosively, through the combination of two effects. The first effect is the one we just discussed—the amplification of vorticity through its interaction with the strain rate. The second effect is the unstable growth of the strain rate through a "secondary instability." The simple instabilities, such as RT or KH, tend to produce very ordered two-dimensional or three-dimensional flows. On the jets of Fig. 5.23, for example, the KH instability produces curled structures that wrap around the column of the jet to form loops. (These have vorticity on their surfaces, which soon diffuses into the fluid near these surfaces.) The direction of this vorticity is azimuthal (it wraps around the jet). Initially, the azimuthal strain rate is zero—the system is cylindrically symmetric. However, many symmetric, two-dimensional systems are unstable to fluctuations in the third dimension. In the case of the jets, these fluctuations modulate the surfaces in the azimuthal direction, and these modulations grow exponentially. The key consequence is that these growing modulations create a finite azimuthal strain rate that also grows exponentially. At this point the vorticity amplifies exponentially from the exponentially growing strain. A very turbulent flow develops almost immediately.

The above discussion enables us to understand how the dynamics of the fluid creates a distribution of vortices on all scales down to the Kolmogorov length. The vortices larger than any given scale produce the strain rate that is experienced by the smaller vortices. Thus, vortices at any given scale are both stretched and rotated through the influence of the larger scales. Tennekes and Lumley (1972) show that any given scale is most strongly affected by slightly larger scales. As a result, the flow of energy and vorticity to smaller scales can accurately be described as a *cascade*. Kolmogorov first described these dynamics. For this reason, the spectrum of $E(k)$ observed in turbulent systems is often described as a *Kolmogorov cascade*.

We need to address a few more details in order to conclude this discussion. First, note that vortices are inherently three-dimensional objects. They have structure in all three directions. As a result, vorticity and its effects cannot be captured by one-dimensional or two-dimensional calculations or simulations. This makes the accurate modeling of turbulence a very challenging problem. Second, in the above we have used the notion of fluctuating vorticity. This can be misleading, as we also saw that the vorticity through a given surface is conserved unless viscous diffusion matters. The vorticity in turbulence fluctuates because vortices move and change their shape, not because there is a vorticity oscillation. Another way to put this is that vortex motion is not wave motion and does not involve the oscillation of physical quantities.

Finally, there is the question of how the vortices at small scales begin. We have seen that vortices on a given scale can affect those on a smaller scale, but this assumes that the smaller scale vortices are already present. There is certainly some thermal level of vorticity, but growing this to large amplitude would be a very long process. The global instabilities such as KH do deposit vorticity on surfaces, and perhaps these or other more-complex processes are responsible for the initial production of vorticity at small scales. One may hope and expect that further research will clarify the details of the transition to turbulence.

This takes us to the end of our discussion of unstable hydrodynamic behavior. Given that acceleration, deceleration, shear layers, and shock waves can each produce such behavior, it is no surprise that hydrodynamic instabilities are common in high-energy-density experiments. We proceed now to turn our attention to another aspect of high-energy-density systems. They are often quite hot, and being hot they tend to radiate profusely. We begin to cope with this in the next chapter.

Homework Problems

5.1 Consider a system with water above oil as described in Sect. 5.1.1. Suppose there is an small, sinusoidal ripple on the surface. Find the vertical profile of the force density between the lower and upper boundaries of the ripple for a region of denser fluid and for a region of less-dense fluid. Discuss the comparison of the two fluids and the shape of the force density profile.

5.2 The final relation in (5.37) is significant for our specific applications, in which one needs to integrate, across an interface, equations that contain discontinuous quantities along with derivatives of discontinuous quantities. By treating the delta function and the step function as limits of appropriate functions (see a mathematical methods book), prove this relation.

5.3 Find the solution for the velocity profiles and the growth rate for the RT instability for two uniform, constant density fluids that are confined by two planar surfaces each a distance d from the interface, which is accelerated at constant g.

5.4 The discussion above (5.55) shows that $\tilde{n} = (n/\sqrt{kg})\sqrt{\tilde{k}}$. This would suggest that it might make more sense to separate the meaning of the axes more cleanly by using $\tilde{\delta} = (n/\sqrt{kg})$ and $\tilde{k} = [(k^2\nu)/\sqrt{gk}]^{2/3}$ as the two variables. Recast this equation in terms of these new variables, solve it, and plot the real roots from $\tilde{k} = 0$ to 2. Discuss the results and compare them to $n = \sqrt{A_n gk}$.

5.5 In the derivation of the dispersion relation for the Rayleigh-Taylor instability with viscosity, some steps were skipped. Derive (5.56) and (5.57) from (5.50). Comment on the nature of the terms that have been dropped.

5.6 Explore the global RT mode in arbitrary directions. Find the plane-wave solutions in x, y and z to (5.60) and discuss their behavior.

5.7 Consider an exponential density profile that decreases in the direction of the acceleration, g, as $\rho = \rho_o e^{-z/L}$, and thus is the opposite of the case analyzed in Sect. 5.2.3. Apply the RT instability analysis to find n for this case. Discuss the results.

5.8 Carry out the calculation described in Sect. 5.3 and find (5.78). Then find the limits when (a) $k_p \to 0$ and $k_x L \gg 1$ and (b) when $A_n = 0$ and $L_p = 0$. Compare these with previous results in the chapter.

5.9 Work out the linear theory of the Rayleigh-Taylor instability to find an expression for the growth rate for the case of an exponential density gradient that extends for a finite distance between two layers of constant density.

5.10 This problem relates to the derivation of the mode-coupling results. By operating on (5.95) and (5.97), create two scalar differential equations that can be subtracted to eliminate terms involving p. Compare the resulting differential equation to (5.21) and discuss.

5.11 If we take the point of view that the modulations of interest in Kelvin-Helmholtz instabilities are proportional to e^{int}, then we would insist on finding negative imaginary n in order to have growth of the modulations, as opposed to damping, in time. However, this should give us pause because the complex representation is only a mathematical convenience while the physical quantities are real. Considering the real, physical quantities, what is the significance of finding positive or negative imaginary n? (The chapter in Jackson (1999) that introduces waves may be of some help regarding the connection of real physical quantities and a complex representation.)

5.12 Suppose β is small enough that terms involving β in (5.140) can be dropped. Determine whether the two boundaries seen in Fig. 5.11 ever meet, completely eliminating the instability.

5.13 Analyze the shock conditions for a small-amplitude ripple and show that the change in the $\hat{z}$ component of $\boldsymbol{u}$ due to the ripple, relative to the $\hat{z}$ component of $\boldsymbol{u}$ produced by a planar shock, is second order in the ripple amplitude [i.e., generalize (5.143)].

5.14 Delve into the origins of the response to a rippled shock wave. Develop (5.159) and (5.160) from the equations in Chap. 2.

5.15 Explore further the effects of a rippled shock wave. Solve (5.146) through (5.148) to find the ratio of α, η, and χ to β. Plot the results for various values of γ and comment on what you observe.

5.16 Evaluate the small-angle limit of the equations for a shock at an oblique, rippled interface with a density decrease, and produce a plot similar to Fig. 5.20 for this case.

5.17 Consider the qualitative behavior of the postshock interface when there is a rarefaction but $\chi < 0$. Redraw Fig. 5.20 for this case. Discuss the evolution of the interface.

5.18 Work out the steady state, mean flow equations from the Reynolds decomposition. Derive (5.162) through (5.164). Comment on the meaning of each term.

5.19 To be more precise about the frozen-in property of vorticity, one should recognize that what moves with the fluid is the vorticity passing through a surface S. Prove this by taking the time derivative of the integral of $\boldsymbol{\omega} \cdot d\boldsymbol{S}$ over a surface S that moves with the fluid and may change its shape in time. Relate the result to (5.169). Hint: The key here is the evaluation of the partial derivative in time of the surface as a contour integral involving the edge of the surface.

5.20 Obtain the various equations describing the behavior of vorticity, (5.168) through (5.170), from the momentum equation. Discuss the point of each one.

References

Bychkov VV, Liberman MA, Velikovich AL (1990) Analytic solutions for Rayleigh-Taylor growth-rates in smooth density gradients – comment. Phys Rev A 42(8):5031–5032

Chandrasekhar S (1961) Hydrodynamic and hydromagnetic stability. Dover, New York

Dimonte G (2000) Spanwise homogeneous buoyancy-drag model for Rayleigh-Taylor mixing and experimental evaluation. Phys Plasmas 7(6):2255–2269

Dimotakis PE (2000) The mixing transition in turbulent flows. J Fluid Mech 409:69–98

Dimotakis PE (2005) Turbulent mixing. Annu Rev Fluid Mech 37:329–356

Duff RE, Harlow FH, Hirt CW (1962) Effects of diffusion on interface instability between gases. Phys Fluids 5(4):417–425. https://doi.org/10.1063/1.1706634

Haan SW (1991) Weakly nonlinear hydrodynamic instabilities in inertial fusion. Phys Fluids B-Plasma Phys 3(8):2349–2355. https://doi.org/10.1063/1.859603

Hinze JO (1959) Turbulence. McGraw-Hill, New York

Ishizaki R, Nishihara K, Sakagami H, Ueshima Y (1996) Instability of a contact surface driven by a nonuniform shock wave. Phys Rev E 53:R5592–R5595

Jackson JD (1999) Classical electrodynamics. Wiley, New York

Landau LD, Lifshitz EM (1987) Fluid mechanics, course in theoretical physics, vol 6, 2nd edn. Pergamon, Oxford

Oron D, Arazi L, Kartoon D, Rikanati A, Alon U, Shvarts D (2001) Dimensionality dependence of the Rayleigh-Taylor and Richtmyer–Meshkov instability late-time scaling laws. Phys Plasmas 8(6):2883–2890

Richtmyer RD (1960) Taylor instability in shock acceleration of compressible fluids. Commun Pure Appl Math 13:297

Tennekes H, Lumley JL (1972) A first course in turbulence. MIT Press, Cambridge

Tritton DJ (1988) Physical fluid dynamics, 2nd edn. Clarendon Press, Oxford

Velikovich AL (1996) Analytic theory of Richtmyer–Meshkov instability for the case of reflected rarefaction wave. Phys Fluids 8(6):1666–1679

Velikovich A, Phillips L (1996) Instability of a plane centered rarefaction wave. Phys Fluids 8(4):1107–1118

Wouchuk JG, Cobos-Campos F (2017) Linear theory of Richtmyer–Meshkov like flows. Plasma Phys Control Fusion 59(1). https://doi.org/10.1088/0741-3335/59/1/014033

Chapter 6
Radiation Transfer and Atomic Processes

Abstract This chapter begins by discussing the new concepts necessary to account for the transfer of energy by radiation, and develops the equations for radiation intensity, radiative energy density, radiative energy flux, and radiation pressure. It then discusses the interaction of radiation and matter. This includes both the way we account for the sum of all the effects that increase or reduce the radiation intensity and the specific atomic processes that contribute. These elements provide the basis of the radiation transfer equation, whose solutions are discussed. This includes a discussion of simplifications that become possible when scattering is isotropic and elastic, and when the behavior of the radiation becomes diffusive. The chapter concludes with a discussion of some simple aspects of radiation transfer in relativistic systems.

Thus far we have focused on systems that are purely hydrodynamic. In so doing, we have ignored a major aspect of many high-energy-density systems: radiation. It is easy to see why radiation often matters. At any given pressure, the temperature increases as density decreases, and there is some density below which radiation fluxes will exceed material fluxes. If we suppose that $Z + 1 = A$, for simplicity, then the temperature is given by $T = m_p p/(\rho k_B)$. The characteristic radiation flux is σT^4, which can be compared to a characteristic material energy flux $\rho \epsilon c_s$. The actual material energy flux may differ from this by some factor, but the threshold density below which radiation fluxes exceed thermal fluxes depends only on the one-fourth power of this factor. Figure 6.1 shows the density at which σT^4 equals $\rho \epsilon c_s$, for pressures from 1 Mbar to 1000 Mbar. Radiative effects matter in gases and foams toward the low end of this regime, and in solid-density materials toward the high end. We consider some other comparisons based on temperature, and when radiation pressures matter, at the beginning of Chap. 7.

To understand radiative effects one must first understand radiation transfer, which is the transport of energy and momentum through a physical system by radiation, including the interactions with matter. Radiation transfer is familiar to us in everyday life. For example, suppose radiation from the sun, with its spectral peak at frequencies we perceive as green, is transmitted through the atmosphere, heating a black asphalt driveway. On a hot, bright day we can feel the radiation emitted from this black surface, if we are smart enough to put our hand or foot near

© Springer International Publishing AG 2018
R. P. Drake, *High-Energy-Density Physics*, Graduate Texts in Physics,
https://doi.org/10.1007/978-3-319-67711-8_6

Fig. 6.1 Energy flux regimes. For any given density, the radiation energy flux exceeds the material energy flux at a high enough material pressure

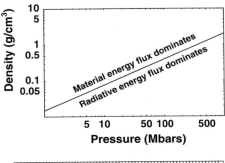

Fig. 6.2 Luminosity variations in several Cepheid variable stars. Adapted from the Michigan Math and Science Scholars Summer Program

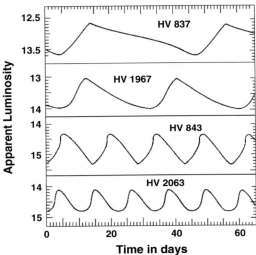

it before we step on it. We experience another aspect of radiation transfer, namely scattering, when looking through a fog bank at a bright light.

The emission, absorption, and transfer of radiation are central to much of astrophysics. Shu (1992), for example, devoted one entire volume of his two-volume set, *The Physics of Astrophysics*, to radiation. Fusion generates heat in the core of a star that in the end must be radiated from its surface. The stellar structure depends strongly on the absorption and emission of radiative energy. In some variable stars, structure in the absorption causes the luminosity of the star to oscillate. This occurs for example in Cepheid variable stars, which are often used as distance indicators because in order to oscillate their luminosity must remain in a narrow and well-known range. Figure 6.2 shows an example of luminosity oscillations from four Cepheid variable stars. The period of oscillation varies, and can be used to establish their absolute luminosity based on the empirically established relation between period and luminosity. The luminosity oscillations are a consequence of oscillations in the temperature dependence of the absorption length for radiation within the star, which leads the surface of the star to regularly become hotter and cooler. The inverse of the absorption length is known as the *opacity*, and is discussed at length below. The period–luminosity relation for Cepheid variable stars is now understood, thanks to more-sophisticated versions of the calculations described in this chapter, verified by laboratory measurements of the opacity, some of which are shown in Sect. 6.2.3.

To develop an understanding of radiation transfer, we will first consider the description of radiation in isolation, which is analogous to our discussion of matter in isolation in Chap. 2. We will then discuss the atomic processes that occur in high-energy-density systems, and how we summarize mathematically their net impact on the radiation. These two discussions will have prepared us to take on the radiation transfer problem: how radiation evolves as it interacts with matter. We will conclude with some discussion of relativistic effects in radiation transfer.

6.1 Basic Concepts for Radiation

In order to work with radiation, its transport, and its effects on matter, one must first face the task of describing the radiation itself. This is a task at least as complicated as the description of the particles. Like particles, the radiation can fill space, vary in time, and propagate in any direction. The radiation does have a unique velocity, which is simpler than the situation with particles, but this is more than offset by the fact that the radiation can have any frequency and can interact through several mechanisms, some of which would be described in mechanics as "inelastic." In addition, since radiation does move at the speed of light, it is a bit easier for relativistic effects to matter. Nonetheless, by proceeding step by step we can develop useful descriptions. As we do so, we will use primarily a subset of the notation in the book by Mihalas and Weibel-Mihalas (1984), where one can find a much longer and much more complete discussion, especially in the area of relativistic effects.

6.1.1 Properties and Description of Radiation

Here we will build up our description from fundamentals. This corresponds to the development of plasma theories beginning with the Boltzmann equation. The analog of the distribution function, for radiation, is the *spectral radiation intensity*, I_ν, which has units of energy per unit area per unit time per unit solid angle per unit frequency, or ergs $cm^{-2} s^{-1} sr^{-1} Hz^{-1}$ in cgs units. Thus, within differential elements of area (perpendicular to the direction of propagation), time, solid angle, and frequency are given by $dA, dt, d\Omega$, and $d\nu$, respectively, the increment of energy delivered is

$$\Delta\text{energy} = I_\nu dA dt d\Omega d\nu. \tag{6.1}$$

It should be evident that the spectral intensity is fundamentally related to the Poynting vector. Working out the relationship is an interesting problem, but not one that we need to take up here. A fundamental and complete description of radiation would have to describe the variations in I_ν as a function of all these variables and distance of propagation. Fortunately, much can be understood by working with less-complete models. We will identify the *radiation intensity*, I_R, as the energy per unit area per unit time per unit solid angle in the radiation. That is,

$$I_R(\boldsymbol{x}, t, \boldsymbol{\Omega}) = \int_0^\infty I_\nu(\boldsymbol{x}, t, \boldsymbol{\Omega}, \nu)d\nu, \tag{6.2}$$

with cgs units of $\mathrm{ergs\,cm^{-2}\,s^{-1}\,sr^{-1}}$. The total intensity emitted by a black body is the *thermal intensity*, $B(T)$, given by

$$B(T) = \sigma T^4/\pi \;\mathrm{ergs\,cm^{-2}\,s^{-1}\,sr^{-1}}. \tag{6.3}$$

Useful values of σ are $1.03 \times 10^{12}\,\mathrm{ergs\,cm^{-2}\,s^{-1}\,eV^{-4}}$ or $1.03 \times 10^5\,\mathrm{J\,cm^{-2}\,s^{-1}\,eV^{-4}}$.

The radiation intensity is not necessarily uniform in direction, as hotter regions generally emit more thermal radiation (though not necessarily more radiation in atomic line emission, as we discuss below). It turns out that quantities known as the *mean spectral intensity*, J_ν, and *mean intensity*, J_R, are quite useful. These are defined as

$$J_\nu(\boldsymbol{x}, t, \nu) = \frac{1}{4\pi} \int_{4\pi} I_\nu(\boldsymbol{x}, t, \boldsymbol{\Omega}, \nu)d\Omega \tag{6.4}$$

and

$$J_R(\boldsymbol{x}, t) = \frac{1}{4\pi} \int_{4\pi} I_R(\boldsymbol{x}, t, \boldsymbol{\Omega})d\Omega. \tag{6.5}$$

We will see how these quantities are important later in this chapter when we discuss radiation energy transport.

No matter what the distribution of the radiation in angle may be, its energy density is an important property that appears in some of the dynamic equations. In general the density of something is a ratio of flux to velocity, but in particular the mathematics depends upon the details. When material particles move in many directions, as in a gas, their total energy density is much larger than the directed energy of motion of the gas viewed as a fluid. Similarly, the energy density of the radiation is not the net radiation flux divided by the propagation speed. Instead, the *radiation energy density*, E_R, is the integral over solid angle of the radiation intensity divided by the group velocity. Thus,

$$E_R(\boldsymbol{x}, t) = \frac{1}{c} \int_{4\pi} I_R(\boldsymbol{x}, t, \boldsymbol{\Omega})d\Omega = \frac{4\pi}{c} J_R, \tag{6.6}$$

in which we have made the nearly-always-valid assumption that the group velocity is isotropic and is equal to c. There is of course a corresponding *spectral radiation energy density*, given by

$$E_\nu(x, t, \nu) = \frac{1}{c} \int_{4\pi} I_\nu(x, t, \Omega, \nu) d\Omega = \frac{4\pi}{c} J_\nu. \qquad (6.7)$$

The above quantities are similar to integrals of a particle distribution function to find total density, distributions of speeds, and so on. There is a choice in both cases regarding what variables to use to define the distribution. Particle distributions could be defined by their energy density in a space of position, direction, and energy, but it is generally more intuitive to use the number density in a space of position and velocity. Similarly, one can treat photons in terms of their number density, by dividing I_ν by $ch\nu$, and this is at times useful. But it is generally more intuitive to work with their energy density in a space of position, direction, and energy.

Continuing the analogy with distributions of particles, we next discuss the moments of the distribution of photons. By direct analogy, one would say that the photon flux is

$$\int_{4\pi} \frac{I_\nu(x, t, \Omega, \nu)}{ch\nu} \boldsymbol{v} d\Omega, \qquad (6.8)$$

in which $\boldsymbol{v}$ is the photon velocity vector. But the *radiation energy flux*, $\boldsymbol{F}_R$, is generally more useful, and we know that the speed is c, so we write

$$\boldsymbol{F}_R(x, t) = \int_{4\pi} I_R(x, t, \Omega) \boldsymbol{n} d\Omega, \qquad (6.9)$$

in which $\boldsymbol{n}$ is a unit vector in the direction of propagation for any value of Ω. Thus it varies as one integrates. The z-component of $\boldsymbol{n}$, for example, is $\cos\theta$ in a standard spherical coordinate system. The cgs units of radiation energy flux are ergs s^{-1} cm^{-2}. It will come as no surprise that the *spectral radiation energy flux* is

$$\boldsymbol{F}_\nu(x, t, \nu) = \int_{4\pi} I_\nu(x, t, \Omega, \nu) \boldsymbol{n} d\Omega, \qquad (6.10)$$

with units ergs s^{-1} cm^{-2} Hz^{-1}. The radiation flux is a particularly important quantity, because there is a large and important regime in which transport of energy by radiation is crucial even though the energy density and pressure of the radiation are negligible. The radiation flux is related to the radiation momentum density. The total radiation momentum density is $\boldsymbol{F}_R/c^2$, and the spectral radiation momentum density is $\boldsymbol{F}_\nu/c^2$. Thus, the total radiation momentum transport across an element of area, $d\boldsymbol{A}$, is $\boldsymbol{F}_R \cdot d\boldsymbol{A}/c$.

As is the case with particle distributions, one can define further moments of the radiation distribution function indefinitely. In practice, the second moment is as far as one typically needs to go. The *spectral radiation pressure tensor*, $\underline{P}_\nu$, is defined in dyadic notation by

$$\underline{P}_\nu(\boldsymbol{x}, t, \nu) = \frac{1}{c} \int_{4\pi} I_\nu(\boldsymbol{x}, t, \boldsymbol{\Omega}, \nu) \boldsymbol{n}\boldsymbol{n} \mathrm{d}\Omega. \tag{6.11}$$

This is clearly a symmetric tensor since reversing the order of the components of $\boldsymbol{n}$ does not change the integral. The integral of (6.11) over frequency is the *total radiation pressure tensor*, $\underline{P}_R$. The transport of momentum by radiation, in the absence of matter, is fundamentally described by

$$\frac{1}{c^2} \frac{\partial \boldsymbol{F}_\nu}{\partial t} = -\nabla \cdot \underline{P}_\nu. \tag{6.12}$$

This is again perfectly natural, as one can see by integrating over a finite volume and using Gauss' theorem. It is worth noting that the left-hand side of this equation is nearly always negligible in systems involving both radiation and matter. The radiative contribution to the energy flux in a system of radiation and matter is often large, but the material momentum nearly always dominates over the radiation momentum. Even so, the radiative momentum source (the right-hand side of this equation) can be the dominant momentum source for the matter.

The radiation field often is symmetric in one of three ways that produce simpler results for the radiation pressure. In general, the *scalar spectral radiation pressure* p_ν is defined by

$$\begin{aligned}
p_\nu(\boldsymbol{x}, t, \nu) &= \frac{2\pi}{c} \int_0^\pi I_\nu(\boldsymbol{x}, t, \theta, \nu) \cos^2(\theta) \sin(\theta) \mathrm{d}\theta \\
&= \frac{2\pi}{c} \int_{-1}^1 I_\nu(\boldsymbol{x}, t, \mu, \nu) \mu^2 \mathrm{d}\mu,
\end{aligned} \tag{6.13}$$

where $\mu = \cos\theta$ is a very convenient way to represent the variation with polar angle.

Here we have evaluated the zz component of $\underline{P}_\nu$. If the radiation field is isotropic, then $\underline{P}_\nu$ is evidently diagonal with three equal, nonzero elements. One then has

$$p_\nu(\boldsymbol{x}, t, \nu) = \frac{2\pi I_\nu(\boldsymbol{x}, t, \nu)}{c} \int_{-1}^1 \mu^2 \mathrm{d}\mu = \frac{1}{3} E_\nu, \tag{6.14}$$

which is the simplest example in which one can see this relation between pressure and energy density. In the isotropic case the divergence of the pressure tensor in the momentum equation becomes the gradient of the scalar pressure, $\nabla \cdot \underline{P}_\nu = \nabla p_\nu$, just as occurs with material pressures. (One might protest that a truly isotropic radiation field cannot have pressure gradients, because local isotropy cannot be maintained without having a spatially uniform radiation intensity. This is mathematically true. However, in practice, significant pressure gradients can correspond to negligible anisotropy.) Treating the radiation field as isotropic is for example justified in the diffusion regime, which is of great importance and which we will discuss at length below.

The integral of (6.13) over frequency gives the (total) *scalar radiation pressure* p_R. In the isotropic limit, one then has $p_R = E_R/3$.

A second useful, symmetric case is the planar case, in which I_ν varies only in angle relative to one direction, and is isotropic in the two orthogonal directions. In this case we choose the direction of variation as the z axis and write $I_\nu = I_\nu(z, t, \mu, \nu)$. In this case the zz element of $\underline{P}_\nu$ is again p_ν, and the xx element, P_{xx}, is

$$P_{xx} = \frac{1}{c} \int_0^{2\pi} \int_{-1}^1 I_\nu(z, t, \mu, \nu)(1 - \mu^2)\mathrm{d}\mu \cos^2 \phi \mathrm{d}\phi$$

$$= \frac{1}{2}E_\nu - \frac{1}{2}p_\nu = p_\nu - \frac{1}{2}(3p_\nu - E_\nu),$$

$$(6.15)$$

which is also equal to P_{yy}. Note that for an isotropic intensity or for any angular distribution that yields $E_\nu = 3p_\nu$, the pressure again reduces to a scalar. In the planar case, the only nonzero derivatives are in the z direction, so

$$\nabla \cdot \underline{P}_\nu = \nabla p_\nu = (\partial p_\nu / \partial z)\, \hat{z}, \qquad (6.16)$$

where $\hat{z}$ is a unit vector in the z direction.

A third useful case is that of spherical symmetry. In this case the diagonal components corresponding to the polar and azimuthal angles equal P_{xx} from (6.15), and the radial component is again p_ν from (6.14). In this case the only nonzero derivatives are in the radial direction, so

$$\nabla \cdot \underline{P}_\nu = [\partial p_\nu / \partial r + (3p_\nu - E_\nu)/r]\, \hat{r}, \qquad (6.17)$$

in which $\hat{r}$ is a unit vector in the radial direction.

Now consider in general terms the ratio of p_ν to E_ν. This ratio is known as the *Eddington factor*, $f_\nu = p_\nu/E_\nu$. The Eddington factor depends on the angular variation of I_ν, as is clear in (6.13). In the limit of a plane wave at frequency ν, with energy flux (power per unit area) I, one would have $E_\nu = I/c$ and $p_\nu = (I/c^2)c = I/c$ so in this case $p_\nu = E_\nu$ and $f_\nu = 1$. A sufficiently beam-like intensity distribution can have $f_\nu \sim 1$. The limit where the radiation propagates freely with little interaction is known as the *free-streaming limit*. In the free-streaming limit, f_ν will approach 1 as the distance from the source increases. It is also clear from (6.13) that f_ν decreases as the distribution spreads in angle, reaching 1/3 when the distribution becomes isotropic. In natural systems, f_ν typically varies between 1/3 and 1. A "pancake-like" distribution, in which most of the energy is transverse to the symmetry axis, produces an Eddington factor below 1/3. This occurs in some radiative shocks, as is discussed in Chap. 8. It also might occur in nature if an extended, hot, source region were sandwiched between two strongly absorbing regions.

It should be evident that one can define a total radiation pressure tensor, $\underline{P}_R$, and a scalar radiation pressure, p_R, by integrating the relations above over frequency. This permits one to define an overall Eddington factor, p_R/E_R. Some computational approaches to radiation transport are formulated in terms of an Eddington factor, which can be an effective way to improve the accuracy of a calculation without always dealing explicitly with all the possible directions of radiation propagation.

6.1.2   Thermal Radiation

Thermal radiation is very important even in systems with very nonequilibrium radiation. The reason is that the electrons are responsible for the emission of radiation, and the electrons very often develop a Maxwellian or piecewise-Maxwellian distribution. In such cases, the spectral intensity of the emitted radiation in some frequency range is proportional to the equilibrium spectral radiation intensity at the temperature of the electrons that are responsible for the emission. Nearly all texts on modern physics or statistical mechanics derive the properties of thermal equilibrium (or *blackbody*) radiation, so we need not repeat this here. By considering the relative probability that a state will be occupied, the density of states in phase space, and the two possible polarizations, one can show that the *spectral thermal radiation intensity*, $B_\nu(T)$, is,

$$B_\nu(T) = \frac{2h\nu^3}{c^2} \frac{1}{e^{h\nu/(k_B T)} - 1}, \tag{6.18}$$

in which h is Planck's constant and the units are those of spectral intensity: energy per unit area per unit time per unit solid angle per unit frequency. Here the subscript indicates that $B_\nu(T)$ is frequency dependent. Integrating over ν, one finds that the *total thermal radiation intensity*, $B(T)$, is

$$B(T) = \sigma T^4/\pi, \tag{6.19}$$

in which σ is the Stefan–Boltzmann constant. The energy density of the radiation is

$$E_R(T) = \frac{4\pi}{c} B(T) = \frac{4}{c}\sigma T^4. \tag{6.20}$$

Thermal radiation must be isotropic, so the pressure is a scalar and $p(T) = E_R(T)/3$.

6.2 Atomic Processes in High Energy Density Systems

Having defined the variables necessary to describe the radiation in isolation, we now are ready to ask how radiation and matter interact. This will prepare us to consider the combined problem, known as the *radiation transfer problem*. This will also lead us to discuss the detailed processes that determine the relative populations of various ionization states. Two books provide a next level of detail regarding the processes discussed here. These are Salzman (1998), a very readable basic description, and Griem (1997), which provides more detail on some issues but is much harder to read. Sobel'man et al. (2013) is additional book with some helpful material.

6.2.1 Types of Interaction Between Radiation and Matter

The only significant processes involved in the absorption and emission of radiation involve electrons, because they accelerate much more readily than nuclei. One can identify three fundamental types of interaction between radiation and matter involving electrons. We survey these here, before discussing some of them in more detail in later sections.

The first of these involves *bound–bound transitions*, which one encounters in elementary physics upon being introduced to the Bohr atom. The electron in a Bohr atom is bound to the atom, but can transition among the energy levels of the atom, known as "states." The lowest energy state is the *ground state* and all bound, higher-energy states are *excited states*. Radiation can be emitted when the electron "decays" from a higher-energy state to a lower-energy one. This is *bound–bound emission*, which is responsible for the familiar Lyman and Balmer spectral series. The inverse process, compared to decay, is excitation, in which an electron is given energy and moves from a lower-energy state to a higher-energy one. In general, both excitation and decay require the involvement of at least one additional particle, other than the electron and the atom or ion, to conserve energy and momentum. This additional particle can be any type of particle including a photon. In practice, certain particles tend to dominate the rate at which a specific decay or excitation occurs. In low-density plasmas, a decay is nearly always radiative and an excitation nearly always is produced by an electron collision. In high-energy-density plasmas, both collisional and radiative decay often matter, collisions may move the electron from one excited state to another, and radiative excitation will also be important if the radiation is near equilibrium with the matter.

Bound–bound emission produces line radiation, whose frequency ν is given by $E_\Delta = h\nu$, where E_Δ is the energy difference between the levels and h is Planck's constant. The spectral width is quite narrow. Energy emitted from one particle can be absorbed by a second particle, but only if the frequency of the radiation as seen by the second particle overlaps a bound–bound transition in the second particle. Since any Doppler shift changes the frequency seen by the second particle, the transfer of energy by repeated absorption and emission involving bound–bound transitions can become complex.

The second category of radiation–matter interaction involves *free–free transitions*. These transitions move an electron from one continuum state to another. The interaction of a free electron with any other particle (including a photon) produces a free–free transition. Such transitions often produce photon emission or absorption. Two of the most common and important free–free interaction processes are *bremsstrahlung* emission and *inverse bremsstrahlung* absorption. In bremsstrahlung, a particle (in practice, an electron) is accelerated by interaction with another charged particle (in practice, a nucleus), and this results in the emission of photons. This is the primary source of continuum emission from hot dense matter. In inverse bremsstrahlung, a photon (or light wave) moves an electron past a nucleus. The interaction with the nucleus randomizes the motion of the electron, which has the effect of extracting energy from the light. The absorption coefficient for inverse bremsstrahlung is discussed in Sect. 11.2. The high-energy limit of inverse bremsstrahlung is Compton scattering, in which the photon–particle energy exchange is quantized. Another free–free emission mechanism, important in magnetized plasmas, is *synchrotron emission*.

The third fundamental type of radiation–matter interaction involves the *bound–free transition*. The limiting energy of the bound states is the ionization energy, given in the Bohr model in the limit that the principal quantum number goes to infinity. In plasmas, ionization potential depression may reduce this, as was discussed in Chap. 3. When an electron is given energy that moves it above the ionization energy, it becomes a *free electron*. It is then said to be in *the continuum*, so called because the energy of the allowed states can vary continuously. Just as in the case of bound–bound transitions, an electron can decay from or be excited to a continuum state as a result of an interaction with any particle including photons.

Thus, a photon can be absorbed by an atom or ion, releasing one of the electrons through a transition from a bound state to a free state, a process known as photoionization. This is a major contribution to the absorption of X-ray photons by materials. Figure 6.3 shows the transmission of a thin layer of titanium. There is an abrupt decrease in transmission at 4.7 keV, which is the lowest energy at which the X-ray

Fig. 6.3 The transmission of a 5 μm thick titanium slab to X-rays

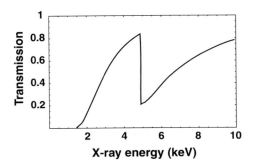

photon can pull an electron from the bound state whose principal quantum number is $n = 1$ (the K-*shell*) into the continuum. This absorption feature is known as the K-*edge*. As the photon energy increases above the K-edge, any absorbed photon will place the electron into a higher-energy continuum state. The cross section for this process decreases as the energy of the continuum state increases, which leads to the increase in transmission as energy increases above the K-edge. The energy of the K-edge increases with increasing atomic number. The next edge, with lower ionization energy, is the *L-edge*. It corresponds to the extraction of an electron from the $n = 2$ shell. Higher-Z materials such as tin or rhodium have an L-edge in the 3–4 keV energy range. All solid materials become strongly absorbing at X-ray energies below 1 keV, where bremsstrahlung (free–free) absorption becomes large.

Alternatively, a photon can be emitted when an electron in the continuum recombines with an ion, a process known as radiative recombination. This produces an X-ray line just below the K-edge, as it is a much stronger process for electrons in continuum states with near zero energy. On the low-energy side of this line, one may observe structure resulting from electrons that recombine into an excited state and then decay into the ground state. On the high-energy side of this line, one may observe a continuous feature from electrons that make free–bound transitions from higher-energy states in the continuum. The spectrum of free–bound radiation, at energies just above the X-ray line, can in some cases be used as a temperature diagnostic.

There are analogues of photoionization and radiative recombination involving electrons rather than photons, known as electron-impact ionization and three-body (or collisional) recombination. The bound–free transition is ionization of ions by electron impact. In this type of transition, a free electron impacts the ion and ejects a bound electron from the ion, losing the corresponding amount of energy in the process. Electron-impact ionization is central to the determination of ionization state distributions at all densities. The free-bound transition is three-body recombination, in which an electron rejoins an ion, with the involvement of another electron that enables conservation of momentum.

6.2.2 The Net Interaction of Radiation and Matter

We will further discuss most of the processes just mentioned below. But fortunately, one often need not explicitly account for every distinct interaction of radiation and matter. Instead, one can obtain an adequate description of many systems by considering the net total emission, absorption, and scattering. We develop such a description here. Plasmas emit radiation, both directly through the interactions of the particles, such as bremsstrahlung, and indirectly by scattering radiation in angle and/or energy. We will write the *spectral emissivity*, η_ν, as

$$\eta_\nu(\mathbf{x}, t, \mathbf{n}, \nu) = \eta_{\nu\text{th}}(\mathbf{x}, t, \mathbf{n}, \nu) + \eta_{\nu\text{sc}}(\mathbf{x}, t, \mathbf{n}, \nu), \tag{6.21}$$

which has (cgs) units $\text{ergs}\,\text{cm}^{-3}\,\text{s}^{-1}\,\text{sr}^{-1}\,\text{Hz}^{-1}$. In some writings, the term *spectral emission coefficient* is used rather than spectral emissivity. Here the *spectral thermal emissivity* is $\eta_{v\text{th}}$, which is an approximation assuming that the particles have a single, Maxwellian energy distribution. A more general and complete expression would explicitly include all the processes by which all the particles in the system can emit radiation, including for example line emission following collisional excitation and bremsstrahlung emission by high-energy tails on the electron distribution. Note that the integral of this term over frequency and solid angle gives the power loss rate of the matter in the plasma due to radiation. The *spectral scattering emissivity* is $\eta_{v\text{sc}}$, which includes all processes that scatter radiation in angle or energy. We will not pursue this in any depth here, but in general this emissivity at a given angle or energy depends on an integral over the radiation intensity present at other angles or energies. Unlike the quantities discussed previously in this section, this implies that the integral of (6.21) over frequency or angle is not straightforward, unless one can simply approximate the scattering term or ignore it for some reason.

The rate of attenuation of energy from radiation depends inherently on the radiation intensity, so the energy attenuated must be an expression involving the radiation intensity. We express the energy attenuation per unit volume per unit time per unit solid angle per unit frequency as $\chi_v I_v$, in which χ_v is the *spectral total opacity* in units of cm^{-1}, also at times known as the *spectral extinction coefficient*. In analogy with the case of emission, there is a *spectral absorption opacity*, κ_v, for absorption by the particles, which contributes to heating of the matter, and there is a *spectral scattering opacity*, σ_v, for scattering by the particles, which changes the direction of the radiation. Thus

$$\chi_v(x, t, n, v) = \kappa_v(x, t, n, v) + \sigma_v(x, t, n, v). \tag{6.22}$$

Here again, the general interaction of radiation and matter may be much more complex. In principle it may involve dielectric tensors and powers of the electric field of the light waves. But for nearly all problems in high-energy-density physics, it is sufficient to take the energy absorption rate to be linearly proportional to the radiation intensity, as we do here.

In the following we will not specifically treat the emission, absorption, and scattering of radiation by spectral lines. In some contexts, these can be approximated as an overall effective emissivity and opacity. In other contexts, they must be treated explicitly. The methods will be similar to those discussed here, but it will be necessary to treat each line discretely and to associate Doppler shifts with all relative material motions.

A centrally important case for high-energy-density physics is that of thermal emission. Kirchoff's law states that emission and absorption of radiation must be equal in equilibrium, which today we would view as an application of the principle of detailed balance to radiation emission and absorption. Mathematically, we would write this as

$$\eta_{v\text{th}}(\boldsymbol{x}, t) = \kappa_v(\boldsymbol{x}, t)B_v(T). \qquad (6.23)$$

Local thermodynamic equilibrium (LTE) is a state in which each species in the plasma, including the radiation, has an equilibrium distribution of energies, and in which the temperatures of these distributions are all equal. Plasmas are in LTE if the photon mean free path for absorption (and the collisional mean free paths) are very small compared to the gradient scale length of the temperature and if any variations in time are slow compared to the time required for an equilibrium distribution of ionization and excitation to be established. Thus, a system must be quite dense or very large to establish LTE. Some high-energy-density systems can be accurately described as LTE. In some experiments, for example, a low-Z envelope is used to confine a higher-Z material of interest, and this entire sample is maintained within an equilibrium radiation environment long enough to be uniformly heated. This places the material of interest in LTE. In contrast, in most astrophysical systems outside of stars, photons are not confined; such systems will nearly always be far from LTE.

Relation (6.23) is also useful even when the entire system is not in LTE. Collisions are usually rapid enough that the distribution of electrons in a high-energy-density system is nearly Maxwellian. The thermal radiation emitted by these electrons is accurately described by (6.23). Also the emission from any excited states whose populations are maintained at their equilibrium (Saha) values by electron collisions will be described by (6.23). However, the excited states that interact with the tail of the electron distribution, and with photons having long mean free paths, typically are not in their equilibrium distributions. These states are not in LTE, and (6.23) does not accurately describe their emission. A useful side note is that the intensity produced by atomic line emission cannot rise above $B_v(T)$, assuming the electron distribution is Maxwellian at the relevant energies. This is because the thermal intensity corresponds to the thermodynamically correct occupation of energy states by the photon distribution at that temperature. No emission process can rise above this value.

6.2.3 Opacities in Astrophysics and the Laboratory

Opacities are often tabulated as specific opacities, which can often be approximated as power laws in density and temperature. The density dependence of the specific opacity is typically weak at densities of interest for high-energy-density physics. Figure 6.4 shows the Planck mean opacity of Aluminum from one standard set of tables (this is LANL SESAME table 13710). The Planck mean is an average over frequency, discussed in detail in Sect. 6.3. There are a number of interesting features in this figure. Note the regions with relatively straight line segments, at low and high temperature. In these regions free-free (inverse bremsstrahlung) absorption dominates the behavior. In such regions the specific opacity becomes proportional to density. The slope of these line segments steepens once the Al becomes fully ionized at some fraction of a keV in temperature. Under most conditions shown,

Fig. 6.4 The specific Planck
mean opacity of Al versus
electron temperature, with the
curves from bottom to top
showing densities in g/cm^3 of
10^{-6}, 10^{-4}, 10^{-2}, 1, 10^2, and
10^4 (based on LANL
SESAME table 13710)

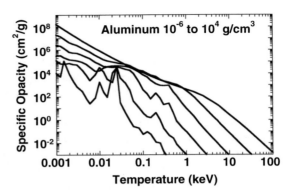

though, transitions involving bound states increase the opacity above that due only
to inverse bremsstrahlung. These effects are so strong that, in the range of density
(1–100 g/cm^3) and temperature (10 eV to 1 keV) typically of interest in high-energy-
density experiments, the density dependence of the specific opacity is very weak.
This is the origin of the standard formula given below. The absorption peak near
30 eV in temperature, associated with the second ionization (producing neon-like
aluminum), is strong enough to keep the opacity large even at very low densities.

Here are some approximations to the specific Planck mean opacity, κ_m, based on
a standard set of tables (the SESAME tables):

$$\kappa_m \approx 2 \times 10^5 \, T_{eV}^{-1} \, \text{cm}^2/\text{g} \quad \text{for CH}$$

$$\kappa_m \approx 3 \times 10^6 \, T_{eV}^{-1} \, \text{cm}^2/\text{g} \quad \text{for Al} \quad\quad (6.24)$$

$$\kappa_m \approx 3 \times 10^9 \, T_{eV}^{-2} \, \text{cm}^2/\text{g} \quad \text{for Xe.}$$

The scaling of the opacity is different in low-density systems, where the absorption
and emission may be dominated by bound–bound transitions and line radiation.
The net emission from low-density astrophysical plasmas is often described using
a *cooling function*. The cooling function Λ is the power loss per unit volume per
unit electron density per unit ion density. Thus, the power loss per unit volume is
$n_e n_i \Lambda$, and one can see that Λ has units of ergs-cm^3/s or equivalent. The discussion
in Sect. 7.2.2.2 shows that the relation of the absorption opacity to the cooling
function is $\kappa = n_e n_i \Lambda / (2\sigma T^4)$. Using typical numbers of $n_e \sim n_i \sim 10$ cm^{-3},
$\Lambda \sim 10^{-22}$ ergs cm^3/s, and $T = 10$ eV, one finds $\kappa \sim 5 \times 10^{-37}$ cm^{-1}, or about
10^{-19} parsec^{-1}. Figure 6.5 shows typical astrophysical cooling functions, based
on results in Sutherland and Dopita (1993). These particular results correspond
to a model assuming that the distribution of ionization states is in an equilibrium
determined by collisions and radiative recombination.

It often proves useful to have a calculation of the contribution to the opacity and
emissivity by inverse bremsstrahlung, working from (6.23). The relevant thermal
energies correspond to X-ray photons, for which the densities present are always

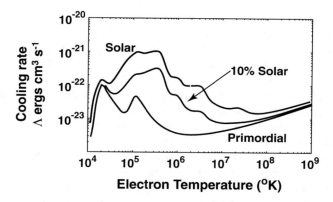

Fig. 6.5 Typical astrophysical cooling functions. The overall shape does not vary, but the location of the minimum depends upon the density of elements above He. The primordial case has only H and He while the other two cases have solar or 10% solar concentrations of such heavier elements, as indicated. Adapted from Sutherland and Dopita (1993)

a small fraction of their critical density. In this limit the absorption rate, (9.15) [discussed in Chap. 9]. simplifies to

$$\kappa_{EM} = \frac{\nu_{ei}\omega_{pe}^2}{8\pi^2 c\nu^2} = \frac{\nu_{ei}e^2 n_e}{2\pi c\nu^2}, \tag{6.25}$$

in which ν is the frequency of the photon, the rightmost term is in Gaussian cgs units, and the other quantities are as discussed in Sect. 9.1.3. Using the expressions for ν_{ei} in (2.59), we can write this as

$$\kappa_{EM} = \frac{1}{3(2\pi)^{5/2}} \frac{e^2 Z \omega_{pe}^4}{c\nu^2 v_e^3} \ln \Lambda = 4.03 \times 10^{-9} \ln \Lambda \frac{n_e^2 Z}{\nu^2 T_{eV}^{3/2}}, \tag{6.26}$$

in Gaussian cgs units and with T_{eV} as the electron temperature in eV.

One can do two useful calculations using (6.26). The first is to integrate $\kappa_{EM} B_\nu$ over a Planckian thermal photon distribution to find the total thermal emissivity. One obtains

$$\eta_{th} = 5.68 \times 10^{-27} \ln \Lambda \, n_e^2 \sqrt{T_{eV}} Z \text{ ergs cm}^{-3} \text{ s}^{-1}, \tag{6.27}$$

with n_e in cm^{-3}. The other is to find the absorption mean free path, $\lambda_x = 1/\kappa_{EM}$ for frequency corresponding to the mean photon energy in a thermal distribution, which is $\nu = 2.7 k_B T_e / h$. One finds from (6.26), still with ρ in cgs units,

$$\lambda_x = \frac{1.05 \times 10^{38} T_{eV}^{7/2}}{n_e^2 Z \ln \Lambda} = \frac{2.95 \times 10^{-10} A^2 T_{eV}^{7/2}}{\rho^2 Z^3 \ln \Lambda} \approx \frac{10^{-9} A^2 T_{eV}^2}{\rho^2} \text{ cm}, \tag{6.28}$$

where the final term is under the approximations that $Z \sim 0.63\sqrt{T_{eV}}$ and $\ln \Lambda \sim 1$.

Laboratory measurements can determine opacities under material conditions that are the same as those present in some astrophysical systems. Indeed, they are essential for this. We saw in the case of EOS that such measurements under identical conditions were all one might hope for. The situation with regard to opacity is somewhat better, because opacities depend fundamentally upon quantum-mechanical processes within atoms. (In contrast, the EOS depends in part upon chemical interactions among groups of atoms.) The quantum-mechanical processes can be reliably scaled from one atom to another in computations. This is especially true along *isoelectronic sequences*, for which the number of electrons attached to the nucleus is the same. Such scalings may break down when the difference in nuclear charge becomes too large, introducing new issues such as relativistic effects into the calculation. The net effect is that laboratory measurements are essential in determining those opacities that can be measured. In addition, these results can validate computational approaches to calculating opacity of other elements, when such calculations can be scaled to the experiment.

The radiative transfer of energy through a star or a supernova is an example of a process that is complicated and three-dimensional, that is difficult to model, and that cannot be evaluated in a static experiment. Exploding stars create a homologous expansion, with velocity, v, radial distance, r, and time, t, related by $v = r/t$. As a result, each radiating region resides in a velocity gradient and sees plasma receding from it in all directions. In other words, the absorbing regions are always red shifted relative to the emitting regions. The relative motion of any two locations creates Doppler shifts that move any specific emission line out of resonance with itself and (perhaps) into resonance with other lines. For photons emitted in one region to escape the star, they have to pass through "windows" in opacity, where the absorption probability is low. An adequate radiation transfer calculation must include the effects of the Doppler shifts in the opacity line and edge locations, due to the expansion. In due course, laboratory observations may prove to be of great value because of the near-impossibility of incorporating a fully correct treatment of radiation transfer into a computer simulation of an entire system. Experimental examples will be needed to validate (or invalidate) various possible approaches.

At this writing a number of experiments have been conducted to measure the LTE opacities of a variety of materials (e.g., Fe, Ge, Na, Al) at temperatures in the range of 10–200 eV and electron densities of 10^{18} to 10^{22} cm^{-3}, using either lasers or Z-pinches as the energy source. We discussed above the difficulty of obtaining LTE conditions and this is a key issue for all of these experiments. The most common approach uses hohlraums (see Sect. 9.3) to provide an equilibrium radiation environment free of energetic electrons or strong nonthermal emission, either of which could alter the conditions of the sample. Measurement of this temperature is an essential detail. The sample is "tamped" by surrounding it with a low-Z material. This prevents rarefactions from reaching the sample and constrains it so that its density remains uniform and changes slowly. At a chosen moment, one produces a source of high-energy radiation that enters the hohlraum through a small hole, irradiates the sample, and exits the hohlraum through another hole. By making

Fig. 6.6 Iron transmission spectra. Comparison of data (gray line) and calculations for a plasma of 80.2% Fe plus NaF, of density 0.0113 g cm^{-3} and areal density 339 μg cm^{-2} at a temperature of 59 eV. The various models are described in Springer et al. (1992). Reproduced with permission

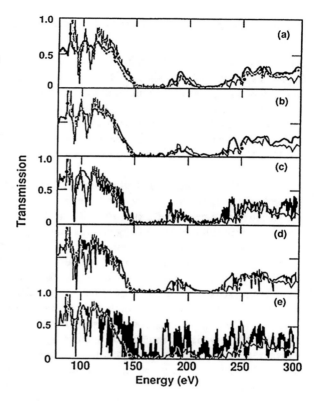

spectral measurements of this radiation, both through the sample and around the hohlraum, one can determine the spectral transmission through the sample and thus the total spectral opacity.

As an example we show in Fig. 6.6 results, from Springer et al. (1992), of this type of measurement of the opacity of Fe at $T_e = 59$ eV and $\rho = 11$ mg/cm^3. The electron temperature was measured using the spectrum from a Na dopant. The 1D radiographic spatial imaging gave the sample density of the thin Fe foil (sandwiched between tamping layers). Hence, the opacity of Fe was measured for known conditions of T_e and ρ. The experimental results shown in Fig. 6.6 were compared with several different opacity calculations employing different approximations. The conclusion of this work was an unambiguous demonstration of the need to include quantum-mechanical term splitting in the opacity calculations. Models that neglect this, such as DCA (panel e), significantly underpredict the opacity.

6.2.4   Non-LTE Models and Atomic Processes

At lower densities, typical of the plasmas penetrated by laser light (or produced in Z pinches away from any wires or imploded cores), neither the excited-state nor the ionization-state populations will be in their LTE distribution. In this case one must evaluate numerically the effect of all the relevant atomic processes to determine the populations. Such calculations are known as *collisional-radiative* calculations, and models performing such calculations are often described as *non-LTE (NLTE)* models. Here we discuss how such models work and then discuss the processes involved.

In a complete collisional-radiative model, one would account for all the processes through which electrons and photons interact with ions. The calculation of rates for the various processes makes extensive use of statistical detailed-balance arguments, since in equilibrium the rate of any process and its inverse must be equal. Table 6.1 lists the sets of inverse processes, now described. The first two lines of the table show collisional mechanisms that change the state of one electron, with the participation of another electron that is free. The simplest is electron-impact ionization, in which a free electron strikes an ion and knocks out one of the bound electrons. The inverse of this process is three-body recombination, in which a free electron becomes attached to an ion, reducing its ionization state, with the participation of another free electron to enable conservation of momentum and energy. The second line of the table shows the corresponding pair of processes for excitation and de-excitation, in which a bound electron changes its state, and a free electron either gives or takes energy equal to the change in the binding energy of the bound electron.

The third and fourth lines in Table 6.1 show the analogous processes in which the third particle is a photon, and the quantum state of one electron is changed. In photoionization, a photon knocks a bound electron out of an ion, changing its energy and making it free. At energies relevant here, the process is photo-electric, so that the photon is fully absorbed and the freed electron carries the energy of the photon less the ionization energy. (At some tens of keV photon energy, Compton scattering becomes dominant over photo-electric absorption.) In radiative recombination, an initially free electron becomes attached to an ion, with the emission of a photon to conserve momentum and energy. The left image in Fig. 6.7 illustrates this process. Similarly, in photoexcitation and radiative de-excitation, a bound electron changes its state, with absorption or emission of a photon.

Table 6.1 Table of inverse atomic processes

Process	Inverse process
Electron-impact ionization	Three-body recombination
Electron-impact excitation	Electron-impact de-excitation
Photoionization	Radiative recombination
Photoexcitation	Radiative de-excitation
Autoionization	Di-electronic capture

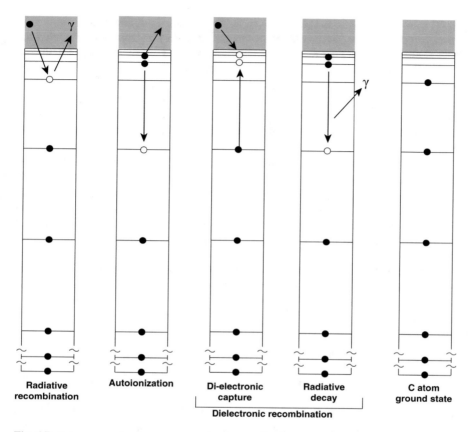

Fig. 6.7 Processes and sub-processes related to radiative and dielectronic recombination of singly ionized carbon. The energy levels are distributed vertically, with the gray representing the continuum of free electronic states. The solid dots show the electrons in neutral or singly ionized carbon, including the two, deeply trapped 1s electrons, the two 2s electrons, and one or two 2p electrons. Dielectronic recombination occurs in the fraction of cases where di-electronic capture is followed by radiative decay rather than by autoionization

   The above four sets of processes all involve a single bound electron. But there is no rule of physics that says that only one bound electron can be in an excited state. And there is one process involving a doubly excited ion that is of practical importance, found on the fifth line of Table 6.1 and shown in the second image of Fig. 6.7. Autoionization begins with a doubly excited ion, in which the ionization energy of the electron in a higher quantum state is smaller than the de-excitation energy of the electron in a lower quantum state. Such doubly excited states can be described as autoionizing states. They can then decay by autoionization: the emission of the higher electron and the de-excitation of the lower one. Autoionization competes with the radiative decay of the lower electron, so that statistically some fraction of autoionizing states become singly-excited ions and the remainder become ionized.

The inverse process associated with autoionization is di-electronic capture. In di-electronic capture, a free electron joins an ion with the excitation of one of the bound electrons. This is one way to create an autoionizing state. In the fraction of cases where the autoionizing state relaxes by radiative decay, the two-step sequence has the net effect of attaching a free electron to an ion, reducing the ionization state. This process is known as dielectronic recombination, and is of substantial practical importance. The third and fourth images in Fig. 6.7 show this case. The outer electron then decays into its ground state to produce, in the case shown in the fifth image, a neutral C atom.

We ignore another process in which a given ion acquires an electron, known as charge exchange. In charge exchange, one particle gives an electron to a charged particle and becomes more ionized in turn. The rates for this process are relatively high only for the interaction of neutral atoms with ions, for which the Coulomb forces are small. But for charge exchange to be important, there must be a substantial density of neutral atoms available. This is typically not the case in high-energy-density systems.

To construct a collisional-radiative model, one solves simultaneously a set of rate equations that incorporate all the processes that change the population of any given state. The rate equation describing the evolution in time of the density of ionization state j is

$$\frac{dn_j}{dt} = n_{(j-1)}cN_\gamma\bar{\sigma}_{(j-1)} - n_j cN_\gamma\bar{\sigma}_j + n_{(j-1)}n_e\langle\sigma_{ei}v\rangle_{(j-1)}$$

$$-n_j n_e\langle\sigma_{ei}v\rangle_{(j)} + n_{(j+1)}n_e\mathcal{R}_{(j+1)} - n_j n_e\mathcal{R}_j, \qquad (6.29)$$

in which the subscripts involving j, $(j-1)$, and $(j+1)$ designate the ionization state, the next-lower, the next-upper states, respectively, the photon density is N_γ and the appropriately averaged photoionization or photo-excitation cross section is $\bar{\sigma}$, discussed further below. The electron density is n_e, and the rate coefficient for electron-impact ionization is $\langle\sigma_{ei}v\rangle$. The total rate coefficient for lowering the ionization state by recombination is shown as $\mathcal{R}$, having in principle contributions from collisional (three-body) recombination, radiative recombination, and dielectronic recombination, all discussed below.

One can write a similar equation for the population of a given excited state within some ionization state. However, one will need in principle to consider transitions to all other accessible excited states. In most cases these will be negligible, but decay into the ground state, when possible, is a strong process, and excitation from the ground state may be large if the ground-state population is large. One figures out which states matter by looking at the distribution of population and at the specific rates. Codes that do calculations of state distributions, such as FLYCHK or FAC, typically include all the interactions among some range of excited states, such as those up to some principal quantum number. Codes that evaluate opacities, typically assuming equilibrium conditions, evaluate the contributions of many more states (at times millions). We proceed now to consider the processes that enter into (6.29).

The fundamental approach will be to show standard fits to the rate coefficients, using actual ionization energies for the ionization states and, when needed, a hydrogenic model for the excited states. This is a common practice in both models and codes.

6.2.4.1   Collisional Rate Coefficients

Here we discuss two sets of inverse processes for changes created by collisions. The first set is collisional ionization (by electron impact) and collisional recombination. The second set is collisional excitation and collisional de-excitation, in which the state of the electron (described by its quantum numbers) is changed without a change of ionization level.

Electron collisions are rapid, and so electrons typically thermalize quite rapidly in dense, laboratory plasmas. Here we assume that the electron distribution has become Maxwellian. In the case of electron-impact ionization by Maxwellian distributions of electrons, the relevant basic parameter is the rate coefficient, which is an average of the cross-section σ_{ei} times the electron velocity v, denoted here by $\langle \sigma_{ei} v \rangle$. The contribution to the rate of ionization from state j to $j + 1$ by this process is then $n_e \langle \sigma_{ei} v \rangle_j$, in which n_e is the electron density. The number of ionizations per unit volume per unit time is $n_e n_j \langle \sigma_{ei} v \rangle_j$.

Early historic formulae had $\langle \sigma_{ei} v \rangle$ proportional to e^{-U}/U, where $U = E_{th}/(k_B T_e)$, where E_{th} is again the ionization energy. In his iconic formula for electron-impact ionization, Lotz (1967) replaced the exponential with an exponential integral function, obtaining:

$$\langle \sigma_{ei} v \rangle = 3 \times 10^{-6} \frac{\xi_{Znl}}{T_{eV}^{3/2}} \frac{\text{ExpIntegralE}[1, U]}{U} \text{ cm}^3/\text{s}, \tag{6.30}$$

in which the number of electrons in the shell being ionized is ξ_{Znl} and ExpIntegralE$[1, U] = \int_U^\infty (e^{-t}/t)dt$. This formula remains reasonably accurate. For the excited states, a hydrogenic model is reasonably accurate. The Lotz formula then applies to these states by replacing E_{th} with $E_{th,n} = E_{th}/n^2$ in (6.30).

Figure 6.8 shows the rate coefficients for ionization of carbon, and Table 6.2 shows the values of E_{th} for its ionization states. Note that the four, p-shell electrons ionize more readily than one would expect from a hydrogenic model, while the two $1s$ electrons have E_{th} near the value given by the hydrogenic model. In principle, there is a finite rate of ionization at any finite temperature, because the electron distribution includes some particles above the ionization threshold. In practice, the rate coefficient becomes significant once T_e reaches about $E_{th}/3$. (More precisely, it reaches 1% and 10% of its maximum value at 0.23 E_{th} and 0.45 E_{th}, respectively.) It then continues to rise with T_e until eventually reaching a peak when $T_e \sim 10 E_{th}$, above which it slowly decreases. The behavior at higher temperature is often unimportant, because the rate coefficient for the next higher ionization state becomes significant as T_e reaches $\sim 1/3$ the ionization energy of this next state.

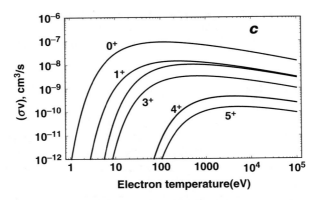

Fig. 6.8 Rate coefficients for electron-impact ionization from the ground states of C ions, for Maxwellian electrons, based on Lotz (1967). The leftmost curve is neutral C, and the rightmost curve is C 5^+. Note the large gap between the curves for Li-like C 3^+ and He-like C 4^+, reflecting the large increase in ionization energy

Table 6.2 Table of threshold ionization energies for carbon

C ionization state	0^+	1^+	2^+	3^+	4^+	5^+
Ionization threshold (eV)	11.3	24.4	47.9	64.5	392	490

Note that the ionization rates can be very fast. For a density near 10^{21} cm^{-3} and $\langle \sigma_{ei} v \rangle \sim 10^{-9}$ cm^3/s, the ionization timescale is 1 ps. As a result, a given species often ionizes rapidly up to its steady-state value. If the system is dense enough, this will be the equilibrium value. Our crude estimate of the equilibrium value in Chap. 3 corresponds to $E_{th} \sim 5T_e$. When radiative effects and dielectronic recombination are negligible, one tends to find a value more like $3T_e$.

Turning to collisional, three-body recombination, one obtains rate coefficients by detailed-balance arguments relative to collisional ionization. Defining the three-body recombination rate coefficient into state n by α_n, the detailed balance relation is expressed as

$$n_e n_{(j,n)} \langle \sigma_{ei} v \rangle_{(j,n)} = \alpha_n n_e^2 n_{j+1}. \qquad (6.31)$$

One applies the Saha equation to express the relation between the level populations in detailed balance. The numerical result is often expressed in effective two-body form, so that

$$\alpha_{TBR} = (\alpha_n n_e) = 1.65 \times 10^{-22} \langle \sigma_{ei} v \rangle_{j,n} \frac{n_e n^2}{T_{eV}^{3/2}} \exp\left(\frac{E_{th,n}}{T_{eV}}\right). \qquad (6.32)$$

One can substitute the Lotz formula into this equation. Then the factor of U in the denominator of $\langle \sigma_{ei} v \rangle_{j,n}$ implies that the fundamental scaling of $(\alpha_n n_e)$ is as n^4. One obtains

Fig. 6.9 Rate coefficients for three-body recombination into the ground ($n = 1$) states of C ions, for $n_e = 10^{21}$ cm^{-3}. These coefficients are proportional to n_e and to n^4. The uppermost curve is neutral C, and the bottom curve is C 5$^+$. Note the large gap between the curves for Li-like C 3$^+$ and He-like C 4$^+$, reflecting the large increase in ionization energy

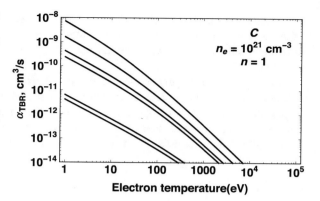

$$\alpha_{TBR} = 4.95 \times 10^{-8} \frac{n_{e20} n^4 \xi_{Zn\ell}}{E_{th} T_{eV}^2} \exp\left(\frac{E_{th,n}}{T_{eV}}\right) \text{ExpIntegral}\left[1, \frac{E_{th,n}}{T_e}\right], \tag{6.33}$$

where $n_{e20} = n_e(\text{cm}^{-3})/10^{20}$. Figure 6.9 shows results for carbon ions in the ground state. These appear low in comparison to the ionization rate coefficients of Fig. 6.8, but note that at low enough temperatures they exceed the ionization rate coefficients. Whenever the recombination rate into a given state exceeds the ionization rate for that state, the state above it will depopulate and the average ionization will decrease. Because α_{TBR} increases with n_e while $\langle \sigma_{ei} v \rangle$ does not, increasing n_e can make three-body recombination dominant and lead to a decrease in ionization. In contrast, decreasing n_e makes three-body recombination unimportant in comparison with radiative and dielectronic recombination (discussed below). In the next section, we discuss the implications of the fact that all the collisional rates increase rapidly with n.

Collisional excitation and de-excitation describe collisionally induced transitions of electrons between specific excited states $n\ell \to n'\ell'$. The overall behavior of these rates is similar to those of ionization. Salzman (1998) attributes to van Regemorter a commonly used formula,

$$1.6 \times 10^{-6} \frac{1}{T_{eV}^{3/2}} \frac{e^{-U}}{U} G(U) f_{(n\ell \to n'\ell')} \text{ cm}^3/\text{s}, \tag{6.34}$$

for the rate coefficient. Here $U = \Delta E/(k_B T_e)$, where ΔE is the energy difference of the transition. As a result, U becomes rapidly smaller for transitions among highly excited states. In addition, $G(U)$ is a Gaunt factor, accounting for some quantum mechanical corrections and having a maximum value below 1, and $f_{(n\ell \to n'\ell')}$ is the quantum-mechanical oscillator strength, whose maximum value approaches unity but which is often much smaller. The author is not aware of a simple parameterization of the oscillator strength, and engaging the quantum mechanical calculations is beyond the scope of this book. The reader can begin with the other books mentioned, if there is a need to explore these issues. Comparing (6.34) with (6.30), one can see that small values of $f_{(n\ell \to n'\ell')}$ will tend to make this coefficient smaller than that for ionization while small values of U will have the reverse effect.

All the collisional processes just discussed become much more rapid as n increases, reflecting the decreasing value of the transition energies. This drives the upper excited states toward Saha equilibrium with the ions in the next-higher ionization state, as we discuss in the next section. In dense plasmas, though, these effects are limited as continuum lowering eliminates many of the most-excited states.

6.2.4.2 States in Saha Equilibrium

For excited states, radiative decay rates decline rapidly with increasing principal quantum number, n, while those for ionization into the ground state of the next-higher ionization state increase, as do three-body recombination rates from it. As a result, states above some principal quantum number n_o come into collisional equilibrium with the population of the ground state of the next higher ionization state. In effect, the ground state population of any given ionization state is shared with some range of excited states of the next lower ionization state.

One can estimate n_o as follows, for the excited states of an ion of charge $Z_u - 1$. Sobel'man et al. (2013) give a formula for the Einstein A coefficient,

$$A = 7.89 \times 10^9 \frac{Z_u^4}{n^5} \tau(n), \qquad (6.35)$$

in which, for the range of n that is relevant here, their tabulated values of $\tau(n)$ are reasonably fit by $\tau(n) = 4\sqrt{n/2} - 2$. For collisional, electron-impact ionization, use the formula of Lotz (6.30). Use the actual ionization energies along with the hydrogenic model discussed above for the excited states, and note that $\xi_{Znl} = 2n^2$ if one does not discriminate by orbital angular momentum. Thus we replace U by U/n^2 in (6.30). We denote the rate coefficient for ionization state j and principal quantum number n by $\langle \sigma_{ei} v \rangle_{j,n}$. One sets $n_e \langle \sigma_{ei} v \rangle_{j,n} = A$ from (6.35) and solves for n_o, finding the value above which the states will be in Saha equilibrium, obtaining the results shown in Fig. 6.10 for C at selected densities.

The consequence is that, for most conditions of interest at high energy density, only recombination into the ground state or perhaps the $n = 2$ state contributes meaningfully to the determination of the ionization state distribution under typical conditions of interest. The other states, all of which are within $E_{th}/4$ of the ground state energy of the next ionization state, are in Saha equilibrium.

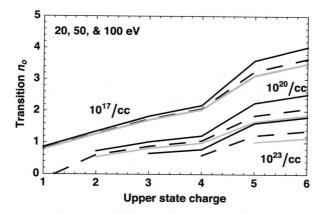

Fig. 6.10 Value of principal quantum number above which states are in Saha equilibrium, n_o, for carbon for three ion densities as labeled. This figure shows three sets of three curves, one for each density as indicated. For each density the lower curve is for $T_e = 100$ eV, the middle curve is for $T_e = 50$ eV, and the upper curve is for $T_e = 20$ eV

6.2.4.3   Photoionization

The inverse processes that involve radiation and changes of ionization state are photoionization and radiative recombination. The principle of detailed balance is used to relate the cross sections for these two processes. In practice, however, researchers have developed fitting formulae that independently capture the behavior of each process. We will consider them in turn.

Photoionization cross sections for the outer shells of atoms and ions turn out to be fit well as $\sigma(E) = \sigma_o G(E)$, where σ_o is in Mb and

$$G(E) = \left(\left[(x-1)^2 + y_w^2 \right] y^{(0.5P-5.5)} \left[1 + \sqrt{y/y_a} \right]^{-P} \right) \mathcal{H}(E - E_{th}), \qquad (6.36)$$

where E_{th} is a threshold energy, $\mathcal{H}$ is the Heaviside step function, $y = \sqrt{x^2 + y_1^2}$, and $x = (E/E_o) - y_o$. Here the tabulated fit parameters are $\sigma_o, E_{th}, E_o, P, y_w, y_a, y_o$ and y_1. For the specific cases of C, N, and O, this gives the total cross section for ionization from the 2p and 2s shells. There is an additional tabulated parameter E_{max} that gives the energy where photoionization of the first inner shell reaches threshold, at which point this becomes the dominant mechanism and the formula described next must be used. These fits are discussed in Verner et al. (1996) and the parameters are available at http://www.pa.uky.edu/~verner/atom.html.

The inner shell photoionization cross sections are fit by a similar but different formula, described in Verner and Yakovlev (1995), with fit parameters available at the same website. One has $\sigma(E, n, \ell) = \sigma_o K(E, n, \ell)$, where σ_o is in Mb and

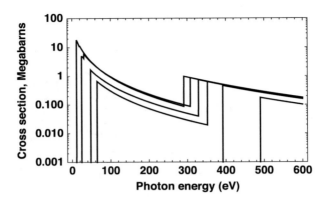

Fig. 6.11 Total cross sections for photoionization of the ionization states of carbon. The lower ionization states have progressively higher cross sections and lower thresholds

$$K(E, n, \ell) = \left(\left[(y - 1)^2 + y_w^2 \right] y^{-Q} \left[1 + \sqrt{y/y_a} \right]^{-P} \right) \mathcal{H}(E - E_{th}), \qquad (6.37)$$

where E_{th} is a threshold energy, $\mathcal{H}$ is the Heaviside step function, $y = (E/E_o)$, and $Q = 5.5 + \ell - 0.5P$, with ℓ being the quantum number for orbital angular momentum. Here the tabulated fit parameters are $\sigma_o, E_{th}, E_o, P, y_w$, and y_a and fits are given for each available set of quantum numbers n, ℓ. Figure 6.11 shows the total photoionization cross sections for carbon.

For the rate equation above, it is the spectral average of the photoionization cross section over the source spectrum that is important. One defines this cross section, $\bar{\sigma}$, by means of the following equation, denoting the total photon density as a function of photon energy E_γ as $N_\gamma = \int_0^\infty N_E(E_\gamma) dE_\gamma$,

$$N_\gamma \bar{\sigma} = \int_{E_{th}}^\infty \sigma(E_\gamma) N_E(E_\gamma) dE_\gamma, \qquad (6.38)$$

in which $\sigma(E_\gamma)$ is the sum of the value from Eq. (6.36) for energies up to E_{max} and the value from Eq. (6.37) for the 1s shell above E_{max}. If needed, one can define averaged subshell cross sections for subshells using the same equation, replacing $\sigma(E)$ by $\sigma(E, n, \ell)$ in the integral. As a result, the subshell values add to equal $\bar{\sigma}$.

6.2.4.4 Radiative and Dielectronic Recombination

Here we must discuss radiative recombination and dielectronic recombination. In both cases the recombination rate is $n_e \alpha$, where the rate coefficient is α, usually given in cm^3 per second. Dielectronic recombination is not possible for recombination producing H-like species. As Verner and Ferland (1996) summarize, radiative recombination also dominates at low temperature for recombination producing He-like and Li-like species, since the only possible dielectronic recombination pathways involve excitation of the tightly bound 1s electrons. It also turns out that radiative recombination is dominant for recombination to Na-like species, creating one 3s electron outside the closed $n = 2$ shell. Dielectronic recombination dominates otherwise.

The radiative recombination coefficients are fit to find

$$\alpha_R(T) = a \left[\sqrt{T/T_o} \left(1 + \sqrt{T/T_o} \right)^{(1-b)} \left(1 + \sqrt{T/T_1} \right)^{(1+b)} \right]^{-1}, \qquad (6.39)$$

where tabulated fitting parameters are T_o, T_1, a, and b. Table 6.3 shows these parameters for some materials. Note that Si IV is Na-like, and has only one 3s electron outside a closed 2p shell.

Turning to dielectronic recombination, the total rate coefficient is the sum over the rates for each allowed pathway. As a result, the dielectronic recombination rate coefficients are something of a mess. We tabulate the relevant formulae in Table 6.4. The formulae are also shown in Salzman (1998). Here N is the number of electrons on the target ion before the capture increases this number, and Z_c is the nuclear charge. Thus the ionization state before capture is $Z = Z_c - N$.

Table 6.3 Examples of fitting parameters for radiative recombination, from Verner and Ferland (1996)

State	Z	N_V	a (cm³/s)	b	T_o (eV)	T_1 (eV)
C IV	6	3	8.540×10^{-11}	0.525	0.043	1275
C V	6	2	2.765×10^{-10}	0.686	0.013	2200
C VI	6	1	6.556×10^{-10}	0.757	0.006	2109
N V	7	3	1.169×10^{-10}	0.547	0.059	1420
N VI	7	2	3.910×10^{-10}	0.699	0.014	2820
N VII	7	1	7.586×10^{-10}	0.756	0.008	2880
O VI	8	3	2.053×10^{-10}	0.602	0.041	1480
O VII	8	2	4.897×10^{-10}	0.705	0.016	3530
O VIII	8	1	8.616×10^{-10}	0.756	0.010	3750
Ne VIII	10	3	3.200×10^{-10}	0.620	0.055	2260
Ne IX	10	2	6.161×10^{-10}	0.703	0.028	5380
Ne X	10	1	1.085×10^{-09}	0.757	0.016	5840
Si IV	14	11	5.942×10^{-11}	0.393	0.077	1050
Si XII	14	3	5.373×10^{-10}	0.634	0.100	4030
Si XIII	14	2	8.722×10^{-10}	0.700	0.061	9960
Si XIV	14	1	1.517×10^{-09}	0.757	0.031	11,500

The state shown is that after the recombination, and N_V is the number of bound electrons after the recombination. The states are labeled using the spectroscopic convention, in which the neutral state is I, the 1+ state is II, and etc.

Table 6.4 Fitting formulae for dielectronic recombination rate coefficients, in units of 10^{-13} cm^3/s, with T in keV

Excited electron	Formulae
1s	$\alpha_R = \left[A_1 e^{-A_2/T} T^{-3/2}\right] \exp\left[-A_3(N-2)^2\right]\left[6/(4+N)\right]^{0.9}$ $A_1 = 1230 \exp\left[-44/(Z_c + 2.86)\right] Z_c^{-0.14}$ $A_2 = 0.0075(Z_c + 1/N)^2$ $A_3 = 0.0222 Z_c$
2s	$\alpha_R = \left[B_1 e^{-B_2/T} T^{-3/2}\right](N-2)(10-N)(N+B_3)^{-2.5}(1 + 0.3/T^{0.21})$ $B_1 = 52 \exp\left[-18/(Z_c - 1)\right]\left[Z_c(10 + 0.011 Z_c^2)^{-1}\right]^{0.65}$ $B_2 = 0.0023(Z_c - 2)(1 + 0.0015(Z_c - 2)^2)^{-1}$ and $B_3 = 0.8$
2p	$\alpha_R = C_1(F_c T^{-3/2}) \exp\left[-C_2\left(1 + 0.0001/(N+1)\right)/T\right]$ $C_1 = 2.15 \exp\left[-0.004(Z_c - 37)^2\right](Z_c - 2)^{1.8}$ $C_2 = 0.00115(Z_c - 2)^2(1 - 0.003(Z_c - 2))$ and $C_3 = 0.17$ $F_c = (10/(N+1))^{1/2} \exp\left[-C_3 \lvert N - 9.6\rvert\right]$
3s	$\alpha_R = (D_1 T^{-3/2}) \exp\left[-(D_2/T)(1 - 0.15(N-10)^{-1.5})\right] \times$ $(N-10)(Z_c - N)(N - D_3)^{-1}$ $D_1 = 0.16 Z_m^2 \exp\left[-0.11 Z_m\right]$ $D_2 = 0.0024 Z_m(1 - 0.01 Z_m)$ and $D_3 = 6$ $Z_m = Z_c - 10$
3p	$\alpha_R = (E_1 T^{-3/2}) \exp\left[-E_2/T\right] \exp\left[-E_3(N-12)^2\right](N-10)(Z_c - 10)/8$ $E_1 = (0.45/Z_m) \exp\left[Z_m/(4 + 0.02 Z_m)\right]$ $E_2 = 0.0003(Z_c - 7)^2(1 - 0.003(Z_c - 7))$ and $E_3 = 0.02$

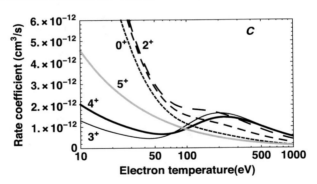

Fig. 6.12 Comparison of radiative plus dielectronic recombination rate coefficients for C. The curves are labeled by the ionization state following recombination. The medium-dashed curve shows the result for the 1^+ state, which lies between those for 0^+ and 2^+. The text discusses the details

Figure 6.12 shows the sum of the rate coefficients for radiative and dielectronic recombination of C ions. These rates are independent of density. The gray curve labeled 5^+ shows the purely radiative recombination of the bare nucleus, for which dielectronic recombination is impossible. Radiative recombination becomes slower as the charge of the ion decreases, which accounts for the low-temperature behavior of the solid black curves labeled 3^+ and 4^+. At higher temperatures, the recombination rate for these states increases due to the emergence of dielectronic recombination involving the 1s electron. At low temperatures, the recombination rate into the 0^+, 1^+ and 2^+ states is dominated by dielectronic recombination involving the 2s and/or 2p electrons. The latter two of these also show some impact of the 1s state at higher temperatures.

At sufficiently high densities, the rates of dielectronic recombination decrease because continuum lowering progressively removes the excited states that can participate. This effect is roughly co-incident with the emergence of three-body recombination to become the dominant process. We will leave the question of whether there is a zone in parameter space where the total rate of recombination is reduced to the atomic physicists.

For the dominant ionization states, the magnitude of the total recombination rate coefficient will be in the range of 1 to a few $\times 10^{-12}$ cm^3/s, so the timescale for radiative and dielectronic recombination at ion densities $\sim 10^{21}$ cm^{-3} will be within a factor of a few of 1 ns.

6.3 Radiation Transfer

To account for radiation transfer, we return to the notation introduced at the beginning of the previous section. We will assume that someone has evaluated (or could evaluate) all the detailed mechanisms that produce, absorb, and scatter radiation, and has produced values of emissivity η and total opacity χ as functions of the plasma parameters. This enables us to consider the absorption and emission of radiation along a ray as it propagates. Since we are concerned with thermal radiation at energies of tens of eV or more, we develop the corresponding equations in the geometric-optics limit, assuming that the rays move in straight lines. This is generally a good assumption except when it matters that rays propagate along a path nearly orthogonal to a density gradient, as in some soft-X-ray-laser designs.

6.3.1 The Radiation Transfer Equation

The radiation transfer equation is no more than an accounting of the change in radiation intensity due to sources and losses of radiation in a specific element of phase space. In this case the phase space includes an element of solid angle, $d\Omega$, about a direction, n, an element of frequency, $d\nu$, about a frequency, ν, and an element of volume of length ds and cross-sectional area $dA \perp n$ beginning at position x. The net rate that energy from this element of phase space adds to the radiation intensity in some direction is the difference between the rate that energy enters the element at x and t and the rate that energy leaves it at $x + nds, t + \Delta t$, where $\Delta t = ds/c$. Mathematically, we can write this as

$$\left[I_\nu \left(x + n ds, t + \frac{ds}{c}, n, \nu \right) - I_\nu(x, t, n, \nu) \right] dA d\Omega d\nu$$

$$\rightarrow \left[\frac{1}{c} \frac{\partial I_\nu}{\partial t} + \frac{\partial I_\nu}{\partial s} \right] ds dA d\Omega d\nu, \tag{6.40}$$

which has units of energy per unit time. What causes this increase in intensity is the difference between the total rate of energy emission from the medium into the element of phase space, which includes scattering from other elements of phase space, and the total rate of energy removal from the radiation by absorption or by scattering into other elements of phase space. In terms of the total spectral emissivity, η_ν, and the total opacity, χ_ν, we can express this conceptual description mathematically as

$$\left[\frac{1}{c} \frac{\partial}{\partial t} + \frac{\partial}{\partial s} \right] I_\nu(x, t, n, \nu) = \eta_\nu(x, t, n, \nu) - \chi_\nu(x, t, n, \nu) I_\nu(x, t, n, \nu). \tag{6.41}$$

This is the "classical" equation of radiation transfer. It is perhaps most accurately thought of as an equation for photons treated as particles, which is an excellent approximation for the X-ray photons that carry the thermal energy in high-temperature plasmas. Wave effects, including diffraction, refraction, and polarization are ignored. Note that, unlike most of the equations above, one cannot obtain an equation for the total intensity by integrating this equation without making a severe simplifying assumption about the frequency dependence of χ_ν. Also note that this form of the equation will be most useful when scattering, including absorption and reemission, is not a central feature of the radiation dynamics. If, for example, the emissivity includes significant scattering from other angles (or other frequencies), then the emissivity involves an integral over I_ν times a scattering coefficient and (6.41) becomes an integro-differential equation. For example, the problem of radiation transport through an expanding envelope, as in a supernova, introduces just this sort of complexity.

It is worthwhile for ease of applications to elaborate on $\partial/\partial s$. From the chain rule we can write

$$\frac{\partial I_\nu}{\partial s} = \frac{\partial I_\nu}{\partial x} \cdot \frac{\partial x}{\partial s} = n \cdot \nabla I_\nu + \frac{\partial n}{\partial s} \cdot \frac{\partial I_\nu}{\partial n}, \tag{6.42}$$

in which $\partial/\partial x$ is equivalent to the gradient operator. The second equality follows because the jth component of the position vector, x_j, along n is $x_j = s \cos \alpha_j$, where the direction cosine for the jth component of x is $\cos \alpha_j$. Of course, n is the unit vector composed of these direction cosines. In a Cartesian coordinate system, α_j is fixed and only the first term on the right is nonzero. In curvilinear coordinate systems, α_j varies along s. This can lead to very complicated expressions in general cases. In a standard spherical coordinate system, for example, one needs three variables (r, θ, ϕ) to specify the location of a point on the ray and in addition two variables (a polar angle Θ and an azimuthal angle Φ) to specify the direction of the ray with respect to the local radial direction. In a spherically symmetric system, such as a star treated as a symmetric object, the location is fully specified by r. At any specific point, the radiation intensity varies with direction, but it is symmetric about the local radius vector. As a result, one needs a single angle, Θ, to specify the local direction of the ray. Defining $\mu = \cos \Theta$, one can show that

$$\left[\frac{1}{c} \frac{\partial}{\partial t} + \frac{\partial}{\partial s} \right] I_\nu(\boldsymbol{x}, t, \boldsymbol{n}, \nu) = \left[\frac{1}{c} \frac{\partial}{\partial t} + \mu \frac{\partial}{\partial r} + \frac{(1 - \mu^2)}{r} \frac{\partial}{\partial \mu} \right] I_\nu(r, t, \mu, \nu).$$

(6.43)

6.3.2   Radiation Transfer Calculations

We have at last come to the end of our first task relating to radiation transfer. We have defined the properties of radiation and have developed an Eq. (6.41) accounting for its interactions with matter. We now face the problem of actually describing this interaction, and of developing applied equations that will prove useful in various limits. In many circumstances, the time derivative in (6.41) can be neglected, as the motion of the radiation is effectively instantaneous in comparison with that of the matter. We now develop a sequence of models that we will use in the next two chapters to describe radiation transfer under various conditions.

6.3.2.1   Direct Solutions of the Radiation Transfer Equation

To solve (6.41), it is often useful to normalize (6.41) by the opacity. To do so, we introduce a new variable known as the *optical depth*, τ_ν, which is inherently a function of frequency. We define an infinitesimal increment of optical depth as

$$d\tau_\nu = \chi_\nu ds.$$

(6.44)

Thus, the optical depth, at frequency ν, between point s and point s_o, is

$$\tau_\nu = \int_s^{s_o} \chi_\nu(s') ds'.$$

(6.45)

In applications, sign conventions are usually chosen so that the optical depth is a positive quantity. This often takes care of itself. For example, if radiation is propagating in the $-z$ direction then $ds = -dz$ and τ_ν becomes the integral of the opacity from smaller to larger z. A layer of material is said to be *optically thick* (at some frequency) if $\tau_\nu \gg 1$, and *optically thin* if $\tau_\nu \ll 1$. All materials are optically thick at long enough wavelengths and optically thin at short enough wavelengths. (We should also mention that the use of optical depth and related terms is not always consistent in the literature and can be misleading. In some experimental work, the transmission of a layer may be given as its "optical depth." In contrast, the optical depth of (6.45) is the natural logarithm of the transmission. In optics, one encounters filters designated by "O.D." However, these initials stand for "optical density", not optical depth. Optical density is consistently defined as the logarithm to the base 10 of the transmission.)

To complete the normalization of (6.41), we define the source function, S_ν, as

$$S_\nu = \eta_\nu / \chi_\nu. \tag{6.46}$$

As an application of Kirchoff's law, discussed above, the source function due to a Maxwellian distribution of electrons is

$$S_\nu = B_\nu(T). \tag{6.47}$$

The resulting simplified version of (6.41), for steady radiation, is

$$\frac{\partial I_\nu}{\partial \tau_\nu} = I_\nu - S_\nu. \tag{6.48}$$

One can see by normalizing (6.41) that the spectral intensity will be driven toward S_ν as optical depth becomes large. We will frequently constrain the electron distribution to be Maxwellian, as opposed to specifying that the system be in equilibrium. For (6.47) to describe the emission, it is a necessary and sufficient condition that the electron distribution be Maxwellian. This often occurs, and (6.47) often describes the source function, even in systems that are far from complete equilibrium.

Solutions to the time-independent radiation transfer equation are often useful. First, note that (6.48) has an integrating factor, which is just $\exp(-\tau_\nu)$. With this realization, integrating and simplifying this equation gives

$$I_\nu(x_o + ns) = e^{-\tau_\nu(0,s)} I_\nu(x_o) + \int_0^s S_\nu(x_o + ns') e^{-\tau_\nu(s',s)} ds', \tag{6.49}$$

where the optical depth $\tau_\nu(a, b)$ is the integral of χ_ν along n from $s = a$ to $s = b$. This shows how the intensity at some propagation distance s from an initial point x_o (where $s = 0$) is the transmitted intensity from x_o plus the attenuated contribution from the emission at each intervening point.

For planar systems, or systems that otherwise have a preferred z axis, the definition of τ_ν in (6.45) becomes impractical, because it gives a different value of τ_ν for each direction of propagation through a layer of some thickness. In this case, it is practical to define $d\tau_\nu = -\chi_\nu dz$, so that $d\tau_\nu = -\mu\chi_\nu ds$, where $\mu = \cos\theta$, with θ being the polar angle relative to the z axis. Then (6.41) becomes

$$\mu \frac{\partial I_\nu}{\partial \tau_\nu} = I_\nu - S_\nu. \tag{6.50}$$

In this case, the more oblique a ray is the smaller μ is, and the more quickly I_ν is driven to S_ν as z increases.

The solution of (6.50), with $ds = dz$ and $d\tau_\nu = \chi_\nu dz$, and $\tau_\nu = 0$ at a specified value of z, is

$$I_\nu(\tau_\nu) = e^{-(|\tau_\nu - \tau_o|/\mu)} I_\nu(\tau_o) + \int_{\tau_\nu}^{\tau_o} S_\nu(\tau') e^{-(|\tau' - \tau_\nu|/\mu)} \frac{d\tau'}{\mu}. \tag{6.51}$$

This solution is essentially identical to the previous one, except that it accounts for the variation in optical depth with angle through the introduction of $\mu = \cos\theta$. For applications in which an approximate treatment of the total intensity is relevant, one can define $d\tau = \kappa dz$ and integrate (6.57) to find

$$I_R(\tau) = e^{-|\tau - \tau_o|/\mu} I_R(\tau_o) + \int_{\tau_o}^{\tau} B(\tau') e^{-|\tau - \tau'|/\mu} \frac{d\tau'}{\mu}. \tag{6.52}$$

We will use this in the next chapter.

6.3.2.2   The Radiation Transport Regime

The behavior of the matter does not generally depend directly on the radiation intensity, but rather on the divergence of the net radiation flux. In order to use the radiation transfer equation to find the net radiation flux, one must solve for the rays in all directions and do the integral as shown in (6.9). This is complex.

Simpler results can often be obtained by making some assumptions that are often correct. Suppose that the thermal emission is given by (6.23), and that the scattering is isotropic and elastic (or "coherent"), so that it does not change the photon energies. Then we have

$$\eta_\nu = \kappa_\nu B_\nu + \sigma_\nu J_\nu \tag{6.53}$$

so

$$S_\nu = (\kappa_\nu B_\nu + \sigma_\nu J_\nu)/(\kappa_\nu + \sigma_\nu). \tag{6.54}$$

Using this definition of η_ν, one can find from (6.41),

$$\frac{1}{c}\frac{\partial I_\nu}{\partial t} + \frac{\partial I_\nu}{\partial s} = \kappa_\nu(B_\nu - I_\nu) + \sigma_\nu(J_\nu - I_\nu). \tag{6.55}$$

Integrating this equation in frequency one obtains

$$\frac{1}{c}\frac{\partial I_R}{\partial t} + \frac{\partial I_R}{\partial s} = \int \kappa_\nu(B_\nu - I_\nu)d\nu + \int \sigma_\nu(J_\nu - I_\nu)d\nu. \tag{6.56}$$

It is often sensible to approximate the second term on the right-hand side in this equation as negligible, either because the intensity distribution is nearly isotropic, so $I_\nu = J_\nu$, or because the system is optically thin so that it changes I_R primarily by emission, or because the scattering is small relative to the absorption. (However, in some astrophysical systems scattering is large relative to absorption, especially when the radiation is dominantly line radiation.) In the static limit, and with this approximation, we have

$$\frac{\partial I_R}{\partial s} = \kappa \left[B(T) - I_R \right], \tag{6.57}$$

in which κ is a nonlinearly averaged absorption coefficient, approximately equal to the Planck mean opacity, defined and discussed just below. Equation (6.57) may be needed to determine the angular variation in (and integral of) the radiation intensity reaching some surface of interest in an application. One important example is the calculation of the radiation intensity emerging from an optically thin layer of material.

If instead one integrates (6.56) over all solid angle, then the second term on the right vanishes identically. Note that this is equivalent to taking the zeroth moment over angle of (6.41). In Cartesian coordinates, we find

$$\frac{\partial E_R}{\partial t} + \nabla \cdot \boldsymbol{F}_R = 4\pi \int_0^\infty \kappa_\nu \left(B_\nu - J_\nu \right) d\nu \equiv 4\pi \kappa \left(B - J_R \right), \tag{6.58}$$

which relates the overall absorption of radiation to changes in the radiation flux and energy density. Note that $\nabla \cdot \boldsymbol{F}_R = \int (\partial I_R / \partial s)\, d\Omega$. Equation (6.58) is an equation for the radiation energy-density, although conceptually it is the analogue of the continuity equation for mass, in which the rate of change of a density is related to the divergence of a flux and to sources. It is convenient, but also intuitively sensible, that the scattering terms cancel out of this result. The energy density is very often negligible, in which case this becomes a fairly simple equation for $\nabla \cdot \boldsymbol{F}_R$. The regime where this model applies is described as the "transport regime". It is also important to note that (6.58) is only valid for a static medium. In a moving medium, there are both a convective energy flux and pdV work associated with the radiation, as discussed in Chap. 7.

Equation (6.58) also defines an *absorption opacity*, κ. One sees that κ is again a nonlinear average over frequency of κ_ν. This is one of several similar averages that one encounters in simple calculations. The average in (6.57) is also labeled κ, even though it represents a nonlinear average over $B_\nu - I_\nu$ rather than $B_\nu - J_\nu$. The reason we make no distinction here is that in practice one seldom calculates either average. Unfortunately, one cannot evaluate κ as defined unless one has already solved the problem of the radiation transport and knows I_ν and J_ν. So to obtain practical solutions one must somehow approximate κ. It turns out that, for systems to which it makes sense to apply (6.57) or (6.58),

$$\kappa \approx \kappa_P \equiv \frac{1}{B(T)} \int_0^\infty \kappa_\nu B_\nu \mathrm{d}\nu. \tag{6.59}$$

Here κ_P is the *Planck mean opacity,* which depends only on the equilibrium properties of a material, is often tabulated, and may be available in an approximate functional form. Of course, one could develop iterative solutions to either (6.57) or (6.58), in which one determined κ from an initial solution assuming $\kappa = \kappa_P$. But if one actually needed to do this, one might be better advised to employ a more-sophisticated radiation transport calculation from the start.

6.3.2.3 Radiation Diffusion Models

An even simpler model for $\nabla \cdot \boldsymbol{F}_R$ can be found from the first moment in angle of (6.41). This gives

$$\frac{1}{c^2} \frac{\partial \boldsymbol{F}_\nu}{\partial t} + \nabla \cdot \underline{\boldsymbol{P}}_\nu = \frac{1}{c} \int_{4\pi} (\eta_\nu - \chi_\nu I_\nu) \, \boldsymbol{n} \mathrm{d}\Omega. \tag{6.60}$$

This equation greatly simplifies in most situations, as the first term on the left is only significant for relativistic motions and as the emission nearly always is isotropic and averages to zero. One then has

$$\nabla \cdot \underline{\boldsymbol{P}}_\nu = \frac{-\chi_\nu}{c} \int_{4\pi} I_\nu \boldsymbol{n} \mathrm{d}\Omega \mathrm{d}\nu = \frac{-\chi_\nu}{c} \boldsymbol{F}_\nu. \tag{6.61}$$

In planar systems, with only one direction of inhomogeneity, and using the Eddington approximation, this becomes

$$\nabla p_\nu = \nabla (f_\nu E_\nu) = -\frac{\chi_\nu}{c} \boldsymbol{F}_\nu, \tag{6.62}$$

providing a diffusive model connecting $\boldsymbol{F}_\nu$ and E_ν:

$$\boldsymbol{F}_\nu = -\frac{c}{\chi_\nu} \nabla (f_\nu E_\nu). \tag{6.63}$$

For near-equilibrium conditions, one has $f_\nu = 1/3$ and $E_\nu \to 4\pi B_\nu/c$. If we recognize that B_ν is a function of T only and that T varies with position, then

$$\boldsymbol{F}_\nu = -\frac{4\pi}{3\chi_\nu} \frac{\partial B_\nu}{\partial T} \nabla T. \tag{6.64}$$

Note that the spectral radiation flux is the first derivative of B_ν, which is small compared to B_ν. Even so, the radiation flux can become important at much lower temperatures than those at which the radiation pressure does.

Integrating over frequency, we obtain for F_R

$$F_R = -\int_0^\infty \frac{4\pi}{3\chi_\nu} \frac{\partial B_\nu}{\partial T} d\nu \nabla T = \frac{4\pi}{3\chi_R} \frac{\partial B}{\partial T} \nabla T, \qquad (6.65)$$

in which the Rosseland mean opacity χ_R is defined by

$$\frac{1}{\chi_R} = \int_0^\infty \frac{1}{\chi_\nu} \frac{\partial B_\nu}{\partial T} d\nu \Big/ \int_0^\infty \frac{\partial B_\nu}{\partial T} d\nu. \qquad (6.66)$$

We can rewrite the expression for F_R as

$$F_R = -\frac{4\pi}{3} \frac{1}{\chi_R} \frac{\partial B}{\partial T} \nabla T = -\frac{16\sigma T^3}{3\chi_R} \nabla T \equiv -\kappa_{rad} \nabla T. \qquad (6.67)$$

We will call this the *Rosseland heat flux*. This equation defines the coefficient of radiative heat transport, κ_{rad}, valid only in the equilibrium diffusion limit. It is worth noting that the Rosseland heat flux is the equilibrium heat flux, σT^4, multiplied by a small quantity. The small quantity is the fractional change in temperature per unit optical depth, times (16/3). We can write this as $16 / (3\ \chi_R L_T)$, where L_T is the temperature scale length.

Some characteristic values of χ_R/ρ, in cm^2/g, for near-solid densities, are

$$\chi_R/\rho \approx 2 \times 10^6 \rho^{1/7} T_{eV}^{-2} \quad \text{for CH}$$

$$\chi_R/\rho \approx 3 \times 10^6 T_{eV}^{-4/3} \quad \text{for Al}$$

$$\chi_R/\rho \approx 2 \times 10^8 T_{eV}^{-2} \quad \text{for Xe} \qquad (6.68)$$

$$\chi_R/\rho \approx 6 \times 10^6 \rho^{0.3} T_{eV}^{-3/2} \quad \text{for Au.}$$

Again these are from the SESAME tables (but from Lindl (1995) for Au). Note that with these values κ_{rad} scales as T^4 to T^5. In astrophysical regimes where the cooling function has a negative slope, κ_{rad} scales somewhat more rapidly, as T^6 to T^7.

The thermal diffusive limit is actually more restrictive than it would appear from the above derivation. We can evaluate $\nabla \cdot F_R$ from (6.58), which assumes only that the scattering is isotropic, given an expression for J_ν. We find this as follows. Consider the radiation properties to be a function of optical depth. Take the point of view that the temperature can vary slowly, but only so slowly that the temperature change is negligible over a distance of one radiation mean free path (i.e., χ_ν^{-1}). In LTE, as in other cases with Maxwellian electron distributions, the radiation source is the thermal source, $S_\nu(\tau_\nu) = B_\nu(\tau_\nu)$. If the temperature were constant, then the solution to (6.50) would be that $I_\nu(\tau_\nu) = B_\nu(\tau_\nu)$. As a result $J_\nu(\tau_\nu) = B_\nu(\tau_\nu)$ and (6.55) would imply that there is no radiation flux. But suppose instead that there is a temperature gradient, small in the sense described above so that S_ν can be described by a Taylor expansion relative to some initial location, τ_ν, as

$$S_\nu(\tau'_\nu) = \sum_{n=0}^{\infty} \frac{\partial^n B_\nu(\tau_\nu)}{\partial \tau_\nu^n} \frac{(\tau'_\nu - \tau_\nu)^n}{n!}. \tag{6.69}$$

Equation (6.51) can then be integrated with $\tau_o \to \infty$ to find $I_\nu(\tau, \mu)$, again with $\mu = \cos\theta$, giving

$$I_\nu(\tau_\nu, \mu) = B_\nu(\tau_\nu) - \mu \frac{\partial B_\nu(\tau_\nu)}{\partial \tau_\nu} + \mu^2 \frac{\partial^2 B_\nu(\tau_\nu)}{\partial \tau_\nu^2} + \cdots. \tag{6.70}$$

Note that for forward-going radiation ($\mu > 0$), if $\partial B_\nu/\partial \tau_\nu > 0$ then I_ν is smaller than the local value of B_ν. This is as it should be. One can average this in solid angle to obtain

$$J_\nu(\tau_\nu) = B_\nu(\tau_\nu) + \frac{1}{3} \frac{\partial^2 B_\nu(\tau_\nu)}{\partial \tau_\nu^2} + \cdots, \tag{6.71}$$

If we substitute from (6.71) for J_ν into (6.58), and if we further assume that the variation in the opacities on the scale of the temperature gradient is negligible, and convert to a more general vector notation, we find

$$\nabla \cdot \boldsymbol{F_R} = -\frac{4\pi}{3} \nabla \cdot \left[\int_0^\infty d\nu \frac{\kappa_\nu}{\chi_\nu^2} \frac{\partial B_\nu}{\partial T} \right] \nabla T. \tag{6.72}$$

This result is only consistent with (6.65) if the scattering opacity is much smaller than the absorption opacity, so that $\kappa_\nu/\chi_\nu \sim 1$.

Pause a moment here. This is truly a bizarre result, because the derivation of (6.65) seems completely general. Yet the calculation leading to (6.72) is more fundamental, and only produces the same result in the case of small scattering. The solution to this dilemma lies in the assumption that $E_\nu \to 4\pi B_\nu/c$. Near-LTE conditions can only exist, in the presence of a temperature gradient, if scattering is much smaller than absorption. Otherwise the photons are transported down the density gradient much faster than the material is heated, which will drive the radiation "temperature" out of equilibrium with the material temperature. Absorption may be larger than scattering if free–free transitions, notably bremsstrahlung and inverse bremsstrahlung, dominate the radiation–matter interactions. This is often the case in high-energy-density plasmas. But when bound–bound transitions dominate the radiation–matter interactions then scattering will dominate and LTE will be much less likely to occur.

6.4 Relativistic Considerations for Radiation Transfer

While there are some systems in the universe, such as pulsar envelopes or experiments with lasers at 10^{21} W/cm^2, that are manifestly relativistic, most laboratory and astrophysical systems seem at first glance manifestly nonrelativistic. This, however, is often not true in at least two senses. First, because spectral lines are very narrow, often having a normalized line width of order 10^{-4}, Doppler shifts can complicate the transport of energy by line radiation at velocities as small as $10^{-4}c$. Second, the Lorentz transformation between an observer and a fluid, or between different parts of a fluid, introduces terms in all orders of v/c, where v is a velocity difference between frames of reference. But the moments of the nonrelativistic transfer equations already contain terms that differ from one another by v/c, and the terms that are first order in v/c at times become the dominant ones (for example if $F_R/E_R < v/c$). As a result the leading relativistic terms, which are of order v/c, may contribute as much to the radiation transfer as the nonrelativistic terms in the equations. It turns out that the terms which matter, to this order, are just those one would find from Galilean relativity.

We will not attempt a derivation of all the relations among relativistic radiation transfer equations here. This is a large project, carried out for example in Mihalas and Weibel-Mihalas (1984) and references therein. Our goal, instead, is to introduce the relativistic effects, discuss their origin, and discuss the equations that result.

In discussing relativistic effects, we will write equations that relate quantities in two frames of reference. The first frame, designated by the subscript o, is the frame that is at rest locally within the fluid. This is the frame in which the microscopic interactions of radiation and matter are correctly described by nonrelativistic equations, sometimes known as the *proper frame*. For example, emission from random processes is isotropic only in this frame. The second frame, designated by no special subscript, is in motion with velocity $\boldsymbol{v}$ relative to the first frame. Recalling that the phase space of radiation intensity involves radiation (or photons) of frequency ν and direction $\boldsymbol{n}$, the Doppler shift and aberration are given by

$$\nu_o = \gamma_r \nu (1 - \boldsymbol{n} \cdot \boldsymbol{v}/c) \tag{6.73}$$

and

$$\boldsymbol{n}_o = \frac{\nu}{\nu_o} \left[\boldsymbol{n} - \gamma_r \frac{\boldsymbol{v}}{c} \left(1 - \frac{\gamma_r \boldsymbol{n} \cdot \boldsymbol{v}/c}{\gamma_r + 1} \right) \right] \tag{6.74}$$

in which γ_r is the relativistic contraction factor, $\gamma_r = 1/\sqrt{1 - v^2/c^2}$. The inverse transformations of these quantities are

$$\nu = \gamma_r \nu_o (1 + \boldsymbol{n} \cdot \boldsymbol{v}/c) \tag{6.75}$$

and

$$n = \frac{\nu_o}{\nu}\left[n_o + \gamma_r\frac{v}{c}\left(1 + \frac{\gamma_r n_o \cdot v /c}{\gamma_r + 1}\right)\right]. \tag{6.76}$$

There are some aspects of these equations worth noting from the standpoint of radiation transport. First, the Doppler shift depends on direction, so that local emission in some frequency band appears distributed into a range of frequencies when viewed in a moving frame of reference. Second, the measured direction depends on the frame of reference, so that isotropic emission in the local frame of reference does not appear isotropic in another frame of reference. The well-known "beaming effect" is present in (6.76)—as γ_r becomes large, the radiation is all observed to lie near the direction of v in the moving frame.

It is very helpful in working with relativistic systems to identify which quantities are Lorentz invariant, as this greatly facilitates the conversion between frames of reference. From the above equations, one can show that

$$\nu d\nu d\Omega = \nu_o d\nu_o d\Omega_o. \tag{6.77}$$

In addition, photon number in a given volume must be independent of frame of reference, from which one can obtain several relations of use in radiation transfer. Specifically,

$$I_\nu(\mu, \nu)/\nu^3 = I_{\nu o}(\mu_o, \nu_o)/\nu_o^3, \tag{6.78}$$

$$\eta(\mu, \nu)/\nu^2 = \eta_o(\mu_o, \nu_o)/\nu_o^2, \tag{6.79}$$

and

$$\nu\chi(\mu, \nu) = \nu_o\chi_o(\mu_o, \nu_o). \tag{6.80}$$

It may also be worthwhile for reference to provide here the result of the Lorentz transformation from the frame moving with the material to the frame of an observer moving at velocity v relative to the material, in planar geometry, for the moments of I_R. These are

$$E_R = \gamma_r^2\left(E_{Ro} + 2\frac{v}{c}\frac{F_{Ro}}{c} + \frac{v^2}{c^2}p_{Ro}\right) \approx E_{Ro} + 2\frac{v}{c}\frac{F_{Ro}}{c}$$

$$F_R = \gamma_r^2\left[\left(1 + \frac{v^2}{c^2}\right)F_{Ro} + vE_{Ro} + vp_{Ro}\right] \approx F_{Ro} + vE_{Ro} + vp_{Ro} \tag{6.81}$$

$$p_R = \gamma_r^2\left(p_{Ro} + 2\frac{v}{c}\frac{F_{Ro}}{c} + \frac{v^2}{c^2}E_{Ro}\right) \approx p_{Ro} + 2\frac{v}{c}\frac{F_{Ro}}{c},$$

in which the second approximate equality gives the result to order v/c.

In the context discussed above, our interest here is in the relativistic effects that are first-order in v/c. Dealing with strong relativistic effects is beyond our scope. The fact that the radiation transfer equation is relativistically invariant has less utility than we might wish for, because the emission and opacity are very inconvenient in frames of reference in which they are not isotropic in angle and frequency. This leads us to always want to evaluate these quantities as functions in the local frame. So our first task is to obtain a relation between the spectral intensity in an arbitrary inertial frame and the plasma properties in a local frame, to first order in v/c. Accordingly, we take

$$\nu = \nu_o(1 + \boldsymbol{n} \cdot \boldsymbol{v}/c), \tag{6.82}$$

from which one can relate the emission or opacity in the moving frame to the corresponding quantity in the local frame, evaluated at the same frequency, as

$$\chi(\boldsymbol{n}, \nu) = \chi_o(\nu) - (\boldsymbol{n} \cdot \boldsymbol{v}/c)\left[\chi_o(\nu) + \nu_o(\partial\chi_o/\partial\nu)\right] \tag{6.83}$$

and

$$\eta(\boldsymbol{n}, \nu) = \eta_o(\nu) + (\boldsymbol{n} \cdot \boldsymbol{v}/c)\left[2\eta_o(\nu) - \nu_o(\partial\eta_o/\partial\nu)\right]. \tag{6.84}$$

From these, one can obtain the radiation transfer equation, in Cartesian coordinates, for an inertial frame, as

$$\frac{1}{c}\frac{\partial I_\nu(\boldsymbol{n}, \nu)}{\partial t} + \boldsymbol{n}\cdot\nabla I_\nu(\boldsymbol{n}, \nu) = \eta_o(\nu) - \chi_o(\nu)I_\nu(\boldsymbol{n}, \nu) +$$
$$\left(\frac{\boldsymbol{n} \cdot \boldsymbol{v}}{c}\right)\left[2\eta_o(\nu) - \nu\frac{\partial\eta_o}{\partial\nu} + \left(\chi_o(\nu) + \nu\frac{\partial\chi_o}{\partial\nu}\right)I_\nu(\boldsymbol{n}, \nu)\right]. \tag{6.85}$$

In (6.85) we have achieved our goal of relating the spectral intensity in a moving, inertial frame to the plasma properties in a local frame where they are angularly symmetric. By taking moments of this equation, as above, one can obtain the following equations for the radiation energy and momentum:

$$\frac{\partial E_R}{\partial t} + \nabla \cdot \boldsymbol{F}_R = \int_0^\infty \left[4\pi\eta_o(\nu) - c\chi_o(\nu)E_\nu\right]d\nu +$$
$$\frac{\boldsymbol{v}}{c} \cdot \int_0^\infty \left[\chi_o(\nu) + \nu\frac{\partial\chi_o}{\partial\nu}\right]\boldsymbol{F}_\nu d\nu \tag{6.86}$$

and

$$\frac{1}{c^2}\frac{\partial \boldsymbol{F}_R}{\partial t} + \nabla \cdot \underline{\boldsymbol{P}}_R = \frac{-1}{c}\int_0^\infty \chi_o(\nu)\boldsymbol{F}_\nu d\nu + \frac{4\pi\boldsymbol{v}}{c^2}\int_0^\infty \eta_o(\nu)d\nu$$

$$+\frac{\boldsymbol{v}}{c}\cdot\int_0^\infty\left[\chi_o(\nu)+\nu\frac{\partial\chi_o}{\partial\nu}\right]\underline{\boldsymbol{P}}_\nu d\nu. \tag{6.87}$$

In the first of these equations, for example, the second term on the right-hand side is new by comparison with (6.58). It can be essential. In the diffusion regime, for example the two components of the integral in the first term on the right-hand side are nearly in balance, differing only because of the relativistic shift of E_ν between the reference frame and the local frame. The net remaining value of the first term is of the same order as the second term.

The ability to work with radiation in an inertial frame of reference moving at a fixed velocity relative to a specific volume of plasma may be of use, but is in fact insufficient for typical radiation hydrodynamic problems. What one actually needs is the ability to *always* treat the radiation in the local frame of reference, so the emission and opacity are isotropic, even as the plasma velocity changes from place to place. An analysis in which one continuously transforms the frame of reference as the radiation moves through the plasma is described as an analysis in the *comoving frame*. This name is rather misleading however, as it represents no fixed frame of reference and as any given frame of reference may be accelerating and therefore not an inertial frame. Instead, the "comoving frame" represents a continuously varying sequence of frames of reference that are always at rest in the local plasma. This has the effect of introducing terms into the radiation transfer equation that depend upon the local fluid velocity, $\boldsymbol{u}$, derivatives of the local fluid velocity, and the local acceleration, $\boldsymbol{a}$. We will not work through this derivation here (see Mihalas and Weibel-Mihalas 1984) but will provide the resulting first two moments of the radiation transfer equation, which are

$$\rho\frac{D}{Dt}\left(\frac{E_{Ro}}{\rho}\right)+\nabla\cdot\boldsymbol{F}_{Ro}+\underline{\boldsymbol{P}}_{Ro}:\nabla\boldsymbol{u}+\frac{2}{c^2}\boldsymbol{a}\cdot\boldsymbol{F}_{Ro}$$

$$=\int_0^\infty\left[4\pi\eta_o(\nu_o)-c\chi_o(\nu_o)E_{\nu o}\right]d\nu_o \tag{6.88}$$

and

$$\frac{\rho}{c^2}\frac{D}{Dt}\left(\frac{\boldsymbol{F}_{Ro}}{\rho}\right)+\nabla\cdot\underline{\boldsymbol{P}}_{Ro}+\frac{1}{c^2}\boldsymbol{F}_{Ro}\cdot\nabla\boldsymbol{u}+\frac{1}{c^2}(E_{Ro}\boldsymbol{a}+\boldsymbol{a}\cdot\underline{\boldsymbol{P}}_{Ro})$$

$$=-\frac{1}{c}\cdot\int_0^\infty\chi_o(\nu_o)\boldsymbol{F}_{\nu o}d\nu_o. \tag{6.89}$$

In these equations the velocity terms mentioned above have been compactly expressed through the division by ρ in the convective derivative and the terms involving the tensor $\nabla\boldsymbol{u}$. Also, $\underline{\boldsymbol{P}}_{Ro}:\nabla\boldsymbol{u}$ is dyadic notation and could also be written as $(\underline{\boldsymbol{P}}_{Ro}\cdot\nabla)\cdot\boldsymbol{u}$. Recall that these equations are only accurate to order u/c. Mihalas and Weibel-Mihalas (1984) give more general results in the nonaccelerating limit of special relativity.

Comparing (6.88) with (6.58) and its derivation, one can see that one would recover the latter from the former for a stationary medium with isotropic scattering. The terms involving ρ and $\boldsymbol{u}$ in these equations arise from the flux of radiative energy due to fluid motion and from adiabatic work done by the radiation and the fluid upon one another. We could have obtained these terms from a nonrelativistic derivation, but would have been unsure of their correctness. The good news is that Galilean intuition suffices for systems in which $u \ll c$. As mentioned above, these effects may not be negligible. In particular uE_{Ro} can exceed F_{Ro} under some circumstances. To properly treat radiation in higher-velocity systems would require that one revisit the derivation of radiation transfer under the conditions of interest.

Homework Problems

6.1 Integrate the thermal intensity B_ν over 2π steradians to find the total radiation power per unit area from a surface at temperature T.

6.2 Using the particle treatment of the radiation, derive an expression for the total radiation momentum density, and show that it equals $\boldsymbol{F}_R/c^2$.

6.3 Derive the relation between radiation pressure and energy, (6.14).

6.4 Graduate students frequently struggle with units, and in particular with the problem posed here. First, integrate B_ν, (6.18), symbolically, over frequency to obtain (6.19). Second, evaluate the coefficients in the integral independently for cgs and mks units, and show that you obtain equivalent results. Third, convert (6.18) to have units of energy per unit area per unit time per unit solid angle per unit photon energy. Integrate this new expression over photon energy and show that you obtain the same result. You would be well-advised to do all of this within a computational mathematics program, either with excellent comments your work file, or with an independent document (in LaTeX at the time of writing this) that describes the calculations and their results.

6.5 From the uncertainty principle, the spectral width in frequency, $\Delta\nu$, of an emission line is roughly the inverse of the decay time. For a typical decay time of 1 ns, find the normalized spectral width $\Delta\nu/\nu$, for emission lines in the visible and in the soft X-ray with a photon energy of 100 eV. Discuss the significance of this result.

6.6 Derive the radiative transfer equation for a spherically symmetric system, (6.43).

6.7 Take moments of the radiation transfer equation to derive the equations for radiation energy density and radiation pressure, (6.58) and (6.61).

6.8 Demonstrate that $\nu d\nu d\Omega$ is Lorentz invariant, i.e. (6.77).

6.9 Given relations (6.78) through (6.80), show that the radiation transfer equation is relativistically invariant.

6.10 Derive the relativistic transformations of opacity and emissivity (6.83) and (6.84), and the implied radiative transfer equation, (6.85). Discuss the limits on v/c for this specific description if the emission and absorption are dominated by (a) continuum emission or (b) line emission.

6.11 Rework the relativistic equation for radiative energy density (6.88) into the form of a conservation equation. Discuss the meaning of the terms that result.

References

Griem H (1997) Principles of plasma spectroscopy. Cambridge University Press, Cambridge

Lindl JD (1995) Development of the indirect-drive approach to inertial confinement fusion and the target physics basis for ignition and gain. Phys Plasmas 2(11):3933–4024

Lotz W (1967) Electron-impact ionization cross sections and ionization rate coefficients for atoms and ions. Astrophys J Suppl S 14:207

Mihalas D, Weibel-Mihalas B (1984) Foundations of radiation hydrodynamics, vol 1, Dover (1999) edn. Oxford University Press, Oxford

Salzman D (1998) Atomic physics in hot plasmas. Oxford University Press, New York

Shu FH (1992) The physics of astrophysics: radiation, vol I. University Science Books, Mill Valley, CA

Sobel'man II, Vainshtein L, Yukov EA (2013) Excitation of atoms and broadening of spectral lines. Springer Series on atomic, optical, and plasma physics, 2nd edn. Springer, New York

Springer PT, Fields DJ, Wilson BG, Nash JK, Goldstein WH, Iglesias CA, Rogers FJ, Swenson JK, Chen MH, Bar-Shalom A, Stewart RE (1992) Spectroscopic absorption measurements of an iron plasma. Phys Rev Lett 69:3735–3738

Sutherland RS, Dopita MA (1993) Cooling functions for low-density astrophysical plasmas. Astrophys J 88(September):253–327

Verner DA, Ferland GJ (1996) Atomic data for astrophysics. 1. Radiative recombination rates for H-like, He-like, Li-like, and Na-like ions over a broad range of temperature. Astrophys J Suppl Ser 103(2):467–473

Verner D, Yakovlev D (1995) Analytic fits for partial photoionization cross sections. Astron Astrophys Suppl Ser 109:125–133

Verner DA, Ferland GJ, Korista KT, Yakovlev DG (1996) Atomic data for astrophysics. 2. New analytic fits for photoionization cross sections of atoms and ions. Astrophys J 465(1):487–498

Chapter 7
Radiation Hydrodynamics

Abstract This chapter is concerned with the basic description of dynamic systems in which radiation transport and hydrodynamics both may matter. It first presents the fundamental equations of radiation hydrodynamics and some related thermodynamic considerations. The next topic is radiation and fluctuations, including acoustic waves with radiation in systems of various optical depth and the radiative thermal cooling instability. It then discusses radiation diffusion waves, notably including Marshak waves

It is fair to say that we never directly experience radiation hydrodynamic phenomena—that is, phenomena in which the radiation directly participates in the hydrodynamic evolution of a system. We do experience consequences of radiative heat transport, as for example when heating by solar irradiation produces wisps of fog above a wet road. And we are aware of some systems, such as solar sails, in which radiation directly causes material motion. But as we shall see, radiation hydrodynamic phenomena require temperatures of millions of degrees, more or less, so they are outside the realm of our direct experience.

One would like to know when radiation affects hydrodynamics in important ways. This requires either that the radiation flux becomes comparable with the material energy flux or that the radiation pressure becomes comparable with the material pressure. Thus our first goal is to see when radiation hydrodynamics matters.

To find the conditions under which radiation affects hydrodynamics by a simple calculation, one must make some assumptions. Assume that the electron temperature and the radiation temperature are comparable and equal to T. If the ion temperature matters, which it often does not, we also assume it to be not too far from T. Further assume that the systems of interest are optically thick, which matters in determining the radiation flux but also in keeping the temperatures comparable. Under these conditions, the maximum radiation flux is σT^4. We can evaluate the material energy flux by examining (2.4), finding it to be $\rho u[\epsilon + (u^2/2) + (p/\rho)]$, where u is some characteristic velocity. Here we take u to equal the sound speed c_s—any other reasonable number will be within a small enough multiple of c_s that the results will not be affected. For the specific energy of the material, ϵ, we use the hydrogenic model discussed in Chap. 3. For specific conditions, we can then identify

© Springer International Publishing AG 2018

R P. Drake, *High-Energy-Density Physics*, Graduate Texts in Physics,

https://doi.org/10.1007/978-3-319-67711-8_7

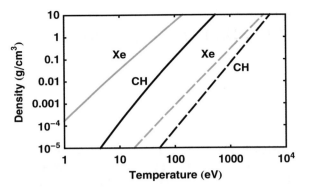

Fig. 7.1 Radiation hydrodynamic regimes. Dashed curves show where P_R equals the material pressure, and solid curves show where F_R equals the material energy flux. The gray curves are for Xe and the black curves are for CH

the boundary in temperature above which the radiation flux dominates. Similarly, taking the radiation pressure from (6.20) and the material pressure from (3.1), we can identify the boundary in temperature above which the radiation pressure dominates.

Figure 7.1 shows the results of such a calculation for C_1H_1, assumed to be fully ionized, and for Xe, assumed to be ionizing with $Z = 0.63\sqrt{T}$. In the region to the right of a solid line, the radiation flux exceeds the material energy flux. In the region to the right of a dashed line, the radiation pressure exceeds the material pressure. This is the radiation-dominated regime discussed briefly in Chap. 3. One sees that Xe is more radiative than C_1H_1; the Xe curves are displaced to lower temperature. This is no surprise. Xe has $A = 131.3$ and $Z_{nuc} = 54$. Over this range of temperatures it still has electrons in s, p, d, and f shells, allowing very many X-ray transitions, while the C has only six electrons. The C_1H_1 is assumed to be fully ionized; accounting for the internal energy of the C properly would move the solid black curve slightly to the right. Leaving aside the fine details, one can see that radiation fluxes become important at temperatures of tens of eV, and that the exact value depends on details and especially density. The plasma becomes radiation-dominated at temperatures of hundreds of eV, again with exact values depending on details. In the important case of plasmas near 1 g/cm^3 in density, it takes temperatures of about 2 keV to make the plasma radiation-dominated.

Stellar interiors include regions in which radiation affects the hydrodynamics. This is not surprising as the essential behavior of stars is to release energy in their cores and then to radiate it away from their surfaces. As a result, radiative fluxes must exceed material energy fluxes, and they do, for example, in the sun. Figure 7.2 shows results of a simulation of the sun. A convective zone exists because diffusion of the radiation is not fast enough to transport the energy generated by nuclear fusion to the solar surface. Larger, hotter, stars include regions in which the radiation pressure exceeds the material pressure. In addition, all supernovae heat the stellar interior into the radiation-dominated regime, where it stays until it cools sufficiently through volumetric expansion. Some dense astrophysical environments, such as neutron stars and black hole regions, can be strongly radiation-dominated,

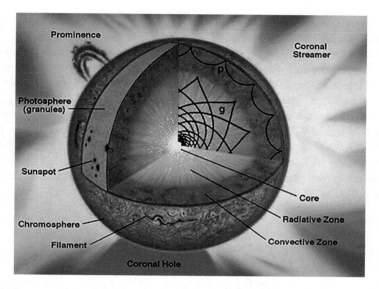

Fig. 7.2 Cut-away view of the Sun. Energy is released by fusion in the core, propagates by radiative transfer in the radiative zone, and drives radiation-hydrodynamic convection in the convection zone. From the entry "Sun", in the *Encyclopedia of Planetary Sciences*, published by Chapman & Hall in 1997, used with permission

and even relativistic. Here we will not consider the relativistic cases, as laboratory experiments are a long way from accessing them.

Interstellar astrophysical systems, including the interstellar medium, interstellar shocks, and molecular clouds, have densities more than 15 orders of magnitude below those shown in Fig. 7.1. The curves there make it appear that such systems would be in a radiative regime. However, such interstellar astrophysical systems the radiative flux never approaches σT^4, because they are optically thin. The radiative flux from an optically thin system, for thermal emission, equals $\kappa d \sigma T^4$, where κ is the absorption opacity of the system and d is its size. However, many optically thin systems, especially in astrophysics, produce primarily line emission, in which case κd would be an appropriate average over the spectral variation of the optical depth and thermal spectrum. (The relation of κ and the cooling function, Λ, is discussed in Sect. 6.2.3.) Moreover, at low density the opacity decreases as density squared while the material energy fluxes decrease linearly with density. In this regime, the temperature required to enter the radiative regime must increase as density decreases. Curiously, this increases the shock velocity required for radiation fluxes to be significant into the range of >100 km/s, which is just where it is for laboratory experiments with foams or dense gas. The small optical depth and lack of sources or boundaries also implies that the radiation "temperature" remains decoupled from and much smaller than the electron temperature. As a result, the radiation pressure never exceeds the material pressure in interstellar astrophysical systems and the radiation-dominated regime is genuinely difficult to reach in experiments.

However, the radiation flux may at times be essential to the dynamics of such systems even at densities below 10^{-20} g/cm^3. This can occur when the energy input from the interactions of matter, such as pdV work, becomes smaller than the radiative cooling. An astrophysical shock wave compresses and heats the material it shocks, but after that there is often no further energy input. The material cools by radiation, however long this may take. The shocked layer of material produced by a supernova is driven for a long time, gaining energy from the pressure of the hot "bubble" created by the explosion. But eventually this ends, and the layer subsequently cools by radiation, eventually decreasing in temperature and increasing in density by orders of magnitude. We discuss the dynamics of these astrophysical phenomena in the next chapter, when we discuss optically thin radiative shocks.

7.1   Radiation Hydrodynamic Equations

Our first task is to develop the equations we will need to account for radiation hydrodynamic phenomena in the high-energy-density regime. The simple equations of Chap. 2 ((2.27) and (2.28)) are useful for considering the relative contributions of the various processes that may occur in an energetic fluid or plasma, all of which are included, for example, in many simulation codes. However, these equations are rarely practical for simple calculations. In the context of the present chapter we will ignore viscous effects and usually heat conduction—these play a very limited role in most systems hot enough for radiative effects to matter. We will also necessarily work with only the simplest models of radiation transport. However, there certainly are cases in which more-complicated models are needed to obtain an accurate description. A wider range of such models is discussed in Mihalas and Weibel-Mihalas (1984).

7.1.1   The Fundamental Equations of Radiation Hydrodynamics

We need equations to describe the evolution of the system when radiation is important. In the simplest cases, which we discuss first, it is productive to work with equations for the combined, total energy density and momentum density. These equations simplify easily when the energy density of either the matter or the radiation is dominant. This transition is fairly abrupt, since the material energy density scales as T while the radiation energy density scales as T^4. When the energy density of both the matter and the radiation is important, these equations are most useful near LTE, when the radiation intensity is Planckian and has the same temperature as the matter. Yet often in applications the system is not optically thick enough for the radiation and matter to come into equilibrium. In this case it can be helpful to treat the radiation and the matter as distinct fluid species, coupled to one another. Nearly all codes employ such nonequilibrium models. We discuss them second.

7.1.1.1 The Equations for Total Energy and Momentum

Here we focus on nonrelativistic systems with isotropic radiation fields. In this case, the momentum equation remains quite simple. The momentum of the radiation is negligible compared to that of the fluid in this limit. It will prove useful to define variables relating total pressure and internal energy, as follows:

$$\widetilde{p} = p + p_R, \ \rho\widetilde{\epsilon} = \rho\epsilon + E_R, \ \text{and} \ \rho\widetilde{\epsilon} = \frac{\widetilde{p}}{\widetilde{\gamma} - 1}, \ \text{so} \ \widetilde{p} + \rho\widetilde{\epsilon} = \widetilde{\gamma}\rho\widetilde{\epsilon} = \frac{\widetilde{\gamma}}{\widetilde{\gamma} - 1}\widetilde{p}. \tag{7.1}$$

The effect of the isotropic radiation field on the matter can be expressed as a gradient in the scalar radiation pressure, p_R. The momentum equation and mechanical-energy equations then become

$$\rho\frac{D}{Dt}\boldsymbol{u} = -\nabla\widetilde{p} \ \text{and} \ \rho\frac{D}{Dt}\frac{u^2}{2} = -\boldsymbol{u}\cdot\nabla\widetilde{p} \ \text{or} \ -\boldsymbol{u}\cdot\nabla\left[(\widetilde{\gamma} - 1)\rho\widetilde{\epsilon}\right], \tag{7.2}$$

respectively, in which p_R is the scalar radiation pressure. There are usually no other significant forces in laboratory systems, but there might be other forces such as gravitation in an astrophysical problem. In this case the right-hand side will have additional terms. The total-energy equation, found from Eq. (2.28), dropping the terms involving viscosity and heat conduction, and using our combined variables, is

$$\frac{\partial}{\partial t}\left(\frac{\rho u^2}{2} + \rho\widetilde{\epsilon}\right) + \nabla\cdot\left[\rho\boldsymbol{u}\left(\frac{u^2}{2} + \widetilde{\gamma}\,\widetilde{\epsilon}\right)\right] = -\nabla\cdot\boldsymbol{F}_R. \tag{7.3}$$

If there are additional forces such as gravitation, they will introduce terms relating to both potential energy and work into these equations. Note that these equations involve fundamentally a Galilean treatment of the radiation energy density. In particular, it convects with velocity $\boldsymbol{u}$. This is certainly not valid for strongly relativistic fluid velocities (or frames of reference), but is accurate to order u/c, as we found in Sect. 6.4. We also note that formally the radiation quantities in these equations should be evaluated in a frame of reference moving with the fluid—the comoving frame discussed in Sect. 6.4. For problems that can be addressed with the above equations, one can generally ignore this distinction at least until it is necessary to consider what will be detected by an external observer.

The most useful energy equations are gas-energy equations, like (2.31). To obtain such equations, we rewrite (7.3) as

$$\rho \frac{D}{Dt} \frac{u^2}{2} + \frac{\partial}{\partial t}(\rho \widetilde{\epsilon}) + \nabla \cdot [\boldsymbol{u} \widetilde{\gamma} \rho \widetilde{\epsilon}] = -\nabla \cdot \boldsymbol{F}_R. \tag{7.4}$$

We then subtract the mechanical-energy equation above to find

$$\frac{D}{Dt}(\rho \widetilde{\epsilon}) + \widetilde{\gamma} \rho \widetilde{\epsilon} \nabla \cdot \boldsymbol{u} = -\nabla \cdot \boldsymbol{F}_R. \tag{7.5}$$

One can replace $\nabla \cdot \boldsymbol{u}$ from the continuity equation to find other useful forms,

$$\rho \frac{D \widetilde{\epsilon}}{Dt} - \frac{\widetilde{p}}{\rho} \frac{D\rho}{Dt} = -\nabla \cdot \boldsymbol{F}_R \text{ and} \tag{7.6}$$

$$\frac{D \widetilde{p}}{Dt} - \frac{\widetilde{\gamma} \widetilde{p}}{\rho} \frac{D\rho}{Dt} = -(\widetilde{\gamma} - 1)\nabla \cdot \boldsymbol{F}_R, \tag{7.7}$$

in which $\widetilde{\gamma}$ is the value appropriate to heat conduction.

Equation (7.6) can be expressed more simply when there is a simple relation between pressure and energy density. Let us explore this. On the one hand, if the radiation pressure is negligible and the matter is a polytropic gas, then $\rho \widetilde{\epsilon} = p/(\gamma - 1)$ and the left-hand side of (7.6) reduces to that of (2.31) with an appropriate value of γ. On the other hand, if the material pressure is negligible and the radiation pressure is dominant, then we showed in Chap. 3 that the same relation applies, so one also recovers the left-hand side of (2.31) with $\gamma = 4/3$. One can also see from (7.2) and (7.3) that, when radiation is dominant, the momentum and energy equations can be placed in the form of the Euler equations, with radiation pressure replacing the material pressure.

Note that up to this point, we have assumed only that the system is not relativistic. We have made no assumptions about the equation of state ($\widetilde{\gamma}$ could have any arbitrary dependence on other parameters). We can find a form of the gas-energy equation that will prove useful below for systems in which E_R and p_R are negligible, for a polytropic gas with constant c_V and R. (Recall also that $R = (\gamma - 1)c_V$.) Then, using the polytropic equation of state and the continuity equation, one can rework (7.6) to obtain

$$\rho c_V \frac{D}{Dt} T + \rho(\gamma - 1)c_V T \nabla \cdot \mathbf{u} = -\nabla \cdot \mathbf{F}_R. \tag{7.8}$$

This form of the gas-energy equation is particularly useful when one can express F_R as a function of T.

It is important to observe that the primary way in which radiation transfer enters into these equations is through the term $\nabla \cdot \boldsymbol{F}_R$. In the previous chapter we identified three approaches to evaluating $\nabla \cdot \boldsymbol{F}_R$. One can solve for I_R and integrate as in (6.10).

This is often necessary in optically thin systems. Alternatively, when the scattering is elastic and one can evaluate J_R, one can use (6.58) to obtain $\nabla \cdot \boldsymbol{F}_R = 4\pi \kappa_p (B - J_R)$, a model known as the "transport regime". Here the absorption coefficient will only be close to the Planck mean opacity when J_R is either negligible or close enough to a Planckian of the same temperature as B. The third approach is valid when the system is optically thick and near equilibrium and a radiation diffusion model describes the radiation transport. In these cases on most often uses $\nabla \cdot \boldsymbol{F}_R = -\nabla \cdot (\kappa_{\text{rad}} \nabla T)$, as we will later in the chapter. In addition, when one finds the cooling rate by one of these methods, it often can be expressed as a rate ν_c times some heat content. It can then be convenient to take $-\nabla \cdot \boldsymbol{F}_R = -\nu_c \rho c_V T_c$, in which T_c might be a temperature or a temperature deviation, depending on the context.

7.1.1.2 Equations for Radiation and Matter Out of Equilibrium

It very often is necessary to treat the matter and radiation separately, and to explicitly include the coupling between the two fluids. To obtain an equation for the radiation in this case, we extract the terms for the radiation energy, use the continuity equation, and add terms describing the emission and absorption of radiation to (7.5), obtaining

$$\frac{DE_R}{Dt} - \frac{4E_R}{3} \frac{1}{\rho} \frac{D\rho}{Dt} = -\nabla \cdot \boldsymbol{F}_R - c\kappa_E E_R + 4\kappa_p \sigma T_e^4, \qquad (7.9)$$

in which the electron temperature is T_e, the Planck opacity remains κ_p, and the absorption opacity is another averaged value, in this case being

$$\kappa_E = \frac{\int_0^\infty \kappa_\nu E_\nu d\nu}{\int_0^\infty E_\nu d\nu} \sim \kappa_p. \qquad (7.10)$$

The second term on the left hand side of (7.9) is the change in radiation energy density due to changes of volume. The corresponding gas-energy equation for the matter is

$$\rho \frac{D\epsilon}{Dt} - \frac{p}{\rho} \frac{D\rho}{Dt} = c\kappa_E E_R - 4\kappa_p \sigma T_e^4. \qquad (7.11)$$

Note that if we add (7.9) and (7.11), the coupling terms cancel and we regain an equation equivalent to the gas-energy equations above. Also note that if $E_R \sim 0$ then $\nabla \cdot \boldsymbol{F}_R = -c\kappa_E E_R + 4\kappa_p \sigma T_e^4$ and we recover (7.6) in the same limit.

One most straightforwardly creates a *nonequilibrium diffusion model* by writing $\nabla \cdot \boldsymbol{F}_R = -\nabla \cdot (c\nabla E_R/(3\bar\chi))$ in (7.9), so that with an equation of state for the matter one has a closed set of equations. Simple diffusion models like the one written here fail in optically thin regions, and fail so as to transport too much energy too rapidly. The computational implementation of such models generally involves a radiation flux limiter, which limits the radiation fluxes when the medium becomes optically thin. Such flux-limiters are beyond the scope of our discussion here, but will become important to you if you engage in modeling of radiation hydrodynamic systems. A nonequilibrium-diffusion model in which the radiation is treated as a single fluid is also known as a "gray" or "gray diffusion" model, because it does not distinguish in any way the spectral behavior of the radiation. This failure is the largest source of inaccuracy in such models, because in fact the more energetic photons have much smaller opacities and so penetrate and heat the matter much more deeply. A very common way to address this is to split the radiation spectrum into energy groups (typically some tens of groups), evaluate the absorption opacities for each group, and solve an equation like (7.9) for each energy group. Then one modifies (7.11) so that the radiation absorption and emission terms are sums over all groups. Such models are known as *multigroup diffusion models*. These are the most commonly used models in codes that model laboratory experiments.

7.1.2 Thermodynamic Relations

As discussed above, one can often use the EOS for either the matter or the radiation, as any given system tends to be in one regime or the other. It is of some interest to see what the transition looks like, which we seek to do here. Our approach is to take the results from the simplest EOS of Chap. 3 and to add the quantities describing the contributions of the radiation. Here we assume that the plasmas in question are near equilibrium, to that the matter and radiation can be described using a common temperature. The total pressure, $\widetilde{p}$, is the sum of the plasma pressure from (3.88) and the radiation pressure, giving

$$\widetilde{p} = n_i(1 + Z)k_B T - \frac{3}{10}\frac{n_i Z^2 e^2}{R_o} + \frac{4\sigma}{3c}T^4. \tag{7.12}$$

The total specific energy density, $\widetilde{\epsilon}$, is the sum of the contributions from thermal particles, radiation, and internal energy, again designated as $R(T)$.

$$\rho\widetilde{\epsilon} = \frac{3}{2}n_i(1 + Z)k_B T - \frac{9}{10}\frac{n_i Z^2 e^2}{R_o} + n_i\frac{E_H}{6}Z(1 + Z)(1 + 2Z) + \frac{4\sigma}{c}T^4. \tag{7.13}$$

Here the third term on the right-hand side is the energy density of ionization, evaluated using a hydrogenic model, and could be replaced with some other, more-accurate evaluation.

One then can proceed to evaluate the various types of γ, in the generalization of polytropic indices introduced in Sect. 3.6. One has for the shock gamma

$$\widetilde{\gamma} = 1 + \frac{\widetilde{p}}{\rho\widetilde{\epsilon}}, \tag{7.14}$$

as above. To find the value of γ_h, required for heat transport calculations, we need the specific heat at constant volume,

$$c_V = \left(\frac{\partial \widetilde{\epsilon}}{\partial T}\right)_\rho = \frac{3k_B}{2Am_p}\left[1 + Z + T\left(\frac{\partial Z}{\partial T}\right)_\rho\right] + \tag{7.15}$$

$$\frac{E_H(1 + 4Z + 6Z^2)}{Am_p}\left(\frac{\partial Z}{\partial T}\right)_\rho + \frac{16\sigma}{\rho c}T^3.$$

The partial derivative of $\widetilde{p}$ with respect to T at constant density, is

$$\left(\frac{\partial \widetilde{p}}{\partial T}\right)_\rho = n_i\left[1 + Z + T\left(\frac{\partial Z}{\partial T}\right)_\rho\right]k_B - \frac{6}{10}\frac{n_iZe^2}{2}\left(\frac{\partial Z}{\partial T}\right)_\rho + \frac{16\sigma}{3c}T^3. \tag{7.16}$$

This then gives

$$\frac{1}{(\gamma_h - 1)} = \frac{\rho c_V}{(\partial\widetilde{p}/\partial T)_\rho} \tag{7.17}$$

$$= \frac{(3/2)\left[1 + Z + T(\partial Z/\partial T)_\rho\right]}{\left[1 + Z + T(\partial Z/\partial T)_\rho\right] - [6Ze^2/(R_ok_B)](\partial Z/\partial T)_\rho + 16\sigma T^3/(3n_ik_Bc)}$$

$$+ \frac{(E_H/k_B)(1 + 4Z + 6Z^2)(\partial Z/\partial T)_\rho + 16\sigma T^3/(n_ik_Bc)}{\left[1 + Z + T(\partial Z/\partial T)_\rho\right] - [6Ze^2/(R_ok_B)](\partial Z/\partial T)_\rho + 16\sigma T^3/(3n_ik_Bc)}$$

for γ_h. Note that γ_h depends on $(\partial Z/\partial T)_\rho$, which drops abruptly to 0 when the plasma is fully ionized. Note also that for $Z \propto \sqrt{T}, T(\partial Z/\partial T)_\rho = Z/2$. One can rearrange this last equation to have

$$c_V = \frac{1}{(\gamma_h - 1)}\frac{1}{\rho}\left(\frac{\partial \widetilde{p}}{\partial T}\right)_\rho, \tag{7.18}$$

a form that is useful for applications when one already has a sufficiently accurate value for γ_h.

We also need the sound speed, from (3.140), for calculations involving acoustic waves. To evaluate this, we need

$$\left(\frac{\partial \widetilde{p}}{\partial \rho}\right)_T = \frac{(1 + Z)k_BT}{(Am_p)} - \frac{2e^2Z^2}{5Am_p}\left(\frac{4\pi\rho}{3Am_p}\right)^{1/3} \tag{7.19}$$

and

$$\left(\frac{\partial \widetilde{\epsilon}}{\partial \rho}\right)_T = -\frac{4\sigma T^4}{c\rho^2} - \frac{e^2Z^2}{5Am_p}\left(\frac{9\pi}{2Am_p\rho^2}\right)^{1/3}. \tag{7.20}$$

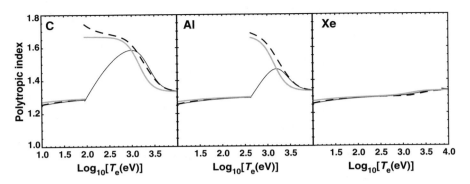

Fig. 7.3 Values of γ (solid curve), γ_s (dashed curve), and γ_h (thick gray curve) for carbon (left), aluminum (middle), and xenon (right), all evaluated at $\rho = 2 \, \mathrm{g/cm^3}$

The expression for γ_s is even messier than the one for γ_h, and so not shown here. It is easily evaluated by computer.

Figure 7.3 shows values of γ, γ_s, and γ_h for near-equilibrium radiating plasmas composed of three different elements. Here we have assumed $Z = 0.63 \sqrt{T}$ with T in eV until $T = 2.5 Z_{\mathrm{nuc}}^2$, above which we have taken $Z = Z_{\mathrm{nuc}}$, the nuclear charge. One can see in the figure the abrupt change in γ_s and γ_h, but not γ, when the plasma becomes fully ionized. Note that all three indices remain in the range of 1.25–1.75 throughout, which is less than a 50% variation. It is not a terrible assumption to take $\gamma_s = \gamma_h = \gamma$, except for temperatures within a factor of several above the temperature corresponding to full ionization. We ignore this regime in the applications. It is worth noting, though, that using a single value of γ does not permit one to use the simple equations of Sect. 3.1.1. One still must evaluate quantities such as the specific heat properly for the radiative regime.

To go beyond equilibrium models, one has the same sorts of choices we explored in Chap. 2 (and Sect. 6.3.2) regarding how to treat the entire physical system. Just as we saw there that there is a choice between a single-fluid and a multiple-fluid treatment of the particles, there is a similar choice regarding the radiation, especially with regard to the energy equation. A common type of model in simulations is the single-fluid, three-temperature model discussed in Chap. 2. Such models use a single continuity equation and a single momentum equation, but make the pressure the sum of the electron, ion, and radiation pressures. Then one works with separate energy equations for the electrons, the ions, and the (often multigroup) radiation, keeping track of the energy exchange between these species. Such models often employ a generalized nonequilibrium diffusion treatment of the radiative energy transport, along the lines discussed in Sect. 7.1.1.2. Alternatively, they may employ more-sophisticated radiation transport methods in connection with a single-fluid, two-temperature (electron and ion) treatment of the matter. Such models, even with a diffusive treatment of radiation heat transport, are comparatively tractable and often give qualitatively correct results.

7.2 Radiation and Fluctuations

Strong radiation, when present, affects every hydrodynamic process that occurs in a medium. Much of the material in this chapter and the next involves the examination of how radiation can alter the behavior of phenomena we have already explored, such as acoustic waves and shocks. Some additional phenomena, including thermal instabilities and diffusive heat waves, have analogues in systems with heat conduction.

7.2.1 Radiative Acoustic Waves: Optically Thick Case

To see the effect of radiation on acoustic waves, we will begin by examining what would happen deep within an optically thick, near-equilibrium, radiating medium such as a stellar interior or an experimental volume that is hot enough and dense enough. We begin with the limit in which the system is so near equilibrium and so optically thick to the radiation that $B \sim J_R$. In this limit the radiative flux is given by (6.67). In this case, (7.7) becomes

$$\frac{D\widetilde{p}}{Dt} - \frac{\gamma_s \widetilde{p}}{\rho}\frac{D\rho}{Dt} = (\gamma_h - 1)\nabla \cdot (\widetilde{\kappa}\nabla T), \qquad (7.21)$$

in which $\widetilde{\kappa} = \kappa_{th} + \kappa_{\text{rad}}$, the sum of the thermal and radiative coefficients of heat conduction, and (formally) we take γ_s and γ_h from Sect. 7.1.2. We linearize this equation, taking $\widetilde{p} = \widetilde{p}_o + \widetilde{p}_1, \rho = \rho_o + \rho_1$, and $T = T_o + T_1$, working in the comoving frame and assuming that the zeroth-order gradients of temperature, velocity, and density are zero, to obtain

$$\frac{\partial\widetilde{p}_1}{\partial t} - \frac{\gamma_s \widetilde{p}_o}{\rho_o}\frac{\partial\rho_1}{\partial t} = (\gamma_h - 1)\widetilde{\kappa}\nabla^2 T_1. \qquad (7.22)$$

We then need to eliminate one variable by linearizing the equation of state (7.12). We choose to eliminate T_1, but to do so we have to evaluate the variation in Z. Here we choose $Z \propto \sqrt{T}$ in the ionizing regime. Then we find

$$\widetilde{p}_1 - p_o\frac{\rho_1}{\rho_o} = \alpha\frac{T_1}{T_o}, \text{ where} \qquad (7.23)$$

$$\alpha = \frac{16\sigma T_o^4}{3c} + \frac{\rho_o(1 + \beta Z_o)k_B T_o}{Am_p}, \qquad (7.24)$$

with $\beta = 1$ in a fully ionized plasma or 3/2 in an ionizing plasma. We note that p_o in (7.23) is the particle pressure and that α has units of pressure. Substituting for T_1 in (7.22), we have

$$\frac{\partial \widetilde{p}_1}{\partial t} - \frac{(\gamma_h - 1)\widetilde{\kappa}T_o}{\alpha}\nabla^2\widetilde{p}_1 = \frac{\gamma_s\widetilde{p}_o}{\rho_o}\frac{\partial \rho_1}{\partial t} - \frac{(\gamma_h - 1)\widetilde{\kappa}T_o}{\alpha}\frac{p_o}{\rho_o}\nabla^2\rho_1. \tag{7.25}$$

Note that the coefficient in the second term on the left-hand side has units of cm^2/s, making it some sort of generalized kinematic diffusion coefficient (see Chap. 2). We can arrive at an equation that includes acoustic waves by noting that the linearized versions of the momentum equation (7.2) and the continuity equation imply that $\partial^2 \rho_1/\partial t^2 = \nabla^2\widetilde{p}_1$, which enables us to differentiate (7.25) twice with respect to t to find

$$\left[\frac{\partial^2}{\partial t^2} - \frac{\gamma_s\widetilde{p}_o}{\rho_o}\nabla^2\right]\frac{\partial \widetilde{p}_1}{\partial t} = \frac{(\gamma_h - 1)\widetilde{\kappa}T_o}{\alpha}\left[\frac{\partial^2}{\partial t^2} - \frac{p_o}{\rho_o}\nabla^2\right]\nabla^2\widetilde{p}_1. \tag{7.26}$$

This is a fourth-order wave equation describing radiation-modified acoustic waves and related waves. We perform a plane-wave analysis as in Chap. 2 to find the dispersion relation, which is

$$\left[1 - \frac{\gamma_s\widetilde{p}_o}{\rho_o}\frac{k^2}{\omega^2}\right] = -i\left(\frac{(\gamma_h - 1)\widetilde{\kappa}T_o}{\alpha}\omega\frac{k^2}{\omega^2}\right)\left[1 - \frac{p_o}{\rho_o}\frac{k^2}{\omega^2}\right]. \tag{7.27}$$

One can simplify this by defining $\eta = 4\sigma T^4/(3cp_o)$ to capture the relative importance of radiation and $\gamma_s v_n^2 = (\omega/k)^2(\rho_o/p_o)$ giving $v_n = 1$ when the phase velocity equals the usual, isentropic sound speed, so that (7.27) can be written

$$\left[1 - \frac{(1 + \eta)}{v_n^2}\right] = -i\left[\frac{\omega_n}{v_n^2}\right]\left[1 - \frac{1}{\gamma_s v_n^2}\right], \tag{7.28}$$

in which the normalized frequency, ω_n, is given, assuming $\widetilde{\kappa} = \kappa_{\text{rad}}$, by

$$\omega_n = \left(\frac{(\gamma_h - 1)}{\left[1 + \frac{1}{4\eta}\left(\frac{1 + \beta Z_o}{1 + Z_o}\right)\right]}\frac{\omega}{v_e(c_s^2/c^2)}\right), \tag{7.29}$$

where we have introduced an extinction rate $v_e = c\chi_R$ to clarify the normalization. It is evident that as radiation becomes negligible so $\eta \to 0$, $\omega_n \to 0$ and the phase velocity assumes its usual value ($v_n = 1$). When η is not small, the behavior depends on ω_n. At any given η, ω_n defines a frequency scale, so that "low" frequencies make this term small and "high" frequencies make it large. The high-frequency regime is the one in which thermal conductivity is very effective on the scale of the wavelength, smoothing the temperature fluctuations that acoustic waves otherwise produce. The waves in this regime have phase velocity $(c_s/\sqrt{\gamma})$. Such waves are

Fig. 7.4 Acoustic waves in
the radiation diffusion limit.
(**a**) The phase velocity
normalized by the
nonradiative, isentropic sound
speed. (**b**) The spatial
damping length normalized
by the wavenumber.
Parameters were $\gamma_h \sim \gamma_s \sim$
$1.3, \beta = 1$, and $\eta = 0.01$

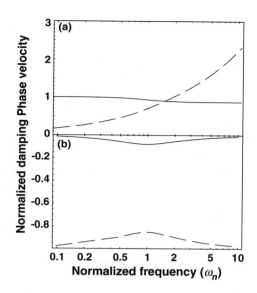

known as *isothermal acoustic waves,* as this is the phase velocity found from the
Euler equations by assuming the plasma to be isothermal.

In general, (7.28) has four roots, corresponding to two pairs of oppositely
propagating waves. The roots are complex, giving the phase velocity and the
damping, both normalized to the nonradiative, isentropic sound speed. [The spatial
damping rate, normalized by the wavenumber, is the ratio of the imaginary root
of (7.28) to the real root.] Figure 7.4 shows the normalized phase velocity and
spatial damping rate, as a function of ω_n, for values of the parameters shown in
the caption. The weakly damped root with a normalized phase velocity near 1
is the acoustic wave. The other root, often described as a *thermal wave,* is very
strongly damped except at very low frequency, where its phase velocity becomes
negligible. It corresponds to a strongly damped perturbation in temperature and the
other quantities.

For typical laboratory values (at approximately $1 \, \text{g/cm}^3$) of $\chi_R \sim 10^4 \, \text{cm}^{-1}$,
$c_s \sim 30 \, \text{km/s}$, and $\gamma_h \sim \gamma_s \sim 1.3, \omega_n = 1$ when $\omega \sim (3/\eta) \times 10^6 \, \text{rad/s}$, which
corresponds to wavelengths of order $2\pi \eta \, \text{cm}$. Thus, once η decreases below about
10^{-5} (at approximately $100 \, \text{eV}$), so the wavelength becomes shorter than $1/\chi_R$, the
acoustic waves will be in the "high-frequency regime." For stellar interiors, one
might have $\chi_R \sim 10 \, \text{cm}^{-1}$, $c_s \sim 100 \, \text{km/s}$, and $\gamma_h \sim \gamma_s \sim 1.3$, so $\omega_n = 1$ when
$\omega \sim (3/\eta) \times 10^4 \, \text{rad/s}$, which corresponds to wavelengths of order $600 \, \eta \, \text{cm}$. Thus,
only very short wavelengths, by comparison to the stellar radius, will be in the "high-
frequency" regime.

7.2.2 Cooling When Transport Matters

We next turn to systems that are not quite so optically thick, so that $B \neq J_R$ and radiation transport is significant. It is not immediately clear what may happen in this case. The question is where the radiation goes, how is it absorbed, and what are the consequences. We assume that the radiation is in steady state, as it equilibrates rapidly on the timescale of material motion. We further assume that the plasma particles are in near-equilibrium distributions, so they emit at the equilibrium rate. Under these assumptions, the divergence of the radiative flux is $4\pi\kappa(B - J_R)$ as given by (6.58) for steady state, and in applications we will approximate $\kappa \sim \kappa_p$, the Planck mean opacity. Before we consider the dynamics of fluctuations in this context, we need to know the rate of cooling produced by a plane-wave fluctuation in temperature and thus thermal emission. In the next section, we calculate this.

7.2.2.1 Cooling of Temperature Fluctuations

We consider a system with no zeroth-order gradients. Such a system will be in a steady-state (or an equilibrium) in which the sources and losses of thermal radiation are in balance. Our goal is to determine the radiative cooling produced by a plane-wave fluctuation in the emission. Formally we write $B = B_o + B_1$ and $B_1 = \hat{B}\exp[ik(z - z_1)]$, which defines the z-axis to lie along the wave vector $\boldsymbol{k}$ of the fluctuation. Our convention will be that the physical quantity represented by any variable, a, is $(a + a^*)/2$. We then seek the radiation intensity due to B_1, ignoring the steady-state contribution due to B_o. Equation (6.44) then describes the incremental radiation intensity, I_1. We assume the medium is uniform and refraction is negligible, and we evaluate the intensity by integrating in the $+z$ direction along some ray, so we have

$$I_1(z) = \int_{z_o}^{z} \frac{\mathrm{d}z'}{\mu} \kappa\hat{B}e^{ik(z'-z_1)}e^{-\kappa(z-z')/\mu} + I(z_o)e^{-\kappa(z-z_o)/\mu}, \qquad (7.30)$$

where μ is the cosine of the angle of the ray relative to the z-axis. It is consistent with our context to take z_o to be a large negative number, so that $|\kappa z_o| \gg 10$. Then, after integrating, (7.30) becomes

$$I_1(z, \mu) = \frac{\kappa\hat{B}}{(i\mu k + \kappa)}e^{ik(z-z_1)} = \left[\frac{(\kappa - i\mu k)}{(\mu^2 k^2 + \kappa^2)}\right]\kappa\hat{B}e^{ik(z-z_1)}. \qquad (7.31)$$

When we integrate I_1 to find J_1 the imaginary term, which is odd in μ, integrates to zero, and we are left with

$$J_1(z) = \frac{1}{2}\int_{-1}^{1} \mathrm{d}\mu I_1(z, \mu) = \hat{B}e^{ik(z-z_1)}\frac{\kappa}{k}\mathrm{Cot}^{-1}\left(\frac{\kappa}{k}\right). \qquad (7.32)$$

This is the result we need to evaluate the heat input to the matter in the plasma, from the steady-state limit of (6.58), as the negative of the input to the radiation.

For systems in which the radiation pressure is negligible, the impact of the radiation is through the heat it transports. It is then useful to express this result in terms of the fluctuation of the temperature, T_1. We treat the incremental energy loss rate, which is $-\nabla \cdot \mathbf{F}_R$, as a damping rate on the incremental local energy density, $\rho c_V T_1$. Thus we have, again in the comoving frame of a uniform medium,

$$\frac{\partial T_1}{\partial t} = -\nu_1 T_1 = \frac{4\pi\kappa}{\rho c_V}(J_R - B). \tag{7.33}$$

Just as we have expanded the other physical parameters, we expand J_R as $J_R = J_o + J_1$, where J_1 is the deviation in J_R due to the temperature fluctuation from (7.32). Since $J_o = B_o$ the first-order expression for the right-hand side is $\kappa(J_1 - B_1)$, where κ is the zeroth-order value of κ. In addition, it is clear from (7.33) that T_1 must have the same plane-wave dependence as J_1 and B_1, so we have

$$\nu_1 = \frac{4\pi\kappa\hat{B}_1}{\rho c_V \hat{T}_1}\left[1 - \frac{\kappa}{k}\mathrm{Cot}^{-1}\left(\frac{\kappa}{k}\right)\right] = \frac{16\sigma\kappa T_o^3}{\rho c_V}\left[1 - \frac{\kappa}{k}\mathrm{Cot}^{-1}\left(\frac{\kappa}{k}\right)\right]. \tag{7.34}$$

This result is plotted in Fig. 7.5. At small κ/k, ν_1 becomes independent of k and can be designated as

$$\nu = \frac{16\sigma\kappa T_o^3}{\rho c_V} = 11.5\frac{A}{(Z+1)}T^3\kappa_m, \tag{7.35}$$

in which the second equality gives ν in s^{-1} for T in eV and κ_m in cm^2/g. As κ/k increases, the damping is smaller. Consider this further. At small κ/k, which is the optically thin limit for the perturbation, the radiation travels many wavelengths before it is absorbed. The emission is small and the absorption is spatially uniform. (Thus, if one writes the equation for T_o, it will have a higher-order heating term.) As κ/k becomes larger, the absorption becomes increasingly local so the energy from

Fig. 7.5 Radiative cooling of optically thin fluctuations

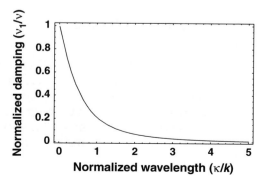

any given temperature maximum stays increasingly near that maximum, slowing the cooling. At very large κ/k, there is a net flow of energy from hot regions to cool regions within each wavelength. However, in this limit J_R also approaches B and the transition to a diffusive regime. The frequency of a fluctuating perturbation also must be considered. At high enough frequency, ignoring the first term in (6.58) is no longer justified. One can estimate when this might be important by taking $E_R \sim F_R/c$, $\nabla \to k$, and $\partial/\partial t \to \omega$ in the two terms on the left-hand side of (6.58). They become comparable when $\omega/k \sim c$. This is not a surprising result. We are now ready to apply the cooling rate of (7.34) to two cases of interest.

7.2.2.2   Cooling of Thin Layers

We have seen that faster shock waves produce higher postshock temperatures, and also that the opacity decreases as the temperature increases. As a result, experiments may produce optically thin shocked layers that endure for some time. The shocked layers produced by interstellar shocks in astrophysics are optically thin as well. Such layers may radiate so strongly that they cool substantially, their density profiles change, and more complicated dynamics also become possible. We will explore the profiles of the resulting cooling layers in our discussion of radiative shocks. It is often helpful, though, to evaluate the cooling rate of a hot, thin layer. If the corresponding cooling time is long compared to the duration of the system, then changes in layer structure due to the radiation will be minimal.

The cooling of a thin, planar layer of infinite lateral extent also provides a nice application of the radiative transfer equation. For a planar layer of thickness d, we use (6.52), writing

$$I_R = \int_0^d \rho \kappa_m B e^{-(z\rho\kappa_m/\mu)} \frac{\mathrm{d}z}{\mu} = B\left(1 - e^{-(d\rho\kappa_m/\mu)}\right). \qquad (7.36)$$

We use this to find the radiation flux emerging from one surface of the layer, as it was defined in (6.9), but integrating over only the one hemisphere of outgoing radiation. This gives

$$F_R = 2\pi \int_0^1 I_R \mu \mathrm{d}\mu = \pi B\left[1 + e^{-\tau}(\tau - 1) - \tau^2 E_1(\tau)\right], \qquad (7.37)$$

in which $\tau = d\rho\kappa_m$ is the optical depth of the shocked layer based on its thickness and $E_1(\tau)$ is the exponential integral function with $n = 1$. The total energy flux removed from the layer is twice this value, as radiation leaves from both sides. Figure 7.6 shows the radiative flux, normalized to σT^4, as a function of optical depth. One sees that the flux increases linearly from zero with optical depth at small optical depth but soon saturates and approaches σT^4.

Fig. 7.6 Radiative flux from optically thin layer. The ordinate shows the radiative flux leaving one surface of the layer, normalized by $\pi B = \sigma T^4$. The flux increases as the optical depth increases

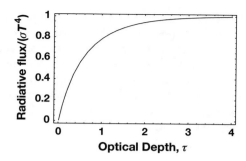

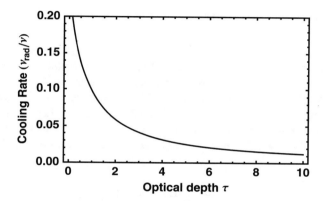

Fig. 7.7 The normalized cooling rate, ν_{rad}/ν decreases as the optical depth increases

From Fig. 7.6, one might expect cooling to be fastest when a layer is of order one optical depth thick, because one then extracts radiation from the entire volume but at a fairly high rate. This expectation would be false, however, because the amount of material to be cooled increases faster than the radiation flux. We can see this as follows. The radiation cooling rate ν_{rad} is the ratio of twice the radiative flux from a single surface to the energy content per unit area of the layer

$$\nu_{\text{rad}} = \frac{2F_R}{\rho d c_V T} = \nu \frac{\left[1 + e^{-\tau}(\tau - 1) - \tau^2 E_1(\tau)\right]}{8\tau}, \tag{7.38}$$

in which the normalizing factor ν is that of the previous section. Figure 7.7 shows the corresponding normalized cooling rate. Its limiting value as the optical depth becomes very small is 0.25. It decreases as optical depth increases, approaching 0.01 for an optical depth of 10. Beyond that point the normalized cooling rate changes slowly, scaling as $1/\tau$ at large τ. Thus, if $T \sim 10\,\text{eV}$ and $\kappa_m \sim 10^6\,\text{cm}^2\,\text{g}^{-1}$, one has $\nu \sim 3 \times 10^{10}\,\text{s}^{-1}$ giving sub-ns cooling times for $\tau \lesssim 4$. The cooling times will be shorter for smaller optical depth, and longer for larger optical depth.

It may be useful to develop a comparison of the cooling rate just found and the standard expression for the astrophysical cooling function, Λ. The cooling function Λ is the power loss per unit volume per unit electron density per unit ion density, in ergs-cm^3/s or equivalent. Thus, the power loss per unit volume is $n_e n_i \Lambda$. The corresponding cooling rate, ν_{astro}, is $n_e n_i \Lambda/(\rho c_V T)$. Setting this equal to ν_{rad}^*, we find the opacity $\kappa_{astro} = \rho \kappa_m$ corresponding to the optically thin astrophysical case, as $\kappa_{astro} = n_e n_i \Lambda/(2\sigma T^4)$. Note also that one can identify the emissivity of the thin layer as $\kappa_{astro} d$.

7.2.3 Optically Thin Acoustic Waves

The diffusive, near-equilibrium regime of Sect. 7.2.1 is not easily achieved in the laboratory or often encountered in astrophysics except within stars. Much more common are systems hot enough that radiative cooling matters, but optically thin, or at least not very thick, so that (7.33) and (7.34) describe the cooling. These systems also have negligible radiation pressure and energy density. Let us consider how acoustic waves behave in this regime, in the limit that the system is so optically thin that $\nu_1 = \nu$ from (7.35). Then, informed by the discussion relating to (3.141), (7.7) becomes

$$\frac{Dp}{Dt} - \frac{\gamma_s p}{\rho}\frac{D\rho}{Dt} = -(\gamma_h - 1)\nu\rho c_V T_1, \tag{7.39}$$

We once again linearize for an initially uniform plasma, using (7.23) and (7.24) for negligible radiation, to find

$$\frac{T_1}{T_o} = \frac{(1+Z)}{(1+\beta Z)}\left[\frac{p_1}{p_o} - \frac{\rho_1}{\rho_o}\right], \tag{7.40}$$

in which, as in (7.24), $\beta = 1$ for an ionized plasma, 3/2 for an ionizing plasma with $Z \propto \sqrt{T}$, and some other value or function in more general cases. From this we obtain, from (7.39) in the comoving frame, after linearizing and redefining ν to absorb the factor of $(\gamma_h - 1)$, the equation

$$\frac{\partial p_1}{\partial t} - \frac{\gamma_s p_o}{\rho_o}\frac{\partial \rho_1}{\partial t} = -\nu\left(p_1 - \frac{p_o}{\rho_o}\rho_1\right). \tag{7.41}$$

Here we again use the result from the continuity and fluid momentum equations that $\partial^2 \rho_1/\partial t^2 = \nabla^2 p_1$, to find

$$\left[\frac{\partial^2}{\partial t^2} - \frac{\gamma_s p_o}{\rho_o}\nabla^2\right]\frac{\partial p_1}{\partial t} = -\nu\left[\frac{\partial^2}{\partial t^2} - \frac{p_o}{\rho_o}\nabla^2\right]p_1. \tag{7.42}$$

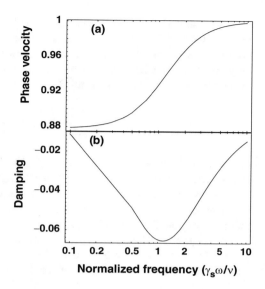

Fig. 7.8 Acoustic waves in the optically thin limit. The phase velocity, normalized to the isentropic sound speed, c_s, and the spatial damping rate, normalized to the wavenumber k vs. normalized frequency

The corresponding dispersion relation is

$$-i\omega\left[-\omega^2 + \frac{\gamma_s p_o}{\rho_o}k^2\right] = -\nu\left[-\omega^2 + \frac{p_o}{\rho_o}k^2\right],\tag{7.43}$$

which can be solved for the normalized inverse phase velocity, $c_s k/\omega$, to obtain

$$\frac{c_s^2 k^2}{\omega^2} = \frac{1}{1 + \nu^2/(\gamma_s^2 \omega^2)}\left[1 + \gamma_s\left(\frac{\nu}{\gamma_s \omega}\right)^2 + i(\gamma_s - 1)\frac{\nu}{\gamma_s \omega}\right].\tag{7.44}$$

One sees in (7.44) that one recovers ordinary isentropic acoustic waves as the cooling rate goes to zero, and damped, isothermal acoustic waves as the cooling rate becomes large. Figure 7.8 shows how the phase velocity and damping rate implied by this dispersion relation depends on the natural normalized frequency for (7.44), $\gamma_s \omega/\nu$. The phase velocity increases from the isothermal sound speed at low normalized frequency to the isentropic sound speed at high normalized frequency. In both these limits, the spatial damping rate is small. The spatial damping rate increases somewhat during the transition.

Returning to (7.44), the limiting behavior in this regime is easy to recover and merits discussion. In the limit of very high frequency or very small damping, we evidently recover ordinary acoustic waves. For a high frequency with $\nu/\omega \ll 1$, we find

$$k = \frac{\omega}{c_s}\left[1 + i\frac{1}{2}\frac{\nu}{\omega}\frac{(\gamma_s - 1)}{\gamma_s}\right],\tag{7.45}$$

corresponding to damped sound waves with a characteristic, very-long damping length of $(2\gamma_s c_s/v)/(\gamma_s - 1)$. The opposite limit, in which $v/\omega \gg 1$, is more complicated because at a low enough frequency the waves will experience an optically thick medium and the present calculation will not apply. Assuming that over some range of frequencies this limit does make sense, one can see that k will be

$$k = \frac{\sqrt{\gamma_s}\omega}{c_s}\left[1 + i\frac{1}{2}\frac{\omega}{v}(\gamma_s - 1)\right], \tag{7.46}$$

and that the damping length is $[2c_s/(v\sqrt{\gamma_s})](v^2/\omega^2)/(\gamma_s - 1)$. This damping length is also quite long, compared to the wavelength of the fluctuations. In this limit we have weakly damped, isothermal acoustic waves with phase velocity $(c_s/\sqrt{\gamma})$. Physically, in this case the radiation damps out the temperature fluctuations at a rate much faster than the wave frequency.

To see the implications of this, consider acoustic waves in a CH plasma at $T = 100\,\text{eV}$. With $\kappa_P = 2 \times 10^5 \rho/T_{eV}, \rho = 1\,\text{g/cm}^3$, the radiative damping rate from (7.35) is just over $10^9\,\text{s}^{-1}$. If the sound speed is about $10^6\,\text{cm/s}$ ($10\,\text{km/s}$), then acoustic waves with wavenumbers of $0.001\,\text{cm}^{-1}$ will be the most damped. The corresponding wavelength of order $100\,\text{m}$ is large compared to experiments, so this result is most relevant to the gradual damping of large-scale structures in the plasma. The waves inside the plasma, having larger wavenumbers, will be isentropic acoustic waves.

Our two treatments of radiative acoustic waves, in this section and Sect. 7.2.1, show rather different behavior, because they apply to different regimes. The lowest frequencies, in any medium, are optically thick, in the sense that the absorption will occur in a very small fraction of a wavelength, so the description of Sect. 7.2.1 will apply. The highest frequencies are optically thin, so the description of the present section will apply. As a result, Fig. 7.8 connects naturally to Fig. 7.4, because increasing the frequency also takes one from an optically thick to an optically thin regime. Overall, acoustic waves progress from isentropic to isothermal and back to isentropic as frequency increases. The transition between the two regimes of optical depth, and the even-more-complicated case of frequencies so high that the propagation time of the radiation matters, are discussed in detail in Mihalas and Weibel-Mihalas (1984).

7.2.4 Radiative Thermal Cooling Instability

An important application of cooling by radiation is the *radiative thermal cooling instability*. This regulates the pulsations of Cepheid variable stars and creates structures within high-energy-density plasmas.

As we shall see, this instability occurs when a finite region of material cools by radiating, and when the derivative of the opacity with temperature is sufficiently negative. Consider a system with spatially constant initial density ρ and temperature $T(t)$, gradually cooling. For this problem we want to consider the interplay of density and temperature fluctuations. The key equations are the gas-energy equation (7.6) and the basic momentum equation, which can be written as

$$\rho c_V \frac{\partial}{\partial t} T + \rho c_V \mathbf{u} \cdot \nabla T + \rho R T \nabla \cdot \mathbf{u} = -\nabla \cdot \mathbf{F}_R, \text{ and} \tag{7.47}$$

$$\rho \frac{\partial}{\partial t} \mathbf{u} + \rho \mathbf{u} \cdot \nabla \mathbf{u} = -\nabla p = -R T \nabla \rho - \rho R \nabla T. \tag{7.48}$$

Linearizing with non-subscripted zeroth order quantities and first-order $\mathbf{u}$, keeping zeroth order and first order terms, gives

$$\rho c_V \frac{\partial}{\partial t} (T + T_1) + \rho R T \nabla \cdot \mathbf{u} = -\nabla \cdot \mathbf{F}_R, \tag{7.49}$$

and

$$\rho \frac{\partial}{\partial t} \mathbf{u} = -R T \nabla \rho_1 - \rho R \nabla T_1, \tag{7.50}$$

with the continuity equation adding

$$\frac{\partial}{\partial t} \rho_1 = -\rho \nabla \cdot \mathbf{u}. \tag{7.51}$$

Eliminating $\mathbf{u}$ gives

$$\rho c_V \frac{\partial}{\partial t} (T + T_1) - R T \frac{\partial}{\partial t} \rho_1 = -\nabla \cdot \mathbf{F}_R, \tag{7.52}$$

and

$$\frac{\partial^2}{\partial t^2} \rho_1 = R T \nabla^2 \rho_1 + \rho R \nabla^2 T_1. \tag{7.53}$$

Suppose our system has an initial temperature fluctuation, so that $T_1 = \hat{T} e^{ikx}$ and to first order $B[T + T_1] = B_o(1 + 4T_1/T)$ so $B_1 = \hat{B} e^{ikx}$ with $\hat{B} = 4B_o \hat{T}/T$. Suppose also that the system is optically thin enough that $J \sim 0$. Then we have

$$-\nabla \cdot \mathbf{F}_R = 4\pi \kappa (J - B) = -4\pi \left(\kappa + \frac{\partial \kappa}{\partial T} T_1 \right) (B_o + B_1). \tag{7.54}$$

Keeping zeroth and first order terms gives

$$-\nabla \cdot \mathbf{F}_R = -4\pi \left(\kappa B_o + \kappa B_1 + B_o \frac{\partial \kappa}{\partial T} T_1 \right)$$

$$= -4\pi \kappa B_o - 4\pi B_o \left(4\kappa \frac{T_1}{T} + \frac{\partial \kappa}{\partial T} T_1 \right) \tag{7.55}$$

Now we can drop the first order terms in Eq. (7.49) to see that the zeroth order equation is

$$\rho c_V \frac{\partial}{\partial t} T = -4\pi \kappa B_o \tag{7.56}$$

Since $B_o \propto T^4$ and κ is more or less some power of T, this corresponds to cooling of the system as a power law in time. This can be relatively slow or rapid, depending on the temperature dependence of κ. In contrast, the first-order behavior offers the possibility of having exponentially growing fluctuations, as follows. Keeping only the first order terms, Eq. (7.49) becomes

$$\rho c_V \frac{\partial}{\partial t} T_1 - RT \frac{\partial}{\partial t} \rho_1 = -4\pi B_o \left(4\kappa \frac{T_1}{T} + \frac{\partial \kappa}{\partial T} T_1 \right), \tag{7.57}$$

Now look for growing modulations proportional to $\exp[ikx + nt]$. Then Eq. (7.50) gives

$$(n^2 + RTk^2) \frac{\rho_1}{\rho} = -RTk^2 \frac{T_1}{T}. \tag{7.58}$$

Substituting in Eq. (7.57) then yields

$$n\rho c_V T \frac{T_1}{T} + n\rho RT \frac{RTk^2}{(n^2 + RTk^2)} \frac{T_1}{T} = -4\pi B_o \left(4\kappa \frac{T_1}{T} + \frac{\partial \kappa}{\partial T} T_1 \right), \tag{7.59}$$

which can be written

$$n\rho RT \left(\frac{1}{(\gamma - 1)} + \frac{RTk^2}{(n^2 + RTk^2)} \right) = -4\pi B_o \left(4\kappa + \frac{\partial \kappa}{\partial T} T \right). \tag{7.60}$$

Note that every symbol in this equation is positive except possibly $\partial \kappa / \partial T$ and n. As a result n is negative and the modulations decay unless

$$\frac{\partial \kappa}{\partial T} < -4 \frac{\kappa}{T}, \tag{7.61}$$

in which case they grow exponentially. For $\kappa \propto T^{-m}$ this requires $m > 4$. When this occurs, it is an example of a radiative cooling instability. Such steep cooling behavior does occur. For example, consider the cooling curves shown in Chap. 6. These are proportional to the product κB, and so any cooling curve with a negative slope corresponds to $m > 4$. The large-scale structure of κB is to decrease with T out to some small number of keV, and then to increase as bremsstrahlung and perhaps synchrotron radiation become significant. The cooling instability will thus be active across a broad range of temperatures. Structure in the long-term evolution of conductively heated, radiatively cooled high-Z plasmas has been attributed to this instability. Also, one can sometimes see a variant of it in the simulated evolution of initially uniform plasmas that are cooling after being heated by radiation. Since the radiative damping is smallest at the largest k, this will tend to create fluctuations on the scale of the zones in a simulation.

Beyond the thermal instability, structure in κ can lead to other effects in the plasma evolution. For example, nonlinear oscillations in temperature can arise in a system that is steadily heated. This occurs in the visible layers in Cepheid-variable stars, steadily heated from within, which operate in a range of temperatures where the opacity of Fe has maxima and minima. When the visible layer of the star is at a temperature where $\partial \kappa / \partial T > 0$, the stronger radiative cooling can cause the temperature to decrease, overshooting the minimum in κ, so that $\partial \kappa / \partial T < 0$, in response to which the temperature increases, again causing κ_o to overshoot the minimum. The result is a steady oscillation in temperature and luminosity. Figure 7.9 shows the experimentally measured transmission through a sample of Fe at a temperature near 100 eV. High transmission corresponds to low opacity. One can see that the opacity fluctuates with temperature.

As another example, the presence of a region where $\partial \kappa_o / \partial T > 0$ over some narrow range of temperatures in the plasma expanding from a laser-heated surface can lead to a local density maximum. The pressure of the adjacent regions compresses the region where the radiation losses are larger. In the context of laser fusion, such structures have been designated (Hazak et al. 1999) *radiative plasma structures*.

Fig. 7.9 Opacity of Fe. The plot shows transmission through an Fe sample, as reported by DaSilva et al. (1992)

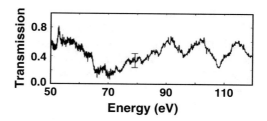

7.3 Radiation Diffusion and Marshak Waves

When a cool region warms by the transport of radiative heat from a hot region, this transport is often diffusive. The mean-free path of the radiation in the cool material can be quite short. However, the mean-free path often increases rapidly with temperature, so that diffusive heat transport is more complicated than simple diffusion. Diffusion in the presence of a variable diffusion coefficient is often referred to as *nonlinear diffusion*. Nonlinear diffusion is fundamental to high-energy-density plasmas, because they are ionizing. The opacity of each ionization state is different. In general, as material ionizes the spectral regions of largest opacity shift to higher energy, because more energy is needed to access the bound-free and bound–bound transitions of the more highly ionized state. In addition, even ordinary bremsstrahlung absorption is nonlinear. Its opacity decreases with increasing temperature as $1/T_e^{3/2}$ for otherwise fixed conditions. Two kinds of nonlinear-diffusion problems merit our attention here. In the first, a constant-temperature source drives a radiative heat wave, known as a Marshak wave, into a cooler material. In the second, a finite amount of energy is spread through the material by radiative diffusion. We consider these in turn.

7.3.1 Marshak Waves

The Marshak wave describes the solution to a simple problem that nonetheless has great relevance to many real situations. The simple problem is the near-equilibrium diffusion of radiative energy into an initially cold material, through an initially sharp boundary, from a constant-temperature energy source. Marshak was the first to show that this problem admits self-similar solutions. The medium is assumed to be at rest and to remain at rest. We will revisit this assumption later, but note here that it would be a poor assumption if the temperature were high enough that the radiation pressure was the dominant pressure. Thus, Marshak waves are relevant to the common situation that the radiative heat transport is essential but the radiative pressure is small.

An essential example of this is the heating of a high-Z material wall by a sustained radiation source, such as the emission of thermal X-rays from laser-heated regions on the wall. An enclosed structure within which this occurs is known as a hohlraum. We will discuss hohlraums further in Chap. 9. They are of real importance for inertial fusion and for other experiments that require a sustained radiation environment.

To obtain a solvable description of the Marshak-wave problem, we assume that the radiative coefficient of heat conductivity scales as $\kappa_{rad} \propto T^n$. This is reasonable, with $n \sim 4$ to 5 for typical materials in high-energy-density systems (in contrast, $n \sim 6$ to 7 in typical astrophysical systems). We further assume constant density and specific heat. Given these assumptions, the gas-energy equation (7.5) describing the temperature structure becomes

$$\rho \frac{\partial \epsilon}{\partial t} = \rho c_V \frac{\partial T}{\partial t} = \nabla \cdot \kappa_{\mathrm{rad}} \nabla T = \frac{T_o}{n+1} (\kappa_{\mathrm{rad}})_{T_o} \nabla^2 \left(\frac{T}{T_o} \right)^{n+1}. \tag{7.62}$$

With $f = T/T_o$ and $W = (\kappa_{\mathrm{rad}})_{T_o}/[\rho c_V (n+1)]$, this can be written as

$$\frac{\partial f}{\partial t} = W \nabla^2 f^{n+1}. \tag{7.63}$$

This equation has only the one, dimensional parameter, W, so recalling Chap. 4 we can expect to find a planar similarity solution with similarity variable $\xi = x/\sqrt{Wt}$. This gives

$$-\frac{\xi}{2} \frac{df}{d\xi} = \frac{d^2 f^{n+1}}{d\xi^2}. \tag{7.64}$$

Note that $f = 1$ at $\xi = 0$. One can show that the second derivative of f remains negative so that f eventually reaches zero at some $\xi = \xi_o$. This makes possible a simple, approximate calculation. One can assume that the radiative heat flux must be constant from the source location to the end of the heat wave. This must be approximately true; otherwise, the temperature somewhere would increase above the source temperature or decrease below that of its surroundings. The radiative flux, in terms of the variables just defined, is

$$F_R = -\kappa_{\mathrm{rad}} \frac{\partial T}{\partial x} = -\frac{T_o}{n+1} (\kappa_{\mathrm{rad}})_{T_o} \frac{\partial f^{n+1}}{\partial x} = -\rho c_V T_o \sqrt{\frac{W}{t}} \frac{df^{n+1}}{d\xi}. \tag{7.65}$$

Even though $f(\xi)$ retains the same shape, the flux decreases with time as the physical temperature gradient at the boundary decreases. If we assume that the flux at any given time is constant throughout the wave, then recalling that $f = 0$ at $\xi = \xi_o$ we find

$$f(\xi)^{n+1} = C (\xi_o - \xi) = (1 - \xi/\xi_o), \tag{7.66}$$

with C a constant equal to $1/\xi_o$ because $f = 1$ at $\xi = 0$. This is equivalent to

$$T = T_o (1 - \xi/\xi_o)^{1/(n+1)}. \tag{7.67}$$

Figure 7.10 shows the temperature profile from the constant-flux model and from the solution of the more exact Eq. (7.64), for some values of n. The constant-flux model is sufficiently accurate for nearly all purposes, as the other assumptions in the Marshak-wave model are certainly not exact (e.g., see the next section for a discussion of ionization).

Fig. 7.10 Marshak-wave
temperature profiles. The
normalized temperature,
$f = T/T_o$, decreases
nonlinearly as ξ/ξ_o
approaches 1. The gray
curves give the temperature
profiles from the
constant-flux model for
$n = 3, 4.5,$ and 7 from
bottom to top. The black
curves give the corresponding
numerical solutions

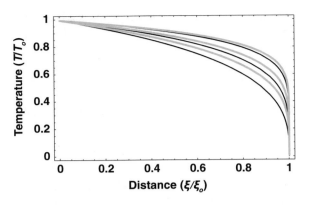

Fig. 7.11 The lines show the
value of ξ_o from the
constant-flux model (grey
curve) or a numerical solution
(black curve) for traditional
Marshak waves. The dashed
line shows the result for an
ionizing radiation wave

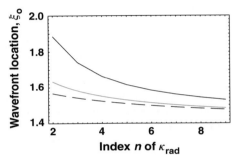

Continuing with the constant-flux calculation, we can find ξ_o by realizing that the flux through the initial boundary must equal the rate of increase of energy, E_w, in the wave, or

$$\frac{\partial E_w}{\partial t} = \frac{\partial}{\partial t}\left[\frac{dx}{d\xi}\int_o^{\xi_o}\rho c_V T d\xi\right] = \frac{\rho c_V T_o}{2}\sqrt{\frac{W}{t}}\left(\frac{n+1}{n+2}\right)\xi_o. \qquad (7.68)$$

Setting this equal to the flux at $x = 0$, and knowing from (7.66) that $\partial f^{(n+1)}/\partial\xi = -1/\xi_o$, we find

$$\xi_o = \sqrt{2}\sqrt{(n+2)/(n+1)}. \qquad (7.69)$$

Figure 7.11 compares this value of ξ_o with a more exact solution. One sees that the constant-flux model underestimates the extent of the heat front by roughly 10%. Here again, this is a small effect compared with other probable differences between a real situation and a Marshak-wave model.

The position of the radiation wavefront, x_o, from the constant flux model, is

$$x_o = \frac{\sqrt{(n+2)}}{(n+1)}\sqrt{\frac{(\kappa_{rad})_{T_o}}{\rho c_V}}\sqrt{2t}, \qquad (7.70)$$

and the front velocity u_o is

$$u_o = \frac{\sqrt{(n+2)}}{(n+1)} \sqrt{\frac{(\kappa_{\text{rad}})_{T_o}}{\rho c_V}} \frac{1}{\sqrt{2t}}. \tag{7.71}$$

Note that this velocity decreases from infinity to very small values as time increases. Of course, the physical velocity is never infinite, because the assumptions of the model break down as t approaches zero.

Equation (7.64) can be solved numerically but to do so one must find workable boundary conditions. One knows that $f = 1$ at $\xi = 0$ and that $f = 0$ at $\xi = \xi_o$, but one does not know ξ_o, the value of ξ corresponding to the head of the wave. One does know that the flux through $\xi = 0$, from (7.65) evaluated at $\xi = 0$, must equal the time rate of increase in the energy content of the radiation wave, E_w. Generalizing (7.68) and using (7.65), one has

$$\left(\frac{df}{du}\right)_{u=0} = \frac{-\xi_o^2}{2(n+1)} \int_0^1 f(u)du, \tag{7.72}$$

in which $u = \xi/\xi_o$. One can then solve (7.64) and (7.72), with independent variable u, iteratively for ξ_o and for the value of the integral in (7.72), by seeking conditions such that $f = 0$ at $\xi = \xi_o$. This procedure produced the numerical curves shown in Figs. 7.10 and 7.11.

The decrease with time of the velocity of the radiation wave has important consequences for real systems. At first, the velocity of the radiation front far exceeds any other velocity in the system. In this regime, the wave is known as a *supersonic radiation wave*. During this period, the radiation wave reaches any location in the medium first and is affected only by changes in the radiation source.

Two competing effects, not included in the simple Marshak model, alter the structure of any real radiation wave as it slows. The first effect is the launching of a shock by the radiation-heated matter. The important parameter is the downstream Mach number of the radiation wave, which is the ratio of its speed to the sound speed in the heated matter. The higher-pressure, heated matter would have launched a shock wave, except that at early times the radiation wave is faster. It turns out that the shock is launched once the wave speed decreases to Mach 2. We discuss this effect further in Sect. 8.2.1. Once this shock is launched, the radiation-heated region expands behind it, which will tend to reduce the pressure and to weaken and slow the shock.

The second effect is the overtaking of the radiation wave by an ablatively driven shock wave. In a real system, the advent of a radiation flux is rarely if ever the only process to occur at the boundary. Whether one considers the birth of a star or any other release of energy within an optically thick environment, or the initiation of an X-ray source within a high-Z container, the inner boundary of the affected material is also disturbed. Very often, the absorption of radiation produces ablation at this boundary, launching a shock wave into the material. The downstream Mach number of this shock wave varies from just below 2 to several, depending on γ. The location of the shock wave is initially proportional to t, and it slows very gradually. Eventually it must overtake the radiation wave. The timing of the ablatively driven shock relative to that of the shock launched by the radiation wave depends on details. The velocity of the radiation wave eventually drops below the sound speed in the radiation-heated medium, at which time it becomes a *subsonic radiation wave*.

The shock waves alter the density structure in the system. The extent to which this alters the shape of the radiation wave will depend on the density dependence of the opacity. The shock heating, in contrast, is typically not significant in real cases where there is enough radiation to sustain a radiation wave.

7.3.2   Ionizing Radiation Wave

The largest error in the Marshak-wave model, especially in a laboratory environment, is the assumption that the specific heat at constant volume, c_V, is constant. This is very much not true, as c_V depends on Z, through both the thermal energy and the ionization energy, and Z is not constant. We can describe a wave in which c_V and Z change through ionization as an *ionizing radiation wave*. (This should not be confused with an *ionization front*, discussed Sect. 8.2.3.) To obtain an evaluation of the difference between such a wave and a Marshak wave, we can revisit the analysis of the previous section. Assuming $Z \propto \sqrt{T}$ and a hydrogenic model of the ion, we have

$$c_V = \frac{3}{2} \frac{(1 + (3/2)Z)k_B}{Am_p} + \frac{k_B E_H}{12T} \frac{(Z + 6Z^2 + 12Z^3)}{Am_p}. \tag{7.73}$$

This would not admit a self-similar solution if all the terms in c_V were important. However, for $T_e > 10\,\text{eV}$, the terms of highest order in Z dominate. In this regime we can take $c_V = c_{Vo}\sqrt{T/T_o}$, where c_{Vo} is the value of c_V when $T = T_o$. One then can show, just as in (7.62)–(7.64), that

$$-\frac{\xi}{2}\frac{df}{d\xi} = \frac{1}{\sqrt{f}}\frac{d^2 f^{n+1}}{d\xi^2}, \tag{7.74}$$

with the same definitions of f and ξ. Once again, this can be integrated numerically.

Alternatively, one can develop a constant-flux description of this system. The flux becomes

$$F_R = -\rho c_{Vo} T_o \sqrt{\frac{W}{t}} \sqrt{f}\frac{df^{n+1}}{d\xi}. \tag{7.75}$$

Holding this constant and integrating with $f = 1$ at $\xi = 0$ gives

$$T = T_o \left(1 - \xi/\xi_o\right)^{2/(2n+3)}, \tag{7.76}$$

from which we integrate to get the increase in energy

$$\frac{\partial E_w}{\partial t} = \frac{\partial}{\partial t}\left[\frac{dx}{d\xi}\int_o^{\xi_o} \rho c_V T d\xi\right] = \frac{\rho c_{V_o} T_o}{2}\sqrt{\frac{W}{t}}\left(\frac{2n+3}{2n+6}\right)\xi_o. \tag{7.77}$$

Setting this equal to the flux, we find

$$\xi_o = \frac{2\sqrt{(2n+6)(n+1)}}{2n+3} \quad \text{and} \tag{7.78}$$

$$x_o = \frac{2\sqrt{(2n+6)}}{(2n+3)}\sqrt{\frac{(\kappa_{\mathrm{rad}})_{T_o}}{\rho c_{V_o}}}\sqrt{t}. \tag{7.79}$$

Figure 7.11 shows the value of ξ_o at the front from (7.78). The Marshak wave will be shorter in an ionizing system than in a system with constant specific heat. Figure 7.12 compares the shape of this wave (from 7.76) with the shape of a traditional Marshak wave. One sees no dramatic differences.

7.3.3   Constant-Energy Radiation Diffusion Wave

Now we turn to the second case of common interest. A finite event, such as a laser pulse, a Z-pinch implosion, or an astrophysical burst, may produce a definite

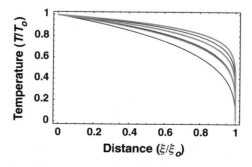

Fig. 7.12 Ionizing radiation wave profiles. The normalized temperature $f = T/T_o$ is somewhat flatter in an ionizing medium than it is in the fixed-Z case of the traditional Marshak wave. Here the gray curves give the temperature profiles from the ionizing model for $n = 3$, 4.5, and 7 from bottom to top. The thin black curves give the corresponding numerical solutions for the traditional Marshak wave

amount of radiation. If the radiation is released into a uniform near-equilibrium medium in which the radiation pressure is negligible and the radiation transport is diffusive, then the same fundamental equations apply as in the Marshak-wave case. We consider the planar case here. (Zel'dovich and Razier (1966) discuss the spherical case.) One still has, from (7.62),

$$\frac{\partial T}{\partial t} = \frac{(\kappa_{\text{rad}}/T^n)}{\rho c_V (n+1)} \nabla^2 T^{n+1} = Y \nabla^2 T^{n+1}, \tag{7.80}$$

thus defining Y where, if T is in energy units, then Y has units of $cm^2 s^{-1}$ energy^{-n}. As κ_{rad}/T^n does not depend on T, Y is independent of T. In contrast to the Marshak-wave case, we need a time-dependent normalization for T since the maximum temperature must decrease with time as energy is carried outward. This normalization is found below. Note also that $\rho c_V T$ is the energy per unit volume so T is the energy per unit volume per unit ρc_V. However, the total energy per unit area is also fixed so this is a problem with two independent dimensional parameters. Defining the energy per unit area per unit ρc_V as $Q = \int T dx$, which has units of cm^1 energy^1, the quantity $Q^n Y t$ has units of $cm^{(n+2)}$. Thus, an effective dimensionless similarity variable is

$$\xi = x/ (Q^n Y t)^{1/(n+2)}. \tag{7.81}$$

As a result, we expect that the position of any point on the heat wave, where for example the temperature is some fraction of the maximum temperature, will be $\propto t^{1/(n+2)}$. Since n is typically 4 or 5, such diffusion waves propagate *much* more slowly than Marshak waves. The normalization of T need not be spatially dependent, as all the spatial dependence can be in the evolution with ξ. To see what normalization makes sense, we consider the spatial derivative of a normalized function, f, finding

$$\frac{\partial f}{\partial x} = \frac{1}{(Q^n Y t)^{1/(n+2)}} \frac{df}{d\xi}. \tag{7.82}$$

We can also see that the right-hand side (RHS) of (7.80), in terms of ξ, becomes

$$\text{RHS} = \frac{Y}{(Q^n Y t)^{2/(n+2)}} \frac{d^2}{d\xi^2} T^{n+1}. \tag{7.83}$$

Since the time derivative on the left-hand side of (7.80) will introduce a factor of $1/t$, it makes sense to multiply (7.83) by t, from which we can find that an effective normalization for T, with consistent units, is $[Q^2/(Yt)]^{1/(n+2)}$, so $f = T/[Q^2/(Yt)]^{1/(n+2)}$. It is important to note that the denominator in the definition of f is not the central temperature $T(0)$. Rather, one obtains $T(0)$ by multiplying

$[Q^2/(Yt)]^{1/(n+2)}$ times the value of f that we will find at $\xi = 0$. Because f is *not* time-dependent but depends only on ξ, to develop the equation for f we must observe that

$$\frac{1}{[Q^2/(Yt)]^{1/(n+2)}} \frac{\partial T}{\partial t} = \frac{\partial f}{\partial t} - \frac{f}{(n+2)t}. \tag{7.84}$$

We also have the usual type of relation between derivatives in t and ξ,

$$\frac{\partial f}{\partial t} = \frac{-1}{n+2} \frac{\xi}{t} \frac{df}{d\xi}, \tag{7.85}$$

so (7.80) becomes

$$f + \xi \frac{df}{d\xi} + (n+2) \frac{d^2}{d\xi^2} f^{n+1} = 0. \tag{7.86}$$

The solution to this equation is

$$f(\xi) = \left[\frac{n\xi_o^2}{2(n+2)(n+1)} \right]^{1/n} \left[1 - \left(\frac{\xi}{\xi_o} \right)^2 \right]^{1/n}. \tag{7.87}$$

This has a fixed shape in ξ, as expected. The time dependence is entirely included in the normalization of T, as it should be, so

$$T(0) = \left[\frac{n\xi_o^2}{2(n+2)(n+1)} \right]^{1/n} \left(\frac{Q}{Y} \right)^{1/(n+2)} t^{-1/(n+2)}. \tag{7.88}$$

Figure 7.13 shows the shape of the constant-energy, radiation-diffusion wave and compares it to that of the Marshak wave. The constant-energy wave produces a much flatter temperature profile. Both waves have the very steep front that is characteristic of nonlinear diffusion waves.

Fig. 7.13 Radiation diffusion wave profiles. The shape function, $f(\xi)$, is shown against the normalized similarity variable, ξ/ξ_o, for both constant-energy radiation diffusion waves (gray) and Marshak waves (black). In each case, from bottom to top, the curves correspond to $n = 3, 4.5,$ and 7

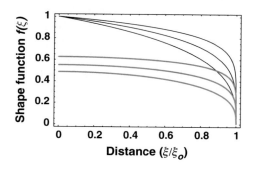

Fig. 7.14 Wavefront location
for a constant-energy
diffusion wave

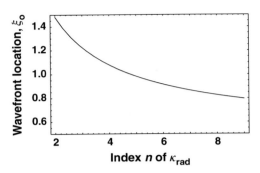

The value of ξ_o must be determined from an integral of the total energy. The definitions of Q and f imply that the correct normalization is

$$1 = \int_{-\infty}^{\infty} f(\xi)d\xi = \left[\frac{n}{2(n+2)(n+1)}\right]^{1/n} \xi_o^{(2+n)/n} \sqrt{\pi} \frac{\Gamma\left(1+\frac{1}{n}\right)}{\Gamma\left(\frac{3}{2}+\frac{1}{n}\right)}. \tag{7.89}$$

Figure 7.14 shows the value of ξ_o as a function of n. One sees that $\xi_o \sim 1$ to within 10% over the range of interest for radiation waves in high-energy-density plasmas. We can use this value to evaluate the location of the heat front, obtaining (7.90). Note that in a real problem one probably knows Q and does not know T_o, but does know $(\kappa_{rad})_{T_o}/T_o^n$. The heat front location is

$$x_o = \xi_o \left(Q^n Yt\right)^{1/(n+2)} \approx \left[\left(\frac{Q}{T_o}\right)^n \frac{(\kappa_{rad})_{T_o}}{\rho c_V(n+1)}\right]^{\frac{1}{(n+2)}} t^{1/(n+2)}. \tag{7.90}$$

The constant-energy radiation wave slows very rapidly. Its speed is $\propto$ $t^{-(n+1)/(n+2)}$ as compared to $1/\sqrt{t}$ for the Marshak wave. Yet the same phenomena occur as it slows as do with the Marshak wave. It will launch a shock when it slows enough and it will be overtaken by ablative shocks, if they exist. Because its speed decreases more rapidly, these effects will occur sooner. One may be challenged in any experiment at high energy density to discern the period when a constant-energy radiation wave is the dominant feature. Even so, this problem is a good example of a type of self-similar system, where the scale of the shape function evolves in time, and it can be a good test problem for simulation codes.

This concludes our discussion of radiation diffusion waves. We discuss the related topic of ionization fronts in the next chapter.

Homework Problems

7.1 Carry out the calculations of radiation and material energy fluxes and pressures and compare the behavior of pure hydrogen as opposed to C_1H_1 (used in Fig. 7.1).

7.2 Derive the dispersion relation for isothermal acoustic waves from the Euler equations. That is, demand constant temperature and see what happens.

7.3 Figure 7.4 shows the wave properties as ω varies for fixed η. Consider how the wave properties vary with η for $\beta = 1$ and fixed $\omega/(v_e c_s^2/c^2)$. Plot the normalized phase velocity and damping length for $0.01 \leq \eta \leq 10$ and discuss the results.

7.4 There should be a sensible connection between the present calculation and the optically thick one as the system becomes optically thick. For the limit in which $\kappa >> k$, seek to reconcile (7.21) and (7.33).

7.5 We did not explore the angular variation in the contributions to (7.37). One might imagine that the largest contributions could come at grazing angles, where μ is very small and the optical depth along a line of sight becomes large. The model used here would be less realistic if most of the emission came at grazing angles, because real systems will have layers that are not truly planar and certainly are not infinite in extent. Use a computational mathematics program to derive (7.37). Then modify the calculation to explore how large the contribution is from such grazing angles. Conclude whether or not the results above might be reasonable estimates for real layers.

7.6 It is curious that (7.41) and (7.43) do not depend on β, so that these waves seem not to care whether the system is fully ionized. Beginning with (7.39), derive (7.43) and discuss why there is no β dependence.

7.7 Beginning with $\rho(\partial \epsilon/\partial t) = \nabla \cdot (\kappa_{rad} \nabla T)$ derive (7.64).

7.8 Work through the constant-flux model for Marshak waves, providing all the missing mathematical steps. Then plot the positions vs. time of the radiation wave and of a disturbance (in the radiation-heated material) moving at Mach 1 or Mach 5, for a wave in Au foam with $T_o = 200\,\text{eV}$, $\rho = 0.1\,\text{g/cm}^{-3}$, and $Z = 40$. Discuss the results.

7.9 For the constant-energy, radiation-diffusion wave, show that (7.87) is a solution to (7.86). Clearly annotated work with a computational mathematics program is preferred.

7.10 Consider a gold container shaped so that a planar approximation is reasonable, having planar walls spaced 1 mm apart in vacuum. Assume $\rho = 20\,\text{g/cm}^3$ and treat $c_V = 10^{12}\,\text{ergs/(g\,eV)}$ as constant. Use other parameters from Chap. 6 as appropriate. Suppose $100\,\text{kJ/cm}^2$ is the initial energy content of the vacuum between the walls and that the initial wall temperature is negligible. Assume that the gold material does not move. Apply the self-similar model of the constant-energy radiation diffusion wave to this system, on the assumption that the two walls are touching but contain the specified energy. In doing so, approximate ξ_o from Fig. 7.14. From zero to 10 ns, find the position of the heat front and the temperature of the surface as a function of time. Realizing that the walls are in equilibrium with the temperature of the radiation in the vacuum, plot the ratio of the energy content of the walls to the energy content of the vacuum. Discuss the meaning of this result from the standpoint of the accuracy of an intermediate asymptotic model (see Chap. 4).

7.11 Develop the equivalent of (7.90) describing the radius of a spherically symmetric, constant-energy, radiation-diffusion wave.

References

DaSilva LB, MacGowan MJ, Kania DR, Hammel BA, Back CA, Hsieh E, Doyas R, Iglesias CA, Rogers FJ, Lee RW (1992) Absorption measurements demonstrating the importance of delta n = 0 transitions in the opacity of iron. Phys Rev Lett 69:493–496

Hazak G, Velikovich AL, Klapisch M, Schmitt AJ, Dahlburg JP, Colombant D, Gardner JH, Phillips L (1999) Study of radiative plasma structures in laser driven ablating plasmas. Phys Plasmas 6(10):4015–4021

Mihalas D, Weibel-Mihalas B (1984) Foundations of radiation hydrodynamics, vol 1, Dover (1999) edn. Oxford University Press, Oxford

Zel'dovich YB, Razier YP (1966) Physics of shock waves and high-temperature hydrodynamic phenomena, vol 1, 2002nd edn. Dover, New York

Chapter 8
Radiative Shocks and Heat Waves

Abstract This chapter is concerned with radiative shocks and heat waves. There are several types of radiative shocks, depending primarily on the optical depth of upstream and downstream regions. Despite this multiplicity, the fluid dynamics sharply constrains the possible structures of such systems. The discussion covers the behavior of radiative shocks in optically thin systems, optically thick systems, and some mixed systems relevant to experiments. The chapter then turns to radiative heat fronts, in which incoming radiation is deposited in a relatively narrow zone across which the plasma parameters change. The detailed behavior depends upon the boundary conditions imposed by the adjacent regions. Specifically considered are Marshak-like heat fronts, expansion heat fronts, and photoionization fronts.

8.1 Radiative Shocks

A *radiative shock* is one in which the structure of the density and temperature is affected by radiation from the shock-heated matter. This simple definition covers an enormous range of phenomena, all at a high enough temperature that we are fortunate not to encounter them in ordinary life. Yet radiative shocks can readily be produced in high-energy-density experiments, and they are frequently encountered in astrophysics. An astrophysical example is found in the supernova remnant developing from SN 1987A, shown in Fig. 8.1. The bright spots in this image are produced by the collision between the ejecta from the star and matter at the edges of the inner ring that encircled the star. Analysis of spectra has shown that the shock waves being driven into the ring are radiative, at least in places. The presence of bright spots, rather than a continuous ring of emission, indicates that there are spikes of dense material at the inner edge of the ring. These might be a result of the Rayleigh–Taylor instability during ring formation.

8.1.1 Regimes of Radiative Shocks

Here we discuss the conditions under which a radiative shock occurs, and the physical conditions that determine its structure. In the introduction to the previous

© Springer International Publishing AG 2018
R. P. Drake, *High-Energy-Density Physics*, Graduate Texts in Physics,
https://doi.org/10.1007/978-3-319-67711-8_8

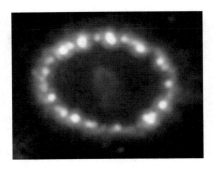

chapter (see Fig. 7.1) we found, depending in detail on density and material, that the radiative flux and pressure became important at temperatures of tens of eV and hundreds of eV, respectively. Shock waves can provide the heating that pushes a plasma into a radiative regime, or they can occur within a plasma that is already in a radiative regime. In order for a shock to push a plasma into a radiative regime, it must at minimum be fast enough that the radiative fluxes, which scale as the fourth power of the temperature and thus the eighth power of shock velocity, exceed the material energy fluxes, which scale as the third power of shock velocity.

In the nonradiative regime, the immediate postshock temperature T_s is given by (4.20), which we rewrite here as

$$RT_s = \frac{2(\gamma - 1)}{(\gamma + 1)^2} u_s^2, \tag{8.1}$$

in which u_s is the shock velocity and it will be useful at times below to work with the gas "constant" $R = k_B(Z + 1)/(A m_p)$, which in general is temperature dependent. Note that RT_s has units of energy per unit mass. For $\gamma = 4/3$ and $Z + 1 = A/2$, T_s is 6.4 eV at $u_s = 100$ km/s, which is one reason why radiative effects are rarely important for shock velocities much below 100 km/s.

The average number of electrons that share energy with each ion is Z, but this can be a source of difficulty in shock waves. We have already discussed how Z can vary with temperature, in Chap. 3. In addition, the shock heats the ions and then the electrons and ions equilibrate, so that in sufficiently low-density matter Z would be zero immediately following the density jump. Thus, in general, one may need to allow separate temperatures for ions and electrons, a point we return to in Sect. 8.1.8 (which was also discussed previously in Sect. 2.3.2 with reference to Fig. 2.2). It is the electrons, though, that couple significantly to the radiation. Here for simplicity we assume immediate equilibration of ions and electrons. In practice, this means that the equilibration zone just behind the shock (the jump in density and ion temperature) where ions and electrons equilibrate is ignored. The radiation from this equilibration zone increases as the fourth power of the electron temperature, so that most of the equilibration zone is not a significant contributor to the radiation dynamics. In addition, as we will see in Sect. 8.1.8, the equilibration zone is quite small.

Fig. 8.2 Postshock
temperatures, for xenon
(dashed) and C_1H_1 (gray).
This figure ignores the role of
radiation pressure at high
velocity

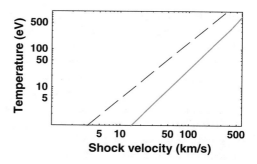

Figure 8.2 shows the temperature implied by (8.1), for xenon and C_1H_1. For a plasma of C and H, one replaces A by $(1 + 12) = 13$ and takes $Z = 1 + Z_C$, where H is assumed to be ionized and Z_C is the average ionization of the carbon. This modifies the result at low temperature, but makes little difference on a log–log plot. This equation only applies while the radiation pressure remains negligible. When radiation pressure matters, a more careful calculation based on the fundamental equations would be needed. But the figure suffices to indicate the conditions required to reach the radiative regime. In round numbers, one needs shock velocities of tens of km/s to reach temperatures of tens of eV where radiative fluxes matter, and of hundreds of km/s to reach temperatures of hundreds of eV where radiative pressure matters. The velocities required with xenon are smaller than those required with CH, by a factor of a few.

By the time that radiative fluxes exceed material energy fluxes, the radiation will have affected the medium ahead of the density increase produced by the shock. The affected region is a *radiative precursor*, which we will discuss in more detail in later sections. To connect our discussion with other usage, we should begin by identifying two possible types of "radiative precursors." The first we will call a *transmissive precursor*. The most familiar example is lightning. One sees a precursor—the lightning flash—before the resulting thunder, which has evolved from the shock wave, arrives. In this case the precursor is created by the explosion that drives the shock and not by the shock itself. A second example would be an explosion in the atmosphere strong enough to drive a radiative shock. In this case, some of the radiation from the shock itself could be seen far beyond the volume directly affected by the radiative shock. Thus, one would say that a transmissive precursor is radiation from a shock front or its source that is weakly absorbed while propagating. Thus, it can be seen at a long distance. This type of precursor is not of much interest to us, although we will discuss it briefly below.

The second type of precursor is of much more interest to us and we call it an *absorptive precursor*. In this type of precursor, the radiation is absorbed and is intense enough to affect the upstream medium, principally by increasing its temperature. Unless we specify otherwise, when we write of a "precursor" or "radiative precursor" in the following, we refer to this type of precursor. An important issue for precursors is that of geometry. In order for the precursor to remain planar, a real experiment would need a radiation source whose lateral size substantially exceeded the steady-state precursor length. This is a very demanding constraint. Spherical experiments can avoid this constraint but suffer severely from the dilution of their energy in the three-dimensional expansion.

The concept of a shock inherently involves some separation of physical scales, as we discussed in Chap. 4. In ordinary hydrodynamic shocks, the scale on which viscous diffusion matters must be much smaller than the global scale of the flow. Radiation introduces another scale into the problem, fundamentally related to optical depth. (Likewise, electron-ion energy exchange (Sect. 8.1.8), heat conduction, or magnetohydrodynamic effects also introduce additional scales under various circumstances.) Yet the physical scale over which the radiation matters is much larger than the scale of any viscous effects. As a result, there are two ways to think about the entire system including the effects of radiation both upstream and downstream of the density increase associated with the shock.

On the one hand, if one views the medium as infinite (measured in optical depths), then one may take the point of view that the radiation alters the structure of the shock transition, extending it in space over a (potentially large) number of radiation mean-free paths. In this case one will speak of the "shock" as the entire region between a distant, undisturbed upstream region and a distant, steady-state downstream region. One would then speak of the comparatively localized density increase as the "density jump" or the "viscous shock transition." This is the viewpoint taken in much prior literature, including Zel'dovich and Razier (1966) and Mihalas and Weibel-Mihalas (1984).

On the other hand, and as is discussed in the Introduction to this chapter, the system may be optically thin. It may be thin in the upstream direction, the downstream direction, or both. What specifically this means is that the sum of radiation from distant sources and radiation returning to the shock from any matter it has heated is negligible. Whenever the entire region affected by radiation from the shock is not well isolated from other influences, it seems more natural to speak of the "shock" as the region across which the rapid density increase takes place. This use of "shock" is more common in discussions of optically thin astrophysical shocks, as for example in Shu (1992). In this case, the interactions of the radiation and the surrounding medium may affect both the upstream and the downstream conditions.

Optical depth provides an effective way to classify radiative shocks. We saw in Chap. 6 that the treatment of radiation transport depends on the structure of the medium within which the transport occurs, and in particular on optical depth. The different regimes of radiation transport correspond to major differences in shock behavior. In one limit—that of very small optical depth, where the radiation serves only to cool the shocked layer—the shocked layer can evolve to become orders of magnitude denser than the preshock medium. In another limit—that of an optically thick and radiation-dominated plasma—the increase in density is limited to a total of a factor of 7. (Recall from Chap. 3 that the radiation-dominated plasma behaves like a polytropic gas with $\gamma = 4/3$.) An effective way to categorize radiative shocks and their behavior is to plot them in a space defined by the optical depth of the

Fig. 8.3 Radiative shock regimes, identified in a space based on optical depth. The four regimes corresponding to the corners of this plot are discussed in the text. The curve shows the qualitative trajectory of a supernova blast wave

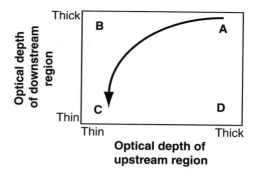

upstream and downstream regions. Figure 8.3 shows a qualitative depiction of this space. We next briefly discuss each of the four labeled regions.

8.1.1.1 Thick-Thick Shocks

In regime A, both the downstream and the upstream regions are optically thick. This is the realm in which it makes the most sense to treat the viscous density increase and all the radiative effects as part of a single, extended, shock structure. Many of the features of this structure can be found from a theory that assumes the medium to be in LTE everywhere. We discuss this regime in Sect. 8.1.6. For typical ideal gases with $\gamma \geq 4/3$, the density ratio never exceeds 7. In addition, under some circumstances the density transition is continuous, with no localized jump. Astrophysical environments in which such shocks exist are necessarily both hot and dense. Shocks in stellar interiors are of this type, as is the blast wave within the exploding star in a supernova. Such shocks may also exist within some astrophysical "compact objects," such as pulsars, but their treatment would have to be relativistic. It is difficult, however, to imagine planar laboratory experiments in this regime other than transiently and in special cases. One difficulty is that the precursor length increases so strongly with shock velocity (see Sect. 8.1.4) that one could not produce a measurable precursor of finite length for realistic variations of the experimental parameters. There may be more potential for experiments in spherical geometry, but the challenge of producing a system many optical depths in scale will remain substantial.

8.1.1.2 Thick-Thin Shocks

In regime B the downstream region is optically thick but the upstream region is thin. We discuss this regime in Sect. 8.1.6. There is a cooling layer downstream of the viscous shock transition, followed by a steady downstream final state. This regime is common in experiments, in which an optically thick piston (and in some cases optically thick shocked material) drives a radiative shock into a medium whose depth is small compared to the steady-state precursor length. The upstream medium is then quickly heated so that it becomes optically thin. Astrophysical examples of such systems include the blast wave in a supernova as it emerges from the star and the accretion shocks produced in some binary systems.

8.1.1.3 Thin-Thin Shocks

In regime C, discussed in Sect. 8.1.5, both downstream and upstream regions are optically thin. Such shocks are the most-commonly observed in astrophysics, in part because they are easy to see (as the radiation escapes). Supernova remnant (SNR) shocks in dense enough environments are of this type—it is thought that Type II supernovae from red-supergiant precursor stars produce such conditions. Many shock-cloud interactions, including some of those driven by SNR shocks, are also of this type. Shocks that propagate up jets (or are driven by clumps propagating up jets) may be of this type. In such shocks, the entire downstream region is a radiative cooling layer, and it ends (in large enough systems) when the downstream temperature reaches a value determined by local sources and losses of energy rather than by the shock. The density increase associated with such shocks is formally unbounded in the sense that it is limited only by external factors, such as the compression of an initially negligible magnetic field or the presence of a limiting temperature due to other energy sources. Some experiments, with shocks in sufficiently low-density gases, may produce these conditions.

8.1.1.4 Thin-Thick Shocks

Regime D is not trivial to produce in steady state, as it would require that the shocked material become optically thin when it is shocked while simultaneously remaining optically thick in the upstream region, over a sufficient distance to sustain a steady precursor. Such a change in optical depth can be produced in ionization fronts driven by radiation, discussed in Sect. 8.2.3 and in Sect. 9.3.1. Obtaining this response in a shock involving flowing material is more difficult. It might occur, for example, if a very-high-velocity, low-density incoming flow impacted a comparatively dense material. If such a system could be produced, it would have a very dense shocked layer as energy continued to be lost in the downstream direction. Two transient examples are certain shock-cloud collisions and certain experiments. A shock-cloud collision in which the cloud was dense enough and large enough to be optically thick for some time would be of this type. The collision of SNR 1987A with its inner "ring" may be of this type. An experiment might be in this regime while a hot, thin layer of gas drives a shock through a much larger volume of gas. All these cases seem likely to transition to the thin–thin regime if driven harder or longer, and they may never develop a thick upstream region in the sense discussed above.

8.1.2 Fluid Dynamics of Radiative Shocks

It turns out that several important properties of radiative shocks are independent of the details of the radiation transport. In this section, for a polytropic gas, we consider how the fluid properties must vary within radiative precursors and radiative shocks. That is, we will consider what things must be true independent of the details of the radiative transport. Our conclusions here will apply even to shocks whose transport is far more complex than are the models we use later. One example of such a complex situation would be the transport of energy by line radiation in the presence of significant Doppler shifts. We will frame most of this discussion in the radiative flux regime, assuming the radiation pressure and energy density to be negligible. We analyze a planar system in steady state, working as usual in the shock frame. In this case the divergence of the flux terms in the conservative form of the mass, momentum, and energy equations must be zero.

We will work in our usual shock frame as described in Chap. 4 so that the incoming fluid has a negative velocity. The mass flux ρu must be constant everywhere and equal to its value in the region beyond the precursor, $-\rho_o u_s$, where the far upstream density is ρ_o and the shock velocity is u_s. The constancy of momentum flux gives

$$p + \rho u^2 = \rho_o u_s^2 + p_o, \tag{8.2}$$

in which the initial upstream pressure is p_o. The continuity and momentum relations thus give

$$\frac{p}{\rho_o u_s^2} = 1 - \frac{\rho_o}{\rho} + \frac{p_o}{\rho_o u_s^2} \text{ and} \tag{8.3}$$

$$\frac{RT}{u_s^2} = \frac{\rho_o}{\rho}\left(1 + \frac{p_o}{\rho_o u_s^2}\right) - \left(\frac{\rho_o}{\rho}\right)^2, \tag{8.4}$$

where we have used $p = \rho RT$ so RT is proportional to the thermal energy per unit mass of the plasma at temperature T and $R = (1 + Z)k_B/(Am_p)$, when the Coulomb and radiative contributions to p are small. These results are shown in Figs. 8.4 and 8.5. We use as the independent variable the *inverse compression*, ρ_o/ρ. It is

Fig. 8.4 Postshock material pressure, normalized to $\rho_o u_s^2$, against inverse compression, for normalized upstream pressure of zero (black), Mach $3/\sqrt{\gamma}$ (dashed), and Mach $2/\sqrt{\gamma}$ (gray)

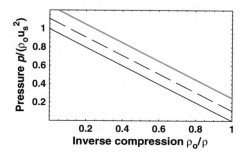

Fig. 8.5 Postshock
temperature. The thermal
energy of the matter per unit
mass, RT, normalized to u_s^2 is
shown against the inverse
compression, ρ_o/ρ, for
normalized upstream pressure
of zero (black), Mach $3/\sqrt{\gamma}$
(dashed), and Mach $2/\sqrt{\gamma}$
(gray)

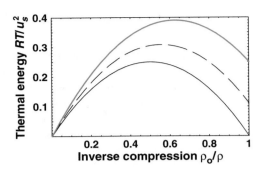

worth noting that radiation does not enter into these relations (so long as $p_R \ll p$),
and that these figures apply both to nonradiative shocks and to shocks in the
radiative flux regime. Whether in the precursor or across the shock jump, a change
in compression corresponds to an increase in pressure and a change in temperature
as shown. The pressure increases continuously as ρ_o/ρ decreases, and so places no
constraints on the shock transition. In contrast, the competition between heating and
pdV work of compression creates the maximum in the thermal energy in Fig. 8.5.
A formal discontinuity in density occurs only if the initial and final states of the
viscous density transition are on opposite sides of the temperature maximum seen
in this figure. (The temperature cannot increase and then decrease across the density
transition, without unphysical consequences for the radiation flux.) In sufficiently
weak shocks, the inverse compression can remain to the right of this maximum,
producing a continuous transition. This also can occur under certain conditions
when the radiation pressure becomes significant. Figure 8.5 shows curves for three
values of the normalized pressure. Note that this normalized pressure, $p/(\rho_o u_s^2)$, can
also be written as $1/(\gamma M^2)$ for upstream Mach number M.

The radiation flux enters into the energy flux equation, which for a polytropic gas
gives

$$\left(\frac{\gamma p}{\gamma - 1} + \frac{\rho u^2}{2}\right) u + F_R = -\frac{\rho_o u_s^3}{2}\left(1 + \frac{2\gamma}{\gamma - 1}\frac{p_o}{\rho_o u_s^2}\right) + F_o, \qquad (8.5)$$

in which ρ_o, p_o, and F_o are the density, pressure, and radiation flux, respectively,
in some presumably steady upstream state. Note that the negative sign on the first
term on the right-hand side is the consequence of the flow velocity being negative
(according to our standard conventions for shocks throughout this book) and of
taking the "shock velocity," u_s, as a positive quantity.

One must understand the physical context in order to set F_o. There are three
limiting cases in which F_o has some specific value for specific reasons. First, in a
shock wave that is fully contained within an optically thick medium, it is sensible
to consider the upstream state to be beyond the reach of any radiation, so one takes
$F_o = 0$. In this case, the energy flux into the system is the (negative) value of the
material energy influx. Second, in a planar shock of infinite lateral extent with an

upstream region that has limited optical depth, F_o will correspond to the upstream radiation flux as some location beyond which heating by absorption of this flux is negligible. We can take the parameters at this location to be the steady, initial state. In this case, the values of density and temperature at this location may have evolved to this state (from some other values) during the initiation of the shock. When $F_o > 0$, the net energy influx to sustain the shock is reduced by comparison with the $F_o = 0$ case. The net energy influx is then the difference between the incoming material energy flux and the outgoing radiation flux. Third, one may have a (more or less) planar shock of finite lateral extent. This is relevant to various experiments that produce optically thin upstream regions. In this case F_o represents the sum of the energy lost beyond some designated axial position and the energy lost radially before reaching that position. This is the case because (8.5) keeps track of energy conservation, so that any energy removed from the radiation flux, as represented in this equation, must be absorbed in the matter. As a result, when absorption decreases the upstream radiative energy flux, the returning material energy flux (from further upstream) must also decrease. In contrast, when lateral losses decrease the upstream radiative energy flux, the material energy flux does not decrease.

To further clarify what the radiation is doing, it is worthwhile to discuss the recycling of energy that occurs in radiative shocks. The flow of energy through the system involves the following sequence as the shock is established. One begins with incoming mechanical and internal energy from far upstream. In the shock and the shocked matter, the plasma converts the mechanical energy to additional internal energy and to radiation. Some of the radiation flows away downstream with the material. The rest of the radiation flows upstream. If absorbed there (reducing F_R where the absorption occurs) the radiation adds internal energy to the incoming material. In a steady state, the mechanical and internal energy incoming to the shock is larger than its initial value before the shock was established. Thus, the elements of the shock as a system include incoming material energy, recycling of energy by upstream absorption of radiation, the escape of radiation upstream including perhaps radially, and the escape of radiation and material energy downstream.

Returning to (8.5), in the absence of a heat flux (F_R and F_o here, but this could be any heat flux), this equation provides a second, redundant condition for the pressure. The simultaneous solution of (8.3) and (8.5) then determines the only possible compression at the shock. The presence of the heat flux opens a larger range of possibilities. Equations (8.3) and (8.5) imply

$$F_R - F_o = \frac{\rho_o u_s^3}{2} \left[\frac{2\gamma}{\gamma - 1} \frac{\rho_o}{\rho} - \frac{\gamma + 1}{\gamma - 1} \left(\frac{\rho_o}{\rho} \right)^2 - 1 - \frac{p_o}{\rho_o u_s^2} \frac{2\gamma}{\gamma - 1} \left(1 - \frac{\rho_o}{\rho} \right) \right].$$
(8.6)

We will use this equation extensively in what follows, choosing F_o according to the discussion above.

Figure 8.6 shows the dependence of the net radiation flux ($F_R - F_o$) on the inverse compression, for $p_o = 0$. The flux is normalized to the incoming kinetic energy flux, $\rho_o u_s^3 / 2$. Note that this curve depends on γ while the previous figures for pressure and

Fig. 8.6 Radiative flux in
radiative shocks, normalized
to $\rho_o u_s^3/2$, against the inverse
compression, ρ_o/ρ, for
$\gamma = 4/3$. The gray line
shows a characteristic shock
trajectory

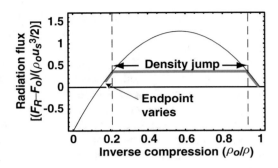

the specific thermal energy RT do not. The net, normalized radiation flux rises above zero only in the presence of energy recycling in the plasma, when shock-heated matter emits radiation that warms upstream matter which then carries the energy back to the shock. A rather magical physical system, with multiple independent recycling loops, would be needed to drive the flux above 1. The net, normalized radiation flux reaches –1 if all the energy entering the system is radiated away, corresponding to a state of formally infinite compression and zero temperature.

Figure 8.6 is tremendously important for understanding the properties of radiative shocks. We discuss some aspects of this now, but we will return to this type of figure repeatedly in what follows. The general behavior of a radiative shock is shown as a trajectory on this figure. There is some compression in the precursor region, as $(F_R - F_o)$ increases from zero to a maximum that equals the value of $(F_R - F_o)$ entering the precursor from the shocked region. Because $(F_R - F_o)$ is continuous across the viscous shock transition (the density jump), this value of $(F_R - F_o)$ then fixes the compression produced by this transition. Further evolution may also occur after the density jump. The nature of this evolution depends on the downstream boundary condition. This boundary condition could correspond to (a) a positive radiation flux if there is a bright source downstream of the shock, (b) a radiation flux of zero if the downstream region is optically thick, or (c) a negative radiation flux if the shocked matter also cools by emitting radiation that is lost from the system. The value of the inverse compression where $(F_R - F_o) = 0$ corresponds to the final state one would reach in a nonradiative shock, and is $(\gamma + 1)/(\gamma - 1)$. This may or may not correspond to the final density in an actual radiative shock. We will discuss some specific cases in later sections.

When we reach the point of discussing the structure within the shocked material in a radiative shock, we will need an equation describing the spatial evolution. We will limit our discussion to steady-state shocks. Our approach will be to assume that either the final state or the immediate postshock state can be determined on the basis of the fundamental conservation equations. From this starting point we will integrate the energy equation to determine the profiles. We again start with (7.3), but now we are interested in the spatial derivatives. We assume a polytropic gas, work in the radiative flux regime, and substitute from (8.2) and the continuity relation to obtain

$$\frac{\rho_o u_s^3}{2} \frac{\partial}{\partial x} \left[-\frac{2\gamma}{\gamma - 1} \left(\frac{\rho_o}{\rho} \right) + \frac{\gamma + 1}{\gamma - 1} \left(\frac{\rho_o}{\rho} \right)^2 \right] = -\frac{\partial F_R}{\partial x}. \tag{8.7}$$

This is the equation we will use to explore the spatial profiles.

8.1.3 Transmissive Radiative Precursors

Before turning to specific models, we first discuss one simple threshold for radiative effects. One can say that a radiative precursor will be present when the flux of ionizing photons radiated ahead of the shock equals the flux of neutral atoms incident on the shock. This point of view is that one will certainly see heating and a change of state of the upstream medium when all (or most) of the incoming atoms are ionized. To be precise, the threshold would be when the upstream flux of ionizing photons times the fraction that are absorbed in the upstream region equals the flux of incoming atoms. One way to express the flux of ionizing photons is as the flux of photons emitted by a black body at the postshock temperature, which is $2.3 \times 10^{23} T_s^3$, with T_s in eV, times the fraction of these photons that are emitted and are ionizing. This fraction is the product of the emissivity of the downstream region, ϵ_d, and the fraction α_i of all photons that are ionizing. Recall that ϵ_d is equal to the optical depth if the downstream region is optically thin. The fraction α_i is near unity for shock velocities above 50 km/s.

The fraction of ionizing photons that is absorbed in the upstream region is equal to the upstream emissivity, ϵ_u. One can assemble the last few lines of material into an equation for the threshold:

$$2.3 \times 10^{23} \epsilon_u \epsilon_d \alpha_i T_s^3 > \rho u_s / (A m_p). \tag{8.8}$$

Using (8.1), one can convert this into a threshold for the shock velocity, given by

$$u_s > 270 \left[\rho / (\epsilon_u \epsilon_d \alpha_i) \right]^{1/5} \text{ km/s}. \tag{8.9}$$

In laboratory experiments with dense gases or foams, the quantities in square brackets may all be of order unity. For low-density astrophysical systems, (8.9) is correct but not very useful. With ρ of order 10^{-24} g/cm^3, obtaining a radiative precursor will require first of all that the postshock temperature be high enough to obtain a significant fraction of ionizing photons. Beyond that it will depend on the optical depth of the system.

Let us return to the properties of the precursor region. Assuming zero initial upstream pressure p_o, (8.4) and (8.5) can be solved for the normalized temperature in the precursor, RT_p/u_s^2, which turns out to depend only on the normalized net radiation flux, $F_{Rn} = 2(F_R - F_o)/(\rho_o u_s^3)$. One finds

$$\frac{RT_p}{u_s^2} = \frac{1}{(\gamma + 1)^2} \left(1 - \sqrt{1 - (\gamma^2 - 1)F_{Rn}} \right) \left(\gamma + \sqrt{1 - (\gamma^2 - 1)F_{Rn}} \right). \tag{8.10}$$

This temperature goes to zero as F_{Rn} goes to zero, as it should. F_{Rn} rises above zero and the precursor heats when there is recycling. As F_{Rn} increases, T_p can approach but cannot exceed the temperature of the radiation from the shocked matter. (We discuss this point further below.)

The precursor can include a diffusive region, in which absorption and emission of radiation is a significant part of the dynamics. This cannot occur unless the upstream matter is optically thick, and so we discuss this case when we discuss such shocks. There are also transmissive regions in precursors, in which emission is negligible and the emerging radiation flux attenuates only by absorption (or, in actual applications, lateral divergence). Some precursors are entirely transmissive, and one can show as follows that all precursors must included a transmissive region once F_{Rn} attenuates sufficiently. In the limit that the precursor region is optically thin, the precursor is formally transmissive and may be heated by the radiation, while the fraction of the radiation flux that is absorbed may be negligible.

When the radiation and matter are in equilibrium in the precursor, the energy density of the radiation will be $4\sigma T_p^4/c$. This is the maximum possible energy density of radiation from the precursor. The actual value is near this limit when the precursor is optically very thick and T_p is changing slowly. In contrast, the minimum possible energy density of the radiation from the shocked matter is F_R/c. This is larger than $(F_R - F_o)/c$, although F_o is often small or zero when the present discussion is relevant. The actual value is near F_R/c once the radiation is far enough from the shock front that absorption (or escape) has attenuated the oblique rays. When the first of these energy densities falls below the second, then the precursor must be transmissive. We can formulate the relationship as follows. The ratio of the energy densities is

$$\frac{4\sigma T_p^4}{(F_R - F_o)} = 4\frac{\sigma u_s^5}{\rho_o R^4 F_{Rn}}\left(\frac{RT_p}{u_s^2}\right)^4 = 4\frac{Q}{F_{Rn}}\left(\frac{RT_p}{u_s^2}\right)^4, \qquad (8.11)$$

in which $Q = 2\sigma u_s^5/(R^4\rho_o)$ is a shock strength parameter that we will also find useful below. It has a typical value of 10^4 to 10^5 in laboratory radiative shocks. Here Q is nondimensional but must be evaluated using consistent units, such as cgs units with σ in ergs/s^{-1} cm^2/eV^4 and with R in ergs/g^1/eV^1. Figure 8.7 shows the relation

Fig. 8.7 As the radiation flux decreases, all precursors eventually become transmissive

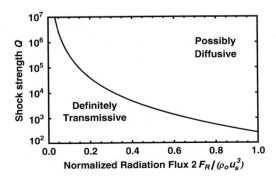

between Q and F_{Rn} for which (8.11) equals 1. For any given Q, values to the left of the curve have an energy density of thermal precursor radiation that is necessarily below that of the radiation from the shock. Beyond the location where this occurs, as the radiation flux further decreases, the thermal radiation density becomes rapidly even more negligible. Given actual radiation transport, the actual boundary of the transmissive regime will usually be to the right of the curve shown in the figure.

Because all precursors include a transmissive regime, it makes sense to analyze this aspect of the behavior now. For any transmissive precursor that extends over many absorption lengths, one can model the precursor using the nonequilibrium diffusion theory of Sect. 7.1.1.2. One has from (6.67) and (7.9), again under the assumptions that the radiation reaches steady state on hydrodynamic time scales and that the system is in the radiative flux regime, that

$$F_R = -\frac{4}{3\bar{\chi}} \nabla \sigma T_R^4, \text{ and} \tag{8.12}$$

$$\nabla \cdot F_R = 4\kappa_P \sigma T_p^4 - 4\kappa_E \sigma T_R^4, \tag{8.13}$$

in which $\bar{\chi}$ and κ_E are averaged opacities defined in that section and κ_P is the Planck mean opacity. Although there are three distinct opacities in (8.13), it seems common in the literature to assume without comment that these are all equal. While this has the virtue of being consistent with the treatment in some computer codes employing nonequilibrium diffusion, it is not numerically correct, and may introduce significant errors. In the case of a precursor with T_p less than some large fraction of T_R, T_p can be ignored in (8.13). In this case the radiation flux is just attenuated. Solving (8.12) and (8.13), with the optical depth τ defined here as the magnitude of the distance times the "opacity," $\tau = \sqrt{\bar{\chi}\kappa_E}z$, gives

$$F_R = F_o e^{-\sqrt{3}\tau}, \tag{8.14}$$

in which F_o is the radiation flux emerging from the shock. If the absorption is dominated by bremsstrahlung, this result may be accurate. However, only a fairly sophisticated computer code will treat the opacities here correctly and thus evaluate the exponential scale length accurately.

8.1.4 Optically Thin Radiative Shocks

In the present section we consider shocks that are optically thin throughout, so that radiation freely escapes in both directions. We also assume that nearby radiation sources are negligible, so that the shock exists in isolation. In this case all of the incoming energy eventually leaves the system as radiation. It is important that an optically thin system is *not* energy conserving. A limiting case is to assume that the pressure in the precursor is negligible compared to $\rho_o u_s^2$. In this case, the ordinate

Fig. 8.8 Temperature and
radiation flux properties for
optically thin shocks. Note
$p_o = 0$

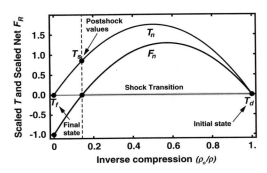

of Fig. 8.6, $(F_R - F_o)/(\rho_o u_s^3/2)$, is zero at the shock but F_o is finite and represents
energy escaping in the upstream direction. Figure 8.8 shows the temperature and
radiation flux curves for this type of shock. We use the subscript d to designate the
precursor properties at the density jump. Here by assumption $T_d = 0$, and because
the absorption in the precursor is assumed to be zero, so is $F_R - F_o$ at the shock. As a
result, the density jump across the shock is $(\gamma+1)/(\gamma-1)$ and T_s equals the value for
a non-radiative shock. After this, radiative cooling moves the inverse compression
toward zero and the normalized radiation flux will decreases toward -1, as all the
incoming energy is converted to radiation. The final density will be formally infinite,
limited only by factors outside this analysis, such as increasing magnetic field,
increasing optical depth of the downstream plasma, or external radiation sources.
The question we address here is what profiles develop during this cooling.

To work with (8.7) to find the structure within the shocked material, we consider
that J_R is negligible compared to B, so in the transport regime $\partial F_R/\partial x = -4\pi\kappa B$.
We now want to simplify (8.7), to find the essential parameters that control the
behavior. We use the subscript s to designate the immediate postshock state. Then
we can take $\kappa = \kappa_s(\rho/\rho_s)^m(T/T_s)^{-n}$ so that

$$\frac{\partial F_R}{\partial x} = -4\pi\kappa_s B_s \left(\frac{\gamma+1}{\gamma-1}\frac{\rho_o}{\rho}\right)^{-m}\left(\frac{T}{T_s}\right)^{4-n}. \tag{8.15}$$

With this definition, the natural normalization of (8.7) is to create a radiation
parameter R_r, defined by

$$R_r = \frac{\gamma+1}{\gamma}\frac{4\pi B_s}{\rho_o u_s^3} = \frac{\gamma+1}{\gamma}\frac{4\sigma T_s^4}{\rho_o u_s^3}, \tag{8.16}$$

which is approximately the ratio of the radiative flux from an optically thick shocked
layer at temperature T_s to the incoming energy flux. When $R_r > 1$, the radiative
fluxes from an optically thick postshock layer would exceed the material energy
fluxes, which would violate the conservation of energy. Thus, when $R_r \geq 1$ a
cooling layer will develop, in which the plasma temperature decreases rapidly to
some final value T_f, such that the entire system conserves energy. In the limit we are

considering $T_s \propto u_s^2$, so R_r increases as u_s^5. We also define an optical-depth variable $\tau = -\kappa_i z$ In addition, (8.4) (the equation of state) implies

$$\frac{T}{T_s} = \frac{(\gamma + 1)^2}{(\gamma - 1)\left(2 + (\gamma + 1)\frac{p_o}{\rho_o u_s^2}\right)} \frac{\rho_o}{\rho} \left(1 + \frac{p_o}{\rho_o u_s^2} - \frac{\rho_o}{\rho}\right), \tag{8.17}$$

where we keep the terms proportional to p_o (which will be needed later). With these definitions, (8.7) can be rewritten as

$$\frac{\gamma + 1}{\gamma} \left[-\frac{\gamma(1 + \frac{p_o}{\rho_o u_s^2})}{\gamma - 1} + \frac{\gamma + 1}{\gamma - 1}\left(\frac{\rho_o}{\rho}\right) \right] \frac{\partial}{\partial \tau}\left(\frac{\rho_o}{\rho}\right) \tag{8.18}$$

$$= R_r \left(\frac{\gamma + 1}{\gamma - 1}\right)^{-m} \left(\frac{\rho_o}{\rho}\right)^{-m} \left[\frac{(\gamma + 1)^2}{2(\gamma - 1)}\left(1 - \frac{\rho_o}{\rho}\right)\frac{\rho_o}{\rho}\right]^{4-n}.$$

Given boundary conditions at the shock, along with R_r, p_o, and the parameters describing the material in the system (γ, m, n), one can integrate this equation to find the density profile. Note that if there is absorption of radiation in the precursor, then p_o is not the pressure at the shock, and ($F_R - F_o$) is no longer zero at the shock. In this case p_o is the pressure at a point beyond which there is negligible heating by radiation. The pressure on the upstream side of the shock has been accounted for through both p_o and its relation to compression.

We will see below that R_r is within a few orders of magnitude of unity for typical laboratory experiments. Under interstellar astrophysical conditions, however, R_r is enormous. Using $\gamma = 5/3$, $k_B(1 + Z)/Am_p = 1.5 \times 10^{12}$ ergs/g/eV, and $\rho_o = 10^{-22}$ g/cm^3, one finds $R_r \sim 10^{18}$ for $u_s \sim 100$ km/s. Figure 8.9 shows the resulting density profiles for relevant parameters with $R_r = 10^{18}$ or 10^{19}. Note that the optical depth required for the evolution of the profile is approximately $1/R_r$. This remains

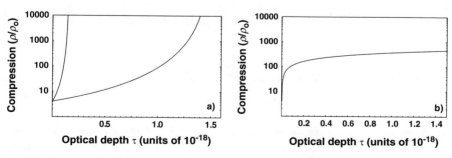

Fig. 8.9 Density profiles for optically thin shocks. (**a**) Cool plasmas. Here $\gamma = 5/3$, $n = 4/3$, and $m = 2$. Profiles are shown for $R_r = 10^{18}$ (on the right) and $R_r = 10^{19}$ (on the left). In this model, the density increases without limit as all the energy goes into radiation. (**b**) Hot plasmas. Here $\gamma = 5/3$, $n = -1$, and $m = 2$, and $R_r = 10^{21}$, corresponding qualitatively to the behavior at temperatures above the minimum of the cooling function near 100 eV

true even as R_r becomes much smaller (~ 10). In the astrophysical case, recall (from Sect. 6.3.2) that we had $\kappa \sim 10^{-38}$ cm. Thus the cooling distance $1/(\kappa R_r)$ for these parameters is of order 10^{20} cm or 100 light years. This distance becomes smaller as the shock velocity increases, producing more radiation.

As we remarked above, it will be some factor not in this model that stops the increase in density. The increase will slow or stop, for example, if the slope of κ with T changes or if the magnetic pressure becomes significant. Interstellar astrophysical shocks often cool by radiation until the temperature of the downstream, shocked material equals that of the nearby environment both beyond and ahead of the shock. These are sometimes known by the horribly unphysical designation "isothermal shocks." (The term *isothermal shock* is also sometimes used to describe the very idealized limit in which a shock in the presence of heat conduction may have no jump in temperature where the density jump occurs. The presence of a lighter particle species, such as photons, that transports heat eliminates this solution except as a limiting case. One example is the "supercritical shock" as discussed in Sect. 8.1.6.)

One sees in Fig. 8.9a a density increase that becomes increasingly rapid with increasing optical depth. This type of cooling is sometimes known as *catastrophic cooling*. This occurs, for example, when *old supernova remnants* cool sufficiently. What is required to produce catastrophic cooling is that $\partial \kappa / \partial T < 0$. Then as cooling for some time interval leads to a density increase and a temperature decrease, the rate of cooling increases so there is more cooling in the next time interval. However, the behavior seen in Fig. 8.9a is not universal in all interstellar astrophysical shocks. As Fig. 6.5 showed, the astrophysical cooling function reverses slope at some temperature above a few hundred eV, corresponding to a shock velocity of approximately 300 km/s. Figure 8.9b shows the cooling that occurs for $n = -1$ and $R_r = 10^{21}$. Some rapid cooling again takes place over a small distance, but following this there is gradual cooling over a much larger distance. Under these conditions, a shocked layer will cool slowly until κ reaches its minimum and $\partial \kappa / \partial T < 0$. Then rapid cooling to very low temperature will ensue. The rapid density increase may be described as a *density collapse* or *collapse of the shock*, because the thickness of the shocked layer decreases in inverse proportion to the increase of density.

8.1.5 Radiative Shocks That Are Thick Downstream and Thin Upstream

We turn now to a circumstance that is common in laboratory experiments with radiative shocks. The upstream medium, being limited in extent, may quickly become ionized and after that will be optically thin. Alternatively, for example, in a spherical experiment in gas, the upstream medium may quickly become optically thin out to a heat front (or an ionization front; see Sect. 8.2.3) where the radiation is absorbed. However, the optical depth of the heated upstream medium is likely to be small, as is the radiation flux from the precursor region back toward the shock. Correspondingly, for steady-state calculations we will take the view that the precursor region is a uniform plasma with some initial temperature, but that essentially none of the radiation entering the precursor returns across the density jump. We also will treat the electron and ion temperatures as equal throughout. Immediately behind the shock front, this is not correct, as ions are heated much more than electrons at any shock transition (see Chap. 4). But the ion and electron temperatures equilibrate much faster than the radiative properties change, and so we ignore this detail as part of the microstructure of the shock.

Before considering the steady-state case, let us qualitatively analyze the evolution of such a shock from an initial, optically thin limit as a laboratory experiment begins. One has $\partial\kappa/\partial T < 0$ under many conditions of interest, so the initial behavior will involve a density collapse like that discussed in the previous section. This could be limited by the maximum in κ at low temperature but is more likely to be limited by the transition at some temperature (and after some time) to an optically thick shocked layer. The resulting density and temperature of the shocked layer, making it optically thick, might be said to represent the initial attempt of the system to establish a steady state. However, the radiation from the cooling layer controls the ultimate steady state, as we discuss next. This radiation may heat the shocked matter to a final temperature above the initial value, establishing a steady state only when the shocked matter becomes optically thick at the temperature necessary for self-consistency.

Figure 8.10 shows the behavior of the temperature and radiation flux for this case. The initial temperature is non-zero because the precursor has been heated. The curve shown is for $p_{on} = 1$, shown below to be an overestimate. Because the upstream region is optically thin, $F_R = F_o$ at the shock and the net radiation flux across it is zero. Because p_{on} is finite, the initial density jump is smaller than $(\gamma + 1)/(\gamma - 1)$. The radiative cooling then reduces the temperature below T_s until it reaches a

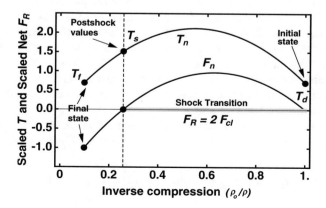

Fig. 8.10 Temperature and radiation flux properties for shocks that are optically thin upstream and optically thick downstream. Note $p_o = 1$ here

final value, T_f, consistent with energy balance. The cooling layer emits a flux of magnitude F_{cl} in both directions, and for a semi-infinite, optically thick downstream region, we show below that $F_R = F_o = 2F_{cl}$ across the shock transition. We will first discuss the properties of the radiative precursor and then the energy balance and its implications. We will find that, because σT_s^4 is typically much greater than $\rho u_s^3/2$, the optical depth of the cooling layer must be quite small to conserve energy. As a result, in physical space the spike in temperature downstream of the shock is quite narrow, and the temperature rapidly decreases to its final value. This type of temperature structure was first discovered, for radiative shocks in optically thick media, by Zel'dovich in the 1950s, and is known as a *Zel'dovich spike*.

8.1.5.1 Radiative Precursors in the Transport Regime

In many real situations the radiative transfer within the precursor (and the shocked matter too) may be in the transport regime. The transport regime is the relevant one in the case that the upstream plasma is optically thin or is limited in extent. Moreover, in this case the radiative flux may approach the full flux from the shocked region, σT_{eff}^4, while in the diffusion regime the Rosseland flux is much smaller, being a blackbody flux σT_R^4 multiplied by the (small) fractional change in temperature per unit (reduced) optical depth. We assume first that the radiation pressure is negligible, that the radiation and matter temperatures are the same, and that the density and velocity in the precursor region are not changed by the precursor (just as is the case for the Marshak wave), so that $\nabla \cdot \boldsymbol{u} = 0$. Under these assumptions and for a planar precursor (7.5) becomes

$$\rho \frac{\partial \epsilon}{\partial t} + \rho u \frac{\partial \epsilon}{\partial z} = -\frac{\partial F_R}{\partial z} = 4\pi \kappa (J_R - B). \tag{8.19}$$

This equation allows useful estimates of the steady-state plasma temperature and the time required to reach steady state. First consider $(J_R - B)$. The average intensity J_R has three components. These are the contribution from the shocked matter (J_1), the contribution from the region between the shock and a given location (J_2), and the contribution from the region upstream of the given location (J_3). To make a simple analysis tractable, we suppose that the upstream plasma has a characteristic extent D in the upstream direction and is infinite laterally. We then examine the plasma in a location a distance d from the shock.

In calculating the contribution from the shocked matter, we take the radiation intensity (power per unit area per unit solid angle) to be $\sigma T_{\text{eff}}^4/\pi$. Then we have

$$J_1 = \frac{\sigma T_{\text{eff}}^4}{4\pi^2} \int e^{-\kappa d/\mu} d\Omega = \frac{\sigma T_{\text{eff}}^4}{2\pi} \int_0^1 e^{-\kappa d/\mu} d\mu, \tag{8.20}$$

in which we integrate over the hemisphere facing upstream. This integral evaluates to

$$J_1 = \frac{\sigma T_{\text{eff}}^4}{2\pi} \left[e^{-\kappa d} - \kappa d \Gamma(0, \kappa d) \right] \approx \frac{\sigma T_{\text{eff}}^4}{2\pi}, \tag{8.21}$$

in which Γ is the incomplete Gamma function and the second approximate equality requires that we stay where the optical depth to the shock (κd) is small.

To find J_2 and J_3, we will have to integrate B over space. We will designate the plasma temperature as T_p so that $B = \sigma T_p^4/\pi$. Our point of view here is that we are analyzing a very thin system, so that measured in optical depths, the precursor of interest is very near the shock. This allows us to assume that T_p is constant for the purpose of evaluating this integral. We then have

$$J_2 = \frac{1}{4\pi} \int d\Omega \int_0^d \kappa B e^{-\kappa z/\mu} dz/\mu, \tag{8.22}$$

in which the integral over distance evaluates the radiation intensity at a polar angle corresponding to μ and the solid-angle integral is over the forward hemisphere. One finds

$$J_2 = \frac{B}{2} \left[1 - e^{-\kappa d} + \kappa d \left(\Gamma(0, \kappa d) \right) \right]. \tag{8.23}$$

For small κd, $J_2 \sim \kappa d[1 + \Gamma(0, \kappa d)]$. Proceeding to J_3, one has

$$J_3 = \frac{1}{4\pi} \int d\Omega \int_0^{D-d} \frac{\kappa}{\mu} B e^{-\kappa z/\mu} dz = \frac{B}{2} \left[1 - e^{-\kappa D} + \kappa D \Gamma(0, \kappa D) \right], \tag{8.24}$$

in which the solid-angle integral is now over the entire hemisphere in the downstream direction and in writing the rightmost expression we have assumed $d \ll D$. We can now rewrite (8.19), realizing that $\epsilon = RT_p/(\gamma - 1)$, as

$$2\kappa \left[\sigma T_{\text{eff}}^4 - \sigma T_p^4 \left(1 + e^{-\kappa D} - \kappa D \Gamma(0, \kappa D) \right) \right] = -\frac{\rho_o u_s RT_p'}{\gamma - 1}. \tag{8.25}$$

The term on the right-hand side of this equation should be small, by our assumptions. One can check this, using (8.10) to evaluate the derivative of RT_p, taking the derivative of F_{Rn} to be $-\kappa F_{Rn}$. One finds this term to be small for any strength parameter $Q > 10$. Thus one finds for T_p:

$$T_p = \frac{T_{\text{eff}}}{[1 + e^{-\kappa D} - \kappa D \Gamma(0, \kappa D)]^{1/4}}. \tag{8.26}$$

For small κD, T_p is 84% of T_{eff}. In this equation, T_p increases as κD does, approaching T_{eff} at large κD. However, the calculation of J_R becomes invalid as κD

increases. One knows from the flux balance equation that T_p will decrease as the net radiation flux decreases with increasing κD. Note that this system is optically thin by assumption and that much of the radiation flux crossing the shock is in the end lost from the system and so increases F_o. We do not know how much is lost a priori, as is indeed the case in real systems of this type. As we discussed in Sect. 8.1.2, the flux balance equation demands only a relation between the temperature profile in the precursor and the radiation flux that is actually absorbed there.

The relevance of (8.26) depends on how readily the steady state is achieved and whether there is in fact time for the precursor plasma to be heated as the shock approaches it. To evaluate this, we use (8.4) and take $\partial \epsilon / \partial t \sim \epsilon(T_{\text{eff}})/t_{ss}$, defining the time we seek as t_{ss}, and we take $J_R \sim J_1$ and $B \sim 0$. This gives

$$t_{ss} = \frac{\rho \epsilon(T_{\text{eff}})}{2\kappa \sigma T_{\text{eff}}^4}. \tag{8.27}$$

This turns out to be remarkably fast. For a laboratory plasma, we can take $\epsilon \sim 10^{12} T_{\text{eff}}$ ergs/g and $\kappa \sim 10^6 / T_{\text{eff}}$ cm^{-1} to find $t_{ss} \sim 10^{-6} \rho / T_{\text{eff}}^2$, which is 1 ns for $T_{\text{eff}} \sim 10$ eV and $\rho \sim 0.1$ g/cm^3. Radiation hydrodynamic experiments usually have timescales of at least several nanoseconds. Thus we conclude that the precursor plasma may approach its steady-state temperature reasonably quickly in laboratory experiments. For an interstellar astrophysical plasma with $\kappa \sim 5 \times 10^{-37}$ cm^{-1}, $\rho \sim 10^{-23}$ g/cm^{-3}, and $T_{\text{eff}} \sim 10$ eV one finds $t_{ss} \sim 10^{10}$ s ~ 300 years. This too is very fast, but it will turn out that T_{eff} is typically not large enough to be significant in such astrophysical plasmas. In both these examples, there will be an initial transition period, during which the shock processes precursor material that has not yet reached this steady state.

Thus, the precursor may approach steady state and the plasma temperature T_p will be close to T_{eff} in a shock with an optically thin upstream layer. Note that this result has no explicit dependence on the value of T_{eff}. Instead, the temperature in the precursor will depend primarily on the optical depth of the (downstream) shocked layer. The temperature near the shock will tend to increase as the optical depth of the precursor increases, but a more complete calculation would be needed to assess how much. The precursor will be determined by T_{eff}, which in turn is determined by details of the downstream layer discussed in the next two sections.

8.1.5.2 Structure of Thick-Thin Shocks

Figure 8.11 illustrates the energy balance in a steady shock of this type. One can see how the cooling layer controls the final state. The net flux at the downstream boundary of the cooling layer must be zero, so the final temperature must increase until the thermal flux from the steady downstream layer equals the flux from the cooling layer. When this occurs, the net upstream radiation flux, lost from the system in our description, but perhaps in reality having the effect of extending the length of

Fig. 8.11 Energy fluxes in thick-downstream, thin-upstream shocks. Note $p_o = 0$

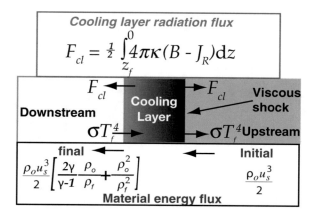

a precursor region, is $2\sigma T_f^4$. One self-consistency test for calculation of profiles as is described here is that the integrated radiation source, and the integrated change in material energy should both equal $2\sigma T_f^4$.

We can develop solutions for the structure of the cooling layer in such shocks, as follows. Based on our discussion of optically thin precursors, the upstream temperature will be quite close to T_f. Here we will take it to be equal to T_f, so the normalized upstream pressure is $p_{on} = \rho_o RT_f/(\rho_o u_s^2) = RT_f/u_s^2$. From (8.4) this implies that $p_{on} = RT_f/(u_s^2) = \rho_o/\rho_f$. Knowing p_{on}, we can evaluate the flux balance equation (8.6) to find the conditions at the boundaries of the cooling layer. At the shock transition $F_R = F_o = 2\sigma T_f^4$, so given $p_{on} = \rho_o/\rho_f$ one finds the initial inverse compression in terms of the final inverse compression, from (8.6),

$$\frac{\rho_o}{\rho_s} = \frac{\gamma - 1}{\gamma + 1} + \frac{2\gamma}{\gamma + 1}\frac{\rho_o}{\rho_f}. \tag{8.28}$$

Figure 8.12 shows the implications of this equation, for $\gamma = 4/3$. One can see that a very large final compression will be required before the initial compression becomes as large as the nonradiative, strong-shock value of 7. The initial compression is smaller than this because of the finite pressure p_o in the precursor.

In the final state, $F_R = 0$, so using (8.6) with $\rho_o/\rho = \rho_o/\rho_f = p_{on}$ one finds the net (normalized) radiation flux at the final density to be

$$\frac{-2F_o}{\rho_o u_s^3} = -1 + \left(\frac{\rho_o}{\rho_f}\right)^2, \tag{8.29}$$

which turns out to be independent of γ other than through the final density. Note that the limiting value of F_o, as the final density becomes very large, equals $\rho_o u_s^3/2$ as it should (per Fig. 8.10).

Fig. 8.12 Compression in
thick–thin shocks. The final
compression (gray) and
immediate postshock
compression are shown as a
function of the final
compression, for $\gamma = 4/3$

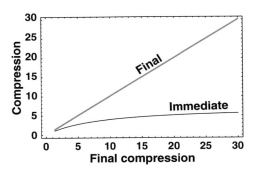

To pin down the final state, we observe that the energy supplying the radiation
flux comes from the shocked material, so the net radiation flux must equal the net
energy flux lost from the material between the immediate, postshock state, and the
final state. The behavior of the radiation flux in the shock thus has three differences
from that of the optically thin case, although these are not dramatic on a plot like
Fig. 8.10. The shape of the curve including the value of its maximum are altered
by the finite value of p_{on}, the value of the postshock inverse compression is not the
value for a nonradiative shock, and the final normalized flux is close to but larger
than -1. By setting the net flux $(F_R - F_o)$ across the shock equal to $2\sigma T_f^4$ (see
Fig. 8.11), we can solve (8.6) to find the final inverse compression. This turns out to
depend on the radiation strength parameter we defined above as $Q = 2u_s^5\sigma/(R^4\rho_o)$,
in terms of which

$$\frac{\rho_o}{\rho_f} = \sqrt{\frac{\sqrt{1+8Q}-1}{4Q}}. \tag{8.30}$$

This is independent of γ, although the detailed structure is not. For $R = 10^{12}$
ergs/g/eV, u_s in km/s $= u_{kms}$, and ρ_o in g/cm^3, one finds $Q = 4 \times 10^{-11}u_{kms}^5/\rho_o$.
At large enough shock velocity and thus large Q the final inverse compression
approaches zero, although by that point the system may be entering the radiation-
dominated regime. This result may seem strange, given that the radiation gets
stronger as velocity increases. But recall that a smaller final inverse compression
corresponds to a decreasing fraction of the incident energy ending up as thermal
energy, as more and more energy is radiated away. At small velocity (8.30) would
take the inverse compression to 1. However, we require $u_s^2 > c_{so}^2$, where c_{so} is the
upstream sound speed, to have a shock. Evaluating u_s^2/c_{so}^2 we find

$$\frac{u_s^2}{c_{so}^2} = \frac{u_s^2\rho_o}{\gamma p_o} = \frac{u_s^2}{\gamma RT_f} = \frac{\rho_f/\rho_o}{\gamma}. \tag{8.31}$$

Thus, the final compression must exceed γ, or equivalently the inverse compression must be smaller than $1/\gamma$, in order to have a shock at all. Correspondingly,

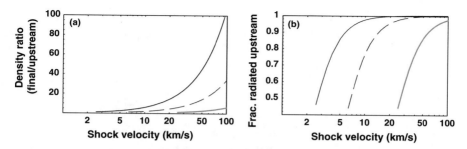

Fig. 8.13 Final state conditions for thick–thin shocks, for $\gamma = 4/3$, $A = 130$, and $Z = 17$. The curves show results for 1 g/cm^3 (gray), 1 mg/cm^3, (dashed), and $10 \, \mu\text{g/cm}^3$ (solid). (**a**) The ratio of final density to upstream density as a function of shock velocity. (**b**) The fraction of the incoming energy flux carried away upstream by radiation

one can show that $Q > \gamma^2(\gamma^2 - 1)/2$ in order to have a shock. The parameter Q depends primarily on u_s^5/ρ_o, which is the same as the ratio we found in our first, preliminary discussion of precursors. The temperature dependence of R, if included, would introduce additional complications in the solution. Here, to see the main qualitative behavior, we will assume R (and thus Z) to be constant.

Figure 8.13 shows (a) the dependence of the final compression on shock velocity for three densities and (b) the fraction of the energy radiated away under the same conditions. This figure uses $Z = 17$ and $A = 130$, corresponding to the use of a high-Z material to maximize the radiative effects. One sees that the radiation indeed carries away most of the energy as the system becomes more radiative. One sees that the compression can indeed become very high as shock velocity increases, but that this requires radiation of very nearly all of the incoming energy flux.

We are now ready to determine the profiles. To do so, we note that R_r evaluates to

$$R_r = Q \frac{\gamma + 1}{\gamma} \left[\frac{2(\gamma - 1)}{(\gamma + 1)^2} - \frac{(1 - 6\gamma + \gamma^2)}{(\gamma + 1)^2} \frac{\rho_o}{\rho_f} - \frac{2\gamma(\gamma - 1)}{(\gamma + 1)^2} \left(\frac{\rho_o}{\rho_f} \right)^2 \right]^4. \qquad (8.32)$$

By integrating (8.7) as represented by (8.18), beginning at the shock transition, one can obtain the profiles shown in Fig. 8.14. (Recall that $\kappa = \kappa_s(\rho/\rho_s)^m(T/T_s)^{-n}$.) The two cases shown correspond to $R_r \sim 20$ and $R_r \sim 1200$, so one can see that here again the distance required for the cooling decreases as the shock velocity and hence R_r increase. One also sees that the optical depth of the cooling layer is indeed quite small, as we anticipated above.

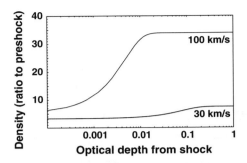

Fig. 8.14 Density within the cooling layer, plotted as the ratio to the preshock density and against the optical depth from the shock transition. The curves show results for a shock velocity of 30 km/s (lower) and 100 km/s (upper). These are evaluated for $\rho = 1\,\text{mg/cm}^3$, $\gamma = 4/3$, $n = 4/3$, $m = 0$, $A = 130$, and $Z = 17$

Fig. 8.15 Energy balance in optically thick (thick–thick) radiative shocks

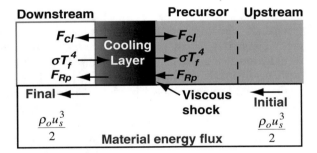

8.1.6 Optically Thick Radiative Shocks: Radiative Flux Regime

We now turn to a type of radiative shock that can exist only within an extensive system, in which both the upstream and the downstream media are optically thick but yet the shock is in steady state. The historical literature considered only this case, and the case of optically thin shocks in an astrophysical context. One might be inclined in this case to consider the system to be in LTE, and to use a diffusion model to describe the dynamics. However, this is not strictly valid because the shock itself drives the plasma out of equilibrium with the radiation. We will take the point of view that the precursor region, far enough away from the shock, can perhaps be described by a modified Marshak-wave model, and that the downstream region away from the shock is also in LTE. To start with, there are some general things we can say about this system. We first discuss the implications of energy flux balance, with reference to Fig. 8.15, and then the implications of the fluid dynamics, with reference to Fig. 8.16.

There are two places where the net radiation flux must be zero. These are at the head of the precursor, where all the net upstream radiation from the shock has been converted to heat, and at the boundary of the downstream region, where a new

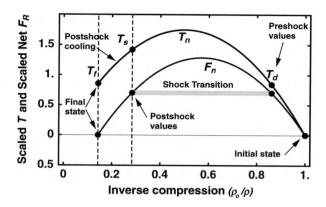

Fig. 8.16 Fluxes and temperatures in thick–thick shocks

postshock steady state is established. In the radiation-dominated regime, radiation energy and pressure are convected through these boundaries, but there is still no net radiation flux through them. If the absolute value of the radiation flux generated in the cooling layer is F_{cl}, then at the downstream boundary of the cooling layer one has $F_{cl} + F_{Rp} = \sigma T_f^4$, where F_{Rp} is the radiation from the precursor plasma. (The cooling layer differs only in details from that discussed in the previous section. Recall that the cooling layer is optically very thin, so the fluxes from adjacent regions are fully transmitted.) The net radiation flux moving upstream through the shock transition must balance the increase in convected energy flux in the precursor, which is negative and which we can designate as F_p. Thus $F_{cl} + \sigma T_f^4 - F_{Rp} + F_p = 0$, so that

$$2F_{cl} = 2\sigma T_f^4 - F_{Rp} = -F_p. \tag{8.33}$$

This also equals the net radiative flux across the shock.

We can take F_{Rp} to be approximately σT_d^4, where T_d is the temperature of the first few optical depths of the precursor. As T_d approaches T_f, both F_{cl} and F_p become a smaller and smaller fraction of the radiation flux in either direction. However, these cannot become zero, and in fact we will find below the limiting value of F_p from the fluid dynamics. The implication is that $T_d < T_f$ always.

This casts some doubt on the traditional definition of a supercritical shock, which in fact exists only as a limiting case. In the prior literature of optically thick radiative shocks, the distinction between a *subcritical shock,* having $T_d < T_f$, and a *critical* or *supercritical shock,* having $T_d = T_f$, is emphasized. Here T_f is the steady-state temperature of the downstream region. The traditional viewpoint can be summarized as follows. The radiative flux must be continuous when one crosses the shock, which sets the immediate postshock density as can be seen in Fig. 8.16. Because the temperature curve is shifted to the left relative to the flux curve, the immediate postshock temperature T_s is *always* higher than T_d. T_f is smaller than T_s, but how much smaller depends on the details of the radiation transport and on the radiation strength Q. If Q is small enough, then T_d can never approach T_f. Such a shock is known in the literature as a subcritical shock. The traditional notion is that a strong enough shock will produce $T_d = T_f$ and such a shock is known as a critical or supercritical shock. This is implied by an LTE analysis using equilibrium radiation diffusion. One might imagine that the temperature could become continuous as the diffusion limit is approached. However, this is not an accurate conclusion because the shock and cooling layer is always an out-of-equilibrium, nondiffusive structure. We will see below that T_d may approach T_f in various realistic circumstances, but that this is a limiting case rather than the threshold of a regime. As a result, the present author disagrees with the notion that a shock of any specific shock strength can unambiguously be identified as a critical shock.

8.1.6.1   Fluid Dynamics of Optically Thick Shocks: Radiative-Flux Regime

It is next worthwhile to consider what the fluid dynamics may imply in this case. The flux balance equation, if necessary including radiative pressure and energy terms, applies to all optically thick radiative shocks. No energy is "lost" from the system (except perhaps laterally in a laterally limited planar system, but we do not consider this here). This calculation naturally separates into flux-dominated and radiation-dominated cases. We take these up in turn.

Figure 8.16 shows the fluid-dynamics trajectory in the flux-dominated regime. The plasma is heated and compressed in the precursor, undergoes the shock transition at $F_R = -F_p$, and then cools to a final state with zero flux. For negligible initial upstream pressure, the final state thus has $\rho_f/\rho_o = (\gamma + 1)/(\gamma - 1)$ and has a normalized temperature given by (8.4) as

$$RT_f/u_s^2 = 2(\gamma - 1)/(\gamma + 1)^2, \tag{8.34}$$

which is 0.12 for $\gamma = 4/3$. For the same reasons as in the thick-thin case of the previous section, the postshock temperature drops rapidly from T_s toward T_f, the cooling layer is physically very thin, and thus there is a Zel'dovich spike.

The fluid dynamics also implies F_p as a function of T_d, as the difference between the net material energy flux reaching the shock from upstream (which includes recycled energy) and the incident material energy flux, $\rho_o u_s^3/2$. If one evaluates (8.6) in the precursor just before the density transition, taking $p_o = 0$ and using (8.4) (which originates from the continuity equation, momentum equation, and equation of state), one finds that

$$F_p = -\frac{\rho_o u_s^3}{4(\gamma - 1)}\left[(1 - \gamma) + 2(\gamma + 1)\frac{RT_d}{u_s^2} + (\gamma - 1)\sqrt{1 - 4\frac{RT_d}{u_s^2}}\,\right]. \tag{8.35}$$

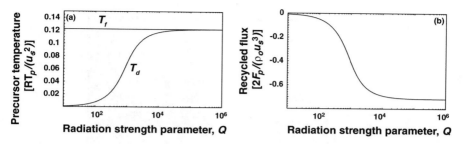

Fig. 8.17 Precursor properties in optically thick radiative shocks, for $\gamma = 4/3$. (a) Normalized temperature and (b) recycled flux

This goes to zero as $T_d \to 0$, as it should, but also goes to zero as $T_d \to T_f$. Thus $T_d \to T_f$ cannot be a solution, because recycled energy is present for any $T_d \neq 0$.

Given F_p from (8.35) T_f from (8.34), and $F_{Rp} = \sigma T_d^4$, one can express (8.33) as an implicit equation for T_d:

$$\left(\frac{RT_d}{u_s^2}\right)^4 + \frac{1}{2Q}\frac{RT_d}{u_s^2}\frac{(\gamma+1)}{(\gamma-1)} + \frac{1}{4Q}\left(\sqrt{1 - 4\frac{RT_d}{u_s^2}} - 1\right) - 16\frac{(\gamma-1)^4}{(\gamma+1)^8} = 0,$$

(8.36)

in which we again encounter the radiation strength parameter, $Q = 2u_s^5\sigma/(R^4\rho_o)$. Noting from (8.35) that the final term on the left-hand side in this equation is RT_f/u_s^2, one sees that T_d reaches T_f only in the limit that $Q \to \infty$, as expected from the discussion above.

To understand what is happening in this system, we can examine Fig. 8.17, which plots in (a) the solution of (8.36) for RT_d/u_s^2 and in (b) the solution of (8.35) for $2F_p/(\rho_o u_s^3)$, which is the recycled flux as a fraction of the energy flux incident on the shock, both for $\gamma = 4/3$. We see that as Q increases above 1000, T_d becomes a large fraction of T_f. We also see that, as this occurs, the magnitude of the recycled energy flux asymptotes to about 70% of the incident energy flux. As the incident shock velocity (and thus Q) increases further, the net recycled flux (which is the net radiative flux across the shock, and also twice the flux from the cooling layer) remains a fixed fraction of the incident flux while the radiative fluxes in each direction, σT_d^4 and σT_f^4, increase much more rapidly.

We can now discuss the implications of the fluid dynamics of the precursor. From Fig. 8.16, one can see that there is a nonzero final temperature for any finite inverse compression and thus for any possible final value of F_R. For example, the final normalized temperature is approximately 0.12 for the specific value of the inverse compression (approximately 0.15) corresponding to $(F_R - F_o) \sim 0$ and $\gamma = 4/3$. One can see in Fig. 8.16 that if the precursor temperature reaches this final temperature then the density increase in the precursor will be between 10 and 15%. This is quite small in comparison with the total density increase, of order 10, which justifies somewhat the assumption in some following sections that the density is unchanged in the precursor.

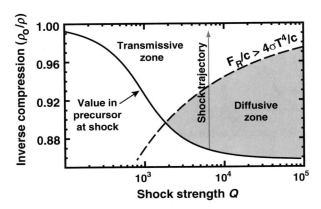

Fig. 8.18 Precursor regimes in thick–thick shocks. The solid curve shows the value at the shock for any given Q. The dashed curve shows the lowest possible density where the precursor becomes transmissive

We can also consider whether the precursor is in a transmissive regime or a diffusive regime, continuing the discussion of Sect. 8.1.3 for the specific case of optically thick shocks. We can set the terms in (8.11) to one, and use the basic fluid dynamics equations to determine at what densities F_R/c must be $> 4\sigma T_P^4/c$ as Q changes. These are shown in Fig. 8.18 as the dashed curve on a plot of inverse compression vs. Q. Also shown is the value of the inverse compression in the precursor at the shock front, obtained from the profile calculations discussed below. The trajectory of the radiation within the shock is to emerge from the shock front, to diffuse through the precursor if a diffusive zone exists, and then to be transmitted through the remainder of the precursor once it becomes transmissive. In the diffusive zone, the temperature profile is concave downward, as is typical of diffusive profiles. In the transmissive zone, the temperature profile is concave upward, corresponding to exponential attenuation.

8.1.6.2   Diffusive Radiative Precursors

In the diffusion regime, one is tempted to model the precursor in a supercritical shock as a Marshak wave, since it is a diffusive radiation wave emanating from a constant-temperature source. In this case, one must deal heuristically with the fact that the source is moving. The Marshak wave has a length-dependent velocity, being very fast when it is short (early in time after its initiation) and slowing down monotonically as its length increases. One can argue that, in steady state, the precursor length ahead of a shock must be such that the diffusion wave velocity equals the shock velocity. In either (7.70) or (7.79) one has the length $z_o = \xi_o \sqrt{Wt}$ with ξ_o a constant near 1.6 and $W = (\kappa_{rad})_{T_o}/[\rho c_V (n+1)]$. Matching the precursor velocity to the shock velocity gives $u_s = (\xi_o/2)\sqrt{W/t}$, which determines the "time" in the Marshak wave evolution at which the length is maintained. Combining these gives the steady-state precursor length as

$$z_o = \xi_o^2 W / (2u_s). \tag{8.37}$$

Note that W, being proportional to the coefficient of radiative heat conduction, κ_{rad}, scales as a large power of the temperature (T^4 to T^7). This is why the precursor length has a *very* strong dependence on shock velocity. However, in the context of real systems one would have difficulty observing such precursors. One can use the steady-state precursor length in the initial relation for x_o to find the time, t_o, required for the diffusion wave to reach this length. One obtains

$$t_o = (z_o / \xi_o \sqrt{W})^2 = \xi_o^2 W / (4u_s^2). \tag{8.38}$$

This time also increases very rapidly with shock velocity. Real experiments in the planar geometry of this analysis will achieve steady-state precursors only over a very narrow range in velocity.

The qualitative situation is better for experiments in spherical geometry, if in fact they can produce large enough systems that a diffusion model can meaningfully apply. Zel'dovich and Razier (1966) consider the case of nonlinear radiative heat diffusion from a point source in spherical geometry. They find that the diffusion wave moves with $r \propto t^{1/(3n+2)}$, so that the velocity is $\propto t^{-(3n+1)/(3n+2)} \sim 1/t$.

It is not too hard to improve the analysis above for a planar system. The gas-energy equation, simplified with the assumptions relevant to a Marshak wave, is (8.19). Applying the analysis of Chap. 4, one can show that this equation does not admit a self-similar solution. But we can hypothesize that a steady precursor may develop, reflecting the balance of upstream diffusion and downstream flow, so we consider this problem in steady state. If we take $\epsilon = c_V T$, approximating c_V as constant, and assume F_R is produced by diffusive heat transport we find

$$-u_s \frac{\partial T}{\partial z} = \frac{\partial}{\partial z} \frac{\kappa_{rad}}{\rho c_V} \frac{\partial}{\partial z} T, \tag{8.39}$$

from which as in Sect. 7.3 we obtain

$$\frac{\partial f}{\partial z} = -\frac{W}{u_s} \frac{\partial^2 f^{n+1}}{\partial z^2} \tag{8.40}$$

in this case, with $f = T/T_d$ (taking the radiation and matter to be in equilibrium in the optically thick system). Now we define a new variable $\zeta = z u_s / W$, so that (8.40) becomes

$$\frac{\partial f}{\partial \zeta} = -\frac{\partial f^{n+1}}{\partial \zeta^2}. \tag{8.41}$$

We can obtain a boundary condition from the fact that the upstream radiative heat flux at the precursor boundary balances the thermal energy brought back to the shock by the incoming flow. (This has some subtle aspects, because the diffusion treatment is not fully self-consistent.) The relevant equation is (7.4), evaluated, for the upstream region, in the frame of a steady shock for constant ρ and u. Here we write the flux balance as

$$\kappa_{\text{rad}} \nabla T \big|_{\text{shock}} = \gamma \rho \epsilon u_s. \tag{8.42}$$

Upon ignoring differences between $T_{\text{eff}}, T_R\big|_{\text{shock}}$, and the material temperature at the shock, which are caught up in the subtleties just mentioned, this becomes

$$\frac{\partial f}{\partial \zeta}\bigg|_{\zeta=0} = \frac{\gamma}{n+1}. \tag{8.43}$$

Figure 8.19 shows the precursor profile for $n = 4$ and $\gamma = 4/3$. This solution finds $\zeta \sim 0.9$ at the leading edge of the precursor. Let us compare the size of this precursor with our simple estimate above. The scaling is the same, as W/u_s. The ratio of the result of the improved calculation to that of the simple estimate is $1.8/(\xi_o^2) \sim 0.7$. One can see that the precursor length from the diffusion model is somewhat shorter than of the precursor length from the Marshak-wave estimate. In other words, the effect of the incoming flow is to reduce the size of the precursor in addition to limiting its expansion.

We can estimate the diffusive precursor length as follows. For typical parameters ($\chi_R \sim 10^6 T_{\text{eV}}^{-4/3}$ cm^2/g, $n = 4, \gamma = 4/3, \rho = 0.1$ g/cm^3, $c_V = 10^{12}$ ergs g^{-1}/eV, $u_s = 2 \times 10^6 T_{\text{eV}}^{1/2}$ cm/s), one finds $W = 10^{-4} T_{\text{eff}}^{13/3}$ and $W/(\gamma u_s) = W/(8 \times 10^6 T_{\text{eff}}^{1/2})$, with T_{eff} in eV. This is 19 μm at 100 eV for this density but would be 1.9 mm for

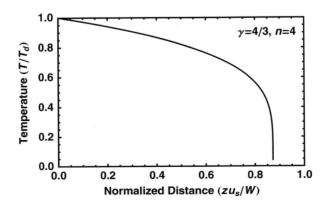

Fig. 8.19 A diffusive precursor profile. A sample solution of (8.39), for $\gamma = 4/3$ and $n = 4$, which gives $f'(0) = -0.26$. The temperature ratio f is shown on the ordinate, with the normalized distance zu_s/W on the abscissa

Fig. 8.20 Diffusive
precursor lengths. The
boundaries show the
indicated values. Experiments
with foams and gasses at
densities below 10 mg/cm^3
would tend to produce very
large precursors, but in actual
experiments lateral energy
losses will limit the precursor
length

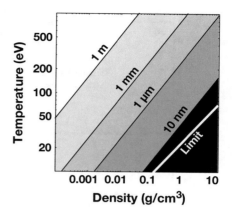

$\rho = 0.01$ g/cm^3. Figure 8.20 shows the steady-state diffusive precursor lengths in
a space of density and temperature. The boundary labeled "limit" in this figure is
where the precursor length decreases to ten interparticle spacings for ionized Be.
The model is certainly not valid beyond that point.

8.1.6.3   The Structure of the Optically Thick Radiative Shock

One can calculate the structure of such a shock using the equations presented above.
Figure 8.21 shows the results of doing so using the three-layer model described
here. One divides the structure into the three regions illustrated in Fig. 8.15: a
steady downstream state, a cooling layer, and a precursor. Note the vast change
in the optical depth scale at the shock location. The profiles show the behavior
expected from Fig. 8.15. The temperature and compression increase throughout the
precursor, with the change in compression remaining small. The density increases
by a factor smaller than 7 at the viscous transition, where the temperature roughly
doubles. (Reducing Q leads to a cooler precursor and larger jump in normalized
temperature.) In the cooling layer, the temperature and compression rapidly change
toward the final, downstream values. The expected Zel'dovich spike is present;
physically, it is extremely small. Eventually, the compression reaches its limiting
value of $(\gamma + 1)/(\gamma - 1) = 7$.

We now discuss the modeling of the various regions. The downstream region is
a steady-state whose properties can be inferred using Fig. 8.16. The precursor is a
very extended region, whose actual length changes very rapidly with u_s, for reasons
discussed in the previous section. As a result, one can treat the precursor using a
diffusion model with an Eddington factor $f_E = 1/3$. One can see in the plot where
the precursor becomes transmissive and the second derivative of the temperature
profile changes sign. In that part of the profile the actual Eddington factor changes
from 1/3 to 1 as the radiation becomes beam-like, and so the spatial scale obtained
by the calculation is not correct. The cooling layer is modeled using a transport
model, taking $\partial F_R/\partial z = 4\pi \kappa (B - J_R)$. This is tractable because it turns out that, to

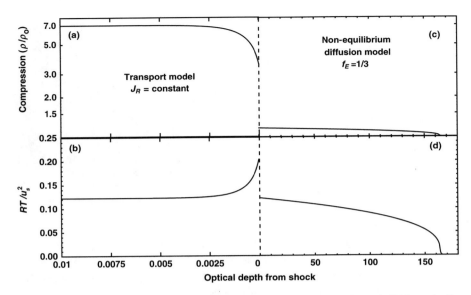

Fig. 8.21 Profiles of compression [in (**a**) and (**c**)] and temperature [in (**b**) and (**d**)] in optically thick shocks. $Q = 3 \times 10^5$, $\gamma = 4/3$, and $f_E = 1/3$. The shock location is at zero. The optical depth of the precursor [shown in (**c**) and (**d**)], increasing to the right, is far larger than that of the cooling region [shown in (**a**) and (**b**)], increasing to the left

quite good accuracy, $J_R = B_f$ throughout the cooling layer. The reason is that J_R is an average over all angles, and the emission is isotropic. As one moves through the cooling layer, rays producing left-going radiation are replaced by an equal number of rays producing right-going radiation. At the left boundary of the cooling layer, the net radiation flux must go to zero, which sets the value of J_R. The completion of the calculation involves satisfying the energy balance at the downstream edge of the cooling layer, where J_R must remain constant and includes contributions from the downstream region, the cooling layer, and the precursor.

The structure of the cooling layer is quite informative in the context of radiation transfer, which shows why it is not accurately described by a diffusion model (single-group or multi-group). First we develop the diffusion model in this context. The fundamental equations of Chap. 6 imply that

$$\frac{\partial p_R}{\partial z} = -\frac{\bar{\chi}}{c} F_R \tag{8.44}$$

for radiation along the z axis, and in which $\bar{\chi}$ approaches the Rosseland mean opacity as the system approaches LTE. This equation is always correct, if one evaluates $\bar{\chi}$ for the actual distribution of radiation intensity. It is the Eddington approximation that enables closed diffusion models, by taking $p_R = f_E E_R = f_E(4\pi/c)J_R$, giving

$$4\pi f_E \frac{\partial J_R}{\partial z} = -\bar{\chi} F_R. \tag{8.45}$$

One obtains a diffusion model in this context by normalizing the intensities by $\rho_o u_s^3/2$ and designating this with an added subscript of n, using (8.45) in (8.6) and a transport model in (8.7), recognizing that the normalized thermal intensity can be expressed in terms of Q and ρ_o/ρ and that $p_o = 0$ for these shocks, to obtain

$$\frac{4\pi f_E}{\bar{\chi}} \frac{\partial J_{Rn}}{\partial z} = \left[1 - \frac{2\gamma}{\gamma-1}\frac{\rho_o}{\rho} + \frac{\gamma+1}{\gamma-1}\left(\frac{\rho_o}{\rho}\right)^2\right] \text{ and} \tag{8.46}$$

$$\left[-\frac{\gamma}{\gamma-1} + \frac{\gamma+1}{\gamma-1}\left(\frac{\rho_o}{\rho}\right)\right] \frac{\partial}{\partial z}\left(\frac{\rho_o}{\rho}\right) = 2\pi\kappa_P \left[J_{Rn} - \frac{Q}{\pi}\left(1 - \frac{\rho_o}{\rho}\right)^4\left(\frac{\rho_o}{\rho}\right)^4\right]. \tag{8.47}$$

Figure 8.21 showed the results of integrating these equations from the leading edge of the precursor toward the shock, to find the spatial structure of the precursor, for the given γ and Q and assuming κ_P to be constant for simplicity.

Returning to the cooling layer, note that the sign in (8.44) is always correct; the direction of the radiation flux opposes that of the radiation-pressure gradient. It is widely assumed that this also applies to (8.45), so that radiation always flows "downhill", toward regions of lower radiation-energy-density. This is why diffusion models are widely assumed to give qualitatively correct answers. However, in cooling layers and similar systems this turns out to be false. This is a consequence of two facts. First, J_R and p_R are different moments of the radiation intensity, I_R. One has $J_R = \int_{-1}^{1} I_R d\mu/2$ while $p_R = (2\pi/c)\int_{-1}^{1} I_R \mu^2 d\mu$. Second, the planar layer always becomes optically thick near enough to $\mu(= \text{Cos } \theta) = 0$. The consequence is that J_R, although nearly constant in this application, develops a maximum within the cooling layer (offset to the right because of the contribution of the warm downstream region), while p_R monotonically decreases toward the shock. To the left of the maximum in J_R, (8.45) forces the radiation flux to be leftward when in actual fact it is large and rightward. A code implementing a diffusion model, if written so as to conserve energy, will generate incorrect profiles of J_R in order to support the overall features necessary to conserve energy. The general point is that a diffusion model will give an explicitly wrong answer whenever a system develops an optically thin hot (or cold) layer that extends far enough laterally to have some optically thick rays. One can find a more detailed discussion of this point, and of optically thick shocks in the radiative flux regime, in several papers involving the present author from 2007 and 2010.

8.1.7   Radiation-Dominated Optically Thick Shocks

For the radiative shocks in optically thick systems that are our subject here, one could hope to evaluate the structure in the radiation-dominated regime, using an approach similar to the one we just described for the radiative flux regime. However, this would be mathematically even more complicated, because the radiation pressure depends on the fourth power of T. We will leave the problem of the structure, and even more so of the structure in transition regimes, to the specialized literature. It is important to note, as Sect. 8.1.6.1 showed, that treatments of the internal structure near the shock that use only the diffusion approximation will be qualitatively wrong. Here we consider only the relation between initial and final states, where the radiative flux is zero.

In this case the continuity equation is unchanged. The momentum and energy equations ((7.2) and (7.3)) become, for steady-state planar shocks,

$$\rho u^2 + p + p_R = \rho_o u_o^2 + p_o + p_{Ro} \text{ and} \tag{8.48}$$

$$u\left(\rho\frac{u^2}{2} + \rho\epsilon + E_R + p + p_R\right) = u_o\left(\rho_o\frac{u_o^2}{2} + \rho_o\epsilon_o + E_{R_o} + p_o + p_{R_o}\right), \tag{8.49}$$

where as usual these equations are in the shock frame. We would like to develop useful relations from this, just as we have done previously. Because $u_o = -u_s$ and $u = -u_s(\rho_o/\rho)$, we find

$$\frac{p + p_R}{\rho_o u_s^2} = \left(1 - \frac{\rho_o}{\rho}\right) + \frac{p_o + p_{Ro}}{\rho_o u_s^2} \text{ and} \tag{8.50}$$

$$\frac{1}{2}\left(1 - \left(\frac{\rho_o}{\rho}\right)^2\right) - \left(\frac{\gamma p/(\gamma-1) + 4p_R}{\rho_o u_s^2}\right)\left(\frac{\rho_o}{\rho}\right) + \left(\frac{\gamma p_o/(\gamma-1) + 4p_{Ro}}{\rho_o u_s^2}\right) = 0. \tag{8.51}$$

Here we have two equations for three unknowns (ρ, p, and p_R). In the case that radiation completely dominates, so p can be neglected, these are readily solved for p_R and ρ. If p and p_R are both known functions of T, one can solve for T and ρ. Alternatively, if the medium is ionizing and has $\gamma = 4/3$, one can solve for the total pressure and ρ. We consider the third case here and leave the first two to homework.

If $\gamma = 4/3$, then one can substitute from (8.50) into (8.51) and express the total pressure as p_t to obtain

$$\frac{1}{2}\left(1 - \frac{\rho_o}{\rho}\right)\left(1 - 7\frac{\rho_o}{\rho} + 8\left(\frac{p_{to}}{\rho_o u_s^2}\right)\right) = 0, \tag{8.52}$$

in which p_{to} is the total pressure in the upstream state. The two solutions of this equation for the inverse compression (ρ_o/ρ) give the total density change across the shock transition. These solutions are 1 (the upstream density) and $(1/7)[1 + 8p_{to}/(\rho_o u_s^2)]$. Thus, with negligible upstream pressure the density increases by a factor of 7 and this density increase gets smaller as the upstream pressure, normalized by the ram pressure, increases. The shock will vanish when the upstream pressure reaches 7/8 of the ram pressure. One can substitute for the inverse compression in (8.50) and solve for the final total pressure, finding $p_t = (6\rho_o u_s^2 - p_{to})/7$. At the most, this can be 6/7 of the ram pressure when the upstream pressure is negligible.

8.1.8 Electron-Ion Coupling in Shocks

To this point we have ignored the equilibration region behind the density jump in radiative shocks. The shock heating of the electrons is small [consider (8.1) for electrons], so that it is the ions that are primarily heated. The ions then heat the electrons by Coulomb collisions, in a region we will designate the *equilibration zone*. At issue here is the slowing of the ions, which would be described in collision theory as the "test particles" in this case. The field particles, which interact with the ions and cause them to slow, are the electrons. The electron-heating coefficient is $\nu_{ie} = 3.2 \times 10^{-9} n_i Z^3 \ln \Lambda/(A T_e^{3/2})$, with T_e in eV and n_e in cm^{-3}. The evolution of the ion temperature is then

$$\frac{\partial T_{ion}}{\partial t} = -\nu_{ie} (T_{ion} - T_e). \tag{8.53}$$

The electron heating also may correspond to increased ionization of the ions. Typically the electron-heating coefficient and the temperature difference both decrease as the electrons heat. But curiously, if $Z \propto \sqrt{T_e}$ as is approximately true in ionizing plasmas, then the electron-heating coefficient remains constant as the electron temperature increases, leading to more rapid equilibration. The exchange of energy between the electrons and the ions, including ionization and the heating of the new electrons, does not change the total postshock pressure. In contrast, radiation can affect the pressure, and also the rate of equilibration through the density. This makes it worthwhile to compare the radiative rates with the electron heating.

First we compare the heating or cooling rates. Then we consider more carefully the structure of the equilibration zone. The rate of energy emission, in power per unit volume, is $2\kappa\sigma T_e^4$ at high density and $n_e n_i \Lambda$ at low density (see Chap. 6). The rate of energy transfer per unit volume per unit fractional temperature difference $(T_{ion}/T_e - 1)$ is $\rho c_V \nu_{ie}$. Figure 8.22 shows the ratio of $\rho c_V \nu_{ie}$ to the radiative cooling rate for both laboratory and astrophysical conditions. For electron temperatures that do not approach keV levels, electron heating clearly dominates, except at low density (0.01 g/cm^3), and using a density-independent scaling for κ (which may not apply in actuality). One concludes that radiative cooling of electrons would become

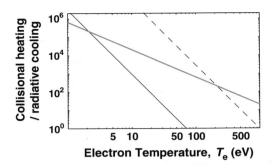

Fig. 8.22 Ratio of electron heating power to radiative emission vs. T_e for cases of interest. In the laboratory regime, with $\rho = 0.01 \, \text{g/cm}^3$ (black solid) and $1 \, \text{g/cm}^3$ (dashed), using the density-independent form of κ for Al from (6.45). In the astrophysical regime (gray), where the ratio depends only on T_e, using $\Lambda = 10^{-22} \, \text{erg} \, \text{cm}^{-3}$

important in shocks producing ion temperatures of many keV, and possibly under some conditions for somewhat lower temperatures.

However, Fig. 8.22 overstates the importance of radiation, because it is the net difference of absorption and emission that heats or cools the electrons. In a shock, the radiation from the final state will at first overwhelm the radiation emission by the electrons and will contribute to their heating. Later, when the electron temperature rises above the final plasma temperature, radiation will have a net cooling effect, in opposition to electron heating by the ions. However, in many cases these differences won't matter because the effect of the radiation on the equilibration zone will be negligible. Assuming that radiation plays no role, let us consider the structure of the equilibration zone produced by electron heating alone.

We can convert (8.53) into a spatial equation in the shock frame by dividing by the postshock fluid velocity, $u = u_s(\gamma - 1)/(\gamma + 1)$. There is a corresponding equation for the electrons in which T_e and T_s are exchanged and the right-hand side is negative. One can use (8.1) for the ions with $Z = 0$ to get a characteristic initial value of T_s, and one can assume $T_e \sim 0$ to start. Figure 8.23 shows the resulting spatial profiles of T_e and T_s for a shock velocity of 100 km/s. One sees that even for very low density, low-Z gas (part a) the equilibration occurs within a few micrometers. For Xe gas at somewhat higher density, the equilibration occurs within a fraction of a micrometer.

Recall that in detail the shock transition itself is not instantaneous, but occurs over a distance of a few ion–ion mean-free paths. Both electron heating and radiation emission do occur simultaneously with the shock transition, so there may be regimes in which all three processes are simultaneously important. It will typically be adequate, though, to assume that the shock transition occurs instantaneously, that the equilibration zone is at most small, and that the radiation becomes important on a larger spatial scale. In simulations, the electron heating occurs on the scale of the shock transition for conditions that produce postshock temperatures of order 10 eV,

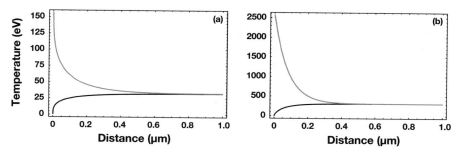

Fig. 8.23 Postshock electron–ion equilibration. Positive distance is downstream from the 100 km/s shock. (**a**) Be at $\rho = 10^{-5}$ g/cm^3. (**b**) Xe at $\rho = 0.01$ g/cm^3, assuming $Z \propto \sqrt{T_e}$

and occurs on a larger (though still small) scale as the postshock ion temperature reaches hundreds of eV.

8.2   Radiative Heat Fronts

In general a front is a region where some parameter describing a system changes abruptly, where "abruptly" means that this happens on a spatial scale small compared to the overall size of the system. The front and whatever structure follows behind it is often called a "wave". Thus far, we have primarily been concerned with shock fronts and shock waves. We have seen that they can have considerable internal structure, but also that on a large enough scale, they can be viewed as transitions whose behavior must conserve mass, momentum, and energy. Radiative shock waves in their simple form produce radiation by emission from shock-heated matter. Section 8.1 discussed the consequences, which we found to be diverse. It is also not uncommon to find that radiation sources cause the heating of matter. If this goes on long enough, then the penetration of the heat into the matter can be described as a "heat wave" and the leading edge of the heated region as a "heat front".

Under most conditions at high energy density, the heat transport by radiation is large compared to that by electrons, and so here we consider formally only radiative heat fronts. However, one can find circumstances where the electron heat transport is dominant, and in these cases one might have an electron heat front. The fluid dynamics below applies equally well to such fronts.

In this section we will analyze steady heat fronts, in the "heat-front frame". We assume that there is a radiation flux incident on the system from the left, and that the radiation flux is not rapidly absorbed until it reaches the front, where the remaining radiation is absorbed over a short distance. (There may be gradual absorption, as we will see below.) There is some similarity to Marshak waves, for which the constant flux model discussed above proved fairly accurate. As a result of these assumptions, we have jump conditions across the front that are similar to those we had for radiative shocks. Once again, and for the same reasons as in most of the discussion of radiative shocks, we consider the radiation-flux regime, neglecting the momentum and pressure of the radiation.

We rewrite the equations here for a polytropic gas of index γ, and state the definitions of the variables, for convenience:

$$\rho_1 u_1 = \rho_2 u_2, \tag{8.54}$$

$$\rho_1 u_1^2 + p_1 = \rho_2 u_2^2 + p_2, \text{ and} \tag{8.55}$$

$$\left(\frac{\rho_1 u_1^2}{2} + \frac{\gamma p_1}{\gamma - 1}\right) u_1 - F_R = \left(\frac{\rho_2 u_2^2}{2} + \frac{\gamma p_2}{\gamma - 1}\right) u_2, \tag{8.56}$$

where the density, velocity, and pressure in the upstream, unheated region are ρ_1, u_2, and p_1, respectively, and the corresponding quantities in the downstream, heated region have the subscript 2. The radiation flux F_R here represents the energy flux absorbed at the heat front, so that we have implicitly ignored any radiation that streams through it. In addition, the radiation might in principle be incident from either side, but more often is incident from region 2. We retain the same basic geometric conventions we used above for shocks, so that in the frame of the heat front, u_1 and u_2 will be negative. Positive F_R then will correspond to the typical case where radiation from the heated region, region 2, is incident on the front. The radiative energy deposited at the heat front then will add to the energy carried in from region 1 to produce the parameters of region 2.

Because steady heat fronts have constant temperatures on each side of the front, it is natural to normalize the pressure in terms of the square of the isothermal sound speed, written as $a^2 = p/\rho$. (a^2 here is RT in our discussion of radiative shocks.) Then the momentum and energy equations become, using (8.54) and (8.55) to simplify (8.56),

$$\rho_1 (u_1^2 + a_1^2) = \rho_2 (u_2^2 + a_2^2), \text{ and} \tag{8.57}$$

$$\frac{(u_2^2 - u_1^2)}{2} + \frac{\gamma}{(\gamma - 1)} (a_2^2 - a_1^2) = -\frac{F_R}{\rho_1 u_1}. \tag{8.58}$$

The differences among various heat fronts relate to the assumptions and boundary conditions used to solve these equations. We will first consider Marshak-like heat fronts, having a definite downstream temperature and heat flux, and will see a new aspect of this type of behavior. A slightly different case of great practical importance is the expansion heat front, which we next explore. Then we will discuss photoionization fronts. The difference between these types of front lies in the boundary conditions. Marshak-like waves involve radiation diffusion from a source, expansion heat fronts must sustain a downstream rarefaction, and photoionization fronts involve radiation that streams through the medium to a region of ionization. The impact of this difference lies in the implied boundary conditions. Marshak-like waves and expansion heat-fronts feature a nearly constant temperature in the downstream region, while photoionization fronts have a front velocity determined by simple kinematics.

8.2.1 Marshak-Like Heat Fronts

When we examined Marshak waves in Sect. 7.3, we assumed that the density remained constant as the wave propagated through the system. This excludes the possibility of seeing behavior in which the fluid dynamic response matters. Here we begin with assumptions appropriate to steady waves that have much in common with Marshak waves, in order to examine the fluid-dynamic possibilities that result. We ignore the radiation transport, which was discussed above. Thus, our calculation complements the previous one.

Marshak waves feature a constant temperature at the downstream boundary. The incoming radiation flux (from downstream but directed toward the front) decreases very rapidly at early times but then decreases more slowly. We will approximate this incoming heat flux as constant, in order to see the hydrodynamic steady state that would correspond to a given heat flux. In effect, we consider the Marshak-like wave to represent an abrupt, steady transition in the properties of the plasma, and we examine the relation between upstream and downstream quantities that is allowed by the conservation of mass, momentum, and energy in steady state. This approach is sensible in a regime where absorption opacity dominates and where the diffusion of energy is nonlinear, so that the heat wave becomes a spatially limited structure. If scattering opacity were dominant, then one would have to think through this problem differently. Based on these assumptions, we consider F_R, a_2, a_1, and ρ_1 to be input parameters that are initially known. This leads us to normalize the above equations by taking $\rho_2 = \rho_1 \widetilde{\rho}$, $a_1 = \widetilde{a}_1 a_2$, $u_1 = \widetilde{u}_1 a_2$, $u_2 = \widetilde{u}_1 a_2 / \widetilde{\rho}$, and $F_R = \widetilde{F}_R \rho_1 a_2^3$, giving

$$\widetilde{\rho}(\widetilde{u}_1^2 + \widetilde{a}_1^2) = (\widetilde{u}_1^2 + \widetilde{\rho}^2), \text{ and} \tag{8.59}$$

$$0 = -\widetilde{F}_R + \frac{\widetilde{u}_1^3}{2}\left(1 - \frac{1}{\widetilde{\rho}^2}\right) + \widetilde{u}_1 \frac{\gamma}{\gamma - 1}(\widetilde{a}_1^2 - 1). \tag{8.60}$$

One can solve Eq. (8.59) for $\widetilde{u}_1$, obtaining

$$\widetilde{u}_1 = -\frac{\sqrt{\widetilde{\rho}(\widetilde{\rho} - \widetilde{a}_1^2)}}{\sqrt{\widetilde{\rho} - 1}}, \tag{8.61}$$

taking the positive values of the square roots. This remains real for all $\widetilde{\rho}$ such that numerator and denominator are both real or both imaginary. Substituting into Eq. (8.60), one obtains

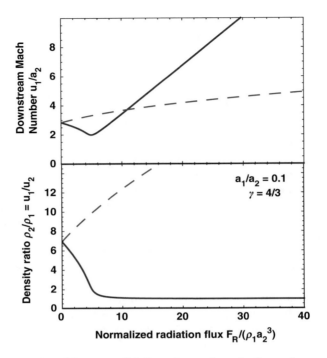

Fig. 8.24 Density ratio and downstream Mach number are shown for the two denser roots based on the jump conditions, assuming a step-function temperature profile. Note that, for some temperature T characteristic of the source, if $F_R \propto T^4$ and $a_2 \propto T$ then the abscissa is proportional to T

$$0 = 2\widetilde{F}_R(\gamma - 1)\widetilde{\rho}\sqrt{\widetilde{\rho} - 1} +$$

$$\left[\widetilde{\rho}\left(\gamma(\widetilde{\rho} - 1) - (\widetilde{\rho} + 1)\right) + \widetilde{a}_1^2\left(\gamma(\widetilde{\rho} - 1) + (\widetilde{\rho} + 1)\right)\right]\sqrt{\widetilde{\rho}(\widetilde{\rho} - \widetilde{a}_1^2)}. \qquad (8.62)$$

From Eq. (8.62) one can solve for the density ratio $\widetilde{\rho}$, obtaining three real roots as functions of γ, $\widetilde{F}_R$, and $\widetilde{a}_1$. Figure 8.24 shows the two denser roots. The solid curves show the Marshak-like heat wave. At large enough radiation flux the density ratio remains clamped at one, consistent with the standard Marshak wave. The difference is that here we have fixed the radiation flux while in the standard Marshak wave it is steadily decreasing. But once the downstream Mach number reaches 2, the solution changes and a shock wave begins to develop at the transition. Since in real heat waves the radiation flux does drop as the wave penetrates further and the Marshak wave continues to slow with time, this emergent shock wave moves out ahead of the heat front. The emergent shock wave has been observed experimentally. It is distinct from another phenomenon that has also been observed. An experiment often launches a shock wave from the surface through which the radiation penetrates to heat the medium. As the Marshak wave slows, this launched shock wave slows much less if at all. It will then overtake the Marshak wave at some point. Ultimately, these systems involve the interplay of the Marshak wave, the emergent shock wave, and the ablative shock wave.

The dashed curve seen in the figure can be identified as a shock wave corresponding to extra energy deposition at the shock front. This might occur if there were radiation flux incident, under conditions such that the radiation was much more strongly absorbed at the front than away from it. Such a flux might be incident from either direction. The third root in the solution of Eq. (8.62) corresponds to a decrease in density. While such a solution, with fixed F_R and a_2, might be relevant under some very special boundary conditions, we do not discuss it further here. Instead we take up the similar but more relevant case of an expansion heat front.

8.2.2 The Expansion Heat Front

Ionizing radiation often encounters a surface where the density and absorption strongly increase, leading to the local deposition of the radiation flux that reaches the boundary. The ablated material then expands away from the surface, while the high pressure in the absorption zone drives a shock into the dense material. This occurs in particular at the surface of an X-ray-driven fuel capsule for inertial confinement fusion, on the surface of a planar ablator when soft X-rays are used to drive other experiments, or at confining walls when such radiation reaches them, as for example in radiative shock experiments. In laboratory systems, there is often enough absorption of the radiation in the outflow to sustain the temperature of the expanding plasma. When astrophysical radiation ablates molecular clouds, this will not be the case. In addition, the heating will then occur at a photoionization front, discussed in the next subsection. Here we consider the laboratory case.

We describe these dynamics with reference to Fig. 8.25. We again treat the evolution in the context of a fluid model, assuming the shock front and the heat front to be infinitesimally thin. For typical conditions, the absorption depth in solid-density materials is in the vicinity of 1 μm, so this defines what corresponds to thin

Fig. 8.25 Schematic of expansion heat front for analysis. Soft X-rays incident from the left sustain an isothermal rarefaction up to the heat-front location. The heat front sustains the pressure that drives a shock wave to the right

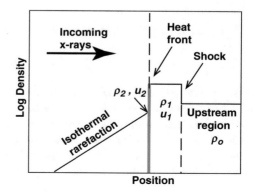

in the present context. The shock front in the dense matter is some small number of orders of magnitude steeper than the heat front. Here we discuss the solutions for the shock wave, for the heat front and expansion, and how to match the two solutions. We use the same fundamental variables we used above, begin with (8.54)– (8.56), and again take $a^2 = p/\rho$. We denote by F_{Ro} the radiation flux that enters the system. Some of this flux is absorbed while penetrating the expanding plasma, and the remainder is absorbed at the heat front. We denote the density of the unshocked matter as ρ_o.

The combination of absorption and heat conduction (at lower density) tends to sustain the temperature of the expansion, and so we approximate its behavior as an isothermal rarefaction. We are seeking a steady solution for the heat front and shock, so the rarefaction must meet the heat front at its sonic point, which is the only position at which the density does not change with time. Take the density at the sonic point to be ρ_2. To examine the energy requirements, we take the rarefaction to extend from $x = 0$ to ∞, with the sonic point at $x = 0$. The speed in the rarefaction is $u_r = a_2(1 + x/(a_2t))$ and the total energy flux required to sustain the rarefaction is

$$F_{tot} = \frac{\partial}{\partial t} \int_0^\infty \rho_2 \left(\frac{u_r^2}{2} + \frac{a_2^2}{\gamma - 1} \right) e^{-x/(a_2t)} dx = \rho_2 a_2^3 \left(\frac{5\gamma - 3}{2(\gamma - 1)} \right). \tag{8.63}$$

We can find another expressions for F_{tot} from the fluid equation for energy conservation, which tells us, in 1D, that

$$\frac{\partial}{\partial t} \left(\frac{\rho u_r^2}{2} + \frac{\rho a_2^2}{\gamma - 1} \right) = -\frac{\partial}{\partial x} \left(\frac{\rho u_r^3}{2} + u_r \rho a_2^2 \frac{\gamma}{\gamma - 1} + F \right), \tag{8.64}$$

in which F may include energy fluxes from conduction or from radiation transport. With $u > 0$ here, one will have $F < 0$. Equation (8.63) then implies that $F_{tot} =$

$$-\int_0^\infty \frac{\partial}{\partial x} \left(\frac{\rho u_r^3}{2} + u_r \rho a_2^2 \frac{\gamma}{\gamma - 1} + F \right) dx = \rho_2 a_2^3 \frac{3\gamma - 1}{2(\gamma - 1)} + (F_2 - F_\infty), \tag{8.65}$$

for $u_2 = a_2$. Taken altogether, the above results imply that

$$F_2 - F_\infty = \rho_2 a_2^3. \tag{8.66}$$

Because $F < 0$, $|F_\infty| > |F_2|$, as it should.

For the overall calculation of the structure we have three known quantities, F_{Ro}, a_2, and ρ_o, and ten unknowns, defined in Table 8.1. We have six equations describing the jump conditions for mass, momentum, and energy in the two frames. We have as equations of state that $p_2 = \rho_2 a_2^2$ and $p_1 = \rho_1 a_1^2$. We know that $u_2 = -a_2$ because the heat front is at the sonic point, and we know that the velocity of the

Table 8.1 Unknowns for the expansion heat front problem

Name	Symbol
Density in rarefaction at heat front	ρ_2
Pressure at sonic point	p_2
Velocity at sonic point	u_2
Shocked matter velocity in heat-front frame	u_1
Density of shocked matter	ρ_1
Pressure of shocked matter	p_1
Sound speed (isothermal) in shocked matter	a_1
Shocked matter velocity in shock frame	u_{1s}
Lab-frame velocity of shock front	u_s
Lab-frame velocity of heat front	u_{hf}

shocked matter in the lab frame, inferred from either the shock frame or the heat-front frame, must be the same, so that $u_{1s} + u_s = u_{hf} + u_1$. We also know that the incoming radiation flux must supply the energy required to sustain the structure in time, and include this fact in formulating the energy equation for the rarefaction. We continue to use the standard sign convention throughout the book, taking $u_s > 0$ and u_{1s}, u_1, and $u_2 < 0$.

On the assumption that the shock is a strong shock, the relations for the variables in the shock frame are simple and familiar:

$$\rho_1 = \frac{\gamma + 1}{\gamma - 1} \rho_o, \tag{8.67}$$

$$p_1 = \frac{2}{\gamma + 1} \rho_o u_s^2, \tag{8.68}$$

$$u_{1s} = -u_s \frac{\gamma - 1}{\gamma + 1}, \text{ and} \tag{8.69}$$

$$a_1 = u_s \frac{\sqrt{2(\gamma - 1)}}{\gamma + 1}. \tag{8.70}$$

In the heat-front frame we normalize (8.54)–(8.56) by taking $\rho_2 = \rho_o \widetilde{\rho}_2$, $u_1 = \widetilde{u}_1 a_2$, $p_1 = \widetilde{p}_1 \rho_o a_2^2$ and $F_{Ro} = \widetilde{F}_{Ro} \rho_o a_2^3$. One can then solve the normalized equations, also using the other known facts relating to the rarefaction from above, to obtain

$$\widetilde{u}_1 = \frac{(\gamma - 1)\widetilde{\rho}_2}{(\gamma + 1)}, \tag{8.71}$$

$$\widetilde{p}_1 = \widetilde{\rho}_2 \frac{[2(\gamma + 1) - \widetilde{\rho}_2(\gamma - 1)]}{(\gamma + 1)}, \text{ and} \tag{8.72}$$

Fig. 8.26 Solutions for the expansion heat front, for $\gamma = 1.5$

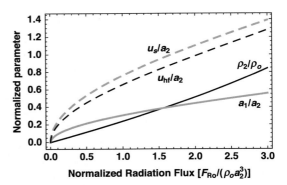

$$\widetilde{F}_{Ro} = \frac{\widetilde{\rho}_2(7\gamma - 5)}{2(\gamma - 1)} - \frac{\widetilde{\rho}_2^2(2\gamma)}{(\gamma + 1)} + \frac{\widetilde{\rho}_2^3(\gamma - 1)}{2(\gamma + 1)}. \tag{8.73}$$

One can solve (8.73) for $\widetilde{\rho}_2$, obtaining a complicated expression that is a function of γ and $\widetilde{F}_{Ro}$ only.

The shock frame and heat-front frame are connected as stated above:

$$u_s + u_{1s} = u_{hf} + u_1, \tag{8.74}$$

and in addition the one must have only one pressure in the shocked matter, so $\widetilde{p}_1$ from (8.72) equals $p_1/(\rho_o a_2^3)$ with p_1 from (8.68). This enables one to connect the two sets of results just given. One can show that, in order to sustain $u_s > u_{hf}$, necessary for an expansion heat front, one must have $\rho_2 < 2\rho_o$. At this point in the calculation, one can find all the variables as functions of F_{Ro} and γ. Figure 8.26 shows the results of this calculation.

This concludes our discussion of the expansion heat front. We have focused on behavior relevant to the irradiation of solid surfaces by soft X-rays, which matters for many applications. Hatchett (1991) provides a more general discussion of the expansion heat front.

8.2.3   Photoionization Fronts

If an ionized medium remains strongly absorbing, then radiative energy proceeds diffusively through it and one has a Marshak wave or related phenomenon. However, the opacity of the medium may decrease strongly after ionization. This is particularly true of ionized hydrogen, which can no longer undergo bound-free transitions once it is ionized. An opacity decrease can also happen under some circumstances with other materials. In these cases, the incident radiation can stream through the ionized medium until it reaches the ionization front, where the ionization occurs over a comparatively short distance. This establishes a *photoionization front*. In

Fig. 8.27 The Horsehead Nebula. Located in an intense UV environment, this is an example of a structure through which an ionization front has passed. Credit: N.A. Sharp/National Optical Astronomy Observatory/Association of Universities for Research in Astronomy/National Science Foundation

the astrophysics literature, photoionization fronts are typically referred to just as ionization fronts. Here we use the longer word, to avoid confusion with shock fronts and diffusive heat fronts, both of which can often be described as ionization fronts, in the sense that they are fronts where the ionization changes.

Examples of photoionization fronts are common in astrophysics. They occur when bright stars emitting primarily in the UV irradiate molecular clouds containing mostly H_2. The Horsehead Nebula, shown in Fig. 8.27, is a structure that has developed, perhaps through hydrodynamic instabilities, in a molecular cloud through which an ionization front is passing.

8.2.3.1 The Kinematics of Photoionization Fronts

The speed of a photoionization front depends on the photon flux that creates it. Suppose the flux of ionizing photons is F_γ, in number per unit area per unit time. This flux can be determined from the properties of the photon source including its geometry. Suppose in turn that the initial number density of atoms or ions to be ionized further is $n_a [= \rho/(A m_p)]$ and that the local density of not-yet-ionized particles is n_o. The velocity of the front will from simple kinematics be the flux divided by the number density, $v_f = F_o/n_a$ where the initial photon flux is F_o. It is straightforward to write equations for the evolution of the front, which are

$$\frac{dF_\gamma}{dx} = -F_\gamma n_o \sigma_o \quad \text{and} \quad \frac{dn_o}{dt} = -F_\gamma n_o \sigma_o. \tag{8.75}$$

These equations can be placed in dimensionless form by taking $f = F_\gamma/F_o$, $z = x/\lambda$, $\eta = n_o/n_a$, and $y = v_f t/\lambda$, where $\lambda = 1/(n_a \sigma_o)$ is the mean-free path in the un-ionized medium. One can substitute for f to obtain the following integro-

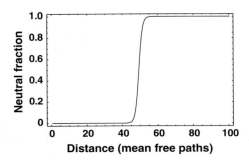

Fig. 8.28 Structure of an ionization front. Neutral particle fraction n_o/n_a vs. distance in particle mean-free paths, in an ionization front that has evolved for a time of $50\lambda/v_f$

differential equation for η:

$$\frac{d\eta}{dy} = -\eta e^{-\int_0^z \eta(z')dz'}. \qquad (8.76)$$

Figure 8.28 shows the resulting profile, from numerical integration of this equation, at a time of $50\ \lambda/v_f$. The full scale in the figure is $100\ \lambda$. One sees that the ionization front has moved very nearly $50\ \lambda$, so it moves at the expected velocity. One also sees that the width of the front (from 10% to 90% ionization) is less than 5 mean-free paths.

Since the speed of a photoionization front is determined by the photon flux, the front can be subsonic or supersonic relative to the sound speed in the upstream medium. It may also change from one of these to the other as the radiation source evolves in time. A supersonic front will encounter an unperturbed upstream medium. As subsonic front, in contrast, has the potential to produce very complicated behavior, for example, if it drives a shock wave that acts to reduce the opacity of the shocked material.

In the photoionization region around a star, the radiation flux and hence u_1 decrease with distance from the star, because of the spherical geometry. In addition, because the rate of recombination is finite, the radiation energy within a given solid angle also decreases. The radiation flux then gets driven to zero at a finite radius, known as the Strömgren radius.

8.2.3.2   Photoionization Fronts as Heat Fronts

In any regime of likely interest for front behavior, photoionization occurs via the photoelectric effect, so that the liberated electron retains the difference in energy between the ionizing photon and the ionization energy. In astrophysics, the radiation sources of interest are typically stars, emitting Planckian spectra corresponding to a temperature of a few eV. As a result, photoionization is driven by the high-energy tail of the photon distribution, except perhaps sometimes for H. The average energy provided to the liberated electrons corresponds to the integral of the product of the (energy dependent) photoionization cross section and the spectral photon flux, both of which drop rapidly above the ionization energy. The temperature of the liberated electrons is then of order the temperature of the Planckian, and often much less than the ionization energy. For this reason, photoionized plasmas in astrophysics are often described as "overionized". They are ionized far beyond the collisional ionization that would correspond to the electron temperature. In the laboratory, in contrast, the temperature of the radiation source likely exceeds the ionization energies, so that the plasma may well at first be underionized. In either case, though, the photoionization front is also a heat front, and we can analyze it accordingly.

The boundary conditions for photoionization heat fronts correspond to the kinematics of photoionization, discussed in the previous subsection. Simply put, the front uses up all the photons. Designating the flux of ionizing photons as F_γ, the front speed as v_f, the density of atoms to be ionized as n_a, and the degree of ionization achieved as ζ, we have

$$F_\gamma = n_a \zeta v_f. \tag{8.77}$$

Using the geometric conventions we established above for heat fronts, we thus have $u_1 = -v_f = -F_\gamma/(\zeta n_a)$. We include the ionization here to allow for the possibility of such fronts, in the laboratory, in gasses other than H.

To consider the photoionization front as a heat front, and in comparison with the astrophysical literature, we need to account for the incoming energy flux that produces ionization. For the simple case of photoionization of H, it can be convenient to subtract the ionization energy of the H, ϵ_H, from the mean energy of the electrons produced by ionization, ϵ_γ, so that the energy flux entering the fluid equations is $(\epsilon_\gamma - \epsilon_H)n_a u_1$. This allows one in principle to otherwise ignore the atomic physics and take $\gamma = 5/3$. This may not always be justified for low-temperature fronts that can produce significant radiation. For the laboratory case, it makes more sense to account for the full incoming, ionizing energy flux and to use a value of γ that includes the internal energy of ionization, consistent with our discussion in Chap. 3. Designating the mean energy of the ionizing photons as $\bar{E}_\gamma$, we can replace F_R in Eq. (8.56) as follows:

$$F_R \to \bar{E}_\gamma F_\gamma = \bar{E}_\gamma n_a \zeta v_f = -\frac{\bar{E}_\gamma \zeta}{\bar{M}} \rho_1 u_1 = -Q^2 \rho_1 u_1, \tag{8.78}$$

where the signs are necessary to support our intention that the energy deposited at the heat front be added to the energy of state 1 to produce state 2, the average mass per atom is $\bar{M}$, and Q is a variable with units of velocity that characterizes the energy deposition. It is significant that the energy flux as such is not important to the energy equation in this context, but rather the energy per atom deposited via the photoionization. Changing the energy flux changes u_1 but does not change the energy deposited per atom as the photoionization front progresses. This is an enormous qualitative difference, compared to the behavior of diffusive heat fronts.

The upstream isothermal sound speed, a_1, is independent of the front and makes a useful parameter for normalization of Eqs. (8.54)–(8.56). So we take $\rho_2/\rho_1 =$

μ, $a_2 = \hat{a}_2 a_1$, $u_1 = \hat{u}_1 a_1$, $u_2 = \hat{u}_1 a_1 / \mu$, $F_R = -Q^2 \rho_1 u_1$, and $Q = \hat{Q} a_1$. The normalized equations are

$$\mu + \hat{u}_1^2(\mu - 1) - \hat{a}_2^2 \mu^2 = 0, \text{ and} \tag{8.79}$$

$$\mu^2 \left((2\hat{Q}^2 + \hat{u}_1^2)(\gamma - 1) + 2\gamma \right) - \mu 2\gamma(\hat{u}_1^2 + 1) + \hat{u}_1^2(\gamma + 1) = 0. \tag{8.80}$$

Here the independent variables, set by the problem, are $\hat{u}_1$ and $\hat{Q}$, so we can solve Eq. (8.79) for $\hat{a}_2$, finding

$$\hat{a}_2 = \frac{\sqrt{\mu + \hat{u}_1^2(\mu - 1)}}{\mu}. \tag{8.81}$$

Meanwhile Eq. (8.80) can be solved to give two real, positive roots for μ, as follows

$$\mu = \frac{\rho_2}{\rho_1} = \frac{\gamma(\hat{u}_1^2 + 1) \pm \sqrt{(\hat{u}_1^2 - \gamma)^2 - 2\hat{Q}^2 \hat{u}_1^2(\gamma^2 - 1)}}{2\hat{Q}^2(\gamma - 1) + \hat{u}_1^2(\gamma - 1) + 2\gamma}. \tag{8.82}$$

Note that in these last two equations the density ratio depends only upon known quantities—the upstream Mach number of the heat front, $\hat{u}_1 = u_1/a_1$, γ, and $\hat{Q}$— and that knowing these quantities one can infer everything else.

It may be worth mentioning that some of the literature emphasises only the momentum relations or the energy relations, but solving both as is done here is necessary to find an un-ambiguous picture. Figure 8.29 shows the solutions for $\hat{Q} = 1$ and $\gamma = 4/3$. The labels show the standard jargon as defined by Kahn (1954) and Axford (1961). For finite radiation flux, there is a gap in u_1/a_1 between the two sets of roots. This gap vanishes as $\hat{F}_R$ decreases, until the two weak solutions for the density ratio join in a single line corresponding to no perturbation at $\rho_2 = \rho_1$. Perhaps this is why they are known as the "weak" fronts.

One can ask what will happen when some physical system produces a value of u_1/a_1 that lies in the gap seen in Fig. 8.29. The system will have to adjust to create a flow pattern that conserves mass, momentum, and energy. In this case what happens is as follows. The initial heating launches a shock wave, which heats and compresses the upstream matter. This increases the value of a_1 and decreases the value of u_1 corresponding to the radiative heat front, so that it now exists in the allowed, subsonic zone on these plots. The heat front will typically connect to an isothermal rarefaction, carrying the ablated matter off to the left.

We can say a bit more regarding the various solutions shown in Fig. 8.29. The Weak R-Type front is the classic, simple ionization front described in the previous subsection, in which photons stream through an ionized medium and are stopped over a short distance where they ionize the neutral particles. The Strong R-Type front is a modified shock wave; the effect of the radiation is to reduce the compression. This solution requires an incoming flow velocity to support the shock

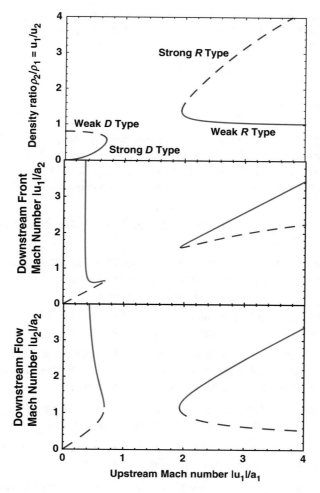

Fig. 8.29 Density ratio, sound speed ratio, and downstream internal Mach number are shown against the upstream Mach number for $\hat{Q} = 1$ and $\gamma = 4/3$. Labels show the standard identification of the roots

structure (or equivalently, a piston driving the flow from a distance). The Weak D-Type front represents a subsonic expansion with modest heating. This would require that the boundary conditions on the outflow support this as a steady-state solution. The Strong D-Type front is a supersonic rarefaction, in which the density drops by some amount at the front. In practice such behavior is often connected to an isothermal rarefaction at lower densities.

This concludes our discussion of the fundamental phenomena of radiation hydrodynamics. We have seen here that radiation hydrodynamic phenomena appear frequently in astrophysics and can readily be produced in the laboratory. We will see in the following that some radiation hydrodynamic effects are essential to the production of high-energy-density conditions and specifically to inertial confinement fusion.

Homework Problems

8.1 Demonstrate that the material energy flux coming into a radiative shock does not decrease when lateral losses decrease the upstream energy flux, by considering a system having a planar flow of material within a cylinder of some diameter and of finite length yet losing radiation both radially and axially, and integrating over the cylinder.

8.2 Derive (8.6), relating the radiation flux to the material properties, and discuss the origin and significance of each term.

8.3 Working with the Planck description of blackbody radiation, find and plot the fraction of photons that are ionizing as a function of temperature. You will need a computational mathematics program to generate the plot.

8.4 Evaluate the net radiation flux $(F_R - F_o)$ for an optically thin precursor using a radiative-transfer calculation similar to that done in (8.20) and (8.24).

8.5 Explore further the behavior of radiative shocks that are optically thin upstream and thick downstream. Beginning with (8.4)–(8.6), derive the final inverse compression (8.30) under the assumptions of the present section.

8.6 Determine whether the equation for energy flow in a radiative precursor, (8.19), admits a self-similar solution, assuming a diffusive model for F_R.

8.7 Examine the behavior of diffusive precursors. Solve (8.41) numerically, for several relevant values of n. Comment on the results.

8.8 We saw in Sect. 7.3 that Marshak waves are inherently unsteady, yet in Sect. 8.2.1 we analyzed them using steady jump conditions. Develop a condition for the validity of the use of steady jump conditions for Marshak waves, and determine when this is realistic.

8.9 Consider a truly radiation-dominated case, so p can be neglected in (8.50) and (8.51). Solve these equations for p_R and ρ. Find the dependence of the postshock T on the shock velocity, and compare it to the dependence of a non-radiative shock.

8.10 Express p and p_R as reasonable functions of T and solve (8.50) and (8.51) to find T and ρ in the postshock state. This may be a numerical solution, for which you should make reasonable choices about the parameters and show a few cases. Provide at least one graph based on these equations as part of the analysis.

References

Axford W (1961) Ionization fronts in interstellar gas: the structure of ionization fronts. Philos Trans R Soc Lond A 253:301–333

Hatchett SP (1991) Ablation gas dynamics of low-z materials illuminated by soft X-rays. Tech. Rep. UCRL-JC-108348, Lawrence Livermore National Laboratory

Kahn F (1954) The acceleration of interstellar clouds. Bull Astron Inst Neth XII:187–200

Mihalas D, Weibel-Mihalas B (1984) Foundations of radiation hydrodynamics, vol 1, Dover (1999) edn. Oxford University Press, Oxford

Shu FH (1992) The physics of astrophysics: radiation, vol I. University Science Books, Mill Valley, CA

Zel'dovich YB, Razier YP (1966) Physics of shock waves and high-temperature hydrodynamic phenomena, vol 1, 2002nd edn. Dover, New York

Chapter 9
Creating High-Energy-Density Conditions

Abstract This chapter discusses a sequence of physical mechanisms and practical issues associated with the production of high-energy-conditions, primarily using lasers. After a brief introduction to the lasers themselves, the first topic is the effect of the laser beams on the plasma and of the plasma on the laser beams. This includes absorption, refraction, reflection, laser scattering, and laser–plasma instabilities. The next topic is the transfer of energy by electron heat conduction, which enables a discussion of the laser heating of the plasma, of the consequent ablation of dense matter, and of the eventual resulting, rocket-like acceleration of the target. This is followed by a consideration of the dynamics of mid-Z and high-Z targets, in which radiation plays a larger role, and specifically of the use of gold targets as an X-ray source. The final topic is hohlraums, including discussions of the soft-X-ray energy fluxes they generate, the ablation of low-Z matter by such X-rays, and the problems that experiments using hohlraums encounter.

Nature often creates high-energy-density conditions. At root, the cause is always gravity. At the center of the Earth, for example, the pressure is 3.6 Mbar, almost entirely due to gravity. Jupiter is similar, with a pressure of $\sim$80 Mbar at its core. In stars, the gravitational assembly of the stellar mass leads to heating by nuclear fusion, which produces much larger pressures—the pressure at the core of the Sun is roughly 0.2 terabars. Once fusion creates conditions that lead to supernova explosions, even larger pressures occur. For example, some supernovae produce neutron stars, and the magnetic field at the surface of a typical neutron star is near 1 teraGauss. The pressure of such a magnetic field is about 40 petabars.

Nature also creates conditions whose laboratory analog must be in the high-energy-density regime. One can consider, for example, shock waves that are fast enough to ionize matter and perhaps to cause radiative effects. A 1000 km/s shock wave in the interstellar medium, where the density of particles is of order one per cubic centimeter, has a ram pressure of tens of femtobars. This pressure approaches nanobars when the density becomes large enough that radiative losses become important. In contrast, a laboratory system that is a good analog of this astrophysical shock wave might involve a shock wave at 10 km/s in a material that is $\sim$1 g/cc, for which the ram pressure is of order 1 Mbar.

In this chapter we take up the problem of producing high-energy-density conditions. It would have been convenient to take this up much sooner, but in fact many of the concepts we have already introduced are needed to understand how to do this. We will discuss the technology that makes this possible in the early twenty-first century. We will also discuss the conceptual, physical, and mathematical models that are necessary to understand how such conditions are produced. These topics subdivide naturally into five areas: direct laser irradiation, laser-driven hohlraums, ultrafast lasers, high-energy-density beams, and Z-pinches. The first two of these are covered here; the next two are covered in Chap. 13. The next chapter addresses Z-pinches and related systems.

9.1 Direct Laser Irradiation

All *lasers*, from laser pointers to megaJoule systems, have certain features in common. They involve the preparation of a medium that can reach a lower-energy state by giving energy to a light wave. This often involves the excitation of a specific atom, such as neodymium (Nd), so that more electrons are present in the upper state than in the lower state of an atomic transition. All lasers also involve the initiation of a light wave within such a system, sometimes by thermal emission of radiation and sometimes from an external source. This light wave is then amplified coherently, as the medium gives energy to it. The resulting beam of light may be well collimated, but this depends in part on the geometry of the lasing system. In the present book we will not discuss these aspects of lasers in further detail, leaving this subject to other books. Instead, we will discuss the aspects of lasers that are specific to high-energy-density physics—high energies in this section and high powers in Chap. 13.

Before discussing the specifics of lasers, some discussion of the relevant units in common use is a good idea. The common units, also used here, are somewhat mixed. The energy of a laser pulse is typically given in Joules (J), or in related units such as kJ or MJ. Correspondingly, and considering that the timescale of the pulses is ns, the power is given in W, gigawatts (GW), terawatts (TW), or petawatts (PW). However, the practical unit of distance for real lasers is the cm, not the meter, so the power per unit area is typically given in W/cm^2, much to the horror of SI units purists. Perhaps more important is that the high-energy-density community has a habit of describing this power per unit area as an intensity, so one will see for example in the literature that the laser intensity in our experiment was 10^{14} W/cm^2. This spawns confusion across disciplines, because the word "intensity" is used to mean various physical quantities in various disciplines. It horrifies the conceptual purists in optics, as the general meaning of intensity involves power per unit area per unit solid angle (as in Chap. 6). The correct term in optics for power per unit area is irradiance. Alternatively, these units in simple language represent an energy flux, and there is a lot to be said for using such basic terms in order to improve communication and clarity, so we will use this terminology here.

9.1.1 Laser Technology

Now we turn to the specific issues involved in the lasers that produce high-energy-density conditions. A number of high-energy lasers have been constructed over the past few decades, motivated by the challenge of achieving inertial confinement fusion. The long-term goal is to create miniature fusion explosions with an energy gain of about 100. Such a laser system begins with a very high-quality laser beam, initially of low energy, which propagates through and extracts energy from Nd-doped laser glass. (We do not discuss here lasers based on gasses such as CO_2 or KrF. No high-energy CO_2 lasers remain in operation. The KrF lasers represent an important alternative to Nd-glass lasers for high-energy-density physics.) The first challenge for glass-based lasers is to prevent defects and diffraction from damaging the laser components as the energy per unit area of the laser beam reaches high levels. The second is to extract a large fraction of the stored energy.

The first challenge was met in the 1970s. The key inventions here were *image relaying* and *spatial filters*. A laser system that uses image relaying first creates a high-quality, low-energy laser beam at a specific position (the object plane). It then designs the optics in the laser system so that the object plane is imaged onto or near the planes where the highest-irradiance laser light penetrates optical materials. These locations are often the lenses that inject the light into spatial filters. Spatial filters are also essential to reduce the structure in a laser beam. They do so by placing a pinhole at the focus of an input lens. This clips most of the energy in hot spots or other structures in the incident laser beam, as these small structures are focused to a much larger spot than the uniform beam is. The output lens then recollimates the remaining laser light into a smoother, more-uniform beam. These inventions led to several high-energy lasers capable of delivering >1 kJ of laser energy to a target.

Figure 9.1 shows one example of such a laser system, the Omega facility (Boehly et al. 1995), which can deliver 30 kJ to a target. The laser occupies approximately the area of one (American) football field. The capacitors in the basement accumulate

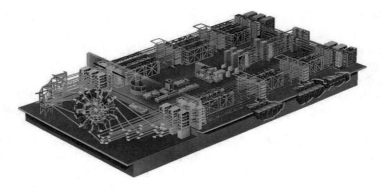

Fig. 9.1 A drawing of the Omega laser system. Credit: Laboratory for Laser Energetics

energy for several minutes before delivering it to flashlamps in the laser amplifiers, preparing the Nd glass to amplify light. The initial laser beam, formed and amplified at the center of the laser bay, is split, amplified further, and eventually feeds the 60 amplifier chains that proceed down the sides of the laser bay toward the output end. Frequency conversion crystals then triple the frequency of these laser beams, decreasing their wavelength from 1.05 to 0.35 μm. Mirrors then direct the laser beams toward the center of the target chamber.

The second challenge cited above, of using the stored energy more efficiently, was met by large lasers constructed in the early twenty-first century. For the laser beam to extract more of the stored energy it must pass through the Nd-doped glass several times, without destroying the quality or the focusability of the laser beam. To accomplish this it is necessary to clean up the laser beam between passes, using spatial filters, and/or to compensate for the phase differences across the amplifying optics, using *adaptive optics*. (An adaptive optic deforms an optical surface either continually or in small segments, allowing local adjustment of the distance that the light travels.)

9.1.2   Laser Focusing

It would seem that a simple lens would be sufficient to focus a high-energy beam to high energy flux for experiments, just as a simple magnifier can focus sunlight to start a fire. Unfortunately, a typical laser beam, especially when focused to a spot that is larger than the smallest (or best focus) spot that a lens can produce, creates a very irregular spot. This has a variety of adverse consequences, some of which we will touch on later. Inertial fusion, for example, requires the irradiation of a target by a very smooth laser beam. Small lasers often use a *Gaussian beam* to produce a high-quality laser spot. A Gaussian beam has a profile of energy flux that is approximately Gaussian as a function of radius (proportional to $\exp(-r^2/a^2)$, where r is radius and a is a distance). Such a beam can be image-relayed through an optical system to maintain high quality. This type of laser is comparatively inefficient, however, as most of the beam is at low intensity and does not extract much of the stored energy from the laser glass. High-energy beams must extract as much as possible of this stored energy, and thus must use much flatter energy flux profiles. Unfortunately, thorough studies proved that no practical optics could produce laser beams with flat energy flux profiles whose phase fronts were uniform enough that they would focus to smooth spots without some sort of extra processing. This has led to the invention of a number of techniques for so-called beam-smoothing. We discuss some of these here.

These techniques typically rely on the diffractive behavior of laser optics. When a light wave passes through a circular aperture, diffraction of the light by the aperture is well known to produce an Airy pattern (see Fig. 9.2). If such a light wave is focused, then diameter, d, of the first zero of the Airy pattern is given by

$$d = 2.44\lambda f, \tag{9.1}$$

Fig. 9.2 An Airy pattern

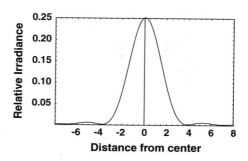

in which the wavelength of the light is λ and the *f number*, the ratio of length to aperture, of the focusing system is f. The central maximum of the Airy pattern contains about 88% of the energy in the light wave. Thus, for example, for a 30-cm dia. lens with a 3-m focal length, $f = 10$ so for 0.35 μm light $d = 8.5$ μm. This is also called the diffraction-limited spot size of the laser system. Typical high-energy laser systems, using only a focusing lens, produce best-focus spots that are larger than ten times the diffraction-limited spot size. The size of these best-focus spots and their structure is due to the gradual variation of the phase of the light across the aperture of the lens, and to the interference of the beam from different portions of the aperture.

The simplest of the beam-smoothing systems, no longer in much use, is the *random phase plate* (RPP). A random phase plate passes the laser beam through an array of hundreds or thousands of adjacent optical elements, of randomly varying thickness, thus dividing the beam into small beamlets. The elements are typically hexagonal in shape. The elements are sized so that the diffraction-limited spot of each element is the size of the desired laser spot. Thus, to obtain a 1 mm spot with 0.35-μm light at a distance of 3 m, (9.1) implies the aperture of an element must be 2.6 mm. This determines the overall size of the laser spot. (In actual experiments, one is most often concerned with the size of the spot that contains half the laser energy, or at the edges of which the energy flux is half its peak value. These numbers are somewhat smaller than the value from (9.1) and for real systems must be determined numerically or experimentally.) In addition, the beamlets from different elements interfere with one another, typically producing small, local maxima in the energy flux pattern, known as *speckles*. The minimum speckle diameter is produced by interfering beamlets that originate from opposing edges of the laser beam, and is given by (9.1) using the aperture of the entire laser beam to determine the f number. These speckles are actually very long and narrow structures. Their length, L_s, is

$$L_s = 7\lambda_a f^2, \tag{9.2}$$

so the ratio of length to width is roughly $3f$.

What has largely replaced the random phase plate in practice is the *distributed phase plate*, or DPP, in which the phase of each small element is controlled by

Fig. 9.3 Pattern of energy
flux from a distributed phase
plate. The graph shows a line
through the image, shown in
the inset. Credit: Laboratory
for Laser Energetics

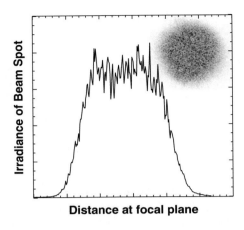

design in order to determine the shape and structure of the resulting laser spot. In
particular, this allows one to produce laser beams with flatter overall profiles of
energy flux and with less energy in the wings of the laser spot. Figure 9.3 shows
the typical energy flux pattern produced by a distributed phase plate. One sees a
smooth overall profile modulated by many speckles. A related type of optic, often
used in combination with a DPP, is a *distributed polarization rotator* or DPR. A
DPR uses birefringent optical elements to rotate the polarization of each beamlet
by a controlled amount. Since only the components of two beamlets with parallel
polarizations interfere with one another, this provides a further dimension of control
when designing the shape of a laser spot. It also in principle allows one to tailor the
polarization of the laser beamlets as they interact with the target.

Plasmas conduct heat easily, which will tend to smooth the effect of small-scale
spikes like those seen in Fig. 9.3. This smoothing, combined with the tendency of
shock waves to anneal as discussed in Chap. 5, implies that for some experiments
the use of an RPP or DPP is sufficient to obtain high-quality results. However,
for inertial fusion, at least in cases where the laser directly illuminates the fusion
capsule, such smoothing techniques are not sufficient. The fixed location of each
speckle creates lasting effects at a high spatial frequency. An improvement on this
is to cause the speckles to move around, so that the profile of time-averaged energy
flux across the overall laser spot is very smooth. There are two approaches to this.

The first approach is *induced spatial incoherence*, or ISI. One again breaks the
laser beam into beamlets, but now one arranges that the difference in optical path
length between the beamlets exceeds the distance over which the laser is coherent.
This is only feasible using a comparatively broadband laser. The result is that the
phase difference between beamlets varies in time, causing the speckles to move
around on a timescale comparable to the coherence time of the laser. ISI was
demonstrated on Nd-glass lasers, and is particularly well suited for implementation
in KrF lasers. where it can be integrated into the laser design.

The second approach is *smoothing by spectral dispersion* (SSD), which has
proven to be more practical for large glass lasers. In this approach one produces

a broadband laser pulse, disperses it in angle using a diffraction grating, and then collimates it to produce a laser beam whose frequency varies in the direction that was dispersed. One can also use two gratings (or complicated optics) to disperse the beam in two directions, producing 2D SSD. When such a beam is focused through a distributed phase plate, the phase difference between beamlets varies in time because the beamlets have different frequencies so their individual phases vary at different rates. This again causes the speckles to move around with a timescale determined by how large the differences in frequency are.

9.1.3 Propagation and Absorption of Electromagnetic Waves

Now that we have some idea how to irradiate a target with a high-energy laser beam that has been smoothed, we are ready to examine what happens when such a laser beam actually strikes a target. At the energy fluxes of interest, which are typically 10^{12}–10^{16} W/cm^2, the laser light immediately produces a plasma at the surface of the target. At the higher energy fluxes in this range, the electric field of the laser light is sufficient to directly ionize the atoms. At the lower energy fluxes, the process is more complicated but nonetheless a plasma is quickly produced. Figure 9.4 illustrates the three fundamental processes that occur when laser light penetrates a plasma. The laser light is refracted as it propagates through the density gradient in the plasma, it is reflected when it reaches a high enough density, and it is absorbed along its entire path of propagation. We will analyze these processes by examining the fundamental behavior of light in plasma.

To understand the behavior of laser light as it penetrates a plasma, we begin with Ampere's law, (2.18), in Gaussian cgs units, and use the standard vector and scalar potentials in the Coulomb gauge. This gives

$$\nabla \times \nabla \times \boldsymbol{A} = \frac{-1}{c^2} \frac{\partial^2 \boldsymbol{A}}{\partial t^2} + \frac{4\pi \boldsymbol{J}}{c} - \frac{1}{c} \frac{\partial \nabla \Phi}{\partial t}. \tag{9.3}$$

Fig. 9.4 The simple processes that occur when a light wave enters a plasma

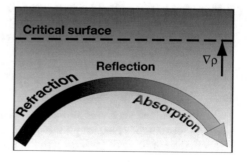

This equation has two parts, and it turns out that they separate completely. Any vector can be decomposed into a *transverse* (or rotational) part, the divergence of which is zero, and a *longitudinal* (or compressive, or irrotational) part, the curl of which is zero. By definition, A is purely transverse. Also, any gradient has zero curl, so $\nabla\Phi$ is purely longitudinal. Taking the divergence of (9.3) yields a continuity equation for the charge in the plasma, in which the variation with time charge density (proportional to $\nabla^2\Phi$ via the Poisson equation) is balanced by the divergence of a flux of charge (i.e., a current).

One can expand the left-hand of (9.3) and then subtract the longitudinal terms from the equation, to obtain a fundamental equation for light wave propagation:

$$\frac{\partial^2 A}{\partial t^2} - c^2\nabla^2 A = 4\pi c J_t, \tag{9.4}$$

in which J_t refers to the transverse part of J. (One can construct J_t from J if needed. This is discussed for example in Jackson (1999).)

For any specific plasma environment, the behavior of light waves is thus determined by J_t. It is also true that the frequencies of the lasers of interest are so large that the ion motion is negligible on the timescale of the laser propagation. (Ion motion can have important consequences on longer timescales, some of which are discussed in Sect. 9.1.4.) The net current carried by the electrons is $-en_e u_e$; we are seeking the transverse part of this current.

To find this transverse current, we work with the continuity and momentum equations for the electron fluid, which are (2.42) and (2.43). The momentum equation, written again here for electrons,

$$m_e n_e \frac{\partial u_e}{\partial t} + m_e n_e u_e \cdot \nabla u_e = -n_e e\left(E + \frac{u_e}{c}\times B\right) - \nabla p_e - \sum_{j=\text{ions}} m_e n_j (u_e - u_j)\nu_{ej}, \tag{9.5}$$

is key. In applying this equation here we can make several observations and simplifications. The velocity of any ions is negligible and can be neglected. In addition, the pressure gradient is an inherently longitudinal vector, so we can drop it as we are seeking the transverse velocity. We can also ignore $\nabla\Phi$ after we again use the scalar and vector potentials. Finally, we can divide each term by $n_e m_e$. After these adjustments, we have an equation for the transverse electron velocity,

$$\frac{\partial u_e}{\partial t} + u_e \cdot \nabla u_e = \frac{-e}{m_e}\left(-\frac{1}{c}\frac{\partial A}{\partial t} + \frac{u_e}{c}\times\nabla\times A\right) - u_e\nu_{ei}. \tag{9.6}$$

From standard vector identities, $u_e \cdot \nabla u_e = -u_e \times \nabla \times u_e + \nabla u_e^2/2$, but the gradient term here is also longitudinal and can also be dropped. (This term will play a role in the coupling of laser light to longitudinal waves in the plasma.) Substituting and rearranging, we have

$$\left(\frac{\partial}{\partial t} + \nu_{ei}\right)u_e - u_e\times\nabla\times u_e = \left(\frac{\partial v_{os}}{\partial t} - u_e\times\nabla\times v_{os}\right), \tag{9.7}$$

where we have defined $v_{os} = eA/(m_e c)$, which we will identify as the oscillating velocity of the electron within the light wave. One can see from (9.7) that in the absence of any collisional energy loss to the ions, one would have precisely $u_e = v_{os}$. The presence of v_{ei} introduces spatial damping of the electromagnetic wave and a phase variation between u_e and v_{os}. There are typically two simplifying aspects. The first, dealt with below, is that the spatial scale of the variation in A is large compared to the wavelength of the light. The second is that u_e is a small fraction of the speed of light. (This second is true for typical high-energy lasers, having pulses of order a ns, but not for sufficiently intense, high-power lasers, discussed in Chap. 13. To treat the relativistic regime one must use modified equations.) So one can divide (9.7) by c^2 and note that the terms involving the curl are much smaller than the other terms because $|\partial/\partial(ct)| \sim \omega/c \sim k \sim |\nabla|$. This justifies the use of linearization and the dropping of the terms involving the curl. Then, assuming that A and thus u_e vary as $e^{-i\omega_o t}$, we find

$$u_e = v_{os}\frac{1}{1 + iv_{ei}/\omega_o}. \qquad (9.8)$$

Here the imaginary term produces the phase shift mentioned above, but this is small so long as $v_{ei} \ll \omega_o$. Also note that u_e is purely transverse, because A and E for an electromagnetic wave are purely transverse.

It will be helpful for applications to connect v_{os} and v_{ei} with practical units. The direction of v_{os} is the direction of the electric field of the laser, typically described as the direction of polarization of the laser. The magnitude, v_{os}, can be related to the energy flux of the laser in vacuum, I_L, as follows. Because the energy density of the electromagnetic field is $E^2/(8\pi)$, and because this propagates with a group velocity in vacuum of c, one has $I_L = cE^2/(8\pi)$. The magnitude of this is $I_L = \omega_o^2 A^2/(8\pi c) = v_{os}^2 \omega_o^2 m_e^2 c/(8\pi e^2)$, from which

$$v_{os}/c = \sqrt{I_{14}\lambda_\mu^2/117}, \qquad (9.9)$$

in which I_{14} is I_L in units of 10^{14} W/cm^2 and λ_μ is the wavelength of the light in μm. We discussed v_{ei} in Chap. 2. In practical units it is

$$v_{ei} = 3 \times 10^{-6} \ln\Lambda \frac{n_e Z}{T_e^{3/2}}, \qquad (9.10)$$

with n_e in cm^{-3}, T_e in eV, and where $\ln\Lambda = \max\left(1, [24 - \ln(\sqrt{n_e}/T_e)]\right)$.

We then take $J_t = -en_e v_{os}/(1 + iv_{ei}/\omega_o)$ in (9.4). This is completely accurate if n_e is constant. (However, if a variation in n_e is designated as n_{e1}, then the part of J_t proportional to $n_{e1}v_{os}$ may have transverse and longitudinal components, depending on the direction of the gradient in n_e. This detail matters for wave coupling calculations, but we ignore it here.) Substituting into (9.4), simplifying, and rearranging, we obtain

$$\left(\frac{\partial^2}{\partial t^2} + \frac{\omega_{pe}^2}{(1 + i\nu_{ei}/\omega_o)} - c^2\nabla^2 \right) A = 0, \tag{9.11}$$

in which ω_{pe} is the plasma frequency defined in Chap. 2. This is the wave equation for a damped electromagnetic wave in a plasma.

It is worthwhile to examine and discuss the dispersion relation implied by (9.11), which is

$$\omega_o^2 - \frac{\omega_{pe}^2}{\left[1 + (\nu_{ei}/\omega_o)^2 \right]} - c^2 k^2 + i \frac{\nu_{ei}/\omega_o}{\left[1 + (\nu_{ei}/\omega_o)^2 \right]} \omega_{pe}^2 = 0, \tag{9.12}$$

in which k is the wavenumber of the light wave. The imaginary term here introduces an imaginary component to the phase of the wave, which may be expressed as an imaginary part of either ω_o or k. It causes damping as the wave propagates. We will examine this damping below. The real component of (9.12) describes the propagation of the wave. The reader may recall that a propagating light wave traveling through a stationary, unchanging medium experiences variations in k and not in ω_o. Furthermore, when k is driven to zero the wave cannot propagate further and must be reflected and/or absorbed. Equation (9.12) implies that k is driven to zero approximately when $\omega_o = \omega_{pe}$. Physically, when $\omega_o = \omega_{pe}$ the electrons resonantly oscillate at the frequency of the light wave, creating a reflecting surface like a mirror. This surface is known as the *critical surface* and the density there is the *critical density*, n_c. From $\omega_o = \omega_{pe}$ one can show that

$$n_c(\text{cm}^{-3}) = 1.1 \times 10^{21}/\lambda_\mu^2. \tag{9.13}$$

Thus, for visible and UV lasers, $n_c < 10^{23} \text{ cm}^{-3}$. If one now looks again at Fig. 9.5, one can draw some implications for the absorption of light in plasmas. The critical density for visible and UV lasers is typically small enough that $\nu_{ei} < \omega_{pe}$ there. Such laser beams propagate through the plasma and reflect, perhaps having been substantially absorbed in the process. In contrast, the critical density for X-rays is typically above any density present in the plasma. Thus, in the absence of

Fig. 9.5 Refraction at an interface, with solid lines showing phase fronts

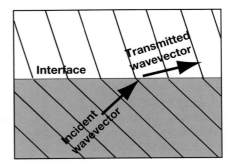

absorption the X-rays would penetrate freely through the target. But absorption can be strong: ν_{ei} becomes quite large at densities near or above solid density. The result is that soft X-rays, with energies below roughly 1 keV, are very strongly absorbed by collisions. Harder X-rays usually are not strongly absorbed by collisions, but the atomic absorption of these X-rays can be significant as discussed in Chap. 6.

Let us focus now on the *absorption of laser light* and assume that $(\nu_{ei}/\omega_o)^2$ is small enough to be ignored. We are interested in the spatial rate of absorption, so we assume ω_o to be real and take $k = k_r + i\kappa_{EM}/2$, with real and imaginary parts k_r and $\kappa_{EM}/2$, respectively. Here κ_{EM} is the spatial rate of absorption of the laser energy, proportional to E^2, and $\kappa_{EM}/2$ is the spatial rate of change of E (or A). We then solve (9.12) for k_r and κ_{EM}, ignoring the term involving κ_{EM}^2 subject to verifying our assumption that the light is absorbed slowly as it propagates (i.e., $\kappa_{EM} \ll k$). We obtain

$$k_r = (\omega_o/c)\sqrt{1 - n_e/n_c} \tag{9.14}$$

and

$$\kappa_{EM} = \nu_{ei}\frac{\omega_{pe}^2}{\omega_o^2}\frac{1}{c\sqrt{1 - n_e/n_c}} = \frac{\nu_{ei}}{\nu_g}\frac{n_e}{n_c}, \tag{9.15}$$

in which ν_g is the group velocity of the light wave in the plasma. Equation (9.15) is easy to understand—only that fraction of the energy in the light wave that participates in electron oscillations, n_e/n_c, can be affected by electron–ion collisions, and the spatial rate at which this effect occurs is the temporal collision rate divided by the rate at which energy propagates in space (i.e., the group velocity). Also note that our previous assumption that $\nu_{ei} \ll \omega_o$ assures that $\kappa_{EM} \ll k$.

The above discussion implies that the fraction of the incident laser light transmitted through a uniform plasma of length D is $\exp[-\kappa_{EM}D]$. Kruer (1988) shows how to determine the absorption in more complicated circumstances. Two of his results are worth quoting here. Let the electron-ion collision rate at the critical density be ν_{ei}^*, let z be a spatial variable and let L be a scale length. Then, for a laser beam normally incident on a plasma with a linear density profile, so $n_e = n_c z/L$, the absorption f_A is

$$f_A = 1 - \exp\left(\frac{-8\nu_{ei}^*L}{3c}\right), \tag{9.16}$$

while for a laser beam incident at an angle θ from the normal on a plasma with an exponential density profile, so $n_e = n_o e^{-z/L}$, the absorption is

$$f_A = 1 - \exp\left(\frac{-8\nu_{ei}^*L}{3c}\cos^3\theta\right). \tag{9.17}$$

The third important process that occurs during penetration of a plasma by laser light is *refraction*. Refraction refers to the bending of rays of light as they propagate. The concepts of light rays and refraction are valid in the geometric-optics limit, when variations in the medium occur on scales large compared to the wavelength of the light. In high-energy-density systems, refraction is a sensible concept when light, incident at some angle from normal incidence, penetrates a plasma that has been expanding from an initial solid surface for 100 ps or more. In contrast, if the light penetrates to the critical density or if the plasma is only a few wavelengths in extent, one must analyze the light as a wave. Kruer (1988) provides a discussion of laser light reflection at the critical surface, where some energy can be absorbed by *resonance absorption*.

Continuing with the discussion of refraction, the variation of k_r in a plasma is given by (9.14). The variation in the wave vector $\mathbf{k}$ is given by the equations of ray propagation (see Landau and Lifshitz 1987) as

$$\frac{d\mathbf{k}}{dt} = -\nabla\omega, \tag{9.18}$$

which is the analogy in geometric optics of the relation between the rate of change of momentum and the gradient of the Hamiltonian in mechanics. It is sensible to identify the components both parallel and perpendicular to the density gradient as $k_{||}$ and $k_{\perp}$, respectively. Thus, the component perpendicular to the density gradient does not change. Intuitively, this may be easiest to see by recalling the refractive behavior at a sharp interface, illustrated in Fig. 9.5. The phase velocity (ω/k) of the wave changes at the interface, implying that the distance between the phase fronts must change. However, the distance between the phase fronts along the interface is determined by the incident wave. As a result the component of the phase velocity (and $\mathbf{k}$) that is perpendicular to the interface is what changes. One can view propagation up a gradient as the limit of propagation up a series of steps as the number of steps becomes large and the step size becomes small.

Also, because the boundary condition is that the fields in the wave must vary continuously across the boundary of the plasma, one has $k_{\perp} = \sin(\theta)\omega_o/c$, in which θ is the angle of incidence, measured with respect to a normal vector. This implies

$$k_{||} = (\omega_o/c)\sqrt{\cos^2\theta - n_e/n_c}. \tag{9.19}$$

This equation has an obvious interpretation: an obliquely incident light wave (in a planar plasma) reflects at a density such that $n_e/n_c = \cos^2\theta$. Thus, when the angle of incidence is 45°, the laser light is reflected at roughly $n_c/2$, and at 60° this decreases to $n_c/4$.

9.1.4   Laser Scattering and Laser–Plasma Instabilities

Most of the applications in high-energy-density physics would be simpler and easier if laser beams did no more than propagate, refract, and absorb in plasmas. Even laser scattering from small fluctuations, discussed just below, would not disturb these applications. From this point of view, it is unfortunate that the underdense plasma produced by the laser is host to a variety of waves, and that these waves can couple unstably to the laser light wave. The unstable waves can become large and can have large effects, scattering large amounts of laser light, producing substantial populations of energetic electrons or ions, and even causing modulations in the laser absorption dynamics. From the point of view of such dynamics, *laser–plasma interactions* (LPI) is a tremendously exciting field. It even has a few astrophysical applications, relating for example to the dynamics in certain solar bursts and in the turbulence within stellar winds. One such application—the scattering of pulsar radiation by the plasma of a binary companion—might at times occur at a high energy density. But LPI is not our primary topic here. Our goal here is to cover as much of LPI as our reader should know to work intelligently in high-energy-density physics. Those interested in more details should start with the book on LPI by Kruer (1988).

We will now discuss the dynamics involved in *laser scattering* from density fluctuations in the plasma. We will consider only uniform plasmas. This is an oversimplification, as most scattering and instabilities actually occur in plasmas that are nonuniform. Here we ignore nonuniformity because it introduces complications without introducing many new ideas. Collisions may at times be important but we will ignore them too for the same reason. Thus, we take $A = A_L + A_s$, in which s refers to a scattered light wave, and we take $n_e = n_{eo} + \delta n_p$. We will use L, s, and p as subscripts for the laser light wave, the scattered light wave, and the wave in the plasma, respectively. It is worth noting that the laser light interacts primarily with the electrons, because the ions move so much more slowly and thus carry far less current. However, the electron-density fluctuation δn_p may be produced by any wave in the plasma. On the one hand, it could be produced by an electron–plasma wave in which the ion density is effectively fixed. On the other hand, it could be produced by an acoustic wave in which the ion density fluctuates and the electrons are forced by the ion-charge variations to move with the ions. We also assume that proper normalization would show that terms involving A_L alone are large in comparison with those involving only A_s or δn_p. We then substitute into (9.11) and linearize. We find that we can drop the terms involving only A_L because they cancel one another, so we obtain

$$\left(\frac{\partial^2}{\partial t^2} + \omega_{pe}^2 - c^2 \nabla^2 \right) A_s = -\omega_{pe}^2 \frac{\delta n_p}{n_{eo}} A_L. \qquad (9.20)$$

With reference to (9.4), the interpretation of this equation is simple. The interaction of the laser light with the density fluctuation produces an additional transverse current proportional to the right-hand side of (9.20). Another aspect of

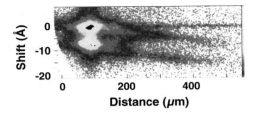

Fig. 9.6 This spectral image from Thomson scattering in the collective regime, with wavelength varying in the vertical direction shows two peaks due to oppositely propagating acoustic waves. The feature near zero shows the wavelength of the probe laser. The colors on this image cycle from white to black twice as the intensity of the signal increases. Adapted from Montgomery et al. (1999)

(9.20) is that it is fundamentally an equation describing wave beating. The laser light wave and the electron density fluctuation beat together to drive scattered light waves. The reader should also recall, from elementary physics, that this will produce beat waves having both the sum and the difference of the frequencies and wavenumbers of the two driven waves. One may wonder where the energy in the scattered-light wave comes from. It is obtained from the laser light wave through a second-order term that slowly reduces A_L.

Indeed, when scattering measurements are used to characterize the fluctuations in the plasma, a technique known as *Thomson scattering*, one sees both of the resulting scattered-light waves. Figure 9.6 shows an example of such data. This figure is a gated imaging spectrum, meaning: (a) the measurement was limited to a brief period (it was *gated*); (b) the instrumentation provides an image along a line through the object, horizontal in the figure; and (c) the signal from this line was resolved spectrally in the orthogonal (vertical) direction. Thus, a vertical cut through the image shown in the figure provides a spectrum at that specific location. The plasma motion at any location shifts the entire spectrum, by the Doppler effect, providing a measurement of the local velocity. One can see that this average frequency shift increases with distance in the figure, which implies that the plasma flow velocity is increasing with distance. This is sensible because this image is from a plasma expanding from a laser-irradiated surface, as an isothermal rarefaction. The fact that the rarefaction is isothermal is confirmed by the constant spacing between the two peaks in the spectrum. The spacing is proportional to the frequency of the acoustic waves causing the thermal density fluctuation; thus it measures the sound speed in the plasma, which depends mainly on the electron temperature. The weakening of the signal from left to right in the image is due to decreasing plasma density in the rarefaction. Studies of the ionosphere employ similar Thomson-scattering techniques, also for the purpose of diagnosing density and temperature. Both the laser–plasma and the ionospheric applications take place in what is known as the collective regime of Thomson scattering, to which (9.20) applies. In contrast, the use of Thomson scattering in magnetic-fusion plasmas takes place in the single-particle regime, in which the scattering is effectively from individual particles. The different Doppler shifts of these particles produces a scattered-light spectrum from which temperature can be inferred. Sheffield et al. (2010) treat the fundamental theory of Thomson scattering.

We now turn to cases in which one of the scattered-light waves can participate in an instability, producing much stronger scattering. We consider explicitly the case of *stimulated Raman scattering* (SRS). This process involves scattered-light waves, which we have just discussed, and electron–plasma waves, discussed in Chap. 2. To develop an initial understanding, SRS and the other instabilities are most profitably described in uniform plasmas irradiated by laser pulses that are not depleted by the action of the instability. In the simplest form of these instabilities, only the difference frequency participates, and we will selectively include here only those terms in our discussion here.

Before proceeding with this selective analysis, some words on notation are needed. All the complex quantities in any derivation must be related to real quantities by some convention, such as $\mathbb{R}(A_L) = (A_L + A_L^*)/2$, in which $*$ designates the complex conjugate. We will assume that the amplitude is real (an approximation that retains the essential physics) and will use a caret ^ to designate the amplitude of each wave, so that for example $A_s = \hat{A}_s e^{i(k_s \cdot x - \omega_s t)}$. When doing theory with no nonlinear terms, we might typically write the time-and-space dependence of the scattered light as $e^{i(k_s \cdot x - \omega_s t)}$. When working accurately with equations that have nonlinear terms, one would have to write $A_s = (\hat{A}_s e^{i(k_s \cdot x - \omega_s t)} + \hat{A}_s^* e^{-i(k_s \cdot x - \omega_s t)})/2$. The interaction of the two real physical quantities in the nonlinear terms involves cross terms, and these are essential to accounting for the beat waves.

For a cursory analysis we will work with selected terms. Consider first the beat term in which the phase of the light waves varies as $e^{i(k_s \cdot x - \omega_s t)}$ and that the phase of the plasma wave varies as $e^{-i(k_p \cdot x - \omega_p t)}$. One then has from (9.20)

$$
\left(-\omega_s^2 + \omega_{pe}^2 + c^2 k_s^2\right)\hat{A}_s = \frac{-\omega_{pe}^2}{2} \frac{\delta \hat{n}_p}{n_{eo}} \hat{A}_L \times
$$
$$
\exp i\left[\left(k_L - k_s - k_p\right) \cdot x - \left(\omega_L - \omega_s - \omega_p\right) t\right],
\tag{9.21}
$$

in which the exponential term is a phase-matching term. Its argument must be zero to obtain a nonzero averaged response. This imposes the beating condition that we expect—the frequency and wavevector of the driven wave must equal the difference of the values for the laser and for the electron–plasma wave.

To see the unstable behavior, we need to reconsider the derivation of the electron–plasma wave from (2.42) to (2.50), in the presence of light waves and informed by the discussion earlier in this chapter. We are looking for the ways in which light waves can affect plasma waves. The electron plasma wave is a purely longitudinal wave, so one can write $u_e = v_{os} + v_p$, in which v_{os} is purely transverse but may involve the sum of contributions from more than one light wave and v_p is the purely longitudinal vector describing the motion of the electron fluid in the plasma wave. Substituting into (9.5) and again using scalar and vector potentials, one obtains an equation from which (9.6) can be subtracted. After linearizing in $|v_p|$ and ignoring collisions we find

$$\frac{\partial \boldsymbol{v}_p}{\partial t} - \frac{e}{m_e}\nabla\Phi + \frac{\nabla p_e}{n_e m_e} = -\nabla\frac{v_{os}^2}{2}, \tag{9.22}$$

in which we have also dropped a term proportional to $\boldsymbol{v}_p \times \nabla \times \boldsymbol{v}_{os}$, whose direction is orthogonal to $\boldsymbol{v}_p$.

As in Chap. 2, one obtains a wave equation by taking the divergence of this equation, then using continuity and the equation of state to simplify all the terms on the left-hand side. The right-hand side of this equation represents the force known as the *ponderomotive force*. This term is the gradient of v_{os}^2, which is equivalent to the gradient in the energy density (or pressure) of the electromagnetic waves. The ponderomotive force can be important in other contexts, but here we focus on its role in instabilities. If the light waves present have vector potentials $\boldsymbol{A}_L$ and $\boldsymbol{A}_s$, then v_{os}^2 has three terms. However, only the cross term involves the beating of two waves. Keeping only the cross term, one obtains

$$\left(\frac{\partial^2}{\partial t^2} + \omega_{pe}^2 - 3\frac{k_B T_e}{m_e}\nabla^2\right)\frac{\delta n_p}{n_{eo}} = \frac{e^2}{m_e^2 c^2}\nabla^2\left(\boldsymbol{A}_L \cdot \boldsymbol{A}_s\right). \tag{9.23}$$

This equation describes the driving of electron–plasma waves by the beating of light waves in the plasma. Once again the wave beating produces source terms that are upshifted or downshifted in frequency relative to the laser frequency, and once again only the downshifted source term is significant for the simple instability. To see the unstable coupling, consider the beat term in which the phases of the laser-light wave, the scattered-light wave, and the plasma wave vary as $e^{-i(\boldsymbol{k}_L \cdot \boldsymbol{x} - \omega_L t)}$, $e^{i(\boldsymbol{k}_s \cdot \boldsymbol{x} - \omega_s t)}$, and $e^{-i(\boldsymbol{k}_p \cdot \boldsymbol{x} - \omega_p t)}$, respectively, and again use a caret ˆ to designate the amplitude of each wave (assumed to be real). One then finds

$$\left(-\omega_p^2 + \omega_{pe}^2 + 3\frac{k_B T_e}{m_e}k_p^2\right)\frac{\delta\hat{n}_p}{n_{eo}} = \frac{-e^2 k_p^2}{2m_e^2 c^2}\left(\hat{A}_L \cdot \hat{A}_s\right)$$
$$\times \exp i\left[\left(-\boldsymbol{k}_L + \boldsymbol{k}_s + \boldsymbol{k}_p\right)\cdot\boldsymbol{x} - \left(-\omega_L + \omega_s + \omega_p\right)t\right]. \tag{9.24}$$

One sees here the same sort of phase-matching term we encountered in (9.21). If one now multiplies (9.21) by $\hat{A}_L$, and substitutes for $\delta\hat{n}_p$ from (9.24), one finds

$$\left(-\omega_s^2 + \omega_{pe}^2 + c^2 k^2\right)\left(-\omega_p^2 + \omega_{pe}^2 + 3\frac{k_B T_e}{m_e}k_p^2\right) = \omega_{pe}^2\frac{k_p^2 v_{os}^2}{4}. \tag{9.25}$$

This coupled dispersion relation describes the growth of an instability. Physically, the laser light beats with density fluctuations to drive the scattered light and beats with the scattered light to drive density fluctuations. When phase-matching is satisfied—that is, when $\boldsymbol{k}_L = \boldsymbol{k}_s + \boldsymbol{k}_p$ and $\omega_L = \omega_s + \omega_p$—the process is resonantly reinforcing. Note that the two sets of parentheses on the left-hand side each enclose the dispersion relation of one of the normal modes of oscillation of the plasma. In the absence of driving or coupling and for any given wavenumber, each set of parentheses would determine the frequency for each mode independently. The coupling represented by the right-hand side, in the presence of the phase matching that connects the mode frequencies and wavevectors, leads to instability growth.

Equation (9.25) implies an exponential growth rate for the instability. For the uniform-plasma case considered here it makes sense to look for temporal growth. Mathematically, the two driven waves grow as $e^{\gamma t}$ while the amplitudes are small. One finds γ by identifying the real part of each frequency with the subscript r and the imaginary part of both of them as γ_o, with the sign corresponding to growth in time. For the components we have chosen to consider, this gives us $\omega_s = \omega_{sr} + i\gamma_o$ and $\omega_p = \omega_{pr} - i\gamma_o$. We also assume here that the real part of each frequency is the normal-mode frequency, and thus cancels the other real terms in its part of the equation. If γ_o is much smaller than the wave frequencies, as is nearly always the case for SRS (and is sometimes the case for other instabilities), then one finds the growth rate for SRS in a homogeneous plasma,

$$\gamma_o = \sqrt{\frac{\omega_{pe}^2}{\omega_{sr}\omega_{pr}}\frac{k_p v_{os}}{4}}. \tag{9.26}$$

The growth of SRS in other more complicated situations can be usefully expressed in ways involving this growth rate. SRS can occur in principle at densities up to $n_c/4$, where ω_{pr} and ω_{sr} are both $\sim \omega_o/2$. It more typically occurs near $n_c/10$, where $\omega_{pr} \sim \omega_o/3$, $\omega_{sr} \sim 2\omega_o/3$, and $k_p \sim 1.5\omega_o/c$. The growth rate, for $I_{14}\lambda_\mu^2 \sim 1$, is $\gamma_o \sim 0.002\omega_o$. Thus, SRS indeed grows slowly on the scale of the wave cycles. But note that $1/\omega_o \sim 1$ fs, so $1/\gamma_o < 1$ ps. The instability, when present, thus grows extremely rapidly on the ns scale of the typical laser pulse.

We should confess that the derivation just provided involves cheating at several levels, in order to most simply make its physical point. Strictly speaking, one should express all the wave amplitudes as real quantities and follow through with all the wave-beating terms that arise. This is the only way one can obtain the factor of 2 that mysteriously appeared in (9.21) and (9.24). One also should not assume that the light wave and the electron plasma wave, which are normal modes of an undisturbed plasma, will be unchanged by the instability. Doing all this properly would involve several more pages, however, and in the end would produce the same result with many nuances. One could then proceed to consider other complications such as nonuniform plasmas and depletion of the laser pulse. A first level of improved analysis can be found in the book by Kruer (1988). Doing better than that requires submersion in the archival literature.

Of greater importance than the details of the SRS theory is to understand that once energy is given to electron–plasma waves (by any instability), it tends to be converted to electron energy in a tail on the distribution function. (In such a tail, the density of the electrons in phase space at energies above the thermal energy is increased by comparison to the density present in a Maxwellian.) The electrons in the tail are known as *suprathermal electrons*. They are in most cases a terrible nuisance. They readily penetrate target materials, especially low-Z materials, and so can preheat them and alter the initial conditions for later evolution of the target. This is a crucial issue for laser fusion, as we discuss in Chap. 11. In addition, because laser irradiation produces large-scale magnetic fields that wrap around the target, the energetic electrons can very easily travel around almost any shielding to penetrate and heat surfaces that are distant from the laser spot. This can affect both the physics of an experiment and the signals seen by diagnostics.

Landau damping produces the energetic electrons. Landau damping typically dominates over collisional damping of electron–plasma waves. (If collisions are strong, the waves will not be driven, and if collisions are weak, then the waves are Landau damped.) Some readers will recall that Landau damping operates by accelerating electrons in the wave, so that the electrons produced have an energy of order the phase velocity of the wave. This energy, for the typical SRS conditions given above, is then

$$\frac{1}{2}m_e v^2 \sim \frac{1}{2}m_e \left(\frac{\omega_{pr}^2}{k_p}\right)^2 \sim \frac{m_e c^2}{40} \sim 13 \text{ keV}. \tag{9.27}$$

This increases and becomes closer to 30 keV as the density approaches $n_c/4$. Two-plasmon decay, discussed below, can produce smaller wavenumbers, higher phase velocities, and higher-energy electrons. The penetration of such energetic electrons into materials is discussed at the end of Sect. 9.1.5.

Table 9.1 summarizes the instabilities driven by the laser beam in laser plasmas. In all cases, each of the two driven waves is coupled to the laser light wave so as to drive the other driven wave. The density where each instability occurs is given, as is the growth rate in a homogeneous plasma, with the two driven waves indicated by subscripts 1 and 2. In the column giving the growth rates, ω_{pi} is the ion plasma frequency, $\omega_{pi}^2 = 4\pi Z^2 e^2 n_i/m_i$, with ion density n_i and mass m_i, and c_s is the sound speed, derived for plasmas in Chap. 2. As shown in (2.53) it is given approximately for a two-fluid plasma by $c_s^2 = Zk_BT_e/m_i + 3k_BT_i/m_i$. We now briefly discuss the instabilities shown in the table.

Table 9.1 Laser-driven instabilities

Name	Driven wave 1	Driven wave 2	Where	Growth rate
Stimulated Raman scattering	Scattered light	Electron plasma	$\lesssim n_c/4$	$\sqrt{\frac{\omega_{pe}^2}{\omega_1\omega_2}}\frac{k_2 v_{os}}{4}$
Stimulated Brillouin scattering	Scattered light	Acoustic	$\lesssim n_c$	$\sqrt{\frac{\omega_{pi}^2}{\omega_1 k_2 c_s}}\frac{k_2 v_{os}}{4}$
Two-plasmon decay	Electron plasma	Electron plasma	$\sim n_c/4$	$\frac{k_2 v_{os}}{4}$
Parametric decay	Electron plasma	Acoustic	$\sim n_c$	$\sqrt{\frac{\omega_{pi}^2}{\omega_1 k_2 c_s}}\frac{k_2 v_{os}}{4}$
Filamentation	Modulated light	Zero-frequency acoustic	$\lesssim n_c$	$\frac{v_{os}^2}{8(T_e/m_i)}\frac{\omega_{pe}^2}{\omega_o}$

SRS is strongly reduced by collisional effects for short laser wavelengths and by the creation of smooth plasmas with SSD or other methods. *Stimulated Brillouin scattering* (SBS) is a direct analog of SRS in which the second-driven wave is an acoustic wave. It can be strongly reduced by the introduction of bandwidth in the laser and also saturates fairly easily under many conditions. However, both SRS and SBS have at times been observed to convert more than 50% of the laser energy, so they can be enormous.

In addition, there are two *decay instabilities* in which the laser light wave drives two waves within the plasma. These instabilities directly produce no scattered light. *Two-plasmon decay*, being localized at a single density surface in the plasma, saturates fairly easily. However, at this writing it appears to be the largest potential threat to laser fusion by direct laser irradiation. In most modern experiments, collisional absorption of the laser light prevents the occurrence of the *parametric decay instability* near n_c, except perhaps for a brief period at the start of the laser pulse. (During this early period, the laser electric field may also penetrate to n_c, where it may excite resonance absorption—see Kruer (1988).) There is a variant of this instability involving a zero-frequency acoustic wave, sometimes described as the *oscillating two-stream instability*. The variant of SBS with a zero-frequency acoustic wave is *filamentation*, which in the nonlinear limit will break the laser beam into discrete intense beamlets. At present it appears that strong filamentation near the leading edge of the plasma may smooth the illumination of the denser regions, by producing beamlets that focus strongly and then spray their energy into a wide angular range. There are actually several types of filamentation. The growth rate shown in Table 9.1 is for ponderomotive filamentation, in which the ponderomotive force causes the plasma motion just as it does in SBS. In colder plasmas, thermal filamentation, involving differential heating, can be important. When v_{os} becomes relativistic, relativistic self-focusing can arise; this process involves variations in the effective electron mass.

One is led to wonder why all the laser energy is not consumed by SRS or some other instability (most of them grow very quickly on the scale of the laser pulse). The answer to this question is twofold. On the one hand, all the laser energy (more or less) is consumed by these instabilities if the laser wavelength is too long— roughly 1 μm or longer. This nearly led to the death of the laser fusion program, which initially attempted to use laser wavelengths of 1–10 μm. Such lasers produced some spectacular phenomena but not much progress toward fusion. Lasers with wavelengths in the visible and UV have two advantages—v_{os} is smaller for a given laser energy flux and collisional effects begin to play a role. On the other hand, so long as v_{os} is not too large, some of the instabilities saturate at low values and they all are strongly affected by plasma nonuniformity or laser bandwidth. It is worth emphasizing here that a very common mistake, among those doing experiments not focused on laser-plasma instabilities, to fail to consider and test the ways in which they may adversely affect the experiment.

9.1.5   *Electron Heat Transport*

We have emphasized the mobility of electrons by comparison with ions, so that the electrons dominate for example the direct interactions of the laser with the plasma. This might lead one to expect that the thermal electrons would play a dominant role in transporting energy throughout all plasma systems. This, however, is not the case in the systems of interest here. Laboratory systems in the high-energy-density regime are typically so collisional that the electrons cannot manage to escape the ions and do not manage to affect the dynamics very strongly. We saw this quantitatively in Chap. 3. It is also remarkable that there are very few astrophysical systems in which the electrons carry significant heat. The electrons, because of their small mass, are very tightly bound to the magnetic field, and the magnetic field is typically tangled enough to keep them from accomplishing any large-scale heat transport. Systems involving instabilities in loops of magnetic field, which occur for example near the Sun, or involving magnetic reconnection, which occurs in many places, produce bursts of energetic electrons. These electrons in turn radiate, so that the electron radiation can be an important diagnostic of the phenomena. However, the electrons do not dominate the overall dynamics of reconnecting systems. Likewise, the radiation from electrons has become an important indicator of cosmic-ray acceleration in supernova remnants, but the cosmic rays that actually reach the Earth are almost all ions. So electrons are important. However, with two crucial exceptions, they rarely carry much heat anywhere that matters.

One crucial exception is in the delivery of energy from a laser beam to dense material. This is an essential aspect of inertial fusion and any other high-energy-density experiment using lasers. A second exception is in the loss of energy from the burning region in inertial fusion. For these reasons, it is worthwhile to have some understanding of electron heat transport. Figure 9.7 shows the profile of a laser-irradiated plasma, taken from a computer simulation. The plasma expands but absorbs little laser light in the expansion zone while absorption takes place in the absorption zone, over some range of densities below the critical density. The entire region below critical density is often designated the corona, by analogy with

Fig. 9.7 The electron density profile from a computer simulation, with various regions indicated. The simulation corresponds to a laser wavelength of 0.35 μm and an energy flux of 10^{15} W/cm^2

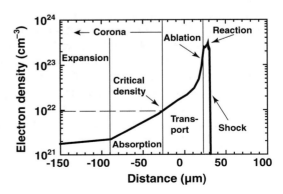

the solar corona. In the transport zone, electron heat transport carries the energy to higher densities above the critical density. Ablation occurs in this high-density material. In reaction to the ablation (or equivalently, in response to the ablation pressure), a shock wave propagates into the material, creating a region of high electron density.

To gain an understanding of electron heat transport, we will discuss here the classic derivation by Spitzer and Harm, after which we discuss the limit when transport becomes too strong for this derivation to be valid. Because this process involves the behavior of individual particles, we use the kinetic description of Sect. 2.5 and discuss the behavior of the distribution function $f(v)$, normalized for this purpose so that its integral over velocity space gives n_e.

Suppose we have a plasma with a gentle temperature gradient, a condition we would express mathematically as $\lambda_{mfp} \ll T_e/|\nabla T_e|$, where λ_{mfp} is the collisional mean free path of the electrons. The heat flux, Q, in a direction z within such a plasma is found by integrating the energy carried by each particle over the distribution:

$$Q = \int \left(\frac{1}{2} m_e v^2 \right) v_z f(v) \mathrm{d}^3 v. \tag{9.28}$$

One can see that $Q = 0$ if $f(v)$ is Maxwellian or any symmetric function in v_z. Actual distributions are often nearly Maxwellian but are seldom fully symmetric, so that heat is usually carried by the electrons. The source of asymmetry can be found by thinking about what would happen if the plasma initially consisted of Maxwellian distributions with a slow spatial variation of the temperature. As temperature increases, the number of hot particles with energies above $k_B T_e$ increases while the number of cold particles, with lower energies, decreases. As a result, the flow of particles through some point from a Maxwellian distribution in a warmer region will include a surplus of hot particles and a deficit of cold ones. From the opposite, cooler, direction, there will be a deficit of hot particles and a surplus of cold ones. Thus, if the distribution functions were initially Maxwellian but had a varying temperature, they would almost immediately develop non-Maxwellian structure, asymmetric in velocity, and thus able to carry heat.

To find an equation for the heat flow, we assume a plasma of constant density and slowly varying temperature. Note that the temperature gradient defines a unique direction within the plasma, and that the effects of interest involve motions in that direction, which we will call z. This motivates the definition of a polar angle θ with respect to that direction and the expansion of the distribution function by means of Legendre polynomials. Keeping only the first term in this expansion, we have

$$f(v) = f_o(v) + f_1(v) \cos \theta, \tag{9.29}$$

in which $f_o(v)$ is a Maxwellian and $f_1(v)$ must be small to justify using only the first two terms in the expansion. In some experiments, such as those of Liu et al. (1994), two or three coefficients of this expansion have been directly measured.

In addition, it is shown by Shkarofsky et al. (1966) that one can accurately treat the effects of collisions in this problem using a simple relaxation rate. Expressing this rate in terms of the electron–ion collision rate ν_{ei} defined in Sect. 2.4 gives

$$
\left(\frac{\delta f}{\delta t}\right)_C = \frac{-3}{4\pi}\left(\frac{2\pi k_B T_e}{m_e}\right)^{3/2}\frac{\nu_{ei}}{v^3}f_1(v)\cos\theta = \frac{-W}{v^3}f_1(v)\cos\theta,
\tag{9.30}
$$

which defines the coefficient W strictly for convenience in what follows. We can substitute (9.29) and (9.30) into (2.63) and keep only the terms proportional to $\cos\theta$ to obtain

$$
\frac{\partial f_1}{\partial t} + v_z\frac{\partial f_o}{\partial z} - \frac{eE}{m_e}\frac{\partial f_o}{\partial v} = \frac{-W}{v^3}f_1(v),
\tag{9.31}
$$

in which the electric field must be in the z direction from the symmetry of the problem. We ignore B here, which is justified in the dense target material near the center of the laser spot, where collisions are large and the magnetic field produced by the laser is weak. In steady state, this implies

$$
f_1(v) = \frac{-v^3}{W}\left(v_z\frac{\partial f_o}{\partial z} - \frac{eE}{m_e}\frac{\partial f_o}{\partial v}\right).
\tag{9.32}
$$

To find E, one can note that any net flow of charge would cause an electric potential to develop that would then shut off the flow of charge, so in steady state the net current in the z direction, J_z, must be zero. This gives

$$
J_z = 0 = -e\int_0^\infty v_z f(\boldsymbol{v})d\boldsymbol{v} = -2\pi e\int(v\cos\theta)f_1(v)\cos\theta v^2\sin\theta d\theta dv,
\tag{9.33}
$$

from which

$$
0 = \int_0^\infty v^3 f_1(v)dv = \int_0^\infty v^6\left(v\frac{\partial f_o}{\partial z} - \frac{eE}{m_e}\frac{\partial f_o}{\partial v}\right)dv.
\tag{9.34}
$$

Integrating this, solving for E, and substituting for the derivatives of f_o, assumed to be Maxwellian, one finds $eE = -4k_B\partial T_e/\partial z$, so

$$
f_1(v) = f_o(v)\frac{v^4}{2Wk_B T_e}\left[8 - \frac{m_e v^2}{k_B T_e}\right]k_B\frac{\partial T_e}{\partial z}.
\tag{9.35}
$$

The heat flux per $d^3 v$ is proportional to $v^3 f_1(v)\cos^2\theta$ per (9.27) and (9.28). Also, the contribution at a large velocity reverses sign compared to that at a small velocity, as we expected from our qualitative analysis. Integrating (9.28) one finds the *Spitzer–Harm heat flux*, Q_{SH}, as

$$Q_{SH} = \frac{-128}{3\pi} \frac{n_e k_B T_e}{m_e \nu_{ei}} k_B \frac{\partial T_e}{\partial z} = -\kappa_{th} \frac{\partial k_B T_e}{\partial z}, \tag{9.36}$$

in which the heat transport coefficient is κ_{th}. Here k_B must convert T_e to the energy units used in Q_{SH}, and if T_e is expressed independently in some other units then the heat transport coefficient is $\kappa_{th} k_B$. Note that, through ν_{ei}, κ_{th} is proportional to $T_e^{5/2}$, so that the heat transported increases very rapidly with temperature. Remarkably, the heat transported is independent of density. The increase in the flux of particles with increasing density is precisely balanced by the increase in collision rate. This model is also known as a diffusion model or a description of diffusive heat transport, because the electrons carry the heat through a diffusive process. Indeed, when one uses (9.36) in a fluid equation for the electron energy, one obtains a diffusion equation (in simple limits). Note that $k_B T_e/(m_e \nu_{ei})$ has units of a kinematic diffusion coefficient (e.g., cm^2/s). Kruer (1988) points out that (9.36) overestimates the transport because the derivation ignores electron–electron collisions, and that this can be approximately corrected for by multiplying κ_{th} by $g(Z) = (1+3.3/Z)^{-1}$.

The Spitzer–Harm transport model, adjusted as just described, gives accurate results regarding the heat transport within limits we are about to define, despite the fact that it has some fundamental problems. These relate to the use of the expansion in (9.29), which is only valid if $f_1/f_o \ll 1$. One can show from (9.35) that this ratio is

$$\frac{f_1}{f_o} = \frac{1}{\sqrt{2\pi}} \left(8 \frac{v^4}{v_{th}^4} - \frac{v^6}{v_{th}^6} \right) \frac{\lambda_{mfp}}{L_T}, \tag{9.37}$$

in which the electron mean-free-path is $\lambda_{mfp} = v_{th}/\nu_{ei}$, the temperature scale length is $L_T = T_e/|\nabla T_e|$, and $v_{th}^2 = k_B T_e/m_e$. This particular definition of a thermal velocity v_{th} is common in laser–plasma interactions. (In various other areas of physics there is some numerical factor multiplying the right-hand side of this definition.) Figure 9.8 shows this ratio, normalized by λ_{mfp}/L_T. Two points are important with regard to this figure. First, the expansion of $f(v)$ always breaks down at a large velocity. The only reason the Spitzer-Harm model ever gives accurate results is that negligible heat is carried by the high-velocity electrons.

Fig. 9.8 The normalized perturbation to the distribution function from the Spitzer–Harm theory

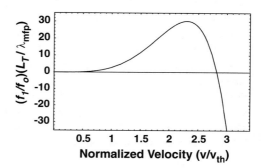

Fig. 9.9 Ratio of temperature scale length to electron mean free path, for $n_e = 6 \times 10^{21}$ per cc and $T_e = 500$ eV. This ratio scales with T_e^{-2}, Z, and n_e. The heat transport should be flux-limited for values not $\gg 30$

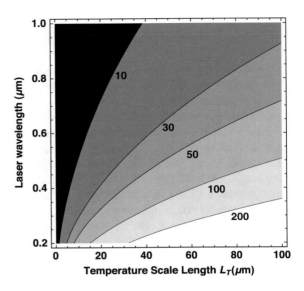

Mathematically, f_o decreases much more rapidly than $|f_1/f_o|$ increases. Second, the heat is carried primarily by the electrons with energies of about three times $k_B T_e$, corresponding to the maximum in Fig. 9.8 for v/v_{th} between 2 and 2.5. In this range, $f_1/f_o \sim 30\lambda_{mfp}/L_T$, so the Spitzer–Harm model will be accurate if

$$L_T/\lambda_{mfp} \gg 30. \qquad (9.38)$$

Unfortunately, condition (9.38) is only sometimes satisfied in laser-irradiated plasmas. As a rule of thumb, this condition is likely to be satisfied for UV lasers, may or may not be satisfied for visible lasers, and is not satisfied for infrared lasers. Figure 9.9 shows this ratio for relevant plasma parameters and laser wavelengths.

The proper way to proceed when (9.38) is not satisfied is to find a better solution to the Boltzmann equation, which may also require finding a more sophisticated expression of the collision term. The standard, more-sophisticated version of the Boltzmann equation is known as the Fokker–Planck equation, but this equation must be solved numerically for all cases of interest. In addition, finding the heat flux from the Fokker–Planck equation is difficult enough that it cannot readily be included in radiation-hydrodynamic simulations of the overall dynamics of a laser target. This difficulty strongly motivates the search for simple models that can be of some use.

There is such a model, very crude but very widely used, known as *flux-limited heat transport*. The maximum possible flow of energy would occur if the thermal energy density in the plasma were transported as some characteristic thermal velocity, producing a *free-streaming heat flux*, equal to $n_e k_B T v_{th}$. In real physical systems, the maximum heat transported only approaches some fraction of this limit. This is described by introducing a *flux limiter*, f, so that

$$Q_{FS} = f \times n_e k_B T_e v_{th}. \qquad (9.39)$$

Under typical conditions, matching the heat flux found in Fokker–Planck calculations requires taking $f \sim 0.1$. Figure 9.9 can be used to determine how likely this is. The ratio Q_{SH}/Q_{FS} scales with T_e^2.

However, in some historical experiments, especially using infrared lasers and high intensities, the observations could only be explained using f as small as 0.01. This indicated that additional, noncollisional processes were impeding the heat transport in these cases. Possible explanations of these observations include the effects of laser-generated magnetic fields and of intense acoustic fluctuations produced by plasma instabilities. A very common model in simulations is to set f at some value and to take Q to be the smaller of the values given by (9.39) and (9.36). Some more sophisticated computer simulations employ a technique known as multigroup, flux-limited diffusion. In this technique, the electrons are divided into a number of groups, with each group being treated as either diffusive or flux-limited as appropriate.

The models just described are adequate for the calculation of the global influence of the laser on the material it irradiates. For example, they can be used, after some tuning, to calculate accurately the production of pressure and shock waves by the laser. In practice, most tuned models of experiments use $f = 0.05$ or 0.06. However, when it comes to the detailed structure of the target plasma near the irradiated surface, these models give results that are not correct. Unfortunately, although flux-limited transport can provide a quantitative estimate of the local flow of heat, it fails fundamentally to capture the dynamics of heat transport in strong temperature gradients. The reason is that such transport is at root *nonlocal*. The heat deposited at a given point is not determined only by the local conditions, but rather involves particles transported from a range of distances. As a result, the long-term evolution of the plasma differs from what would occur if heat transport were local. In particular, because the more energetic particles have smaller Coulomb cross sections and longer mean free paths, they tend to penetrate deeper into the target and to produce a warm foot ahead of the main heat front. There has been a great deal of work over several decades aimed at producing a model of nonlocal, electron heat transport that would be practical to implement within radiation-hydrodynamic codes. One of these, by Schurtz et al. (2000), has recently emerged and has been implemented in some of these codes.

In the presence of laser–plasma instabilities, the reality becomes even more complex than this. As we discussed in Sect. 9.1.4, these instabilities produce electrons with energies of tens of keV. Such electrons penetrate far deeper into materials than the electrons from the thermal population that transport heat inward. As a result, they can *preheat* the initially cold material in an experimental system, altering the initial conditions for the subsequent evolution. The distributions of electrons produced by instabilities are often observed to be exponential in energy, even though they do not arise from waves with a wide distribution of phase velocities. Thus, this is another of the many cases in which a plasma anomalously produces an exponential particle distribution. (The first was Langmuir's paradox, from the early 1900s, relating to the behavior of low density plasma in evacuated chambers.) The exponential energy distribution allows one to assign a temperature to the energetic electrons. These are most often described as *suprathermal electrons* or *hot electrons*, with a temperature designated by T_{hot}.

The penetration of such electrons into materials is complex, as they are strongly scattered by any nucleus they get close to and also by the cold, local electrons. The result is that their penetration is diffusive, with a step size of an electron mean free path and a collision time of the mean free path divided by the velocity. However, the electron velocity decreases steadily as it loses energy in successive collisions and by drag on the electrons in the material. The net effect is that of diffusive penetration with a steadily decreasing diffusion coefficient. Rosen et al. (1987) provide a resulting mean electron range for carbon. This range should be proportional to the density of atoms in the matter. Scaling their result accordingly, one finds a mean range as an areal mass density, x_o, in g/cm^2, of

$$x_o = 5 \times 10^{-6} \left(A/Z_{nuc}^{3/2} \right) T_{hot}^2, \tag{9.40}$$

where as before Z_{nuc} is the nuclear charge and T_{hot} is in keV. For $A = 2Z_{nuc}$, $Z_{nuc} \sim 4$, and $T_{hot} \sim 30$ keV, one finds $x_o \sim 5 \times 10^{-3}$ g/cm^2. For plastic at ~ 1 g/cm^3 this is a 50 μm mean penetration depth. This is enough to affect many experiments, and the penetration increases strongly as T_{hot} increases. For example, one might irradiate a 1 mm spot with 10 kJ of laser energy for 1 ns, producing an energy flux of 1.3×10^{15} W/cm^2. If 1% of the laser energy were converted to hot electrons, at 30 keV temperature, that were deposited in a 30 μm layer of plastic, the temperature of this material would increase to ~ 30 eV and its pressure would be ~ 10 Mbar. This is why preheat is very often a concern in laser experiments. In addition, since the penetration of energy is a diffusive process, noticeable heating may occur at depths well beyond x_o. For their case, Rosen et al. (1987) find the heating to scale as $\eta^{0.1} \exp(-1.65\eta^{0.4})$ with $\eta = x/x_o$. (For different reasons, preheat is also a concern in Z-pinch experiments.)

9.1.6 Laser Heating and Ablation Pressure

Although much of what we have discussed in the preceding two sections is confined primarily to laboratory environments, the ablation of matter by irradiation is found much more widely. Figure 9.10 shows an image of the Eagle Nebula, justly famous for its dramatic structures, referred to as Elephant Trunks. The Eagle Nebula exists within a star-forming region—a zone with many dense molecular clouds that can provide mass for very large new stars. These new stars are very massive and very bright, with much of their radiation in the deep UV. These photons have energies large enough to directly ionize the material they encounter, creating an ionization front that is also a region of comparatively high pressure. One hypothesis regarding the origin of the structures is that they might have been produced by Rayleigh–Taylor instabilities that developed when the hot, low-density, ablated plasma began pushing on the cooler, denser plasma behind it.

Fig. 9.10 The Eagle Nebula.
From Hubble Space
Telescope, WFPC2. Credit:
NASA, Jeff Hester and Paul
Scowen Arizona State
University

The intense lasers used in high-energy-density experiments also substantially affect the material they encounter. The photons in these lasers cannot individually ionize the material, but in combination they can and do ionize it. At most relevant energy fluxes the electric field of the laser can directly ionize the atoms. At lower energy fluxes the interaction with the target quickly produces plasma, although the mechanisms are more complex. (We leave it as an exercise for the student to find the threshold for direct ionization.) In this section we explore the ablation of matter and discuss its effects.

9.1.6.1   Generation of Ablation Pressure

Even technically informed people often first imagine that the influence of light on a material is primarily due to reflection, as is the case for example in a solar sail. The magnitude of this effect is straightforward to estimate. The pressure produced by reflection, P_{ref}, is the rate of change of momentum by the reflection. The momentum of a single photon is $\hbar k$ and the change of momentum upon its reflection is $2\hbar k$. The flux of photons, F, is the energy flux I_L divided by the energy per photon, $F = I_L/(\hbar\omega)$. The pressure is thus $P_{ref} = 2\hbar k F = 2(k/\omega)I_L = 2I_L/c$. If the photon is absorbed and not reflected, then the light pressure is half this value. If one chooses instead to take a microscopic view of reflection, one can do so by evaluating the ponderomotive pressure at the reflection surface. A first estimate of this, based on the relations given above, gives $n_e m_e v_{os}^2/2 = I_L/c$. One recovers $2I_L/c$ by taking into account the doubling of the laser electric field during reflection and by taking the time-averaged value of v_{os}. In practical units, $P_{ref} = 0.067 I_{14}$ Mbar. However, reflection competes with ablation, which nearly always dominates the effect of the laser on the target material at solid density in experiments using lasers whose pulse duration exceeds a few hundred ps. Let us consider this.

Ablation is the process in which material is heated and then flows away from a surface. By Newton's third law, the surface experiences a reaction force equal to the rate at which momentum is carried away. In detail, sustained laser heating produces a high pressure in the laser-heated region, beyond which the electrons carry heat into the material, causing it to ablate and flow outward. These two perspectives are equivalent—in rockets driven by thermal release of fuel, it is the pressure of the hot fuel that drives the rocket forward. Let us evaluate the laser heating.

We will continue to use I_L for the laser energy flux, although our calculation really involves the absorbed energy flux. In effect, we are assuming the laser energy is completely absorbed. This is quite accurate for UV lasers and becomes less and less accurate at longer laser wavelength. The energy deposited by the laser in the absorption region supports three phenomena. It sustains the rarefaction of material away from the target surface, it sustains an electron heat flux into the denser matter above critical density, and it sustains radiative losses from the matter below critical density. The third of these is negligible for conditions of interest. For an initial calculation, we assume that the laser energy divides equally between the first two.

We thus suppose that half the energy is transported through the critical surface. Using a flux-limited heat transport model, we then have

$$0.5I_L = fn_c k_B T_e \sqrt{k_B T_e/m_e}, \tag{9.41}$$

from which for $f = 0.05$

$$T_e = 2.6 \left(I_{14}\lambda_\mu^2\right)^{2/3} \text{ keV}, \tag{9.42}$$

in which I_{14} is I_L in units of 10^{14} W/cm^2 and λ_μ is the wavelength of the light in μm. The ablation pressure is then the corresponding pressure at the critical density. If $T_e = T_i$ at this density, this is

$$P_{abl} = n_c(1 + 1/Z)k_B T_e = 6.1 I_{14}^{2/3}\lambda_\mu^{-2/3} \text{ Mbar}, \tag{9.43}$$

A standard value (Lindl 1995) of the coefficient which is 6.1 here would be 8.6, based on computer simulations of the detailed energy transport, using a flux-limited, multigroup, diffusive electron-heat-transport model. But in actuality, as of 2017, systematic studies that would firmly ground these theoretical estimates in experimental data remain to be undertaken.

Continuing the comparison with the pressure produced by reflection, the formulae we just obtained imply that the laser energy flux would have to reach about 10^{21} W/cm^2 before reflection became dominant over ablation. This is not correct in detail because relativistic effects become important at 10^{18} W/cm^2, as we will discuss in Chap. 13. An accurate statement is that throughout the non-relativistic regime ablation is more important than reflection in the acceleration of material at or near solid density. This does not, however, imply that reflection and the ponderomotive force have no effects. The ponderomotive pressure grows to equal the plasma pressure at the critical density by the time the laser energy flux reaches 10^{16} W/cm^2. Above this intensity, the ponderomotive pressure steepens the density profile by pushing the critical surface inward. Even so, this does not prevent the outward flow of material that corresponds to the ablation pressure produced by the heating and removal of solid matter at higher density.

Before proceeding, take a moment to note that the ablation pressure produced by these systems is an amazing number. Using late-twentieth century laser facilities, it is straightforward to irradiate a large (mm^2) area with 0.35 μm laser light at 10^{15} W/cm^2. This produces an ablation pressure of 75 Mbar, which is near the pressure at the core of Jupiter! The idea that one can produce pressures of many millions of atmospheres in an Earth-bound laboratory is pretty exciting. Indeed, this is what has made the work described in this book as experimental astrophysics possible.

Let us return to laser heating, and explore some of its aspects further. For comparison with the pressure evaluated above, we can quantify the outward flux of momentum through the critical surface. This flow is well approximated as an isothermal rarefaction, although simulations often find that there is a density drop across the critical surface. Because the density at this location remains steady, the represents the centered location for the rarefaction, which occurs at the sonic point. If the density there is ρ_s, and is a fraction g is the mass density at the critical surface, then the pressure inferred from the momentum flux, P_{flux} is

$$P_{flux} = \rho_s c_s^2 = \frac{gn_c}{Z} \left[Zk_B T_e + 3k_B T_i \right], \qquad (9.44)$$

with c_s from (2.53). Out in the underdense corona, simulations typically find $T_i \sim T_e/3$. If this applied near the critical surface, the result here would equal that of (9.43) for $g = 1$. The models here are not accurate enough to make a further exploration of these differences worthwhile.

As to the value of the temperature, we can improve the calculation above by explicitly including the energy flux required to sustain the coronal plasma, treated as an isothermal rarefaction. This energy flux is $\rho_s c_s^3 (5\gamma - 3)/(2(\gamma - 1))$. This is the total energy flux required to support a rarefaction that starts at the sonic point. Thus we write

$$I_L = f n_c k_B T_e \sqrt{k_B T_e / m_e} + \rho_s c_s^3 \frac{(5\gamma - 3)}{2(\gamma - 1)}, \qquad (9.45)$$

from which

$$k_B T_e = \left(\frac{I_L}{n_c} \right)^{2/3} \left(\frac{f}{\sqrt{m_e}} + \frac{g(1 + Z)^{3/2}(5\gamma - 3)}{2\sqrt{Am_p}Z(\gamma - 1)} \right)^{-2/3}, \qquad (9.46)$$

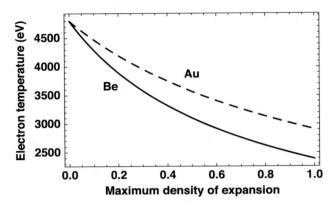

Fig. 9.11 Electron temperature from a model in which the laser energy sustains an isothermal rarefaction and flux-limited, free-streaming inward heat transport. The maximum density of expansion is the fraction of critical density at which the rarefaction starts. The upper curve is for $A = 200, Z = 35$, and $\gamma = 4/3$ and the lower curve is for $A = 8, Z = 4$, and $\gamma = 5/3$

which implies

$$T_e = 4.19 \left(I_{14}\lambda_\mu^2\right)^{2/3} \left(20f + \frac{0.233g(1+Z)^{3/2}(5\gamma-3)}{\sqrt{AZ}(\gamma-1)}\right)^{-2/3} \text{keV}. \qquad (9.47)$$

Figure 9.11 shows the results for $I_{14} = 10$ and $\lambda_\mu = 0.35$, using parameters that correspond to a low-Z (Be) plasma and a high-Z (Au) plasma. Supporting the plasma expansion takes about half the laser energy flux when the rarefaction begins near n_c (so $g \sim 1$), and progressively less as g decreases. This leads to larger temperatures. Higher-Z targets tend to produce larger temperatures, a trend qualitatively seen in data.

The exact value of the electron temperatures and the ablation pressure can be affected by several details. The temperature may differ from that given by the models here, as the heat-transport model used is only approximate. The temperature can also be reduced by the lateral transport of energy, which is not considered here. Current computer simulations of laser absorption may not necessarily give correct results. One reason for this is that they often do not include the nonlocal effects of heat transport. Other phenomena that inhibit the transport of heat, including magnetization of the electrons and turbulence in the plasma, tend to increase the coronal temperature. Simulations of such effects, using crude models, find that they also decrease g. For any specific type of experiment, it is often necessary to adjust the simulation parameters based on (direct or indirect) measurements of the ablation pressure, in order to obtain realistic results.

9.1.6.2 Mass Ablation and the Rocket Effect

Next we discuss the effects of ablation on the target itself. First, the ablation removes mass from the target, creating a loss of mass per unit area often designated $\dot{m}$. This mass flows through the critical surface into the rarefaction, so the mass flux is $\rho_s c_s$, ignoring any motion of the critical surface. In terms of the above models and evaluating T_e from (9.42), this gives

$$\dot{m} = g \frac{Am_p n_c}{Z} \sqrt{\frac{Zk_B T_e + 3k_B T_i}{Am_p}} = 1.5 \times 10^5 I_{14}^{1/3} \lambda_\mu^{-4/3} \text{ g cm}^{-2} \text{ s}^{-1}, \tag{9.48}$$

for $g = 1$, $T_i = T_e/3$, $Z = 4$, and $A = 8$. The standard scaling from Lindl (1995), which is based on simulations, has a coefficient of 2.6 rather than 1.5 here. The reality is that this coefficient depends on details that are beyond both this calculation and the standard simulations, so that it must be measured in any case where its precise value matters.

Second, the ablation pressure pushes on the target. The immediate effect of the ablation pressure on the target is to launch a shock into it, with consequences that were discussed in Chaps. 4 and 5. A short time after the shock wave has traversed the target, the entire target begins to accelerate. This is essential to laser fusion, for example, which needs initially to deposit as much kinetic energy as possible within moving material. To assess the acceleration of the target, which is like the acceleration of a rocket, we work with equations describing the conservation of momentum for such an object.

We take the object to have an initial mass, m_o, and an instantaneous remaining mass $m_r = m_o - m_a$, where m_a is the total ablated mass. (Equivalently, all the masses may be replaced by mass per unit area.) One can derive the resulting behavior using the conservation of momentum, which must apply to the combined system of rocket and exhaust. Working in the lab frame, when an element of mass dm is ejected, the exhaust carries away (in the opposite direction) a momentum $dm(V_{ex} - V)$, in which the rocket velocity is V and the exhaust velocity is V_{ex}. The increase in momentum of the remaining rocket mass is $(m_r - dm)(V + dV) - m_r V$. Setting these equal and taking the limit that dm and dV are infinitesimal (so the product $dmdV$ is negligible), one finds

$$V_{ex} dm = m_r dV = (m_o - m_a)dV, \tag{9.49}$$

where m_a is the variable whose increase is measured by dm. Integrating from $m_a = 0$ to some value, and taking $V = 0$ initially, one finds

$$V = V_{ex} \ln (m_o/m_r). \tag{9.50}$$

This shows that the velocity of the rocket increases rapidly at first and then more slowly, reaching V_{ex} when about 2/3 of the mass has been ablated. It would seem that one could reach an arbitrarily high velocity by ablating nearly all the mass. Unfortunately, the hydrodynamic instabilities discussed in Chap. 5 place a limit on the amount of mass that can be ablated without breaking up the target.

The *ablation efficiency* of such a rocket, ϵ_R, is defined as the ratio of kinetic energy of the remaining mass to the total kinetic energy of the rocket plus the exhaust. This is the efficiency of an ideal system in which all the energy was kinetic and no energy went to heat. Thus $\epsilon_R = m_r V^2/(m_r V^2 + 2K_{ex})$, where K_{ex} is the kinetic energy of the exhaust. To find this one must determine the total kinetic energy of the exhaust, from

$$
\begin{aligned}
K_{ex} &= \int_0^{m_a} \frac{1}{2} \left(V_{ex} - V \right)^2 \, \mathrm{d}m_a \\
&= \int_0^{m_a} \frac{1}{2} V_{ex}^2 \left[1 + \ln\left(1 - \frac{m_a}{m_o} \right) \right]^2 \, \mathrm{d}m_a \\
&= \frac{1}{2} m_o V_{ex}^2 \int_{m_r/m_o}^1 \left[1 + \ln\left(\frac{m_r}{m_o} \right) \right]^2 \, \mathrm{d}\left(\frac{m_r}{m_o} \right).
\end{aligned}
\tag{9.51}
$$

This integral evaluates to $1 - (m_r/m_o)[1 + \ln(m_r/m_o)]$. Evaluating the efficiency, one finds

$$
\epsilon_R = \frac{(m_r/m_o) \ln^2(m_r/m_o)}{1 - (m_r/m_o)},
\tag{9.52}
$$

which is plotted in Fig. 9.12. When a small fraction of the mass has been ablated, one can expand the logarithm in (9.51), using the fact that $m_a \ll m_o$. Doing this, one finds to the lowest order $\epsilon_R = m_a/m_o$, so that the efficiency is equal to the fraction of the initial mass that has been ablated. *This is a very useful result.* One can see that it is reasonably accurate up to about 70% ablated mass fraction, which is far beyond its formal range of validity. The observed efficiency, in laser–plasma experiments and computer simulations, is roughly half this value. This reflects the fact, discussed above, that only about half of the energy provided by the laser energy flux actually reaches the densities above the absorption zone, where the ablation occurs.

Fig. 9.12 Ideal efficiency of rocket-based acceleration. The dashed line shows the simple estimate for small ablated mass

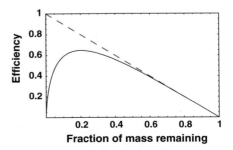

9.2   Dynamics of Mid-Z to High-Z Targets

Mid-Z and high-Z targets exhibit some common behavior upon irradiation with energy fluxes relevant to inertial fusion, and do so whether the energy source is laser light or X-rays. We discuss this behavior here.

For low-Z targets, heating near the surface ionizes the matter fully, or at least into a He-like state. This reduces the opacity to a value not much above the free-free, inverse-bremsstrahlung value. We saw in Chap. 6 that this opacity is much smaller than the opacity when more electrons are attached. As a result the radiation emission is relatively small and radiation plays at most a minor role in the dynamics. For the case of laser irradiation, the electrons carry heat into the denser matter, sustaining its pressure and driving a shock forward into the matter. The resulting profile is like that shown in Fig. 9.7.

As the nuclear charge Z_{nuc} of an irradiated target increases, the emission of thermal radiation by matter at densities near 1 g/cm^3 plays an increasing role in the dynamics. The ionization there becomes less complete and the opacity increases, so that the matter in this region emits significantly more radiation than one sees in low-Z targets. The target evolution then involves the interplay of radiation emission and transport, shock-wave generation, and hydrodynamic expansion. We will discuss this interplay here.

Begin with heating of a layer at the surface of a target, either by heat conduction from a laser-heated corona, or by absorption of an external X-ray flux. When the radiation emitted by this hot layer is sufficient, three effects ensue. First, the radiation launches a Marshak-like heat wave into the target. As we saw in Chap. 7, such diffusive waves initially have a high velocity but slow with time. Second, the high pressure of the heated matter at the target surface launches a shock wave. This shock wave penetrates the target at a relatively steady velocity, and so eventually overtakes the heat wave. The shock launched as the heat-wave slows (see Sects. 7.3.1 and 8.2.1) may also play a role. For practical experimental parameters, and for all materials of interest, this all occurs within small fractions of a nanosecond. Third, the heated matter at the target surface expands, forming a nearly isothermal rarefaction at a density near 1 g/cm^{-3}.

The temperature in the rarefaction is small hundreds of eV, for reasons discussed below, so the sound speed associated with the rarefaction is many tens of μm/ns and its spatial scale quickly becomes tens of μm. So long as the radiation mean-free-path is a few μ or less, the radiation transport will be diffusive. This requires that the specific opacity be of order 10^4 cm^2/g or larger. This requirement is satisfied at such temperatures for all mid-Z and high-Z materials. As heating continues, the radiation penetrates diffusively through the rarefaction, sustaining its temperature and eventually heating and ablating the denser matter. The structure of the dense matter evolves into an expansion heat front (Sect. 8.2.2), although the evolution of the densest matter will also be affected by the returning rarefactions and consequent reshocks (Sect. 4.6.3).

At the outer surface of the optically thick layer, the hot matter emits radiation outward. It turns out that this process soon becomes dominant in the energy balance. Suppose the incident energy flux, in units of 10^{14} W/cm^2, is F_{14}, and that the fraction radiated outward is β. Then the effective temperature of the radiation is

$$T_{eff} = 177(\beta F_{14})^{1/4} \text{ eV.} \qquad (9.53)$$

When the incident energy flux is X-rays, $\beta \sim 1$ after the initial transients, and so the re-emission becomes nearly 100% efficient. One says the material has an albedo of ~ 1.

The story for laser-irradiated targets is more complex. Additional phenomena that are important in these targets include laser absorption below the critical density and electron heat transport to higher density, where the dynamics just described applies. The heating by laser absorption drives an isothermal rarefaction in the corona below critical density. The heat transported inward by the electrons acts to ablate the lower-density end of the rarefaction driven by the X-rays. The resulting structure, featuring two ablation fronts and two rarefactions, has become known as a *double ablation front* (Sanz et al. 2009; Drean et al. 2010). The value of β in this case does vary strongly with Z. It depends on how much of the energy flux is expended in the low-density matter and does not reach the strongly radiative zone. We discuss the specific case of high-Z targets such as Au in the next section.

Figure 9.13 shows profiles of plasma parameters for Cu and Ag irradiated by laser light, under conditions described in the caption. In both materials, one can see three distinct physical regions, separated by relatively abrupt transitions. There is the underdense corona, where the ionization is nearly complete and T_e is a few keV. The steepness and extent of the transition to the middle region is affected by electron heat transport. If the transport is more inhibited then the transition is steeper and extends to lower density [lowering g in (9.44) through (9.47)]. In the middle, one sees the optically thick rarefaction produced by expansion of the denser matter heated by X-ray diffusion. Here Z is reduced, the density is relatively flat, and T_e

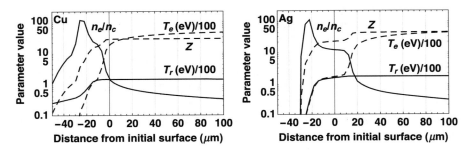

Fig. 9.13 Axial profiles of plasma parameters, as labeled, for Cu and Ag targets irradiated with 0.35 μm light at 10^{15} W/cm^2, as evaluated after 1 ns of laser irradiation. The profiles were taken near the axis of symmetry from simulations using a two-dimensional radiation hydrodynamics code with accurate laser absorption. The simulation used a 680 μm laser spot

and T_r are both somewhat above 100 eV. The extent of this region is relatively small in Cu at 1 ns; it would increase with time. Finally there is a denser region of shocked target material, in which Z, T_e, and T_r are further reduced. To the left is the shock front, which in reality would be much steeper than can be resolved in this simulation. The target material extends much further to the left. For Ag, as Z drops to zero so does n_e. For Cu, the opacity is small enough that some radiative heating penetrates beyond the shock-wave, so there is some ionization and n_e remains finite there. In detail, one sees in Cu the separation of T_r from T_e discussed for radiative-shock precursors in Chap. 8.

9.2.1 X-Ray Conversion of Laser Energy

Experiments using thermal X-rays to drive matter require an initial X-ray source. In many cases, this X-ray source is produced by irradiating a high-Z surface with a laser. The most common material used is gold, because it produces X-rays with high efficiency, it is easy to work with for target fabrication, and it is chemically inactive, so that oxidation or other processes are not a concern. In the present section, we specifically discuss laser-irradiated gold.

Below the critical density, collisional absorption, which scales as Z^2, is much larger than it is in low-Z plasmas. The gold becomes ionized to a Z of several tens. There are two limiting types of behavior relating to the heat transport. In the first case, the heat transport is the Spitzer-Harm value and small enough that flux-limiting is not needed, and the absorption occurs over an extended volume below critical density. The radiation from the lower-density, multi-keV, gold plasma is not dominant, but may include noticeable radiation at higher energies, as we see below. In the second limiting case, the heat transport is well below the Spitzer-Harm value because some mechanism interferes with it. In response, the plasma below critical density becomes significantly hotter and the density profile across the critical density becomes much steeper. The heat is still carried by the electrons, but the behavior near n_c then resembles the expansion heat front discussed in Sect. 8.2.2. In this limit, laser absorption near critical density by resonance absorption or other mechanisms may become important.

At least two mechanisms can potentially interfere with the electron heat transport through the critical surface. One of these is self-generated magnetic fields, which inhibit transport by trapping electrons locally, penetrating only as collisions displace their orbits. Such fields mainly have been thought to occur near the edges of the plasma, but some recent evidence suggests that they may not be so simple. The second such mechanism is the excitation of ion-acoustic turbulence. The Landau damping of ion-acoustic waves is small in high-Z plasmas. When heat transport is large enough, simulations and theory indicate that they can be driven into a turbulent state. The initiating mechanism is a two-stream instability, developing because the entire electron distribution drifts outward to balance the current carried inward by the heat-transport process. When this drift speed exceeds the sound speed, ion-acoustic waves become unstable. At this writing, both of these mechanisms remain active areas of research.

The plasma above the critical density strongly radiates the energy that reaches it. As we saw above, the energies being radiated are hundreds of eV. The mechanisms involved are a combination of bremsstrahlung emission and emission from the very many transitions that involve the N, O, and P shells in the Au. There are so many such transitions, that each shell is associated with a band of emission rather than with discrete lines. Accurate atomic physics modeling of them requires accounting for millions or more of such transitions. As we discussed in Chap. 6, the limiting intensities of any such line (band) emission are the Planckian thermal spectrum, and these intensities are reached when any given energy becomes optically thick. The outcome is that the spectrum from the laser spots is similar enough to a Planckian that we can treat it that way for our purposes here. When the emission from laser spots irradiates a gold wall, the re-emission from that wall is even closer to Planckian.

Returning to the gold plasma, the dense material above the critical density is heated by the energy transported to it, primarily by electron heat conduction. The consequences were discussed in the previous section. There is an initial competition between a Marshak-like heat wave and a shock waves. Before long an expansion heat front develops. It includes a nearly isothermal rarefaction, heated by radiation diffusion. The expansion within the rarefaction, and the ablation of material at its high-density end, enables the radiation to penetrate the material more quickly than it would in a pure Marshak wave. A standard estimate by M.D. Rosen, reported in Lindl (1995), for the penetration depth x_M into the dense matter is

$$x_M = 0.53 T_{\text{heV}}^{1.86} t_{ns}^{0.75} \mu\text{m}, \qquad (9.54)$$

in which T_{heV} is the temperature of the material at 1 ns, measured in hundreds of eV, and t is as usual the time, measured in ns here. Note that T_{heV} is the temperature in the rarefaction above critical density, and is not the electron temperature in the absorption region, which is much larger. The power required to heat this layer is

$$P_M = \rho R T_e \frac{dx_M}{dt}, \qquad (9.55)$$

after converting the units of x_M as needed. Here ρ is the density of the solid material but R and T_e correspond to the properties of the material within the rarefaction. For temperatures of interest, below a few hundred eV, this power remains less than 10% of the power incident on the target. The shock wave moves a few times faster into the dense matter, but heats it much less (to $\sim$10 eV), and so adds only a small fraction to the required power.

In addition, even without additional inhibition of heat conduction, simulations typically find the temperature in laser-heated Au plasmas to be a few keV and find a significant drop in density through the critical surface, both of which correspond

Fig. 9.14 A typical spectrum from an irradiated gold surface. Here J_x is the X-ray energy into 2π steradians and J_L is the laser energy. The energy flux was $\sim 5 \times 10^{14}$ W/cm^2, at 0.35 μm wavelength. The data below 2 keV is from a 10-channel X-ray diode detector system; above 2 keV it is from two crystal spectrometers. Credit: Robert L. Kauffman

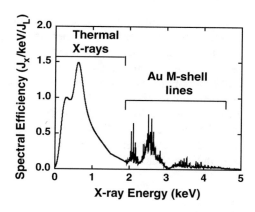

(see Fig. 9.11 and the related discussion) to having a majority of the laser energy penetrate to higher density. And from the above discussion, nearly all of this energy should be re-radiated as soft X-rays. Experiments using irradiation of spherical targets have observed conversion efficiencies from laser energy to soft-X-ray energy above 70% for Au targets and above 90% for targets using a mixture of Au, U, and Dy. There are other complexities. Targets with small laser spots produce much lower conversion efficiencies, an effect attributed to lateral heat conduction. Some evidence suggests that laser spots within the high-temperature environment of hohlraums have higher conversion efficiencies than laser spots in isolation do. The effect on nominal effective temperature is modest. For conversion efficiencies of 50%, 70%, and 100%, the coefficient in (9.53) becomes 148, 162, and 177, respectively. In the discussion of hohlraums that follows, we will take the conversion efficiency to be 70%.

Figure 9.14 shows a typical spectrum of the X-rays emitted by a gold laser spot. The total radiation has soft component dominated by emission from the N, O, and P bands in the Au, in addition to inverse bremsstrahlung. This is often approximated as a blackbody spectrum having a temperature of order the estimates just given. It is often described as the "thermal" emission. The emission from the hotter, underdense, absorption region has a more-energetic, but nonequilibrium spectrum, dominated by radiation from Au M-band transitions (involving lower-state electrons whose principal quantum number is $n = 3$). These have an energy near 2 keV and typically contain from a few % to more than 10% of the laser energy. The M-band radiation can pose challenges for inertial fusion or other experiments, as it penetrates much more deeply into low-Z surfaces than the thermal radiation does.

9.2.2 X-Ray Production by Ion Beams

Another approach to heating hohlraums is to use beams of heavy ions. These can in principle be produced with a high efficiency and at a high repetition rate, making heavy ions a plausible source of energy for a power plant based on inertial confinement fusion. But the ion beams are not focusable to the degree that a laser is. An ion-beam-heated hohlraum will be irradiated from a single direction or from two opposing directions. The ion beams will deposit their energy within the outer wall of a hohlraum, or within a beam target placed within the hohlraum. This will produce a hot source that will irradiate those walls exposed to it, beginning the same process of absorption and reemission that occurs in a laser-heated hohlraum. For applications needing uniform X-ray heating of some object, the ion-beam-heated spot will typically be hidden from the object, which will be heated only by radiation from the hot walls.

9.3 Hohlraums

When Max Planck was exploring the fundamental nature of thermal radiation, in the late nineteenth century, he found it useful to conceive of a completely enclosed volume, within which the radiation field would reach equilibrium with the material in the walls of the volume. He designated such a volume a *hohlraum*. Placing a very small hole in such an enclosure disturbs it negligibly, and the image of an enclosed volume whose radiation emerges from a small hole has become a standard one in the study of blackbody radiation in courses on statistical and thermal physics. Evidently by heating such an enclosure one can increase the temperature of the radiation field. This is the notion behind laser-driven hohlraums, which allow one to produce radiation fields having very nearly the spectrum of blackbody radiation (described as Planckian) at temperatures of millions of degrees. In this section, we explore these devices and some of their behavior.

Figure 9.15 shows an image of a typical hohlraum. The overall purpose of the hohlraum is to create a useful radiation environment. They are often cylindrical, as this is easy to manufacture, but they can be made and have been made with a very wide variety of shapes. Their typical dimensions are a few mm or less, for experiments using kJ-class lasers. Hohlraums are composed of some high-Z material. Gold is often used because good methods have been developed for

Fig. 9.15 Image of a typical hohlraum. The hohlraum is the cylindrical object, with a visible laser entrance hole to the right. The laser beams strike the interior walls of the hohlraum, as indicated in one case. An experimental package may be located within the hohlraum or on the wall as shown

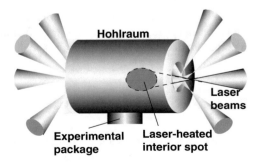

producing gold hohlraums. Every so often, someone does an experiment in which the laser beams heat a (thin-walled) hohlraum by striking it on the outside. But it is more typical to do as shown in Fig. 9.15—to provide one or more laser entrance holes, through which some number of laser beams enter the volume. In many cases beams enter the hohlraum from two directions, which makes it easier to create symmetric conditions inside the hohlraum. Some experiments, however, use only one laser entrance hole and a comparatively short cylinder. Such targets are known as halfraums, which is something of a pun as it confuses hohl with whole. (Another common mis-use of language is to describe an enclosed target made of a low-Z material as a "low-Z hohlraum".) Hohlraums also can be heated by ion beams (Sect. 9.2.2) or Z-pinches, discussed in Sect. 10.10. Here for laser-heated hohlraums, we discuss the establishment of a radiation environment, the application of the radiation to the ablation of matter, and what can go wrong.

At times the radiation is used to heat a spherical capsule in the center of the hohlraum, with the important example being inertial confinement fusion. Sometimes the radiation is used to heat a sample rather than to ablate it, for example in order to measure the structure of the X-ray opacity. At other times the object to be heated is mounded on the wall of the capsule, as in Fig. 9.15.

Consider the energy balance for a hohlraum within which there is an energy source, in the form of X-rays emitted from laser-heated spots. (One can adjust the calculation for other initial energy sources.) This radiation from the laser spots illuminates the interior walls of the hohlraum. This is similar to the irradiation of the laser spots, although the X-rays penetrate to higher densities, where they contribute to heating, driving Marshak and shock waves, and reemission. But an arbitrary point on the wall of the hohlraum is not just irradiated by the laser spots. It is also irradiated by the other walls of the hohlraum that it can see, which further contributes to the local heating. Because the transit and reemission times are short compared to the timescale for the evolution of the system, we can express the total energy flux reaching the wall, I_w, as a series:

$$I_w = I_o \left(1 + \alpha f + \alpha^2 f^2 + \cdots \right) = \frac{I_o}{(1 - \alpha f)} = \frac{\eta f A_L / A_w}{(1 - \alpha f)} I_L = \xi I_L, \qquad (9.56)$$

in which $I_o = \eta f I_L A_L / A_w$ is the average energy flux of the walls due to the laser spots. Here η is the X-ray conversion efficiency, A_L and A_w are the areas of laser irradiation and of the wall, respectively, and f is the fraction of the emission from the laser spots that reaches other walls. The fraction f is less than 1 because energy is lost through the laser-entrance hole and to capsules or other objects. The fraction of the radiation incident on a point on the wall that is reemitted by the wall is the albedo α; so the fraction of the energy flux of the walls by the laser spots that is reemitted and is then absorbed again by other walls is αf. There is in turn reemission of this radiation, and so on. One thus constructs the power series seen in (9.56). This series converges and the final result is that the wall energy flux is proportional to the laser energy flux. The constant of proportionality is $\xi = \eta f (A_L / A_w) / (1 - \alpha f)$. We assume here that this applies uniformly and in an averaged sense. In actual hohlraums, the wall temperature varies because the transport of energy is not uniform. We will return to this point in Chap. 11 Typical values of the parameters are $\eta \sim 0.7, f \sim 0.9$, which is basically the ratio of entrance hole and package areas to the total area, $\alpha \sim 1$, because the reemission is much larger than the energy penetration into the walls, and $A_L \sim 0.1 A_w$, because one wants A_L to be as large as feasible but must also inject the laser beam into the hohlraum and place the laser spot where the experiment demands. Taken together, one finds $I_w \sim 0.6 I_L$. Thus, ξ can be near unity.

Take note of this result. Hohlraums are amazingly efficient devices. They can absorb the input energy and keep most of it bouncing around from wall to wall for many bounces. They can irradiate a capsule or an experimental package with an energy flux of soft X-rays that approaches the energy flux of the laser beams themselves. We will return to this point soon. In addition, it is now clear that we were not really justified in treating the laser spot in isolation, because the laser spot is also illuminated by the hohlraum walls and by other laser spots. Indeed, there is some evidence (Lindl 1995) that the effective conversion efficiency is higher within hohlraums than outside them. But the evidence is complicated and not entirely conclusive. We will leave its exploration to the interested reader.

Solving the approximate solution to the energy transport equation, $I_w \approx \sigma T_w^4$, where T_w is the wall temperature, we obtain an equation for the wall temperature (often called the hohlraum temperature or the radiation temperature, T_r):

$$T_w = \left[\frac{\eta f A_L / A_w}{\sigma (1 - \alpha f)} I_L \right]^{1/4} = 177 \left[\xi I_{14} \right]^{1/4} \text{ eV.} \qquad (9.57)$$

The units of T_w are determined by those of σ, which must be consistent with those of I_L.

Next suppose we place a capsule or other experimental package, of area A_c, within or on the wall of the hohlraum. The energy flux experienced by such an object is I_w. The material is usually low-Z as the goal is often to produce ablation pressure, and even targets irradiated for other purposes very often use a low-Z outer layer to prevent motion of the interior materials (such a layer is called a *tamper*). The X-rays readily penetrate through the ionized, low-Z material that has already been heated. (This material may be fully ionized but in any event has few atomic transitions to absorb the X-rays. It also has a low enough collision rate to be weakly absorbing.) The X-rays are then absorbed in a short distance once they reach the cooler material that has not yet been heated. In most cases the object absorbs nearly all the energy incident upon it and thus contributes to f. The fraction of the laser power that is delivered to the capsule is

$$\frac{I_w A_c}{I_L A_L} = \frac{\eta f}{(1 - \alpha f)} \frac{A_c}{A_w}. \qquad (9.58)$$

This can be a large fraction but cannot exceed 1 because $\eta < 1, \alpha < 1$, and $f < (1 - A_c/A_w)$. One sees that the larger one makes the capsule, relative to the hohlraum, the more efficiently one will deliver energy to it. Unfortunately, this comes at a cost because larger capsules experience less uniform irradiation, as we will discuss further in Chap. 11.

9.3.1 X-Ray Ablation

As is mentioned above, many applications of hohlraums involve the ablation of low-Z materials by the X-rays produced in the hohlraum. This drives a shock-wave into the matter, and then acts to accelerate samples that are sufficiently thin, as discussed in Sect. 9.1.6.2. During the period of shock propagation, the structure is that of an expansion heat front, as discussed in Sect. 8.2.2 and illustrated in Fig. 8.25.

A first estimate of the ablation pressure (on the downstream side of the heat front) can be obtained as follows. Some fraction of the power delivered to the ablating object by the soft X-rays is deposited at the heat front, where it provides the flow of energy into the rarefaction at speed c_s and the energy necessary to sustain the pressure behind the shock. Balancing these gives the energy density at the heat front, ϵ_{hf}, and the ablation pressure is $(\gamma - 1)\epsilon_{hf}$. The incident X-ray energy flux must also provide the required downstream heating and kinetic energy of the rarefaction and the energy for pdV work, so we assume that 50% of the incident X-ray energy contributes to the ablation pressure, which gives

$$P_{abl} = 0.50(\gamma - 1)\sigma T_w^4 \sqrt{\frac{Am_p}{(Z+1)k_B T_w}}. \tag{9.59}$$

Note that this is formally independent of density. However, based on the above discussion one will only find the conditions necessary to create the structure depicted in Fig. 8.25 with certain materials and over a certain density range. If we evaluate this for $\gamma = 5/3$ and for Be, we find

$$P_{abl} = 4.4 \left(\frac{T_w}{100 \text{ eV}}\right)^{3.5} = 33 \, [\xi I_{14}]^{7/8} \text{ in Mbar}. \tag{9.60}$$

The standard estimate, based on simulations and reported in Lindl (1995), would replace 4.4 by 3 in the first part of this equation or would correspond to $\xi = 0.68$ in the second part. In detail, the ablation pressure depends on materials and varies in time. It is also useful to have an equation for c_s subject to the same assumptions. This is

$$c_s = \sqrt{\frac{(Z+1)k_B T_w}{Am_p}} = 7.3 \times 10^6 \sqrt{\frac{T_w}{100 \text{ eV}}} = 0.97 \times 10^7 \, [\xi I_{14}]^{1/8} \text{ cm/s}. \tag{9.61}$$

For comparison, we can also evaluate these quantities using the solutions from Sect. 8.2.2 for a self-similar expansion heat front when the source temperature determines the temperature in the rarefaction. The expressions are algebraically complex, but the result scales quite closely to $T_w^{3.5}$, as does the simple estimate above. One finds a coefficient of about 7.5, for comparison with the values of 4.4 and 3 discussed above. We will return to X-ray ablation in Chap. 11, where we discuss its applications to inertial fusion. What is important at present is to see from (9.60) that it is also straightforward to obtain pressures of order 100 Mbar by using lasers to heat a hohlraum. Although the target is more complicated in this case, the resulting irradiation is inherently uniform, unlike that obtained even with smoothed laser beams. As a result, hohlraums are an option worth evaluating for any experiment that would benefit from highly uniform irradiation.

9.3.2 Problems with Hohlraums

In addition to their complexity, hohlraums have other limitations that affect their usefulness in certain experiments. We will discuss here four of these: asymmetry, LPI, plasma pressure pulses, and crossed-beam effects.

9.3.2.1 Hohlraums: Asymmetry

For laser-driven fusion and for some other applications, it is important to irradiate a capsule symmetrically. In practical experiments hohlraums do not produce isotropic radiation fields. Using the equations of this chapter, one would be led in designing almost any experiment to maximize A_L/A_w, to maximize the energy delivered to a capsule or other target. But a given point on an experimental package sees larger X-ray emission (a hotter environment) from the laser spots and little X-ray emission (a colder environment) from the laser entrance holes. As one moves around the surface of a laser capsule, or even of a planar target, this can produce significant variations in the X-ray flux. This in turn can produce asymmetric pressures on the surface of the irradiated object. Efficient target designs require the use of viewfactor codes, which integrate a specified X-ray source distribution over solid angle at each point on a target, to assess quantitatively the radiation uniformity.

Unfortunately, as the hohlraum walls expand the X-ray sources at the laser spots move. The result is that one cannot obtain symmetric irradiation of a capsule at all times, from a fixed set of laser beams. It is necessary to design the experiment so as to produce a desired symmetric outcome (such as a spherical capsule implosion) by using asymmetric irradiation that varies in time. The design must be such that the effect of these time variations averages out. If one has enough independent laser beams available, then one can also vary their properties in a way to assist with this. Note that to succeed at such a design one must successfully model both the motion of the X-ray sources and the response of the capsule. In practice this has not proven feasible. To obtain a desired pattern of irradiation, it has proven necessary to measure the time-dependent structure of the X-ray irradiation in three dimensions and to adjust the experimental setup in response. The issue of the effect of asymmetry on fusion capsules is discussed in Chap. 11.

9.3.2.2 Laser-Plasma Instabilities

Any hohlraum very quickly develops a volume of plasma that is very large compared with the scale that matters for laser-plasma instabilities (LPI), which is of order the laser wavelength. Hohlraum plasmas also are often much more uniform than those emerging from open targets. Because the strongest limitation on LPI is nonuniformity, the implication is that hohlraums have the potential to produce much larger levels of LPI than open targets do. These problems are very much larger in experiments with infrared lasers, because they have higher oscillating velocities and smaller collisional effects at a given energy flux. Managers of research programs seeking to produce fusion using hohlraums have never liked the fact that LPI threatens their success. Unfortunately, they have often chosen to ignore it and have failed to prepare for the potential adverse effects. This unwise prioritization has nearly killed the laser-driven-fusion program in the US at least twice.

The first time was in the late 1970s, at a time when attempts to produce fusion in hohlraums were centered on "vacuum hohlraums", having only the fusion capsule within them. Time and again, the neutron yields from fusion experiments fell orders of magnitude below those predicted by the computer simulations that were used to define the experiments. In that era almost none of the details of target performance were checked by measurement, and so many relevant issues were not controlled. But the dominant explanation of the observations turned out to be plasma filling.

Plasma filling is jargon, as any hohlraum will fill with plasma at some density almost immediately. Beyond this, plasma filling refers to the establishment of conditions that permit the excitation of the laser–plasma instabilities discussed in Sect. 9.1.3. In particular, stimulated Raman scattering can become very strong in plasmas with a large volume near $0.1n_c$ in density.

We can gain a qualitative understanding of the main aspects of plasma filling from some simple calculations. Both the laser-heated spots and the X-ray heated walls contribute to the filling of the hohlraum. The X-ray heating takes time to develop, so that early in time the laser-heated spots dominate. Based on (2.53), the sound speed $\sim \sqrt{ZT_e/M}$ is above 3×10^7 cm/s, and the hohlraum radius is below 1 mm, so the timescale for filling is less than 3 ns. This is the regime in which vacuum hohlraums irradiated with infrared laser light operate.

The production of hot electrons as a result of plasma filling was both a severe and an unknown problem in early laser fusion experiments. The U.S. laser fusion effort was nearly canceled before the scientists involved discovered and demonstrated the effects of plasma filling. Data from such experiments were eventually well explained by a model that assumed that all the laser energy was lost to LPI as soon as the hohlraum filled to a density of $\sim 0.25n_c$. A next level of detail regarding these observations is provided by Lindl (1995). Some observers, including the present author, believe that this problem was not found sooner because the fact that the phenomena responsible for it, though known to exist, were not incorporated in the primary computer code used to model the experiments.

For hohlraums irradiated with optical and UV lasers, the expansion of the heated walls becomes more important. Once the walls of the hohlraum heat sufficiently, the sound speed of their plasma is $\sim 10^7$ cm/s. However, because the plasma expands from a much higher initial density, it needs to expand for a only a short time to produce effects throughout the hohlraum. Describing this expansion as an isothermal rarefaction (Sect. 4.4.1), so that the density is $n = n_s e^{-x/(c_s t)}$, where x is the distance from the high-density edge of the rarefaction, n_s is the electron density near the solid material, which we will take to be 10^{24} cm^{-3}, and we take c_s from (9.61). We ask how long it will take for the plasma density at the center of the hohlraum from one segment of wall to reach $0.001n_c$, This is about 14 e-foldings below solid density, and when this occurs there is a large volume in which the density is near $0.1n_c$, which is about eight e-foldings below solid density. For a hohlraum radius R_{mm} in mm we find

$$t = 0.61 \frac{R_{mm}}{\sqrt{T_w/100 \text{ eV}}} \text{ ns,} \qquad (9.62)$$

with ρ in g/cm^3 and T_w in eV. For typical parameters, this gives times below 2 ns. Thus, plasma filling can also be an issue in hohlraums irradiated by optical and UV lasers.

The second time that failures involving LPI, among other problems, nearly killed the laser fusion program in the US was in the early years of operation of the National Ignition Facility (NIF). The NIF, a ~ 2 MJ laser, has been spectacularly successful as a research facility. In the several decades during which the author has seen the completion of many major research facilities, NIF is the only one to have been ready when "completed" to operate at its specifications, without needing more time for system integration or needing to soon make major repairs. Unfortunately, the initial research program whose goal was to use NIF to achieve fusion was a failure, despite the involvement of a large team of outstanding scientists fielding mind-blowing new diagnostics. Much could be said about the combination of rationalistic thinking, hubris, and just plain bad management that brought this about, but here we focus only on the LPI aspect.

The initial targets for NIF employed gas-filled hohlraums. The gas was introduced to reduce the expansion of the gold walls and the resulting time variation in the symmetry of capsule irradiation. The evident danger was that the gas immediately would produce uniform plasmas within which LPI could become large. The plan was to produce conditions that would limit the unstable gain of the instabilities (though control of density, temperature, and gradients) by using computer models of the plasma conditions. Unfortunately, these computer models never have been able to model, with high accuracy, the plasma conditions produced even by simple planar targets. It is no wonder that they failed horribly to model the conditions produced in the NIF hohlraums. To make matters worse, the managers in control chose not to prioritize any ability to actually measure the plasma conditions in the hohlraums. In the event, some of the NIF laser beams lost more than half their energy to LPI, and specifically SRS. The NIF scientists developed clever ways to compensate for this that seemed to work in 2D computer models, but that did not work in reality. In other, non-fusion experiments, levels of Stimulated Brillouin backscattering in one case became large enough to damage some of the laser optics.

It is clear that MJ-class UV lasers have entered a regime where LPI effects can readily spoil an experiment and even damage the laser itself. This increases the challenge of doing successful experiments with hohlraums. This challenge has been met since the early failures on NIF. As of 2017, more recent target designs have produced comparable and in some ways better results with much less LPI.

It is worthwhile to add an aside here relating to a very bad habit in the high-energy-density community. Unless one had made a very attentive study of the literature, one could be forgiven for being surprised that the computer models failed so badly at predicting the plasma conditions on NIF. Contributing to this blindness is the entrenched tendency to model experiments using a multiphysics code, to in some way tune the code to obtain results that look somewhat like the data, and then to write in the paper that the code achieved "good agreement with the data". Managers strongly prefer this as it seems to make their program look competent and successful. In contrast, the author has come to conclude that this practice is at best non-scientific and perhaps even anti-scientific. (Alas, too many of the author's past papers do include such statements.) Unless a computer model is truly an implementation of first-principles science that accurately models the actual case of the experiment, to say that the code agrees with the data has no scientific significance. What may be scientifically significant is the specific ways in which the code result differs from the data, and the specifics of the tuning necessary to obtain some level of agreement between the code and the data, as these provide clues to potential missing important physics and to the improvement of the code. In the specific case of multiphysics modeling of plasma hydrodynamics, claims of "good agreement" in many papers might lead one to believe the models are quite accurate, but examination of the specifics of the disagreements would lead to the opposite point of view.

The overall point here is that both attention during experiment design and measurements during the experiments are important to avoid doing experiments that are compromised or worse as a result of LPI in the hohlraum. Typical experiments using hohlraums last at least a few ns. So it should be no surprise that plasma filling can be a significant factor. Both the energetic electrons and the laser scattering that laser–plasma instabilities can produce are of concern for experiments and especially for laser fusion. We discuss some of this further in Chap. 11. Finally, it is very easy to mistake a limited computer model for reality. This is a cautionary note for students of this or any other science.

9.3.2.3 Plasma Pressure

The third phenomenon worth discussing is the development of plasma pressure. A planar isothermal rarefaction, like that described above, has a velocity $v = c_s \zeta$ with $\zeta = x/(c_s t)$ and a mass density $\rho = \rho_o e^{-\zeta}$. The ram pressure, ρv^2, of such a plasma is $\rho_o c_s^2 \zeta^2 e^{-\zeta}$, which has a maximum of $\sim \rho_o c_s^2/2$ when $\zeta \sim 2$. Thus, for plasma from a gold wall with $\rho_o \sim 19$ g/cm^3 and c_s as given above, the maximum ram pressure is

$$P_{\text{ram}} \sim \rho_o c_s^2/2 = 1.3(T_w/100 \text{ eV}) \text{ Gbar}, \tag{9.63}$$

in which T_w is in eV. This is already an enormous pressure, but if the hohlraum wall stayed hot long enough, it would be an underestimate, because the pressure at the center of the hohlraum would be increased by convergence effects. The nominal time required for the point with $\zeta = 2$ to reach the center of the hohlraum would be $t_{\text{max}p} = R/(2c_s)$, which is

$$t_{\text{max}p} = 8.4 R_{\text{mm}}/\sqrt{T_w/100 \text{ eV}} \text{ ns}. \tag{9.64}$$

This typically would be a few ns, but the heating pulse for hohlraums less than 1 mm in radius is typically of order 1 ns. So the pressure actually produced when the gold plasma converges is smaller than that indicated by (9.63). Nonetheless, it can be very large compared to the ablation pressure on a package irradiated by the hohlraum. What happens on this timescale in an actual hohlraum, as elucidated by Hurricane et al. (2001), is that a large pressure pulse propagates outward from the center of the hohlraum. The implication is that any experiment using X-ray ablation to drive a package has only a limited potential duration before the effects of the ablation are overwhelmed by the pressure pulse that will eventually follow. This places a real limitation on the design of experiments to examine the long-term evolution of hydrodynamic phenomena.

A combination of practical considerations make this issue more dangerous than it might seem. It is not practical to set up and run a multiphysics model that adequately describes the behavior of the hohlraum and also the detailed behavior of whatever experiment is produced using the X-rays from the hohlraum. Modeling the hohlraum alone is a major challenge, and is in some ways not very accurate, as discussed in the previous section. In practice, one uses a hohlraum model to define a "radiation source", and one models the physics experiment by driving it with this radiation source. However, the eventual stagnation shock is not included in the radiation source. So it is also important, and easily overlooked, to run the hohlraum model long enough, along with a low-resolution representation of the physics experiment, to get an estimate of the stagnation effects. Unfortunately, the models of such effects may not be accurate. So making relevant measurements also matters.

9.3.2.4 Crossed-Beam Effects

There is some potential for LPI to develop in the plasma that fills the entrance hole through which the lasers penetrate the hohlraum. Both the heating by these lasers and their combined ponderomotive force tend to resist this such filling, but even so one ends up with a finite plasma density there. In the laser entrance hole, many laser beams cross. If one considers any two such beams, their wavevectors can be connected by ion waves having the correct k to satisfy the matching condition required for one of the beams to give energy to the other, discussed in Sect. 9.1.4. In a stationary plasma, this process would only be resonant if the two laser beams also differed in frequency by the frequency of the acoustic wave. In a moving plasma, the Doppler shift of the acoustic wave can bring the transfer into resonance for any difference, or no difference, in laser-beam frequencies. As a result, one can produce resonant, crossed-beam energy transfer either by careful (or accidental) selection of plasma conditions or by tuning the difference in laser-beam frequencies. This process can be quite large. In the early experiments on NIF, it was cleverly used to move energy into the beams who were suffering large losses from LPI.

In general, and in both hohlraums and directly-irradiated targets, one must be aware of crossed-beam effects. These include the transfer of energy just mentioned and also the potential, local plasma heating via damping of the ion waves involved.

Homework Problems

9.1 Derive the general electromagnetic wave equation (9.3) from Maxwell's equations.

9.2 Derive an equation for the conservation of charge from (9.3).

9.3 Using the equation of motion for the electron fluid in the fields of an electromagnetic wave in a plasma of constant density, determine the time-averaged distribution of energy among the electric field, the magnetic field, and the kinetic energy of the electrons. Discuss how this varies with density.

9.4 Derive the wave equation for scattering of laser light from density fluctuations (9.20).

9.5 Derive (9.22) for the longitudinal plasma velocity. Calculate the energy density of the laser light wave and show how this is related to the source term on the right-hand side.

9.6 Develop an energy equation for the electron fluid including a Spitzer–Harm heat flux, and show that it is a diffusion equation.

9.7 Determine the range of electron velocities that contribute significantly to the heat flux, by plotting the first-order contribution to the argument of the heat-flux integral (9.28).

9.8 Find the approximate expression for the ablation efficiency ϵ_R of a rocket, to second order in the quantity m_a/m_o. Plot the corresponding rocket efficiency and the value of (9.52). Discuss the comparison.

9.9 Examine the energy distribution of the dynamics at the surface of a laser-irradiated, mid-Z target, where in the upper rarefaction the plasma temperature is T. Model this region as a material at T and at a density of 1 g/cm^3, growing at the sound speed. Assume the pressure throughout the dense material, up to the shock driven inward, equals that at 1 g/cm^3, and assume the shock is driven into matter whose initial solid density is 10 g/cm^3. Evaluate and plot, as a function of T, the energy fluxes that are radiated outward, that sustain the shock wave driven into the solid material, that sustain the growing region near 1 g/cm^3. Discuss the results.

9.10 Assume that a hohlraum of 1 mm radius is heated for 1 ns at a temperature of 200 eV. Estimate the pressure produced at the center of the hohlraum when the plasma expanding from the gold walls reaches the axis. (Note: this is not an application of (9.63). Instead, you will need to think about the rarefaction produced during the heating pulse.)

References

Boehly TR, Craxton RS, Hinterman TH, Kelly JH, Kessler TJ, Kumpman SA, Letzring SA, McCrory RL, Morse SFB, Seka W, Skupsky S, Soures JM, Verdon CP (1995) The upgrade to the omega laser system. Rev Sci Instrum 66(1):508–510

Drean V, Olazabal-Loume M, Sanz J, Tikhonchuk VT (2010) Dynamics and stability of radiation-driven double ablation front structures. Phys Plasmas 17(12):122701

Hurricane O, Glendinning SG, Remington BA, Drake RP, Dannenberg KK (2001) Late-time hohlraum pressure dynamics in supernova remnant experiments. Phys Plasmas 8:2609–2612

Jackson JD (1999) Classical electrodynamics. Wiley, New York

Kruer WL (1988) The physics of laser plasma interactions. Addison-Wesley, Redwood City

Landau LD, Lifshitz EM (1987) The classical theory of fields, course in theoretical physics, vol 2, 2nd edn. Pergamon Press, Oxford

Lindl JD (1995) Development of the indirect-drive approach to inertial confinement fusion and the target physics basis for ignition and gain. Phys Plasmas 2(11):3933–4024

Liu JM, Groot JSD, Matte JP, Johnston TW, Drake RP (1994) Measurements of inverse bremsstrahlung absorption and non-Maxwellian electron velocity distributions. Phys Rev Lett 72:2717–2720

Montgomery DS, Johnson RP, Cobble JA, Fernandez JC, Lindman EL, Rose HA, Estabrook KG (1999) Characterization of plasma and laser conditions for single hot spot experiments. Laser Part Beams 17(3):349–359, 0263-0346

Rosen MD, Price RH, Campbell EM, Phillion DW, Estabrook KG, Lasinski BF, Auerbach JM, Obenschain SP, McLean EA, Whitlock RR, Ripin BH (1987) Analysis of laser-plasma coupling and hydrodynamic phenomena in long-pulse, long-scale-length plasmas. Phys Rev A 36(1):247–260

Sanz J, Betti R, Smalyuk V, Olazabal-Loume M, Drean V, Tikhonchuk VT, Ribeyere X, Feugeas J (2009) Radiation hydrodynamic theory of double ablation fronts in direct-drive inertial confinement fusion. Phys Plasmas 16:082704

Schurtz GP, Nicolai PD, Busquet M (2000) A nonlocal electron conduction model for multidimensional radiation hydrodynamics codes. Phys Plasmas 7(10):4238–4249

Sheffield J, Froula D, Glenzer S, Luhmann NC (2010) Plasma scattering of electromagnetic radiation: theory and measurement techniques. Academic, Burlington

Shkarofsky IP, Johnston TW, Bachynski MP (1966) The particle kinetics of the plasmas. Addison-Wesley, Reading

Chapter 10
Magnetized Flows and Pulsed-Power Devices

Abstract This chapter concerns the phenomena that occur within the magnetized flows one can produce using high-energy-density facilities, and the generation of high-energy-density conditions that involve strong magnetic fields. It begins with a brief summary of the motions of charged particles in magnetic fields. The bulk of the chapter concerns the magnetohydrodynamic (MHD) equations and their consequences, including magnetic-field generation by various mechanisms, field transport by the Nernst effect, waves and shock waves in MHD plasmas, the magnetized Rayleigh-Taylor instability, and magnetic reconnection. Pulsed magnetic coils, Z-pinches, and flyer plates are discussed as experimental systems involving strong magnetic effects.

The early twenty-first century has seen the maturing of several methods of using high magnetic fields to enable the study magnetized, flowing plasmas or for other purposes including equation-of-state research and inertial confinement fusion. The first is the use of pulsed-power devices to drive currents so intense that they create both strong magnetic fields and plasma, most often generating the plasma by ionizing metals or ionizing and compressing gasses. These devices also can be energy sources for a wide range of experiments and applications. The second is the use of small, pulsed-current drivers (or lasers) to create magnetic fields of up to tens of Tesla within which lasers can produce plasmas to study. The third is to exploit the "thermo-electric" or "Biermann-battery" field long known to be produced near the edges of laser-heated spots. Beyond these, pulsed-power devices can magnetically launch slabs of matter, producing collisions that create high-energy-density conditions. We discuss some of the fundamental dynamics that occur in this chapter.

We begin with a review of single-particle motions in magnetic fields, which can be skipped by those with a background in plasma physics. Then we turn to the simplest model that may apply to magnetized plasma systems, magnetohydrodynamics (MHD), introducing its equations and discussing some of the consequences. Much of the material in the present chapter is covered in more depth and at more length in the many available books on plasma physics, although some of the dynamical aspects that apply most strongly at high energy density, such as magnetized shock waves, are often not covered. One book that covers quite of few topics of interest

© Springer International Publishing AG 2018
R P. Drake, *High-Energy-Density Physics*, Graduate Texts in Physics,
https://doi.org/10.1007/978-3-319-67711-8_10

here is the book, *Plasma Physics for Astrophysics*, by Kulsrud (2005). We finish
with a discussion of some common means of producing magnetic fields and using
them to produce experimental systems of interest.

10.1   Single-Particle Motions

Magnetic fields influence high-energy-density systems primarily by affecting transport properties like heat conduction and electrical resistivity. This is especially
true at the boundaries of such systems. They play a small role in the dynamics
of most nonrelativistic high-energy-density systems. In addition, they are important
in some astrophysical systems that one might hope to understand with the aid of
high-energy-density experiments. Moreover, by using high-energy-density tools and
working at lower mass density, one can seek to examine the behavior of energetic
plasmas in which magnetic fields are dynamically important. And finally, relativistic
high-energy-density systems often inherently involve magnetic fields. For all these
reasons, understanding some simple aspects of particle motion in magnetic fields
is important for those who work with high-energy-density systems. Those readers
who have studied plasma physics have this knowledge. For those readers with no
background in plasma physics, this section is included. In Chap. 13, some aspects
of relativistic motion are discussed. Figure 10.1 illustrates the basic motions derived
and discussed in this section.

The motion of a single particle is in general described by Newton's second law,
which in nonrelativistic form reads

$$m\frac{d\boldsymbol{v}}{dt} = q\left(\boldsymbol{E} + \frac{\boldsymbol{v}}{c} \times \boldsymbol{B}\right) + \boldsymbol{F}, \tag{10.1}$$

in which the particle velocity, mass, and charge are $\boldsymbol{v}$, m, and q, respectively, The
fields are in Gaussian cgs units, and the nonelectromagnetic forces are designated
by $\boldsymbol{F}$. Because of the cross product in the Lorentz force, it makes sense to write $\boldsymbol{v}$ as
a sum of components perpendicular and parallel to the magnetic field, $\boldsymbol{B}$, as

$$\boldsymbol{v} = \boldsymbol{v}_\perp + \boldsymbol{v}_{||}. \tag{10.2}$$

The equation of motion along $\boldsymbol{B}$ is then

$$m\frac{d\boldsymbol{v}_{||}}{dt} = q\boldsymbol{E}_{||} + \boldsymbol{F}_{||}, \tag{10.3}$$

in which the $||$ subscript designates the component of a vector that is parallel to $\boldsymbol{B}$.
Likewise the $\perp$ subscript designates components perpendicular to $\boldsymbol{B}$.

The motion perpendicular to $\boldsymbol{B}$ is more complex but fortunately can be separated
into distinct elements. We find the first of these by assuming $\boldsymbol{E} = \boldsymbol{F} = 0$. The
equation of motion perpendicular to $\boldsymbol{B}$ is then

Fig. 10.1 The motions and drifts of charged particles. This work is licensed under a Creative Commons Attribution 2.5 Generic License

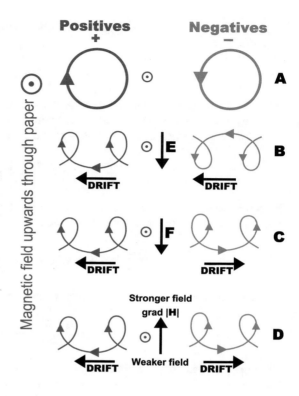

$$m\frac{\mathrm{d}\boldsymbol{v}_\perp}{\mathrm{d}t} = q\frac{\boldsymbol{v}_\perp}{c} \times \boldsymbol{B}. \tag{10.4}$$

One sees that the derivative of $\boldsymbol{v}_\perp$ is inherently perpendicular to $\boldsymbol{v}_\perp$, and is constant in magnitude. This circumstance describes circular motion. One traditionally identifies the center of the circle as a line of magnetic field, and says that the particle "orbits" this field line. The radius of this *orbit* is known as the Larmor radius, and is given by an equation that sets the force involved in circular motion equal to that from (10.4), as follows:

$$\frac{mv_\perp^2}{r_L} = q\frac{v_\perp B}{c}, \tag{10.5}$$

which gives

$$r_L = \frac{mv_\perp}{qB/c} \text{ (cgs)} = \frac{mv_\perp}{qB} \text{ (SI)}. \tag{10.6}$$

The frequency of the orbit is known as the gyrofrequency and (in radians per second) is given in cgs units by $qB/(mc)$ or in SI units by qB/m.

To find the next element of the particle motion, suppose $F = 0, E \neq 0$, and write the particle velocity as

$$v = v_g + v_d + v_{||},$$ (10.7)

in which we understand that the gyromotion is fully included in v_g, and that the vector v_d describes the new, "drift" motion that is due to the electric field; this drift motion is also perpendicular to B. In this case the equation of motion becomes

$$m \frac{dv_d}{dt} = q \left(E + \frac{v_d}{c} \times B \right).$$ (10.8)

To find the effect of E in the direction perpendicular to B, we cross this equation with B, obtaining with the aid of a vector identity

$$\frac{m}{q} \frac{dv_d}{dt} \times B = (E \times B) + \frac{v_d}{c} B^2.$$ (10.9)

The solution of this equation, for constant fields so v_d is constant in time, is

$$v_d = c \frac{(E \times B)}{B^2} \quad \text{cgs.}$$ (10.10)

Thus, the particles drift in a direction perpendicular to both E and B. An interesting aspect of this behavior is that both positive and negative particles drift in the same direction. A similar derivation shows that the drift velocity associated with an arbitrary force F is

$$v_d = \frac{c}{q} \frac{(F \times B)}{B^2} \quad \text{cgs.}$$ (10.11)

If such a force is not proportional to an odd power of q, then the positive and negative charges will drift in opposite directions, driving a current.

In the common event that the magnetic field has a spatial gradient, this also produces a drift, known as the *grad B drift*. In the typical case the gradient is small along B, so that

$$\nabla B \cdot B = 0.$$ (10.12)

Identifying the magnetic field at the center of the orbit as B_o, the field locally experienced by the particle is then

$$B = B_o + r_L \cdot \nabla B,$$ (10.13)

so that the equation of motion perpendicular to $\boldsymbol{B}$ for $\boldsymbol{E} = \boldsymbol{F} = 0$, in Gaussian cgs units, is

$$\frac{d\boldsymbol{v}_\perp}{dt} = \frac{q}{m} \frac{\boldsymbol{v}_\perp}{c} \times \boldsymbol{B}_o \left(1 + \frac{\boldsymbol{r}_L \cdot \nabla B}{B_o}\right). \tag{10.14}$$

If we now write $\boldsymbol{v}_\perp$ as

$$\boldsymbol{v}_\perp = \boldsymbol{v}_d + \boldsymbol{v}_g = \boldsymbol{v}_d + \frac{q}{mc} \boldsymbol{r}_L \times \boldsymbol{B}_o, \tag{10.15}$$

where we have written the orbital velocity in terms of $\boldsymbol{r}_L$, then (10.14) becomes, by substitution and using a vector identity,

$$\frac{d\boldsymbol{v}_\perp}{dt} = -\left(\frac{qB_o}{m}\right)^2 \boldsymbol{r}_L \left(1 + \frac{\boldsymbol{r}_L \cdot \nabla B}{B_o}\right) + \frac{q}{m} \boldsymbol{v}_d \times \boldsymbol{B}_o. \tag{10.16}$$

This is still an instantaneous equation, but we are in fact interested in the average behavior over many particle orbits. We average this equation over an orbit, noting that

$$\langle \boldsymbol{r}_L (\boldsymbol{r}_L \cdot \nabla B) \rangle = \frac{1}{2} r_L^2 \nabla B. \tag{10.17}$$

Then by taking the cross product of the averaged equation with $\boldsymbol{B}$, as before, we find the grad-B drift velocity,

$$\boldsymbol{v}_d = \frac{1}{2} \frac{m v_\perp^2}{q} \frac{\boldsymbol{B} \times \nabla B}{B^3}. \tag{10.18}$$

Note that this drift also drives a current in the plasma. This concludes our brief summary of charged-particle drifts in magnetic fields.

One other aspect of particle motion is worth mentioning. Because $\nabla \cdot \boldsymbol{B} = 0$, any change in the magnitude of $\boldsymbol{B}$ along the direction of the initial field line is also accompanied by a change in some other component of $\boldsymbol{B}$. The simplest example occurs when initially straight magnetic field lines are squeezed together, for example by a magnetic coil, producing an inward radial component to $\boldsymbol{B}$. The Lorentz force due to this second component of $\boldsymbol{B}$ either accelerates or decelerates the particle in its orbit and in its motion along $\boldsymbol{B}$. Since magnetic forces do no work, this does not change the total energy of the particle, but it does redistribute the energy between motion along $\boldsymbol{B}$ and orbital motion.

One can analyze the behavior in more than one way. On the one hand, one can consider explicitly the forces on the particle and determine the particle motion. On the other hand, one can note that the particle has a magnetic moment because its motion represents the flow of current around a circular loop, and can determine the effects of the electric field induced around the loop due to changes in B. (This second calculation is easier to do.) In either case, one finds that the magnetic moment of the particle remains constant, a result that can be expressed as

$$\frac{1}{2}\frac{mv_\perp^2}{B} = \text{constant}. \tag{10.19}$$

The result of (10.19) is that as B increases, so does $v_\perp$. This continues until all the energy of the particle is carried by the orbital motion, at which point the particle will change directions and begin moving in the direction of decreasing B. A magnetic structure in which magnetic field increases to a maximum, causing the reflection of many of the particles incident upon it, is known as a magnetic mirror.

10.2 Magnetohydrodynamic Equations

One often finds that the dynamics of interest in some case are at low frequency, in the sense that the electrons may be taken to be in steady state to sufficient accuracy. As a result, phenomena at frequencies a great deal larger than acoustic frequencies are unimportant, while the capacity of the plasma to carry current and support magnetic fields remains crucial. In addition, in many systems of interest all the distances that matter are a very large multiple of the ion orbit radius (see Sect. 10.1). In such cases, the fluid equations and the Maxwell equations can be simplified in ways we will soon discuss. There are many cases in astrophysics for which MHD modeling can be highly valuable, including the study of stellar atmospheres, planetary magnetospheres, interplanetary and interstellar space, among others. Furthermore, the relativistic generalization of the MHD equations is effective for the description of pulsar magnetospheres, galactic jets, and other phenomena. In some of these systems the magnetic field is dynamically important and strongly affects the behavior of matter. Even when the magnetic field is not dynamically important, it may greatly alter the flow of heat through the plasma. In addition, weak fields present over large volumes can contain substantial amounts of energy.

To describe phenomena at low frequencies in current-carrying plasma, one can simplify the fundamental equations to obtain *MHD equations*. These come in many detailed variations, some of which we will discuss here. First, we can add the mass continuity equations (2.42) and the momentum equations (2.43) for all the species to obtain, in Gaussian cgs units,

$$\frac{\partial \rho}{\partial t} + \nabla \cdot (\rho \boldsymbol{u}) = 0 \text{ and} \tag{10.20}$$

$$\rho \frac{D\boldsymbol{u}}{Dt} = -\nabla p + \frac{\boldsymbol{J} \times \boldsymbol{B}}{c}, \tag{10.21}$$

in which the total pressure is p and the current density is $\boldsymbol{J}$. In practical cases, ρ is dominated by the ions while $\boldsymbol{J}$ is carried almost entirely by the electrons. There is no term involving $\boldsymbol{E}$, because the net charge density is very nearly zero at low frequencies.

Now consider the equations describing the fields, under these assumptions. Because the charge density is negligibly small, one does not need the Poisson equation. Faraday's law stands as written,

$$\frac{1}{c}\frac{\partial \boldsymbol{B}}{\partial t} = -\nabla \times \boldsymbol{E}, \tag{10.22}$$

but the final Maxwell equation can be simplified to Ampère's law,

$$c\nabla \times \boldsymbol{B} = 4\pi \boldsymbol{J}, \tag{10.23}$$

discarding the term involving the time derivative of $\boldsymbol{E}$ that is present in (2.18). This term was essential to electromagnetic waves, discussed in Chap. 9, and is important in other high-frequency waves, but is negligible for the conditions of interest for MHD. In addition, one knows that $\nabla \cdot \boldsymbol{B} = 0$ and the vanishing charge density implies that $\nabla \cdot \boldsymbol{J} = 0$. These two relations may be used in some derivations but are often not listed as part of the MHD equation set.

If we apply Ampère's Law (10.23)–(10.21) we obtain the MHD equation of motion,

$$\rho \frac{Du}{Dt} = -\nabla p - \frac{\boldsymbol{B} \times (\nabla \times \boldsymbol{B})}{4\pi}. \tag{10.24}$$

Expanding the double cross product, this equation can also be cast in the form

$$\rho \frac{Du}{Dt} = -\nabla p - \nabla \frac{|B|^2}{8\pi} + \frac{(\boldsymbol{B} \cdot \nabla \boldsymbol{B})}{4\pi}. \tag{10.25}$$

This version has intuitive value, as it lets one see that the plasma is affected by gradients in the magnetic pressure, $|B|^2/(8\pi)$ and also by magnetic tension. Here $(\boldsymbol{B} \cdot \nabla \boldsymbol{B})$ describes the restoring force when a field line is bent, often compared to the familiar restoring force on a guitar string.

When we come to shock waves, we will also need an energy equation. This requires inclusion of magnetic energies in an equation like (2.28). Here we drop the radiative terms from that equation, assume that heat fluxes and work done by mechanical forces are negligible, and assume the medium is a polytropic gas, to obtain

$$\frac{\partial}{\partial t}\left(\rho\epsilon + \frac{\rho u^2}{2} + \frac{B^2}{8\pi}\right) \tag{10.26}$$

$$= -\nabla \cdot \left[u\left(\frac{\gamma p}{\gamma - 1} + \rho\frac{u^2}{2}\right) - \frac{(\boldsymbol{u} \times \boldsymbol{B}) \times \boldsymbol{B}}{4\pi}\right] - \boldsymbol{J} \cdot \boldsymbol{E},$$

noting that the Joule heating term on the right hand side may be negligible, though
not if the resistivity is significant.

The equations above are justified more fully in many of the textbooks on plasma
physics. Yet they do not enable a complete solution for the physical behavior. For
this, one needs additional equations. One needs an equation of state connecting p to
other parameters. In the equations above and our applications here, we will take the
plasma to be a barotropic ideal gas having polytropic index γ. One also needs an
equation, known as an Ohm's law, connecting E to other properties of the plasma. If
either the equation of state or the Ohm's law involve the plasma temperature, then
one may need an energy equation and perhaps an ionization-balance model, as well.

Because the electric field is established by the motions of the electrons, under the
influence of forces, the most general form of Ohm's law is the electron-momentum
equation. Haines (1986) discusses this form, and the impact of averaging its high-
frequency terms. Here we drop the terms associated with electron motion and
provide the form first developed by Braginskii (1965):

$$en_e\left(E + \frac{u \times B}{c}\right) = -\nabla \cdot \underline{P_e} + \frac{J \times B}{c} + en_e\underline{\alpha} \cdot J - n_e\underline{\beta} \cdot \nabla T_e, \qquad (10.27)$$

in which the electron pressure tensor is $\underline{P_e}$. Braginskii also provided an equation for
the heat flux,

$$Q = -\underline{\kappa} \cdot \nabla T_e - \underline{\beta} \cdot J/e. \qquad (10.28)$$

Epperlein and Haines (1986) provide the best evaluation of the tensor operators
$\underline{\alpha}, \underline{\beta}$, and $\underline{\kappa}$. They do this using Fokker-Planck calculations and provide fits for wide
variations of ionization Z. These coefficients all depend nonlinearly on the electron
magnetization, $\omega_{ce}\tau$, for electron cyclotron frequency ω_{ce} and electron collision time
$\tau = 1/\nu_{ei}$. They also depend linearly (or inversely) upon m_e, n_e, τ, and T_e, in various
combinations.

For reference below, we write out the final two dot products in (10.27). In each
case, there are three terms, each corresponding to one of three orthogonal directions.
These are the direction of B, the direction perpendicular to B in the plane of B and J
or ∇T_e, and the direction perpendicular to both B and J or ∇T_e. The corresponding
subscripts are $\parallel$, $\perp$, and $\times$, respectively. One has, designating a unit vector in the
direction of B as $b = B/B$,

$$\underline{\alpha} \cdot J = \alpha_{\parallel}(J \cdot b)b + \alpha_{\perp}b \times (J \times b) - \alpha_{\times}(b \times J), \text{ and} \qquad (10.29)$$

$$\underline{\beta} \cdot \nabla T_e/e = \beta_{\parallel}(\nabla T_e \cdot b)b/e + \beta_{\perp}b \times (\nabla T_e \times b)/e + \beta_{\times}(b \times \nabla T_e)/e. \qquad (10.30)$$

There are some aspects of these coefficients worth noting. First, as $\boldsymbol{B} \to 0$, $\boldsymbol{\alpha} \cdot \boldsymbol{J}$ and $\boldsymbol{\beta} \cdot \nabla T_e / e$ are replaced by $\alpha_{\parallel} \boldsymbol{J}$ and $\beta_{\parallel} \nabla T_e / e$, respectively, evaluated for $\omega_{ce} \tau = 0$. Second, the final term in $\boldsymbol{\alpha} \cdot \boldsymbol{J}$ has the effect of augmenting the $\boldsymbol{J} \times \boldsymbol{B}$ term (the Hall term) in (10.27). We ignore the consequences of the other terms for the moment.

To solve for plasma behavior using these equations, one substitutes $\boldsymbol{E}$ from Ohm's law into Faraday's law (10.22). We will consider several cases that are important in research using high-energy-density facilities.

10.2.1 Unmagnetized Plasma: The Biermann Battery

While always present, the source term associated with the pressure gradient has its most dramatic effects in unmagnetized plasma. Since the electrons move quickly, they attempt to flow away from the ions in response to any gradient in the electron pressure, p_e. In response, an electric field arises that prevents their escape from the ions. One has

$$\boldsymbol{E} = -\frac{1}{e}\frac{\nabla p_e}{n_e}, \text{ so} \tag{10.31}$$

$$\frac{\partial \boldsymbol{B}}{\partial t} = -\frac{c}{n_e^2 e}\left(\nabla n_e \times \nabla p_e\right) = -\frac{ck_B}{n_e e}\left(\nabla n_e \times \nabla T_e\right). \tag{10.32}$$

Historically, the laser-plasma community did not have a unique name for this source of field, which was sometimes referred to as "thermoelectric" or the "grad n cross grad T" field. More recently, however, the community has become more aware that Biermann identified this source, in an astrophysical context, in 1950, and so it is usually now called the Biermann battery. In astrophysics, a key role for this process is that it may have been the original source of magnetic field in the universe. Any asphericity in the shock wave from a supernova produces misaligned gradients of density and temperature, and so creates a magnetic field via (10.32).

It is not so easy to intuit why misaligned density and temperature gradients produce the current necessary to create magnetic field. Haines (1986) provides some help, pointing out that the energy extracted from the plasma is $\propto \boldsymbol{J} \cdot \boldsymbol{E}$, and that $\boldsymbol{J}$ involves the difference in electron and ion velocities, so the energy source represents the difference in work done by the pressure gradient on the electrons and the ions.

This source of magnetic field is important near the edges of laser-heated plasmas expanding from surfaces they irradiate. In that case, the density gradient is dominantly axial while the temperature gradient is dominantly radial. As a result, a toroidal magnetic field forms around the outer edges of the laser spot. For typical ICF lasers (ns, $\sim 10^{14-15}$ W/cm^2), its magnitude is ~ 1 MG ~ 100 T. It also operates in structures produced by the Rayleigh-Taylor instability, and under various other circumstances.

These Biermann-battery fields have also been used or present in some applications. If one creates two adjacent laser spots, each of which produces such field, then the lateral expansion of these plasmas brings together plasmas having oppositely directed magnetic fields, causing reconnection. This has been used in several experiments to study reconnection and its effects. In addition, under some circumstances the toroidal field can be convected downstream. This creates additional complications, at minimum for diagnostics, in experiments intending to use such flows to drive instabilities or turbulence.

10.2.2   Ideal and Resistive MHD

It is not so uncommon to find plasmas where any currents flow along B and the pressure and temperature-gradient sources of field are small. In this case the Ohm's law simplifies to

$$E + \frac{u \times B}{c} = \eta J, \tag{10.33}$$

in which η is the resistivity (i.e., the scalar value of $\underline{\alpha}$ in this limit). Then the magnetic-field equation becomes

$$\frac{\partial B}{\partial t} = \nabla \times (u \times B) - c\nabla \times (\eta J) \tag{10.34}$$

When $\eta = 0$, one has *Ideal MHD*, commonly used for low-density, magnetized, laboratory plasmas. In this case (10.34) causes the magnetic field to move precisely as the plasma does. The field is then said to be "frozen in" the plasma. In addition, the plasma flow velocity u is then equal to the $E \times B$ drift speed.

We will use the Ideal MHD model to derive some of the basic waves in MHD systems, but in practice magnetized flows at high energy density most often are significantly resistive. This author is partial to the treatment of resistivity in Krall and Trivelpiece (1986). The resistivity is

$$\eta = \frac{\nu_{ei} m_e}{n_e e^2} = \frac{4\pi \nu_{ei}}{\omega_{pe}^2} \text{ sec}, \tag{10.35}$$

in cgs units; the conversion to mks units is $1\,\text{s} = 9 \times 10^9$ ohm-m. Here the plasma frequency ω_{pe} and the electron–ion collision frequency ν_{ei} are both defined in Sect. 2.4.

The implications of the MHD equations for motion of the magnetic field are important for the behavior of Z-pinches and of many astrophysical systems. If one substitutes for J in (10.34) using Ampère's law, one finds

$$\frac{\partial \boldsymbol{B}}{\partial t} = \frac{\eta c^2}{4\pi}\nabla^2 \boldsymbol{B} + \nabla \times (\boldsymbol{u} \times \boldsymbol{B}). \tag{10.36}$$

For a fluid at rest, with $\boldsymbol{u} = 0$, this is a pure diffusion equation for the magnetic field. (The notion that magnetic field can diffuse is often confusing to students. It may help to note two things. First, the presence of a field implies the presence of current, generally carried by particles. Second, the particles are affected by collisions, which will tend to cause any current-carrying region to broaden.) In this case the magnetic diffusion time τ_B, with a system spatial scale of L, is

$$\tau_B = \frac{4\pi L^2}{\eta c^2} = \frac{\omega_{pe}^2 L^2}{\nu_{ei} c^2} = 1.2 \times 10^{-8}\frac{L_{mm}^2 T_{eV}^{3/2}}{Z \ln \Lambda} \text{ sec}, \tag{10.37}$$

in which L_{mm} is L in mm and T_{eV} is T in eV, and the Coulomb logarithm $\ln \Lambda$ is also defined in Sect. 2.5. The numbers implied by (10.37) are quite interesting and are illustrated in Fig. 10.2. Plasmas with sub-mm scale lengths tend to have 1–10 ns magnetic diffusion times, while plasmas with scale lengths of a few mm and temperatures above 100 eV have µs-scale diffusion times. High-energy-density systems that are magnetized can be found in either of these regimes.

If one uses the normalizing relations of Chap. 2 to evaluate the dimensionless scaling of (10.37), one finds that the relative magnitude of the diffusion term scales with $1/Rm$, where Rm is the *magnetic Reynolds number* defined as $Rm =$

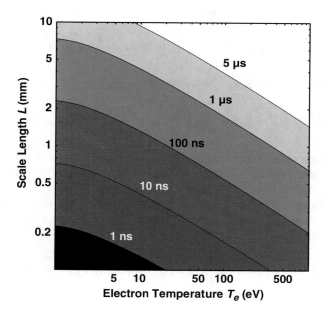

Fig. 10.2 Contours of constant magnetic diffusion time, with contour levels from bottom to top of 1 ns, 10 ns, 100 ns, 1 µs, and 5 µs. For $n_e = 10^{20}$ cm^{-3} and $Z = 0.63\sqrt{T_{eV}}$ (see Chap. 3)

$4\pi UL/(\eta c^2)$. The magnetic Reynold number is thus the ratio of the magnetic diffusion time τ_B to the hydrodynamic timescale $\tau_H \sim (L/U)$.

10.2.3   Hall MHD and the Nernst Effect

When there are cross-field currents, for example corresponding to those particle drifts that drive current, the $J \times B$ term in (10.27) can become significant or dominant. This is known as Hall MHD. We can note that $(J - en_e u)$ is the electron fluid velocity (times en_e); the net effect is that this regime has the magnetic field lines frozen to the electron motion. Hall MHD phenomena are very important in the polar Earth ionosphere and in other heliospheric contexts, and also in the coupling of a pulsed-power energy source to a target. When and whether such effects prove to become significant for the inner working of experiments in high-energy-density physics remains to be seen.

In contrast, the Nernst effect, not very well known in plasma physics, can be very important in high-energy-density systems. This effect has long been thought, from simulations, to be present in some cases. More recently (since 2010), experiments with jet-like flows within imposed magnetic fields have begun to measure its effects. In the Nernst effect, a temperature gradient causes charged particles to move, on average, across a magnetic field. This induces an electric field perpendicular to the field and the gradient, and can in turn induce a magnetic field through Faraday's law.

To find this effect, we recast the final term in (10.30) as

$$\beta_\times(b \times \nabla T_e)/e = -\frac{V_N \times B}{c}, \quad \text{with } V_N = \frac{c}{e}\beta_\times\frac{\nabla T_e}{B}, \qquad (10.38)$$

defining a Nernst velocity, V_N. Then one has

$$\frac{\partial B}{\partial t} = \nabla \times \left[(u + V_N) \times B + \frac{c}{en_e}\left(\nabla \cdot \underline{P_e} - \frac{J \times B}{c} - \text{other terms} \right) \right]. \qquad (10.39)$$

Now the magnetic field is no longer frozen in but rather convects through the flowing plasma at V_N, in consequence of heat transport. In addition, while the $u \times B$ term leads only to convection of magnetic flux with the flow at u, this new term can sustain field amplification during convection. To illustrate this, suppose u and all the terms in the rightmost parentheses in (10.39) are zero, and note that V_N depends on B. Then one has

$$\left(\frac{\partial}{\partial t} + V_N \cdot \nabla \right) B = -B\left[b(\nabla \cdot V_N) - (b \cdot \nabla)(BV_N) + \frac{V_N}{B}(b \cdot \nabla)B \right]. \qquad (10.40)$$

This shows that B can amplify convectively when B has a component along V_N.

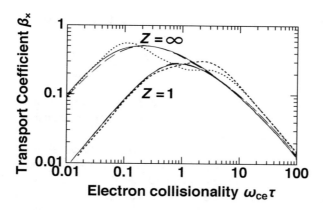

Fig. 10.3 The coefficient of heat conduction that is important to the Nernst effect, $\beta_\times$, depends strongly on the electron collisionality and the charge of the plasma ions. Adapted from Epperlein and Haines (1986). The solid lines are from their calculation, the long-dashed lines are from the fits they provide, and the short-dashed lines are from the earlier work by Braginskii (1965)

Figure 10.3 shows accurate results for the value of $\beta_\times$ in (10.38), from Epperlein and Haines (1986), who also provide fits. One can see that its value is, roughly 0.07, to within a factor of about 7, but that it also can vary by a factor of ~ 50 as the electron collisionality varies. Its value peaks at moderate collisionality.

10.3 Scaling Magnetized Flows

The scaling of magnetized flows is similar to the scaling of hydrodynamic flows discussed in Chap. 2. If we again nondimensionalize the momentum equation (now (10.24)), replacing $\partial/\partial t$ by U/L, replacing ∇ by $1/L$, where U is a characteristic speed and L is a characteristic spatial scale, and then dividing the equation by $\rho U^2/L$, the terms on the right hand side become

$$\frac{L\nabla p}{\rho U^2} \to \frac{1}{M^2} \quad \text{and} \quad \frac{L\boldsymbol{B} \times (\nabla \times \boldsymbol{B})}{4\pi\rho U^2} \to \frac{1}{\beta M^2}, \tag{10.41}$$

with, as usual, Mach number M and ratio of plasma to field pressure β. This captures the reality that magnetic effects become more important as β decreases and that these forces have a much bigger impact on subsonic motion than on supersonic motion.

Turning to the magnetic induction equation (10.36), in a time τ the field diffuses a distance

$$d \sim \sqrt{\frac{\eta c^2 \tau}{4\pi}}, \tag{10.42}$$

for a fluid at rest. As a result, in a time L/U the distance diffused is

$$d \sim \sqrt{\frac{\eta c^2 L}{4\pi U}} = L\sqrt{\frac{\eta c^2}{4\pi UL}} = L\sqrt{\frac{1}{Rm}}. \qquad (10.43)$$

From this we see that the fraction of the system scale over which the fields diffuse in an evolution time L/U is $1/\sqrt{Rm}$. An experiment will be unable to sustain field structures on scales smaller than this.

The behavior of structured, magnetized flows is of interest in contexts ranging from particle acceleration at shock waves to magnetic-field production in turbulence. Sustaining small-scale structure in both the fluid parameters and the magnetic field requires that both Rm and the fluid Reynolds number, $Re = UL/\nu$ (see Chap. 2) be large. Practical expressions for these quantities are as follows:

$$Rm = \frac{4\pi UL}{\eta c^2} = \frac{\omega_{pe}^2 UL}{c^2 \nu_{ei}} =, \text{ and} \qquad (10.44)$$

$$Re = \frac{UL}{\nu}, \qquad (10.45)$$

with the kinematic viscosity ν taken from (2.39). In practical units, one has

$$Rm = \frac{59 T_e^{3/2}}{Z \ln \Lambda} =, \text{ and} \qquad (10.46)$$

$$Re = 5.4 \times 10^{-20} n_e UL \frac{AZ^3 \ln \Lambda}{\sqrt{A}T_i^{5/2} + 0.013 Z^3 T_e^{5/2}}, \qquad (10.47)$$

with temperatures in eV and other quantities in cgs units.

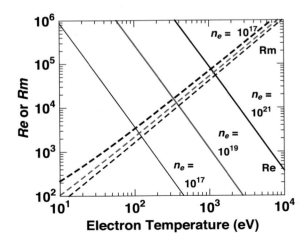

Fig. 10.4 The fluid Reynolds number, Re, shown by solid lines, decreases as T_e increases and increases in proportion to electron density. The magnetic Reynolds number, Rm, shown by dashed lines, increases with T_e and depends weakly on electron density. Rm decreases slightly as density increases; only the density of the uppermost dashed curve is labeled. These curves are for a Be plasma in a system 5 mm in extent with flow velocities of 1000 km/s

Figure 10.4 shows Re and Rm for some parameters of likely interest, evaluated for Be with $A = 9$, $Z = 4$, and $T_i = T_e$. In the abstract, it would be desirable to get Rm above 10^4 so that magnetic structures endure down to 1% of the system scale, and to get Re above 10^4 so that fluid turbulence can be excited. One sees that, for the product UL chosen, which was $5 \times 10^7 \text{ cm}^2 \text{ s}^{-1}$, this can be accomplished only at a temperature near 500 eV and an electron density of 10^{19} cm^{-3}. Such a product of U and L is challenging to achieve in most high-energy-density experiments.

10.4 Alfvén and Magnetosonic Waves

The MHD plasma supports three wave modes, and these do come up in experiments with magnetized plasmas in high-energy-density devices. As for the other waves we considered, in Chaps. 2 and 9, one needs equations for mass and for momentum density, but in the MHD cases we also need an equation for $\boldsymbol{B}$. Drawing from earlier in this chapter, the three relevant equations are

$$\frac{\partial \rho}{\partial t} + \nabla \cdot (\rho \boldsymbol{u}) = 0 \tag{10.48}$$

$$\rho \frac{D\boldsymbol{u}}{Dt} = -\nabla p - \frac{\boldsymbol{B} \times (\nabla \times \boldsymbol{B})}{4\pi}, \text{ and} \tag{10.49}$$

$$\frac{\partial \boldsymbol{B}}{\partial t} = \nabla \times (\boldsymbol{u} \times \boldsymbol{B}) - c\nabla \times (\eta \boldsymbol{J}). \tag{10.50}$$

Here we retain the resistivity to enable calculations of damping. There are several ways to work with these equations, or equivalent ones, and find the dispersion relations. One of them proceeds as follows. We linearize the relevant variables and assume the pressure to be barotropic, so that

$$\rho = \rho_o + \rho_1; \ \nabla p = c_s^2 \nabla \rho; \ \boldsymbol{B} = \boldsymbol{B}_o + \boldsymbol{B}_1, \tag{10.51}$$

and $\boldsymbol{u}$ is first order. The equations become, to first order

$$\frac{\partial \rho_1}{\partial t} + \rho_o \nabla \cdot \boldsymbol{u} = 0, \tag{10.52}$$

$$\rho_o \frac{\partial \boldsymbol{u}}{\partial t} = -c_s^2 \nabla \rho_1 - \frac{\boldsymbol{B}_o \times (\nabla \times \boldsymbol{B}_1)}{4\pi}, \text{ and} \tag{10.53}$$

$$\frac{\partial \boldsymbol{B}_1}{\partial t} = \nabla \times (\boldsymbol{u} \times \boldsymbol{B}_o) + \frac{c^\eta}{4\pi} \nabla^2 \boldsymbol{B}_1. \tag{10.54}$$

We use the first and third of these to simplify the time derivative of the second, which we then rearrange.

$$\rho_o \frac{\partial^2 \boldsymbol{u}}{\partial t^2} = \rho_o c_s^2 \nabla(\nabla \cdot \boldsymbol{u}) - \frac{\boldsymbol{B}_o}{4\pi} \times \left(\nabla \times \frac{\partial \boldsymbol{B}_1}{\partial t} \right), \text{ from which} \tag{10.55}$$

$$\rho_o \frac{\partial^2 \boldsymbol{u}}{\partial t^2} = \rho_o c_s^2 \nabla(\nabla \cdot \boldsymbol{u}) + \frac{(\boldsymbol{B}_o \cdot \nabla)}{4\pi} \frac{\partial \boldsymbol{B}_1}{\partial t} - \nabla \left(\frac{\boldsymbol{B}_o}{4\pi} \cdot \frac{\partial \boldsymbol{B}_1}{\partial t} \right). \tag{10.56}$$

Here we can note that the first term involving the magnetic field represents the restoring force due to field-line tension, which drives Alfvén waves, while the second such term introduces a compressive response. Also, (10.54) can be rewritten as

$$\frac{\partial \boldsymbol{B}_1}{\partial t} = (\boldsymbol{B}_o \cdot \nabla)\boldsymbol{u} - \boldsymbol{B}_o(\nabla \cdot \boldsymbol{u}) + \frac{c^2 \eta}{4\pi} \nabla^2 \boldsymbol{B}_1. \tag{10.57}$$

Combining (10.56) and (10.57) gives

$$\frac{\partial^2 \boldsymbol{u}}{\partial t^2} = c_s^2 \nabla(\nabla \cdot \boldsymbol{u})$$

$$+ \frac{(\boldsymbol{B}_o \cdot \nabla)}{4\pi \rho_o} \left[(\boldsymbol{B}_o \cdot \nabla)\boldsymbol{u} - \boldsymbol{B}_o(\nabla \cdot \boldsymbol{u}) + \frac{c^2 \eta}{4\pi} \nabla^2 \boldsymbol{B}_1 \right]$$

$$- \nabla \left(\frac{\boldsymbol{B}_o}{4\pi \rho_o} \cdot \left[(\boldsymbol{B}_o \cdot \nabla)\boldsymbol{u} - \boldsymbol{B}_o(\nabla \cdot \boldsymbol{u}) + \frac{c^2 \eta}{4\pi} \nabla^2 \boldsymbol{B}_1 \right] \right). \tag{10.58}$$

We define unit vector $\boldsymbol{b} = \boldsymbol{B}_o/B_o$, and define $\boldsymbol{k}_{\parallel} = \boldsymbol{k} \cdot \boldsymbol{b}$ and $\boldsymbol{k}_\perp = \boldsymbol{b} \times (\boldsymbol{k} \times \boldsymbol{b})$. We similarly define $\boldsymbol{u}_{\parallel}$ and $\boldsymbol{u}_\perp$, and the component of velocity in the third orthogonal direction as $\boldsymbol{u}_\times = \boldsymbol{u} \times \boldsymbol{b}$. Also let B_o define the z direction with associated unit vector $\hat{z}$. Define the Alfvén speed as $v_A = \sqrt{B_o^2/(4\pi \rho_o)}$. Seeking plane-wave modulations proportional to $e^{i(\boldsymbol{k} \cdot \boldsymbol{x} - \omega t)}$, and assuming η to be small enough to ignore, we find

$$\omega^2 \boldsymbol{u} = (c_s^2 + v_A^2)\boldsymbol{k}(\boldsymbol{k} \cdot \boldsymbol{u}) + k_{\parallel} v_A^2 \left[k_{\parallel} \boldsymbol{u} - \hat{z}(\boldsymbol{k} \cdot \boldsymbol{u}) \right] - k v_A^2 \left(\left[k_{\parallel}(\hat{z} \cdot \boldsymbol{u}) \right] \right)$$

$$= (c_s^2 + v_A^2)\boldsymbol{k}(\boldsymbol{k} \cdot \boldsymbol{u}) + v_A^2 \left[k_{\parallel}^2 \boldsymbol{u} - k_{\parallel}(\boldsymbol{k} \cdot \boldsymbol{u}) - \boldsymbol{k}(k_{\parallel} \cdot \boldsymbol{u}) \right], \tag{10.59}$$

giving

$$(\omega^2 - k_{\parallel}^2 v_A^2)\boldsymbol{u} = c_s^2 \boldsymbol{k}(\boldsymbol{k} \cdot \boldsymbol{u}) + k v_A^2(\boldsymbol{k}_\perp \cdot \boldsymbol{u}) - k_{\parallel} v_A^2(\boldsymbol{k} \cdot \boldsymbol{u})$$

$$= c_s^2 \boldsymbol{k}(\boldsymbol{k} \cdot \boldsymbol{u}) + k_\perp v_A^2(\boldsymbol{k}_\perp \cdot \boldsymbol{u}) - k_{\parallel} v_A^2(\boldsymbol{k}_{\parallel} \cdot \boldsymbol{u}). \tag{10.60}$$

This is three equations:

$$(\omega^2 - k_{\parallel}^2 c_s^2)u_{\parallel} - c_s^2 k_{\parallel} k_\perp u_\perp = 0, \tag{10.61}$$

$$c_s^2 k_\perp k_{\parallel} u_{\parallel} - (\omega^2 - k^2 v_A^2 - k_\perp^2 c_s^2)u_\perp = 0, \text{ and} \tag{10.62}$$

$$(\omega^2 - k_\parallel^2 v_A^2)u_\times = 0. \tag{10.63}$$

These have the right limits for $k_\parallel = 0$ or $k_\perp = 0$, as one can confirm by doing the homework. The $\times$ direction is not coupled to the others and supports only Alvfen waves having $\omega^2 = k_\parallel^2 v_A^2$. This solution is sometimes called the "intermediate mode", because its phase speed is intermediate to those of the fast and slow modes, found below.

To find the dispersion relation for the other two modes, one constructs the determinant of the coefficients of (10.61) and (10.62), obtaining

$$(\omega^2 - k^2 v_A^2 - k_\perp^2 c_s^2)(\omega^2 - c_s^2 k_\parallel^2) = c_s^4 k_\parallel^2 k_\perp^2. \tag{10.64}$$

Solving for ω^2 gives two roots,

$$\omega^2 = k^2 \frac{(c_s^2 + v_A^2)}{2} \left[1 \pm \sqrt{1 - \frac{4k_\parallel^2 c_s^2 v_A^2}{k^2(c_s^2 + v_A^2)^2}} \right]. \tag{10.65}$$

These roots are the slow and fast magnetosonic waves. Figure 10.5 shows the behavior of the phase velocity for these waves, normalized to $\sqrt{c_s^2 + v_A^2}$. Parallel to the magnetic field, the faster wave moves at the faster of the two speeds, the Alfvén wave is purely transverse, and the sound wave is purely longitudinal. At oblique angles both modes are mixed, including both transverse and longitudinal elements. Perpendicular to the magnetic field, only one mode remains, and it is longitudinal and moves rapidly. There is a connection between c_s, v_A, and the plasma β that can be useful to know. Specifically $\beta = 2c_s^2/v_A^2$. Thus, low-β plasmas are Alfvén-

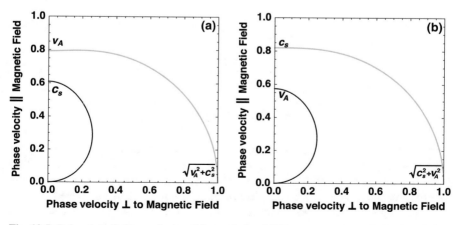

Fig. 10.5 Polar plot of phase velocity of fast and slow MHD waves as the angle of propagation varies. (a) $v_A = 1.3c_s$. (b) $v_A = 0.7c_s$. This type of figure, introduced by K.O. Friedricks in 1957 in a Los Alamos report, is often described as a Friedricks diagram

dominated and high-β plasmas are sound-wave-dominated. In the $\beta \sim 1$ plasmas so often of astrophysical interest, and sought in experiments on magnetized flows, both the sonic and the Alfvénic aspects can be important.

10.5 Magnetized Rayleigh-Taylor Instability

We saw above that magnetic pressure can apply force to an ionized fluid. This makes it relatively easy to create conditions in which a magnetized, lower-density plasma opposes the motion of a higher-density plasma. This in turn produces the opposed gradients of pressure and density that we found in Chap. 5 imply the excitation of the Rayleigh-Taylor instability. But because magnetic field lines resist being bent, we can expect that the magnetic field might in some way alter the behavior of surface modes, including those produced by the Rayleigh-Taylor instability. Here we revisit this instability in the presence of a magnetic field. In the case that arises most often in experiments, the magnetic field is transverse to the direction of the acceleration, g. In addition, the background field may be excluded from the conductive plasma, and may be acting to decelerate it. This is the case we will explore. There are also effects if the magnetic field is aligned with the direction of acceleration; one good treatment can be found in Chandrasekhar (1961).

10.5.1 Differential Equation for Magnetized Rayleigh-Taylor Instability

The useful form of the Ideal-MHD equations for this purpose is

$$\frac{\partial \rho}{\partial t} + \nabla \cdot (\rho \boldsymbol{u}) = 0 \tag{10.66}$$

$$\rho \frac{D\boldsymbol{u}}{Dt} = -\nabla p - \frac{\boldsymbol{B} \times (\nabla \times \boldsymbol{B})}{4\pi} + \rho \boldsymbol{g}, \text{ and} \tag{10.67}$$

$$\frac{\partial \boldsymbol{B}}{\partial t} = (\boldsymbol{B} \cdot \nabla)\boldsymbol{u} - \boldsymbol{B}(\nabla \cdot \boldsymbol{u}) - (\boldsymbol{u} \cdot \nabla)\boldsymbol{B}. \tag{10.68}$$

As usual, we let $\boldsymbol{g}$ define the $-z$ direction. We also let $\boldsymbol{B}_o$ define the x direction but allow the amplitude B_o to vary in z. Thus, $\boldsymbol{k}$ will lie in the x–y plane. This geometry and labeling is shown in Fig. 10.6. Such a geometry is not uncommon in experiments, where expanding plasmas may encounter and be slowed by preexisting magnetic fields. We write $\boldsymbol{u} = (u, v, w)$ as in Chap. 5. We also, as in Chap. 5, take the fluctuations to be incompressible, and assume the medium to initially be stratified so that the background density and pressure vary in z only. The initial hydrostatic balance is then $\rho_o \boldsymbol{g} = \nabla p_o + \nabla(B_o^2/(8\pi))$.

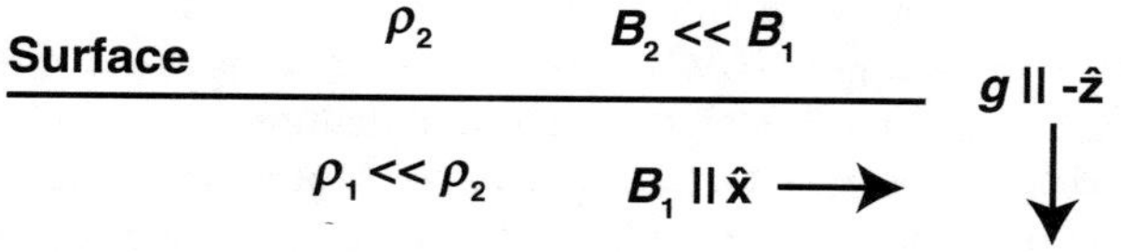

Surface $\qquad \rho_2 \qquad\quad B_2 \ll B_1$ $\qquad\qquad$ $g \parallel -\hat{z}$

$\qquad\qquad\quad \rho_1 \ll \rho_2 \qquad B_1 \parallel \hat{x} \longrightarrow$ $\qquad\qquad$ $\downarrow$

Fig. 10.6 Geometry for analysis of magnetized Rayleigh-Taylor instability. The subscripts correspond to the value at the interface when approaching from the side indicated. k lies in the x–y plane

We need to analyze these equations using a notation that works for surface waves, for which we will need to designate regions below and above the interface as 1 and 2, respectively. Accordingly, we linearize these equations as

$$\rho = \rho_o(z) + \delta\rho; \ p = p_o(z) + \delta p; \ \boldsymbol{B} = \boldsymbol{B}_o(z) + \boldsymbol{h}, \tag{10.69}$$

and take $\boldsymbol{u}$ to be first order. For later convenience, we define

$$\nabla_\perp = \left(\frac{\partial}{\partial x}, \frac{\partial}{\partial y}, 0\right) \text{ and } D = \frac{\partial}{\partial z}. \tag{10.70}$$

The linearized equations become

$$\frac{\partial \delta\rho}{\partial t} + \boldsymbol{u} \cdot \nabla\rho_o = 0 \tag{10.71}$$

$$\rho_o \frac{\partial \boldsymbol{u}}{\partial t} = -\nabla\delta p - \frac{\boldsymbol{B}_o \times (\nabla \times \boldsymbol{h})}{4\pi} + \delta\rho\boldsymbol{g}, \tag{10.72}$$

$$\frac{\partial \boldsymbol{h}}{\partial t} = (\boldsymbol{B}_o \cdot \nabla)\boldsymbol{u}, \text{ and} \tag{10.73}$$

$$\nabla \cdot \boldsymbol{u} = 0. \tag{10.74}$$

Our first problem is just that of previous surface waves: to extract, from these equations, a differential equation that describes the vertical structure. We first work on the x and y components of the momentum equation. To this end, note that the second term on the right hand side of (10.72) has no x component. Taking $\partial/\partial t$ of (10.71) and using (10.69) and (10.72), one finds

$$\rho_o \frac{\partial^2 \boldsymbol{u}}{\partial t^2} = -\nabla\frac{\partial \delta p}{\partial t} - \frac{\boldsymbol{B}_o \times (\boldsymbol{B}_o \cdot \nabla)(\nabla \times \boldsymbol{u})}{4\pi} - gw(D\rho_o), \tag{10.75}$$

If we take the curl of this equation, we find

$$\rho_o \frac{\partial^2 (\nabla \times \boldsymbol{u})}{\partial t^2} = +\frac{(\boldsymbol{B}_o \cdot \nabla)^2(\nabla \times \boldsymbol{u})}{4\pi} + \boldsymbol{g} \times \nabla(wD\rho_o). \tag{10.76}$$

Note that the z component of the final term is zero. The other two terms will support stable, propagating Alfvén waves having a z component of $\nabla \times \boldsymbol{u}$, but not the exponentially growing, Rayleigh-Taylor modulations of interest here. For these modulations we can thus take the z component of $\nabla \times \boldsymbol{u}$, and hence also $\nabla \times \boldsymbol{h}$, to be zero. This implies that the second term of (10.72) has no y component either.

Now (10.74) can be written as $\nabla_\perp \cdot \boldsymbol{u} = -Dw$. With this, one can dot $\nabla_\perp$ with (10.72) to find

$$\rho_o D \frac{\partial w}{\partial t} = \nabla_\perp^2 \delta p. \tag{10.77}$$

This is the first result of primary interest. Next we work on the z component of the momentum equation. Taking the x-derivative of (10.74) and the y-derivative of the z component of the vorticity, we have

$$\frac{\partial}{\partial x} \nabla_\perp \cdot \boldsymbol{u} = -D \frac{\partial w}{\partial x} \quad \text{and} \tag{10.78}$$

$$\frac{\partial}{\partial y} (\nabla \times \boldsymbol{u}) \cdot \hat{z} = \frac{\partial^2}{\partial y \partial x} v - \frac{\partial^2}{\partial y^2} u = 0, \tag{10.79}$$

implying that

$$\nabla_\perp^2 u = -D \frac{\partial w}{\partial x}. \tag{10.80}$$

We also note that

$$\left[\frac{\boldsymbol{B}_o \times (\boldsymbol{B}_o \cdot \nabla)(\nabla \times \boldsymbol{u})}{4\pi \rho_o} \right] \cdot \hat{z} = \frac{B_o^2}{4\pi \rho_o} \frac{\partial}{\partial x} \left(Du - \frac{\partial w}{\partial x} \right) \tag{10.81}$$

Recall that the Alfvén speed is given by $v_A^2 = B_o^2/(4\pi \rho_o)$. The z component of (10.75) is then

$$\rho_o \frac{\partial^2 w}{\partial t^2} = -D \frac{\partial \delta p}{\partial t} - \rho_o v_A^2 \frac{\partial}{\partial x} \left(Du - \frac{\partial w}{\partial x} \right) + gw(D\rho_o). \tag{10.82}$$

We now eliminate u by operating with $\nabla_\perp^2$, to obtain

$$\rho_o \nabla_\perp^2 \frac{\partial^2 w}{\partial t^2} = -D \nabla_\perp^2 \frac{\partial \delta p}{\partial t} + \rho_o v_A^2 \frac{\partial^2}{\partial x^2} \left(D^2 + \nabla_\perp^2 \right) w + g(D\rho_o) \nabla_\perp^2 w, \tag{10.83}$$

We now eliminate δp by operating on (10.77) with $D\partial/(\partial t)$ and substituting, which yields

$$\rho_o \nabla_\perp^2 \frac{\partial^2 w}{\partial t^2} + D \left(\rho_o D \frac{\partial^2 w}{\partial t^2} \right) - \rho_o v_A^2 \frac{\partial^2}{\partial x^2} \left(\nabla_\perp^2 + D^2 \right) w = +g(D\rho_o) \nabla_\perp^2 w. \tag{10.84}$$

This is the differential equation we were seeking, which is a function of only w, derivatives, and known parameters. Seeking plane-wave solutions of the form $e^{nt+i(k\cdot x)}$, we obtain

$$n^2 D\,(\rho_o Dw) - \rho_o k^2 n^2 w + \rho_o v_A^2 k_x^2 \left(D^2 - k^2\right) w = -gk^2 (D\rho_o)w. \tag{10.85}$$

This is the equation we need in order to determine the vertical structure in specific cases.

10.5.2 Magnetized Rayleigh-Taylor Instability When $k \perp B_o$

When one has $k \perp B_o$, $k_x = 0$ and (10.85) reduces to

$$n^2 D\,(\rho_o Dw) - \rho_o k^2 n^2 w = -gk^2 (D\rho_o)w. \tag{10.86}$$

This is identical to (5.21) for zero viscosity, and so the magnetic field has no effect on the instability for modulations in this direction. This may seem strange, but it reflects the fact that flux tubes frozen into plasma do not resist motion. For $k \perp B_o$, the flux tubes move up and down with the fluid but they do not have to bend. It is bending that the field lines actively resist. Thus B does not inherently stabilize Rayleigh-Taylor instability. The boundary conditions also do not change, one finds that the growth rate for this instability is identical to that of the purely hydrodynamic case when $k \perp B_o$.

10.5.3 Magnetized Rayleigh-Taylor Instability When k Is Not $\perp B_o$

Here as a specific example we consider the case of two uniform fluids separated by a surface that is horizontal relative to g. Fluid 2 is the upper fluid and fluid 1 is the lower fluid. When we add a subscript with a number to a given variable, we refer formally to the value of that variable taken in the limit that the surface is approached. The properties of the two fluids may differ.

The basic boundary conditions is that w be continuous across the interface. We take its value at the surface to be w_o. As before, in Chap. 5, we integrate (10.85) across the surface to obtain another boundary condition:

$$n^2 \left(\rho_2 (Dw)_2 - \rho_1 (Dw)_1\right) + \frac{k_x^2}{2} \left[\left(\rho_2 v_{A2}^2 + \rho_1 v_{A1}^2\right) \left((Dw)_2 - (Dw)_1\right)\right]$$
$$= -gk^2 w_o (\rho_2 - \rho_1). \tag{10.87}$$

Electrodynamics also requires that the normal component of $\boldsymbol{B}$ at the surface be continuous. In our initial condition $\boldsymbol{B_o} \cdot \hat{z} = 0$, so there is no initial normal component. Yet (10.73) seems to imply that we generate one, which would not be continuous across the surface if B_o were not continuous. indeed some of the interesting practical cases are ones in which magnetic-field pressure resists the acceleration of a denser medium, and B_o need not be continuous. The solution to this conundrum is that the surface becomes rippled such that $\boldsymbol{B_o} \cdot \hat{q}$ remains zero, where $\hat{q}$ is the surface normal.

Returning to the main problem, the differential equation, (10.85) becomes for this case

$$\rho_o(n^2 + v_A^2 k_x^2)\left(D^2 - k^2\right) w = 0, \tag{10.88}$$

so just as in the simpler, field-free case the solutions for unbounded fluids are

$$w = w_o e^{kz} \text{ for } z < 0 \text{ and } w = w_o e^{-kz} \text{ for } z > 0, \tag{10.89}$$

while the case of finite-depth fluids requires combinations of these functions. This lets us evaluate (10.87), finding

$$-n^2 k w_o \left(\rho_2 + \rho_1\right) - 2k w_o \frac{k_x^2}{2}\left[\left(\rho_2 v_{A2}^2 + \rho_1 v_{A1}^2\right)\right]$$
$$= -gk^2 w_o(\rho_2 - \rho_1), \tag{10.90}$$

so that

$$n^2 = gk\frac{(\rho_2 - \rho_1)}{(\rho_2 + \rho_1)} - k_x^2\frac{\left(\rho_2 v_{A2}^2 + \rho_1 v_{A1}^2\right)}{(\rho_2 + \rho_1)}, \tag{10.91}$$

which also can be written

$$n^2 = gk\frac{(\rho_2 - \rho_1)}{(\rho_2 + \rho_1)} - k_x^2\frac{\left(B_2^2 + B_1^2\right)}{4\pi\left(\rho_2 + \rho_1\right)}. \tag{10.92}$$

As it should, (10.92) reduces to the hydrodynamic growth rate of Chap. 5 for $B_1 = B_2 = 0$, and independently for $k_x = 0$ as was noted above. When these are not zero, though, the magnetic field is stabilizing. The effect is much like that of surface tension, for the same reason. The surfaces of fluids with surface tension resist bending. In practice, this stabilization of Rayleigh-Taylor in some directions matters for experiments. If one hopes to see the modulations produced by Rayleigh-Taylor, one had better be looking along $\boldsymbol{B}$. At small enough k_x (long enough wavelength), the magnetic field becomes ineffective. Consider where this might be. A rough look at the numbers is as follows. For $\rho_1 \ll \rho_2$, $g \sim 10$ km/s/ns, and a 60 μm wavelength, the hydrodynamic growth rate is ~ 1 inverse ns. For a magnetic field of 100 kG and a mass density of $\rho_2 \sim 0.001$ g/cc, v_A is about 10 km/s. Modes with $k = k_x$ will be stabilized for wavelengths below about 60 μm. As ρ_2 decreases, progressively longer wavelengths can be stabilized.

10.6 Magnetized Shocks

High-energy-density systems tend to produce shock waves, in natural consequence of releasing an initial high pressure against matter. This applies just as much to magnetized flows as it does to any others. But, just as the properties of waves in a plasma are affected by the presence of a magnetic field, so is the behavior of shock waves. To explore the behavior, we proceed as we did with oblique shocks in Chap. 4, defining a unit vector $\boldsymbol{n}$, which is normal to the shock front and in the direction of the normal flow into the shock front (thus, $\boldsymbol{n}$ points from right to left in our standard orientation). Then the component of $\boldsymbol{u}$ in the normal direction is $\boldsymbol{u}_n = \boldsymbol{n}(\boldsymbol{u} \cdot \boldsymbol{n})$ while the transverse component of $\boldsymbol{u}$ is $\boldsymbol{u}_\perp = (\boldsymbol{n} \times \boldsymbol{u}) \times \boldsymbol{n}$. If the magnetic field is parallel to $\boldsymbol{n}$, then the shock is said to be a "parallel shock," while if $\boldsymbol{B} \cdot \boldsymbol{n} = 0$, the shock is said to be a "perpendicular shock." Any perpendicular field may have components in the direction of $\boldsymbol{u}_\perp$ and also in the direction of $\boldsymbol{n} \times \boldsymbol{u}$. It is also helpful to work with the conservative form of the momentum equation,

$$\frac{\partial(\rho\boldsymbol{u})}{\partial t} + \nabla \cdot \left[\rho\boldsymbol{u}\boldsymbol{u} + \left(p + \frac{|B|^2}{8\pi} \right) \boldsymbol{I} - \frac{\boldsymbol{B}\boldsymbol{B}}{4\pi} \right] = 0. \tag{10.93}$$

Here for simplicity we assume an isotropic pressure tensor. Using the subscript n for the normal direction, (2.41), (10.93), and (10.26) give us the following jump conditions, obtained by recognizing that the direction these equations must be integrated to fully capture the jump conditions (see Chap. 4) is that of $\boldsymbol{n}$. We have

$$\rho_2 u_{n2} = \rho_1 u_{n1}, \tag{10.94}$$

$$\rho_2 \boldsymbol{u}_2 u_{n2} + \left(p_2 + \frac{|B_2|^2}{8\pi} \right) \boldsymbol{n} - \frac{\boldsymbol{B}_2 B_{n2}}{4\pi}$$

$$= \rho_1 \boldsymbol{u}_1 u_{n1} + \left(p_1 + \frac{|B_1|^2}{8\pi} \right) \boldsymbol{n} - \frac{\boldsymbol{B}_1 B_{n1}}{4\pi}, \tag{10.95}$$

which has components parallel and transverse to $\boldsymbol{n}$, and from the energy equation we have

$$u_{n2} \left(\frac{\gamma p_2}{\gamma - 1} + \rho_2 \frac{u_2^2}{2} + \frac{|B_2|^2}{4\pi} \right) - \frac{B_{n2}(\boldsymbol{u}_2 \cdot \boldsymbol{B}_2)}{4\pi}$$

$$= u_{n1} \left(\frac{\gamma p_1}{\gamma - 1} + \rho_1 \frac{u_1^2}{2} + \frac{|B_1|^2}{4\pi} \right) - \frac{B_{n1}(\boldsymbol{u_1} \cdot \boldsymbol{B_1})}{4\pi}. \tag{10.96}$$

and from $\nabla \cdot \boldsymbol{B} = 0$ we have

$$B_{n2} = B_{n1}. \tag{10.97}$$

Finally, one can integrate Faraday's Law across the shock transition, using the Ohm's Law for ideal MHD. This implies that $-\boldsymbol{n} \times \boldsymbol{E} = \boldsymbol{n} \times (\boldsymbol{u} \times \boldsymbol{B})$ is constant across the shock, or

$$u_{n2}\boldsymbol{B_2} - B_{n2}\boldsymbol{u_2} = u_{n1}\boldsymbol{B_1} - B_{n1}\boldsymbol{u_1}, \tag{10.98}$$

which can also be written

$$u_{n2}(\boldsymbol{B_2} \times \boldsymbol{n}) - B_{n2}\boldsymbol{u_{2\perp}} = u_{n1}(\boldsymbol{B_1} \times \boldsymbol{n}) - B_{n1}\boldsymbol{u_{1\perp}}. \tag{10.99}$$

This expression may have two components if $(\boldsymbol{B} \times \boldsymbol{n})$ is not parallel to $(\boldsymbol{u} \times \boldsymbol{n})$.

Now consider some simple cases. For a pure parallel shock, with $\boldsymbol{B}\|\boldsymbol{n}$, (10.96) implies that the magnetic field does not change across the shock, so the field-related terms drop out of (10.95) and (10.96), while (10.98) and (10.99) are reduced to the condition that the transverse velocity is unchanged.

For a pure perpendicular shock, with $\boldsymbol{B}\perp\boldsymbol{n}$ and $\boldsymbol{u}\|\boldsymbol{n}$, (10.98) implies that the magnetic field increases by the same factor as the density, from (10.94). At the same time (10.99) implies that the direction of $\boldsymbol{B}$ does not change. In this case, there is simply compression of the magnetic field. The energy cost of field compression is high, though, as it scales as $|B|^2$, so when $\beta \lesssim 1$, it will take comparatively more pressure to produce a strong shock.

Shock waves in unmagnetized fluids are associated with an upstream Mach number, and the piston speed must exceed the sound speed in order to drive a shock. This creates some confusion for magnetized plasmas, where there are two "sonic" waves in addition to Alfvén waves. Certainly a piston moving faster than $\sqrt{c_s^2 + v_A^2}$ will drive a shock and this can be strong. But there is also the potential for more complex behavior at slower speeds, including shock waves that switch a component of the magnetic field on or off. Here we leave the pursuit of this subject to other books or articles, but we note that there is the potential for some interesting laboratory experiments, once facilities can access the necessary regime.

There is also a third type of transition that is not a shock, known as a tangential discontinuity. Here a localized current sheet creates a rotation of $\boldsymbol{B}$ transverse to the flow. In the pure case, neither density nor velocity change. These are observed within the Heliosphere.

The discussion above said nothing about the shock transition itself, and produces results that relate the plasma properties on opposite sides of the transition. In dense enough plasmas, the shock is produced by collisions and the shock transition is a few ion-ion mean-free-paths. One can calculate its microstructure by including ion viscosity in the hydrodynamic equations.

However, as the density drops low enough to have $\beta \lesssim 1$ for a magnetic field of hundreds of kG, the collisional mean free path becomes quite large. The relevant densities are below 10^{20} per cc. One can produce supersonic flows at these densities, or lower ones, in the plasmas blown off of laser-irradiated targets or produced by pulsed-power devices. In astrophysics, one encounters supersonic flows at much lower densities. As one example, the collisional mean free path within the solar wind near the Earth is about one astronomical unit. Yet shock waves are still observed in these natural systems. Such shocks are known as *collisionless shocks*. Such shocks have been observed in space, and in some laboratory experiments, since the 1960s. In the early twenty-first century they have seen renewed interest, because of the ability to produce and diagnose them by several new methods.

The shock transition in any collisionless shock is produced by local, fluctuating, electric and/or magnetic fields. These arise via several mechanisms. In unmagnetized plasmas, the simplest is the electrostatic, two-stream instability, which has been observed in both ICF-scale and relativistic experiments. The latter case is discussed further in Chap. 13. More complex, but potentially important in systems such as gamma-ray bursts, is the development of a structured magnetic field via the ion-ion Weibel instability. Ion-ion Weibel is driven by counterstreaming ions, and produces filamentary magnetic structures that eventually become tangled. At this writing (2017), the instability has been observed in experiments; in simulations shocks have been seen to develop on timescales and spatial scales not far from experimental conditions.

In magnetized plasmas, there is a very wide range of phenomena associated with collisionless shocks. Balogh and Treumann (2013) discuss many of them. In pure perpendicular shocks, the particles are trapped by the magnetic field, so the natural scale for structure ahead of the shock is the gyroradius of ions reflected from the shock. These ions are an energy source for a number of instabilities that produce and sustain shock structure. In pure parallel shocks, reflected electrons and ions can stream freely ahead of the shock. One instability, among many, that can contribute to developing shock structure is the firehose instability, discussed next.

10.7 Firehose Instability

When ion reflection or other phenomena create a plasma with a pressure distribution that is anisotropic, the anisotropy can provide the free energy to support an instability. Like the instability of an actual firehose at high pressure, in this case flux tubes will begin to oscillate. This can be a step in leading to the field structures that sustain parallel collisionless shocks. We will briefly visit the theory of this instability here.

Suppose that the pressure tensor can be written as

$$\underline{P} = p_\perp \underline{I} + (p_\parallel - p_\perp)\boldsymbol{bb}, \tag{10.100}$$

in which again $b = B/B$ is a unit vector in the direction of the field B. Note that the rightmost term establishes the pressure as $p_{||}$ in the direction of b. Then we have

$$\nabla \cdot \underline{P} = \nabla p_\perp + (p_{||} - p_\perp)\nabla \cdot bb + b(b \cdot \nabla)(p_{||} - p_\perp)$$

$$= \nabla p_\perp + (p_{||} - p_\perp)\left[(b \cdot \nabla)b + b(\nabla \cdot b)\right] + b(b \cdot \nabla)(p_{||} - p_\perp), \quad (10.101)$$

noting that $\nabla \cdot b \neq 0$.

It will be helpful to note that $\nabla \cdot \underline{P}$ has components both $||$ and $\perp$ B. Using a subscript to designate $||$ and $\perp$ parts, we have

$$(\nabla \cdot \underline{P})_{||} = \nabla_{||}p_{||} + (p_{||} - p_\perp)b(\nabla \cdot b) + b(b \cdot \nabla)(p_{||} - p_\perp) \quad (10.102)$$

$$\text{and } (\nabla \cdot \underline{P})_\perp = \nabla_\perp p_\perp + (p_{||} - p_\perp)(b \cdot \nabla)b. \quad (10.103)$$

A key term for the firehose instability is that which resists field-line bending, $(b \cdot \nabla)b$. This provides a restoring force here, just as it does for Alfvén waves. We then find, from (10.25) for $u_\perp$,

$$\rho\frac{Du_\perp}{Dt} = -\nabla_\perp\left(p_\perp + \frac{|B|^2}{8\pi}\right) + \frac{(B \cdot \nabla B)}{4\pi} + (p_\perp - p_{||})(b \cdot \nabla)b. \quad (10.104)$$

Linearizing with $B = B_o + B_1$, zeroth-order ρ_o, and first-order u gives

$$\rho_o\frac{\partial u_\perp}{\partial t} = -\nabla_\perp\left(p_\perp + \frac{2B_oB_1}{8\pi}\right) + \frac{(B_o \cdot \nabla)B_1}{4\pi}\left[1 + \frac{4\pi(p_\perp - p_{||})}{B_o^2}\right]. \quad (10.105)$$

The other needed equation is (10.57) which, linearized and for zero resistivity, is

$$\frac{\partial B_1}{\partial t} = (B_o \cdot \nabla)u - B_o(\nabla \cdot u). \quad (10.106)$$

We seek modes in which the flux tubes oscillate, and to the simplest case of interest has $k_\perp = 0$ and involves only $u_\perp$, so that these last two equations imply

$$\frac{\partial^2 B_1}{\partial t^2} = \frac{(B_o \cdot \nabla)^2 B_1}{4\pi\rho_o}\left[1 + \frac{4\pi(p_\perp - p_{||})}{B_o^2}\right], \quad (10.107)$$

implying via the usual plane-wave decomposition (B_1 or $u \propto e^{i(k \cdot x - \omega t)}$) that

$$\omega^2 = k^2\left[v_A^2 + \frac{(p_\perp - p_{||})}{\rho_o}\right]. \quad (10.108)$$

One can see that the condition for instability is

$$p_{||} > \rho_o v_A^2 + p_{\perp}. \tag{10.109}$$

If we approximate $p_{\perp} = \rho_o c_s^2$, then the condition is $p_{||} > \rho_o(v_A^2 + c_s^2)$. If $p_{||}$ exceeds the threshold by $\rho_o v_A^2$, then the growth rate will be of order the Alfvén wave frequency $\sim k v_A$, which is quite fast.

10.8   Reconnection

In the case of ideal MHD, field lines are always frozen into the plasma and can never break. In real plasmas, in contrast, field lines always diffuse and can break and reconnect to form new topologies. Reconnection occurs when ions converge on a current sheet, flowing inward from both sides. This flow can be driven or can be in response to $\boldsymbol{J} \times \boldsymbol{B}$ forces, a difference that will not be our focus here. The plasma must then divert the ion flow away from the region where it converges while also reconnecting the field lines so they can escape. Resistive MHD provides a first way to describe this behavior and sets the stage for more complex descriptions when needed.

Figure 10.7 shows the geometry that corresponds to the early descriptions of Sweet (1958) and Parker (1957). One imagines two regions of oppositely directed, dominantly horizontal field lines, separated by a current sheet in which the reconnection occurs. This is the gray rectangle in the figure, of length $2L$ and thickness $2d$. The plasma flows into this region at speed u. Within it, the field lines diffuse and reconnect, becoming vertical where they cross the reconnection layer. The reconnected field lines then move away laterally at the Alfvén speed.

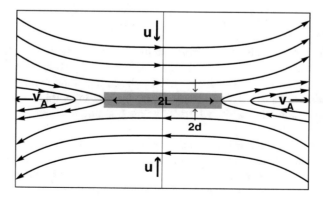

Fig. 10.7 Geometric structure of a reconnection region in the Sweet-Parker description

This geometry is relevant to any system in which a current sheet forms, as happens in any magnetotail including that of the Earth. Through its interaction with the solar wind, the dipole field of the Earth gets dragged into a long tail behind the Earth, in which a current sheet separates regions of oppositely directed magnetic field. The field lines continually (though not steadily) reconnect through the current sheet; the inner ones are drawn back toward the Earth and the outer ones escape downstream.

Suppose the current sheet in Fig. 10.7 extends into and out of the page, so that we can analyze the system in two dimensions. The incoming speed of the field lines, u_{SP}, is often referred to as the "reconnection rate," despite the disparity with what one would expect as the units for a rate. Conservation of mass then implies that

$$u_{SP}L = v_A d. \tag{10.110}$$

Suppose that the inward convection of the field lines balances their diffusive outflow, so that locally B does not change. Then (10.36) implies that

$$\nabla \times (\boldsymbol{u} \times \boldsymbol{B}) = -\frac{c\eta}{4\pi}\nabla^2 \boldsymbol{B}, \tag{10.111}$$

so for a thin, dissipative layer, where $\nabla \sim 1/d$, we approximate (ignoring signs)

$$u = \frac{c\eta}{4\pi d}. \tag{10.112}$$

These equations give

$$u_{SP} = \frac{c\eta v_A}{4\pi u_{SP}L} = v_A \sqrt{\frac{c\eta}{4\pi v_A L}} = \frac{v_A}{\sqrt{S}} \text{ and} \tag{10.113}$$

$$d = \frac{c\eta L}{4\pi v_A d} = L\sqrt{\frac{c\eta}{4\pi v_A L}} = \frac{L}{\sqrt{S}}, \tag{10.114}$$

in which the newly defined *Lundquist number* is $S = 4\pi v_A L/(c\eta)$, comparing Alfvénic convection with resistive dissipation. The Lundquist number is often very large, so that the resistive layer is predicted to be thin and the incoming velocity is a small fraction of v_A. Here d represents the dissipative-layer thickness necessary to produce the reconnection while all the ions flow into and then out of the dissipative layer.

The model as just given is not self-consistent, because along with the resistive field-line diffusion there is resistive dissipation. So one must ask whether the heating might be so large as to cause the expansion of the resistive layer beyond the value just calculated. This was considered and the short answer is no. (See Kulsrud (2005) for further discussion.)

What proved more important is that the timescale, τ_{SP}, for reconnection in the Sweet-Parker model just described is far too slow for many circumstances.

Table 10.1 gives two examples discussed by Kulsrud. The Sweet-Parker reconnection timescale, τ_{SP}, is often a few orders of magnitude faster than resistive decay, of timescale τ_{decay} but yet a few orders of magnitude slower than the observed value τ_{obs}. Even so, the observed timescale is longer than the crossing time for Alfvén waves, τ_A. We discuss the other parameters and timescales below. The resolution of the discrepancy between τ_{SP} and τ_{obs} has several elements.

We can start with a discussion of what has to be true for the Sweet-Parker model to be a valid application of MHD theory using classical resistivity. In the jargon of the field, this case corresponds to "slow reconnection." First, the geometry of Fig. 10.7 has to be accurate. Specifically, the length L must include all the ions whose inflow needs to be redirected. Second, the length d must remain large enough that MHD theory remains accurate. This requires that $d > \delta_i$, the ion skin depth, also shown in Table 10.1. The ion skin depth, $\delta_i = c/\omega_{pi}$ is the scale to which low-frequency external fields can penetrate a plasma, and is also the scale below which electron and ion behavior can differ significantly. (The ion plasma frequency is ω_{pi}.) In laboratory experiments in which $d > \delta_i$, which corresponds to a requirement that the collisional mean-free-path is $\ll \delta_i$—all turn out to break down in various cases of interest. Petschek realized that reconnection events can be local, and need not occur throughout the full extent of a current sheet, and that the actual length of the reconnection layer need not include all the ions needing to be redirected, because the straightening of the field lines after reconnection would sweep along all the ions to the left and right of the reconnection layer. Figure 10.8 shows the resulting, new picture, in which the place that field lines meet is known as an "x-point" and the field lines connected to the x-point are said to form a "separatrix." In real systems, there are often many such x-points, which form and dissipate throughout the current sheet. The reconnection now occurs throughout a volume of much smaller length, L', and the new incoming speed u_P is

$$u_P = u_{SP} \sqrt{\frac{L}{L'}}. \tag{10.115}$$

Table 10.1 The disparity of timescales for reconnection

Parameter	Solar flare	Earth magnetotail	Parameter	Solar flare	Earth magnetotail
B (G)	300	10^{-4}	v_{SP} (cm/s)	20	10^3
Density	$10^{-15}\,\text{g cm}^{-3}$	$10\,\text{cm}^{-3}$	τ_A (s)	37	15
v_A (km/s)	270	70	τ_{SP} (s)	6×10^7	10^5
T_e (eV)	100	100	τ_{decay} (s)	10^{14}	10^9
$\frac{\eta c}{4\pi}$ (cm^2/s)	10^4	10^7	τ_{obs} (s)	10^3-10^4	~ 100
L (km)	10^4	10^3	δ_i/v_A (s)	3×10^{-5}	1
δ_i (cm)	700	5×10^6	S^*	10^8	10^2
S	3×10^{12}	7×10^7			

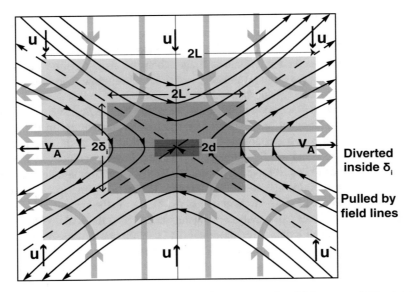

Fig. 10.8 The complicated picture that is two-fluid reconnection. Make sense of this sketch as follows. First, follow the solid black curves. These are lines of magnetic field, which change direction across the separatrix, shown by the dashed line. The largest gray box, of width $2L$, shows the region over which the ion flow must be diverted. Some of this flow, indicated by the outermost, four, thick, gray curves with arrows, is frozen in and is pulled sideways by the straightening magnetic field lines. The middle gray box, of width $2L'$, has a thickness of two ion skin depths ($2\delta_i$). The electrons and ions flow into the top and bottom of this box together, but then go separate ways. The ions are diverted throughout and flow out sideways, as indicated by the innermost, four, thick, gray curves with arrows. The electrons, in contrast, being much more strongly magnetized than the ions, are pulled into a much smaller region, shown by the darkest gray box. This region is of width $2d$, and is where the field lines diffusively reconnect. The lateral outflow is at the Alfvén speed, v_A

The Petschek model thus produces faster reconnection, and the ability for the outgoing ions to flow out in an expanding flow rather than on a narrow sheet is helpful. But the model does not provide a very clear story regarding how to define L', and predicts that shocks will form in the outgoing flow, which have not been observed. Nonetheless, the basic Petschek geometry has been observed when the requirements of the Sweet-Parker model are violated.

Now we turn to the second key requirement, relating to resistivity. Within the context of any model of the plasma, the resistivity can become "anomalous", by which one means the following. The resistance to current flow, averaged over timescales of many ion-acoustic periods, is much larger than the value resulting from Coulomb collisions alone. One way this can occur is when ion-acoustic turbulence develops in consequence of the two-stream instability, excited when the electron drift relative to the ions reaches the ion-acoustic speed. This mechanism was discussed in Chap. 9 in the context of electron heat transport in mid-Z or high-Z targets, and also just above in the context of shock formation. It produces local, fluctuating, electrostatic fields that scatter the electrons and thereby increase the resistivity.

If we suppose that there is such an electron drift, at velocity $\boldsymbol{u}_e$, then the corresponding current is $\boldsymbol{J} = -en_e\boldsymbol{u}_e$ and Ampère's law implies that the corresponding width of the layer containing the current is

$$d = \frac{u_e}{v_A}\delta_i. \tag{10.116}$$

The usual story in such driven turbulence is that the driver is held by the turbulence to a marginally stable value, so that the necessary instability remains excited. This requires an electron drift of order the sound speed c_s, so we have $u_e \sim c_s \sim \sqrt{\beta}v_A$, with β as usual being the ratio of plasma pressure to magnetic-field pressure. Thus, when the thickness of a current sheet drops below $\sim \sqrt{\beta}\delta_i$, anomalous resistivity is likely to develop. Here c_s, v_A and β are defined using the magnetic field just outside the resistive diffusion layer and using the temperatures in the heated plasma within the diffusive layer. With these definitions, one finds typically $\sqrt{\beta} \sim 1$. In other words, when one violates the second requirement given above, one also tends to violate the third requirement. Let us consider the consequences.

The electrons and ions act separately on scales smaller than the ion skin depth, δ_i (see Fig. 10.8). This takes the description out of MHD proper and now the account becomes a two-fluid theory. The ions can always flow laterally in a layer whose thickness is the ion skin depth. (This requires that the ion gyroradius be no larger than δ_i, which implies that the ion speed be no larger than the Alfvén speed, as is generally true.) In addition, the argument of Petschek still applies, so the global scale over which the ions must be redirected, L, may be larger than the scale over which ions and electrons enter the reconnection zone, L'. Figure 10.8 illustrates this case. One can see geometrically that the distance L' depends on the opening angle of the separatrix, which seems not to be determined by simple physics.

The ions now flow out of the reconnection layer in a sheet of thickness $2\delta_i$, so we have

$$u_PL' = v_A\delta_i. \tag{10.117}$$

The electrons flow into the reconnection layer. Recall the discussion of Ohm's law in Sect. 10.2.3: $(\boldsymbol{J} - en_e\boldsymbol{u})$ is $en_e\boldsymbol{u}_e$. When the electrons drift relative to the ions across the field, the magnetic field lines follow the electron motion. This lets us recast (10.112) as

$$u_{ce} = \frac{c\eta^*}{4\pi d} \tag{10.118}$$

for cross-field electron-fluid speed u_{ce} and actual resistivity (perhaps anomalous) η^*. If u_{ce} is of order the marginally stable value $\sim v_A$, as discussed above, then

$$d = L/S^*, \tag{10.119}$$

where S^* is the Lundquist number evaluated using the actual resistivity.

We can analyze this situation a bit more as follows. The electrons are drawn in from the boundary through which they flow with the ions, of length L', to the resistive-diffusion region, of length L''. As a result, one has $u_P L' = u_{ce} L''$. In addition, assuming that the separatrix boundaries are approximately straight lines in this region, and that the width of the resistive-diffusion region is L'', one has

$$d/L'' = \delta_i/L'. \tag{10.120}$$

In combination, and with $u_{ce} \sim v_A$ as discussed above, so that (10.117) implies $d = c\eta^*/(4\pi v_A)$, one has

$$u_P = u_{ce}\frac{L''}{L'} = v_A\frac{d}{\delta_i} = v_A\frac{L}{\delta_i S^*}. \tag{10.121}$$

Although this does not give us a known expression for the reconnection rate, it does indicate that the plasma, by adjusting S^*, can achieve any reconnection rate it needs. The required value can be found by taking

$$\tau_{obs} = \frac{L}{u_P} = S^*\frac{\delta_i}{v_A}. \tag{10.122}$$

For our two examples, Table 10.1 shows the required values of S^*. In each case, the resistivity has to be increased by a few orders of magnitude to produce a self-consistent account. Two-fluid simulations also agree that anomalous resistivity is required to produce Petschek-type reconnection events and to match observations.

Beyond this question of what rate of reconnection the plasma can sustain lie some other issues. One of these is triggering. In some cases, and especially in experiments, reconnection can be driven by relatively steady opposing flows. In other cases, and often in nature, a system exists that can reconnect, reducing the potential energy of the magnetic-field configuration, but that needs a trigger to initiate the reconnection. Two known triggers are a type of instability called an MHD tearing mode, and a modulation instability in current sheets known as the plasmoid instability, which produces isolated plasmas contained within closed magnetic-field lines.

Another issue for reconnection is the energy flow. Some of the energy stored in the magnetic field acts to accelerate the ions, which emerge at more-or-less v_A. But the process of reconnection itself delivers significant energy to suprathermal electrons. These have been detected from reconnection in laser-driven plasmas, via the X-rays they could produce. And it is clear from various data that they produce radio emissions following reconnection in solar flares. Developing a detailed understanding of energy flow is at the forefront at this writing. It seems that a combination of improved experiments and improved simulations will be needed. On the simulation front, researchers are finally now developing codes that can follow the electron physics at the scale of the dissipation layer using Particle-in-Cell techniques, map the output to two-fluid codes that can follow the dynamics on the scale of the ion skin depth, and then map that output to MHD codes that can follow the global-scale behavior.

A summary of the story of reconnection is as follows. Reconnection occurs as plasma flows inward toward a current sheet where the magnetic field changes direction. In order to reconnect the field lines while also diverting the plasma flow, the plasma demands that the reconnection occur within a layer of a specific thickness. If the plasma is sufficiently collisional, this layer remains thicker than two ion skin depths and the ordinary, resistive MHD theory of Sweet and Parker describes the behavior. As the layer becomes thinner than an ion skin depth, three things happen. The length of the reconnection zone contracts laterally, with ions outside that zone being diverted by magnetic forces as field lines straighten. The geometry changes from that of an extended layer to that of an x-point and the outflow becomes more divergent. In addition, the ion flow separates from the electron flow, with the ions escaping from a layer whose thickness is two ion skin depths while the electrons remain magnetized and flow into a layer thin enough to accomplish the needed reconnection. Finally, the cross-field electron currents required by the geometry become large enough to drive turbulence in the plasma, increasing the resistivity and enabling the reconnection in a thicker layer than would otherwise be required. Triggering and energy flow are also issues on which some things are known and much remains to be learned.

10.9 Dynamos

A *dynamo* is a system that causes conducting matter to move across magnetic field lines, producing an electric potential that drives a current which in turn creates magnetic fields. Dynamos are of great importance in nature, as the universe was initially unmagnetized and yet now contains magnetic fields under many circumstances. Examples of dynamos at high energy density include those that produce the magnetic field of the Earth and that of the Sun. Other examples of dynamos include the turbulent dynamos throughout the Universe that produce the observed magnetization of interstellar space and the dynamo that produces the extremely strong magnetic fields near pulsars. Yet the presence of dynamos in traditional plasma-physics texts is irregular. Dynamos have not been relevant to most laboratory plasmas and so have often been overlooked. Today, though, with the increased emphasis on connecting the laboratory and astrophysics, laboratory research focused on plasma dynamos has become common.

Here we will discuss key mechanisms involved in magnetic-field generation in plasmas or other conducting matter—the "α-effect" dynamo and the "Ω-effect" dynamo. A key aspect is that stretching a magnetic field typically strengthens it. Consider a tube of magnetized plasma, bounded laterally by magnetic-field lines and at its ends by two surfaces that are perpendicular to B. For simplicity make this tube cylindrical, although the point we will make here is general. If the plasma is close to incompressible, as is typical, and of density ρ, then the included mass, which must be conserved, is $\rho A \ell$, with tube area A and length ℓ. In the absence of currents on the surface of the flux tube, the magnetic flux, BA, is also conserved. As a result,

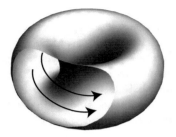 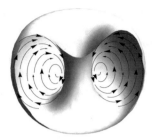

Fig. 10.9 Left: Magnetic field lines for a purely toroidal field. Right: Magnetic field lines for a purely poloidal field

$$\frac{B}{\rho \ell} = \text{constant and } B \propto \ell \tag{10.123}$$

in incompressible fluids where density is constant. The upshot is that stretching magnetic field lines strengthens the magnetic field.

Another necessary element for understanding dynamos is to understand the geometric description of the field relative to an axis of rotation. Key concepts are that the magnetic field can have "toroidal" and "poloidal" components, as illustrated in Fig. 10.9. If the axis of interest defines the z axis in standard spherical coordinates, then a toroidal field lies parallel to the x–y plane and has only a component in the ϕ direction. A poloidal field lies in the z–θ plane. Note that toroidal fields require poloidal currents, and *vice versa*. Dynamo behavior tends to convert one type of field to the other.

The origin of dynamo behavior is found in the magnetic induction equation,

$$\frac{\partial \boldsymbol{B}}{\partial t} = +\nabla \times (\boldsymbol{u} \times \boldsymbol{B}) + \frac{\eta c^2}{4\pi} \nabla^2 \boldsymbol{B}. \tag{10.124}$$

We can see that the term involving the velocity has the potential to cause B to increase. Even so, finding ways that this could actually happen proved challenging. The reason for this became more clear once Cowling, in 1934, proved his theorem that no axisymmetric velocity field could generate new magnetic field.

The α-effect can be understood using a statistical analysis of turbulent flow, similar to the Reynolds decomposition we discussed in Chap. 5 for fluid turbulence. One takes $\boldsymbol{B} = \boldsymbol{B}_o + \tilde{\boldsymbol{B}}$ and $\boldsymbol{u} = \boldsymbol{U} + \tilde{\boldsymbol{u}}$, in which $\boldsymbol{B}_o$ and $\boldsymbol{U}$ are constant averages while $\tilde{\boldsymbol{B}}$ and $\tilde{\boldsymbol{u}}$ vary in time but average to zero. The average in time of (10.124) becomes

$$\frac{\partial \boldsymbol{B}_o}{\partial t} = +\nabla \times (\boldsymbol{U} \times \boldsymbol{B}_o) + \nabla \times \langle \tilde{\boldsymbol{u}} \times \tilde{\boldsymbol{B}} \rangle + \frac{\eta c^2}{4\pi} \nabla^2 \boldsymbol{B}_o, \tag{10.125}$$

where $\langle \rangle$ represents the taking of the average. Under appropriate assumptions (see Kulsrud 2005), one finds

$$\langle \tilde{\boldsymbol{u}} \times \tilde{\boldsymbol{B}} \rangle = \alpha \boldsymbol{B}_o - \beta \nabla \times \boldsymbol{B}_o, \tag{10.126}$$

leading to the mean-field dynamo equation

$$\frac{\partial \boldsymbol{B}_o}{\partial t} = +\nabla \times (\boldsymbol{U} \times \boldsymbol{B}_o) + \nabla \times (\alpha \boldsymbol{B}_o) + \left(\beta + \frac{\eta c^2}{4\pi}\right) \nabla^2 \boldsymbol{B}_o, \tag{10.127}$$

The second term on the right here describes the α-effect dynamo. What is most important is that it can generate poloidal magnetic flux from toroidal magnetic flux. The factor β increases the diffusion of magnetic flux, and so is often described as the turbulent resistivity.

As one example, suppose $\boldsymbol{B}_o$ is along the z axis in Cartesian coordinates. There can be an Alfvén wave having $\boldsymbol{k} \parallel \hat{z}$, $\boldsymbol{B}_1 \parallel \hat{y}$, and $\boldsymbol{u} \parallel \hat{x}$. In ideal MHD, this wave will have $\langle \boldsymbol{u} \times \boldsymbol{B}_1 \rangle = 0$. But in resistive MHD, $\boldsymbol{B}_1$ will be shifted relative to $\boldsymbol{u}$, so that $\langle \boldsymbol{u} \times \boldsymbol{B}_1 \rangle \neq 0$ and it will become possible to generate magnetic field in the $x - y$ plane. This would represent a conversion of poloidal into toroidal flux.

The Ω-effect dynamo describes how differential rotation can convert poloidal flux into toroidal flux. Suppose one begins with a dipole field like that shown on the right in Fig. 10.9, produced by a current loop within a conducting body. For both the Earth and the Sun, this is an accurate global, large-scale description of the magnetic field. Such a field is purely poloidal—it lies entirely in the r-θ plane. If the body in question rotates as a rigid rotor, then the field remains poloidal. But if there is differential rotation, as is the case in the interiors of the Earth and the Sun, then the field lines at some radius are pulled ahead of those at another radius. This creates a toroidal component to $\boldsymbol{B}$. In the limit of long times and small resistivity, most of the magnetic flux can become toroidal. At that point, if not before, the α-effect dynamo can act to produce new poloidal flux from the toroidal flux.

If the signs work out so that the new flux is in the opposite direction of the original flux, this sequence of Ω-effect and then α-effect dynamos can lead to a reversal of the field, as is indeed observed for both the Earth and the Sun. But since the α-effect dynamo depends in detail on the properties of the turbulent flow, the field reversals are not strictly periodic. One also observes this in both the Earth and the Sun.

We now turn to the turbulent dynamo, responsible for the generation of magnetic fields at small scales throughout the Universe. The key mechanism thought to be at work is the α effect, but the geometry is different. Recall that fluid turbulence generates vortices at all scales above the dissipation scale and at all orientations. Also recall that the field is frozen in. So if an eddy develops within a patch of uniform field, it will strengthen the field by winding it into a spiral, as Fig. 10.10 illustrates. Complementary turbulent motions can twist and fold the resulting loop-like magnetic field structures (Schekochihin et al. 2004). In three dimensions (but not two), this can result in amplification of the average field. (Reconnection may also contribute by isolating loops of increased field.) At this writing, simulations have observed such MHD dynamos, and find that the threshold value of R_m for dynamo behavior is ~ 50–200, depending on R_e. Recent experiments have obtained

Fig. 10.10 Magnetic field
lines can be wrapped up and
lengthened when a vortex
develops within a field

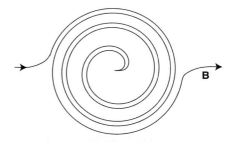

Fig. 10.11 A loop of hot
plasma, held in place by a
magnetic field, near the
surface of the Sun. Credit:
SOHO—EIT Consortium,
ESA, NASA

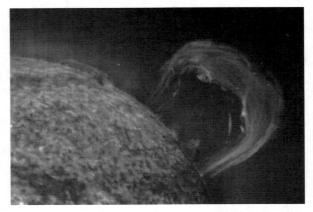

the first observations of magnetic fields attributed to the turbulent dynamo. But
much remains to be learned, and dynamo behavior in collisionless plasma remains
especially challenging to understand.

10.10   Creating Dense Magnetized Plasmas

Figure 10.11 shows a loop of magnetic field near the surface of the sun, visible
in soft X-rays because the plasma it contains is much hotter than the surrounding
plasma. One contributor to such heating is the pinch force, which has the amazing
effect of causing any channel of plasma that carries current to contract. As we will
see, the pinch force is one consequence of the $\boldsymbol{J} \times \boldsymbol{B}$ force. Since most astrophysical
systems include magnetic fields, whose motions induce the flow of current, this
force is present at some level in many circumstances. In the laboratory, modern
pulsed power devices can deliver voltages of >1 MV for >100 ns. As we will see,
this is ample to produce high-energy-density conditions. We will first discuss the
traditional approach to this end, known as the Z-pinch. Later, we will discuss an
alternative use of the same type of pulsed-power system—the magnetically launched
flyer plate. After that, we discuss the direct use of (pulsed) magnets for producing
magnetized plasmas at high-energy-density facilities.

10.10.1 Z-Pinches for High-Energy-Density Physics

André-Marie Ampère showed that current-carrying wires exert forces on one another. We can revisit this briefly to set the stage for the more complicated discussions that follow. Imagine an infinitely long, straight wire carrying a steady current. If one integrates Ampère's law over the area of a surface centered on the wire and bounded by a circle, and applies Stoke's theorem, one finds, using SI units,

$$\mu_o I_1 = \mu_o \int_{A_1} \boldsymbol{J}_1 \cdot \hat{\boldsymbol{n}} dA = \oint_{\ell_1} \boldsymbol{B}_1 \cdot d\boldsymbol{\ell} = 2\pi r_1 B_1, \qquad (10.128)$$

where $\hat{\boldsymbol{n}}$ is a unit vector normal to the surface and the subscript 1 represents the properties produced by the wire. One can show, from symmetry and the absence of magnetic monopoles, that $\boldsymbol{B}_1$ is purely azimuthal. The direction of $\boldsymbol{B}_1$ is given by the right-hand rule because the line integral is by convention always taken in the counterclockwise direction as viewed from the direction toward which $\hat{\boldsymbol{n}}$ points. If there is a thin, parallel wire some distance R from the first wire, the force per unit length on this second wire, $\boldsymbol{F}_2$, due to the magnetic field from the first wire, is

$$\boldsymbol{F}_2 = \int_{A_2} (\boldsymbol{J}_2 \times \boldsymbol{B}_1) \, dA = \mu_o \frac{I_1 I_2}{2\pi R} (-\hat{\boldsymbol{r}}_{12}), \qquad (10.129)$$

where subscript 2 refers to the second wire and $\hat{\boldsymbol{r}}_{12}$ is a unit vector from the first wire to the second wire. The minus sign means that the force is attractive (for $\boldsymbol{J}_1 \| \boldsymbol{J}_2$), as one can verify from the right-hand rule. The general point is that parallel currents attract. This has the implication that any compressible medium carrying current will tend to contract.

This fact enables one to create a type of device known as a Z-pinch. A Z-pinch uses an axial current (in the z direction in a standard Cartesian coordinate system) to create a pinch force, with the aim of producing a high-temperature plasma. Some of the early approaches to magnetic fusion were based on this principle. One can find equilibria in which the inward pinch force, produced by current in a plasma, balances the outward pressure of the plasma. Unfortunately, these equilibria are not stable; if they were stable then we might indeed have had fusion power plants in the 1960s. Most modern Z-pinches are so-called *fast* Z-pinches, in which a rapidly rising current causes the implosive contraction of material. The imploding material is accelerated and then converts the kinetic energy of implosion to heat when the material stagnates on axis. Such implosions can occur with varying relative amounts of heating *versus* acceleration. As we shall see, the implosions of interest to high-energy-density physics are violent indeed. Such implosive pinches avoid the slowly growing instabilities that plague equilibrium pinches. However, they create the transient growth of the Rayleigh–Taylor instability, discussed in Chap. 5, and this imposes some limitations on their operating range. Here we discuss the basic aspects of the implosion of a fast Z-pinch. More extensive discussions of Z-pinch physics can be found in the book, *Physics of High-Density Z-Pinch Plasmas*, by Lieberman et al. (1999), and in the article in *Reviews of Modern Physics* by Ryutov et al. (2000).

We begin by considering the self-consistent behavior of a long cylindrical shell with a uniform current density in the z direction, given initially as a function of radius r by

$$J = J_o \text{ for } r_1 < r < r_2; \ J = 0 \text{ otherwise.} \tag{10.130}$$

Such a current produces no magnetic field in the z direction, and one can show from this and the absence of magnetic monopoles that there is no radial component of magnetic field. Applying Ampère's law to the interior of the cylinder, there is also no azimuthal magnetic field inside the shell, so $B = 0$ there. Within the shell itself, Ampère's law in SI units implies

$$\oint B \cdot d\ell = 2\pi r B = \mu_o \int J \cdot \hat{n} dA = \mu_o J_o \pi \left(r^2 - r_1^2 \right) \approx \mu_o J_o \pi \left(2r \delta r \right), \tag{10.131}$$

where $\delta r = r - r_1$. The equation of motion relates the acceleration of the fluid to the inward force density $\mathcal{F}$, and is

$$\rho \frac{\partial u}{\partial t} = \mathcal{F} = J \times B = -\hat{r} \frac{\mu_o J_o^2}{2} \frac{\left(r^2 - r_1^2 \right)}{r} \approx -\hat{r} \mu_o J_o^2 \delta r, \tag{10.132}$$

where $\hat{r}$ is a unit vector in the radial direction. From this, the equation of motion for the radial acceleration of a fluid element is

$$\rho \ddot{r} = -\frac{\mu_o J_o^2}{2} \frac{\left(r^2 - r_1^2 \right)}{r} \approx \mu_o J_o^2 \delta r, \tag{10.133}$$

in which ρ is the mass density. One can integrate this over the cross-section of the shell to find the total inward force per unit length and thus the approximate equation of motion for the entire shell

$$\hat{m} \ddot{r} = -\frac{\mu_o I^2}{4 \pi r}, \tag{10.134}$$

in which I is the total current and $\hat{m}$ is the mass per unit length, and again this is in SI units. For constant current, one can integrate this equation to obtain

$$u_r^2 = \frac{\mu_o I^2}{4 \pi \hat{m}} 2 \ln \left[\frac{r_o}{r} \right] \text{ (SI)} = \frac{I^2}{c^2 \hat{m}} 2 \ln \left[\frac{r_o}{r} \right] \text{ (cgs)} = u_{\text{Alf}}^2 2 \ln \left[\frac{r_o}{r} \right]. \tag{10.135}$$

in which, u_{Alf} is the velocity of Alfvén waves at the initial outer edge of the pinch. One can in turn integrate (10.135) to find the time t_{imp} at which the implosion reaches a radius r, which is

Fig. 10.12 Behavior of a
constant-current Z-pinch
implosion, showing radius as
a fraction of r_o, and with
inward velocity normalized to
u_{Alf}, versus time, normalized
to r_o/u_{Alf}

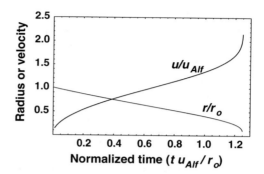

$$t_{imp} = \frac{r_o}{u_{Alf}} \sqrt{\frac{\pi}{2}} \operatorname{erf}\left[\sqrt{\ln(r_o/r)}\right], \tag{10.136}$$

where u_{Alf} is defined by (10.135) and erf is the error function. In practical units,

$$u_{Alf} = 3.3 \times 10^6 I_{MA}/\sqrt{\hat{m}(\text{mg/cm})} \text{ cm/s}, \tag{10.137}$$

where I_{MA} is the current in MA and the units of $\hat{m}$ are shown. The behavior produced
by (10.135) and (10.136) is shown in Fig. 10.12. One can see that the shell of current
moves inward slowly at first, and that only late in the implosion does the acceleration
greatly increase. The development of an actual pinch implosion, which has a slowly
increasing pinch current, is even more gradual. An essential phenomenon, included
in the equation of motion only by the boundary condition that $u_r = 0$ at $r = 0$,
is that the pinch material must stagnate before it reaches the axis of the cylindrical
shell, where the incoming matter will accumulate. In the simplest conception, the
pinch material is accelerated inward, gaining kinetic energy, and is shocked and
compressed as it stagnates when it symmetrically reaches the axis, converting the
kinetic energy into thermal energy and later into radiation and an outward expanding
plasma.

The convergence, r_o/r, enters into these equations. We can estimate a plausible
radius at stagnation r_s in order to determine the maximum convergence. If, for
example, one used a metallic shell whose density was $10\,\text{g/cm}^3$, with an initial mass
of 1 mg/cm, and the imploded and shocked material on the axis had a density of
$40\,\text{g/cm}^3$, all of which are plausible numbers, then the radius of the imploded pinch
would be about $30\,\mu\text{m}$. If the initial radius of the shell were 1 cm, then one would
have $r_o/r_s \sim 300$. In actuality, instabilities limit the degree of implosion, which
typically ends at $r_o/r_s \sim 10\text{–}20$. Note that differences of a factor of 2 in r_o/r_s have
a very small effect on the final implosion velocity because the convergence enters
into the logarithm. Indeed, even increasing the convergence to $r_o/r_s \sim 300$ would
increase u_r by less than 50%. Pinch research in the 1950s was focused on creating
a high-density matter where fusion would occur, for which high convergence is

essential. Modern applications of Z-pinches to high-energy-density physics more often depend primarily on energy (and on its efficient conversion to radiation), and so are less sensitive to convergence.

Returning to the simple model of (10.136), and using $r_o/r_s = 10$, one can find the kinetic energy of the pinch material just at stagnation. Remarkably, this quantity depends only on the pinch current. It is

$$\text{K.E.} = 2.3\hat{m}u_{\text{Alf}}^2 = 2.3\frac{\mu_o I^2}{4\pi} \text{ (SI)} = 2.3\frac{I^2}{c^2} \text{ (cgs)} = 2.3I_{\text{MA}}^2 \text{ kJ/cm.} \tag{10.138}$$

With the currents above 20 MA that are now feasible, this energy can exceed 1 MJ/cm. The total energy of the pinch material may be higher than this, because it has been heated by Joule heating (i.e., $J \cdot E$) and by compression (i.e., pdV work), but it also may lose energy to radiation before the end of the implosion. Assuming that the heated material stagnates symmetrically, all the remaining energy is momentarily converted to heat. The energy of stagnation initially develops in the ions, as they carry the kinetic energy, and is then transferred by collisions to the electrons. Once the temperatures have equalized, the heating produced by the kinetic energy of (10.138) gives a temperature in eV, T_{eV}, of

$$T_{\text{eV}} = \frac{\text{K.E.} \times Am_p}{k_B\hat{m}(Z + 1)} = 20\frac{I_{\text{MA}}^2}{\hat{m} \text{ (mg/cm)}}\left(\frac{A}{Z + 1}\right) \text{ eV.} \tag{10.139}$$

In evaluating this equation, one may have to allow for the dependence of Z on T_e (Chap. 3). Such dense matter, at typical stagnation temperatures above 1 keV, is a very intense radiator. Note that one can adjust this temperature, to seek an optimum for some purpose, by adjusting the mass per unit length.

As was mentioned above, the current in an actual pinch is not constant. In fact, it often has a sinusoidal profile in time. This leads the implosion to develop more gradually than Fig. 10.12 shows. However, there are only limited analytic solutions for the motion of the pinch with more realistic, time-dependent current profiles. To make matters even more complex, the current is not fundamentally independent of the pinch and its dynamics. The pulsed-power machine provides a time-dependent voltage pulse to its load, which in this case is the pinch and the supporting structures for the pinch. The principal limitation on the current through the pinch is the inductance of the pinch itself. Thus, more-accurate pinch modeling specifies the time-dependent voltage supplied to the pinch, determines the current by calculating the instantaneous induction of the pinch and solving a circuit equation, and simultaneously solves an equation like (10.134) for the implosion of the pinch itself. One result of such circuit modeling has been that pinches have become shorter in recent years, and often now have a height that is only a fraction of their initial radius. By reducing the height, one can decrease the inductance and thus increase the current.

This has an application when one considers the duration of the pinch implosion. In actual Z-pinches the implosion time must be matched to the duration of the

Fig. 10.13 Sketch of
structure at the core of a
modern Z-pinch. The part that
matters for the physics is the
wire array. The return current
at the height of the wire array
flows through a canister that
has holes in it for diagnostic
access and for radiation
escape

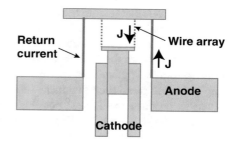

voltage pulse that can be produced by the pulsed-power system. Using (10.136)
and (10.137), for initial radii of a few cm, masses within a factor of five times
1 mg/cm, and currents within a factor of 3 of 10 MA, one can see that the implosion
time is within an order of magnitude of 100 ns. Of course, the reasoning actually
must be done in reverse. Given the ability to deliver a voltage pulse of some duration,
one must choose the mass and radius of the Z-pinch load to obtain an implosion of
the same duration with the current that results. Let us explore this further.

Figure 10.13 shows a sketch, roughly to scale, of the hardware at the core of a
modern Z-pinch. The inductance L of a current-carrying cylinder of height H and
radius r, with the return current carried at some larger radius r_{ret}, is easily found to
be

$$L = \frac{\mu_o}{2\pi} H \ln\left(\frac{r_{\text{ret}}}{r}\right). \tag{10.140}$$

With an available voltage V of duration τ, we solve $V = L dI/dt \sim LI/\tau$ with
$\tau = t_{\text{imp}}$ from (10.136) to find

$$I_{MA} = 10^{-4} \sqrt{\frac{V_{MV}\sqrt{\hat{m}(\text{mg/cm})}}{\sqrt{2}\mu_o^{3/2}}\left(\frac{r_o}{H}\right)\frac{\text{Erf}\left[\sqrt{\ln(r_o/r_s)}\right]}{\ln(r_{ret}/r)}}$$

$$= 1.47\,(\hat{m}(\text{mg/cm}))^{1/4}\sqrt{V_{MV}\frac{r_o}{H}}, \tag{10.141}$$

in which r_o and H must be in the same units and the second result is obtained using
$r_{\text{ret}}/r = 10$ and $r_o/r_s = 20$ but depends weakly on the exact values of these ratios.
We see that multi-MA currents are straightforward to achieve.

The options for increasing the current are limited. The simplest is to decrease
H. It seems from (10.141) that one could increase r_o. However, in order to keep
$t_{\text{imp}} \sim \tau$, (10.136) implies that one must keep the ratio $I_{MA}/(r_o\sqrt{\hat{m}})$ constant, so
this would require decreasing $\hat{m}$. Z-pinches that drive their currents through puffs
of gas accomplish this, obtaining small $\hat{m}$ and very large r_o. This has produced
implosion velocities near 1000 km/s for the production of intense, K-shell radiation.

Ultimately, one can rebuild the machine to increase V (or more accurately $V\tau$). This is the only path that can produce substantial increases in current.

One application of Z-pinches that is most relevant to high-energy-density physics at present involves the production of radiation. (Ryutov et al. (2000) discuss some other possible applications in their review paper, and Chap. 11 discusses their application to inertial confinement fusion.) The first such application is to use the pinch to produce the largest possible soft X-ray energy by blackbody radiation from hot, dense matter. For this purpose one implodes a high-Z material, typically tungsten (W). After the stagnation, the plasma both radiates and expands. We can evaluate the ratio of the blackbody radiation, σT^4, to the power involved in plasma expansion $(\rho/Am_p)k_BTc_s$ as follows:

$$\frac{\sigma T^4}{[Z\rho/(Am_p)]k_BTc_s} = 35\frac{T_{keV}^{2.5}}{\rho}\left(\frac{A}{Z}\right)^{3/2}, \qquad (10.142)$$

where T_{keV} is the temperature in keV. This implies that radiation will be strongly dominant above some temperature of order 1 keV. Such Z-pinch radiation sources are often produced within high-Z hohlraums, similar to those discussed in Sect. 9.3. These hohlraums can confine the pinch radiation and sustain for some time a high-temperature, thermal radiation environment. They have been used to irradiate packages either mounted on their walls, to study ablatively driven phenomena or radiation flow, or mounted within the hohlraum, to study radiation transport or photoionization effects. They have also been used to irradiate capsules for inertial-confinement-fusion research.

A second radiation-related application of Z-pinches is to use them to produce X-ray line radiation. For this purpose, one uses wires of a material whose K_α X-rays have an energy of a few keV, such as titanium. The radiation balance is not as easy to estimate as it is in the case of blackbody radiation. The efficiency is large enough to produce useful yields for practical applications.

In order to maximize the power radiated by a Z-pinch during stagnation, one must maximize the stagnation power, P_s. Because the plasma expands during the implosion, the duration of the stagnation can be expected to scale with the implosion time, which by design one makes equal to the duration τ of the voltage pulse. Thus, since the kinetic energy is proportional to I_{MA}^2,

$$P_s \propto \frac{HI_{MA}^2}{\tau} \propto \frac{HI_{MA}V_{MV}}{L} \propto \frac{V_{MV}^{3/2}\left(\sqrt{\tilde{m}}r_o\right)^{1/2}}{H^{1/2}} \propto \frac{V_{MV}^2}{H}, \qquad (10.143)$$

so for fixed pulsed-power parameters one can increase the stagnation power only by decreasing H. There are limits to this, as the implosion will be compromised near the ends of the pinch. Nonetheless, at around the turn of the century pinches less than 1 cm high were imploded with good results on the Z-device.

For many years the ability of Z-pinches to actually produce X-ray radiation fell far below the expectations one would have from the scalings discussed above. This changed dramatically during the 1990s, with the development of Z-pinches that use a load composed of hundreds of fine wires (typically ∼10 μm dia.). All the previous approaches, which included wire arrays with fewer, thicker wires, solid cylindrical conducting shells (known as liners), and various schemes involving gas, performed far less well. It is tempting to infer that the use of many wires finally produced a structure resembling the uniform plasma shell of our simple modeling, while all the previous methods produced a less uniform plasma that did not stagnate as effectively. The success with arrays of fine wires led to a large increase in the X-ray yield from such devices.

The observed implosion time of pinches using arrays of many wires, determined from the timing of the X-rays produced at stagnation, is typically in excellent agreement with the time predicted by modeling of a uniform plasma shell. This led some authors to conclude that the wires in such wire arrays do expand and merge so as to produce such a plasma shell. Further support for this conclusion has come from MHD modeling, which can reproduce the stagnation time and the size of the resulting plasma, although to do so one must assume very large initial perturbations to seed the Rayleigh–Taylor instability. However, the issues are not so simple and the evidence is rather complex. There are two ways that an array of wires can develop a comparatively uniform implosion. The simplest notion is that the wires explode into small plasmas and that if the wires are close enough then these plasmas will connect and the current will flow uniformly in azimuth. However, the evidence indicates that the wires, and especially those of materials such as Al and W that perform well, do not initially explode.

Instead the wires ablate because the current flowing on their surfaces heats them so strongly that plasma flows away from their surfaces. This can create a structure in which the plasma and magnetic field have merged but the wire cores remain. The likely behavior of the magnetic field is complex—the field is not frozen in. The magnetic diffusion time, from (10.37) and Fig. 10.2, for distances of fractions of a mm, with electron temperatures that are not so many eV, may be of order 1 ns and certainly is much smaller than the implosion time of order 100 ns. The field is initially strongest near the wire surfaces and will tend to diffuse outward into the developing plasma, where it can merge (via reconnection) to form a more symmetric structure. The diffusion of the field also corresponds to a diffusion of the current, so the plasma experiences a $J \times B$ force that accelerates it inward. In typical cases it appears that of order half the wire mass may be accelerated inward before the final phase of the implosion. Some magnetic field will be carried with such plasma, and more may diffuse into it.

It is unclear at this writing how much force is actually delivered to the wire cores, and whether the cores themselves actually move. On the one hand, if the wire cores eventually become small enough to explode into plasma, then they probably do move. On the other hand, there is some evidence that the wire ablation ceases once the wires develop gaps, which are likely the result of MHD instabilities in the wires themselves. After that, there is no longer a source of plasma to sustain the current and magnetic field at the edge of the array, and the outer edge of the plasma will implode inward, sweeping up the interior mass in what is usually described as a snowplow implosion. There may be a fundamental underlying cause, but at the moment it seems amazing and fortuitous that modeling of this more complicated process produces implosion times that are nearly identical to those produced by a uniform plasma shell. Whatever these details turn out to be, the important consequence is that wire-array implosions can be efficient sources of X-rays for high-energy-density experiments.

10.10.2 Magnetically Driven Flyer Plates

As another application of the pulsed-power technology that drives Z-pinches, we consider how such a source of current and voltage can be used to isentropically compress and/or accelerate samples. The traditional Z-pinch uses the fact that nearby conductors carrying parallel currents attract, as we discussed in the beginning of Sect. 10.10. By running parallel currents through an array of low-mass conductors, one can make them implode. There is a return current in a Z-pinch, but it is placed at a large radius so that it has little effect on the implosion, as we discussed. However, it is also true that conductors carrying opposing currents repel one another. By placing the opposing currents in close proximity, one can create a large force that drives them apart. If one makes one conductor quite massive and gives the adjacent conductor a much-lower mass, then the low-mass conductor will be preferentially accelerated. This is the key to what is sometimes known as *magnetic drive*.

The reason for this name becomes more clear if one thinks about the magnetic fields that these currents generate. The two conductors generate a magnetic field between them, perpendicular to the currents and with a direction given by the right-hand rule. One way to think of the resulting drive is to consider that each conductor experiences a $J \times B$ force, just as we did when we discussed the Z-pinch. A second way to think about magnetic drive is to note that the magnetic pressure drops to zero across a thin layer at the surface of the conductor where the current flows, so that one can say that the magnetic pressure is applied to the conductors. This magnetic field can be enormous.

Thus far this description shows how to apply a large force to accelerate an object. There is an additional aspect of this possibility that gives it more value. By adjusting the increase of the current with time, one can control the time dependence of the force. In particular, one can increase it slowly enough to avoid launching a shock into the driven material. Observing the response of the material to such an isentropic compression can provide substantial insight into the equation of state. Beyond this, by isentropic compression and acceleration one can launch a cold flyer plate at a higher velocity than can be produced by traditional flyer-plate launchers such as gas guns. At this writing, Al flyer plates have been launched isentropically at velocities above 30 km/s. See Chap. 4 for discussions of the hydrodynamics of flyer plates and of their use in making equation-of-state measurements.

10.10.3   Direct Use of Magnets

Another way to study plasma in a magnetic field is to use some energy source
to create a magnetic field and then to produce the plasma within it. There is a
conceptual limitation to such experiments, though, not always well understood. The
plasma is very nearly a superconductor, so it cannot be penetrated by new magnetic
field on timescales shorter than the diffusion time discussed above, unless the Nernst
effect becomes substantial. If the flow becomes turbulent then the diffusion time
might be shortened, but there is scant evidence for this, or reason to expect it, in
simple plasmas blown off of surfaces. If one injects plasma along the magnetic field,
then one can get the plasma into the field. If one creates the plasma from matter,
such as a gas, already in the field, then it will contain the field. But if one tries
to inject matter across the field lines, then the simplest physics says the field will
not penetrate the plasma, and the full physics is complex enough that knowledge of
the field penetration must be based on measurements. Despite this limitation, much
physics of interest can be done and has been done using a combination of magnetic-
field sources and plasma sources. There are at least three approaches in common use.

The first approach is to build a coil or wire, not much larger in scale than the laser
target, to use some source of pulsed power to drive a current through it, producing
the field, and then to use a laser (or other source) to produce a plasma that interacts
with the field. This approach is necessary in laser facilities with limited access to the
central area of the target chamber, such as Omega. The Omega facility has built a
device they call MIFEDS, containing powerful small capacitors that can be carried
in close to the chamber center, within a diagnostic insertion tube. It can produce
fields approaching 10 T, depending on the application. The facility is supporting the
upgrade of this device, seeking to enable fields of several times this value. Variations
on this approach have been common also at the Magpie pulsed-power machine at
Imperial College, with some structures used to produce fields and others to launch
plasma at or into such fields. The Z machine operates a larger set of coils that can
also bring its central volume to fields of order 10 T, for a variety of applications.

The second approach, applicable to smaller laser facilities where one can access
the volume near the focus of the beam, is to devise a small magnetic coil with holes
in the structure that provide laser and diagnostic access to a central volume, usually
a few mm in diameter. Such magnets can produce magnetic fields of some tens of T
over this volume. They have been constructed in both the US and Europe, and have
been used in experiments on lasers having hundreds of Joules of available energy.

The third approach is to use one set of laser beams to drive a current loop and
another set to create the plasma of interest, as Fig. 10.14 illustrates. One places two
conductors close enough to one another, and then irradiates one of them (perhaps
through a hole in the other). Some of the electrons escaping the irradiated conductor
are captured by the second conductor, creating a voltage difference between the
two of them. Connecting the two conductors to a current loop enables this voltage
difference to produce a magnetic field. This technique has been measured to produce
fields of tens of T near the center of the coil.

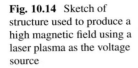

Fig. 10.14 Sketch of structure used to produce a high magnetic field using a laser plasma as the voltage source

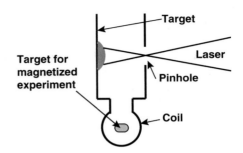

Finally, some experiments have emerged, working at lower energy density, that inject plasma blobs from laser targets or other sources into large plasmas created in kG-scale fields, produced by steady solenoids. This has begun to enable experiments on collisionless shocks and other phenomena that can complement the science that can be done at higher energy density.

Homework Problems

10.1 Find the sizes and directions of the orbits of protons and electrons. Explain from fundamental laws of electromagnetics why their direction is as it is. Show pictorially why the $E \times B$ drift moves particles in the same direction.

10.2 The MHD equations assume charge neutrality, yet MHD plasmas may contain electric fields. Explore this seeming contradiction by, first, evaluating the electric field and relative charge imbalance for a steady electric field in an isothermal plasma having a density gradient (so that $e n_e E = -\nabla p_e$), for reasonable choices of parameters. Compare this to the electric field produced in an electron plasma oscillation for which the amplitude of the electron-density fluctuations is 10%. Express the magnitude of the electric force as $eV/\mu m$.

10.3 Begin with (10.48)–(10.50), keeping the resistivity. Derive the dispersion relation for damped Alfvén waves, starting with the assumptions that k is parallel to B and that u is purely transverse. (You should find this much simpler than the general case just discussed.) For reasonable choices of plasma parameters, plot the ratio of damping rate to real frequency as a function of electron temperature.

10.4 Begin with (10.48)–(10.50), assuming small resistivity. Derive the dispersion relation for cross-field sound waves, starting with the assumptions that k is perpendicular to B and parallel to u. (You should find this much simpler than the general case just discussed.)

10.5 Show that when the field evolves as (10.73) describes, the quantity $B \cdot \hat{q}$ remains zero to the first-order accuracy of the present model.

10.6 While one can vary the properties of a Z-pinch load from one experiment to the next, one can modify the pulsed-power device itself on a somewhat longer timescale. Such devices are typically characterized by the number of Volt-Seconds they can produce, and operate so that $V\tau$ = constant. First, consider and then explain why Volt-Seconds is a reasonable way to characterize a pulsed-power device. Second, using the scaling relations developed in Sect. 10.10.1, discuss how to optimize the stagnation power for a device with $V\tau$ = constant.

10.7 Revisit the derivation at the beginning of Sect. 10.10. Consider two infinitely wide, plane parallel conductors carrying opposing currents. Find the force per unit area between them and express it in terms of the magnetic field magnitude. Discuss how the force per unit area compares to the energy density of the magnetic field.

References

Balogh A, Treumann RA (2013) Physics of collisionless shocks. Springer, New York

Braginskii SI (1965) Transport processes in a plasma. Rev Plasma Phys 1:205

Chandrasekhar S (1961) Hydrodynamic and hydromagnetic stability. Dover, New York

Epperlein EM, Haines MG (1986) Plasma transport-coefficients in a magnetic-field by direct numerical-solution of the Fokker-Planck equation. Phys Fluids 29(4):1029–1041. https://doi.org/10.1063/1.865901

Haines MG (1986) Magnetic-field generation in laser fusion and hot-electron transport. Can J Phys 64(8):912–919

Krall NA, Trivelpiece AW (1986) Principles of plasma physics. San Francisco Press, San Francisco

Kulsrud RM (2005) Plasma physics for astrophysics. Princeton University Press, Princeton

Lieberman MA, De Groot JS, Toor A, Spielman RB (1999) Physics of high-density Z-pinch plasmas. Springer, Berlin/Heidelberg/New York

Parker EN (1957) Magnetic reconnection. J Geophys Res 62:509

Ryutov DD, Derzon MS, Matzen MK (2000) The physics of fast z pinches. Rev Mod Phys 72(1):167–223

Schekochihin AA, Cowley SC, Taylor SF, Maron JL, McWilliams JC (2004) Simulations of the small-scale turbulent dynamo. Astrophys J 612(1):276–307. https://doi.org/10.1086/422547

Sweet PA (1958) Magnetic reconnection. In: Electromagnetic phenomena in cosmical physics, vol 6. Cambridge University Press, Cambridge, p 123

Chapter 11
Inertial Confinement Fusion

Abstract This chapter begins by discussing the nuclear physics that makes possible laboratory fusion, and considering on basic grounds the energy gain that might be possible and what is required for a power plant. Some exploration of the properties of DT fuel follows, enabling one to identify its behavior under compression and its response to entropy deposition. The next section describes the capsule implosions that put kinetic energy into the fuel. After that comes consideration of the stopping of the incoming fuel capsule and the requirements for ignition, whether via a central hot spot or via a spark. A discussion of some of the challenges that must be overcome for targets to actually ignite follows. The final section considers how the combination of all the constraints affects the potential for Inertial Confinement Fusion using laser energy sources to succeed.

The early chapters in this book were focused on the physical fundamentals of high-energy-density physics. Chapters 9 and 10 showed how we could create such conditions, which in turn makes possible the application of high-energy-density systems to the pursuit of various goals. This and the next chapter are much more focused on these goals. The goal that has been and remains dominant in high-energy-density research is the development of *inertial confinement fusion*, or ICF. This is our topic in the present chapter.

Fusion is the joining of two nuclei. This leads to the production of various reaction products, which often carry significant kinetic energy. Whether or not nuclear fusion releases energy depends on the masses of the nuclei involved. If the total mass of the reaction products is less than the mass of the initial nuclei by an amount Δm, then the net energy released by the reaction is Δmc^2. It is by fusion that all the elements beyond the very lightest few were created. However, not all fusion events release energy. Figure 11.1 shows a plot of the nuclear binding energy versus atomic number. The *binding energy* is the energy one must invest to disassemble the nucleus into its component protons and neutrons. This is proportional to the mass difference between the mass of its constituent protons and neutrons and the mass of the nucleus. The most-stable nucleus is iron, with an atomic mass of 56. As a result, energy can be released by combining elements lighter than iron, or by dividing (by fission) elements heavier than iron. One can see that some light

© Springer International Publishing AG 2018
R P. Drake, *High-Energy-Density Physics*, Graduate Texts in Physics,
https://doi.org/10.1007/978-3-319-67711-8_11

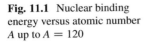

Fig. 11.1 Nuclear binding energy versus atomic number A up to $A = 120$

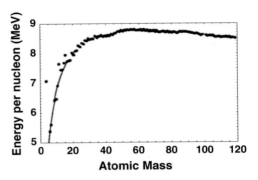

elements have relatively large binding energies—these are those with closed nuclear shells and correspond to elements that accumulate in stars.

Stars begin by assembling a very large mass of light elements. Through gravitational compression, their cores become dense and hot enough to initiate fusion burning. Through fusion, they begin to convert their light elements to heavier elements. Heavier elements require progressively higher temperatures to fuse, as the heavier nuclei have higher nuclear charge, so that it takes more energy to overcome the Coulomb repulsion. Low-mass stars like the sun accumulate He by burning H and eventually are able to ignite He, producing cores of C and O, but they cannot go further than this. High-mass stars (larger than about eight solar masses) can create all the elements up to Fe, and accumulate significant amounts of Si in the process. The Fe proves to be the death of these stars, as it cannot burn, so it cannot sustain the pressure necessary to resist the gravitational contraction. The eventual collapse of the Fe core triggers some types of supernova explosions.

All this has much to do with elemental abundances in the universe. Elements up to iron can be created by stars during their lifetime, and the most abundant ones are those that represent stable endpoints during stellar evolution. The eventual stellar explosion creates an environment rich in neutrons and neutrinos, which rapidly process the material that exists, producing the elements heavier than Fe and altering the populations of the lighter elements. Arnett (1996) is a good first source for more details on this.

This context leads to natural questions. We can create conditions of high pressure and high temperature, if only briefly, using high-energy-density devices. Any concentrations of matter and energy we produce are confined inertially, not gravitationally. (That is to say, they blow apart in roughly one sonic transit time.) Even so, can we perhaps do this in a way that causes light elements to fuse, releasing energy? Can we perhaps release useful amounts of energy? Let us see. We will proceed from asking what conditions we have to end up with, moving to how we might get there and then to what might go wrong. Our approach here will be to use simple arguments to identify the important issues and resolve them. This will get us into the ballpark of actual ICF parameters. But real designs for producing ICF must consider every issue that can be identified, and not just the most important ones.

Computer simulations are an important tool for including many of these details. A next level of detail may be found in Lindl (1995) and in Atzeni and Meyer-Ter-Vehn (2004).

11.1   The Final State

To answer our question about the possibility of inertial fusion, we will proceed from the end toward the beginning. We start by asking what we might use for fuel. Then we will ask what physical conditions are required to make this fusion fuel burn and provide an energy yield. This will lead to the question of how we can produce these conditions, and what the pitfalls might be in attempting to do so.

11.1.1   What Fuel, Under What Conditions?

A first question, determining much that follows, is what elements we might use for fuel. This would lead us to examine tables of nuclear reactions, from which we could find the following few that offer some potential for the easiest laboratory fusion systems:

$$D + T \rightarrow {}^4He \ (3.5 \ MeV) + n \ (14.1 \ MeV), \tag{11.1}$$

$$D + D \rightarrow {}^3He \ (0.82 \ MeV) + n \ (2.45 \ MeV), \tag{11.2}$$

$$D + D \rightarrow T \ (1.01 \ MeV) + H \ (3.02 \ MeV), \tag{11.3}$$

$$D + {}^3He \rightarrow {}^4He \ (3.6 \ MeV) + H \ (14.7 \ MeV), \tag{11.4}$$

and

$$p + B^{11} \rightarrow 3 \ {}^4He \ (8.68 \ MeV \ each). \tag{11.5}$$

The first of these reactions (known as DT) is the focus of nearly all fusion research at this writing. Any plasma producing these reactions will produce the next three as well. We will see the advantage of this focus in a moment. The disadvantage, for applications such as the production of electricity, is that the energy emerges primarily as neutrons (designated n in the equations). One can only manipulate energy from neutrons by first converting the energy to heat, and heat cycles have limited efficiency. (The heat cycle efficiency is $\sim$40%, which applies after the conversion of neutron kinetic energy to heat.) This leads one to look toward *advanced fuels*, such as the reaction of p and B^{11}, that produce only charged-particle reaction products. In the longest run, these offer the potential to escape the need for heat cycles and to eliminate all the associated hardware from fusion power plants.

Fig. 11.2 Rate coefficients
for the DT, DD, D–He^3, and
p–B^{11} reactions

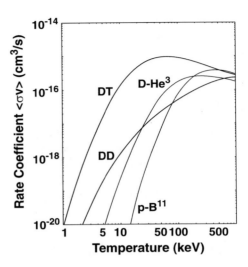

In the shorter run, fusion systems that produce a lot of neutrons may prove more useful as breeders of fuel for power plants using nuclear fission. But we are not yet at the point of thinking about power production, so let us return to the question of whether one could do this at all.

Figure 11.2 shows the dependence on energy of the rate coefficients for these four fusion reactions, found by averaging the reaction cross section over Maxwellian distributions of interacting particles, just as we discussed in Chap. 2 for collisional processes and Chap. 6 for atomic ones. One sees that the rate coefficient for the DT reaction becomes large at temperatures far below those required for the other processes. A star may not care much about this. It can keep the material in place for a long time. But to attempt ICF we do care. We have to burn the fuel before it blows apart—getting the rate coefficient up near its maximum matters. One can see that DT is clearly the fuel of choice for initial attempts to achieve ICF.

Next suppose we have brought a clump of DT fuel into a final state, with conditions that encourage it to burn. How much of it burns? To answer this suppose that the density of deuterium nuclei in the clump is N_D, the density of tritium nuclei is N_T, and the density of pairs of reaction products is n. Also, ignore the DD reactions as we are seeking a simple estimate rather than a complete account. The rate equation describing the accumulation of reaction products is

$$\frac{dn}{dt} = N_D N_T \langle \sigma v \rangle_{DT}, \tag{11.6}$$

in which the rate coefficient for the reaction is $\langle \sigma v \rangle_{DT}$. Next suppose that $N_D = N_T = 0.5 N_o - n$, and define the burn fraction ϕ, given by $2n/N_o$. Then (11.6) becomes

$$\frac{d\phi}{dt} = \frac{N_o}{2}(1 - \phi)^2 \langle \sigma v \rangle_{DT}, \tag{11.7}$$

Now if we make the approximation that $\langle \sigma v \rangle_{\mathrm{DT}}$ is constant as the fuel burns, and integrate over a time τ during which the fuel burns, we find

$$\frac{\phi}{(1 - \phi)} = \frac{N_o \tau}{2} \langle \sigma v \rangle_{\mathrm{DT}}. \tag{11.8}$$

Next we need to estimate how long the fuel burns. This time should be proportional to the fuel radius r and inversely proportional to the sound speed c_s, but is clearly less that the ratio r/c_s. We account for this by taking $\tau = r/(3c_s)$, where the factor of 3 would be only a guess without more-advanced knowledge. We take the temperature to be 30 keV in order to evaluate the sound speed and rate coefficient. (This amounts to assuming that self-heating by the burning fuel will push the temperature to the value corresponding to the maximum of the rate coefficient.) We also convert N_o to obtain

$$\phi = \frac{\rho r}{\rho r + 6 \ \mathrm{g/cm^2}}. \tag{11.9}$$

Here we meet the quantity ρr for the first time. This is the mass per unit area, which is the *areal mass density*. We see that this quantity controls the burn fraction. The transmission of particles or photons through the fuel also depends on ρr. Within the context of the approximations above, one can see that when ρr is 3 g/cm² the burn fraction is 33%, while ρr increases to 6 g/cm², the burn fraction increases only to 50%. The returns are clearly diminishing, and the cost of ρr is high, so let us assume that our final state before burning has $\rho r = 3$ g/cm², producing a burn fraction of 33%.

The discussion just above applies most closely to uniform burning of an entire volume of fuel. We will see that actual fusion designs involve a *propagating burn*. A propagating burn is like a forest fire, in which the fuel begins burning at one location, after which the heating of nearby locations causes them to burn too.

We can pin down the final fuel conditions further by thinking about the total fusion energy released (the *yield*). The range of a 3.5 MeV α particle (a ^{4}He nucleus) in DT is $\rho r = 0.3$ g/cm², so the α particles do not typically escape the fuel. Instead, they contribute to self-heating. This means that the energy released is the energy carried out of the fuel by the 14.1 MeV neutrons. One easily finds that the neutron yield, Y, is

$$Y = \frac{\phi}{2} \frac{m_F}{A m_p} \times 14.1 \ \mathrm{MeV} \approx 5.6 \times 10^{20} \ m_{\mathrm{mg}} \ \mathrm{MeV} \approx 90 \ m_{\mathrm{mg}} \ \mathrm{MJ}, \tag{11.10}$$

in which we have taken $\phi = 0.33$ $A = 2.5$ for DT, and the total mass of the fuel is m_F, also expressed in mg as m_{mg}. Since we are talking about the abrupt release of substantial energy, we need to place this in context. One ton of TNT is 4.2 GJ. The nuclear device exploded at Hiroshima in 1945 released about 10 kilotons of explosive energy. In a large laboratory device, we need to keep the yield small enough that it can be easily contained. For the present calculation, we will seek to produce ~0.05 ton, or about 250 MJ, which will require 3 mg of fuel mass according to (11.10). Designs in this range appear to be compatible with laser-based energy sources. Designs for other energy sources, such as Z pinches, often seem to be more compatible with significantly larger yields. Even so, one can't go too far without blowing up the laboratory.

Thus, we have found that $\rho r = 3 \, \text{g/cm}^2$ and that the mass of the fuel is $4\pi \rho r^3/3 = 3 \, \text{mg}$. We assume that the fuel volume is spherical, as this is most efficient for both the assembly of the dense fuel and for its burning. Substituting for ρr, we find $4\pi r^2 = 0.003 \, \text{cm}^2$, from which we find $r = 0.015 \, \text{cm} = 150 \, \mu\text{m}$. This in turn implies that $\rho = 190 \, \text{g/cm}^3$. However, it turns out that we cannot produce this state.

11.1.2   Implosions and Energy Gain: Is This Worth Doing?

We just described an ideal clump of fuel as a sphere of density $190 \, \text{g/cm}^3$ and radius $150 \, \mu\text{m}$. At $190 \, \text{g/cm}^3$, and considering that the average A for DT is 2.5 while $Z = 1$, the electron density is $n_e = 190 Z/(A m_p) = 4.5 \times 10^{25} \, \text{cm}^{-3}$. In our discussion in Sect. 3.1 of Fermi degenerate systems, we saw that the Fermi energy is $7.9(n_e/10^{23})^{2/3}$ with n_e in cm^{-3}. This is 464 eV for the compressed DT fuel. The corresponding pressure p is 13.5 Gbar. Unfortunately, we have no means to directly create and apply such a pressure. Laser-plasma instabilities limit laser-driven systems to a pressure of ~100 Mbar. Z-pinches and other known energy sources cannot produce pressures even this high. This leaves us with the quandary of how to convert pressures well below 1 Gbar into pressures of many Gbars.

One solution to this quandary is a *capsule implosion*. One concentrates the fuel in a thin layer at the inner boundary of a thin, spherical capsule. Current designs make the fuel a layer of DT ice and the capsule that contains it of some low-Z material. The low-Z material is the *ablator*, intended to be ablated in the creation of the ablation pressure. The idea is that the capsule behaves as a "spherical rocket", so that the ablation pressure accelerates the fuel to high kinetic energy. The fuel also compresses via convergence as it moves inward. Then, when the incoming fuel stagnates on axis, the kinetic energy is converted to the internal energy to produce the required high density. We will focus on several of the details of this process below. Here we connect the implosion process to the energy gain.

There are several steps involved in converting the energy from a source (or a "driver") to energy of implosion. No known driver can deliver its energy directly to the ablating surface. Instead, the energy is first absorbed, and then in some way it is transported to the ablation layer, and then the acceleration to the implosion velocity v_{imp} occurs. As a result, there is an *implosion efficiency*, η, of conversion of the driver energy E_d to kinetic energy of the fuel, and this efficiency is smaller than the rocket efficiency discussed in Chap. 9. Thus one has

$$m_f v_{\text{imp}}^2/2 = \eta E_d. \tag{11.11}$$

This lets us write the energy gain of the implosion, G, as

$$G = \frac{Y}{E_d} = \frac{\eta\phi}{Am_p v_{\text{imp}}^2}, \tag{11.12}$$

in which A is again the average atomic weight of the fuel (2.5 for DT). This result assumes that the fuel is ignited somehow, a topic we take up below.

The implosion efficiency depends on details of the driver and how it is used. For all known uses of laser drivers, though, its value and scalings are similar. Here we will take a scaling from a 1D simulation study by Zhou and Betti (2007), who have

$$\eta = 7.4 \times 10^{-4} v_{\text{kps}}^{3/4} \left(I_{14}\lambda_\mu^2\right)^{-1/4}, \tag{11.13}$$

where v_{kps} is the implosion velocity in km/s and, as in Chap. 9, I_{14} is the laser energy flux in units of 10^{14} W/cm^2 and λ_μ is the laser wavelength in μm. Unfortunately, η is quite a bit below the rocket efficiency, being typically below 10%. We then have, for DT fuel,

$$G = \frac{Y}{E_d} = \frac{4.0 \times 10^5 \phi}{\left(I_{14}\lambda_\mu^2\right)^{1/4} v_{\text{kps}}^{5/4}}. \tag{11.14}$$

Figure 11.3 shows the corresponding curve, along with some regimes we will discuss later. One can achieve an energy gain near 100 with an implosion velocity near 300 km/s. This enables us to assess whether power production based on laser-driven fusion is worth pursuing.

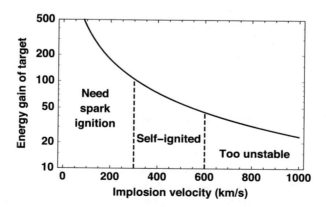

Fig. 11.3 Gain curve from (11.14), for $I_{14}\lambda_\mu^2 = 1$. Regimes to be discussed are indicated, although the precise location of the boundaries (the dashed lines) is not well-known

Fig. 11.4 Schematic of power flow in an electric plant. The numbers here imply that one must have $G \gtrsim 250$

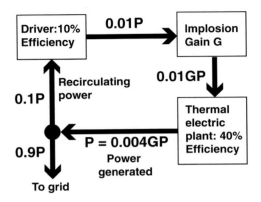

Figure 11.4 shows the essential elements of an electric plant, generating a total electric power P, and the associated numbers. We can use it to discuss the key issues. A plant needs to sell most of the power it generates, and a generally accepted number for the maximum allowable recirculating power is 10%. Existing viable drivers are not 10% efficient, but some (KrF lasers) are close. Diode pumped solid state lasers can be more efficient, and so can fiber lasers. Other driver technology, using pulsed power or ion accelerators, is potentially more efficient but much less developed. The question for all these potential drivers is whether they can be scaled to the required energy and power. If one can make the driver more efficient, without introducing other losses, then the required gain is reduced proportionally.

Once the driver delivers energy to a target, there is some gain G and then energy extraction. Using any fusion approach that generates neutrons requires having them heat matter and then producing electricity via a thermal cycle, whose maximum efficiency is 40%. If one could use an aneutronic fuel cycle, such as p–^{11}B, one could replace the thermal cycle with more-efficient, direct extraction of energy and reduce the required gain by another factor of ∼2. The net result is that a power plant requires $G \sim 250$ with straightforward extensions of present technology, and that the required G might be reduced by a factor of 2–10 by means of improvements in technology and performance. To work in the regime of Fig. 11.3 labeled "self-ignited" would require a several-fold improvement in driver efficiency beyond that available today.

11.1.3   Energetic Considerations for Ignition

The discussion above did not focus on how to assure that the fuel becomes hot enough to ignite. Here we consider ignition from an energetics and conceptual perspective. Later, in Sect. 11.3, we will address the more detailed physics.

The simplest way to assure ignition would be to compress all the fuel so that its final temperature was ∼5 keV. Suppose that the fuel at high temperatures behaves as an ideal gas with $\gamma = 5/3$. And suppose that we decide, from more detailed calculations, that a temperature of 5 keV will suffice for ignition. The specific energy of DT fuel at 5 keV temperature is 5.7×10^8 J/g. If we assume 100% conversion of kinetic energy to internal energy, which is optimistic, the required implosion velocity would be >1100 km/s. The resulting energy gain would be < 20. The discussion above showed this to be impractical for power production. The minimum required energy and implosion velocity would be for Fermi-degenerate fuel. The actual compressed fuel density may be higher than the value of 190 g/cm^3 we found above, as a result of the behavior of the implosion that we will explore later, but will end up in the range of 200–1000 g/cm^3. The specific energy of Fermi-degenerate DT fuel, for this density range, is $(1–3) \times 10^7$ J/g. If we assume 100% conversion of kinetic energy to internal energy, which is optimistic, the required implosion velocity would be 140–250 km/s, corresponding to significantly larger energy gains.

We seem to be in another quandary here, as we have to ignite the fuel but apparently cannot afford to heat it. All the solutions to this quandary revolve around the range of the alpha particles in compressed DT fuel. We mentioned above that the alpha particles do not escape the fuel. In fact, their range is about 0.3 g/cm^2, or 10% of the ρr of the entire compressed fuel. This corresponds to 0.1% of the fuel volume. If one could magically heat only such a volume to 5 keV, then it would begin to create fusion products, the alpha particles would heat the surrounding, cold fuel, and the fusion burn would propagate. The energy cost would still be 5.7×10^8 J/g in the fuel that ignited, but averaged over the entire capsule the extra cost of ignition would be only 5.7×10^5 J/g, which is much less than the cost of compression. Thus, some sort of *hot-spot ignition* is the key to obtaining enough gain from fusion to make fusion-powered electricity viable. In consequence, the fuel will not burn throughout its entire volume all at once, but instead one will have a propagating burn.

Several approaches to producing such a hot spot have been proposed. The simplest and most thoroughly explored is to tailor an implosion so as to create the hot spot at the center of a fuel capsule. We will call this *self ignition*. In other approaches, which we will call *spark ignition*, some source of energy is deposited in a small volume of fuel after compression. When this is done via local delivery of energy, it is known as *fast ignition*. The most studied option for doing this, usually called the *fast ignitor,* involves using a short-pulse laser of very high energy flux to create relativistic particles that penetrate the compressed matter and heat it. Spark ignition can also be accomplished by driving a shockwave through the compressed fuel so as to ignite a small region when it converges. This is known as *shock ignition.*

11.1.4 Properties of Compressed DT Fuel

We have seen that the compressed DT fuel will have a pressure of many Gbars and a density above 1000 times the density of solid DT. Creating the necessary pressure costs money, and any increase in the required final pressure will increase the cost or decrease the performance of an inertial fusion system. For this reason we need to understand the relation of the pressure in DT fuel to the heating that may occur during compression. It is specifically helpful to understand the relation of pressure to entropy, because in practice compression by a factor of 1000 must involve shock waves, and shock waves increase entropy (see Sect. 4.1.4). This leads us to explore further the fundamental properties of DT fuel. It is straightforward to consider the fuel as a collection of fundamental particles. We do this here, ignoring collective effects such as ionization and dissociation.

We will examine the properties of DT fuel with equal numbers of deuterons and tritons. Following Lindl (1995), we consider the initial state of the fuel to be solid DT at $11\,^\circ$K and $0.25\,g/cm^3$. The electrons and the tritons are fermions while the deuterons are bosons. Applying (3.15) to the initial state, we find that the Fermi energy for the electrons is $5.6\,eV$. This is more than three orders of magnitude above the initial temperature, with the implication that the electrons must be treated as fermions until conditions change greatly. In contrast, the initial Fermi energy of the tritons corresponds to a temperature of $7.4\,^\circ$K. This implies that the tritons may be treated as a classical gas throughout the compression and heating, based on the discussion in Sect. 3.1.3.

As bosons, the behavior of the deuterons is not among the topics we discussed in Chap. 3. Upon referring to a statistical physics text such as Landau and Lifshitz (1987), we find that the behavior of bosons varies across three temperature regimes. In the lowest temperature regime, particles accumulate in the lowest-energy state, which must be treated separately from the other states. The temperature, T_o, below which this occurs is

$$T_o = \frac{3.31}{g^{2/3}}\frac{h^2}{4\pi^2}\frac{n_D^{2/3}}{2m_p},\tag{11.15}$$

in which h is the Planck constant, m_p is the proton mass, n_D is the number density of the deuterons, and g is their degeneracy (equal to 3, as they have spin 1). Evaluating T_o for the conditions given above, one finds that it is $3.7\,K$. The implication is that the deuterons in fusion fuel do not collect in the lowest energy quantum state but are instead distributed across many energy states. They are in the second temperature regime, in which the deuterons must be treated as bosons and an analysis involving integrals similar to those of Sect. 3.1.3 is valid. At some higher temperature, whose value depends on the accuracy one needs, the behavior of the deuterons becomes like that of a classical gas. Thus, to determine how to treat the deuterons for fusion fuel, we do need to evaluate their behavior as bosons.

The properties of bosons can be conveniently expressed in terms of integrals similar to those used for fermions. We define $G_n(\phi) = \int_0^\infty x^n[\exp(x-\phi)-1]^{-1}dx$, in which $\phi = \mu/(k_BT)$. Then we have

$$n_D = \frac{12\sqrt{2}\pi}{h^3}\left(2m_pk_BT\right)^{3/2}G_{1/2},\tag{11.16}$$

from which

$$\Theta = 1/(6305G_{1/2}^{2/3}), \tag{11.17}$$

where for convenience we again define $\Theta = T/T_d$, in which T_d is the degeneracy temperature of the electrons given by $k_B T_d = \epsilon_F$. Equation (11.17) defines the relation between the chemical potential (which is negative) and Θ. The deuteron pressure, p_D, is given by

$$p_D = \frac{8\sqrt{2}\pi}{h^3} \left(2m_p\right)^{3/2} \left(k_B T\right)^{5/2} G_{3/2}, \tag{11.18}$$

while the specific entropy of the D in DT fuel, s_D, is given by

$$s_D = \frac{k_B}{5m_p} \left(\frac{5}{3}\frac{G_{3/2}}{G_{1/2}} - \phi\right). \tag{11.19}$$

Note that to obtain the specific entropy for the DT fuel we divide the entropy per particle by the average mass per D particle in the DT fuel ($5m_p$). One can compare the results of these calculations with the classical partial pressure of D in DT fuel,

$$p_{Dcl}/p_F = 5\Theta/4, \tag{11.20}$$

and the classical entropy

$$s_{Dcl} = \frac{k_B}{5m_p} \left[\frac{5}{2} + \ln\left(\frac{5m_p}{\rho}\right) + \frac{3}{2}\ln\left(k_B T\right) + \frac{3}{2}\ln\left(\frac{2\pi 2m_p}{h^2}\right)\right]$$
$$= 0.191 \times 10^8 \left[15.1 + \frac{3}{2}\ln(\Theta)\right]. \tag{11.21}$$

One finds that the classical pressure and the boson pressure are identical to high accuracy for any Θ above 0.001, while the classical entropy remains 5–20% below the boson entropy for all Θ of interest here.

One can put this all together as follows. The total pressure of the DT fuel, normalized to the Fermi pressure of the electrons, is the sum of the electron pressure from Sect. 3.1.3 and the classical pressures of the deuterons and the tritons (each equal to $5\Theta/4$). Figure 11.5a shows the resulting pressures. The total specific entropy is the sum of the specific entropy of the electrons, based on (3.40), the specific entropy of the deuterons, from (11.19), and the classical triton specific entropy, given by

$$s_{Tcl} = \frac{k_B}{5m_p} \left[\frac{5}{2} + \ln\left(\frac{5m_p}{\rho}\right) + \frac{3}{2}\ln\left(k_B T\right) + \frac{3}{2}\ln\left(\frac{2\pi 3m_p}{h^2}\right)\right]$$
$$= 0.191 \times 10^8 \left[15.7 + \frac{3}{2}\ln(\Theta)\right]. \tag{11.22}$$

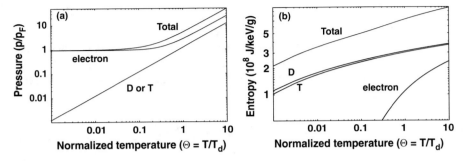

Fig. 11.5 Origins of pressure and entropy in DT fuel. The pressure (**a**) is normalized to the Fermi pressure of the electrons, p_F. The specific entropy (**b**) is for DT fuel

Fig. 11.6 The pressure increases only after enough entropy generation. The initial entropy of the fuel, in this model, is 1.1×10^8 J/keV/g

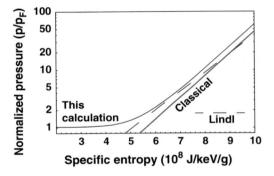

Figure 11.5b shows the contributions to the entropy. The comparison of the two parts of Fig. 11.5 leads to the following conclusion. If one begins with cold, Fermi-degenerate DT fuel, one can heat this fuel until $\Theta \sim 0.1$ before the pressure begins to increase. This is thanks to the degenerate electrons. Such heating corresponds to an increase in entropy. Equivalently, one could say that one can increase the entropy of the fuel, which is dominated by the ions, by some amount before the pressure begins to increase. This second point of view is useful if the entropy will be increased primarily by shocks.

By combining the results shown in the two parts of Fig. 11.5 one can obtain Fig. 11.6. The solid black curve in this figure shows the dependence of the normalized pressure on the entropy of the fuel. One sees that the pressure is constant up to some value of the entropy and increases exponentially with entropy at high entropy. It is a fortunate development for fusion that adding a certain amount of entropy imposes no cost.

The solid gray line in Fig. 11.6 shows the classical result, whose derivation is left as a homework exercise. The dashed curve shows the equation given in Lindl (1995), which is

Table 11.1 Some parameters of a model high-gain ICF system

Parameter	Value
Fuel mass	3 mg
Burn fraction	33%
Fuel ρr	3 g/cm^2
Final fuel density	200–1000 g/cc
Maximum specific entropy	4×10^8 J/keV/g

$$p_{DT} = p_F \exp\left[0.75 \left(\frac{\Delta s}{10^8} - 4 \right) \right], \qquad (11.23)$$

in which Δs is the difference in specific entropy from the initial state. We can see that this expression is a reasonable fit to the result of our calculation in the important regime where the pressure is a few times the Fermi pressure. The entropy of the initial state in the present calculation is 1.1×10^8 J/keV/g. Equation (11.23) describes the behavior of the pressure and the entropy, for densities above 5 g/cm^3, in QEOS or tabular EOS descriptions of DT that include atomic and molecular binding effects that we have ignored here. These effects do alter the equation of state at low densities. Because our calculation matches the results of more-sophisticated models, summarized in (11.23), one concludes that the behavior of highly compressed DT is dominated by the behavior of the individual particles (electrons, deuterons, and tritons). One also concludes that it is acceptable to increase the entropy during the creation of the final fuel state, but that ideally the increase (Δs) should be kept no larger than 4×10^8 J/keV/g. Above that level, the pressure required to obtain a desired state increases exponentially with increasing entropy.

We have thus defined our task. Take 3 mg of DT fuel, make a layer of it within a capsule, and implode the capsule at some hundreds of km/s without adding too much entropy, and then start it burning. Table 11.1 summarizes the properties of this final state to the extent we have determined them. This set of parameters poses two difficulties. First, the initial density of solid DT is 0.25 g/cm^3. Thus, to achieve fusion energy by ICF without blowing up the lab, we must implode the DT fuel so that it reaches thousands of times liquid density. Second, the problem of igniting the fuel is nontrivial. We will take up these issues in turn.

11.2   The Physics of Capsule Implosions

At this point we have some knowledge of the final fuel state, but without considering how we will create it, how we will ignite it, or how the final state might be modified so it can ignite. Here we discuss how one can create such a state. In the next section we will consider what is required to ignite it, and what is required for self ignition or spark ignition. Figure 11.7 illustrates the evolution of the density and pressure profiles during a typical implosion of a CH capsule containing DT fuel. The DT/CH

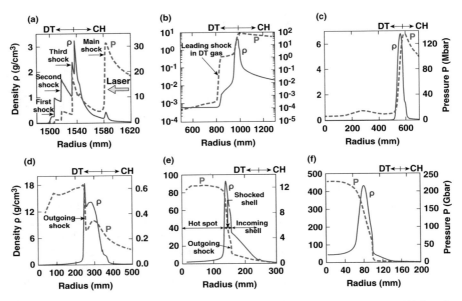

Fig. 11.7 Properties of an imploding capsule, with fusion reactions suppressed (which only matters in the final frame). This design, relevant to the National Ignition Facility, is illustrative of how implosions work but includes less fuel (1.1 mg) than would be needed for electric power production. This is from a 1D simulation using 1.5 MJ of laser energy at a 0.35 μm wavelength. Credit: Steve Craxton. Adapted from Craxton et al. (2015). The scale on part (**e**) has been corrected. (**a**) $t = 5.2$ ns; (**b**) $t = 8.9$ ns; (**c**) $t = 10.1$ ns; (**d**) $t = 10.8$ ns; (**e**) $t = 11.05$ ns; (**f**) $t = 11.4$ ns

boundary is indicated in each panel. We will refer to this figure as we discuss the details.

To be able to accelerate a capsule to the highest possible momentum, we want the largest feasible ablation pressure, P_{abl}. As discussed above, for laser drivers this is about 100 Mbar. Applying this pressure to a surface will launch a shock. If one were to use a single shock to take DT from solid density to 100 Mbar, the entropy generated (Lindl 1995) would be $\sim 9 \times 10^8$ J/keV/g. This would then represent the minimum entropy of the fuel as it implodes. The consequence, shown in Fig. 11.6, would be that the pressure was about ten times the Fermi pressure. Since a given driver and target can produce only a limited pressure, this implies that the density would be reduced by a factor of $10^{3/5}$. This is why all ICF target designs use multiple shocks, which reduces the entropy generation as discussed in Chap. 4. One could also attempt to achieve a shockless compression, using a precisely tailored pressure approach, but slight variations from the ideal pressure profile then launch unintended shocks at uncontrolled times. It appears that the use of multiple, controlled shocks is more reliable.

In practice the first shock is generally chosen to be ~ 1 Mbar. One might hope that a side benefit would be to push the target material into an ionized state, making the equation of state simpler and more knowable. Unfortunately, for the plastic capsule material used first on the National Ignition Facility, this first-shock pressure was not sufficient to make the equation of state tractable. The entropy generation is $\sim 3 \times 10^8$ J/keV/g. The increase from 1 to 100 Mbar is then accomplished in two to three more steps by shocks that are not in the strong-shock limit. The density increases to a value near 6 g/cm^3 once the final pressure is reached. Panels (a) and (b) of Fig. 11.7 show these shocks. Laser pulses for ICF are generally described as having a foot followed by a main pulse. The foot is what launches the shocks that precede the final shock. In the design of the target for Fig. 11.7, the foot included three laser pulses of short duration. As a result, the shock strength decayed as it moved through the capsule, producing the telltale decrease in density behind each shock. This approach had the advantage that there was more entropy generated near the ablator surface, where it helped limit the Rayleigh-Taylor instability (see Sect. 11.5.1), and less generated within the fuel. Most current designs also use three shocks in the foot.

The use of a sequence of shocks introduces a new issue known as *shock timing*. If one is going to use a sequence of shocks to increase the ablation pressure to 100 Mbar, then one must time these shocks so that the later, stronger, faster shocks do not overtake the earlier, weaker shocks too soon. Otherwise the resulting, stronger shock produced when two shocks coalesce would produce too much entropy (and would also produce an internal rarefaction). In order to compress the entire fuel layer, without placing any of it into a high-entropy state, it needs to be compressed by all the shocks and yet they must not coalesce within the fuel. The implication is that the shocks must be timed so that they coalesce just as they reach the inner surface of the fuel. Yet the shock behavior is extremely sensitive to the equation of state of the capsule and the fuel. As a result, the shock behavior must be observed and the driver pulse must be adjusted in order to achieve the required shock timing. Diagnostics and techniques for doing this were implemented on the National Ignition Facility, and were one of the major early successes of that research team. On panel (b) of Fig. 11.7, one can see the shock that emerged into the interior gas in the capsule after coalescence, followed by the rarefaction from the inner surface of the capsule. Remarkably, most of the laser duration is involved in pushing the shocks through the capsule. In the case shown, it requires nearly 9 ns from the time of the first picket (and about 4 ns from the start of the main pulse) to do this, and then less than 3 ns to finish the implosion.

Once the entire capsule has been shocked, the acceleration begins, by the rocket effect discussed in Chap. 9. The specific driver and target produce some exhaust speed, V_{ex}, corresponding to a mass ablation rate $\dot{m} = P_{abl}/V_{ex}$. During the acceleration, the rest of the ablator material is removed (ideally). The final remaining mass fraction for the rocket equation is

$$f_m = \frac{m_F}{m_F + m_{abl}} = e^{-v_{imp}/V_{ex}}, \tag{11.24}$$

in which m_{abl} is the ablator mass present at high density at the start of the acceleration, and not at the start of the laser pulse. In the case of Fig. 11.7, one can see in panels (b) and (c) that $f_m \sim 0.5$ or somewhat more. For the model below, we assume a laser energy flux of 10^{15} W/cm^2 and a laser wavelength of 0.35 μm. Using the standard scalings, this gives $V_{ex} = 740$ km/s for laser irradiation of the capsule and $V_{ex} = 180$ km/s for X-ray irradiation using a hohlraum. We use these values in the scalings that follow.

We now know that we will ablatively accelerate the fuel, using up about half the total mass. This must be accomplished before the capsule has moved too far, or it will begin to decelerate before we have injected the energy. We will suppose that we can accelerate it over a distance of $r_s/2$. Here r_s is the initial inner radius of the capsule. Assume this acceleration occurs at a constant pressure P_{abl}. Also assume that the mass per unit area of the fuel plus the ablator is $m_o = \rho_i \Delta r_i / f_m$, where the initial compressed fuel density when the acceleration starts is ρ_i and the initial thickness of the compressed shell of fuel is Δr_i. (The thin-shell approximation for the geometry is valid for the compressed shell but not for the initial, uncompressed capsule.) Then we integrate the rocket equation to find

$$\frac{r_s}{2} = V_{ex} \int_0^{t_a} \ln\left[\frac{m_o}{m}\right] dt = -V_{ex} \int_0^{t_a} \ln\left[1 - \frac{\dot{m}t}{m_o}\right] dt, \qquad (11.25)$$

where the integral proceeds until $\dot{m}t_a = (1 - f_m)m_o$. Defining $\zeta = \dot{m}t/m_o$, one has $\zeta_a = (1 - f_m)$ and finds

$$\frac{r_s}{2} = \frac{-m_o V_{ex}^2}{P_{abl}} \int_0^{\zeta_a} \ln[1 - \zeta]\, d\zeta = \frac{m_f v_{imp}^2}{4\pi r_s^2 P_{abl}} g(f_m) \quad \text{where} \qquad (11.26)$$

$$g(f_m) = \left[\frac{1 - f_m + f_m \ln(f_m)}{\ln^2(f_m)}\right].$$

The function $g(f_m)$ has a value of ~ 0.3 at $f_m = 0.5$ and increases from ~ 0.2 to ~ 0.4 as f_m increases from 0.2 to 0.8. In the plots below, we evaluate f_m from (11.24).

Solving (11.26) for r_s gives

$$r_s = \left(\frac{1}{2\pi} \frac{m_F v_{imp}^2}{P_{abl}} g(f_m)\right)^{1/3}. \qquad (11.27)$$

The implosion time total implosion time t_{tot} is

$$t_{tot} = t_a + \frac{r_s}{2v_{imp}} = \frac{m_f}{4\pi r_s^2 P_{abl} \ln(f_m)} \frac{-v_{imp}}{} + \frac{r_s}{2v_{imp}}. \qquad (11.28)$$

Note that t_{tot} is the time from start of the main pulse, not of the foot. Figure 11.8 shows t_{tot} for $m_f = 3$ mg and $\rho_i = 6$ g/cm^3. With these constraints, laser drivers need longer pulses than X-ray drivers.

One can determine the initial thickness of the unshocked fuel layer, Δ_o by solving

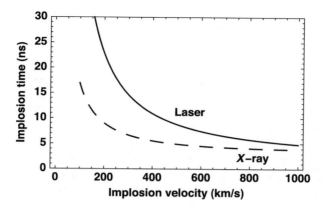

Fig. 11.8 Implosion time from (11.28), for laser or hohlraum drive as indicated

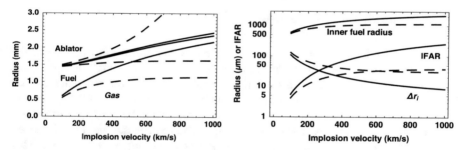

Fig. 11.9 Properties before acceleration begins. Left from bottom to top: inner fuel radius, outer fuel radius, outer ablator radius. Right: properties of the implosion as indicated. Solid curves are for laser irradiation and dashed curves are for hohlraum irradiation

$$m_f = \frac{4\pi}{3}\rho_o r_s^3 \left[\left(1 + \frac{\Delta_o}{r_s}\right)^3 - 1\right]. \qquad (11.29)$$

for the ratio Δ_o/r_s. One can then use the same solution to find the thickness of the ablator. The left panel in Fig. 11.9 shows the inner and outer radii of the fuel and the ablator prior to the laser pulse, for f_m as indicated.

The initial thickness of the shocked shell of DT fuel is

$$\Delta r_i = \frac{m_f}{4\pi r_s^2 \rho_i} = \frac{m_f}{4\pi \rho_i}\left(\frac{1}{2\pi}\frac{m_F v_{imp}^2}{P_{abl}}g(f_m)\right)^{-2/3}. \qquad (11.30)$$

The right panel in Fig. 11.9 shows r_s and Δr_i for $m_f = 3$ mg and $\rho_i = 6$ g/cm³. One sees that the inner fuel radius is near 1 mm and the shell thickness is tens of μm. We will see in Sect. 11.5.1 that the key parameter determining the growth of Rayleigh-Taylor instabilities is the *in-flight aspect ratio*, known as IFAR. The instability growth is larger when IFAR is. One has

$$IFAR = \frac{r_s}{\Delta r_i} = \frac{2\rho_i v_{\mathrm{imp}}^2}{P_{\mathrm{abl}}} g(f_m). \tag{11.31}$$

Figure 11.9 also shows IFAR, which is in the range of several tens. One would like IFAR to be smaller, but can see that this is a challenge. To make a given capsule design work, one needs some value of v_{imp}. One already plans to operate at the maximum feasible value of P_{abl}; increasing this would require changing the driver in some way. A given driver, producing a given P_{abl}, also produces a specific V_{ex}. Obtaining the required v_{imp} for that V_{ex} sets f_m. Decreasing ρ_i requires that the fuel converge further to reach the required density, which would impose stricter requirements on the symmetry of the implosion. The bottom line is that decreasing IFAR will reduce the yield from a given driver and will not come cheap in a new one.

11.3   Stagnation and Ignition

We have next to consider how the implosion comes to an end, and what the consequences are. Every ICF capsule develops a central hot spot, whether or not it is used for ignition. At minimum, the material released from the inner surface of the fuel after the shocks coalesce is compressed and heated as the implosion continues. If one desires to have more interior density than this, one can arrange for the capsule to contain gas. The capsule design used to generate Fig. 11.7, intended to produce self-ignition, included interior gas at a density of $0.6\,\mathrm{mg/cm^3}$. This interior material is compressed and heated in two ways. When the leading edge of the hot-spot material, or the leading shock in the interior gas, converges at the center, a return shock processes and heats the hot-spot material, and this shock may reverberate more than once between the center and the incoming fuel. But its heating of the matter in the hot spot turns out to remain modest. This shock is present and indicated in panels (d) and (e) of Fig. 11.7. In that case, it slows the incoming dense fuel somewhat, but the amount of slowing will vary with details of the design.

The larger source of compression and heating of the hot spot is pdV work from the incoming fuel. The energy for this comes from the imploding shell of cold fuel. In the process, the shell converges and compresses further, until the shell comes to a stop. At that point the fuel must in some way be caused to ignite. We first consider the general requirements for ignition and then the specifics of possible approaches.

11.3.1   Hot Spot Power Balance

The challenge for ignition is to keep a volume of fuel hot enough to sustain fusion despite the energy losses that are present. As a result, our problem boils down to a question of power balance. The net heating of the hot spot must be positive. We discuss this here. This subsection and the next two follow closely the treatment of Atzeni and Meyer-Ter-Vehn (2004). We use the subscript h for the properties of the hot spot. The equation for the internal energy of the hot spot, ϵ_h, assumed to remain at constant density is

$$V_h \rho_h \frac{d\epsilon_h}{dt} = P_{dep} V_h - Q S_h - F_R S_h - p_h u_h S_h, \tag{11.32}$$

in which the hot-spot volume is V_h, the hot-spot surface area is S_h, and the hot-spot density and pressure are ρ_h and p_h, respectively. This equation assumes that the kinetic energy of the hot spot material remains negligible. The rate at which fusion deposits energy in the hot spot is P_{dep}, while the radiation energy flux and thermal heat flux leaving the hot spot are F_R and Q, respectively. As the hot spot expands with radial velocity u_h, it does pdV work on the cold fuel, and this is represented by the final term. We now consider these terms in turn, assuming that in this dense matter $T_e = T_i = T_h$.

The fusion energy deposition is

$$P_{dep} = \frac{dn}{dt} E_{fus} f_{dep} = N_d N_T \langle \sigma v \rangle_{DT} E_{fus} f_{dep}$$

$$\approx C_\alpha \rho_h^2 T_h^2 \left(1 - \frac{0.3}{4 \rho_h R_h} \right), \tag{11.33}$$

where dn/dt is from (11.6) and R_h is the hot-spot radius. To obtain the numerical expression for P_{dep}, we use here a standard fit to $\langle \sigma v \rangle_{DT}$ for the range of 8–25 keV,

$$\langle \sigma v \rangle_{DT} \cong 1.1 \times 10^{-24} T_h^2 \ cm^3/s, \tag{11.34}$$

with T_h in eV; one notes that in hot-spot ignition only the alpha particles provide significant heating, so $E_{fus} = 3.5 \, MeV$; and one defines f_{dep} as the fraction of the alpha particle energy deposited in the hot spot. Equation (11.34) overestimates the rate coefficient for temperatures below 8 keV. For a first calculation, we could assume that we must make the hot spot large enough to absorb nearly all the alpha-particle energy, and take $f_{dep} = 1$. In more detail, energy deposition by charged particles in matter is complicated and is not represented by simple exponential functions. We will leave that as a detail, but do represent f_{dep} by the quantity in curved brackets, which accounts for the first departure from full absorption as ρr decreases. Then with $N_D = N_T$ we find $C_\alpha = 3.8 \times 10^{11} \ ergs \, cm^3 \, s^{-1} \, g^{-2} \, eV^{-2}$. Atzeni and Meyer-Ter-Vehn (2004) provide more detail.

The heat flux Q is the Spitzer–Harm heat flux, which we derived in Chap. 9. One has

$$Q = -\kappa_{th} \nabla T_h \approx \kappa_{th} \frac{T_h}{R_h} \cong C_Q \frac{T_h^{7/2}}{R_h}. \tag{11.35}$$

Using (9.10) and (9.36), and including the factor of $(1 + 3.3/Z)^{-1}$ due to electron–electron collisions, one can evaluate C_Q, finding, with T_h in eV and R_h in cm, $C_Q = 5.9 \times 10^8 \, \text{ergs} \, \text{eV}^{-7/2} \, \text{cm}^{-1} \, \text{s}^{-1}$.

The radiation flux is the result of bremsstrahlung emission from the plasma in the hot spot. If the hot spot were optically thick, this would evidently be σT_h^4. However, the hot spot is typically not optically thick. In the optically thin limit one has

$$F_R \cong \kappa_B R_h \sigma T_e^4 = C_B \rho_h^2 \sqrt{T_h} R_h, \tag{11.36}$$

in which κ_B is the spectrally averaged absorption coefficient [i.e., the Planck mean opacity of (6.59)]. Properly, one would determine the characteristic distance by integrating over the solid angle, accounting for the path length through the source volume. Here we assume that this gives a distance close to R_h. One finds $C_B = 3.2 \times 10^{21} \, \text{ergs} \, \text{cm}^3 \, \text{g}^{-2} \, \text{s}^{-1} \, \text{eV}^{-1/2}$. During the implosion, Zhou and Betti (2007) argue that much of the energy removed from the hot spot by conduction and radiation come back into it within ablated fuel. During the ignition phase, however, these losses represent real potential limitations.

The approach to ignition determines whether and how much the hot spot expands. On the one hand, the implosion might be designed to produce a central hot spot. In this case the cold fuel and the hot spot are all at the same pressure at the moment of stagnation. This is described as an *isobaric* configuration. In this case, u_h is initially zero in (11.32). The hot spot will expand as it heats up, but because the rate of fusion increases strongly with temperature it is the initial heating that matters. So we can use (11.32) with $u_h = 0$ to find an ignition threshold for an isobaric configuration. On the other hand, the implosion might be designed to assemble all the fuel, at the end of which a hot spot will be created by some other means. In this case the cold fuel and the hot spot will initially have the same density. Such a configuration is described as *isochoric*. The hot spot will expand as it heats, and it is a reasonable estimate to take u_h to be the velocity of the fluid behind a strong shock entering the cold fuel. Then from (4.23) one has $p_h = (\gamma - 1)\rho_s u_h^2/2$, which determines u_h. Here ρ_s is the density of the shocked, cold fuel $\rho_s = \rho_c(\gamma + 1)/(\gamma - 1)$ with ρ_c being the cold-fuel density. In addition, $p_h = \rho_h T_h (Z + 1) k_B/(A m_p)$. In this case with $\rho_h = \rho_c$ one has

$$p_h u_h = C_h \rho_h T_h^{3/2}, \tag{11.37}$$

in which $C_h = 5.8 \times 10^{17} \, \text{ergs} \, \text{cm} \, \text{g}^{-1} \, \text{eV}^{-3/2}$ for DT.

Returning now to (11.32), one can show that none of the four terms is always small throughout the regime of interest. As a result, the power of T_h of a given term determines where it will have its impact. One sees that the bremsstrahlung cooling term scales as the lowest power of T_h, so this will dominate at low temperatures. The pdV work term is next, scaling as $T_h^{3/2}$, so when significant this term will increase the minimum temperature needed for ignition. Next is the fusion energy production term, scaling as T_h^2, so that eventually fusion energy production can

Fig. 11.10 Thresholds for ignition. Hot spots in the gray area will produce net heating and will ignite

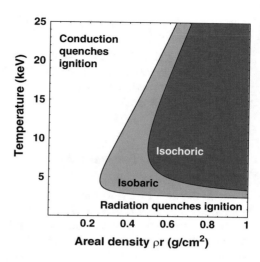

overcome these first two losses. However, lurking at high temperatures is the heat-conduction term, scaling as $T_h^{7/2}$. At high enough temperatures, heat conduction will quench the ignition.

The ignition threshold occurs when the right-hand side of (11.32) equals zero. Figure 11.10 shows this condition for isobaric and isochoric configurations. The qualitative shape of the boundaries shown is correct. However, the lower boundaries of the ignition regimes are too low in this figure because (11.34) overestimates $\langle \sigma v \rangle_{DT}$ at these temperatures, the actual lower boundaries straddle 5 keV. For the isobaric case, one indeed finds that $\rho r \sim 0.3$ g/cm^2 and $T_h \sim 5$ keV is the minimum ignition condition. For the isochoric case, relevant to spark ignition, one sees that one will need twice this ρr and a bit higher temperature. This may change the optimum properties of the cold, dense, fuel, because doubling the ρr of the hot spot at fixed density requires eight times the energy invested in the hot spot. Depending on the cost of this energy, it might or might not make more sense to compress the cold fuel further.

It turns out that capsules that do not satisfy the threshold condition may ignite, if they are at temperatures where heat conduction quenches ignition initially. In this case the hot spot may cool, heating a surrounding region, and in effect creating a modified hot spot with smaller temperature and larger ρr. This modified hot spot may then ignite. In addition, the fact that ignition occurs does not guarantee in principle that the resulting burn will propagate. If the heated region expanded too quickly relative to fusion energy production, then expansion cooling could quench the ignition. However, for the regime relevant to ICF the parameters work out favorably, and ignited capsules typically continue to burn. These last two effects are discussed further by Atzeni and Meyer-Ter-Vehn (2004). They can be summarized as a condition for successful ignition and burn, given by

$$\rho_h R_h T_h > 6\sqrt{\rho_h/\rho_c} \text{ g cm}^{-2} \text{ keV},\qquad\qquad(11.38)$$

in which ρ_c is again the density of the compressed, cold fuel. (A typical value of ρ_h/ρ_c for an isobaric case is 1/16.) This condition is the analog for ICF of the well-known Lawson criterion for MFE, expressed in that case as a threshold value of density times confinement time.

11.3.2   Igniting From a Central Hot Spot

It seems natural to try to get some benefit from the work done stopping the imploding capsule, by making the central gas become the hot spot that initiates the fusion burn. This places a constraint on the final ρr of the hot spot, which must be ~0.3 g/cm^2 to localize the alpha particles. It also places a constraint on the final temperature of the hot spot, which needs to be above 5 keV. If this were too low, the fuel in the hot spot would not ignite, but if it were too high the hot-spot fuel would begin burning too soon.

During the implosion, the hot spot obtains energy from pdV work and eventually from fusion burning, while losing energy through electron heat conduction and radiation. One can analyze the energy balance during the implosion using an energy balance equation similar to (11.32), except that ρ_h is no longer constant. Just as in the case of ignition, relatively cool hot spots lose more energy to radiation while relatively warm hot spots lose more energy to heat conduction. There is an optimum path in density and temperature to minimize the energy cost of assembling the hot spot. The further the hot spot is from that path during the implosion, the more energy will be required. This can affect the details of optimizing the shock timing and initial gas fill of the capsule. We will not examine this level of detail here.

Even so, the formation of the hot spot is complicated, as it involves whatever gas is inside the capsule, the material blown off of the fuel layer, compression, and reverberating shocks. Here, in order to have a tractable model that can show us some main trends, we will ignore internal shocks and the specific sources of matter. We also assume a strictly isobaric model, to that the pressure in the compressed fuel equals that in the hot spot. (Figure 11.7 shows that this is not quite correct.) We assume that the energy of the incoming fuel goes entirely to pdV work and to the internal energy of compression of the cold shell. We also assume the material at radii smaller than r_s begins at some initial, uniform density ρ_o and pressure p_o.

We do need to account in some way for the structure of the central gas (the matter interior to the inner radius of the cold fuel, R_c), because the initial fusion burning is not uniform. The final panel in Fig. 11.7 shows that the pressure in the interior peaks in a broad central region while the density has a minimum at the center. As a result, the temperature is peaked on axis. In Fig. 11.2 we can also see that $\langle\sigma v\rangle_{DT}$ increases more rapidly than T^2 at low temperatures. As a result the hot spot, defined as the region where neutron production is within some reasonable fraction of its maximum, is significantly smaller than R_c. Here we approximate $R_h = R_c/2$, which compares well enough with results of simulations. Similarly, we approximate $T_h = 2\bar{T}$, where $\bar{T}$ is the temperature implied by the stagnation pressure p_h and the averaged density of the interior matter, $\bar{\rho}$. We also account crudely for the shape of the density profile by converting the requirement, $\rho_h R_h = 0.3$ to $\bar{\rho} R_c = 0.75$.

Then we have three equations that we can work with, which must be satisfied to achieve ignition. They are

$$-\int p dV + \mathcal{E}_{\text{fuel}} = m_f v_{\text{imp}}^2/2, \tag{11.39}$$

$$\rho_c \Delta R_c = 3 \text{ g/cm}^2, \text{ and} \tag{11.40}$$

$$\bar{\rho} R_c = 0.75, \tag{11.41}$$

where the compressed cold fuel is in a shell of density ρ_c and thickness ΔR_c, and has an internal energy $\mathcal{E}_{\text{fuel}}$. We will see that these relations imply ρ_o, p_o, and the shell convergence, C_s, and that from these and known information we can infer $\bar{\rho}, p_h$, and T_h. Then we will see whether T_h is high enough to ignite and whether $\bar{\rho}$ is in a plausible range. We proceed now to carry through the calculation.

The convergence for some central-gas density ρ is given by $C = (\rho/\rho_o)^{1/3}$, and has a final value C_s. We can find the density of the cold, stagnated fuel because $p_h = p_o C_s^5$ also equals the pressure in the stagnated, Fermi-degenerate, cold shell, assuming the stagnated structure to be isobaric. The density is

$$\rho_c = \frac{A}{Z} \left(\frac{p_h}{9.9 \text{ Mbar}} \right)^{3/5} = 1.25 C_s^3 \left(\frac{p_o}{9.9 \text{ Mbar}} \right)^{3/5} \tag{11.42}$$

$$= 2.0 \times 10^{-8} C_s^3 [p_o(\text{cgs})]^{3/5} \text{ g/cm}^3,$$

where we ignore any departure from Fermi degeneracy (again in the spirit of doing a model that captures the main physics). The internal energy of the dense fuel is

$$\mathcal{E}_{\text{fuel}} = 3 \times 10^{12} \rho_c^{2/3} m_f = 2.2 \times 10^7 m_f C_s^2 [p_o]^{2/5} \text{ in cgs units,} \tag{11.43}$$

so we have

$$p = p_o \left(\frac{\rho}{\rho_o} \right)^{5/3} = p_o C^5 \text{ and } dV = -\frac{M_h d\rho}{\rho^2} = -\frac{M_h dC}{\rho_o C^3} \text{ so}$$

$$-\int p dV = p_o V_s \int_1^{C_s} C^2 dC = \frac{1}{3} p_o V_s \left(C_s^3 - 1 \right) = \frac{m_f v_{\text{imp}}^2}{2} - \mathcal{E}_{\text{fuel}},$$

$$\text{so } \frac{p_o V_s \left(C_s^3 - 1 \right)}{3 m_f} + 2.2 \times 10^7 C_s^2 [p_o(\text{cgs})]^{2/5} = \frac{v_{\text{imp}}^2}{2}, \tag{11.44}$$

in which the mass at radii smaller than R_c is M_h and the initial volume at radii smaller than r_s is $V_s = M_h/\rho_o = 4\pi r_s^3/3$.

The shell thickness relates to the fuel mass as

$$m_f = \frac{4\pi}{3}\rho_c\left[(R_c + \Delta R_c)^3 - R_c^3\right] = \frac{4\pi}{3}\rho_c R_c^3\left[\left(1 + \frac{\Delta R_c}{R_c}\right)^3 - 1\right],\qquad (11.45)$$

from which

$$\Delta R_c = \frac{r_s}{C_s}\left[\frac{1}{\chi}\left(3\chi^2 + \chi^3\right)^{1/3} - 1\right],\text{ in which}\qquad (11.46)$$

$$\chi = \frac{4\pi\rho_c R_c^3}{m_f} = \frac{4\pi\rho_c r_s^3}{m_f C_s^3} = 2.5\times 10^{-7}\frac{p_o^{3/5}r_s^3}{m_f},$$

where quantities in the final term are in cgs units. Now (11.39) becomes

$$\rho_c\Delta R_c = 3 = 2.0\times 10^{-8}r_s C_s^2\left[p_o(\text{cgs})\right]^{3/5}\times\left[\frac{1}{\chi}\left(3\chi^2 + \chi^3\right)^{1/3} - 1\right],\qquad (11.47)$$

and (11.47) and (11.44) imply p_o and C_s.

The requirement (11.41) on central-gas ρR turns out to determine only the initial density and final temperature, based on C_s and p_o. Using (11.41) and $R_c = r_s/C_s$ gives

$$\bar{\rho}R_c = 0.75 = \rho_o r_s C_s^2\text{ so }\rho_o = \frac{0.75}{C_s^2 r_s},\qquad (11.48)$$

from which $\bar{\rho} = \rho_o C_s^3$. Then $T_h = p_h A m_p/[(1 + Z)\bar{\rho}k_B]$.

Solving these equations by computer produces the results shown in Figs. 11.11, 11.12, and 11.13 for the parameters indicated. One can see in Fig. 11.11a that the final fuel density is hundreds of g/cm^3, the central, stagnation pressure is a few hundred Gbars, and the final hot-spot density (plotting $\bar{\rho}$ from the calculation) is of order 100 g/cm^3. In this model, the final hot-spot density rises above the final fuel density as v_{imp} drops below about 250 km/s. There is no inherent physical problem with this, but this circumstance corresponds to very high convergence and so the implosion is less likely to remain spherical. If pdV work fails to stop the imploding fuel, then upon stagnation additional shocks will be launched within the cold fuel. This might not cause any problems if ignition occurs quickly enough. In Fig. 11.11b one sees the values of initial density ρ_o and initial pressure p_o from the model, but recall that these are approximations to the state of the matter that would be equivalent to the processing of the gas by the initial shock wave combined with the injection of matter from the inner surface of the cold fuel. The main point here is that the values shown are not obviously crazy.

The second figure, Fig. 11.12, has more to tell us. The figure shows the convergence of the dense, cold shell of fuel. The convergence of the hot spot

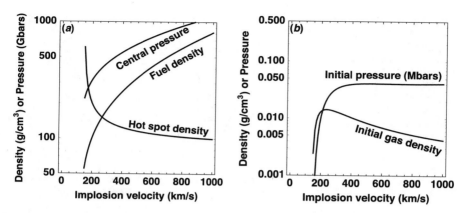

Fig. 11.11 Parameters, as labeled, for ICF capsules intended for power production, having a fuel mass of 3 mg, $P_{abl} = 100$ Mbar, and $V_{ex} = 740$ km/s. The magnitude and shape of these parameters are not strongly sensitive to V_{ex}, except that initial pressure increases as V_{ex} decreases

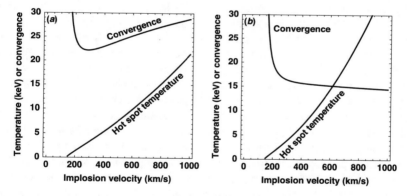

Fig. 11.12 Convergence and hot-spot temperature for ICF capsules intended for power production, having a fuel mass of 3 mg and $P_{abl} = 100$ Mbar. (a) Laser drive, $V_{ex} = 740$ km/s. (b) Hohlraum drive, $V_{ex} = 180$ km/s

is roughly twice this value. We find in Sect. 11.5.2 that increased convergence demands increasing uniformity of the drive on the capsule. This might be costly. But the significant question for a power plant is the total cost, which involves many factors of which this is only one. The important question about the hot-spot temperature is when it reaches ∼5 keV, to enable ignition. For both curves shown here, this occurs at about $v_{imp} = 400$ km/s, Our evaluation of the hot-spot temperature is approximate, and we have included no effect at all of heating by α particles during the implosion (so called α-heating), which can contribute some heating during the final phases of the implosion, before ignition. The consensus appears to be that ignition by a central hot spot becomes unfeasible below about 300 km/s. This is the boundary shown above in Fig. 11.3. It also should be mentioned that the discussion here has ignored the consequences of entropy production in the

Fig. 11.13 Capsule energies,
as labeled, for ICF capsules
intended for power
production, having a fuel
mass of 3 mg, P_{abl} =
100 Mbar, and
V_{ex} = 740 km/s. The
magnitude and shape of these
parameters are not strongly
sensitive to V_{ex}

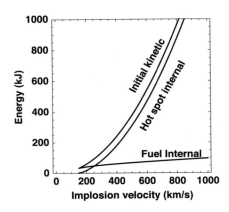

fuel, whether by shock heating or in other ways. One can hope these will be small
but they will not be zero.

Figure 11.13 shows the energetics of the capsule. On can see that the internal
energy of the central gas is usually larger than that of the compressed fuel, and
becomes much larger at high implosion velocities. From this perspective, central
hot spots represent an inefficient way to ignite fuel. In contrast, at low implosion
velocities the energy in the central gas becomes negligible but the convergence
required to get 3 mg to fuel to a state with $\rho R_c > 3$ becomes very large. People who
design implosions for drivers that naturally produce lower ablation pressures than
lasers do tend to decide to work with smaller v_{imp} and increased yield. This choice
creates some difficulties in power-plant design, but as for other issues the tradeoff
might be worth it. Table 11.2 shows numerical values of a number of parameters for
the conditions corresponding to the figures, for an implosion velocity of 400 km/s.

The cases evaluated here are relevant to power production. In the context of
the early twenty-first century, some remarks on the differences of these cases with
those for the National Ignition Facility (NIF) are in order. NIF has only 1.5–2 MJ of
energy, which is not enough to ignite an ICF implosion using 3 mg of fuel. Various
designs for NIF use from 0.2 mg to ~1 mg. (The design of Fig. 11.7 used 1.1 mg.)
At lower fuel masses, one requires a significantly higher convergence, leading to
higher final pressures and higher compressed-fuel densities, than would be needed
for a power plant. Thus, producing ignition on NIF is significantly more demanding
than producing ignition using the larger driver one might need for a power plant.
The driver for a power plant that used ignition by a central hot spot would need to
be in the range of 5–10 MJ, depending on details. Some studies suggest that one
might be able to pull this down some for the case of laser drive. But at least in its
early years, NIF has done only X-ray-driven experiments.

(This brings up a historical aside on the real limitations that major research
programs must deal with. The author happened to be sitting, in the mid 1980s,
in a room where a panel doing a review for the National Academy of Sciences
was discussing the sensible path forward for ICF. There was a strong consensus
that the thing to do would be to build a 10 MJ driver, since this would be virtually

Table 11.2 A model high-gain ICF system ignited by a central hot spot

Parameter	Laser drive	X-ray drive
Fuel mass	3 mg	3 mg
Burn fraction	33%	33%
Fuel ρr	3 g/cm^2	3 g/cm^2
Hot spot ρr	0.3 g/cm^2	0.3 g/cm^2
Implosion velocity	400 km/s	400 km/s
Max. cold-fuel specific entropy	4×10^8 J/keV/g	4×10^8 J/keV/g
Exhaust velocity	740 km/s	180 km/s
Initial fuel Inner radius r_s	1390 μm	1000 μm
Final fuel inner radius R_c	61 μm	64 μm
Fuel shell convergence	23	16
Final fuel density	550 g/cc	620 g/cc
Central gas final density	124 g/cc	118 g/cc
Hot spot temperature	5.3 keV	6.8 keV
Fuel pressure	250 Gbar	310 Gbar
Fermi energy	350 eV	340 eV
Fuel kinetic energy	240 kJ	240 kJ
Central gas energy	180 kJ	175 kJ

assured of being able to ignite capsules and would let one then try to optimize their performance so as to drive down the size of driver required for a power plant. Yet in the era when NIF was sold to the US Congress, the largest project one could sell was for a billion dollars. NIF, at 1.8 MJ, was funded at that budget. It turned out to cost a few times that, which is not unusual for novel projects of any kind. Beyond that, NIF pushed laser technology very far forward. Doing a 10 MJ driver might not have been feasible anyway. NIF is accomplishing a great deal, but at the end of the day may or may not prove able to ignite a capsule based on present knowledge and technology. The author is firmly convinced that, had it proven feasible to fund and build a 10 MJ driver, we would be discussing the results of fusion ignition here rather than its prospects.)

Implosions are the only known way to compress fuel to conditions that enable fusion, but they are costly. The internal pressure of 3 mg of DT fuel, compressed to 200 g/cm^3 and $\rho R_c = 3$ g/cm^2, is 13.5 Gbar, and its energy content is 30 kJ. Compressing this much fuel using an implosion at 200 km/s takes about twice the energy and produces about four times the pressure. Doing it at 400 km/s takes 8 times the energy and produces about 40 times the pressure. The implication is the driver needed to produce implosions that ignite with a central hot spot will need to be two to ten times larger than a driver that assembles fuel for spark ignition. The implication is that a lot of resources are available to produce a fusion ignitor for spark ignition.

11.3.3 Spark Ignition

Spark ignition in general is a much more complicated problem than is ignition by a central hot spot. There are numerous ways one might do the heating to ignite a spark. One might use a beam of electrons, or ions, or photons. Or one might use a slug of dense matter to shock-heat the dense fuel. One also might use a converging shock to accomplish the same effect. It has even been suggested (by Jim Hammer) that one might use a bubble of magnetic field. Here we first discuss the hot-spot properties required to exceed the threshold for ignition discussed in Sect. 11.3, and then some of the options for producing these properties. The first published discussion of fast ignition was that of Tabak et al. (1994). Here we draw from work by Atzeni (1999).

Since fast ignition begins from a nearly isochoric state, we draw from the analysis that produced Fig. 11.10, but using more-precise numbers from more detailed studies. The minimum value of ρR_h required for fast ignition is $0.5 \, \text{g/cm}^2$, at a temperature of $12 \, \text{keV}$. If we consider this to be the optimum hot spot, we can estimate the amount of energy (deposited in the hot spot) that is required for ignition, E_{ig}, assuming that the hot spot is a cylinder of aspect ratio unity, as $E_{\text{ig}} = \pi R_h^3 \rho (1 + Z) k_B T_h / (A m_p)$, which is $36 (100 \, \text{g/cm}^3 / \rho)^2 \, \text{kJ}$ for this case. This seems quite hopeful, since achieving ρ above $200 \, \text{g/cm}^3$ seems plausible and since our simple scaling calculation indicated that the cost in fuel energy of producing a central hot spot was above $60 \, \text{kJ}$. However, energy losses due to radiation, heat conduction, and hydrodynamic motion act to increase E_{ig} substantially.

In addition, the assumed parameters are not necessarily the optimum hot spot. The required temperature decreases as ρR_h increases, reaching $4 \, \text{keV}$ at $\rho R_h = 1.2 \, \text{g/cm}^2$. If, for example, one can readily and efficiently produce a beam whose radius is smaller than $(0.5/\rho) \, \text{cm}$ and whose absorption depth at the desired final density is roughly equal to its radius, then one would want to produce the high-temperature hot spot with the minimum ρR_h. In contrast, if the beam one could efficiently produce had an absorption depth above $1 \, \text{cm}^2/\text{g}$, then the optimum design would heat a larger-radius hot spot to a lower temperature. To make matters even more complex, the duration of the heating beam may be constrained or may be variable, and the beam may be limited in its maximum available power. Moreover we have yet to mention the question of how efficiently the heating energy can be produced and deposited, yet this is very likely the key technical issue that will determine the viability of fast ignition for fusion.

One approach to addressing this complexity is to separate the problem into components. Atzeni did so, asking what the conditions for ignition are for a set of rather general assumptions. He assumed that a beam of radius r_B, power W_B, and energy flux I_B, related by $W_B = I_B \pi r_B^2$, and with an absorption depth R_B, irradiates a constant-density fuel. He then used two-dimensional simulations to assess the parameters required for ignition. For $0.15 \leq R_B \leq 1.2 \, \text{g/cm}^2$, ignition required

$$E_{\text{ig}} = 140 \left(\frac{\rho}{100 \, \text{g/cm}^3} \right)^{-1.85} \, \text{kJ}, \tag{11.49}$$

$$W_B = 2.6 \times 10^{15} \left(\frac{\rho}{100 \text{ g/cm}^3} \right)^{-1} \text{W}, \tag{11.50}$$

and

$$I_B = 2.4 \times 10^{19} \left(\frac{\rho}{100 \text{ g/cm}^3} \right)^{0.95} \text{W/cm}^2. \tag{11.51}$$

The second and third of these relations imply a beam radius of $60 \, \mu$m $\times$ $(100 \text{ g/cm}^3/\rho)^{0.975}$, which may or may not prove feasible in practice. The broad range of absorption depth over which these parameters apply reflects the rough balance between the fact that the overheated hot spots produced when the absorption depth is small ignite only after producing a larger volume at a lower temperature by heat conduction and that increased heating energy is required when the heating is dispersed because of large absorption depth. The net impact of this more-realistic analysis is to increase the optimum fuel density for fast ignition, perhaps to of order 300 g/cm^3, so that the deposited energy must be <20 kJ. At the present level of the discussion this is a guess, since the important question in the end is how much energy must be expended to produce a given amount of deposited energy. The smaller the efficiency of energy deposition, the larger one will need to make the hot spot density. Then the question will be whether the beam radius can be as small as is required. One can see that fast ignition is a challenging problem involving the interplay of numerous constraints and of both physics and technology.

Now let us consider some possible methods for delivering the required energy to the dense fuel, at the required very large energy flux. A simple approach is to direct a sufficiently intense laser beam at the compressed fuel and hope that this leads to enough energy coupling. The most likely source of such coupling would be the generation of relativistic electrons by the intense laser beam. Indeed, beams of relativistic electrons are generated by intense, ultrafast lasers (see Chap. 13). Detailed questions follow. Can one generate enough electrons? Can one do so at useful energies? Can one do so close enough to the dense fuel that they couple efficiently? The dense fuel is surrounded by a formidable quantity of blown-off plasma. This leads one to consider various options. One might use a preliminary laser pulse to drill a hole in this plasma before one introduces the more-intense laser pulse that does the heating. Alternatively, one might implode a capsule that includes a high-Z cone, whose purpose is to provide a region free of such blow-off plasma through which one can introduce the heating beam. Even so, electrons tend to have large absorption depths (though not as large as those of photons at relevant energies). Beams of electrons are also subject to disruption by filamentation or other instabilities. This leads one to consider using heavier particles.

Broadly speaking, the heavier particles might be protons, light ions, or heavy ions, and one might try to accelerate relatively few particles to higher energy or relatively more particles to lower energy. In the high-energy limit, beams of protons have also been observed in experiments with ultrafast lasers (see Chap. 13). If one can put enough energy into these beams, and if one can focus them, this might provide an alternative to the use of electrons. In the low-energy limit, one could try to drive a slug of solid material toward the target with enough energy to cause ignition by shock heating. If protons are still not absorbed readily enough, one could work on devising schemes to deliver a sufficient energy flux of light ions or heavy ions to the target. At this writing, it is clear that there are many options but not yet so clear which may prevail.

In 2007, Betti et al. (2007) reported the invention of another alternative approach to spark ignition, known as *shock ignition*. In shock ignition, the original driver creates an additional pulse during the implosion, timed so that the shock arrives at the center of the capsule just at stagnation. This shock wave is made strong enough to ignite the fuel where it converges. Since its invention, shock ignition has been actively explored. Both theory and experiment indicate that it has potential.

Finally, we should not leave the topic of spark ignition without mentioning one of its major qualitative advantages. We address below the need for ICF implosions to be adequately spherical, both with regard to the symmetry of the ablation pressure and to the impact of the Rayleigh–Taylor instability. When using ignition from a central hot spot, small departures from a spherical implosion can permit the hot spot to be too cool or too convoluted to ignite. Likewise, the injection of cold fuel into the hot spot via the Rayleigh-Taylor instability can quench ignition. In contrast, for fast ignition one typically does not care what the shape of the gas within the imploded fuel capsule may be. Nor does one necessarily care if the implosion is asymmetric to some degree. So long as one can deliver the energy where it is needed to ignite some of the fuel, the fusion burn should be able to proceed.

11.4   Alternative Drivers

Here we briefly discuss some of the alternative possible energy sources for inertial fusion. Z pinches would require some impressive technology development to be able to fire at a steady, high repetition rate, but this is not out of the question. They offer several potential approaches to fusion. By imploding the pinch onto a low-density foam, the Z device has produced a radiative shock that creates an intense X-ray burst along the axis of the pinch. This might be developed into a driver for fusion capsules having a central hot spot (Slutz et al. 2003). The pinch, however, naturally produces a much longer pulse of lower-intensity X-rays than does a laser-driven hohlraum. This could enable low-velocity implosions that compress cold fuel, which is then ignited by some sort of fast ignitor (Vesey et al. 2006).

A more clever approach to ICF using a pulsed-power device as a driver, recently invented at this writing, is known as *magnetized liner inertial fusion*, referred to as MagLIF (Slutz and Vesey 2012). The $\boldsymbol{J} \times \boldsymbol{B}$ force is used, as discussed above in Sect. 10.10.2, but rather than driving a flat plat the driver is configured to implode a conducting cylinder. Approaches to inertial fusion in which an imposed magnetic field reduces heat conduction while an imploding material compresses a plasma toward fusion conditions have been considered at least since the paper by Lindemuth and Widner (1981), but have been thought to be unable to achieve high gain. Two new elements have changed this picture. The first is the addition of a cryogenic DT layer within a metal cylinder, which then implodes at 100–200 km/s. Although cylindrical in shape, this is similar to the low-velocity regime of capsule implosions discussed above. However, rather than somehow engaging in fast ignition after the implosion, the extra energy needed by the hot spot is now deposited before the implosion, by laser heating the gas near the axis of the cylinder. This would not work without the use of a magnetic field to reduce radial heat conduction. The initial design studies showed gains above 1000 by this approach. As mentioned previously, these designs tend to use more fuel and have a higher fusion yield than do designs using higher implosion velocities. Experimental work is underway.

Heavy-ion accelerators have also received significant attention as potential drivers for ICF, beginning in the late 1970s (Arnold 1978). They, like other particle accelerators, seem likely to be operable at high repetition rate. Even so, meeting the requirements for ICF would require some advances in accelerator technology. It has long been imagined that relativistic heavy ions could drive ICF as follows. They would be stopped in high-Z material, such as gold, heating it to some hundreds of eV. The X-rays emitted by the gold would then be transported in a hohlraum and would irradiate a capsule. This would represent X-ray drive, just using an X-ray source other than a laser. It might potentially turn out to have an advantage in efficiency and/or development cost. (Historically, there were at some point also programs aimed at driving ICF with electron beams or with light ions, but these approaches were eventually abandoned.)

A more recent approach using heavy ions (Henestroza et al. 2011), not yet much explored thanks to the vicissitudes of research funding, uses a configuration known as the "X target." This has in common with MagLIF the use of a cylindrical geometry, slow implosions, and high yields. A sequence of ion beams, focused to cylindrical shells, irradiated a low-Z outer shell filled with cryogenic DT. The explosion of the outer shell compresses the DT, and is sustained by the sequence of ion beams until a small, intense beam on axis finally ignites the fuel. One appealing aspect of this novel idea is that it relies on using heavy ion beams for what they are good at—depositing energy throughout large volumes of matter, rather than trying to fit them into an imitation of a miniature fusion bomb.

11.5 Pitfalls and Problems

A pressure of 100 Mbar is fairly easy to produce. One can see from (9.42) that this requires a laser energy flux of 1.6×10^{15} W/cm^2 of 0.35 μm light. The corresponding laser energy during the acceleration, for a capsule of 2 mm radius, accelerated for 10 ns, would be about 2.7 MJ. The laser energy required during the formation and propagation of the shocks would increase the total energy by some factor. Alternatively, one would need an X-ray temperature of 220 eV in a hohlraum to create a pressure of 100 Mbar. This again does not seem very difficult (see Chap. 9). Indeed, achieving such a pressure is easy enough that in the absence of limitations one would seek to use a larger pressure. However, there are several major problems that make ICF a challenge. We discuss three of them here, and then provide a summary.

11.5.1 Rayleigh Taylor

An ICF implosion is Rayleigh–Taylor (RT) unstable during most of its evolution. Early on, during what is often called the acceleration phase, the low-density, hot ablated plasma is at a higher pressure than the cooler, higher-density layer being accelerated. This creates the condition of opposed density and pressure gradients that excites the RT instability (see Chap. 5). One could say that the low-density plasma is pushing on the higher-density plasma. The acceleration phase ends but before long the deceleration phase begins, when the low-density gas within the capsule pushes against and decelerates the denser, incoming fuel layer. Here again the density gradient and pressure gradient are opposed, so one has an RT unstable system. This necessitates understanding what limits RT may place on ICF implosions.

The number of e-foldings of RT growth, $\gamma_{RT}t$, is straightforward to estimate, assuming that the growth rate is the value for an abrupt, embedded interface, which is $\gamma_{RT} = \sqrt{A_n kg}$. The density changes are large so we take $A_n \sim 1$. The most-damaging wavelength is related to the thickness of the capsule during the RT growth, which we designated Δr_i above. Wavelengths that are short compared to Δr_i will grow and saturate without creating large perturbations in the structure. Wavelengths that are long compared to the capsule thickness will distort the capsule and may decrease the compactness of the implosion, but they have less impact and a slower growth rate than wavelengths of order the capsule thickness, which can break up the capsule if they grow large enough. So we take $k = 2\pi/\Delta r_i$. If we approximate the acceleration and deceleration as constant, then we have $r_s/2 = (1/2)g(t/2)^2$, from which $t\sqrt{g} = 2\sqrt{r_s}$. Altogether, this implies a value for the number of e-foldings, GF, of

$$GF = \gamma_{RT}t = \sqrt{\frac{8\pi r_s}{\Delta r_i}}. \tag{11.52}$$

For a variable acceleration, one can generalize GF by integrating γ_{RT} over time. One can see the relevance of the in-flight-aspect ratio, or IFAR, which we defined in Sect. 11.2 as $r_s/\Delta r_i$, and whose value is given by (11.31). As Fig. 11.9 showed, its value is typically several tens. Taking the IFAR to be 30, we would expect nearly 30 e-foldings of RT growth from this calculation. If the noise at such wavelengths corresponded to atomic displacements ($\sim$1 Å), and there were no nonlinear saturation and no spherical effects, the implied amplitude would be many meters. The acceptable growth at such wavelengths is of order several, and certainly not more than 10, e-foldings. One concludes that ICF would not be feasible in the presence of RT growth at such a rate.

Fortunately, the RT growth rate is actually smaller than $\sqrt{A_n k g}$, because the interfaces where the growth occurs are not sharp but rather have some scale length L. The growth is reduced much further during the acceleration phase because ablation carries away the material in which the modulations are growing. Suppose the ablation is carrying away material with some characteristic ablation velocity, V_A (this is not the exhaust velocity, but rather the velocity at which material flows away from the dense interface). Then recall that RT modes are surface waves with an exponential penetration depth of $1/k$. Given that the growth rate without ablation, γ_{RT}, sets a timescale of $1/\gamma_{RT}$, we would expect to see ablation quench the RT growth when $1/k \sim V_A/\gamma_{RT}$ or $kV_A \sim \gamma_{RT}$. This is indeed what is seen, within small numerical factors, in simulations and in detailed analytic theory. A standard relation expressing the net growth rate is

$$\gamma_A = \sqrt{\frac{kg}{1+kL}} - \beta k V_A. \tag{11.53}$$

Here the coefficient β is ~ 1 for X-ray ablation and ~ 3 for laser ablation. The ablation velocity V_A is the velocity at which material leaves the RT-unstable region and equals the mass ablation rate divided by the density in the ablation layer, $\dot{m}/\rho_2$, where ρ_2 is the density of the shocked ablator. The ablation velocity can be estimated from the discussion in Chap. 9. An order-of-magnitude value of the product βV_A is 5×10^5 cm/s, while the order-of-magnitude of g is 3×10^7 cm/s per ns, which is 3×10^{16} cm/s^2. For a steep interface, the maximum unstable wavenumber is $k = g/(\beta V_A)^2 = 10^5$ cm^{-1}, corresponding to a wavelength of about 1 μm. Since the fuel shell is only a few μm thick during the acceleration phase (after compression by the sequence of shocks), the wavelengths whose thickness is of order the shell thickness are strongly stabilized by ablation.

It should be clear that a careful design must consider all possible RT modes in order to assure control of RT during the ablation phase. If the stabilizing effect of ablation alone is not enough, one can consider trying to increase the scalelength L at the ablation surface. Increasing L to a few μm can have an important effect. One way to try to increase L is to design the outer surfaces in the target so that they produce preheat that penetrates the ablator but is not energetic enough to reach and heat the fuel layer. This can be attempted in principle using either electrons or X-rays. Design work at this writing has focused on using X-rays. An alternative approach to increasing L is to launch the first shock into the fuel by using a brief impulsive load that rapidly evolves into a blast wave, as in Fig. 11.7 above. The blast wave will decay as it moves into the capsule, so that the amount of entropy it produces will be larger in the outer layers of the ablator and smaller by the time it reaches the fuel. Either of these approaches, or perhaps another one, could be very helpful in the context of fusion by direct laser ablation.

During the deceleration phase there is little ablation, so only L can act to reduce the RT growth. Fortunately, L tends to be large enough. This happens because the interior of the capsule has been heated by the coalescing shock waves and is much hotter than the cold fuel, in response to which the electrons transport heat into the inner layer of cold fuel, which in turn expands and lengthens the scale length at the inner fuel boundary. A typical estimate, from Lindl (1995), is that at the inner fuel boundary L is roughly 15% of the final fuel radius. This has the consequence that the RT growth during deceleration is limited to about three e-foldings. The initial amplitudes that grow by this amount are determined both by the roughness of the inner surface of the fuel and by the perturbation of the inner surface due to the RT growth during ablation. This coupling of the outer surface to the inner surface is known as *feedthrough*.

Taking together the growth during acceleration and deceleration with the stabilizing effects of ablation and gradients, an approximate limit on the implosion velocity is $\sim$600 km/s, before IFAR becomes too large and RT prevents high gain. The rightmost dashed line in Fig. 11.3 shows this limit. To the extent that the model underlying Fig. 11.11 is accurate, one might be able to push higher with X-ray driven capsules. However, as Fig. 11.3 showed, the gain at high velocities becomes small enough that this may not be of much use.

During the construction of NIF, researchers used extensive 2D and 3D simulations in an attempt to understand the limitations that RT would impose on the design of capsules intended to ignite. It turned out that capsules predicted to be safe from quenching by RT were not safe. Instead, the first few years of NIF experiments produced unambiguous evidence of the mixing of ablator material into the hot spot when GF exceeded about 7, corresponding to a linear-theory amplification of 1000. This was significantly smaller than had been expected based on the simulation studies. The need to keep GF this small limited the neutron yield that could be achieved. The origin of the discrepancy is not yet clear, but the present author suspects that the dissipation of vorticity in the computer simulations is at least partly responsible.

The net result is that it appears that RT can be limited to a low enough level that fusion can succeed, but that doing so places difficult constraints on the initial roughness of the target surfaces and on the smoothness of the ablation pressure. As was mentioned in Chap. 9, the drive to invent ways to smooth the irradiation of surfaces irradiated by lasers came from ICF. Specifically, one needed to reduce the seeding of RT instabilities due to structure in the laser ablation. The precise constraints can be estimated now, but there is a potential to make them less severe through improvements in design of targets and in understanding of RT. The RT instability evidently is present in any approach to fusion, whether initiated with lasers, with a Z-pinch, or with sunlight.

11.5.2 Symmetry

In any analysis of structure in spherical coordinates, it is natural to describe the structure in terms of spherical harmonics. These harmonics form a complete basis set that describes a system by breaking the structure into modes that correspond to having an integer number of wavelengths over 360° in azimuthal angle or 90° in polar angle. The RT modes that grow to the largest amplitude have mode numbers of several tens. (Without any stabilization, the most-unstable mode number would be much larger.) The low mode numbers, such as 2 or 4, correspond to variations in the structure that can be produced if the ablation pressure is not uniform over larger distances. For example, if two ends of a target (the poles) are driven more strongly than the middle of the target (the equator), then the imploded target will be flattened like a pancake. Such an imploded target is indeed described as pancake-like, or may be said to have pancaked. In contrast, if the equator is driven more strongly than the poles, the imploded target will be a long thin tube, for which the common metaphor is a sausage. In either case, the ρr will be smaller than intended over a significant range of solid angle, and one will obtain less burning than one had hoped. One could say that ICF, in order to succeed, must not feed one breakfast.

A simple estimate of the required uniformity is straightforward to make. The fuel radius decreases by a factor of the convergence, C, as the fuel implodes, from an initial radius r_s to a final radius R_c. We can ask what difference in velocity ΔV would cause the fuel at one angle to be at twice the final radius when the fuel at another angle has reached its final radius. To do so, we estimate the implosion time for the fully imploded fuel as r_s/V and we ask what ΔV will satisfy

$$r_s \Delta V / V = R_c, \text{ which is } \Delta V = V R_c / R_s = V / C. \tag{11.54}$$

Thus, to keep the asymmetry of the final state smaller than a factor of 2 requires the variation in implosion velocity to be less than $R_c/r_s \sim 5\%$ since the convergence is of order 20. Since the implosion velocity is proportional to the ablation pressure we require the ablation pressure to be uniform to the same level. In reality, an asymmetry of a factor of 2 is far too large. A more realistic limit is that the ablation pressure must be uniform to 1% accuracy, at least in a time-averaged sense. This is a demanding constraint. With fusion by direct laser ablation, one can use many beams but even so only a few beams overlap on any given point on the target. With fusion within a hohlraum, entrance holes for laser beams or for the flow of energy into the hohlraum from a Z-pinch create a significant asymmetry in the radiation from X-ray heated walls. The asymmetry produced by the energy source must be compensated for by careful design. In all approaches, as the target begins to move (and as the plasma struck by the laser also moves), the irradiation symmetry may change. As in the case of RT instabilities, the requirement of symmetry places difficult constraints on ICF. These constraints are unlikely to be met without measuring the symmetry and fine-tuning the irradiation of the capsule to produce adequately spherical implosions.

A historical and cautionary note is worth making here. During the 1970s portions of the inertial fusion program in the USA attempted to create fusion within hohlraums without making any measurements of the implosion symmetry. This effort failed, as evidenced by neutron yields that were typically 100 times smaller than predicted. Only on the Nova laser, when implosion symmetry and of the other essential aspects of the fusion system were measured, was it possible to first achieve fuel densities above 100 times liquid density and to begin to see neutron yields not very far below those predicted by one-dimensional simulations. This required a close collaboration between experiment and simulation, measuring the details and using simulations to assess the implications of what was seen. Prior to that era, there was far too much reliance on simulation codes without verifying measurements, and far too much belief that the codes were the reality. A good scientist who does simulations understands that the simulation is an essential tool but cannot fully represent reality. In the view of the present author the fact that much of ICF was classified also contributed to its failures in the 1970s. A major benefit of presenting results at open scientific meetings and of publishing in the refereed literature is that these activities force an improvement in the quality of the science being done.

11.5.3   Laser–Plasma Instabilities

We saw in Chap. 9 that laser–plasma instabilities can scatter laser light and can also produce populations of high-energy (suprathermal) electrons. (In Z-pinch-driven fusion, these instabilities are not present but there is also some potential to produce populations of energetic electrons.) We saw that these instabilities, in particular stimulated Raman scattering (SRS) and stimulated Brillouin scattering (SBS), can in some circumstances convert most of the laser light into scattered light or energetic electrons. We also saw that another instability, filamentation, can break the light into filaments, which can potentially change where the light goes or can trigger other instabilities. These processes have three types of adverse consequence. We discuss these first, and then the question of how to control the amount of such scattering.

First, stimulated scattering processes reduce the efficiency of the laser-fusion target. If half the laser light is scattered back toward the source, then one will need to start with twice the energy to produce fusion. The energy sources (or drivers) are expensive enough that this would be a major problem.

Second, stimulated scattering and filamentation both may alter where the light is deposited, both in a directly driven capsule and in a hohlraum. This can affect the symmetry of the implosion, since ablation pressure or soft X-rays would then be produced in unintended places. If these processes were consistent in the amount and direction of the scattering that they produced, then one could tune out the resulting variations through a sequence of experiments that measured the symmetry. This would be fine, but these processes are observed to behave reproducibly in some regimes and irregularly in others. So one cannot count on being able to tune them out.

Third, SRS and the decay instabilities produce energetic electrons. Some of the energy of these electrons reaches the fusion fuel, which heats it and adds to its entropy. We can recall that the entropy limit is then 4×10^8 J/keV/g, but that most of this limit must be used in the process of compression. For purposes of estimates, let us suppose that the limit on the entropy from preheat is 10^8 J/keV/g. The amount of entropy produced depends on the temperature when the preheat occurs, since $ds = dq/T$. Thus, a serious design must address preheat in a time-dependent context. For purposes of a crude estimate, suppose that the preheat occurs when the final shock is being produced, so that the pressure is already $\sim$30 Mbar, the fuel density is $\sim$5 g/cc, and the Fermi energy is $\sim$40 eV, and also suppose that the fuel temperature is $\sim$40 eV. Under these assumptions, an entropy of 10^8 J/keV/g develops if the 3 mg of fuel absorbs 12 kJ of energy. If the final capsule energy is 140 kJ, the rocket efficiency is 10%, and the delivery of energy to the rocket fuel is 50% efficient, then the driver energy is about 3 MJ. Thus, an estimate is that if of order 0.3% of the driver energy is deposited in the fuel as preheat, the fuel entropy would increase above the desired value, the total compression would be less than desired, and the gain would decrease.

This seems like a rather difficult constraint, but it in fact is less severe than it seems. The only energy that does damage is the energy that heats the fuel. But to get to the fuel the electrons must first penetrate the ablator. Those that penetrate the ablator without much attenuation (at energies well above 30 keV) will also penetrate through the (lower-Z) fuel without depositing much energy. Those that cannot penetrate the ablator cannot heat the fuel. A suprathermal electron distribution with a 30 keV temperature has a mean range of about 3 mg/cm^2, which is of order the initial areal density of the ablator. Thus, electrons below this energy will not tend to reach the fuel. In addition, because the electrons transport diffusively in the dense matter and can scatter out of the vicinity of the capsule, many of them may deposit their energy somewhere else, especially in fusion using hohlraums. Depending on details, one would expect that an ICF high-gain target could succeed even if the fraction of the driver energy in suprathermal electrons were of order 1% or perhaps more. By the time this ratio reaches 10%, efficiency is becoming as much of a concern as preheat. In summary, the production of suprathermal electrons must be limited but need not be completely quenched.

It is clear that, if ICF is to succeed, then none of the laser–plasma instabilities can be allowed to grow to a large amplitude. There are at root two approaches to control these instabilities. The first approach is to use short-wavelength (UV) lasers for fusion. This helps in two ways. The growth rate for all the instabilities is proportional to the oscillation velocity of the electrons in the laser light wave, as shown in Table 8.1. The square of this oscillation velocity is proportional to the laser energy flux times the square of the laser wavelength (9.9). Thus, one reduces the growth rate by shortening the laser wavelength. In addition, shorter-wavelength laser light makes collisional effects more important; because critical density increases, all processes occur at higher densities. An added benefit of higher densities is that the laser energy is shared by more particles so the plasma is cooler. This further increases the importance of collisions. Furthermore, the strong collisionality of plasmas made with UV lasers leads to the absorption of the laser light at densities well below the critical density, so that processes at critical density become unimportant. The need for short-wavelength lasers was determined the hard way. Fusion programs using infrared lasers (at $\sim$1 and $\sim$10 μm wavelength) experienced severe problems with energetic electrons, leading, in the long run, away from such wavelengths as serious candidates for ICF.

The second approach to control laser–plasma instabilities is to actively suppress them or at least to reduce their saturation level. If high density and collisions are effective enough, this will not prove necessary. The myriad options for control are beyond our scope here but are also often not realistic possibilities for specific ICF facilities. Fundamentally, one can suppress instabilities involving ion waves, such as stimulated Brillouin scattering, by adding bandwidth to the laser beams. This is not practical for decay instabilities or stimulated Raman scattering. One can limit SRS by creating a smooth, steep density profile in which SRS finds it difficult to grow. Two-plasmon decay, which occurs only near the quarter-critical density, may or may not be a problem. One other option that has recently emerged is the use of spike trains of uneven duration and delay (Afeyan and Huller 2013), which may prevent the instabilities from reaching and sustaining large, damaging levels. Ultimately, laser–plasma instabilities limit the energy flux that can be used for ICF. At this writing, it is unclear whether active control of the laser–plasma instabilities will be needed for ICF.

Early experience on NIF underscored the challenges and importance of limiting the laser–plasma instabilities. A key challenge, especially for hohlraums, is to understand the actual plasma conditions. The hohlraum targets for early experiments on NIF were designed so that the simulated plasma conditions would produce low levels of SRS and SBS. But the program decided to ignore the advice of review committees and did not prioritize the measurement of plasma conditions in the hohlraums or extensive measurements of the instability behavior. In actual fact, we do not at this writing have codes that are predictive regarding plasma conditions, even from flat targets, let alone hohlraums. And indeed, the SRS spectra showed that the plasma conditions were not as predicted. In the experiments, some of the NIF beams converted more than half of their energy into SRS light and energetic electrons. (This is not so easy to detect in the publications, which tend to focus on the total amount of reflected energy, which was near 15% in those experiments.) There are indications that these instabilities were one of the factors limiting the performance of those early experiments. In some other, non-ICF experiments, a small change in the distribution of energy across laser beams, which would have no significant effect at any of the smaller laser facilities, caused SBS to grow large enough to damage some expensive optics. It is unfortunately clear that laser–plasma instabilities on MJ lasers are dangerous to both experiments and the lasers themselves.

Fig. 11.14 Whether ICF
with laser drivers is feasible
depends on whether the gray
window shown here actually
exists

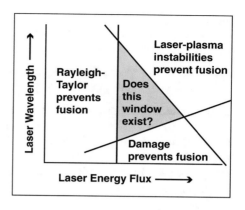

11.5.4 Can Laser-Driven Inertial Confinement Fusion Succeed?

Figure 11.14 illustrates the fundamental constraints on ICF using laser drivers. There is some laser energy flux below which IFAR becomes too big and Rayleigh Taylor prevents fusion from succeeding. There is some laser wavelength, increasing somewhat with laser energy flux, below which damage to the laser optics prevents fusion from succeeding. And there is some laser energy flux, decreasing as laser wavelength increases, above which laser plasma instabilities prevent fusion from succeeding. The question, for some combination of technology and target design, is whether the gray area exists or not. If it does, then perhaps one can produce implosions that will generate high gain.

Homework Problems

11.1 Plot the burn fraction versus ρr. Discuss the impact of the assumptions made in deriving the burn fraction on this curve, and on the size of a system designed to produce a certain quantity of fusion energy.

11.2 Evaluate the classical pressure and the boson pressure for deuterium as a function of temperature. For pure deuterium at a density of $0.1\,\text{g/cm}^3$, plot the ratio of the pressure for deuterium treated as bosons to that for deuterium treated as a classical gas, as a function of temperature. Discuss the comparison.

11.3 For DT fuel, derive the classical relation between entropy and pressure (normalized by the Fermi pressure of the electrons).

11.4 Suppose that one could apply a pressure p for a time t, using some energy source. With this source, we could accelerate some amount of mass per unit area, $\rho_o \Delta r$, to $v_{\text{imp}} = 300\,\text{km/s}$. Define a fusion capsule using the reflected pressure due to sunlight for 12 h as the pressure source. Approximate sunlight as light with a wavelength of 580 nm and an energy flux of 1 kW/m^2. How long would such a capsule take to implode?

11.5 Derive the spectrally averaged absorption coefficient for bremsstrahlung in DT. Check your value against the value found in the NRL plasma formulary.

11.6 Evaluate the appropriate integral of the radiative transfer equation over solid angle to obtain F_R from a spherical volume of optically thin, spatially uniform DT. Find the value of the characteristic distance. Compare your result to the result in (11.36), which assumes that the integral over solid angle of the distance across the fuel gives πR_h. Extra credit: generalize this calculation to include arbitrary optical depth and discuss the results.

11.7 The Lawson criterion is generally written as $n\tau > 10^{14}$ s/cm^3, with density n and energy confinement time τ. Find a way to relate this to (11.38) and comment on the comparison.

11.8 Because the density in a central hot spot is less than that of the cold fuel, a larger fraction of the total energy must be expended to heat it than was estimated above. Revisit the discussion above of the relative energy content of the fuel and the hot spot, and develop an expression for the scaling of the hot-spot energy content with hot-spot density. Find the result as an absolute energy and as a fraction of the energy used to compress the cold fuel.

11.9 Using the equations of this Sect. 11.3.2 and others as necessary, build yourself a computational model of a fusion target that ignites from a central hot spot. Use it to explore target designs for the National Ignition Facility, using fuel masses of 1 mg or less, a fuel ρR_c of 1.5 or less, and other parameters of your choice. Make some relevant plots and comment on what you find.

11.10 Evaluate the amount of Rayleigh Taylor growth for the sunlight-driven fusion system of the problem 11.4.

References

Afeyan B, Huller S (2013) Optimal control of laser plasma instabilities using spike trains of uneven duration and delay (stud pulses) for ICF and IFE. In: Mora P, Tanaka KA, Moses E (eds) IFSA 2011 – Seventh international conference on inertial fusion sciences and applications, EPJ web of conferences, vol 59. EDP Sciences, Cedex A

Arnett WD (1996) Supernova and nucleosynthesis. Princeton University Press, Princeton

Arnold RC (1978) Heavy-ion beam inertial-confinement fusion. Nature 276(5683):19–23. https://doi.org/10.1038/276019a0

Atzeni S (1999) Inertial fusion fast ignitor: igniting pulse parameter window vs the penetration depth of the heating particles and the density of the precompressed fuel. Phys Plasmas 6(8):3316–3326

Atzeni S, Meyer-Ter-Vehn J (2004) The physics of inertial fusion. Oxford, New York

Betti R, Zhou CD, Anderson KS, Perkins LJ, Theobald W, Solodov AA (2007) Shock ignition of thermonuclear fuel with high areal density. Phys Rev Lett 98(15). https://doi.org/10.1103/PhysRevLett.98.155001

Craxton RS, Anderson KS, Boehly TR, Goncharov VN, Harding DR, Knauer JP, McCrory RL, McKenty PW, Meyerhofer DD, Myatt JF, Schmitt AJ, Sethian JD, Short RW, Skupsky S, Theobald W, Kruer WL, Tanaka K, Betti R, Collins TJB, Delettrez JA, Hu SX, Marozas JA, Maximov AV, Michel DT, Radha PB, Regan SP, Sangster TC, Seka W, Solodov AA, Soures JM, Stoeckl C, Zuegel JD (2015) Direct-drive inertial confinement fusion: a review. Phys Plasmas 22(11). https://doi.org/10.1063/1.4934714

Henestroza E, Logan BG, Perkins LJ (2011) Quasispherical fuel compression and fast ignition in a heavy-ion-driven x-target with one-sided illumination. Phys Plasmas 18(3). https://doi.org/03270210.1063/1.3563589

Landau LD, Lifshitz EM (1987) Statistical physics, course in theoretical physics, vol 5, 2nd edn. Pergamon Press, Oxford

Lindemuth IR, Widner MM (1981) Magneto-hydrodynamic behavior of thermonuclear fuel in a preconditioned electron-beam imploded target. Phys Fluids 24(4):746–753. https://doi.org/10.1063/1.863415

Lindl JD (1995) Development of the indirect-drive approach to inertial confinement fusion and the target physics basis for ignition and gain. Phys Plasmas 2(11):3933–4024

Slutz SA, Vesey RA (2012) High-gain magnetized inertial fusion. Phys Rev Lett 108(2). https://doi.org/10.1103/PhysRevLett.108.025003

Slutz SA, Bailey JE, Chandler GA, Bennett GR, Cooper G, Lash JS, Lazier S, Lake P, Lemke RW, Mehlhorn TA, Nash TJ, Nielson DS, McGurn J, Moore TC, Ruiz CL, Schroen DG, Torres J, Varnum W, Vesey RA (2003) Dynamic hohlraum driven inertial fusion capsules. Phys Plasmas 10(5):1875–1882. https://doi.org/10.1063/1.1565117

Tabak M, Hammer J, Glinsky ME, Kruer WL, Wilks SC, Woodworth J, Campbell EM, Perry MD, Mason RJ (1994) Ignition and high gain with ultrapowerful lasers. Phys Plasmas 1:1626

Vesey RA, Campbell RB, Slutz SA, Hanson DL, Cuneo ME, Mehlhorn TA, Porter JL (2006) Z-pinch-driven fast ignition fusion. Fusion Sci Technol 49(3):384–398

Zhou CD, Betti R (2007) Hydrodynamic relations for direct-drive fast-ignition and conventional inertial confinement fusion implosions. Phys Plasmas 14(7). https://doi.org/10.1063/1.2746812

Chapter 12
Experimental Astrophysics

Abstract This chapter begins with a discussion of the general problem of scaling between laboratory systems and astrophysical ones, identifying issues of physical consistency, of what we call the Ryutov scaling of the key parameters, and of the specific scaling for a given experiment focused on a specific astrophysical process. It provides a thorough illustration of how to address these issues using hydrodynamic systems and applies the results to experiments to study instabilities during supernova explosions and the crushing of clouds. It then considers how to scale radiation hydrodynamic systems, with the example of radiative jets. Next it shows how one can analyze MHD systems, taking an example from jet-launching studies. Finally it considers experiments seeking to observe collisionless shocks, in which kinetic effects are essential but where one is more concerned with observation of a process than with scaling of a complete astrophysical system.

In Chap. 1 we looked forward to potential connections between high-energy-density physics and astrophysics. Some of these connections arise because one can produce in the laboratory circumstances that actually exist in astrophysics and can measure the properties of these systems. We have seen examples of this in the areas of equations of state, in Chaps. 3 and 4, and opacities, in Chap. 6. Other connections arise because high-energy-density experiments can produce dynamic behavior under conditions that are relevant to astrophysical systems. By this means, one can explore dynamic processes in ways that allow precise reasoning from the experimental results to the astrophysical process. Even so, astrophysical phenomena involve spatial and temporal scales that are many orders of magnitude (sometimes 25) greater than the scales encountered in laboratory experiments, so one may wonder whether a valid comparison is possible. The issue of whether and how one can make this comparison is the issue of *scaling*. The focus of the present chapter is to discuss how one can establish scaling that relates laboratory processes to astrophysical ones.

We begin with some historical remarks specific to this area. As applications of lasers were first developed, one of the great minds in plasma physics, John Dawson (1964), suggested that they might be useful in the context of astrophysics. With the lasers and instruments of that era, one could have blown up a speck of matter, taken a picture, and marveled at the exploding "star". But such a picture,

© Springer International Publishing AG 2018
R P. Drake, *High-Energy-Density Physics*, Graduate Texts in Physics,
https://doi.org/10.1007/978-3-319-67711-8_12

then or now, would have no relevance to actual exploding stars. The advances in lasers and experimental technique from that time through the 1980s set the stage for the use of lasers to study the astrophysical properties and processes. Effort in both areas began in the early 1990s. Work relating to equations of state and opacities was discussed in Chaps. 3 and 6. Among the early work in astrophysical processes were laser experiments at the Naval Research Laboratory relevant to blast waves in magnetospheres (Ripin et al. 1987) and in astrophysics (Grun et al. 1991). The first publication to specifically suggest the use of lasers to address hydrodynamic processes in supernovae was by Takabe (1993), in Japan. B.A. Remington simultaneously initiated research on this topic in the US. There are several review articles describing the early years of such work (Drake 1999; Remington et al. 1999, 2000, 2006; Takabe 2001).

Our task here is to discuss the fundamental principles that must guide laboratory experiments seeking to advance our understanding of astrophysical processes. With regard to an astrophysical system and a laboratory system that might in some sense represent a scaled model of it, we ask four questions:

1. Can both systems validly be described by the same type of equation?
2. Are the same terms in such equations important in both cases?
3. Can the two systems have good Ryutov scaling?
4. Can the two systems have good specific scaling, with regard to the dynamics of the process of interest and the differences in structure?

The first two of these address the physical consistency of the two systems. The physics that is of interest in the astrophysical system must be present, in a consistent way, in the laboratory. As one example, a purely hydrodynamic laboratory experiment cannot meaningfully simulate any astrophysical process in which forces from magnetic-field curvature are important. We will evaluate the implications of this as we take up various cases below. What we mean, in the third question, by *Ryutov scaling* will become evident below. In an ideal case, the laboratory experiment would be a perfect model of the astrophysical one. The Ryutov scaling establishes the relationship of the parameters that is a necessary but not sufficient condition for perfect scaling. Some experiments can produce parameters, in their regions of interest, that satisfy this scaling. Unfortunately, a parallel requirement for perfect scaling is that the initial shape (or if you prefer, the boundary conditions) of the laboratory system match that of the astrophysical system *over all space*. No experiment can be perfect in this sense. The limitations of either Ryutov scaling or boundary conditions leads to a need for a third aspect of the comparison. We will designate this the specific scaling. This has two aspects. The appropriate dimensionless parameters that characterize the processes of central interest should have similar values in the two systems. And there are also constraints associated with any geometric differences between the two systems. These questions will be addressed below for specific cases that are purely hydrodynamic, or also radiative, or also magnetohydrodynamic. There have been and are active experiments in all these areas.

Another aspect of the laboratory-astrophysics comparison is that of comparisons with and validation of codes. In some cases, an experiment can produce conditions that are consistent with the assumptions of a computational astrophysical simulation tool. In this case an experiment can provide a useful and meaningful test of the code, whether or not the experiment is well-scaled to a particular astrophysical system. The hydrodynamic experiments discussed immediately below are an example in which the differences among the astrophysical system, the experimental system, and the computational system are limited enough that one would expect the experiment to be a good test of the code and that both the experiment and the code would directly inform one's understanding of the astrophysics. In contrast, for the plasma dynamos discussed in Chap. 10, the simulation, the experiment, and the astrophysical system occupy distinct but complementary regimes with regard to dissipative processes (Schekochihin et al. 2004). All can inform a general understanding of the process in question (such as dynamo behavior), but none will be a precise model of any other.

Before proceeding we should note that the laboratory study of processes in astrophysics is not limited to high-energy-density systems. Processes that do not require high Mach numbers or radiation can be studied at lower energy density. One example, at this writing, is the work of a number of groups who are advancing the study of magnetic reconnection, which is one source of energetic particle production in magnetized interplanetary and interstellar plasmas. Another example is the study of the dynamic behavior of magnetic flux tubes. There will doubtless be more examples in the future.

12.1 Scaling in Hydrodynamic Systems

We first consider the issue of scaling between two systems for processes that can be described by hydrodynamic (the Euler) equations. The reader may wonder why one cares about comparing systems whose behavior is hydrodynamic, as the equations seem so simple. Let us consider this first. It may seem as though we understand hydrodynamics, but it would be more accurate to say that we understand the equations that apply to hydrodynamics. We also know that these equations are nonlinear in ways that produce immediate complexity under many circumstances. This has two consequences. First, it means that there are questions that are too complicated to be addressed successfully in computer simulations. We will see one example of such a question below, in our discussion of supernovae. A second example involves the onset of turbulence, in the context of Sect. 5.8. The Reynolds number Re in astrophysical flows is typically far above the value of 10,000 at which the mixing transition can begin. In contrast, because of numerical viscosity, the computer simulations cannot achieve even $Re \sim 10,000$ when modeling moderately complicated systems. In addition, turbulence models are too uncertain to know which approach to an integrated description of turbulence might be best. Another aspect of the impact of turbulence is that it remains unclear whether fine structures in an evolving system can significantly affect the large-scale evolution of the system, a process described in the literature as "stochastic backscatter" (Leith 1990; Piomelli et al. 1991). To address the presence and importance of these effects in the context of astrophysics, experiments that are both clever and well scaled are required.

The complexity of realistic hydrodynamics implies that code validation is essential even for the processes that may in principle be simulated successfully. It is unclear how well we understand how to model numerically the very nonlinear evolution of hydrodynamic systems. Like a series solution to a differential equation, a computer simulation is only an approximation to the actual solution of the physical equations being solved. Unlike a series solution, it is very difficult to judge the error involved in the approximate computational solution. In addition, independent simulations often produce different results (Glimm et al. 2001), especially on a scale below about 10% of an initial perturbation wavelength (Kane et al. 1997). Because of these difficulties, a code that works well for a certain class of problems may not work well for other problems. Thus, validation that is relevant to the dynamics of interest is important for each specific dynamical process.

We now turn to the elements of scaling discussed above. Our intent here is to demonstrate by example the kind of thinking that one must carry through in order to have an experiment with any significance. The requirement of physical consistency is that these equations be valid and adequately complete for both systems. For the systems to behave as a fluid the particles must be well-localized. This may be accomplished by collisions in laboratory systems, or in dense astrophysical environments such as supernovae. In less-dense "collisionless" environments, a hydrodynamic approach may still be valid if the magnetic and electric fields are sufficiently structured, as one expects in turbulent flows. See Ryutov et al. (1999) for a more extensive discussion of this point. In addition, the Euler equations represent the equation of state using a simple, polytropic index γ. To have absolutely perfect scaling, such a model would need to apply to both systems with the same value of γ. However, one still would see any qualitatively important behavior even if the two systems differed in the value of γ. However, such a polytropic model is only relevant in the absence of phase transitions during the period of interest. So these must be avoided.

The Euler equations are adequately complete if all the other terms in a general fluid equation are small. This issue was discussed at length in Chap. 2 and requires that Re and the Peclet number Pe be large and that radiative effects be small. Not emphasized in that discussion was how to think about magnetic fields. This has some subtle aspects. The existence of magnetic energy in a turbulent field, on a small scale, need not invalidate the Euler equations as a description of large-scale behavior and can, if needed, be included in the internal energy that implies some value of γ. However, the sources and impact of global-scale, coherent magnetic field must be small. The corresponding requirement is that the Biermann number (see Sect. 12.6) must be large and the averaged plasma β must not be small.

It is often the case that an experiment and a similar astrophysical system meet these constraints, but that the value of one of the scaling parameters is quite different between them. For example, high-energy-density experiments tend to have Reynold's numbers in the range of $Re \sim 10^5$–10^7 while for otherwise similar astrophysical systems Re is much larger. In such cases it is worthwhile to ask what the significance of these differences may be. This is not a topic we will explore here at any length. For the specific case of Re, one would expect that the presence of an inertial range would matter (see Sect. 5.8), but that the ratio of the Taylor microscale λ_T to the Kolmogorov microscale η_k would not be important. Thus, we would expect that systems having any value of Re above about 10^4 will behave similarly.

Once we have identified an astrophysical system that can be described by the Euler equations, and have some notion of how to model this in the laboratory, we can address the question of the Ryutov scaling for hydrodynamics. The reader can find a more detailed discussion in Ryutov et al. (1999, 2000) and Remington et al. (2006). We are concerned with systems that obey the Euler equations, (2.1)–(2.3). Now consider the initial value problem for this set of equations. Let us present the initial spatial distributions of the density, pressure, and velocity in the following way:

$$\rho(t=0) = \rho^* f\left(\frac{r}{L^*}\right), \qquad p(t=0) = p^* g\left(\frac{r}{L^*}\right),$$

and

$$u(t=0) = u^* h\left(\frac{r}{L^*}\right), \tag{12.1}$$

respectively, where L^* is the characteristic spatial scale of the problem, and the other quantities marked by the asterisk denote a characteristic value of the corresponding parameter; the dimensionless functions (vectorial functions) f, g, and h have maximum absolute magnitude of order unity. They determine the spatial shape of the initial distribution. We note that there are four, dimensional parameters determining initial conditions: L^*, ρ^*, p^*, and u^*, not necessarily corresponding to the same location. Let us then introduce dimensionless variables (which we denote by the tilde) in the following way:

$$\tilde{r} = \frac{r}{L^*}, \qquad \tilde{t} = \frac{t}{L^*}\sqrt{\frac{p^*}{\rho^*}}, \qquad \tilde{\rho} = \frac{\rho}{\rho^*}, \qquad \tilde{p} = \frac{p}{p^*},$$

and

$$\tilde{u} = u\sqrt{\frac{\rho^*}{p^*}}. \tag{12.2}$$

When one expresses the set of (2.1)–(2.3) in terms of the dimensionless variables, one finds that the equations maintain their form, with all the quantities being replaced by their analogs bearing the tilde sign. The initial conditions presented in the dimensionless variables acquire the form

$$\tilde{\rho}(\tilde{t}=0) = f(\tilde{r}), \qquad \tilde{p}(\tilde{t}=0) = g(\tilde{r}),$$

and

$$\tilde{\boldsymbol{u}}(\tilde{t} = 0) = u^* \sqrt{\frac{\rho^*}{p^*}} \boldsymbol{h}(\tilde{\boldsymbol{r}}).$$
(12.3)

Now consider two different systems, say, an astrophysical system and a laboratory system. One sees that the dimensionless initial conditions for the two systems
are identical if the dimensionless functions f, g, and $\boldsymbol{h}$ are the same, and the single
dimensionless parameter, $u^* \sqrt{\rho^*/p^*}$, remains unchanged. In other words, provided
this parameter is invariant, and the initial states are geometrically similar (i.e.,
the functions f, g, and $\boldsymbol{h}$ are the same), one would have the same dimensionless
equations and the same dimensionless initial conditions for *any two hydrodynamical
systems*. This implies that the systems will evolve identically in a scaled sense.

In the papers of Ryutov et al., the parameter $u^* \sqrt{\rho^*/p^*}$ was named the Euler
number. Subsequently, some fluid dynamicists complained that there are lots of
Euler numbers and another one is not needed. The author here suggests that we
consider this number to be the *Ryutov number*, defined as

$$Ry = u^* \sqrt{\frac{\rho^*}{p^*}}.$$
(12.4)

We will add some additional Ryutov numbers below, when we consider more
complex physical systems. One aspect of Ry worth emphasizing is that while it
equal the Mach number in some cases, at least within a factor of $\sqrt{\gamma}$, it need not be.
The values of u, p, and ρ need not be taken at the same location but instead are best
chosen so that the maximum values of the shape functions are near 1. We will see
below that it can be feasible to produce laboratory systems that have the same value
of Ry as an astrophysical system of interest. And we will say that two systems have
good Ryutov scaling when this is true.

There is only one constraint on the four parameters determining evolution of the
system. For the second system, one can choose the scale length, L^*, arbitrarily. One
can then also choose arbitrarily two of the three parameters, p^*, ρ^*, and u^*. Then
by choosing the magnitude of the remaining parameter so as to maintain $u^* \sqrt{\rho^*/p^*}$
constant, one can obtain a system which behaves similarly to the first one. It is very
important that this Ryutov similarity covers not only smooth solutions of the Euler
equations, but also solutions containing shock waves or multiple shocks. The proof
can be found in Ryutov et al. (2000).

There is a special case, often present in astrophysical objects (such as SN
explosions) and in the corresponding laboratory experiments, that is much simpler.
Assume that there is a system with an arbitrarily distributed initial density, and with
some initial pressure profile and initial velocities of the order of the sound velocities
or less. Assume then that a planar (cylindrical, spherical) piston is moved into the
system with a velocity *much greater* than the initial sound velocity. Considering as
an example a spherical piston, we can describe its motion by the equation

$$r = L^* q_{\mathrm{p}}(t/\tau^*),$$ (12.5)

where τ^* is the characteristic time of the piston motion (the time within which it is displaced by the distance $\sim L^*$); the dimensionless function q_p (with subscript p standing for the piston) and its argument t/τ^* are of the order of unity. The initial density distribution will as before be $\rho(t = 0) = \rho^* f(r/L^*)$, with the function f being of the order of unity.

The strong shock propagating in front of the piston brings the plasma to a new state; the characteristic pressure in this new state is

$$p^* \sim \rho^* L^{*2}/\tau^{*2},$$ (12.6)

and the characteristic velocity is

$$u^* \sim L^*/\tau^*.$$ (12.7)

This state is essentially independent of the pressure and the velocity in front of the very strong shock (see Chap. 4).

If one takes a second system, with the scale factors τ^*, L^*, and ρ^* arbitrarily changed but with the function f in (12.3) and the function q_p in (12.5) remaining the same (i.e., initial density distributions are geometrically similar, as are temporal dependences of the piston position), the two systems will evolve identically in the limit that the shock is strong enough.

For example, if in system 1 characterized by scaling factors τ_1^*, L_1^*, and ρ_1^*, the density is $\rho_1(r, t)$ then the function f is $\rho_1(r, t = 0)/\rho_1^*$. The evolution of the density in system 2 will then be

$$\rho_2 = (\rho_2^*/\rho_1^*) \times \rho_1(rL_1^*/L_2^*, t\tau_1^*/\tau_2^*).$$ (12.8)

In this case (12.6) and (12.7) imply that $u^* \sqrt{\rho^*/p^*}$ has the same value in both systems, so there is no need to impose this as a separate constraint. The implication is that all the characteristic parameters (τ^*, L^*, and ρ^*) can be varied independently, and still the similarity exists. This very broad similarity can be extended to include the case where the piston surface deforms in an arbitrary fashion during the piston motion; to do that, one should just describe a piston by the general equation for a surface evolving with time: $F(r/L^*, t/\tau^*) = 0$.

To have perfectly scaled behavior, the shape functions f, g, and h would have to be identical in the two systems and *over all space*. This would guarantee that any dynamics of interest occurred to the same degree in both systems. But in practice, this is (probably) never possible. So we need to find a way to assess the relevance of an experiment to this specific dynamics of interest. This *specific scaling* has two elements. One generally needs to assure that one or more dimensionless parameters have values that place the experiment in the relevant regime. This can be, for example, the normalized spatial scale for radiative cooling in an astrophysical jet

(Sect. 12.5), or the energy confinement time in a plasma device (as in Connor and Taylor 1977). One also needs to assure that the differences between the two sets of shape functions do not interfere with the dynamics of interest. We will discuss the specific scaling for each example given below.

12.2   A Thorough Example: Interface Instabilities in Type II Supernovae

As an example of a well-scaled experiment, we will now discuss experiments to study interface instabilities in Type II *supernovae* in some detail. Overall, Supernovae (SNe) involve a very broad range of physical processes. (The term supernova is represented by SN, so the plural is SNe.) Their complete description requires the use of areas of science as disparate as particle physics and general relativity on the one side, and hydrodynamic stability and turbulence, on the other. Some aspects of SNe are amenable to study by experiments in the laboratory, while others are not.

A nice description of the SN phenomenology, as well as existing theories of their formation, can be found, e.g., in the book by Arnett (1996). Other papers of general interest include (Bethe 1990; Woosley et al. 2002; Woosley and Eastman 1997). SNe are believed to explode by two fundamental mechanisms: collapse of the core (in large stars) and thermonuclear explosion (in small stars below 8 solar masses at birth). The classification by types is based on spectra and is too involved to discuss here. The Type Ia SNe, which will not be our focus here, are predominantly thermonuclear. They occur when white dwarf stars, composed primarily of C and O, accumulate enough mass to overcome the pressure of their degenerate electrons. They then begin to gravitationally collapse, releasing enough energy to cause explosive fusion burning of their C and O. Our focus will be on the predominant core-collapse SNe of Type II.

12.2.1   The Astrophysical Context for Type II Supernovae

Any large-enough, pre-supernova star develops shells of material around an iron core. Iron accumulates in the core because Fe is the most stable nucleus (see Chap. 11). The inner shell is composed primarily of the elements Si and Ca, the second shell is primarily C and O, the third shell is mainly He, and the outer shell is mainly H. One such star became SN 1987A. Any such star is initially composed primarily of H. It develops the interior shells in succession as the gravitational pressure compresses and ignites the accumulating material that will form the next shell. This process stops with Fe, because the star cannot create further energy by converting the Fe to any other material. But once the star accumulates a

"Chandrasekhar mass" (1.4 solar masses) of Fe, the gravitational pressure on the core overcomes the degeneracy pressure of the electrons and the core collapses. The collapsed core forms a neutron star that might later be detected as a pulsar.

The collapse of the Fe core is accompanied by the generation of a short but very intense burst of neutrinos (carrying away some 99% of the released energy). Some of the neutrino energy is coupled to the remaining matter that did not collapse, primarily near the core. The kinetic energy of the exploding matter in a typical SN event is $\sim 10^{51}$ erg. This brief deposition of energy creates an almost classic point explosion case (except for the small hole in the middle). A blast wave (see Chap. 4) develops and propagates out through the star, blowing it apart. When the blast wave emerges from the star, this gives rise to the observed tremendous increase of luminosity. Only a few percent of the hydrodynamic energy is emitted as visible light. In more detail, all the elements of the description just given must be present, but they appear insufficient to explain the actual generation of the blast wave in models. It may be that some nonsymmetric motions, even perhaps involving the generation of a jet during the process of core collapse, are involved in the generation of the blast wave.

Figure 12.1 shows typical light curves for three types of SNe. One should remember that the light as detected by optical telescopes comes not from the core, where the energy release has occurred, but rather from a photosphere, to which the energy is transported by a complex combination of hydrodynamic flows and radiative transport. Linking the energy release in the SN core to a visible light curve is a substantial challenge. A correct description of the material opacities and of the transport of material and radiation is very important. In the present section we are concerned with the transport of material.

Among the broad array of problems related to SN explosions, we shall concentrate on the laboratory simulation of hydrodynamic phenomena in Type II SN explosions and specifically on the evolution of hydrodynamic instabilities at the interfaces in the star. We choose this topic because, on the one hand, multidimensional hydrodynamic effects are thought to be very important and, on the other hand, there already exist experiments of this type, related to the shock breakout

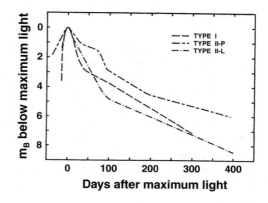

Fig. 12.1 Type I and II curves are normalized to their luminosity at the maximum (reproduced from Dogget and Branch (1985)). Note the presence of two sub-classes of SN Type II, with one of them (P) having a plateau in luminosity, with the other (L) showing a regular decay

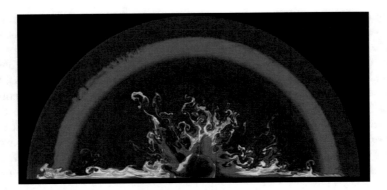

Fig. 12.2 2D simulation of SN 1987A, from Kifonidis et al. (2000)

through the He–H interface. In addition, the analysis of the scalability and other constraints can be nicely illustrated; this example can serve as a template for similar analyses of other problems.

Hydrodynamic instabilities arise as follows during a stellar explosion. At each interface between shells there is a significant density decrease. The interfaces are not smooth but are structured by convection, rotation, and other dynamics. The blast wave is likely to be born with structures resulting from neutrino convection (one model of this leads to the structure seen in Fig. 12.2). In addition, the blast wave is perturbed by the structure at the interfaces as it propagates outward. Even though the shock at the head of the blast wave will tend to anneal as described in Sect. 5.7.1, only small-enough perturbations will anneal completely. As a result, each structured interface will experience in its turn the passage of a structured blast wave. That is, the blast wave communicates between interfaces, so that the structure at inner interfaces produces additional structuring of the outer interfaces. During the deceleration phase that follows the blast wave, the structures at the interface will grow first through the Richtmyer–Meshkov process (Sect. 5.7.3) and then through the Rayleigh–Taylor instability (Sect. 5.1.2). All this had been understood for some time, but the ultimate nonlinear consequences were not clear before the advent of SN 1987A.

In observations of SN 1987A, emissions from the heavy elements, and other indications of their presence in the outer layers of the supernova, were observed only a few months after the explosion. The observed early appearance of heavy elements (like Ni^{56} and Co^{56}) in the photosphere of SN 1987A (see Sutherland 1990, and references therein) is incompatible with a spherically symmetric expansion and seemed to indicate that the instabilities have important and observable effects. Even so, simulations of the explosion in two dimensions (see Arnett et al. 1989; Fryxell et al. 1991; Kifonidis et al. 2000), which was all that was feasible computationally, did not produce rapid enough penetration to explain the observations. Partial simulation studies in three dimensions also indicated that the more-rapid penetration that would be found in three dimensions was not large enough to make up the difference.

In the context of our discussion above, in the introduction to this chapter, these circumstances created two roles for experiments. The first role is code validation. It was worthwhile to determine whether systems of this type behave in some way that existing simulations did not reproduce. A well-scaled experiment with two-dimensional structures would be sufficient for this purpose. We discuss such an experiment next. The second role is the direct observation of cases with realistic complexity. An important question, which cannot be answered by simulations in the current era, is to what extent in three dimensions the coupling between the interfaces in a diverging explosion actually manages to increase the outward penetration of the inner material. This is relevant not only to SN 1987A but also to other cases such as Cassiopeia A. Laser facilities at the MJ scale are required to carry out this second role.

The specific design of an experiment involves the specification of the structure of the target and the parameters of the laser (or other) drive, within the limits of available target-fabrication and laser-system technologies. This typically involves conceptual analysis followed by 1D hydrodynamic simulations to establish a viable approach. 2D or 3D simulations can then evaluate the effects of finite experiment size on the anticipated dynamics.

Figure 12.3 shows the schematic structure of an experiment relevant to the behavior of a single interface in an SN, and to instabilities driven at any decelerating interface. The perspective for the comparison is that the experiment will, for some period of time, evolve like a small patch, whose lateral dimension is a small fraction of its radius, at the interface of the exploding star. Section 4.6.4 discussed the essential, 1D dynamics. The energy input endures long enough to drive a shock wave roughly half way through the denser layer. Once the energy input ends, the rarefaction from the left edge of this material moves rightward and overtakes the shock wave. This forms a blast wave. The shock front moves into the lower-density matter and the interface between the two layers subsequently decelerates, creating a Rayleigh-Taylor-unstable condition. An alternate version of this system, somewhat more relevant to supernova remnants, inserts a gap between the two materials (Drake et al. 2000). The shock wave in the experiment converts material that is initially solid into plasma, enabling fluid equations to describe the dynamics. An actual experimental target is likely to include other structures, including a shock tube to limit lateral expansion, calibration features for spatial and spectral calibration, shields, and other necessary elements.

Fig. 12.3 Schematic of an experiment to study instabilities at decelerating interfaces, including those in supernovae, with some key dimensions d and densities ρ indicated

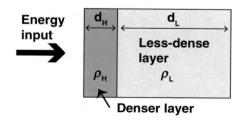

12.2.2   The Scaling Problem for Interface Instabilities in Supernovae

We now turn to the challenge of performing well-scaled experiments to simulate interface instabilities in SN 1987A, specifically at the He–H interface. This problem serves as an example for designing scaled laboratory experiments.

As a representative set of plasma parameters in the He–H transition region (Table 12.1), we consider the set of parameters given in Müller et al. (1991). These are of course parameters from computer simulations, as observations are not possible. However, they should be fairly reliable because of the simplicity of the fundamental blast wave problem. With reference to the table, let h be the density gradient scale-length, u be the characteristic velocity, T be the plasma temperature (the electron and ion temperatures are equal), and ρ be the plasma density. The characteristic deceleration experienced by a given fluid element at the He–H interface in the SN following arrival of the blast wave can be estimated as v/τ, with $\tau \sim h/v$, so $v/\tau \sim v^2/h \sim 5 \times 10^5\,\mathrm{cm/s^2}$. The gravitational acceleration is much smaller and is also negligible in the laboratory experiment to be discussed. Table 12.1 also shows parameters for the laboratory experiment. These are again based on simulations, although in this case several measurements support the detailed numbers given. From the parameters in Table 12.1, we can derive the scaling parameters given in Table 12.2.

Now we proceed to discuss the aspects of scaling described in Sect. 12.1. First consider the issue of physical consistency. One can show that fluid equations, with a polytropic equation of state, are valid for both these systems. The case of the experiment is straightforward—the plasma is quite collisional and radiation is not important. The SN is more complex. To be specific, we discuss properties of the He plasma. At $T = 800\,\mathrm{eV}$ it is fully ionized. Mean free paths with respect to electron–ion (λ_{ei}) and ion–ion (λ_{ii}) collisions are very short, $\lambda_{ei} \sim 10^{-3}\,\mathrm{cm}$ and $\lambda_{ii} \sim 2 \times 10^{-2}\,\mathrm{cm}$. The electron–ion energy equilibration time is less than $10^{-9}\,\mathrm{s}$. Therefore, the electron–ion component behaves as a strongly collisional gas with equal temperatures of electrons and ions. The particle pressure of a helium plasma

Table 12.1 Fundamental hydrodynamic parameters for a supernova experiment

Parameter	Supernova 1987A (2000 s)	Experiment (21 ns)
Length scale (cm)	9×10^{10}	0.0180
Velocity (km/s)	2000	35
Density (g/cm^3)	0.0075	0.4
Pressure (dynes/cm^2)	3.5×10^{13}	5.2×10^{11}
Temperature (eV)	900	7.4
Z_i	2.0	0.6
A	4.0	11.4
Density of nuclei (cm^{-3})	1.1×10^{21}	2.1×10^{22}

Table 12.2 Derived scaling parameters for a supernova experiment

Derived parameter	Supernova 1987A	Experiment
Collisional mfp (cm)	3.5×10^{-3}	7.9×10^{-8}
Kinematic viscosity (cm^2/s)	7.0×10^7	0.334
Reynolds number	2.6×10^{11}	1.9×10^5
Thermal diffusivity (cm^2/s)	1.2×10^6	15
Peclet number	1.5×10^{13}	4.2×10^3
Radiation mfp (cm)	6.8×10^2	2.0×10^{-4}
Radiation Peclet number	10^6	2.5×10^9
Ryutov number $u^*/\sqrt{p^*/\rho^*}$	2.9	3.1

for $\rho = 4 \times 10^{-3}$ g/cm^3, $T = 800$ eV is $p = 2.3 \times 10^{12}$ erg/cm$^3 = 2.3$ Mbar. The radiation pressure for $T = 800$ eV is $p_R = 2 \times 10^{13}$ erg/cm$^3 = 20$ Mbar. In other words, the radiation pressure dominates.

Despite the dominance of radiation pressure, fluid equations apply because the matter entrains the photons. The photon mean free path (mfp) with respect to Compton scattering (Thomson scattering at these low temperatures), ℓ_C, is very short, $\ell_C \sim 10^3$ cm (the mfp for inverse bremsstrahlung is much longer). In other words, the plasma containing the photon gas can be described as a single fluid, whose pressure is the sum of the photon and particle pressures, and which can be characterized by a single velocity of the mass flow u. The energy per unit volume in the case where the pressure is dominated by photons is $E_R = 3p_R$, thereby corresponding to the polytropic gas with $\gamma = 4/3$. Therefore, fluid equations, with a polytropic equation of state, are valid for both these systems.

To determine whether the Euler equations as such are valid for these systems, one must evaluate the dimensionless parameters discussed in Sect. 2.3 and consider their implications. Table 12.2 shows the values of the relevant parameters for the cases of interest. First consider viscous effects. Here all sources of viscosity must be added. Start with the SN. An ordinary (particle) viscosity (from 2.39), for a helium plasma with $T \sim 800$ eV, $\rho \sim 8 \times 10^{-3}$ cm^{-3}, and $\ln \Lambda = 10$, is ~ 2000 cm^2/s. The photon viscosity (from 2.40) is much larger than the ordinary viscosity, $\nu_{rad} \sim 7 \times 10^7$ cm^2/s. Accordingly, the Reynold's number, evaluated for $h \sim 10^{11}$ cm, $u \sim 2 \times 10^8$ cm/s, and $\nu = \nu_{rad}$, is 2.6×10^{11}.

Now consider the experiment. Assuming based on simulations that the average charge of the ions is ~ 0.6, and taking the Coulomb logarithm equal to 1, one finds (from 2.39) that the viscosity of CH$_2$ plasma is ~ 0.3 cm^2/s. Accordingly, the Reynold's number is $Re \sim 2 \times 10^5$. Although smaller than in the supernova, this is higher than a typical critical Reynold's number corresponding to the onset of the instability of the sheared flow ($Re \sim 10^3$) and the mixing transition ($Re \sim 10^4$). Therefore, it is clear that viscous effects are very small in both systems.

The Peclet number evaluated for thermal diffusivity (from 2.34) of the electrons (lines 5 and 6 in the table) is very large for the SN. The electron thermal diffusivity χ in the laser experiment is (from 2.34) ~ 15 cm^2/s, so that the Peclet number corresponding to the particulate heat transfer is high, ~ 4200. (Because the plasma in these experiments could be considered ideal only marginally, the aforementioned estimates of v and χ should be considered as order-of-magnitude estimates.)

Evaluating the magnitude of radiative effects is more complex. For the SN, the photon mean free path ℓ is much less than the characteristic length-scale h. An estimate (Zel'dovich and Razier 1966) for the thermal diffusivity χ of the photons is $\chi \sim c\ell$. For the case under consideration, where ℓ is the mean free path for Compton scattering, ℓ_c, one has $\chi \sim 2 \times 10^{13}$ cm^2/s. The corresponding radiation Peclet number is large, $\sim 10^6$, meaning that the SN is essentially adiabatic.

Turning to the role of radiative losses in the experiment, this subject is complicated by potential effects of optical depth. (The large value of the radiation Peclet number shown in the table implies that diffusive heat conduction by radiation is small but does not preclude large radiative losses from plasma that is not optically thick.) One can give an upper estimate for the effect of radiation as follows. The maximum possible energy loss from the surface of the plasma slab is that corresponding to the blackbody radiation at the plasma temperature, $2\sigma T^4$. On the other hand, plasma energy content per unit area of the plasma surface is $(3/2)h(n_e + n_i)k_B T$. Dividing the second by the first, one finds a lower-bound estimate for the characteristic plasma cooling time. Taking parameters of Table 12.2, we find that it is 1.8 μs, 90 times longer than the characteristic time of the hydrodynamic problem $(h/v \sim 2 \times 10^{-8}$ s$)$. In other words, radiative heat transport also does not affect the plasma dynamics. To conclude this topic: dissipative processes in the problem of the stability of the He–H interface are negligible, and the Euler equations are a legitimate description.

Having determined the validity of fluid models and the applicability of certain equations, the next step in the analysis of scaling is to assess the Ryutov scaling. Ry, also shown in Table 12.2, is nearly identical in the two systems (and could be made so by modest adjustment of the experimental densities). Figure 12.4 shows that the shape functions, along the axis of the experiment or radius of the star, are quite similar. If these were identical, and if the laboratory experiment were spherical, then the experiment would evolve identically to the star. We take up the lateral structure in our discussion of specific scaling, below. In 20 years of practice to date, experimental systems that produce well-matched values of Ry and similar 1D spatial profiles are described as "well-scaled". Here the author suggests being more precise and identifying such cases as having "good Ryutov scaling." In the context of Ryutov scaling, (12.2) implies that 21 ns in the laboratory setting corresponds to 2000 s in the supernova, and 20 μm corresponds to 10^{10} cm. It is then no surprise that, on these scales, the interface velocity profiles are very similar (see Fig. 12.5).

We turn now to the question of specific scaling. Taken literally, the Ryutov papers imply that the shape functions f, g, and $\boldsymbol{h}$ must be *identical* in the two systems in order to have perfect scaling. This is (probably) never true and so the scaling is (probably) never perfect. Addressing this issue is where clever and thoughtful work is needed for any experiment. In general, one can expect this to lead to some additional dimensionless parameters that constrain the experiment or limit its validity. A good guess is that one may have a constraint associated

Fig. 12.4 Experiment and supernova profiles from simulations. (**a**) He–H interface in SN1987A at 2000 s. (**b**) CHBr–Foam interface in experiment at 21 ns

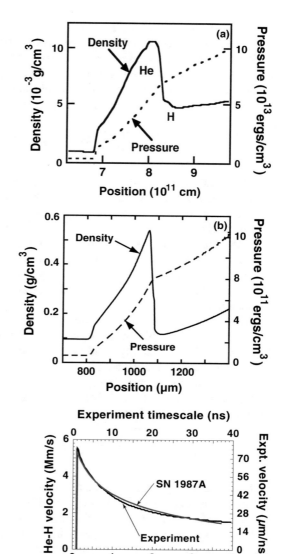

Fig. 12.5 Interface velocity in SN1987A and an experiment, from computer simulations

with each difference between the experiment and the astrophysical system, and also one or more associated with the dynamics of interest. We next discuss several considerations for the present case.

First, the region of interest in the experiment must not be affected by its lateral boundaries. This establishes a maximum experimental time, t_{max}, for which the physics is well-scaled, specified by the condition $t_{max} = R_x/c_x$, where the radius of the experiment is R_x and the sound speed of the shocked matter is c_x. One can write this requirement as a dimensionless parameter, $c_x t / R_x < 1$.

Second, one must assess the consequences of the fact that the stellar explosion is divergent while the experiment is not (prior to t_{max}). The corresponding constraint is that the divergence in the star must not be significant during the evolution time corresponding to t_{max}. The corresponding requirement is

$$\frac{\Delta R}{R_s} = \frac{t_s v_s}{R_s} \sim \frac{v_s}{\sqrt{p_s^* / \rho_s^*}} \frac{h_s}{R_s} \frac{R_x}{h_x} \sim 0.3 R_x \ll 1, \qquad (12.9)$$

where the subscripts s and x refer to the star and the experiment, respectively, the values of p, ρ, and v are from Table 12.2, R_s is the shock radius in the star (from Fig. 12.4), and in the numerical term R_x is in cm. In experiments to date, $R_x \sim 1$ mm, so this condition is well met.

A third difference between the two systems is that the shock wave in the experiment is smooth and has not been structured by crossing prior interfaces. What one can (and should) say about this is that this specific experiment can reveal single-interface dynamics but will not capture any dynamics that may arise from the interaction of multiple interfaces.

We now take up the issue of ensuring the same dynamics in the two systems. The evaluation of this for blast-wave-driven instabilities is as follows. In such systems, the growth of the instability is driven by the deceleration of the interface, almost all of which occurs during an the initial phase before the amount of mass in the shocked, lower-density matter equals the mass of the higher-density matter (Miles 2009). So we desire that t_{max} defined above is later than this time. If the propagation direction is z, this condition becomes

$$t_{max} > \frac{d_L}{v_s} = \frac{d_H}{v_s} \frac{\rho_H}{\rho_L}, \text{ or } R_x > d_H \frac{c_x}{v_s} \frac{\rho_H}{\rho_L} \sim 5 d_H, \qquad (12.10)$$

with reference to Fig. 12.3, experimental shock speed v_s, and the number 5 corresponding to the experimental values with $\rho_H \sim 20 \rho_L$. This condition is an over-estimate, as it ignores ablation of the denser layer. As $d_H \sim 150\,\mu m$, typical values of $R_x \sim 450\,\mu m$ are close to but not quite this large, and so such experiments might not quite capture the full development of the unstable structure before 2D effects become significant. Simulations of the experiments indicate that the experiments manage to squeak by.

Finally, we can consider the initial structure at the unstable interface itself. A limitation of this sort of experiment is that it can only study large-ℓ modes, for which the modulation wavelength λ is a small fraction of R_s. For those modes, the experiment with the best specific scaling would match the relative amplitude, normalized wavelengths, and spectral structure of these modes between the two systems. Knowledge of the structure in the star is limited. Simulations have found that convective effects tend to modulate the stellar interfaces, producing amplitudes of several % of λ or more (Meakin and Arnett 2006) and having a broadband spectral structure. This placed the experiments into a very typical tradeoff. They could do single-mode studies, which could best be measured, would be the most interpretable, and would represent the best test of computer simulations, or they could do studies using broad spectra, which might be most realistic but would also be harder to measure and understand. The actual experiments took a typical, and sensible, approach. They began with the simplest cases and added complexity over time.

12.2.3   Experiments on Interface Instabilities in Type II Supernovae

One of the first attempts to conduct experiments that were a well-scaled study of an astrophysical process (as opposed to measuring a property of astrophysical matter) occurred in the years around the turn of the twenty-first century. A team of researchers conducted the series of experiments whose design and scaling we just discussed, aimed at the problem of hydrodynamic instabilities at the H–He interface.

The first such experiments (Remington et al. 1997; Kane et al. 1997, 1999, 2000) used the Nova laser to examine the RT growth from a single-mode initial perturbation at a planar interface. The instability grew until the distance from the valleys to the peaks in the observed modulations (known as the bubble-to-spike distance) became equal to the initial wavelength of 200 μm. This is very nonlinear (see Sect. 5.7). Simulations, using the astrophysical code PROMETHEUS (Fryxell et al. 2000) and the laboratory code CALE (Barton 1985), reproduced this result, but the details of the structures did not strongly resemble what seemed to be present in the rather poor data obtained in this first attempt. Related experiments were also undertaken during the same period by a French group (Benuzzi-Mounaix et al. 2001).

Subsequent experiments (in a long sequence up to Kuranz et al. 2010) improved the quality of the data. The scaling parameters used in Tables 12.1 and 12.2 were taken from these experiments. Figure 12.4 shows the spatial profiles of SN1987A and these experiments, based on simulations. One can see that these profiles are similar, or in other words that the functions f and g are similar. (Figure 12.5 showed that the interface trajectory is similar. One could view this as showing the similarity of the generalized time-dependence for the strong-shock case or as implying the similarity of the function h, a scalar here.) One concludes from the tables and from these figures that the experiment is a well-scaled (though not perfectly scaled) model of a segment of the exploding star. This will remain true until the interface in the experiment discovers that it is not a small planar patch on a sphere, when the interface is affected by disturbances from the edges.

Figure 12.6 shows data from one such experiment. The image is an X-ray radiograph, taken at 18 ns in the experiment. Darker regions in the image show smaller X-ray intensity, produced primarily by absorption in the Br dopant that

Fig. 12.6 Radiographic data
from a supernova simulation
experiment, at 18 ns

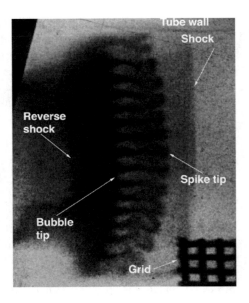

was included in a tracer layer within the denser material. In this case, a single
mode with an initial wavelength of $50\,\mu m$ and an initial peak-to-valley amplitude
of $5\,\mu m$ has grown to a very large amplitude. The tips of the spikes in the image
have been broadened in response to the shear flow and will soon begin to interact
(Miles et al. 2005). The experiments just mentioned set the stage for experiments
that proceeded to take up the challenge developed above—the study of how the
instability would develop in well-scaled experiments that employed more-complex,
three-dimensional, and ultimately realistic initial conditions.

Related experiments have explored the effect of coupling between interfaces,
discussed in Sect. 12.2.1, on the RT instability. These are worth mentioning because
they became the explicit focus of an extensive code validation study. The experiment
to examine this (Kane et al. 2001) produced data that were shown in Fig. 1.5. One
can see the Cu spikes, extending to the right, and the modulations in the second
interface, made visible by the tracer strip in the plastic below the interface. Detailed
simulations of this experiment were carried out as part of a validation study (Calder
et al. 2002) for the astrophysical code FLASH, which included adaptive grids and
other advanced features. A detailed comparison of the simulation results with the
experimental images supported the same conclusion as in the previous case. The
simulations reproduced the qualitative features of the data very well. Quantitatively,
several details were not accurately reproduced, including the exact spike length,
the height of the structuring in the interface, and the behavior at the edges of the
system. In the specific case of the FLASH simulations, the length of the Cu spikes
was found to change with the number of levels of refinement in the simulation but
did not appear to be converging toward the experimental value.

The completes our thorough discussion of this first case. We proceed to a second case in the realm of hydrodynamics and then on to other cases, albeit with a less thorough discussion. The reader is cautioned, however, that the kind of thorough analysis just presented must be carried through to understand the value of any attempt to study astrophysically relevant dynamics in the laboratory.

12.3   A Second Example: Cloud-Crushing Interactions

The previous section discussed a case in which experiments produced a system that had excellent Ryutov scaling and adequate specific scaling to address its question of interest. We also saw in Sect. 12.1 that producing the correct density structure and very strong shocks would be sufficient to obtain good Ryutov scaling in many hydrodynamic systems, but leaves open questions of the specific scaling. We proceed here to examine the second case of cloud crushing, in which blast waves crush and destroy clumps of material. This is common in astrophysics. Blast waves, generally produced by supernovae, propagate through an interstellar medium that is inherently very clumpy. One would like to observe the resulting destruction of the clumps in laboratory experiments and to develop and test the ability to accurately simulate this destruction. In the process, one would like to identify whatever regimes exist, such as regimes in which the dynamics of a given clump may be affected by the presence of other clumps. One would also hope to identify whether any processes develop in a scaled laboratory system that cannot be produced in a computer simulation, such as the onset of turbulence.

In the case we will consider, the experiment and the astrophysical system are both hydrodynamic and the shock wave that induces the destruction of the clump is definitely a strong shock. As a result, any system in which a blast wave encounters a dense clump of some specific shape will have good Ryutov scaling relative to any other such system, subject to the other limitations discussed in Sect. 12.1. Even so, the specific scaling includes several considerations. There is a geometric constraint that the radius of the experimental medium (see Fig. 12.7) must be large enough that

Fig. 12.7 Schematic of experiment on interaction of blast wave with clump

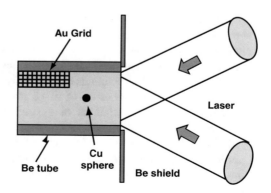

disturbances from its boundaries to not interfere with the behavior of interest. The analysis of this is similar to that discussed above for the previous case.

With regard to the scaling of the dynamics of interest, one can identify two other parameters that characterize the interaction of a blast wave and a clump in a given experiment. Suppose the blast wave propagates at a given velocity u_s through a medium having a density ρ_m and interacts with a clump having a characteristic density ρ_{cl} and radius r_{cl} (and presumed to be spherical). The most-important scaling parameter describing this interaction is the cloud crushing time, defined as $t_{cc} = (r_{cl}/u_s)\sqrt{\rho_{cl}/\rho_m}$. The corresponding dimensionless parameter in the experiment is the ratio of the duration of the experiment (as a well-scaled system) to t_{cc}. This must be large enough to see the dynamics of interest. The second scaling parameter that describes this interaction is the ratio of the width of the blast wave to r_{cl}. To model a specific astrophysical system in detail, one would also have to match this parameter to the astrophysical value. Unfortunately, this issue has been ignored in all work to date at this writing. Yet it is not obviously unimportant.

Figure 12.7 is a sketch of the geometry used in experiments to address the destruction of a single clump by a planar blast wave (Klein et al. 2003; Robey 2002). A number of laser beams irradiated a layer of plastic material, driving a shock into it. After the laser pulses ended, the rarefaction of the front surface overtook this shock, creating a planar blast wave just as described in the previous section. This blast wave eventually encountered a dense (Cu) sphere, whose evolution was observed for several cloud-crushing times. In a fundamental sense, this experiment was not as well scaled as the one described in Sect. 12.2, because the Cu sphere was liquefied but not vaporized by the shock and thus had an equation of state rather different from that of an astrophysical cloud. Later, improved experiments (Hansen et al. 2007a,b) used lower-density materials that could be vaporized by the shock. As a result, this specific experiment is of the type in which the experiment is not completely well scaled but certain key dimensionless parameters are well scaled, so that it can be instructive regarding the dynamics and perhaps useful for code validation. This initial experiment is a relevant model of the incompressible fluid dynamics of cloud destruction.

These experiments observed the evolution of the sphere for several cloud-crushing times. The interaction of the sphere with the postshock flowing plasma produces vortex rings (see Sect. 5.8). This is illustrated in Fig. 12.8. The fluid develops spiral flows around the vortex rings, and may also have shear flow (and hence vorticity) on the surfaces of the spirals. The vortex rings are subject to bending instabilities that produce three-dimensional structure by modulating the rings in the azimuthal direction as described in Widnall and Sullivan (1973, 1974). The development and properties of these structures can be examined in the data and in astrophysical simulations of a similar system. Thus one can use these experiments to test the ability of astrophysical codes to simulate this type of shock–cloud interaction. The results of these experiments were later used by a group of astrophysicists (Hwang et al. 2005) to evaluate the morphological phase of a clump being destroyed by the Puppis A supernova remnant, which they observed with the Chandrasekhar X-ray telescope.

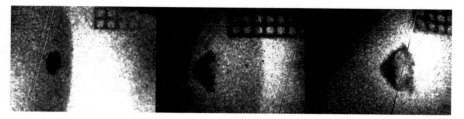

Fig. 12.8 Data from an experiment on interaction of blast wave with clump. These three frames show what has become of the spherical obstacle after the passage of the shock, seen in the first two frames but beyond the region observed in the third frame. One can see the crushing of the sphere followed by the development of vortex rings. The squares in the reference grids are 63 μm on a side. From Robey (2002)

Thus, well-scaled experiments can address significant issues in astrophysical systems that are purely hydrodynamic. The above examples have detailed one case in which initial experiments were useful for code validation and in which eventual experiments will address physical questions that cannot be addressed in simulations of the complete astrophysical dynamics. In addition, the process of doing such experiments initiated a productive interplay of astrophysical data, laboratory experiment, and computer simulation whose ultimate outcome, as of this writing, remains to be seen.

12.4   Scaling in Radiation Hydrodynamic Systems

Radiation hydrodynamic systems are challenging for both theory and experiment. We saw in Sect. 8.1 how the addition of radiation fluxes to shock waves greatly complicates their behavior, even without considering the details of actual radiation emission and transport. We mentioned in Chaps. 7 and 8 a number of examples of astrophysical systems that are radiation hydrodynamic systems. In comparison to hydrodynamic systems, radiation hydrodynamics introduces new difficulties in scaling, in simulation code development, and in experiments. We will consider here only the radiative-flux regime, leaving the scaling of systems having significant p_R to be developed when such experiments exist.

At minimum, such experiments ought to be able to provide benchmarks for the implementation of radiation hydrodynamics in astrophysical codes. Beyond that, one might hope to identify specific processes that matter in astrophysics, that could be produced in the lab, and that were difficult or impossible to simulate with computers. In this section we discuss the issues associated with scaling. While complete scaling is possible in the abstract and perhaps in special cases, it seems more likely that actual experiments will scale the important dimensionless parameters well but will not manage to successfully scale all aspects of the radiating system. In this section we discuss radiation hydrodynamic scaling in general and then discuss radiative jets. We discussed radiative shocks in Chap. 8. For them, the key dimensionless parameters were a shock strength parameter and the optical depth both upstream and downstream of the shock front.

For scaling, one first must consider the two aspects of physical consistency. For the conditions we will consider, fluid equations must apply. This requires that the radiation transport be local. This implies that the system of interest is either optically thin so that radiation is a pure loss term or optically thick so that a diffusion model applies. In these two limits, it is not critical whether the radiation is composed primarily of thermal emission or line radiation. The second aspect of physical consistency—that the same terms matter in the equations—requires here that only hydrodynamics plus radiation are important. This again demands the Re and Rm be large and β not be small. In this context, the discussion of Ryutov et al. (2001) and Cross et al. (2014) are relevant.

Next we must consider how Ryutov scaling applies here. (Our discussion is informed by those of Ryutov et al. (2001) and Falize et al. (2009).) In the radiative-flux regime, the first two Euler equations ((2.1) and (2.2)) remain unchanged. As a result, $Ry = u^* \sqrt{\rho^*/p^*}$ must be equal in the two systems. But now this is not sufficient. When radiative energy fluxes matter, a radiative-heating term must be added to the third Euler equation (2.3), which becomes

$$\frac{\partial p}{\partial t} + \boldsymbol{u} \cdot \nabla p = -\gamma p \nabla \cdot \boldsymbol{u} - (\gamma - 1)\nabla \cdot \boldsymbol{F}_{\mathrm{R}}, \qquad (12.11)$$

in which as usual $\boldsymbol{F}_R$ is the radiation flux.

In the abstract, one may be able to express $\nabla \cdot \boldsymbol{F}_R$ as a power law function of density, pressure, and scale length in some astrophysical system, so that

$$\nabla \cdot \boldsymbol{F}_R = A\rho^{\alpha_1} p^{\alpha_2} r^{\alpha_3}. \qquad (12.12)$$

For an optically thin system, α_3 would be zero, while for an optically thick system it might not be. This case will serve our purposes here, although even more general cases are possible. In principle $\nabla \cdot \boldsymbol{F}_R$ might be a somewhat more complicated function of vectorial position $\boldsymbol{r}$ and might also include a dependence on velocity.

If we now perform the variable transformation described in Sect. 12.1, we obtain

$$\frac{\partial \tilde{p}}{\partial \tilde{t}} + \tilde{\boldsymbol{u}} \cdot \nabla \tilde{p} = -\gamma \tilde{p} \nabla \cdot \tilde{\boldsymbol{u}} - (\gamma - 1) \left[A\rho^{*(\alpha_1 + 1/2)} p^{*(\alpha_2 - 3/2)} L^{*\alpha_3} \right] \tilde{\rho}^{\alpha_1} \tilde{p}^{\alpha_2} \tilde{r}^{\alpha_3}. \qquad (12.13)$$

This equation will be invariant between any two systems, such as an astrophysical system and a laboratory system, if the coefficients α_1, α_2, and α_3 and the quantity in square brackets are the same in both systems. We designate this quantity the *Ryutov radiation number*, Ry_R,

$$Ry_R = A\rho^{*(\alpha_1 + 1/2)} p^{*(\alpha_2 - 3/2)} L^{*\alpha_3}. \qquad (12.14)$$

It will also be helpful below to note that, defining F_R^*/L^* as a characteristic value of $\nabla \cdot \boldsymbol{F}_R$, we have

$$Ry_R = \frac{F_R^*}{p^* \sqrt{p^*/\rho^*}} \tilde{\rho}^{-\alpha_1} \tilde{p}^{-\alpha_2} \tilde{r}^{-\alpha_3} \sim \frac{F_R^*}{p^* \sqrt{p^*/\rho^*}}, \tag{12.15}$$

with the last approximation reflecting the fact that the magnitude of the dimensionless variables is near unity.

In practice the values of ρ^* and L^* (if applicable) will be fairly tightly constrained in an experiment, so the primary adjustment that could be used to obtain a well-scaled experiment would be in increasing p^* to make Ry_R equal in the experiment to its value in the astrophysical system and to adjust the composition of the experimental materials to make the coefficients equal.

As one might expect, one loses some freedom in specifying the parameters by comparison with the purely hydrodynamic case. But all that matters to have a well-scaled experiment is the net cooling rate and its dependences on pressure and density. The microscopic mechanisms are not important. This is significant because the radiation cooling in optically thin astrophysical systems is nearly always due to line radiation while the cooling in high-energy-density experiments is more often dominated by thermal radiation. In an arbitrary experiment modeling some astrophysical system, one seems unlikely to obtain the same, power-law scaling for the radiation in both systems in the sense just described. However, it may happen, and Tikhonchuk et al. (2008) argue that their experiment achieves good Ryutov scaling in this sense. We will discuss this case below when we look more closely at radiative jets.

Beyond attempts to do scaling as described above, we have seen in Chap. 7 that radiation transport is often nonlocal. In such cases, one probably cannot produce an experiment with good Ryutov scaling to a specific astrophysical system. Even so, the behavior of the astrophysical system may depend primarily on certain dimensionless parameters that reflect the processes which control its dynamics. Then one may be able to observe phenomena in the laboratory with the same values of these essential dimensionless parameters. In the case of radiative shocks, we saw in Chap. 8 that the controlling parameters are the optical depth of the upstream and downstream regions. We consider the case of radiative jets shortly.

12.4.1 Perils of the Boltzmann Number: A Detailed Example of a General Point

We use this example to show how the process of determining which terms can be ignored in certain equations has some pitfalls. One often seeks a measure of the relative importance of radiative cooling in the hydrodynamic evolution of some system. One way to do this is to non-dimensionalize the equations as we discussed in Chap. 2. If one divides any energy equation, such as (2.28) or (12.11), by $\rho U^3/L$, one obtains a set of terms that show how important each process is *relative to convective energy transfer*. This helps one assess whether one can use the simpler Euler equations for modeling, but may not tell one what one needs to know about the relative importance of a specific processes in some system of interest. This can be misleading in the case of radiative energy fluxes.

If the fluxes are not dominated by optically thin line radiation, then one always has

$$\nabla \cdot F_R \propto \sigma T_e^4, \tag{12.16}$$

with various other coefficients, such as the optical depth for optically thin systems or $1/(\chi_R L_T)$, with $1/L_T = |\nabla T_e|/T_e$, for optically thick systems. (As we saw in Chap. 6, it is T_e that matters in optically thin cases and $T_e = T_R$ in optically thick cases.) The non-dimensional term is thus proportional to

$$\frac{\sigma T_e^4}{\rho U^3} \sim \frac{\sigma T_e^4}{pU} \sim \frac{1}{Bo}, \tag{12.17}$$

taking $p \sim \rho U^2$ and ignoring factors of order unity in the standard definition of the Boltzmann number Bo as $\rho c_p T U/(\sigma T^4)$. Cross et al. (2014) introduce radiation numbers, also proportional to Bo, to characterize the optically thick and thin cases. Another standard number, the Mihalas number, is $\propto cBo/U$ and describes the ratio of material pressure p to radiation pressure p_R. It can be useful, and relevant to the momentum equation, in the radiation-dominated regime, when $p_R \gtrsim p$. The issue discussed next applies to all these radiation-related dimensionless numbers.

In terms of its intended purpose, to determine whether one can ignore the radiative-flux term in the energy equation, Bo works well. When Bo is very large, radiative effects are small compared to convective energy transport and this term in the equation can be ignored. However, the converse is not true. Radiative effects may not matter much even when Bo is small, and the quantitative smallness of Bo does not necessarily reflect the relative importance of radiative effects. An example of the first point is the case when $U \sim 0$. Blind evaluation of Bo would indicate that radiative effects are important, but they might instead be negligible in comparison with other energy transport processes. In this case, a thoughtful evaluation might replace U with the sound speed for low-speed flows. An example of the second point is the case of radiative shocks. One might imagine that Bo would become very small in strongly radiative shocks, where nearly all the incoming energy flux is converted to radiation. However, in such shocks the ram pressure sustains $p \sim \rho U^2$ so that in fact Bo remains of order 1 and little is learned from its precise value. In these cases, as we saw in Chap. 8, the useful dimensionless parameters depend upon the ratio of the radiative flux that would exist, in the absence of radiative effects on the shock structure, to the incoming mechanical-energy flux.

The point here is that, *in every application*, one must use dimensionless numbers thoughtfully. Just turning the crank on their evaluation, without thinking about the system under study and the processes of interest, can easily produce misleading conclusions. We take up this general point again below, in Sect. 12.7. Before that we discuss radiative astrophysical jets, which illustrate various points discussed above.

12.5   Radiative Astrophysical Jets: Context and Scaling

Many astrophysical jets are purely hydrodynamic, and so the discussion of Sect. 12.3 would apply to simulation experiments aimed at them. Other astrophysical jets are inherently magnetized or involve strong magnetic fields. We discuss these below in Sect. 12.6.2. Our present interest, in the context of radiation hydrodynamics, is in radiative jets. We proceed to discuss these here.

12.5.1   The Context for Jets in Astrophysics

Jets have been a major theme during the first few decades of laboratory astrophysics research using high-energy-density facilities. This is no surprise, given their ubiquity across astrophysics. Galactic and extragalactic jets present some of the most visually intriguing images encountered in astrophysics. One class of such objects are the stellar jets known as Herbig–Haro (HH) objects (Reipurth and Bally 2001), thought to be collimated bipolar outflows emerging from accretion disks during the star formation process. Figure 12.9 shows an image of one such jet, HH 34. The jet shown emanates, at velocities of ~300 km/s, from the pole of a protostar near the bottom of the image. The protostar itself is hidden; one sees reflections through the dust surrounding it. Like other HH jets (Hartigan et al. 2000), HH34 includes

Fig. 12.9 Image of the HH 34 jet. This image shows the hydrogen H_α emission. The protostar is near the left in the center. From Reipurth et al. (2002)

multiple bow shocks, suggesting that the bipolar outflow has been episodic. It is typical that the fastest moving material (at $\sim$200 km/s) occurs on the axis, with slower yet higher luminosity material concentrated at the edges of the jet (Hartigan et al. 1993). This may reflect differences in the launch velocity of the inner and outer material, or perhaps entrainment of ambient material due to the Kelvin–Helmholtz (KH) instability along the edges of the jet, leading to a greater shock excitation but slower velocities at the edges. The HH jets have typical scales of 10^{17} cm, velocities of a few hundred km/s and densities n_{jet} of 10^2–10^3 cm^{-3}. In terms of density contrast, this corresponds to $\eta = n_{jet}/n_{ambient} \gg 1$, where $n_{ambient}$ is the ambient number density. The bow shocks in such jets are often radiative shocks. The internal shocks also tend to be radiative, and to have cooling distances of tens of AU.

Whereas HH jets are thought to be emitted during the formation phase of a star, another category of jet is formed toward the end of the evolution cycle. A star of a certain mass can pass into the asymptotic giant branch (AGB) phase and then to the planetary nebula or proto-planetary nebula (PPN) phase on its way to becoming a white dwarf. During the AGB-to-PPN transition, it appears that bipolar jets can again be emitted, one example being He 3-1475 (Borkowski et al. 1997). The central source for this system is a star at a distance of $\sim$2 kpc, which is in the midst of making the transition from a dust-enshrouded AGB star to a PPN. The star is surrounded by a torus of molecular material expanding at 12 km/s. The most spectacular features of He 3-1475 are the optical jets, and three pairs of symmetric knots, moving in the direction perpendicular to the molecular torus. The knots are located symmetrically with respect to the central star and are moving radially away at velocities of 500 km/s. Closer to the star, the jetlike outflows have velocities as high as 850 km/s. Radiative shocks moving at velocities of $\sim$100 km/s are thought to be the excitation mechanism for the observed emissions.

The basic features of a high-Mach-number jet, present for both radiative and purely hydrodynamic jets, are illustrated schematically in Fig. 12.10 (Hartigan 1989). A source is assumed to exist that creates a beam or jet of material (labeled 1) streaming into the ambient medium (labeled 2). This launches a forward or bow

Fig. 12.10 Schematic of the structure of an astrophysical jet, from Hartigan (1989)

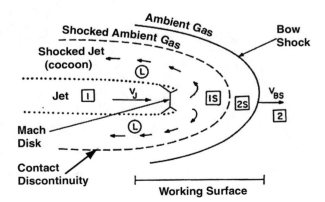

shock into the ambient medium, moving at speed v_{BS}. The presence of the ambient medium causes the jet material to slow down, creating a Mach disk in the jet. Within this description, beam or jet refers to the collimated material streaming from the source to the so-called "Mach disk." Shear along the sides of the jet triggers the Kelvin–Helmholtz (KH) instability, which generates vortices and eddies that churn up mixing along the contact discontinuity. The region of shocked jet material between the contact discontinuity and jet (beam) is referred to as the cocoon. The KH vortices in the cocoon can launch shocks into the jet (beam), which act as a heating mechanism for radiative emissions from within the jet.

12.5.2 Scaling from Radiative Astrophysical Jets to the Laboratory

The astrophysical literature identifies three dimensionless parameters that characterize the properties of a jet and the degree to which radiation is important to its dynamics. The *internal Mach number* (the ratio of the flow velocity to the sound speed within the jet), M_{int}, characterizes the amount of kinetic energy that can potentially be converted to thermal energy through shocks. The *cooling parameter*, χ_j, is defined by $\chi_j = L_{cool}/R_{jet}$, where R_{jet} is the jet radius and L_{cool} corresponds to the length behind the Mach disk beyond which the jet has cooled to some low value. This characterizes the relative scale on which radiation can alter the properties of the matter in the jet. The *density parameter* is η as defined above. It affects the amount of pressure in the shocked ambient medium, which interacts both with the head of the jet in the Mach disk region and with the jet along its length *via* the pressure in the cocoon.

It should not be a surprise that these three parameters connect well with the analysis of scaling above. An experiment with good Ryutov scaling would have Ry and Ry_R be the same in the laboratory experiment and a reference astrophysical jet. We have noted above that Ry has a strong connection to Mach numbers. It can be one although it need not be. Regarding radiation, a natural definition of a cooling time is $\tau_{cool} = (\gamma - 1)2\pi R_{jet}F_R^*/(\pi R_{jet}^2 p^*)$ if we define $c^* = \sqrt{p^*/\rho^*}$, then it is natural to take

$$\frac{L_{cool}}{R_{jet}} = \frac{c^* \tau_{cool}}{R_{jet}} \sim \frac{p^* c^*}{F_R^*} = \frac{1}{Ry_R}. \tag{12.18}$$

The dimensionless parameter η is connected with what we have called the specific scaling, as it identifies which feature of the shape functions is important to the jet dynamics.

The effects of radiative cooling on astrophysical jets can be very large (Blondin et al. 1990; Stone and Norman 1994). Figure 12.11 shows results of simulations that assume a very high Mach number jet, having $M_{int} = 20$, and an equal density with the ambient medium, so $\eta = 1$. The plasma is assumed to be optically thin,

Fig. 12.11 Impact of
radiative cooling on jet
structure, for Mach 20 jets
whose density equals the
initial ambient density. From
Blondin et al. (1990)

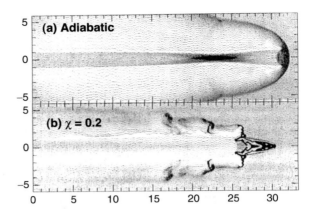

so that photons emitted by hot regions of the jet leave the system. The effects
of radiation are included by means of a time-independent cooling function $\Lambda(T)$,
assuming equilibrium conditions (see Sect. 6.3.2 and Fig. 6.5). Here χ_j is calculated
for cooling to a temperature of 8×10^3 °K. As the magnitude of radiative cooling is
increased from a purely adiabatic jet, $\chi_j \gg 1$ (Fig. 12.11a), to a strongly cooled
jet, $\chi_j = 0.2$ (Fig. 12.11b), the jet morphology changes significantly. Radiative
cooling removes heat from the system, lowering the internal pressure of the jet.
The working surface and the cooling zone behind the shock collapse, just as we
discussed in Chap. 8 for radiative shocks. They contract until pressure equilibrium
is reestablished, making the radiatively cooled zones denser and more compact. In
a more detailed description, radiation should be treated as a nonequilibrium, time-
dependent process, and if the medium is not optically thin, full transport (nonlocally
redepositing the photon energy) may be needed.

For simplicity, and because it has been a focus of experiments, the discussion
here focuses on unmagnetized radiative jets. These, however, may be rare in
astrophysics. If the jet plasma contains magnetic field, then the collapse of the
cooling zone will lead to an increase in magnetic pressure, which will limit the
amount of collapse. Astrophysical observations suggest that this is often significant,
creating a potential direction for experiments beyond those discussed next.

12.5.3 Radiative Jet Experiments

High-energy-density experiments offer the means to create high-Mach-number jets
and to diagnose their subsequent dynamics and evolution (Stone et al. 2000). For
example, the effect on jet dynamics of variations in η, M_{int}, and χ_j can in principle
be directly observed in laboratory experiments. Here we discuss as an example the
first experiment to produce a radiative jet, done on the Nova laser at LLNL (Farley
et al. 1999; Stone et al. 2000). A schematic of the experimental arrangement is

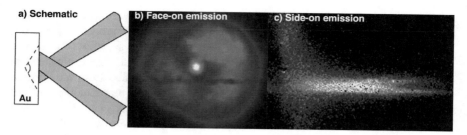

Fig. 12.12 A radiative-jet experiment. Schematic (**a**) and images of self-emission ((**b**) axial and (**c**) side-on) from an experiment producing a radiatively collapsing Au jet. Adapted from Farley et al. (1999)

shown in Fig. 12.12a. A gold disk had a 800 μm diameter cone of 120° full opening angle machined into it. Five 100 ps laser beams from the Nova laser irradiated the inside surface of this cone were symmetric (in azimuth), with an average energy flux of $I_L \approx 3 \times 10^{15}$ W/cm^2. The high-speed, ablated Au plasma thus created expands in a direction normal to the local surface. The radial velocity component causes this plasma to implode and stagnate on the axis of the cone. The axial velocity component brings the imploded plasma out into view as a high-speed jet. Radiation from the hot, stagnated plasma causes the jet to shrink in size and increase in density.

X-ray imaging diagnosed this experiment. One gated X-ray framing camera looked directly face-on at the cone. This is illustrated in Fig. 12.12b with the face-on X-ray image in emission at $t = 0.25$ ns relative to the peak of the Gaussian laser drive. The small bright spot near the center of the image is the imploded Au plasma that has stagnated on the axis, and that is moving at ~750 km/s out of the page, directly toward the recording X-ray camera. Views of this same jet from the side in soft X-ray emission at 1.1 ns (Fig. 12.12c) show that the radiative emissions later in time are on the surface of the Au jet. The reason for the forked nature of the emissions in the side-on image is that the regions of the imploding plasma that stagnate first on the axis (at ~0.5 ns) radiatively cool at first, and appear dark (cool) later in time. By 1.1 ns, this leading tip region stops emitting in the soft X-ray band for which the instrument is sensitive, because the electron temperature T_e has dropped dramatically. A side-on radiograph found the densest part of the jet to be along the cone axis. Simulations of this experiment illustrate the importance of radiation. As the plasma collides on the axis, it heats up to over ~1 keV and has a high ionization state, $Z \sim 40$, but a low density, $\rho \sim 40$ mg/cm^3. In this state, the hot Au plasma cools itself rapidly by radiative emissions, since the plasma is initially optically thin to the keV photons. The temperature was measured by Thomson scattering at 0.6 ns, at which time T_e had already dropped to 250 eV. The radiative cooling leaves a very compact, highly collimated jet moving axially away from the cone at ~750 km/s. In simulations that do not include radiative cooling, the jet is an order of magnitude too broad, since the pressure of the stagnated Au plasma is high, which would stop the implosion (Mizuta et al. 2002).

Researchers using the Gekko-12 laser at the University of Osaka in Japan (Shigemori et al. 2000) continued these experiments. They produced jets from cones of Au, Fe, Al, and CH. The results show a clear correlation: the higher the Z, the shorter the cooling time and the narrower the jet. The experiment at Gekko generating a Au jet reproduced very closely the results from the original experiment, showing that the physics being investigated is reproducible and not facility dependent. In a complementary experiment, Lebedev et al. (2002) used plasma expanding from a conical array of wires to drive radiatively cooled, magnetized, high-Mach-number jets. They did these experiments at the Magpie Z-pinch at Imperial College in London. They observed similar trends to those just described, and continued to examine the collision of this jet with an obstructing object.

These early experiments (on Nova and on Gekko) are a good example of a first attempt to produce a radiation hydrodynamic system that is relevant to astrophysics. However, they are incomplete, because they do not include an ambient medium (thus, $\eta \sim \infty$). In addition, the formation mechanism does not manage to produce an emerging source of material in a consistent initial state. Rather, the first part of the jet to form (that nearest the target) is the trailing portion of the jet that emerges. Each part of this type of jet has a unique history of energy input and cooling dynamics. So if one were seeking a well-scaled experiment that was a direct analog of any astrophysical jet, these first experiments would not achieve this goal. However, the experimenters did manage to vary the radiative cooling parameter χ_j over the range of 0.7–40 and to vary the internal Mach numbers from 2 to 50. Given the paucity of radiation hydrodynamic experimental data, these experiments are in fact of real value for the validation of astrophysical codes.

Some later experiments, using the PALS laser in Prague, produced results that were more complete in some respects. The researchers irradiated a flat surface, with a ring-shaped laser beam, causing the flow inward to converge and form a central jet. They did this within a background gas so that there was an ambient medium. By tilting the target, they proved able to form a jet that moved out of the region of gas affected by the laser beam. This let them scale η from $\sim$0.1 to $\sim$10. In Tikhonchuk et al. (2008), they closely examined the Ryutov scaling and concluded that it was good, even with regard to the values of the α's. This is an impressive result. Unfortunately, despite the control of η, a ring beam also produces a plasma flow in other directions, and this produced a curved shock wave in the ambient gas. This likely prevented them from examining the full range of basic jet behavior. In terms of the analysis above, we would say that the specific scaling was insufficient to allow this. In the future, there is clearly the potential for someone to take radiative jet experiments another step forward.

12.6 Scaling for MHD Systems

To be able to evaluate experiments using magnetized, MHD flows, we need to understand scaling for this case. Here we will ignore radiation, but it enters simply into the energy equation, and so, when relevant, brings requirements that add to those of hydrodynamics and magnetization. As always, we need to begin with the question of validity. For magnetized flows, the primary additional requirement for the MHD equations to be applicable is that the plasma be well-magnetized. This condition also is necessary to justify the application of pure hydrodynamic equations to magnetized plasmas having small current flow. This requirement is sometimes expressed in terms of a magnetization parameter, the ratio of system scale, L, to ion orbit radius, r_{Li}. This parameter, L/r_{Li}, must be large for the MHD equations to be valid. In addition, the requirements for hydrodynamic fluid equations to be valid also apply here.

12.6.1 Validity Considerations and Ryutov Scaling for MHD

To assess the validity of the MHD equations, and which terms in them matter, we need to extend the global scaling arguments of Chap. 2 to the specific context of MHD. Assuming MHD theory to apply, we can work with the momentum and induction equations to find the new, relevant dimensionless parameters. Begin with the momentum equation. If we add the MHD forcing terms from (10.25) to (2.27), we obtain

$$\rho \frac{Du}{Dt} = -\nabla p - \nabla \frac{|B|^2}{8\pi} + \frac{(B \cdot \nabla B)}{4\pi} + \nabla \cdot \underline{\sigma}_\nu + F_{\text{EM}} + F_{\text{other}}. \qquad (12.19)$$

This ignores radiative forcing and quantum effects, both small for our cases of interest. We non-dimensionalize this as before, to see when certain terms are negligible, by taking $\nabla \to 1/L$, $\partial t \to L/U$, and then dividing by $\rho U^2/L$. The $\nabla \cdot \underline{\sigma}_\nu$ term gives $1/Re$, the F_{EM} term gives one of the radiation numbers mentioned above, and the F_{other} term gives other dimensionless numbers (such as the Froude number for gravitational forces). We ignore these here.

Continuing the analysis, the left-hand side and the remaining terms on the right-hand side scale as

$$1, \frac{1}{M^2}, \text{ and } \frac{1}{\beta_{\text{ram}}} \text{ twice}, \qquad (12.20)$$

with M being the Mach number and β_{ram} being the ratio of ram pressure, ρU^2, to magnetic-field pressure, $B^2/(8\pi)$. In the literature of accretion flow, this quantity is known as the "ram β". In some other literature (including Li et al. 2016), this is called a "magnetization" parameter. (It does scale with the magnetization of the ions if they encounter a transverse field of magnitude B, but it is unclear whether the magnetization of ions, defined in this way, will be of specific interest.)

On the one hand, having these two parameters, M and β_{ram} is of some use, indicating when their terms in the momentum equation become unimportant. On the other hand, much greater importance often rests in the ratio of these two terms, which is the usual plasma $\beta = 8\pi p/B^2$. This factor determines whether the dynamic behavior is dominated by plasma pressure or by magnetic forces, or whether both may be significant. This is an example of the point made repeatedly here that there are many cases in which the most relevant dimensionless parameters may not be those found by the global analysis.

Next we examine the general induction equation. If we use (10.27) to determine E and evaluate (10.39) for the case that $\underline{\alpha}$ and P_e are scalars, using the definition of the Nernst velocity in (10.38), and ignoring the other, less-important, heat-transport terms, we obtain

$$\frac{\partial B}{\partial t} = \nabla \times \left[(u_e + V_N) \times B + \frac{c \nabla p_e}{e n_e} \right] + \frac{\eta c^2}{4\pi} \nabla^2 B. \tag{12.21}$$

Dividing this equation by UB/L, we find the relative sizes of the terms as

$$1, \ 1, \ \frac{1}{Ne}, \ \frac{1}{Bi}, \ \text{and} \ \frac{1}{Rm}, \tag{12.22}$$

with magnetic Reynolds number Rm discussed in Sect. 10.2.2. We follow Cross et al. (2014) by defining a Biermann number and Nernst number, though without their normalization of B, obtaining

$$Bi = \frac{eBUL}{ck_BT_e} = \frac{U}{v_{the}} \frac{L}{r_{Le}} \ \text{and} \ Ne = \frac{U}{V_N} = \frac{UL}{\beta_\times T_e} \frac{eB}{c} = \frac{1}{\beta_\times} \frac{U}{v_{the}} \frac{L}{r_{Le}}, \tag{12.23}$$

respectively. Here the electron thermal speed is v_{the} and the electron gyroradius is r_{Le}. We also ignored the difference between u_e and u, which might become important in cases where Hall MHD matters. One sees from (12.23) that the Biermann term and Nernst term are likely to be important under similar conditions. If we anticipate that $U \sim c_s$, then $U/v_{the} \sim \sqrt{m_e/m_i}$. Thus the Biermann effects and Nernst effects will be small once the electrons are well-enough magnetized that $L/r_{Le} \gtrsim 100$. Note, though, that L here reflects mainly the scale of the temperature gradient, which can be quite steep.

We turn now from the question of validity of the MHD equations to the Ryutov scaling for an experiment. If one includes the magnetic-field terms in the momentum equation and the ideal-MHD version of the magnetic induction equation, and carries out the analysis of Sect. 12.1, as done by Ryutov et al. (2001), one finds that there are two dimensionless parameters that must be kept constant in order to have fully well-scaled behavior. These are $Ry = u^* \sqrt{\rho^*/p^*}$, again, and what we will call the magnetic Ryutov number,

$$Ry_M = \frac{B^*}{\sqrt{p^*}}. \tag{12.24}$$

Ry_M is clearly related to the plasma β, so that systems with the same shape, the same β at some common location, and the same value of Ry will have good Ryutov scaling.

The specific scaling will follow as in the above examples. Geometric differences will lead to constraints on the experiments, and the physics of interest will correspond to dimensionless parameters that must lie in the same regime.

12.6.2 Magnetic Jet Launching and Dynamics

A major challenge in understanding astrophysical jets is to understand how they are launched. Most astrophysical models attribute the launching to magnetic effects. It would require unique and unlikely circumstances to launch them hydrodynamically, and the environments where they are launched tend to include dynamically significant magnetic fields. A common theme in models of jet launching is that rotational energy in some system acts to wind up initially poloidal (see Sect. 10.9) magnetic flux into a toroidal field. This requires a surrounding plasma medium through which the current corresponding to the field flows. At the top of this system, the resulting $J \times B$ forces act to drive the field and plasma up along the axis of rotation, creating a structure known as a *magnetic tower*.

Figure 12.13a shows the schematic structure of a magnetic tower. There is a magnetic cavity, containing toroidal magnetic field, necessarily surrounded by a flow of current that must return along the axis. The intense current along the axis produces current-driven MHD instabilities (specifically the so-called "sausage" and "kink" modes). One sees the structures they have produced in part (b) of the figure.

Lebedev et al. (2005) and Ciardi et al. (2007) discuss the astrophysical context more fully, while reporting and analyzing the first experiments to produce and explore the behavior of this type of system. In these experiments, a pulsed-power device drove a current radially inward through an array of wires. The plasma ablating from these wires contained a toroidal magnetic field, and the poloidal loop of current necessary for the field to exist. Thus, the experiments began at the phase postulated in astrophysical models where the rotational energy has created a significant, toroidal magnetic flux. Table 12.3 shows the physical and derived parameters for this experiment. Observed astrophysical jets are typically pulsed, and the experiment just described produced only one such pulse. Later experiments (Suzuki-Vidal et al. 2010) in this sequence found that, by replacing the wires with a flat foil, one could produce episodic jets.

One can see in the table that the experiment is in the regime where ideal MHD, with radiative cooling, is a reasonable model. The Mach number and β imply that compressible effects may matter and that the magnetic field is dynamically important. The large value of Re implies that viscous effects are negligible, while

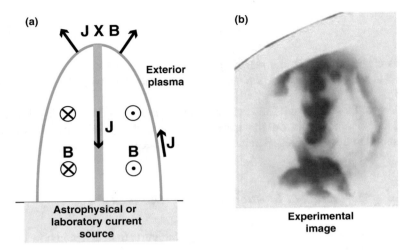

Fig. 12.13 (**a**) Schematic structure of a magnetic-tower jet. (**b**) Self-emission from such a jet, of tungsten, in a pulsed-power experiment. Adapted from Fig. 4 in Lebedev et al. (2005)

Table 12.3 Physical and derived parameters for a magnetic-tower jet experiment

Parameter	Physical	Derived	
Jet parameters			
Velocity (km/s)	100–200	Mach number	3–5
Ion Density (cm^{-3})	10^{18}	Plasma β	~1
Temperature (eV)	120	Reynolds no. Re	~10^4
Ionization Z	20	Magnetic Reynolds No. Rm	~10
A	184 (W)	Peclet number Pe	5–20
Magnetic field B (kG)	>500	Cooling parameter χ	10^{-3} to 10^{-4}
Background plasma			
Density (cm^{-3})	10^{16} to 10^{17}		
Temperature (eV)	<20		
Ionization (Z)	10–15		
Sound speed (km/s)	10–15		
Magnetic field B (kG)	<50		

the modest value of Rm implies that the plasma in the jet would be unable to sustain magnetic structures at a small fraction of the jet diameter. (Inside the magnetic cavity, Rm is much larger.) Fortunately, though, Rm is large enough to allow current-driven MHD instabilities to produce structure on the relative spatial scales that also appear to be important for similar astrophysical jets. As is discussed in Sect. 12.5.2, small values of the cooling parameter χ correspond to conditions in which radiative cooling is rapid compared with hydrodynamic timescales. One concludes that the jet produced in the experiment quickly cools by radiation.

The initial papers in this sequence thus did an excellent job of evaluating the parameters relevant to the Ryutov scaling, and showed the relevance and limitations of an MHD model. They also pointed out that the jet and magnetic bubble could clearly survive for many instability growth times. Eventually (Ciardi et al. 2009) the primary authors also addressed the issues of instability growth times and bubble-ejection times. They noted that the growth time is of order the Alfvén crossing time, which is a few nanoseconds in the experiment and $\sim$ a year in relevant astrophysical jets. We can note that the ratio of Alfvén crossing times, L/v_A, for two jets having the same value of β, is identical to the ratio of timescales implied by the Ryutov analysis. They also found that, in both the experiment and relevant astrophysical jets, the ratio of growth time to bubble-ejection time was $\sim$10. This completed the scaling analysis discussed throughout this chapter, by providing an analysis of dimensionless parameters relevant to the specific process of interest, in addition to having considered those relevant to the global question of modeling the system.

In another experiment, plasmas initiated in jets of gas produced dynamic, MHD structures relevant to jet dynamics and to stellar flares (You et al. 2005). These experiments had the major advantage that they could produce conditions of interest in a relatively simple and inexpensive experimental apparatus. These experiments and associated theory focused more strongly on basic physics aspects of the behavior than on scaled comparisons with astrophysical systems. They helped illuminate the fundamental aspects of MHD dynamics that drive the behavior of both magnetically launched jets and of stellar flares.

12.7 Specific Scaling for Collisionless Shocks

In principle, for any intended laboratory analogue of an astrophysical system, one can address the issues described at the start of this chapter—finding valid equations, keeping the terms that matter, addressing the Ryutov scaling, and identifying the specific scaling. However, this may become less important as the phenomena under study become more complex. For complex phenomena, requiring kinetic plasma theory and involving complex electromagnetic effects, and perhaps nonlocal behavior, there is often considerable uncertainty regarding the fundamental behavior. In these cases it may make the most sense to move directly to specific scaling, in order to focus on the central dynamics for the process of interest. In such cases, one must also address other mechanisms that might interfere with the process of interest. Collisionless shocks are very much a case in point.

It has been argued since the paper by Medvedev and Loeb (1999) that such shock waves may be important to gamma ray bursts, through the specific mechanism of the excitation of the (ion-ion) Weibel instability. The Weibel instability is an electromagnetic instability, in which filaments of magnetic field grow in response to the existence of a particle distribution function with two peaks that are sufficiently separated. The first response of the plasma to such a particle distribution is to produce an electrostatic two-stream instability, which can also matter for electron heat transport (Sect. 9.1.5) and for resistivity (Sect. 10.8). The ion–acoustic turbulence that results from this instability also can cause the formation of shock waves (Forslund and Shonk 1970), and these have been observed (Morita et al. 2010). However, simulations indicate that these shocks saturate at a fairly low level and soon dissipate (Kato and Takabe 2010). If new ions entering some region sustain a two-peaked distribution, then on a longer timescale the Weibel instability can grow, and eventually evolve to sustain a more enduring shock wave containing a modulated magnetic field having a moderate average value.

The equations describing this behavior are those of kinetic plasma theory (see Sect. 2.5) including the collisional effects that may limit the experimental system. Studies of the formation of this type of shock, using collisionless Particle In Cell codes, have identified two dimensionless parameters that are important for this specific process. The growth requires a time of some thousands of ion plasma periods and a few hundred ion skin depths (Kato and Takabe 2008; Spitkovsky 2008). This gives us two relevant dimensionless parameters:

$$\omega_{pi}t > 1000 \text{ and } L > 200c/\omega pi, \tag{12.25}$$

for interpenetrating plasmas, interacting of some distance L, and having ion-plasma frequency ω_{pi}. Knowing this, one can assess whether specific experimental systems will be able to produce such shock waves, as in Drake and Gregori (2012). As matters have developed, researchers using the Omega facility proved able to observe the Weibel mechanism (Fox et al. 2013; Huntington et al. 2015) but not yet to see developed shock waves. Attempts to do so on NIF are ongoing (Ross et al. 2017), at this writing.

The analysis summarized above is not sufficient, however. It fails to address the question of which physical mechanisms might interfere with the process of interest. We discuss one of these here. The design paper mentioned considers several others. For any experiment, one must always think carefully about this question. One can benefit from the fundamental theory in assessing this. The most general equation expressing the kinetic theory, the Boltzmann equation, includes collisional effects. The most likely effect of collisions in the context of Weibel is magnetic diffusion. The global scaling parameter that is relevant to this is the magnetic Reynolds number, Rm. But Rm may not be a good measure of what occurs for a specific microscale process. In Drake and Gregori (2012), we considered the competition between the exponential growth rate γ_W of a magnetic filament at some scale L and its dissipation by magnetic diffusion at a rate $\nu_D = \eta/L^2$. At small scales, this ratio is

$$\frac{\gamma_W}{\nu_D} = \left(\frac{\omega_{pi}}{\nu_{ei}}\frac{V}{c}\right)\left(\frac{L}{c/\omega_{pe}}\right)^2 = \frac{1.5T_e^{3/2}V}{\sqrt{A}n_iZ\Lambda_{ei}}, \tag{12.26}$$

for single-stream speed V, electron-ion collision rate ν_{ei}, and electron plasma frequency ω_{pe}. The rightmost term is in cgs units with T_e in eV, evaluated for the smallest scale length for which structure is seen in simulations, $L = 10c/\omega_{pe}$. (Here

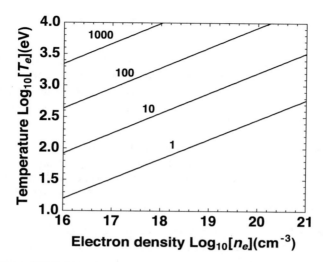

Fig. 12.14 Ratio of Weibel instability growth rate to magnetic diffusion time for a structure size of $10 \, c/\omega_{pe}$, evaluated for a single-stream speed of 1000 km/s. The ratio scales in linear proportion to V

the atomic weight is A, the ion charge is Z, and the Coulomb logarithm is Λ_{ei}.) Figure 12.14 shows curves at specified values of this ratio. One sees that a relatively low density and high temperature is needed to drive this ratio above ~ 10, which is required to make diffusion negligible. Increasing V can help, but getting above 1000 km/s by a big factor is a challenge for known experimental methods. When diffusion is not negligible, a shock might form but its thresholds and behavior could differ from those found in the motivating astrophysical circumstances.

In this section, we provided a specific example of a mechanism for which our usual tools of hydrodynamic, MHD (or even fluid) equations have little applicability. In this as in other circumstances, studies of the mechanism of interest can reveal key dimensionless parameters, and consideration of what might limit an experiment can reveal others.

Homework Problems

12.1 Show that the Euler equations are in fact invariant under the transformations described in Sect. 12.1.

12.2 Design a diverging experiment to address the coupling of two structured, unstable interfaces that are affected by a blast wave. Beyond the basic requirements for hydrodynamic scaling, identify other specific parameters that are important to the dynamics. (Hint: review blast-wave propagation and shock stability as part of your work.)

12.3 Determine why t_{cc} as defined in Sect. 12.3 is the relevant timescale for the crushing of a cloud.

12.4 Suppose that an astrophysical blast wave of interest is produced by a supernova explosion that is a known distance R from a clump of some radius r_{cl}. Determine the properties of an experimental blast wave and the duration of the experiment that would be required to model the shock–clump interaction in this system.

12.5 An approach that has been used to form hydrodynamic jets is to create an adiabatic rarefaction by allowing a shock wave to emerge from a material into an evacuated tube and then to emerge from this tube into an "ambient medium", at a lower density. Using the simple scaling results from this book, develop a design for a similar experiment to produce a radiative jet.

12.6 Magnetized jets must have a ratio of plasma pressure to magnetic field pressure (usually called β in plasma physics) no larger than about 1. For a low-Z plasma with a density of 0.1 g/cm^3 and a temperature of 10 eV, determine how large a magnetic field would be required to satisfy this constraint. How does this compare with the magnetic field of order 1 MGauss that is typically produced in laser–plasma interactions and that might be produced by very clever field-compression experiments?

References

Arnett WD (1996) Supernova and nucleosynthesis. Princeton University Press, Princeton

Arnett WD, Bahcall JN, Kirschner RP, Woolsey SE (1989) Supernova 1987a. Ann Rev Astron Astrophys 27:629–700

Barton RT (1985) The CALE computer code. In: Centrella JM, LeBlanc JM, Bowers RL (eds) Numerical astrophysics. Jones and Bartlett, Boston, pp 482–497

Benuzzi-Mounaix, Bouquet AS, Chieze JP, Mucchielli F, Teyssier R, Thais F (2001) Supernovae Rayleigh-Taylor instability experiments on the CEA-Phebus laser facility. Astrophys Space Sci 277:143–146

Bethe HA (1990) Supernova mechanisms. Rev Mod Phys 62(4):801–866. https://doi.org/10.1103/RevModPhys.62.801

Blondin JM, Fryxell BA, Konigl A (1990) The structure and evolution of radiatively cooling jets. Astrophys J 360:370–386

Borkowski JE, Blondin JM, Harrington JP (1997) Collimation of astrophysical jets: the proto-planetary nebula He 3-1475. Astrophys J 482:L97–L100

Calder A, Fryxell B, Plewa T, Rosner R, Dupont T, Robey HF, Kane JO, Remington BA, Drake RP, Dimonte G, Zingale M, Dursi LJ, Timmes FX, Olson K, Ricker P, MacNeice P, Tufo HM (2002) On validating an astrophysical simulation code. Astrophys J 143:201–229

Ciardi A, Lebedev SV, Frank A, Blackman EG, Chittenden JP, Jennings CJ, Ampleford DJ, Bland SN, Bott SC, Rapley J, Hall GN, Suzuki-Vidal FA, Marocchino A, Lery T, Stehle C (2007) The evolution of magnetic tower jets in the laboratory. Phys Plasmas 14(5):056501

Ciardi A, Lebedev SV, Frank A, Suzuki-Vidal F, Hall GN, Bland SN, Harvey-Thompson A, Blackman EG, Camenzind M (2009) Episodic magnetic bubbles and jets: astrophysical implications from laboratory experiments. Astrophys J Lett 691(2):L147–L150. https://doi.org/10.1088/0004-637x/691/2/l147

Connor JW, Taylor JB (1977) Scaling laws for plasma confinement. Nucl Fusion 17:1067

Cross JE, Reville B, Gregori G (2014) Scaling of magneto-quantum-radiative hydrodynamic equations: from laser-produced plasmas to astrophysics. Astrophys J 795(1). https://doi.org/10.1088/0004-637x/795/1/59

Dawson JM (1964) On the production of plasma by giant lasers. Phys Fluids 7:981–987

Dogget JB, Branch D (1985) A comparative study of supernova light curves. Astron J 90(11):2303–2311

Drake RP (1999) Laboratory experiments to simulate the hydrodynamics of supernova remnants and supernovae. J Geophys Res 104(A7):14505–14515

Drake R, Gregori G (2012) Design considerations for unmagnetized collisionless shock experiments in homologous flows. Astrophys J 749:171

Drake RP, Smith T, Carroll III JJ, Yan Y, Glendinning SG, Estabrook K, Ryutov DD, Remington BA, Wallace R, McCray R (2000) Progress toward the laboratory simulation of young supernova remnants. Astrophys J Suppl 127(2):305–310

Falize E, Bouquet S, Michaut C (2009) Scaling laws for radiating fluids: the pillar of laboratory astrophysics. Astrophys Space Sci 322(1–4):107–111. https://doi.org/10.1007/s10509-009-9983-z

Farley DR, Estabrook KG, Glendinning SG, Glenzer SH, Remington BA, Shigemori K, Stone JM, Wallace RJ, Zimmerman GB, Harte JA (1999) Radiative jet experiments of astrophysical interest using intense lasers. Phys Rev Lett 83(10):1982–1985

Forslund DW, Shonk CR (1970) Formation and structure of electrostatic collisionless shocks. Phys Rev Lett 25(25):1699–1702

Fox W, Fiksel G, Bhattacharjee A, Chang PY, Germaschewski K, Hu SX, Nilson PM (2013) Filamentation instability of counterstreaming laser-driven plasmas. Phys Rev Lett 111(22). https://doi.org/10.1103/PhysRevLett.111.225002

Fryxell B, Muller E, Arnett D (1991) Instabilities and clumping in SN 1987A. I. Early evolution in two dimensions. Astrophys J 367:619–34

Fryxell B, Olson K, Ricker P, Timmes FX, Zingale M, Lamb DQ, MacNeice P, Rosner R, Truran JW, Tufo H (2000) Flash: an adaptive mesh hydrodynamics code for modeling astrophysical thermonuclear flashes. Astrophys J Suppl Ser 131(1):273–334

Glimm J, Grove JW, Li XL, Oh W, Sharp DH (2001) A critical analysis of Rayleigh-Taylor growth rates. J Comput Phys 169:652–677

Grun J, Stamper J, Manka C, Resnick J, Burris R, Crawford J, Ripin BH (1991) Instability of Taylor-Sedov blast waves propagating through a uniform gas. Phys Rev Lett 66(21):2738–2741

Hansen JF, Robey HF, Klein RI, Miles AR (2007a) Experiment on the mass stripping of an interstellar cloud following shock passage. Astrophys J 662(1 Pt 1):379–388

Hansen JF, Robey HF, Klein RI, Miles AR (2007b) Experiment on the mass-stripping of an interstellar cloud in a high Mach number post-shock flow. Phys Plasmas 14(5):056505

Hartigan P (1989) The visibility of the Mach disk and the bow shock of a stellar jet. Astrophys J 339:987–999

Hartigan PJ, Morse JA, Heathcote S, Cecil G (1993) Observations of entrainment and time variability in the HH 47 jet. Astrophys J 414:L121–L124

Hartigan P, Bally J, Reipurth B, Morse JA, Mannings V, Boss AP, Russell SS (2000) Shock structures and momentum transfer in Herbig-Haro jets. In: Protostars and planets 4. The University of Arizona Press, Tucson, pp 841–866

Huntington CM, Fiuza F, Ross JS, Zylstra AB, Drake RP, Froula DH, Gregori G, Kugland NL, Kuranz CC, Levy MC, Li CK, Meinecke J, Morita T, Petrasso R, Plechaty C, Remington BA, Ryutov DD, Sakawa Y, Spitkovsky A, Takabe H, Park HS (2015) Observation of magnetic field generation via the Weibel instability in interpenetrating plasma flows. Nat Phys 11(2):173–176. https://doi.org/10.1038/nphys3178

Hwang U, Flanagan KA, Petre R (2005) Chandra X-ray observation of a mature cloud-shock interaction in the bright eastern knot region of Puppis A. Astrophys J 635(1 Pt 1):355–364

Kane J, Arnett D, Remington BA, Glendinning SG, Castor J, Wallace R, Rubenchik A, Fryxell BA (1997) Supernova-relevant hydrodynamic instability experiments on the nova laser. Astrophys J Lett 478:L75–L78

Kane J, Arnett D, Remington BA, Glendinning SG, Bazan G, Drake RP, Fryxell BA, Teyssier R, Moore K (1999) Scaling supernova hydrodynamics to the laboratory. Phys Plasmas 6(5):2065–2072

Kane J, Arnett D, Remington BA, Glendinning SG, Bazan G, Drake RP, Fryxell BA (2000) Supernova experiments on the nova laser. Astrophys J Suppl 127(2):365–369

Kane JO, Robey HF, Remington BA, Drake RP, Knauer J, Ryutov DD, Louis H, Teyssier R, Hurricane O, Arnett D, Rosner R, Calder A (2001) Interface imprinting by a rippled shock using an intense laser. Phys Rev E 63:055401R

Kato T, Takabe H (2008) Nonrelativistic collisionless shocks in unmagnetized electron-ion plasmas. Astrophys J Lett 681:L93–L96

Kato TN, Takabe H (2010) Electrostatic and electromagnetic instabilities associated with electrostatic shocks: two-dimensional particle-in-cell simulation. Phys Plasmas 17(3):032114. https://doi.org/03211410.1063/1.3372138

Kifonidis K, Plewa T, Janka HT, Muller E (2000) Nucleosynthesis and clump formation in a core-collapse supernova. Astrophys J Lett 531(2):L123–L126

Klein RI, Budil KS, Perry TS, Bach DR (2003) The interaction of supernova remnants with interstellar clouds: experiments on the nova laser. Astrophys J 583(1 Pt 1):245–259

Kuranz CC, Drake RP, Grosskopf MJ, Fryxell B, Budde A, Hansen JF, Miles AR, Plewa T, Hearn NC, Knauer JP (2010) Spike morphology in blast-wave-driven instability experiments. Phys Plasmas 17(5):052709. http://dx.doi.org/10.1063/1.3389135

Lebedev SV, Chittenden JP, Beg FN, Bland SN, Ciardi A, Ampleford D, Hughes S, Haines MG, Frank A, Blackman EG, Gardiner T (2002) Laboratory astrophysics and collimated stellar outflows: the production of radiatively cooled hypersonic plasma jets. Astrophys J 564:113–119

Lebedev SV, Ciardi A, Ampleford DJ, Bland SN, Bott SC, Chittenden JP, Hall GN, Rapley J, Jennings CA, Frank A, Blackman EG, Lery T (2005) Magnetic tower outflows from a radial wire array z-pinch. Mon Not R Astron Soc 361(1):97–108

Leith CE (1990) Stochastic backscatter in a subgrid-scale model: plane shear mixing layer. Phys Fluids A 2(3):297–299

Li CK, Tzeferacos P, Lamb D, Gregori G, Norreys PA, Rosenberg MJ, Follett RK, Froula DH, Koenig M, Seguin FH, Frenje JA, Rinderknecht HG, Sio H, Zylstra AB, Petrasso RD, Amendt PA, Park HS, Remington BA, Ryutov DD, Wilks SC, Betti R, Frank A, Hu SX, Sangster TC, Hartigan P, Drake RP, Kuranz CC, Lebedev SV, Woolsey NC (2016) Scaled laboratory experiments explain the kink behaviour of the crab nebula jet. Nat Commun 7. https://doi.org/10.1038/ncomms13081

Meakin CA, Arnett D (2006) Active carbon and oxygen shell burning hydrodynamics. Astrophys J Lett 637(1 Pt 2):L53–L56

Medvedev MV, Loeb A (1999) Generation of magnetic fields in the relativistic shock of gamma-ray burst sources. Astrophys J 526(2):697–706

Miles AR (2009) The blast-wave-driven instability as a vehicle for understanding supernova explosion structure. Astrophys J 696:498–514

Miles AR, Blue B, Edwards MJ, Greenough JA, Hansen JF, Robey HF, Drake RP, Kuranz C, Leibrandt DR (2005) Transition to turbulence and effect of initial conditions on three-dimensional compressible mixing in planar blast-wave-driven systems. Phys Plasmas 12(5):1–10 (Article No. 056317)

Mizuta A, Yamada S, Takabe H (2002) Numerical analysis of jets produced by intense laser. Astrophys J 567(1 Pt 1):635–642

Morita T, Sakawa Y, Kuramitsu Y, Dono S, Aoki H, Tanji H, Kato TN, Li YT, Zhang Y, Liu X, Zhong JY, Takabe H, Zhang J (2010) Collisionless shock generation in high-speed counterstreaming plasma flows by a high-power laser. Phys Plasmas 17(12):122702. https://doi.org/12270210.1063/1.3524269

Müller E, Fryxell B, Arnett D (1991) Instabilities and clumping in SN 1987A. Astron Astrophys 251:505–514

Piomelli U, Cabot WH, Moin P, Sangsan L (1991) Subgrid-scale backscatter in turbulent and transitional flows. Phys Fluids A 3(7):1766–1771

Reipurth B, Bally J (2001) Herbig-haro flows: probes of early stellar evolution. Ann Rev Astron Astrophys 39:403–455

Reipurth B, Heathcote S, Morse J, Hartigan P, Bally J (2002) Hubble space telescope images of the HH 34 jet and bow shock: structure and proper motions. Astron J 123:362–381

Remington BA, Kane J, Drake RP, Glendinning SG, Estabrook K, London R, Castor J, Wallace RJ, Suter LJ, Muntro DH, Arnett D, Liang E, McCray R, Rubenchik A, Fryxell B (1997) Supernova hydrodynamics experiments on the nova laser. Phys Plasmas 4(5):1994–2003

Remington BA, Arnett D, Drake RP, Takabe H (1999) Modeling astrophysical phenomena in the laboratory with intense lasers. Science 284:1488–1493

Remington BA, Drake RP, Arnett D, Takabe H (2000) A review of astrophysics experiments on intense lasers. Phys Plasmas 7:1641

Remington BA, Drake RP, Ryutov DD (2006) Experimental astrophysics with high power lasers and z pinches. Rev Mod Phys 78:755–807

Ripin BH, McLean EA, Manka CK, Pawley C, Stamper JA, Peyser TA, Mostovych AN, Grun J, Hassam AB, Huba J (1987) Large-Larmor-radius interchange instability. Phys Rev Lett 59(20):2299–2302

Robey HF, Perry TS, Klein RI, Kane J, Greenough JA, Boehly T (2002) Experimental investigation of the three-dimensional interaction of a strong shock with a spherical density inhomogeneity. Phys Rev Lett 89(8):085001-1–4

Ross JS, Higginson DP, Ryutov D, Fiuza F, Hatarik R, Huntington CM, Kalantar DH, Link A, Pollock BB, Remington BA, Rinderknecht HG, Swadling GF, Turnbull DP, Weber S, Wilks S, Froula DH, Rosenberg MJ, Morita T, Sakawa Y, Takabe H, Drake RP, Kuranz C, Gregori G, Meinecke J, Levy MC, Koenig M, Spitkovsky A, Petrasso RD, Li CK, Sio H, Lahmann B, Zylstra AB, Park HS (2017) Transition from collisional to collisionless regimes in interpenetrating plasma flows on the national ignition facility. Phys Rev Lett 118(18). https://doi.org/10.1103/PhysRevLett.118.185003

Ryutov DD, Drake RP, Kane J, Liang E, Remington BA, Wood-Vasey M (1999) Similarity criteria for the laboratory simulation of supernova hydrodynamics. Astrophys J 518(2):821

Ryutov DD, Drake RP, Remington BA (2000) Criteria for scaled laboratory simulations of astrophysical MHD phenomena. Astrophys J Suppl 127(2):465–468

Ryutov DD, Remington BA, Robey HF, Drake RP (2001) Magnetohydrodynamic scaling: from astrophysics to the laboratory. Phys Plasmas 8:1804–1816

Schekochihin AA, Cowley SC, Taylor SF, Maron JL, McWilliams JC (2004) Simulations of the small-scale turbulent dynamo. Astrophys J 612(1):276–307. https://doi.org/10.1086/422547

Shigemori K, Kodama R, Farley DR, Koase T, Estabrook KG, Remington BA, Ryutov DD, Ochi Y, Azechi H, Stone J, Turner N (2000) Experiments on radiative collapse in laser-produced plasmas relevant to astrophysical jets. Phys Rev E 62(6 Pt B):8838–8841

Spitkovsky A (2008) On the structure of relativistic collisionless shocks in electron-ion plasmas. Astrophys J Lett 673(1):L39–L42

Stone JM, Norman ML (1994) Numerical simulations of protostellar jets with nonequilibrium cooling. 3: three-dimensional results. Astrophys J 420:237–246

Stone J, Turner N, Estabrook K, Remington BA, Farley DR, Glendinning SG, Glenzer SH (2000) Testing astrophysical radiation hydrodynamics codes with hypervelocity jet experiments on the nova laser. Astrophys J Suppl 127:497–502

Sutherland PG (1990) Gamma-rays and X-rays from supernovae. In: Petschek AG (ed) Supernovae. Springer, New York, p 111

Suzuki-Vidal F, Lebedev SV, Bland SN, Hall GN, Swadling G, Harvey-Thompson AJ, Chittenden JP, Marocchino A, Ciardi A, Frank A, Blackman EG, Bott SC (2010) Generation of episodic magnetically driven plasma jets in a radial foil z-pinch. Phys Plasmas 17(11):112708. https://doi.org/11270810.1063/1.3504221

Takabe H (1993) Icf and supernova explosions. J Plasma Fusion Res 69:1285–1300

Takabe H (2001) Astrophysics with intense and ultraintense lasers "laser astrophysics". Prog Theor Phys Suppl 143:202–265

Tikhonchuk VT, Nicolai P, Ribeyre X, Stenz C, Schurtz G, Kasperczuk A, Pisarczyk T, Juha L, Krousky E, Masek K, Pfeifer M, Rohlena K, Skala J, Ullschmied J, Kalal M, Klir D, Kravarik J, Kubes P, Pisarczyk P (2008) Laboratory modeling of supersonic radiative jets propagation in plasmas and their scaling to astrophysical conditions. Plasma Phys Control Fusion 50(12). https://doi.org/12405610.1088/0741-3335/50/12/124056

Widnall SE, Sullivan JP (1973) Stability of vortex rings. Proc R Soc Lond Ser A 332(1590):335

Widnall SE, Sullivan JP (1974) The instability of short waves on a vortex ring. J Fluid Mech 66(1):35–47

Woosley SE, Eastman RG (1997) Types 1b and 1c supernovae: models and spectra. In: Ruiz-Lapuente B, Canal R, Iser J (eds) Thermonuclear supernovae. Kluwer, Dordrecht

Woosley SE, Heger A, Weaver TA (2002) The evolution and explosion of massive stars. Rev Mod Phys 74(4):1015–1071. https://doi.org/10.1103/RevModPhys.74.1015

You S, Yun GS, Bellan PM (2005) Dynamic and stagnating plasma flow leading to magnetic-flux-tube collimation. Phys Rev Lett 95(4). https://doi.org/10.1103/PhysRevLett.95.045002

Zel'dovich YB, Razier YP (1966) Physics of shock waves and high-temperature hydrodynamic phenomena, vol 1, 2002nd edn. Dover, New York

Chapter 13
Relativistic High-Energy-Density Systems

Abstract This chapter begins with a discussion of what constitutes a relativistic system at high energy density and how to produce such systems. The primary focus is on lasers intense enough to create relativistic electron motion. The chapter proceeds to discuss the motion of individual electrons in laser pulses and then the problem of producing the interaction of such laser pulses with sold targets, as opposed to with plasma blown off their surfaces. Topics discussed further related to laser irradiation of solids include absorption, harmonic generation, and induced transparency. The discussion then turns to particle acceleration. Wakefield acceleration of electrons is discussed at length. Ion acceleration by target sheaths, by laser pistons, and by Coulomb explosions are discussed next. After that, the chapter analyzes hole drilling by such lasers and the collisionless shocks that may result. The chapter ends with a brief review of several other phenomena, including magnetic-field generation, betatron X-ray production, positron production, nuclear reactions, and phenomena involving intense beams.

In this chapter we address the low-density and high-temperature regime of high-energy-density physics identified in Chap. 1. While phenomena produced in this regime often connect with those discussed in the previous chapters, there are real differences in the underlying physics. A high-energy-density, thermal, relativistic plasma would have a minimum temperature of 511 keV and a density exceeding 10^{18} cm^{-3}. At the turn of the century, such plasmas did not exist in the laboratory. Producing them can be taken as a challenge for the early twenty-first century. However, plasmas did exist at this density with a mean electron energy exceeding 511 keV. Some such plasmas were made relativistic by the electron oscillations caused by intense lasers. We will define a *relativistic laser beam* as one producing a mean electron kinetic energy exceeding 511 keV. Some of these laser-irradiated plasmas produce beams of electrons with characteristic energies of many MeV. In addition, other, denser plasmas existed with a mean electron energy exceeding 511 keV because of the presence of a highly relativistic electron beam in a cold background plasma.

To place these systems in context, we return to the definition of high energy density as corresponding to a pressure exceeding 1 Mbar, or an energy density exceeding 10^{12} ergs/cm^3. Table 13.1 is based on Table 1.1 in the National Research

© Springer International Publishing AG 2018
R P. Drake, *High-Energy-Density Physics*, Graduate Texts in Physics,
https://doi.org/10.1007/978-3-319-67711-8_13

Table 13.1 Quantities corresponding to 10^{12} erg/cm^3

Pressure	1 Mbar = 0.1 TPascal
Energy flux of laser or relativistic particle beam	3×10^{15} W/cm^2
Blackbody radiation temperature	400 eV
Electric field strength	1.5×10^{11} V/m
Magnetic field strength	5 MGauss
Ablation pressure by (1 μm wavelength) laser at	4×10^{12} W/cm^2
Ablation pressure by thermal radiation at	75 eV
Particle density for 511 keV mean kinetic energy	10^{18} cm^{-3}

Council report (Davidson et al. 2003). From the discussion of Chap. 9, it is clear that achieving these conditions is not so difficult. Here we consider three specific relativistic examples.

For lasers, the challenge is to get the kinetic energy of the oscillating electrons up to 511 keV. This requires $I_{18}\lambda_\mu^2 = 1.37$, where I_{18} is the laser energy flux in units of 10^{18} W/cm^2 and λ_μ is the laser-light wavelength in μm. This is not now difficult. It requires a laser of 10^{10}–10^{12} W, assuming the focal spot to be 1–10 μm. This is less than 1 J in 1 ps or 10 mJ in 10 fs. The energy density of the electrons within such a focal spot remains a small fraction of the energy density of the laser beam.

Electron beams at the turn of the century could produce 50 GeV electrons in a 5 μm spot, with bunches of 5 ps duration at a repetition rate of 100 Hz. The bunches contained 150 J each and thus contained 2×10^{10} electrons. The bunches were long and narrow, being more than 1 mm long. Their volume was $\sim 10^{-8}$ cm^3, so the density of these electrons is 2×10^{18} cm^{-3}. When such a beam passes through a solid with an electron density of 2×10^{23} cm^{-3}, the resulting average electron energy is ~ 500 keV. These beams do not deposit their energy very readily, so studies with them primarily involve ways to affect the beam. This includes the important area of wakefield acceleration, discussed in Sect. 13.7.1.

Ion beams at the turn of the century (specifically the Relativistic Heavy Ion Collider) could cause ion bunches to collide at a 50 MHz repetition rate. The ions had an energy of 100 GeV per amu, or 20 TeV for Au ions; the bunches were of ~ 500 ps duration, ~ 200 μm diameter, and ~ 3 kJ energy. Each bunch of such ions has an energy density of about 3×10^{12} ergs/cm^3 shared among about 10^9 ions. Here again, the ion beams do not deposit their energy very readily, so studies with them primarily involve ways to affect the beam. In addition, as we mentioned in Chap. 9, beams of nonrelativistic heavy ions can be used to heat high-Z matter into the high-energy-density regime.

Most of this chapter is devoted to the behavior of matter in the presence of electromagnetic fields strong enough to produce relativistic electron motions. The devices that produce relativistic laser beams are called *ultrafast lasers,* for reasons that will become clear. The emphasis on ultrafast lasers reflects both their broad availability and their potential to produce extremely high electric and magnetic fields. We also discuss relativistic effects that can be produced using high-energy electron beams in the area of particle acceleration. One can expect that more such applications will develop as the twenty-first century proceeds.

As in the previous chapter, this chapter includes many more references to journal articles than most of the book does. This reflects the relative newness of work in this area. However, once again the present chapter is not written as a review. Rather, it is intended as an introduction to the physics and the issues.

13.1 Development of Ultrafast Lasers

Since about 1980, the challenge of producing high-power lasers has become distinct from that of producing high-energy lasers. While the laser systems described earlier can heat cubic millimeters of material to million-degree temperatures, they cannot produce relativistic electrons or distributions of ions with billion-degree temperatures. This requires much more intense laser light. One cannot produce such light by directly amplifying a laser pulse; the amplifying glass would be damaged. The invention by Gerard Mourou of *chirped pulse amplification* (CPA) has allowed laser systems to escape this limitation (Mourou and Umstadter 1992). All intense lasers amplify a laser beam whose spatial area is much larger than the ultimate focused laser spot. CPA goes further, also doing this in the dimension of time. In CPA, one amplifies a laser pulse whose time duration is much longer than the ultimate duration of the pulse reaching the laser spot. This is done as follows.

Despite the notion that lasers are coherent, single-frequency devices, any laser pulse in fact has a finite bandwidth. For a laser pulse that is Gaussian in time, with a frequency bandwidth $\Delta\omega$ and time profile $\exp[-(t/\tau)^2]$, one can show by Fourier transforming the laser pulse that $\Delta\omega\tau \sim 1$. Thus, very short laser pulses may have a significant bandwidth. It is now possible to produce laser pulses of order one cycle in duration ($\sim$1 fs for visible light); such pulses have a very broad bandwidth. There are a number of methods for producing such pulses at low energy and low irradiance. The contribution of CPA is to provide a way to stretch these pulses in time, allowing them to be amplified at low irradiance before they are recompressed to high irradiance, after which they are focused to enormous irradiance.

Figure 13.1 illustrates a simple compressor design that can stretch a pulse in time. The first grating disperses the incoming, broadband, collimated laser pulse in angle, so that the angle of reflection of each frequency is distinct. This is illustrated in the figure by showing a pair of rays one labeled red and one labeled blue. The grating diffracts the longer wavelength, red rays through a larger angle. A lens pair is used to cause the angle of incidence on a second grating to equal the angle of reflection from the first grating. The result is that a collimated laser beam emerges from the second grating, with different frequencies offset in space. A mirror reflects the light back upon itself, so that each frequency retraces its path and one obtains an outgoing beam of the same size as the initial beam. Geometrically nothing has changed, but temporally each frequency has traveled a different distance, and the

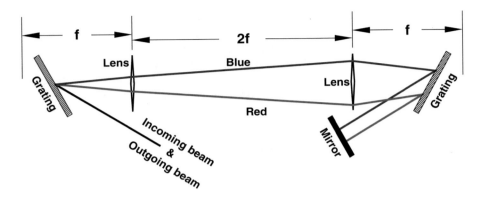

Fig. 13.1 Schematic of pulse stretcher. Credit: Enam Chowdhury

redder frequencies have traveled less distance. One has dispersed the beam in time, producing a *chirped laser pulse* whose frequency varies linearly with distance along the pulse, with redder frequencies at the front.

At this writing some ultrafast lasers have been dedicated systems with a very short laser pulse, an excellent quality laser beam, and comparatively little laser energy. Other systems have been aimed at delivering more laser energy to the targets, which would be necessary for example for inertial fusion using fast ignition (Sect. 11.3.3). These higher-energy systems are often adaptations of a high-energy laser to the task of amplifying short laser pulses. However, none of these lasers yet produces enough energy to create an isolated, thermal, relativistic plasma.

13.2   Single-Electron Motion in Intense Electromagnetic Fields

Many of the fascinating phenomena that ultrafast lasers can produce are a consequence of the relativistic motion of the electrons. To develop some insight into what this makes possible, we consider first the motion of isolated electrons in the fields of such lasers. To do so, we take the electron velocity to be v, the electron momentum to be p_e, the vector and scalar potentials to be A and Φ, and we work as in Chap. 9 in the Coulomb gauge. In Gaussian cgs units, the electromagnetic wave equation is not changed for high velocities, although one must transform the fields properly between inertial frames of reference. To work complicated problems in relativity, four-vector notation becomes very convenient, but we will not invest time in this here. The electromagnetic wave equation is

$$\left(\frac{\partial^2}{\partial t^2} - c^2 \nabla^2 \right) A = 4\pi c J_t, \qquad (13.1)$$

in which as before J_t is the transverse current density. This equation is the same for relativistic and nonrelativistic systems. It is helpful to take note of some aspects of the implied fields, useful in what follows. In a plane-wave decomposition, each spectral component has a distinct k, E, and B, and these are all orthogonal. We let k define the parallel (z) direction. Since $E = (-1/c)(dA/dt)$, so long as the electric potential $\Phi = 0$, and $B = \nabla \times A$, one can see that (for any spectral component) E and B are out of phase with A and in phase with each other. The fact that E and B are in phase and vanish simultaneously may be counterintuitive if you have not considered it previously. We allow A and thus E to define the x direction and B to define the y direction. We define the *pump strength* of the electromagnetic wave as

$$a_o = \frac{eA}{m_e c^2},\tag{13.2}$$

for reasons that will become clear shortly. In practical units,

$$a_o = \sqrt{\frac{I_L \lambda_\mu^2}{1.37 \times 10^{18}\,\mathrm{W}\,\mu m^2/cm^2}}.\tag{13.3}$$

We now consider the motion of an electron in a single plane wave (which may for now include arbitrary spectral components). The equation of motion for an electron is not changed in its fundamental form: the time rate of change of momentum equals the force. But the electron momentum, p_e, is now relativistic. Given our definitions, the y component of the electron momentum is constant in time. The equation of motion for the x and z components of the electron momentum, labeled p_x and p_z, respectively, are

$$\frac{dp_x}{dt} = -\frac{e}{c}\left(cE_x - v_z B_y\right),\tag{13.4}$$

and

$$\frac{dp_z}{dt} = -\frac{e}{c}\left(v_x B_y\right),\tag{13.5}$$

in which $p_e = \gamma_r m_e v$ and in these Gaussian cgs units $E_x = B_y$. To avoid complications involving products of real quantities in complex notation, we here assume that E_x and B_y are real quantities, although (as the particle experiences them) they may vary arbitrarily in time and space. We allow the momenta and velocities to be complex.

If one thinks of a given frequency component of the electric field as seen by the particle, one can see that the detailed motion of the electron has an indefinite number of harmonic components. The electric-field term in (13.4) creates a second-harmonic response in the z motion (13.5), which in turn creates a third-harmonic response in the x motion through the magnetic-field term, which then creates a fourth-harmonic response in the z motion, and so on.

Equations (13.4) and (13.5) and the related conventions enable us to derive a generally useful relation between p_z and p_x, as follows. Defining the total particle energy as $\mathcal{E}_e$, so that with rest mass m_e and rest energy $\mathcal{E}_o = m_e c^2$, one has

$$\mathcal{E}_e = \sqrt{\mathcal{E}_o^2 + p^2 c^2}. \tag{13.6}$$

The rate of change of the particle energy is due entirely to the work done by the electric field, so

$$\frac{d\mathcal{E}_e}{dt} = -e v_x E_x, \tag{13.7}$$

from which by comparison with (13.5) we can see that

$$\mathcal{E}_e - c p_z = \text{const} \equiv \alpha. \tag{13.8}$$

Here α is defined for convenience. It depends on the state of the particle when the field begins. If the particle is at rest at that time, then $\alpha = \mathcal{E}_o$. In general, from (13.6) and (13.8), one finds

$$\mathcal{E}_o^2 = \mathcal{E}_e^2 - p^2 c^2 = (\mathcal{E}_e - p_z c)(\mathcal{E}_e + p_z c) - p_x^2 c^2 = \alpha (\alpha + 2 p_z c) - p_x^2 c^2, \tag{13.9}$$

from which

$$p_z = \frac{1}{2 \alpha c} \left(\mathcal{E}_o^2 - \alpha^2 + p_x^2 c^2 \right). \tag{13.10}$$

If the particle is initially at rest so that $\alpha = \mathcal{E}_o = m_e c^2$, then we obtain the well known result

$$p_z = \frac{p_x^2}{2 m_e c}. \tag{13.11}$$

This equation has some interesting things to tell us. First of all, it says that motions in both x and in z are always part of the response of an electron to a wave. Second, recalling that the momenta are complex in this representation, the motion in z includes a steady drift and an oscillation at twice the frequency of the oscillation in x. This combination creates a path that looks like a figure eight, elongated along E, when p_x is small compared to $m_e c$. Third, isolated electrons cannot be permanently accelerated by light waves. When an electron initially at rest is overtaken by an electromagnetic wave packet, the electron oscillates in x and drifts in z, but as the wave packet passes both these motions cease. The electron ends up displaced but once again stationary. In the absence of collective effects, (isolated) electrons must be created within the light wave, for example by ionization, to end up with significant net energy once the wave has passed. Fourth, in the limit of very large fields, $|p_z| \to \gamma_r m_e c$, in which case $|p_x|$ is smaller, being $\sqrt{2\gamma_r} m_e c$. As a result, v_x decreases, becoming $v_x = p_x/(m_e\gamma_r) \to c\sqrt{2/\gamma_r}$. Finally, the angle of the electron relative to the z-axis, θ, is given by

$$\tan\theta = \sqrt{\frac{2}{\gamma_r - 1}} \tag{13.12}$$

This has sensible limits, going to 90° as γ_r reaches 1 and approaching 0 degrees as γ_r becomes very large.

Thinking strictly in terms of harmonic motion leaves out some aspects of the electron motion, because as the field of the laser increases the electron motion along z soon becomes significant on the scale of the laser wavelength. In more detail, the electric field experienced by an electron within a (z-directed) single-frequency plane wave in vacuum is

$$E_x(\mathbf{x_p}, t) = \hat{E}\cos\left[k(z_p(t) - ct) + \phi_o\right], \tag{13.13}$$

in which

$$z_p(t) = z_o + \int_{t_o}^{t} v_z(t')dt'. \tag{13.14}$$

Here ϕ_o gives the initial phase of the electric field, the z-position of the particle is z_p, and z_p at time t_o is z_o. One can see that the particle will experience this plane wave as a simple harmonic field only if z_p is a linear function of t. Yet we have already seen that this is not the case. Thus, an electron experiences a light wave as a simple harmonic field only in the limit as the motion of the electron in z vanishes. In the other limit, as $z_p \to ct$, the electron will experience a nearly constant field. In this limit the maximum energy the electron can extract from a light wave of finite spot size d is eE_Ld, where the electric field of the focused light wave is E_L.

It is evident that an electron may be introduced into the wave with any phase and thus may experience any field from zero to the maximum when it is born. If the electron is born at rest when the electric field is zero, then the electron returns to rest at the end of each cycle and ends up at rest when the wave has passed. If the electron is produced at rest when the oscillating field is at its maximum, then the electron ends up with the maximum possible energy. When ionization creates the electrons, they may in principle have any phase with respect to the wave. But if the ionization is produced by the wave itself, then the electron will tend to be produced when the electric field is maximum and at near zero velocity. In this case most of the electrons will gain a significant net velocity from their interaction with the wave.

We can explore the electron motion somewhat further as follows. We write the Lorentz factor, which we label as γ_r, as

$$\gamma_r = \left(1 \bigg/ \sqrt{1 - \frac{v^2}{c^2}} \right), \tag{13.15}$$

or

$$\gamma_r = \sqrt{1 + \frac{p_e^2}{m_e^2 c^2}}. \tag{13.16}$$

Now we consider the solutions to (13.4) and (13.5) as the velocity of the electron increases. In the event that v_z is negligible, one has from (13.2) and (13.4)

$$\gamma_r v_x / c = a_o, \tag{13.17}$$

implying

$$v_x / c = \frac{a_o}{1 + a_o}, \tag{13.18}$$

and also

$$\gamma_r = \sqrt{1 + a_o^2} \tag{13.19}$$

in this regime. For small a_o, this is identical to the result we obtained in Chap. 9 for the oscillating velocity of the electron in a light wave. As a_o increases, v_x/c cannot exceed one, as should be the case. Equation (13.19) will fail to be accurate as v_z/c becomes significant (meaning 0.1 for most purposes). To explore this, we can use (13.18) in (13.5), finding

$$\frac{d}{dt} \left(\frac{v_z}{c} \sqrt{1 + a_o^2} \right) = -\frac{a_o}{1 + a_0} \frac{da_o}{dt}. \tag{13.20}$$

In the small a_o limit, one evidently has $v_z/c = a_o^2/2$. One can solve (13.20), using a computational mathematics program, to see when v_z/c approaches 0.1. The additional assumption needed is a specification of a value of v_z (e.g., zero) at some specific phase in the wave (e.g., $\pi/4$). Figure 13.2 plots the maximum value of $\mathbb{R}[v_z]$ in the small a_o approximation and also plots $\mathbb{R}[a_o^2]/4$, against the magnitude of a_o. One sees that the above solutions reach their limits when a_o becomes a few tenths.

Once v_z exceeds about 0.1, the solution to (13.4) and (13.5) for the velocity and the trajectory becomes much more complex for two reasons. First, one must deal with both terms on the left-hand side of (13.4), making their solution a nonlinear mess. Second, one must consider how the variations in v_z affect the phase of the

Fig. 13.2 Electron
oscillating velocity in
direction of $\boldsymbol{k}$ as a_o increases.
The dashed curve shows the
small-a_o limit

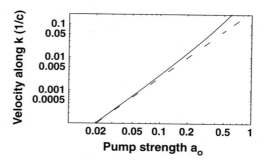

electron in the wave and change its behavior. To see the essential feature that
determines the qualitative behavior, we now pay attention to the phase of the
electron in the wave.

The outcome of the interaction at very large a_o is that the electron moves along $\boldsymbol{k}$
at nearly the speed of light. One can see from (13.4) that in this case the force in the
x-direction will be greatly reduced. To see how this comes about, we can represent
the pump strength as $a_o \sin \phi$, explicitly writing the phase of the wave as ϕ. For
convenience, we treat all the physical quantities as explicitly real for the present
discussion. Now

$$\phi = (\boldsymbol{k} \cdot \boldsymbol{x} - \omega t) = (kz - \omega t) \tag{13.21}$$

in which, without loss of generality for our present purposes, we assume the phase
to be zero when z and t are zero. We also take $\boldsymbol{k}$ to be in the z direction, as assumed
above, and choose the signs so the wave propagates in the forward direction. This
is traditional but not necessary. In the discussion above, we assumed that $z = 0$
throughout. This is reasonable if the electron is essentially stationary in z. However,
as v_z increases, the electron now moves and as a result it no longer experiences
a purely sinusoidal field. The position z_p of an electron is given by (13.14) with
$z_o = t_o = 0$. The phase experienced by the electron, ϕ_e, is

$$\phi_e = \left(k \int_0^t v_z(t')dt' - \omega t \right) = \omega \left(\frac{1}{c} \int_0^t v_z(t')dt' - t \right). \tag{13.22}$$

Recalling that $B_y = E_x = -(1/c)dA/dt$, (13.5) becomes

$$\frac{\mathrm{d}}{\mathrm{d}t}(\gamma_r v_z) = -\omega \left(1 - \frac{v_z}{c} \right) (v_x a_o \cos \phi_e). \tag{13.23}$$

One can see that whether the electron is accelerated or decelerated in z depends
upon the sign of v_x and $\cos \phi_e$. But now consider the impact of v_z on the duration
of the acceleration. One can see from (13.21) and (13.22) that ϕ_e decreases as time

increases. The question now is what time Δt it takes for ϕ_e to change by a certain amount, equal to $-\Delta\phi_e$. Suppose the average velocity during this period is $\overline{v_z}/c$. Then $-\Delta\phi_e = \omega \Delta t(\overline{v_z}/c - 1)$, so

$$\Delta t = \frac{\Delta\phi_e}{\omega}\frac{1}{1 - \overline{v_z}/c}. \tag{13.24}$$

The important point is that periods when $\overline{v_z}$ is larger, as will occur if the electron is being accelerated along $\boldsymbol{k}$, last longer than periods when $\overline{v_z}$ becomes negative. The electron rides the wave when going forward but quickly moves through it when going backward. The result is that the electron is accelerated longer when moving forward, acquiring a high average forward velocity.

The electron eventually will stop ever moving backward and instead will move forward at nearly the speed of light, with the changes occurring primarily in γ_r rather than in v_z. To take an approximate look at this regime, let $v_z/c = 1 - \delta$, where δ is assumed to be small. One finds $v_x/c = \sqrt{2\delta}$ to the lowest order in δ. Then after defining $\eta = \omega t$ for convenience, (13.23) becomes, to the lowest order in δ,

$$\frac{d\delta}{d\eta} = \sqrt{2}\delta^{7/2}a_o \cos\left(\pi + \int_{\eta_o}^{\eta} \delta[\eta']d\eta'\right), \tag{13.25}$$

in which we have added π to the phase to initialize the electrons moving forward.

One can integrate this equation to see how δ behaves as time (η) increases. Figure 13.3 shows the value of δ and the value of the cosine in (13.25), taking $\delta = 0.01$ when $\eta = 0$. One can see that the brief periods of deceleration (which decrease v_z and thus increase δ) are not sufficient to greatly increase δ and (allowing for the logarithmic abscissa) that the duration of a cycle increases as δ decreases.

The simple analysis above is not self-consistent, and it ignores the range of initial conditions produced when an electron is introduced to the light wave with an arbitrary phase. Enam Chowdhury, in his Ph.D. thesis at the University of Delaware (2004), did a numerical treatment of a related problem. He used a tunneling-ionization model to inject electrons into fields corresponding to a model of a focused laser beam. It is helpful to review some of his results here.

Figure 13.4 shows the distribution of final momentum states that result from ionization of various ions in intense fields produced by a focused laser. Even though the initial laser beam is polarized with the electric field in the x direction, focusing the beam introduces finite electric fields in the y and z directions. This produces the finite momentum in y seen in the top set of panels in this figure. The relation between p_z and p_x is dominated by the behavior corresponding to an electron at rest, shown as a dashed line in the bottom panels. This is because most electrons are ionized near the peak of the laser field and are nearly at rest when they are produced. Such electrons form the most intense peaks in Fig. 13.4f, with momenta along the mean electric field of about ± 1.5 MeV/c. The other ionization events produce the range of electron momenta shown.

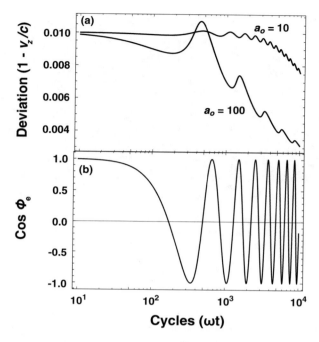

Fig. 13.3 Electron acceleration. (**a**) Deviation from speed of light for $a_o = 10$ (upper curve) and $a_o = 100$ (lower). (**b**) Phase ϕ_e of electron in wave for $a_o = 100$

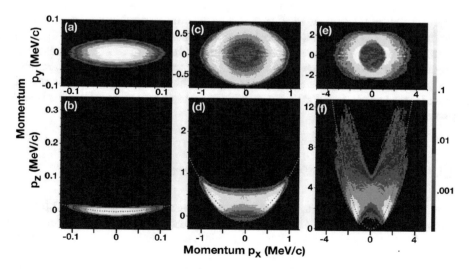

Fig. 13.4 Final state momentum plots for the ionization of atoms in a focus having an f-number of 2.5. The dotted lines represent the relation given in (13.11). (**a**) and (**b**) are for Ne^{7+} at 10^{17} W/cm^2, (**c**) and (**d**) are for Ar^{8+} at 10^{19} W/cm^2, and (**e**) and (**f**) are for Ar^{15+} at 10^{20} W/cm^2. Credit: Enam Chowdhury

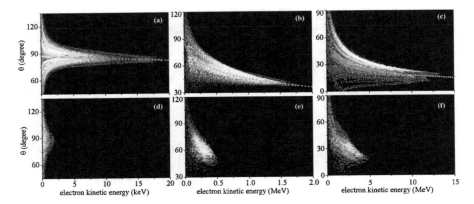

Fig. 13.5 The electron energy spectrum for the same physical conditions as in Fig. 13.4, with the left, center, and right columns corresponding to 10^{17} W/cm^2, 10^{19} W/cm^2, and 10^{20} W/cm^2, respectively. The top panels show the angle of deviation from the z-axis, in the x–z plane. The bottom panels show the angle of deviation from the z-axis, in the y–z plane. Here again the dotted lines represent the relation given in (13.11). Credit: Enam Chowdhury

Figure 13.5 shows the angle of the electrons relative to the z-axis. The top panels show the distribution in the x–z plane. This tends to follow the relation for an electron initially at rest at the maximum field (13.11), shown as a dashed line. The bottom panels show the distribution in the y–z plane. These electrons are influenced by the y-component of the electric field produced by focusing.

Thus, one can produce a beam of energetic electrons, having a distribution of energies like those seen in Figs. 13.4 and 13.5, from ionization produced during the laser pulse. This technique can be used to obtain a brief, energetic electron beam that can be used as a probe or to drive other processes. In contrast, the production of intense, directed beams of electrons for accelerators tends to require the interaction of a high-energy-density source with a plasma. We discuss this at more length in Sect. 13.7.1 after considering some basic aspects of relativistic laser–plasma interactions.

13.3   Initiating Relativistic Laser–Plasma Interactions

One cannot abruptly initiate a laser pulse of high-enough energy flux to produce strongly relativistic effects. The very best one can do is to produce a laser pulse with a Gaussian shape in time, with a pulse width whose characteristic time to reach $1/e$ from its maximum is $\sim 1/\Delta\omega$, where $\Delta\omega$ is the bandwidth. The problem this creates is that any target experiences all energy fluxes from zero to the maximum as the laser pulse arrives. The laser and optical system must have three properties in order to obtain the cleanest-possible laser–solid interactions. First, the laser spot size must be close to the diffraction limit, so that the experimental results will not be confused by structure within the laser spot. Second, most of the laser energy must be present in this spot, to avoid large signals from lower-intensity interactions outside it. Third, the pulse shape in time must be close to Gaussian, without significant structure early in time. Such early structure is known as a prepulse.

These issues have led laser systems devoted to the pure study of interaction phenomena at high fields to produce shorter and shorter laser pulses, to minimize the interactions at lower intensities. Such a laser also needs to use a deformable mirror as part of their focusing system, to produce a nearly diffraction-limited spot. These developments lead in their limit to systems described as λ^3 lasers (λ is the wavelength of the light), whose goal is to obtain a laser pulse of one cycle in the duration of its maximum irradiance (and thus of length λ in space), focused to an area of approximately λ^2. For light at a central wavelength of 800 nm, the duration of one cycle is about 3 fs. In contrast, lasers devoted to delivering large energies to a target, for example to attempt fast ignition (see Chap. 11), end up using much longer pulses, of order picoseconds. Some of the consequences of this are discussed below.

In discussing these consequences, it will prove useful to refer to three laser systems of the early 1990s. The Vulcan laser, at the Rutherford Appleton Laboratory in Britain, produced at that time relatively high energy in a long ($\sim$2 ps) pulse, by doing chirped pulse amplification within a laser system capable of large output energies (Danson et al. 1999). The TITANIA laser in Britain (Chambers et al. 1998) produced a spot that was five times the diffraction limit yet still contained substantial internal structure in addition to a prepulse. The laser facility of that same era at the Center for Ultrafast Optical Sciences (CUOS) at the University of Michigan produced a diffraction-limited spot. It could put high irradiance on target without ever producing a plasma layer whose thickness exceeded a skin depth. (The *skin depth* is the penetration distance of the evanescent laser pulse into a medium past a sharp, reflecting interface. It is the inverse of the imaginary value of the wavenumber in the region where the light cannot penetrate. For a sharp, plasma interface at a very high density, the skin depth is c/ω_{pe}.)

To produce a very thin, plasma layer imposes three requirements on the laser system. First, the irradiance of the laser light that independently makes its way through the laser optics to the target and that arrives nanoseconds before the main pulse, must be kept below 10^8 W/cm^2, to avoid producing vapor or plasma in front of the target (Combis et al. 1991; Lindley et al. 1993; Sauerbrey et al. 1994). Such "prepulses" are often produced by amplified spontaneous emission (ASE) from laser amplifiers, or by leakage through Pockels cells, and can easily be several tens of nanoseconds in extent. ASE prepulses have proven useful for the purpose of maximizing the production of hot electrons (Kmetec et al. 1992) or X-rays (Rousse et al. 1994), but they are harmful if one's goal is to obtain very clear evidence regarding laser interactions with solid matter.

Second, the irradiance of any "pedestal" on the main laser pulse itself must be kept below 10^{12} W/cm^2 to avoid plasma production that is too early. Such pedestals are typically a few ps to a few hundred ps in duration. They are often caused by effects within the laser system that alter the frequency spectrum of the laser light, such as group velocity dispersion in the laser glass or gain-bandwidth narrowing.

Early short-pulse glass lasers typically had a pedestal of order 10^{-3} times the maximum irradiance. Later, with the advent of Ti:sapphire oscillators, this was reduced to less than 10^{-5}. The laser at CUOS in 1992 had a measured pedestal of 2×10^{-6} times the maximum irradiance, next improved to 10^{-8} times the maximum infrared irradiance (Nantel et al. 1998). Doubling the frequency of this pulse kept the pedestal below 10^8 W/cm^2, even as I_L approached 10^{20} W/cm^2. In contrast, the highest-energy, ultrafast, glass lasers of that era had pedestals above 10^{12} W/cm^2 for many ps. More recently, advances in technology such as self-generated plasma mirrors (Dromey et al. 2004; Doumy et al. 2004; Bulanov et al. 2007) have enabled even higher contrast between the pedestal and the main pulse.

The third requirement on the laser system is that the main laser pulse must rise steeply enough at low irradiance. Once I_L exceeds about 10^{12} W/cm^2, plasma forms at the surface of the solid and begins to expand. We give I_L because the onset of preplasma depends primarily on the power delivered. (This does not rule out small effects that depend on wavelength and/or absorption, which we ignore here.) The rate of critical-surface expansion scales as the sound speed, which is proportional to $T_e^{1/2}$, where T_e is the electron temperature. It reaches of order 100 nm/ps at $I_L \sim 10^{15}$ W/cm^2. Once $I_L \lambda_\mu^2$ exceeds 10^{15} W μm^2/cm^2, the ponderomotive pressure of the reflecting light wave becomes large enough to stop the plasma expansion (Liu and Umstadter 1992). At higher energy fluxes, the ponderomotive pressure compresses the plasma and pushes the critical density layer back toward the solid material. A sufficiently intense pulse can first steepen an existing preplasma, so that the local density scale length, L/λ, becomes quite small, and can then push the critical surface inward, decreasing D. We distinguish between L, the local scale length of the density profile at the critical density, and D, the distance from critical density to solid density. L and D are simply connected for a free expansion but may be quite different in plasmas that have expanded and then been compressed. In the end, the only way to retain smooth, thin, planar plasma is to limit the initial plasma expansion so as to keep both L and D small at all times.

The CUOS laser of the early 1990s accomplished this by doubling the frequency of a very clean, Gaussian laser pulse of 400 fs FWHM. The doubling efficiency was saturated at the highest energy flux, so that the FWHM of the converted pulse was also about 400 fs. However, at lower energy flux on the rising edge, the energy flux of the second harmonic, proportional to the square of the first-harmonic energy flux, rose quite steeply. The second-harmonic energy flux, at 527 nm wavelength, increased from 10^{12} to 10^{19} W/cm^2 (16 e-foldings) in 500 fs. Confirmation that this laser system produced negligible preplasma was provided by studies of the high-density plasma using X-ray spectroscopy (Jiang et al. 1995). The increase in I_L from 10^{12} W/cm^2 to 10^{15} W/cm^2 took only 180 fs, which allowed the plasma to expand only about 10 nm. This is less than a skin depth for solid density (17 nm at 10^{23} cm^{-3}). After that, the light pressure prevented the plasma layer from expanding further.

Figure 13.6 compares the nominal profiles of the pulses of our three reference laser systems. A laser system with a clean, Gaussian pulse of 1 ps FWHM, having

Fig. 13.6 Reference laser pulses of lasers in the early 1990s. A schematic CUOS laser pulse is compared to clean, Gaussian pulses having FWHM of 1 and 2.5 ps. The two longer pulses are the best pulses that might have been produced by the TITANIA and Vulcan laser systems of that era, respectively. The longer pulses permit much more time for plasma formation

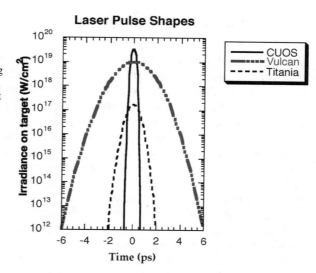

Table 13.2 Comparison of plasma size and excursion distances in planar experiments

Experiment	$I\lambda_\mu^2$ (W µm²/cm²)	Plasma size (D/λ)	Excursion distance (x_{os}/D)
Vulcan (1993)	3×10^{15}	0.2	0.04
Titania (1998)	1×10^{16}	0.2	0.08
CUOS (1999)	7×10^{18}	0.02	7

a peak I_L of 1.6×10^{17} W/cm², requires 700 fs to increase in I_L from 10^{12} to 10^{15} W/cm². This would allow a plasma expansion above 50 nm. This would have been achieved by the TITANIA laser system if it had a clean pulse and no pedestal or prepulses. A laser system with a clean, Gaussian pulse of 2.5 ps FWHM, having a peak I_L of 10^{19} W/cm², requires 1500 fs to increase in I_L from 10^{12} to 10^{15} W/cm². This would allow a plasma expansion above 100 nm. This would have been achieved by the Vulcan laser system if it had no prepulse. The resulting parameters are compared in Table 13.2.

One issue that determines when a plasma layer is thin enough to study laser–solid interactions is whether the interaction dynamics involves the solid-density matter or only involves the lower-density plasma near its surface. A demanding measure of this, and one which determines whether *Brunel electrons,* discussed further below, can participate in the absorption dynamics, is the ratio x_{os}/D. Here $x_{os} = a_o c/(\gamma_r \omega) = (\lambda/2\pi)(a_o/\gamma_r)$ is the excursion distance of the electrons in the electric field of the laser. We compare this parameter in Table 13.2 for the three reference cases.

13.4   Absorption Mechanisms

The absorption mechanisms change as I_L increases in ultrafast laser-target interactions. At low enough irradiance ($<10^{14}$ W/cm^2), the laser cannot heat the dense plasma very much in a short pulse, and collisional absorption of the evanescent laser electric field dominates the interaction. Strong laser absorption is observed at normal incidence and in *s-polarization* (with the laser electric field parallel to the surface of the target) is consistent with that expected from collisions. The absorption is enhanced in *p-polarization* (with the laser electric field parallel to the target normal), for modest scale lengths L, when the evanescent electric field of the laser light penetrating to critical density is resonantly enhanced, leading to *resonance absorption*, discussed in Kruer (1988).

As I_L increases above 10^{15} W/cm^2, the plasma becomes hot enough that collisional effects become small. Then the absorption becomes more complicated, introducing a number of mechanisms that we will not explore here. These include *sheath inverse bremsstrahlung*, the high-frequency skin effect, and the *anomalous skin effect*, all of which are discussed in a common context by Rozmus et al. (1996) and by Gibbon (1996). The simple relation between electron temperature and laser energy flux that we derived in Chap. 9 now breaks down, as the outward convection of hot material is no longer the important energy loss channel for the electrons. Even so, T_e scales as some power of I_L that is not too far from $1/2$. In this regime, resonance absorption is still strong but the energy transfer mechanism may not be collisional. The noncollisional absorption might be due to breaking of the laser-driven electron–plasma waves, but there is also evidence in some simulations of repeated "Langmuir collapse" as in Gibbon (1994). Experimentally, there is clear evidence of resonance absorption in this regime as described in Chaker et al. (1991) and Meyerhofer et al. (1993)

As I_L increases above 10^{18} W/cm^2, the physical excursions of the electrons can become comparable to the thickness of the plasma layer at the surface of a solid. In this case, the excursions of the electrons into the vacuum, accelerated by the laser light wave, can lead to enhanced absorption when they return and enter the solid. This effect, often called the *Brunel effect* as it was first identified theoretically by Brunel (1987) and later replicated in theory and simulation by others including Bonnaud et al. (1991) and Kato et al. (1993), can significantly enhance the absorption for p-polarized light. Resonance absorption gives way to vacuum excursions as I_L increases and the scalelength decreases, but the transition is complex (see Gibbon and Bell 1992). As discussed above with reference to Table 13.2, the condition for significant absorption by vacuum excursions is $x_{os}/D >$ 1. (Here x_{os}/D is the relativistic generalization of the parameter given by Brunel, v_{os}/ω, where v_{os} is the nonrelativistic, electron oscillation velocity in the laser electric field.)

It is worth noting that much earlier experiments produced plasmas that were recompressed, by the ponderomotive force, to have density profiles that were locally very steep. Both some experiments using CO_2 lasers (Bach et al. 1983; Fedosejevs et al. 1990) and more recent ones using 1 μm lasers (Norreys et al. 1996) produced plasmas that, probably, satisfied $x_{os}/L > 1$ at the critical surface. However, in the process of recompression these initially much-thicker plasma layers became rippled. This precluded clean diagnosis of the interaction processes.

In this same irradiance regime the anomalous skin effect begins to become important for normal incidence, if the plasma layer is sufficiently thin. The electrons are heated enough that the normal skin depth becomes smaller than the electron excursion length and the electron mean free path, leading to increased penetration of the laser electric field into the solid. Ruhl and Mulser (1995) used Vlasov simulations to study laser–solid interactions in this irradiance regime. They explain the connection between the Brunel effect and the anomalous skin effect as the natural limits of the force on the electrons through its dependence on the angle of incidence.

The Brunel effect eventually saturates in consequence of $v \times B$ forces (Brunel 1988). As I_L increases into the strongly relativistic regime, further new effects are predicted to arise. An essential aspect of these is that the only way for net heating of the plasma to occur is for the average $\langle J \cdot E \rangle$ to be nonzero. This requires that some physical effect shift the phase of J relative to E as they are perfectly out of phase in the initial laser beam. Mulser et al. (2008) describe the phenomena as *anharmonic resonance*. Gibbon et al. (2012) provide a general model of absorption at very high I_L that is somewhat independent of the specific details. What is of particular interest is that the absorption, directly into energetic electrons, exceeds 50% for $I_L > 10^{20}$ W/cm^2, as observed by Ping et al. (2008). At such high I_L, the $v \times B$ motion of the relativistic electrons in the laser field begins to produce a significant electron velocity directed into the target (Pukhov and Meyer-ter Vehn 1996). This can lead to hole boring and collisionless shocks, discussed in Sect. 13.8.

Despite the variation in specific absorption mechanisms, all the absorption processes involve transferring laser energy to electrons. When laser light is absorbed at a solid surface, this produces a distribution of electrons, often approximately Maxwellian in shape. One can write an energy flux balance equation that qualitatively describes the heating, as

$$f_{\text{abs}} I_L = n_c k_B T_e v_{\text{eff}}, \tag{13.26}$$

in which f_{abs} is the fraction of the laser energy that is absorbed, n_c is the critical density, and v_{eff} is the effective velocity at which energy flows away from the absorption region. In the slow (ns), large-plasma regime of Chap. 9,

$$f_{\text{abs}} \sim 1 \text{ and } v_{\text{eff}} \sim c_s \propto T_e^{1/2} \text{ so } T_e \propto I_L^{2/3}. \tag{13.27}$$

As the temperature increases and the collisional absorption becomes smaller, one enters a regime in which $f_{\text{abs}} \propto v_{ei}/\omega_o$ but v_{eff} is still of order the sound speed. Then, since $v_{ei} \propto T_e^{-3/2}$, one has

$$T_e \propto I_L^{1/3} \text{ in the weakly collisional regime.} \tag{13.28}$$

Heating with this dependence on I_L is often said to have *Beg scaling*, corresponding to data and analysis in the late 1990s (Beg et al. 1997).

Beyond these regimes the excursions of the electrons begin to matter and one finds that f_{abs} depends more weakly on T_e, bringing the increase of T_e with I_L up closer to

$$T_e \propto I_L^{1/2} \text{ in the excursion regime.} \tag{13.29}$$

As the laser irradiance increases further into the strongly relativistic regime, the electrons are observed to have a characteristic energy of order the kinetic energy of the electron oscillations in the laser light. This is a natural result—any process that manages to deflect an electron, changing its velocity in the direction of the laser electric field, will have the effect of converting some fraction of this energy into randomized motion. So the quiver energy must set the energy scale once the absorption loses its temperature dependence and the electrons leave at c. In the relativistic regime, the characteristic energy of these electrons, $\mathcal{E}_{hot}$, is

$$\mathcal{E}_{hot} = \left(\sqrt{1 + a_o^2} - 1 \right) m_e c^2. \tag{13.30}$$

For simple estimates, one often describes the "temperature" of these electrons using the relation $k_B T_h = \mathcal{E}_{hot}$. Once a_o is very large, one has

$$k_B T_h \sim a_o m_e c^2 \propto I_L^{1/2}. \tag{13.31}$$

13.5   Harmonic Generation

As I_L increases above 10^{17} W/cm^2, the interaction of the laser with the overdense target can lead to the emission of many harmonics of the laser light. This phenomenon has utility both as a diagnostic of the interaction mechanisms and as a potential source of coherent, short pulses of soft X-rays for other applications. Early theory on harmonic production from solids (Bezzerides et al. 1982; Grebogi et al. 1983) was motivated by the observation of many harmonics from long-pulse experiments using CO_2 lasers (Burnett et al. 1977; Carman et al. 1981). The important mechanism in this regime is the oscillation of the critical surface in the electric field of the laser light.

If the interaction geometry remains simple and planar, then the pump laser light will be specularly reflected and any harmonics will be emitted within the cone angle of the reflected pump light. It is well established that this requires a plasma layer that is much less than one wavelength in extent. Thus, the angular distribution of harmonic light is an indicator of how planar the surface is. In experiments on the TITANIA laser system, a transition from specular to diffuse harmonic emission was observed when the maximum energy flux of the laser pulse (of 1 ps FWHM for these experiments) exceeded 1.6×10^{17} W/cm^2. The 1 ps FWHM laser pulse shown in Fig. 13.6 corresponds to this maximum. One should note that, with such a pulse, the amount of preplasma produced by the leading edge of the laser pulse increases significantly as I_L increases. In experiments on the Vulcan laser system, whose long laser pulse (illustrated in Fig. 13.6) produced much more preplasma, diffuse harmonic emission was always observed (Norreys et al. 1996). In experiments at Toronto (Zhao 1998), there was a transition from specular to diffuse harmonic emission when a prepulse was introduced 1.5 ns before the main pulse. A consistent interpretation of these data, and of much other data with lower energy fluxes and longer pulse lengths not discussed here, is as follows. Given sufficient preplasma production, the critical surface where the laser light reflects becomes rippled. This is discussed, for example, by Wilks et al. (1992). Possible mechanisms include corrugation through electron clustering, irregular lateral motion of the ions, and Rayleigh–Taylor instabilities. This causes the laser–plasma interaction to occur over a wide range of angles, causing the scattering and reflection of the laser light to become diffuse. Once diffuse scattering sets in, the angular structure of the signal becomes worthless as a diagnostic of the interaction dynamics.

Theoretical work in the 1990s, nicely reviewed by Gibbon (1997), concluded that efficient harmonic production is possible from the interactions of sufficiently intense laser pulses with the overdense plasma at the surface of a solid. Wilks et al. (1993) observed odd harmonics in both 1D and 2D PIC simulations and observed (weaker) even harmonics in the 2D case. Relativistic oscillations of the critical surface introduce harmonic structure into the light reflected from it. One can think of this first in terms of the driven currents that emit the reflected radiation. At all angles of incidence, the ponderomotive force drives density oscillations with even harmonic content, which beat with velocity oscillations at the pump frequency ω to produce a current source with odd harmonic content. (We ignore here the case of circularly polarized light.)

For oblique incidence, the harmonic content also depends on the polarization. In s polarization, a current source develops with only even harmonic content. In p polarization, the density oscillations have both odd and even harmonic content. Those with odd harmonic content beat with velocity oscillations at ω to produce a current source with even harmonic content. In both s and p polarizations, the even harmonics are expected to be very weak near normal incidence and to increase as the angle of incidence increases.

An alternative way to think through the harmonic generation process is to treat the reflection from the critical surface as a reflection from an oscillating mirror, as suggested by Bulanov et al. (1994). Lichters et al. (1996) have done this to produce a cold-plasma model which accounts for the harmonic production observed in their simulations. They note that a correct treatment of the retarded source terms introduces substantial anharmonic content to this oscillation and that including this effect gives quantitative agreement between the model and the PIC simulations in some cases. (Highly resolved calculations for oblique incidence become possible by boosting the frame of a 1D PIC calculation to the Lorentz frame in which the electromagnetic wave appears to be normally incident.) They provide a quantitative model involving few assumptions that can be used to calculate the predicted

Fig. 13.7 The spectrum of
harmonic emission from
simulations by Lichters et al.
(1996), along with predictions
from their model (diamonds).
The crosses are from an ad
hoc model intended to mimic
the effects of surface plasma

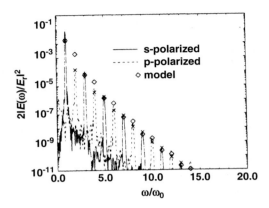

spectrum of the harmonics. They have difficulty with the magnitude of the lowest
even harmonics in oblique incidence with s polarization, as illustrated in Fig. 13.7.
They suggest that the sharpness of the plasma–solid boundary significantly affects
the production of these harmonics.

Another source of harmonic emission is also discussed by Gibbon (1996).
He argues that the particles which undergo vacuum excursions produce a very
anharmonic current source, repeating at the pump frequency, as they reenter the
solid and are strongly decelerated. Gibbon points out that for oblique incidence at
fairly large angles, such as 60°, vacuum excursions will become the dominant source
of harmonic emission.

13.6 Relativistic Self-focusing and Induced Transparency

We now ask what effect the plasma has on the laser light. To do so we return to the
wave equation for the laser light, (13.1), and evaluate $J_t = -en_e v_x = -en_e c a_o / \gamma_r$.
Converting A to a_o, we obtain

$$\left(\frac{\partial^2}{\partial t^2} + \frac{\omega_{pe}^2}{\gamma_r} - c^2 \nabla^2 \right) a_o = 0, \tag{13.32}$$

in which ω_{pe} is the plasma frequency corresponding to the electron density n_e. Upon
first glance, this may appear to be a simple equation, in which the plasma frequency
is reduced by a factor of γ_r. However, since γ_r depends on a_o through v, it also
varies in time on the same timescale as a_o does. One must work out the effects of
this when seeking a detailed solution, but one can formulate the solution in terms of
an appropriately averaged γ_r, which we designate as $\langle \gamma_r \rangle$. One obtains a modified
dispersion relation,

$$\omega^2 - \frac{\omega_{pe}^2}{\langle \gamma_r \rangle} - c^2 k^2 = 0, \tag{13.33}$$

from which the phase velocity is seen to be

$$\frac{\omega}{k} = c \frac{1}{\sqrt{1 - \frac{\omega_{pe}^2}{\langle \gamma_r \rangle \omega^2}}} = c \frac{1}{\sqrt{1 - \frac{n_e}{\langle \gamma_r \rangle n_c}}}. \tag{13.34}$$

Two conclusions follow from these equations. First, the wave travels slower in regions of higher a_o. Since this corresponds to the center of a focused laser beam, the phase fronts become curved and the laser beam tends to self-focus, a phenomenon known as *relativistic self-focusing*. This effect is opposed by diffraction of the laser beam, but above some threshold laser-beam power the beam does self-focus. This threshold laser-beam power, P_{sf}, is

$$P_{sf} = 17.4 \, (n_c/n_e) \text{ GW}, \tag{13.35}$$

in which n_c is the critical electron density as defined in Sect. 9.1. We saw that lasers producing relativistic high-energy-density conditions typically exceed this power. If such a laser-beam interacts with a sufficient volume of plasma, it will self-focus.

The second conclusion is that a relativistic laser beam can penetrate to a higher density than a nonrelativistic one. Taking (13.11) at face value, the density of reflection, where the phase velocity goes to zero, is $n_e = \langle \gamma_r \rangle n_c$. Thus, the laser can penetrate to higher density by a factor of $\langle \gamma_r \rangle$. In detail the effect is more complicated because of the fluctuations in γ_r. Despite this, the net effect is that a relativistic laser beam can penetrate to a higher density than a nonrelativistic beam. This phenomenon is known as *induced transparency*.

13.7 Particle Acceleration

Much of the interest in relativistic laser–plasma interactions revolves around the acceleration of particles by such lasers. It is hoped that beams of particles might be produced for fast ignition. Beams of protons from such lasers may prove useful for proton radiography or proton cancer therapy. Alternatively, beams of electrons may produce γ-rays for radiography or eventually electron beams for high-energy physics. By interacting intense lasers or particle beams with plasmas, researchers are working at this writing to produce advanced particle accelerators. This is possible because one can produce much larger electric fields in plasmas than one can between electrodes in vacuum. For these reasons, it is sensible to discuss the basic mechanisms of particle acceleration here.

13.7.1   Electron Acceleration Within Plasmas

The potential applications of high-energy-density plasmas and beams to electron accelerators nearly all involve the behavior of wakes in plasmas. The mechanism responsible for generating wakes is perhaps best understood by taking some time to throw rocks in a lake. The rock itself pushes some water outward, generating a first circular outgoing wave. But the rock also displaces water, in response to which the remaining water rushes in toward the center and then rebounds outward, with the consequence that a second circular outgoing wave is created. When one proceeds from throwing rocks to watching boats, one can see that these two waves correspond to the bow wave and to the trailing wake. The wakes propagate at an angle to the path of the boat, depending on their phase speed relative to the boat. If one gets onto the water and is able to match speeds with the wake, perhaps on a windsurfer or in a kayak, one can ride the wake, gaining speed and extracting energy from it. At this point one will understand the *wakefield accelerator,* save for a few (i.e. many) technical details. The seminal paper on the application of wakes to acceleration in plasmas was written by Tajima and Dawson (1979).

When one works inside a plasma, the key to producing electron acceleration is to generate some kind of wake. (In some limits, the wake begins to look more like an extended plasma wave.) The fundamental requirement is to create a local source of pressure that moves through the plasma, creating a wake that moves near the speed of light and whose electric field is large. Suppose for a moment that we repeat the experiment with the rock, except by creating a local and brief source of pressure, of some size λ_p, within a uniform cold plasma. (All realistic plasmas are cold when compared to motions at c.) The source is brief in the sense that it moves the electrons but causes little displacement of the ions. The local disturbance of the plasma creates a spherical wake, potentially including structure analogous to the two waves in the water, for the same reasons. The wake propagates because the electron density oscillates in response to a displacement, and because the displacement is local. We know from Sect. 2.4.1 that the electrons will oscillate in response to a charge separation at the electron plasma frequency, ω_{pe}. Since the perturbation moves a distance λ_p in one plasma cycle, the velocity of this wake is

$$v_p = \omega_{pe}/k_p, \tag{13.36}$$

where $k_p = 2\pi/\lambda_p$. If it happens that $v_p \approx c$ and that the wake is strong enough to trap some electrons and carry them along, then our imaginary spherical wake would accelerate these electrons to relativistic velocities.

The simplest application of this idea to real systems is to create a packet of photons or electrons of very-high energy density (and thus high pressure), whose characteristic length is close to $\pi c/\omega_{pe}$, as this will most effectively excite a plasma wake for which $v_p \approx c$. As this high-pressure packet traverses the plasma, it creates a strong plasma wake. The particles in such a wake can be trapped by it and substantially accelerated. The intense field produced immediately behind the high-pressure packet can be particularly effective at producing a beam of accelerated particles having a narrow energy spread and low divergence.

More complicated applications of the wakefield idea abound, because they require less-demanding experimental hardware. A laser beam above the threshold for relativistic self-focusing will tend to focus to produce a small region with very intense fields. When one or more extended laser beams resonantly drives an extended, intense plasma wave with $k \sim \omega_{pe}/c$ the resulting electric field will tend to self-modulate on the scale of ω_{pe}/c through the action of an instability (exactly how depends on details). This will drive in turn a series of wakes in the plasma and can accelerate particles. When two laser beams are used to create such a plasma wave by beating, in a process identical to the beating involved in stimulated Raman scattering (Chap. 9), the accelerator is known as a *beat-wave accelerator.* In the beat-wave accelerator and some of the other approaches, one cannot expect the plasma wake to trap cold particles and instead must inject particles that are then accelerated to higher energy. Overall, there is a veritable zoo of possibilities with an alphabet-soup of acronyms describing various different approaches to the goal of creating plasma wakes to accelerate electrons. We will leave the exploration of this zoo to those who develop a specialized interest.

Much of the interest in the complicated mechanisms just mentioned has been superceded by the discovery of the so-called "bubble regime" by Pukhov and Meyer-ter Vehn (2002). This was an important conceptual development for wakefield accelerators. The limiting idea is that an intense enough, tightly focused pulse of laser light can push the electrons out of a small volume of plasma, creating a "bubble". As the bubble moves through the plasma, it naturally accelerates electrons so as to produce a beam having a narrow spread in energy. One readily sees these effects in PIC simulations (see Fig. 13.8). Several groups actively pursued this approach,

Fig. 13.8 Spatial map of electron density associated with bubble-regime wakefield acceleration, as calculated by a Particle In Cell code. The bubble follows the laser pulse, which is moving to the right here. The movement of the bubble creates a group of accelerated electrons that are well-collimated and have a small spread in energy. Adapted from Geissler et al. (2006); used with permission

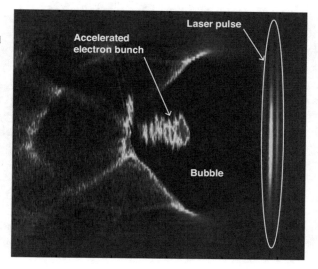

and three of them simultaneously produced results, published in the same issue of *Nature* (Mangles et al. 2004; Geddes et al. 2004; Faure et al. 2004) Actual bubble-regime accelerators do not manage to expel all the electrons from the focal volume. Even so, they work very well to produce electron beams having energies of dozens to hundreds of MeV. Such accelerators are in routine use now in many laboratories around the world. How far in energy they can be pushed remains to be seen.

To determine the potential capabilities of wakefield accelerators, one would like to know the possible electron energy gain. We evaluate this next, following Tajima and Dawson (1979). The maximum energy gain will occur when the wake becomes so large that it breaks. This occurs when the electron excursion distance during a plasma period, x_{os}, reaches $k_p x_{os} \approx 1$. We observe this oscillation in the lab frame, where

$$x_{os} = eE_w/(m_e \omega_{pe}^2). \qquad (13.37)$$

Here E_w is the electric field in the wake, related to the potential Φ by $|E_w| \approx k_p \Phi$. This gives

$$e\Phi = m_e \omega_{pe}^2/k_p^2 \approx m_e c^2 \qquad (13.38)$$

in the lab frame.

The energy gain of the electron is most simply evaluated in the moving frame of the plasma wake, where the electron oscillates in a stationary potential well. Referring to the volume of Landau and Lifshitz (1987) on the theory of fields for the relativistic transformations, we find the depth of the potential well in this moving frame, $\Phi^{(w)}$, in terms of the lab-frame quantities, as

$$e\Phi^{(w)} = \gamma_r e\Phi \approx \gamma_r m_e c^2, \qquad (13.39)$$

in which

$$\gamma_r = 1/\sqrt{1 - v_p^2/c^2}. \qquad (13.40)$$

Thus γ_r is the relativistic factor corresponding to the motion of the plasma wake. We also take $\beta \equiv v_p/c$. The electron has its maximum energy in the lab frame just when it has the maximum energy in the frame of the wake, where the energy is $\gamma_r m_e c^2$. To find the corresponding laboratory-frame energy, $\mathcal{E}_{lab}$, we Lorentz transform the energy back into the lab frame, finding

$$\mathcal{E}_{lab} = \gamma_r[\gamma_r m_e c^2 + \beta c(\gamma_r \beta m_e c)] = \gamma_r^2 m_e c^2(1 + \beta^2) \approx 2\gamma_r^2 m_e c^2. \qquad (13.41)$$

It is helpful to know the relativistic factor γ_{max} of the accelerated electron, seen in the lab frame, that has this much energy. This is evidently

$$\gamma_{\max} = 2\gamma_r^2. \tag{13.42}$$

For the specific case of using a light pulse to create the pressure, the optimum light pulse will have a group velocity, v_g, that equals v_p. As we discussed in Chap. 9,

$$v_g = c\sqrt{1 - \frac{\omega_{pe}^2}{\omega^2}}, \tag{13.43}$$

from which one can infer that

$$\gamma_{\max} = 2\omega^2/\omega_{pe}^2. \tag{13.44}$$

Thus, for example if one uses laser light of wavelength 1 μm to drive a wake in a plasma of density 10^{17} cm^{-3}, one finds $\gamma_{\max} = 2 \times 10^4$ and $\mathcal{E}_{\text{lab}} = 10$ GeV. There are reasons this is not trivial, though. The required laser pulse duration of π/ω_{pe} is 180 fs. The length over which a uniform plasma must be provided so that this acceleration can occur, ℓ_a is

$$\ell_a \approx \mathcal{E}_{\text{lab}}/|eE_w| = 2c\omega^2/\omega_{pe}^3, \tag{13.45}$$

which would be 34 cm. Sustaining a uniform plasma and a focused laser pulse over this distance would be a severe challenge. But if one raises the plasma density (i.e., increases ω_{pe}) to make the distance shorter, the laser pulse length soon approaches one laser cycle. Clever technique will be needed to achieve energies of tens of GeV or more using lasers.

Among other alternatives to lasers, one can use an electron bunch to create the pressure, in which case the plasma wake will accelerate some members of this bunch to higher energy. At this writing, the energy of a ~30 GeV electron beam has been more than doubled by this technique (Corde et al. 2016). In this application, the optimum plasma density is determined by the length of the electron bunch, ℓ_B, which should be

$$\ell_B = \pi c/\omega_{pe} \tag{13.46}$$

for the reasons discussed above. To keep the electron bunch focused as it propagates requires a balance between beam divergence and self-focusing in the plasma. This sets a relation between the plasma density and the beam size at the plasma entrance, r_{bo},

$$\omega_{pe}^2 = 2\gamma_B c^2 \epsilon_B^2/r_{bo}^4, \tag{13.47}$$

in which the relativistic factor for the bunch is γ_B and the emittance of the bunch (related to divergence, and having units of distance times angle) is ϵ_B. Thus, the electron bunch must be carefully shaped for optimum acceleration. It remains true in this case that $\gamma_{\max} = 2\gamma_r^2$ and that $eE_w \approx m_e c\omega_{pe}$, so the acceleration length is

$$\ell_a \approx \gamma_{\max} c/\omega_{pe} = 2\gamma_r^2 c/\omega_{pe}. \qquad (13.48)$$

One can see that keeping ℓ_a from growing while increasing $\gamma_{\max}$ requires increasing ω_{pe}. This in turn requires shortening the bunch and making it smaller in diameter. At this writing, this is also an active area of research.

13.7.2 Ion Acceleration

There are several mechanisms that can produce ion acceleration in targets irradiated by relativistic lasers. We discuss some of them here.

13.7.2.1 Acceleration by Surface Potentials on Solid Targets

The wakefield and beat-wave processes just discussed might perhaps be developed into next-generation particle accelerators. Accelerators place demanding constraints on the dispersion of the accelerated particles in energy and in angle. Some other applications, such as proton radiography, may not be so demanding in these respects. Beams of protons (and other ions) are typically observed when relativistic laser beams strike solid targets. One way that they arise is now known as *Target Normal Sheath Acceleration*, abbreviated as TNSA. The mechanism of electrostatic ion acceleration in the plasma sheath at a target surface is easy to understand. The laser interacts with the electrons, and it removes many of them from the target. The result is that the target becomes positively charged, expelling ions from its surface. Because the surface is planar, the ions are expelled along the target normal. Because the laser produces plasma on the front surface of the target and produces electrons that easily penetrate a target that is not too thick, it is not uncommon to see beams of ions that emerge from both surfaces.

We can analyze this type of acceleration by building on our analysis of the self-similar, isothermal rarefaction in Sect. 4.4.1 and following the discussion of Mora (2003). Assuming the electrons to be Maxwellian, we can write their distribution function, as a function of energy E, as

$$f_e = \frac{n_{eo}}{k_B T_e} e^{-E/(k_B T_e)}, \qquad (13.49)$$

normalized so the integral over all energies gives their density n_{eo}. The sheath plasma expanding from a surface acts as a potential well for electrons. One sets the scale for the plasma potential, Φ, so that $\Phi = 0$ at the target surface and $\Phi < 0$ throughout the expanding plasma. Only those electrons with an energy above the local potential energy in which they are trapped, $-e\Phi$, can reach a given location. The density at that location is then

$$n_e = \int_{-e\Phi}^{\infty} \frac{n_{eo}}{k_B T_e} e^{-E/(k_B T_e)} \, dE = n_{eo} e^{(e\Phi)/(k_B T_e)}, \tag{13.50}$$

in which the initial electron density at the target surface is n_{eo} and one thus has

$$e\Phi = k_B T_e \ln(n_e/n_{eo}). \tag{13.51}$$

This equation is valid so long as electron collisions are fast enough to sustain a Maxwellian distribution, and this remains true even in ultrafast experiments. So we will assume here that the expansion is isothermal, at least while the most energetic ions are produced before the expanded plasma cools significantly.

In a planar, isothermal rarefaction the density is exponential in space. We previously derived the self-similar solution for the ions, finding

$$\rho = \rho_o \exp\left(-\frac{x}{c_s t} - 1\right), \tag{13.52}$$

in which c_s is as usual the sound speed. In an ultrafast experiment, the ions have little time to heat so $c_s = \sqrt{Z k_B T_e/m_i}$ for ion mass m_i. The derivation of this equation assumed quasineutrality of the plasma, so $\rho = n_e A m_p/Z$. One can easily show that the electric field, E_{ss}, corresponding to the self-similar solution, is constant and is given by

$$eE_{ss} = k_B T_e/(c_s t). \tag{13.53}$$

However, in an ultrafast experiment the self-similar model eventually breaks down because the assumption of quasineutrality becomes invalid. Charge separation can always occur on the scale of a Debye length, λ_D. Once λ_D exceeds the local scale length of the ion density profile, $c_s t$, the self-similar model breaks down. Then the electrons are able to stay ahead of the ions, which creates an expansion front and pulls the ions along. This roughly occurs when

$$\lambda_D = \lambda_{Do} \sqrt{n_{eo}/n_e} = \lambda_{Do} \exp\left[\frac{1}{2}\left(1 + \frac{x}{c_s t}\right)\right], \tag{13.54}$$

in which the Debye length at n_{eo} is λ_{Do}. From the self-similar rarefaction, this occurs at $x = (c_s t)(2\ln(\omega_{pi} t) - 1)$, where ω_{pi} is the ion plasma frequency at n_{eo}, equal to $\sqrt{4\pi n_{eo} Z e^2/m_i}$. For fully ionized, solid-density Be, one has $\omega_{pi} \sim 3 \times 10^{14}$ rad/s, so $\omega_{pi} t \sim 100$ in a few hundred fs. In this case the electric field at the ion front, E_{front}, is estimated to be $2E_{ss}$, due to the surplus of electrons at larger x. Simulations of the plasma expansion have shown that to excellent accuracy

$$eE_{\text{front}} = \frac{\omega_{pi}}{c_s} \frac{2k_B T_e}{\sqrt{5.44 + \omega_{pi}^2 t^2}}. \tag{13.55}$$

One can integrate (13.55) (which $= m_i dv_i/dt$) to find the ion velocity v_i and the maximum ion energy, $\mathcal{E}_{max}$. These are

$$v_i = c_s \left[\ln(2\omega_{pi}^2 t^2) - 1 \right],\tag{13.56}$$

and

$$\mathcal{E}_{max} = 2Zk_BT_e \left[\ln(\omega_{pi}t\sqrt{2/2.72}) \right]^2.\tag{13.57}$$

For $\omega_{pi}t \sim 100$, this gives $\mathcal{E}_{max} \sim 40Zk_BT_e$. In the strongly relativistic regime, where $k_BT_e \sim a_o m_e c^2$, this implies $\mathcal{E}_{max} \sim 20Za_o$ MeV. This implies that ion accelerations to many MeV are plausible by sheath acceleration even for modest values of a_o. This process is most likely to be important on the front side of all targets and on the rear side of solid targets that are thick enough that the explosion processes discussed below are not dominant.

13.7.2.2 Acceleration by a Laser Piston

As the laser energy flux increases beyond the levels that drive collisionless shocks, the analysis discussed below in Sect. 13.8 would predict from momentum balance that the laser can produce relativistic ions. In detail, however, the laser pulses push on the electrons and they in turn create an electric field that accelerates the ions. When the electrons are able to accelerate the ions to relativistic velocities during the laser pulse, the system has entered the *laser piston* regime, discussed by Esirkepov et al. (2004). The requirements for this regime can be calculated in simple limits as follows.

We suppose that the laser beam pushes all of the electrons completely out of the initial target, which is a thin layer of thickness d. For simplicity suppose that d is much less than the width of the laser spot, which implies that the electric field is independent of exactly how far the electrons have been pushed. The electric field is produced by both the electron layer and the ion layer, each of which has the same charge. From the Poisson equation one can show that the electric field experienced by the ions closest to the electrons is then

$$E_{||} = 16\pi^2 en_{eo}d,\tag{13.58}$$

which is in cgs units. In SI units the coefficient that is $16\pi^2$ would be 4π. The relativistically correct ion energy is

$$\mathcal{E}_i = \sqrt{m_i^2 c^4 + p_i^2/(2m_i)} = \sqrt{m_i^2 c^4 + (eE_{||}ct)^2},\tag{13.59}$$

in which the rightmost expression is specific to ions that acquire energy while moving through constant electric field at speed c (which is clearly an approximation for early times). The approximate time at which the kinetic energy of the ions reaches their rest mass, thus making them relativistic, can be found from this to be

$$t_{\text{rel}} = \frac{m_i c^2}{e E_\parallel} = \frac{m_i c^2}{16\pi^2 e^2 n_{eo} d} = \frac{Z(c/d)}{4\pi \omega_{pi}^2}. \tag{13.60}$$

If we hope to avoid complications not present in this model, this acceleration should happen within a laser cycle, so that $\omega_o t_{\text{rel}} \sim 2\pi$. We would like to know how intense a laser is required to accomplish this. We find this by recognizing that the electric field in the laser, E_L, must be larger than $E_\parallel$ and using (13.60) to solve for d. We find this implies that the target thickness should be

$$d \sim \frac{Zc\omega_o}{8\pi^2 \omega_{pi}^2} = \frac{c}{8\pi^2 \omega_o} \frac{\omega_o^2}{\omega_{pe}^2} \frac{m_i}{m_e} = \frac{c}{8\pi^2 \omega_o} \frac{n_c}{n_{eo}} \frac{m_i}{m_e}, \tag{13.61}$$

which will typically be some fraction of a laser wavelength. Using this result in (13.58), $E_L \geq E_\parallel$, and $a_o = e E_L / (m_e \omega_o c)$, we find

$$a_o = \frac{1}{2\pi} \frac{m_i}{m_e}, \tag{13.62}$$

which for protons is an $I_L \lambda_\mu^2$ of $\sim 1.2 \times 10^{23}$ W μm^2/cm^2. A major challenge in practice for this and similar schemes is to have the target stay thin enough until a_o reaches the necessary value.

13.7.2.3 Acceleration by Coulomb Explosions

A third important acceleration mechanism, also responsible for high-energy ion production and even for ion beams, is known as a *Coulomb explosion*. Coulomb explosions, created by the irradiation of molecules with photons or particle beams, have been used in chemistry since at least the 1970s. We will see shortly why it took longer to produce them using lasers. With relativistic lasers, they are generally used with gas clusters, which form when a supersonic nozzle releases gas into vacuum under appropriate conditions. They provide a way to absorb a very large fraction of the laser energy, producing a plasma far hotter than one could by irradiating ordinary gas (Ditmire et al. 1997). Our analysis draws in part on Zweiback et al. (2002).

First consider a spherical distribution of ions at constant density n_i, from which the electrons have been magically removed. One can integrate the Poisson equation to find the electric field at location of any ion at radius r inside the cluster, in cgs units, as

$$E = 4\pi n_i Z e r, \tag{13.63}$$

then solve $E = -\nabla \Phi$ to find the potential Φ as

$$\Phi = 2\pi n_i Ze \left[\left(\frac{8\pi}{3}\right) r_o^2 + (r_o^2 - r^2)\right] \text{ for } r < r_o \text{ and}$$

(13.64)

$$\Phi = 2\pi n_i Ze \left(\frac{8\pi}{3}\right) \frac{r_o^3}{r} \text{ for } r > r_o,$$

in which as usual n_i is the ion density, Z is the ionic charge, and e is the charge of an electron. In addition, the initial radius of the sphere is r_o and we have chosen Φ to be zero as $r \to \infty$.

The dynamics of such an ion cloud is simple. All the ions are accelerated outward, with the outermost ions being accelerated most greatly. As r increases for each ion, the acceleration decreases. So no interior ion overtakes any outer ion. As a result, each ion at radius r ends up converting the electrostatic potential energy created by the charge at radii $< r$ into kinetic energy. The maximum ion energy $\mathcal{E}_{max}$ is obtained when an ion at the outer edge of the sphere has been accelerated until the remaining potential is negligible. Then the maximum ion energy is

$$\mathcal{E}_{max} = \frac{8\pi}{3} 2\pi n_i Z^2 e^2 r_o^2.$$

(13.65)

The derivation of the corresponding, normalized, ion energy distribution function $f(\mathcal{E}) = (1/N)dN/d\mathcal{E}$ is left as homework. It is

$$f(\mathcal{E}_i) = \frac{3}{2} \frac{\sqrt{\mathcal{E}_i}}{\mathcal{E}_{max}^{3/2}} \text{ for } \mathcal{E}_i < \mathcal{E}_{max} \text{ and}$$

(13.66)

$$= 0 \text{ for } \mathcal{E}_i > \mathcal{E}_{max}.$$

For $r_o = 10 \ \mu m$, which is a plausible laser-beam spot, $\mathcal{E}_{max}$ evaluates to $0.76 n_i/(10^{18} \text{ cm}^{-3})$ MeV for $Z = 1$. Thus, MeV energies might plausibly be obtained from gasses and much larger energies could be obtained from (thin) solids, if in fact such explosions can be produced. We now consider some aspects of this.

To create a Coulomb explosion, the laser must strip the electrons from the ions and expel them from the cloud. The laser will have to turn on very quickly and reach sufficient irradiance. Two timescales matter. These are the explosion time itself and the sonic expansion of the cloud. Let us consider these in reverse order. The sound speed here again is $c_s = \sqrt{Zk_BT_e/m_i}$, so the sonic disassembly time is

$$\tau_{sonic} = r_o/c_s = \sqrt{r_o^2 m_i/(Zk_BT_e)}.$$

(13.67)

If we crudely estimate $k_B T_e = a_o m_e c^2$ from (13.31), then we find

$$\tau_{\text{sonic}} = 140 \sqrt{\left(\frac{A r_\mu^2}{Z a_o}\right)} \text{ fs} ,$$

(13.68)

in which r_μ is r_o in μm.

The explosion time can be found by solving the equation of motion for an ion at the outer surface of the cloud. Because the charge within the cloud is fixed and remains within the radius of the ion, one can relate the ion velocity to the potential energy. One has, for the ion velocity v_i,

$$\frac{1}{2} m_i v_i^2 = \mathcal{E}_{\text{max}} \left(1 - \frac{r_o}{r}\right) ,$$

(13.69)

which can be written as

$$\left(\frac{v_i}{v_{\text{max}}}\right)^2 = 1 - \frac{r_o}{r} .$$

(13.70)

This allows the equation of motion,

$$\frac{d}{dt} \left(\frac{v_i}{v_{\text{max}}}\right) = \frac{ZeE}{v_{\text{max}}} = \frac{\mathcal{E}_{\text{max}}}{m_i r_o v_{\text{max}}} \left(\frac{r_o}{r}\right)^2 ,$$

(13.71)

to be written as

$$\frac{d}{dt} \left(\frac{v_i}{v_{\text{max}}}\right) = \frac{v_{\text{max}}}{2 r_o} \left(1 - \frac{v_i^2}{v_{\text{max}}^2}\right)^2 .$$

(13.72)

By integrating this equation to find when the ion energy reaches half of $\mathcal{E}_{\text{max}}$, we can obtain a reasonable value for the duration of the explosion, τ_{exp}, which is

$$\tau_{\text{exp}} = 3 \text{ ps} \frac{\sqrt{A}}{Z \sqrt{n_{18}}} ,$$

(13.73)

in which n_{18} is the ion density in units of 10^{18} ions per cm^3. Note that this turns out to be independent of the radius of the cloud. At the densities associated with solids or with clusters in gasses, of order 10^{23} ions per cm^3, this time is of order 10 fs.

The third requirement is that the laser must be able to expel the electrons. The electrons are affected by the ponderomotive force produced by the laser light, discussed in Sect. 9.1. The nonrelativistic expression for this force can be written as the gradient of a ponderomotive potential, defined by $U = m_e \langle v_{os}^2 \rangle / 2$, where the brackets denote an average that introduces a factor of 2. One finds that this potential contributes to the total energy of the electrons and thus can balance other sources of potential energy, such as the Coulomb potential associated with a charged cloud. The relativistically correct ponderomotive potential can be written

$$U = m_e c^2 \left(\sqrt{1 + a_o^2} - 1 \right).$$ (13.74)

Setting this equal to the maximum Coulomb potential, which is $\mathcal{E}_{max}$, we find

$$a_o > 1.2\sqrt{n_{18}}\sqrt{2 + 1.5 n_{18}}$$ (13.75)

to have enough ponderomotive potential to completely remove the electrons. If instead the ponderomotive potential is only large enough to remove some of the electrons, the cloud of ions will still explode. However, it will not produce the maximum possible ion energies.

In summary, to obtain a Coulomb explosion the laser pulse must rise quickly enough to avoid thermal expansion. This time decreases as the laser energy flux increases or the laser spot (or the target cloud, if it is smaller) shrinks. The laser pulse must also rise quickly enough to avoid a premature explosion before the electrons are expelled and must be of long enough duration for the explosion to occur. This duration decreases as density increases, reaching femtoseconds for solid densities. Finally, the laser pulse must be powerful enough to expel the electrons, and the required irradiance increases roughly as the square of the ion density.

13.8 Hole Drilling and Collisionless Shocks

When the laser energy flux onto a high-density target plasma (having $n_e > n_c$) becomes large enough, the ponderomotive pressure no longer merely keeps the density profile steep, but actually drills a hole into the plasma, pushing the ions ahead of it. Under circumstances we will discuss, this can drive a *collisionless shock* into the plasma. The shock reflects plasma ions, but often will not be relativistic. As a result the reflected ions are given roughly twice the velocity of the shock (in the laboratory frame). This is another mechanism, in addition to sheath acceleration and Coulomb explosions, that can produce energetic ions in ultrafast laser–plasma interactions.

The phenomenon of *hole drilling* by relativistic lasers was first discussed by Wilks et al. (1992). One can understand the basic behavior by considering the momentum exchange between the laser beam and the plasma ions. The electrons, of course, carry negligible momentum because of their small mass. The momentum of each photon is $\hbar k$, so that the total incident momentum flux is $\hbar k \times I_L/(\hbar \omega) = I_L/c$ and the total momentum flux delivered to the plasma by the laser is $(1 + \eta)I_L/c$, where the fraction of the laser power reflected is η. This must be balanced by the momentum flux of the ions. If the ions that have already been swept up move into the target at v_i, then the ion flux being swept up is $n_i v_i$ and the momentum flux being delivered to the newly-swept-up ions is $m_i n_i v_i^2$. Thus,

$$\left(\frac{v_i}{c}\right)^2 = \frac{I_L(1+\eta)}{n_i m_i c^3}, \tag{13.76}$$

or if we formulate this in terms of a_o, using also (9.13),

$$\left(\frac{v_i}{c}\right)^2 = \frac{a_o^2(1+\eta)}{m_i c^3} \frac{n_c}{n_i} \frac{1.37 \times 10^{18}}{1.1 \times 10^{21}} = 2.7 \times 10^{-4} a_o^2(1+\eta) \frac{n_c}{n_{eo}} \frac{Z}{A}, \tag{13.77}$$

where the electron density in the target is n_{eo} and the critical density for the laser, from (9.13) is n_c. This equation applies to the regime of non-relativistic ion velocity once a_o is large enough that other pressures are negligible. It implies that a picosecond-timescale laser having $a_o \sim 30$ can drill a hole into a critical-density target that is 50 μm deep. This may prove important in the context of fast ignition (see Chap. 11).

We can compare this ion velocity to the sound speed as follows. For large a_o, the laser-heated electron temperature T_h is given by $k_B T_h \sim a_o m_e c^2$, as discussed above. Thus, the sound speed is

$$\left(\frac{c_s}{c}\right)^2 = \frac{1}{c^2} \frac{Z k_B T_h}{m_i} = a_o \frac{m_e}{m_p} \frac{Z}{A}. \tag{13.78}$$

Thus, the ions penetrate the plasma with a Mach number given by

$$M^2 = \left(\frac{v_i}{c_s}\right)^2 = 0.5 a_o (1+\eta) \frac{n_c}{n_{eo}}. \tag{13.79}$$

This in turn gives a threshold value of a_o to make this Mach number greater than 1, as

$$a_o \geq \left(\frac{2}{1+\eta}\right) \frac{n_{eo}}{n_c}. \tag{13.80}$$

Thus, lasers will drive supersonic ion fluxes into critical-density plasmas beginning at $a_o \sim 1$ and into solid-density plasmas beginning at $a_o \sim 100$.

In the context of hydrodynamics, we would conclude that a shock must form once $M > 1$. However, collisions are far too weak to produce a shock in such a target, so if one is to form it must be by means of collisionless dynamics. This does occur, as follows (for a further description, see the papers by Forslund and Shonk (1970a,b) and Silva et al. (2004)). Interpenetrating plasmas are unstable to an instability known as the *two-stream* instability once the interpenetration velocity exceeds the thermal velocity of one of the species. This instability can resonate between any two species in the plasma, so that there is an ion–ion and an ion–electron two-stream instability. Both these instabilities have a growth rate of some fraction of the ion–plasma frequency, given by $\omega_{pi}^2 = 4\pi n_i Z^2 e^2/m_i$ (in cgs units). This means that they grow to a large amplitude rapidly on the timescale on which the ions move. They are so-called electrostatic instabilities, producing longitudinal electric fields but no magnetic fields.

These instabilities lead to a shock of sorts, as follows. The electrostatic waves become quite large and trap electrons locally where the ions begin to interpenetrate. This takes a time of order $5 \times 2\pi/\omega_{pi}$, as has been confirmed by PIC simulations. The trapped population of electrons prevents further electrons from crossing the interaction zone. As a result, the ions that do cross this zone establish a positive potential and are reflected. The system reaches a steady state. In the shock frame, one would say that the ions coming in from upstream are in large part elastically reflected once they cross the trapped-electron region, while the downstream ions are stationary, being held in place by the ram pressure associated with the reflected ions on one side and the momentum flux delivered by the laser on the other side. In the laboratory frame, one would say that the laser pushes ions into the target, that at their leading edge a group of trapped electrons is produced, and that the resulting positive potential barrier strikes the ions beyond it and sends them ahead of the shock at twice the shock velocity. The density of the ions downstream of the shock, which is effectively compressed from both sides, ends up somewhat larger than the density of the ions ahead of the shock.

Figure 13.9 shows the results of a simulation of such a shock. Like most displays of PIC results, the dots on this plot correspond to the location of a given particle in z velocity and in z position. The initial location of the target was from 500 to 640 distance units. One can see sheath-accelerated particles at the two ends of the target. The shock is at about 550 distance units and is followed by structure that we will not concern ourselves with here. The shock-reflected ions are the highest-energy particles, seen streaming ahead of the shock with momenta above $0.15 m_p c$.

Fig. 13.9 Particle velocities from a PIC simulation of a collisionless shock. A laser pulse having $a_o = 16$ has been incident on a plasma initially having $n_e = 10 n_c$ for a time of $1024/\omega_o$ from Silva et al. (2004)

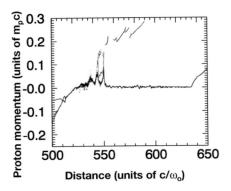

Since few ions join the ions behind the shock, the "piston" velocity at which the laser pulse pushes the ions ahead of it equals the shock velocity. One thus has a reflected-ion velocity, v_{ref}, given by

$$\left(\frac{v_{ref}}{c}\right) = 0.03 a_o \sqrt{\frac{n_c}{n_{eo}} \frac{Z}{A}}, \tag{13.81}$$

which we can also write as

$$\left(\frac{v_{ref}}{c}\right) = \frac{0.06}{(1+\eta)} \sqrt{\frac{n_{eo}}{n_c} \frac{Z}{A}} M^2. \tag{13.82}$$

Even for M not much larger than 1, this can easily reach a few tenths of the speed of light. Under the right conditions, with a solid-density target and a very large a_o, even a shock with $M \sim 2$ might drive relativistic reflected ions.

It would seem from the above equations that relativistic ions would be straightforward to produce by sufficiently increasing M. This, however, does not work in practice because the ability of the fluctuations produced by the two-stream instability to trap electrons is limited, and as a result ions above a certain energy (in the shock frame) cannot be reflected. The limiting Mach number, known as a *critical Mach number*, is $M \sim 3$. Above this Mach number fluctuations are still driven and the interpenetrating ion beams are affected, but there is no longer a shock. Faster ions cannot then be driven by shocks, but they can be driven in the "laser piston" regime, discussed in Sect. 13.7.2.2.

The mechanism of laser-driven collisionless shocks is most effective and most important in comparatively thin targets. This better enables the laser to keep the electrons heated throughout the target so that the shock remains below the critical Mach number and in addition allows the ions to move through the target in time to be further accelerated by the sheath on its rear surface. At this writing, recent experiments have observed 20 MeV beams of protons produced by reflection from such shocks (Haberberger et al. 2012).

A different type of "shock" is produced during any realistic Coulomb explosion of more than a small, isolated cluster. This develops as follows (Kaplan et al. 2003). The density profile of the ions is never entirely uniform, and in addition the electron expulsion may not be complete at the outer edges of the initial ion cloud. This has the consequence that the force on the ions reaches a maximum within the cloud, and that some ions within the cloud are accelerated more strongly than those on its outer edge. As a result, some ions overtake others. If the explosion truly involves cold ions and no electrons, then the Mach number of this interpenetration will be too large to drive the type of shock described above. (Such shocks might develop if the ions were warm enough or some laser-heated electrons were present, to sufficiently increase the sound speed.) But even in the absence of instabilities or other ion interactions, one ends up with ions of three velocities overlapping in a certain regions. These are the slow ions and two groups of faster ions. The two groups of fast ions originate either just inside or just outside the initial radius of the fastest (most strongly accelerated) ions. This leads the ion density to increase in the region where the overlap occurs. In addition, there is a large concentration of ions per unit velocity at the two ends of this region, where in each case two of these groups merge. The resulting structure is often referred to as a *shock shell*.

13.9 Other Phenomena

We have already seen that a host of phenomena can be produced when a laser beam irradiates a target at a relativistic energy flux. We discuss several more here.

Magnetic Field Generation also results from laser-solid dynamics. The beams of electrons produced by the laser carry substantial currents. As a result, they drive substantial magnetic fields. This is easier than it may seem. For example, suppose one wants to produce a 1 GGauss field. From (10.128), supposing that one drives an electron current by sending a beam of electrons, at c, through a circle of diameter 10 μm. The density of this beam must be only $\sim 10^{15}$ cm^{-3}. One can achieve this by accelerating one electron in 10^8 to $\sim c$. Measuring the field, however, is a severe challenge. At his writing, it is believed that GGauss fields have been produced.

Currents can be driven either into the target or along the surface. Ruhl and Mulser (1995) identified the production of a surface magnetic field in their calculations, arising from the ponderomotive force, which generates charge separation and surface currents for oblique incidence. These in turn produce a dc magnetic field, which was found in the simulations to vary strongly with irradiance. This field, which can approach 100 MG, causes dramatic variations in the absorption with I_L and angle of incidence (see their Fig. 6) Wilks et al. (1992) also observed a dc magnetic field in (fully relativistic and electromagnetic) 2D PIC simulations for normal incidence, at intensities larger than those considered by Ruhl and Mulser (1995). They attribute it to electron heating at the light–plasma interface. In addition, other magnetic-field generation mechanisms come into play at relativistic intensities (Bychenkov and Tikhonchuk 1996; Sudan 1993).

Betatron radiation may become an important X-ray source, as an offshoot of wakefield acceleration. Rousse et al. (2004) first demonstrated that a beam of X-ray radiation can be generated by focusing a high-intensity laser pulse into a gas jet, under the usual conditions to produce wakefield acceleration. Within the ion channel produced by the wake, the fields accelerate and wiggle an ultrashort and relativistic electron bunch. The accelerated electrons undergo betatron oscillations, generating a femtosecond pulse of radiation, emitted in a narrow cone angle. The X-ray energies can be in the keV range or higher. This discovery has led to an explosion of research into the production and use of betatron X-rays, which continues at this writing.

Positron production is of potential interest in astrophysical contexts. It requires more than 10^{28} W/cm^2 to produce electron–positron pairs directly from vacuum. However, when the laser light interacts with solid targets, it becomes possible to produce pairs at much lower irradiance. This is discussed, for example, by Liang et al. (1998). This first requires the production of energetic electrons, with the characteristic energy E_{hot} from (13.20). The threshold electron energy for pair production by interaction with a nucleus is $2mc^2$, and the cross-section scales as the square of the nuclear charge, Z_{nuc}^2. The threshold laser energy flux for pair production in steady state is about 10^{20} W/cm^2. For a dynamic, ultrafast interaction, Liang et al. (1998) find a comparable threshold value. At this writing, positron production has been observed (Gahn et al. 2000; Chen et al. 2009, 2010). It remains to be seen whether one can produce a plasma whose energy density is dominated by the pairs and whether a laser–target system can become a practical positron source.

Relativistic laser beams can also produce **nuclear reactions** within a target. The first step is the production of electrons having energies of many MeV. For example, when a beam having an energy flux above 10^{20} W/cm^2 and a total energy of 50 J, with a substantial prepulse, was used by Cowan et al. (2000) to irradiate a Au target, a distribution of electrons was produced with energies up to 100 MeV. These electrons in turn produced bremsstrahlung photons with energies of tens of MeV by interaction with the Au nuclei. These photons exceeded the energy threshold of $\sim$10 MeV for photonuclear reactions in the Au and in Cu located near the target. This caused transmutation of these elements.

Other physics with high-energy beams also can be explored, beyond the use of electron bunches to drive wakefield acceleration in a plasma (Joshi et al. 2002). One can use a tailored plasma to focus, defocus, or steer the beam. One can cause the beam to undulate in the transverse direction in the plasma. On the one hand, this happens naturally if the parameters are such that the beam repeatedly self-focuses and due to the natural tilt of the beam with respect to its propagation direction. On the other hand, one could imagine more active steps to create waves in the plasma that undulate the beams. This would produce tunable, forward-directed radiation with potential applications.

Beyond the above, the collisions involved in high-energy (as distinct from high-energy-density) physics create temporary, relativistic environments of an extremely high energy density. While much of the behavior might be best described in terms of particle physics, one does encounter concepts such as the "quark–gluon plasma" that have much in common with various plasma systems. In addition, some of these systems may have applications to astrophysics (Chen 2003). This potentially includes cosmic acceleration experiments, the spectroscopy of heavy elements, experiments related to event horizons, and experiments with the dynamics of relativistic jets.

The area of relativistic high-energy-density experiments has been developing very rapidly, in recent years, at the time of this writing. One can look forward to exciting further ideas and discoveries in the coming years.

Homework Problems

13.1 Design a pulse stretcher. Suppose you have a laser beam with an 800 nm central wavelength and a bandwidth of 20 nm (corresponding to a 50 fs laser pulse). Use two identical gratings, recalling that for the first diffracted order the scattered wavelength λ is given by $\lambda = d(\sin\alpha + \sin\beta)$, where d is the line spacing on the grating and α and β are angles of incidence and reflection relative to the grating normal. Use two identical lenses, recalling that the object distance, o, image distance, i, and focal length f are related by $o^{-1} + i^{-1} = f^{-1}$. Note that the initial grating must be less than one focal length from the lens to obtain stretching.

13.2 Assuming that the electron motion is due to a plane wave with a single frequency and that the electron movement is small compared to the wavelength of the light wave, solve the equations in Sect. 13.2 to find the electron trajectory. Determine how it changes as the electron velocity increases (while remaining $\ll c$).

13.3 Prove that the definitions (13.15) and (13.16) are equivalent.

13.4 Solve (13.25) to see the electron behavior for a range of values of the initial phase (i.e., change π to various other values, for fixed $a_o = 100$ and $\delta(0) = 0.01$. Comment on the variations in the behavior, in comparison to that seen in Fig. 13.3.

13.5 Find the time required to accelerate the electron to ~ 30 GeV in the example used in discussing (13.43) through (13.45).

13.6 Suppose one has a laser beam that can be focused to 10^{20} W/cm^2 in a 10 μm diameter spot. Would one obtain higher-energy electrons from tunnel ionization (as in Sect. 13.2) or from using the laser for wakefield acceleration, discussed in Sect. 13.7.1?

13.7 Solve for the potential of a spherical cloud of ions having uniform density, and for the energy distribution function of the ions produced by a Coulomb explosion of this cloud.

13.8 Derive the relativistic version of the theory of Sect. 13.8 and find the relativistically correct revision to (13.82).

References

Bach DR, Casperson DE, Forslund DW, Gitomer SJ, Goldstone PD, Hauer A, Kephart JF, Kindel JM, Kristal R, Kyrala GA, Mitchell KB, Hulsteyn DBv, Williams AH (1983) Intensity-dependent absorption in 10.6-μm laser-illuminated spheres. Phys Rev Lett 50(26):2082–2085

Beg FN, Bell AR, Dangor AE, Danson CN, Fews AP, Glinsky ME, Hammel BA, Lee P, Norreys PA, Tatarakis M (1997) A study of picosecond laser-solid interactions up to 10^{19} W/cm^2. Phys Plasmas 4(2):447–457. https://doi.org/10.1063/1.872103

Bezzerides B, Jones RD, Forslund DW (1982) Plasma mechanism for ultraviolet harmonic radiation due to intense CO$_2$ light. Phys Rev Lett 49(3):202–205

Bonnaud G, Gibbon P, Kindel J, Wiliams E (1991) Laser interaction with a sharp-edged overdense plasma. Las Part Beams 9(2):339–354

Brunel F (1987) Not-so-resonant, resonant absorption. Phys Rev Lett 59(1):52–55

Brunel F (1988) Anomalous absorption of high intensity subpicosecond laser pulses. Phys Fluids 31(9):2714–2719

Bulanov SV, Naumova NM, Pegoraro F (1994) Interaction of an ultrashort, relativistically strong laser pulse with an overdense plasma. Phys Plasmas 1(3):745–757

Bulanov SS, Macchi A, Maksimchuk A, Matsuoka T, Nees J, Pegoraro F (2007) Electromagnetic pulse reflection at self-generated plasma mirrors: Laser pulse shaping and high order harmonic generation. Phys Plasmas 14(9). https://doi.org/10.1063/1.2776906

Burnett NH, Baldis HA, Richardson MC, Enright GD (1977) Harmonic generation in CO_2 laser target interaction. Appl Phys Lett 31(3):172–174

Bychenkov VY, Tikhonchuk BT (1996) Magnetic field generation by short ultraintense laser pulse in underdense plasmas. Las Part Beams 14(1):55–62

Carman RL, Forslund DW, Kindel JM (1981) Visible harmonic emission as a way of measuring profile steepening. Phys Rev Lett 46(1):29–32

Chaker M, Kieffer JC, Matte JP, Pepin H, Audebert P, Maine P, Strickland D, Bado P, Mourou G (1991) Interaction of a 1 psec laser pulse with solid matter. Phys Fluids B 3(1):167–175

Chambers DM, Norreys PA, Dangor AE, Marjoribanks RS, Moustaizis S, Neely D, Preston SG, Wark JS, Watts I, Zepf M (1998) Feasibility study of high harmonic generation from short wavelength lasers interacting with solid targets. Opt Commun 148:289–294

Chen P (2003) Laboratory investigations of the extreme universe. Assoc Asia Pac Phys Soc Bull 13(1)

Chen H, Wilks SC, Bonlie JD, Liang EP, Myatt J, Price DF, Meyerhofer DD, Beiersdorfer P (2009) Relativistic positron creation using ultraintense short pulse lasers. Phys Rev Lett 102(10). https://doi.org/10.1103/PhysRevLett.102.105001

Chen H, Wilks SC, Meyerhofer DD, Bonlie J, Chen CD, Chen SN, Courtois C, Elberson L, Gregori G, Kruer W, Landoas O, Mithen J, Myatt J, Murphy CD, Nilson P, Price D, Schneider M, Shepherd R, Stoeckl C, Tabak M, Tommasini R, Beiersdorfer P (2010) Relativistic quasimonoenergetic positron jets from intense laser-solid interactions. Phys Rev Lett 105(1). https://doi.org/10.1103/PhysRevLett.105.015003

Combis P, Cazalis B, David J, Froger A, Louis-Jacquet M, Meyer B, Nierat G, Saleres A, Sibille G, Thiell G, wagon F (1991) Low-fluence laser target coupling. Las Part Beams 9(2):403–420

Corde S, Adli E, Allen JM, An W, Clarke CI, Clausse B, Clayton CE, Delahaye JP, Frederico J, Gessner S, Green SZ, Hogan MJ, Joshi C, Litos M, Lu W, Marsh KA, Mori WB, Vafaei-Najafabadi N, Walz D, Yakimenko V (2016) High-field plasma acceleration in a high-ionization-potential gas. Nat Commun 7. https://doi.org/10.1038/ncomms11898

Cowan TE, Hunt AW, Phillips TW, Wilks SC, al e (2000) Photonuclear fission from high energy electrons from ultraintense laser-solid interactions. Phys Rev Lett 84(5):903–906

Danson CN, Angood S, Booth G, Collier J, Damerell AR, Edwards CB, Flintoff PS, Govans J, Hancock S, Hatton P, Hawkes SJ, Hutchinson MHR, Key MH, Hernandez-Gomez C, Leach J, Lester W, Neely D, Norreys P, Notley M, Pepler DA, Reason CJ, Rodkiss DA, Toss IN, Toner WT, Trentelman M, Walczak JA, Wellstood RA, Winstone TB, Wyatt RWW, Wyborn BE (1999) Focused intensities of 10^{20} W/cm^2 with the upgraded Vulcan CPA interaction facility. Int Soc Opt Eng 3492:82–93

Davidson RC, National, Research, Council, Report (2003) High-energy-density physics: the X-games of contemporary science. National Academies Press, Washington

Ditmire T, Tisch JWG, Springate E, Mason MB, Hay N, Smith RA, Marangos J, Hutchinson MHR (1997) High-energy ions produced in explosions of superheated atomic clusters. Nature 386(6620):54–56. https://doi.org/10.1038/386054a0

Doumy G, Quere F, Gobert O, Perdrix M, Martin P, Audebert P, Gauthier JC, Geindre JP, Wittmann T (2004) Complete characterization of a plasma mirror for the production of high-contrast ultraintense laser pulses. Phys Rev E 69(2). https://doi.org/10.1103/PhysRevE.69.026402

Dromey B, Kar S, Zepf M, Foster P (2004) The plasma mirror – a subpicosecond optical switch for ultrahigh power lasers. Rev Sci Instrum 75(3):645–649. https://doi.org/10.1063/1.1646737

Esirkepov T, Borghesi M, Bulanov SV, Morou G, Tajima T (2004) Highly efficient relativistic-ion generation in the laser-piston regime. Phys Rev Lett 92(17):175003-1–175003-4

Faure J, Glinec Y, Pukhov A, Kiselev S, Gordienko S, Lefebvre E, Rousseau JP, Burgy F, Malka V (2004) A laser-plasma accelerator producing monoenergetic electron beams. Nature 431(7008):541–544

Fedosejevs R, Ottmann R, Sigel R, Kuhnle G, Szatmari S, Schafer FP (1990) Absorption of subpicosecond ultraviolet laser pulses in high-density plasma. Appl Phys B 50:79–99

Forslund DW, Shonk CR (1970a) Formation and structure of electrostatic collisionless shocks. Phys Rev Lett 25(25):1699–1702

Forslund DW, Shonk CR (1970b) Numerical simulation of electrostatic counterstreaming instabilities in ion beams. Phys Rev Lett 25(5):281–284

Gahn C, Tsakiris GD, Pretzler G, Witte KJ, Delfin C, Wahlstrom CG, Habs D (2000) Generating positrons with femtosecond-laser pulses. Appl Phys Lett 77(17):2662–2664. https://doi.org/10.1063/1.1319526

Geddes CGR, Toth C, van Tilborg J, Esarey E, Schroeder CB, Bruhwiler D, Nieter C, Cary J, Leemans WP (2004) High-quality electron beams from a laser wakefield accelerator using plasma-channel guiding. Nature 431(7008):538–541

Geissler M, Schreiber J, Meyer-ter Vehn J (2006) Bubble acceleration of electrons with few-cycle laser pulses. New J Phys 8:186

Gibbon P (1994) Efficient production of fast electrons from femtosecond laser interaction with solid targets. Phys Rev Lett 73(5):664–667

Gibbon P (1996) Harmonic generation by femtosecond laser-solid interaction: a coherent "water-window" light source? Phys Rev Lett 76(1):50–53

Gibbon P (1997) High-order harmonic generation in plasmas. IEEE J Quantum Electron 33(11):1915–1924

Gibbon P, Bell AR (1992) Collisionless absorption in sharp-edged plasmas. Phys Rev Lett 68(10):1535–1538

Gibbon P, Andreev AA, Platonov KY (2012) A kinematic model of relativistic laser absorption in an overdense plasma. Plasma Phys Control Fusion 54(4). https://doi.org/10.1088/0741-3335/54/4/045001

Grebogi C, Tripathi VK, Chen HH (1983) Harmonic generation of radiation in a steep density profile. Phys Fluids 26(7):1904–1908

Haberberger D, Tochitsky S, Fiuza F, Gong C, Fonseca RA, Silva LO, Mori WB, Joshi C (2012) Collisionless shocks in laser-produced plasma generate monoenergetic high-energy proton beams. Nat Phys 8(1):95–99

Jiang Z, Kieffer JC, Matte JP, Chaker M, Peyrusse O, Gilles D, Corn G, Maksimchuk A, Coe S, Mourou G (1995) X-ray spectroscopy of hot solid density plasmas produced by subpicosecond high contrast laser pulses at 10^{18}–10^{19} W/cm^2. Phys Plasmas 2(5):1702–1711

Joshi C, Blue B, Clayton CE, Dodd E, Huang C, Marsh KA, Mori WB, Wang S, Hogan MJ, O'Connell C, Siemann R, Watz D, Muggli P, Katsouleas T, Lee S (2002) High energy density plasma science with an ultrarelativistic electron beam. Phys Plasmas 9(5):1845–1855

Kaplan AE, Dubersky BY, Shkolnikov PL (2003) Shock shells in coulomb explosions of nanoclusters. Phys Rev Lett 91(14):143401-1–143401-4

Kato S, Bhattacharyya B, Nishiguchi A, Mima K (1993) Wave breaking and absorption efficiency for short pulse p-polarized laser light in a very steep density gradient. Phys Fluids B 5(2):564–570

Kmetec JD, Gordon CL, Macklin JJ, Lemoff BE, Brown GS, Harris SE (1992) Mev X-ray generation with a femtosecond laser. Phys Rev Lett 68(10):1527–1530

Kruer WL (1988) The physics of laser plasma interactions. Addison-Wesley Publishing Company, Redwood City

Landau LD, Lifshitz EM (1987) The classical theory of fields, course in theoretical physics, vol 2, 2nd edn. Pergamon Press, Oxford

Liang EP, Wilks SC, Tabak M (1998) Pair production by ultraintense lasers. Phys Rev Lett 81(22):4887–4890

Lichters R, Meyer-ter Vehn J, Pukhov A (1996) Short-pulse laser harmonics from oscillating plasma surfaces driven at relativistic intensity. Phys Plasmas 3(9):3425–3437

Lindley RA, Gilgenbach RM, Ching CH (1993) Resonant holographic interferometry of laser-ablation plumes. Appl Phys Lett 63(7):888–890

Liu X, Umstadter D (1992) Competition between ponderomotive and thermal pressures in short-scale-length laser-plasmas. Phys Rev Lett 69:1935–1938

Mangles SPD, Murphy CD, Najmudin Z, Thomas AGR, Collier JL, Dangor AE, Divall EJ, Foster PS, Gallacher JG, Hooker CJ, Jaroszynski DA, Langley AJ, Mori WB, Norreys PA, Tsung FS, Viskup R, Walton BR, Krushelnick K (2004) Monoenergetic beams of relativistic electrons from intense laser-plasma interactions. Nature 431(7008):535–538

Meyerhofer DD, Chen H, Delettrez JA, Soom B, Uchida S, Yaakobi B (1993) Resonance absorption in high-intensity contrast, picosecond laser-plasma interactions. Phys Fluids B 5(7):2584–2588

Mora P (2003) Plasma expansion into a vacuum. Phys Rev Lett 90:185002

Mourou G, Umstadter D (1992) Development and applications of compact high-intensity lasers. Phys Fluids B 4(7):2315–2325

Mulser P, Bauer D, Ruhl H (2008) Collisionless laser-energy conversion by anharmonic resonance. Phys Rev Lett 101(22). https://doi.org/10.1103/PhysRevLett.101.225002

Nantel M, Ma G, Gu S, Cote CY, Itatani J, Umstadter D (1998) Pressure ionization and line merging in strongly coupled plasmas produced by 100-fs laser pulses. Phys Rev Lett 80(20):4442–4445

Norreys PA, Zepf M, Moustaizis S, Fews AP, Zhang J, Lee P, Bakarezos M, Danson CN, Dyson A, Gibbon P, Loukakos P, Neely D, Walsh FN, Wark JS, Dangor AE (1996) Efficient extreme UV harmonics generated from picosecond laser pulse interactions with solid targets. Phys Rev Lett 76(11):1832–1835

Ping Y, Shepherd R, Lasinski BF, Tabak M, Chen H, Chung HK, Fournier KB, Hansen SB, Kemp A, Liedahl DA, Widmann K, Wilks SC, Rozmus W, Sherlock M (2008) Absorption of short laser pulses on solid targets in the ultrarelativistic regime. Phys Rev Lett 100(8). https://doi.org/10.1103/PhysRevLett.100.085004

Pukhov A, Meyer-ter Vehn J (1996) Relativistic magnetic self-channelings of light in near-critical plasma: three-dimensional particle-in-cell simulation. Phys Rev Lett 76(21):3975–3878

Pukhov A, Meyer-ter Vehn J (2002) Laser wake field acceleration: the highly non-linear broken-wave regime. Appl Phys B-Lasers Opt 74(4–5):355–361

Rousse A, Audebert P, Geinder JP, Failles F, Gauthier JC, Mysrowicz A, Grillon G, Antonetti A (1994) Efficient kalpha X-ray source from femtosecond laser-produced plasmas. Phys Rev E 50(3):2200–2207

Rousse A, Phuoc KT, Shah R, Pukhov A, Lefebvre E, Malka V, Kiselev S, Burgy F, Rousseau JP, Umstadter D, Hulin D (2004) Production of a keV X-ray beam from synchrotron radiation in relativistic laser-plasma interaction. Phys Rev Lett 93(13). https://doi.org/10.1103/PhysRevLett.93.135005

Rozmus W, Tikhonchuk VT, Cauble R (1996) A model of ultrashort laser pulse absorption in solid targets. Phys Plasmas 3(1):360–367

Ruhl H, Mulser P (1995) Relativistic Vlasov simulation of intense fs laser pulse-matter interaction. Phys Lett A 205:388–392

Sauerbrey R, Fure J, LeBlanc SP, van Wonterghem B, Teubner U, Schafer FP (1994) Reflectivity of laser-produced plasmas generated by a high intensity ultrashort pulse. Phys Plasmas 1(5):1635–1642

Silva LO, Marti M, Davies JR, Fonseca RA, Ren C, Tsung FS, Mori WB (2004) Proton shock acceleration in laser-plasma interactions. Phys Rev Lett 92(1):015002-1–015002-4

Sudan RN (1993) Mechanism for the generation of 10^9 g magnetic fields in the interaction of ultraintense short laser pulse with an overdense plasma target. Phys Rev Lett 70(20):3075–3078

Tajima T, Dawson JM (1979) Laser electron accelerator. Phys Rev Lett 43(4):267–270

Wilks SC, Kruer WL, Tabak M, Langdon AB (1992) Absorption of ultra-intense laser pulses. Phys Rev Lett 69(9):1383–1386

Wilks SC, Kruer WL, Mori WB (1993) Odd harmonic generation of ultra-intense laser pulses reflected from an overdense plasma. IEEE Trans Plasma Sci 21(1):120–124

Zhao L (1998) Experimental studies of harmonic generation from solid-density plasmas produced by picosecond ultra-intense laser pulses, Ph.D., University of Toronto

Zweiback J, Cowan TE, Hartley JH, Howell R, Wharton KB, Crane JK, Yanovsky VP, Hays G, Smith RA, Ditmire T (2002) Detailed study of nuclear fusion from femtosecond laser-driven explosions of deuterium clusters. Phys Plasmas 9(7):3108

Appendix A
Constants, Acronyms, and Standard Variables

See Tables A.1, A.2 and A.3.

Table A.1 Table of constants

Name	Symbol	cgs value	Useful alternative	SI mks value
Bohr radius	a_o	5.29×10^{-9} cm	0.529 Å	5.29×10^{-11} m
Speed of light	c	3×10^{10} cm/s		3×10^8 m/s
Electronic charge	e	4.8×10^{-10} statcoul		1.6×10^{-19} Coul
Planck constant	h	6.63×10^{-27} erg s		6.63×10^{-34} J s
Boltzmann constant	k_B	1.6×10^{-12} erg/eV	1.6×10^{-16} J/keV	1.38×10^{-23} J/K
Electron rest mass	m_e	9.11×10^{-28} g		9.11×10^{-31} kg
Proton mass	m_p	1.67×10^{-24} g	$1836 m_e$	1.67×10^{-27} kg
Stefan-Boltzmann	σ	5.67×10^{-5} ergs/(cm^2 s deg^4)	1.03×10^5 W/(cm^2 eV^4)	5.67×10^{-8} W m^{-2} K^{-4}

© Springer International Publishing AG 2018
R P. Drake, *High-Energy-Density Physics*, Graduate Texts in Physics,
https://doi.org/10.1007/978-3-319-67711-8

Table A.2 Table of
acronyms

Acronym	Represents
ASE	Amplified spontaneous emission
CPA	Chirped pulse amplification
DPP	Distributed phase plate
DPR	Distributed polarization rotator
ICF	Inertial confinement fusion
LTE	Local thermodynamic equilibrium
NIF	National Ignition Facility
NLTE	Nonlocal thermodynamic equilibrium
RT	Rayleigh Taylor
KH	Kelvin Helmholtz
RM	Richtmyer Meshkov
RPP	Random phase plate
SBS	Stimulated Brillouin scattering
SN	Supernova
SRS	Stimulated Raman scattering
SSD	Smoothing by spectral dispersion
exp[]	Equivalent to $e^{[]}$

Table A.3 Table of standard variables

Name	Symbol
Atomic weight (average)	A
Vector potential	$\boldsymbol{A}$
Atwood number	A_n
Area of capsule	A_c
Area of laser spots	A_L
Area of walls of hohlraum	A_w
Magnetic field	$\boldsymbol{B}$
Thermal intensity	$B(T)$
Thermal spectral intensity	$B_\nu(T)$
Isentropic sound speed	c_s
Specific heat at const vol	c_V
Small vortex diameter	d
Element of area	$d\boldsymbol{A}$
Critical to solid density distance	D
Electron charge	e
Electric field	$\boldsymbol{E}$
Spectral kinetic energy	$E(k)$
Thermal energy density	E_{BB}
Energy released by fusion	E_{fus}
Hydrogen ionization energy	E_H
Electric field of laser beam	E_L
Total radiation energy density	E_R
Energy in Marshak wave	E_w
Spectral radiation energy density	E_ν
Energy difference between ionization states j and k	E_{jk}
Electron total energy	$\mathcal{E}_e$
Electron rest mass energy	$\mathcal{E}_o$
Ion total energy	$\mathcal{E}_i$
Thermal flux	F_{BB}
Electron free energy	F_e
Electromagnetic force density	$\boldsymbol{F}_{EM}$
Lorentz force	$\boldsymbol{F}_{\bar{L}}$
Radiative energy flux	$\boldsymbol{F}_R$
Total radiation flux	F_R
Photon flux	$\boldsymbol{F}_\gamma$
Spectral radiation flux	F_ν
Eddington factor	$f_\nu = p_\nu/E_\nu$
Distribution function	$f(v)$
"Gravitational" acceleration	g
Laser energy flux	I_L
Total intensity	I_R
Laser energy flux in units of 10^{xx} W/cm^2	I_{xx}
Spectral intensity	I_ν

(continued)

Table A.3 (continued)

Name	Symbol
Current density	$\boldsymbol{J}$
Richardson number	J_r
Total mean intensity	J_R
Transverse current density	$\boldsymbol{J}_t$
Mean spectral intensity	J_ν
Riemann invariants	J_+ or J_-
Wave number	k
Wave vector	$\boldsymbol{k}$
Coefficients in Maxwell's equations	k_1, k_2, k_3
Scale length of a profile	L
Eddy diameter	ℓ
Compton mean free path	ℓ_C
Mach number	M
Upstream Mach number	M_u
Internal Mach number	M_{int}
Fusion fuel mass	m_f
Mass ablation rate	$\dot{m}$
Shock normal	$\boldsymbol{n}$
Electron density	n_e
Ion density	n_i
Critical density	n_c
Scalar fluid pressure	p
Total scalar pressure	$\tilde{p}$
Electron momentum	$\boldsymbol{p}_e$
Fermi-degenerate pressure	p_F
Scalar radiation pressure	p_R
General pressure tensor	$\underline{\boldsymbol{P}}$
Ablation pressure	P_{abl}
Power threshold for relativistic self-focusing	P_{sf}
Turbulent energy dissipation	P_t
Radiation spectral pressure tensor	$\underline{\boldsymbol{P}}_\nu$
Thermal heat flux	$\boldsymbol{Q}$
Radiation strength parameter	Q
Spitzer–Harm heat flux	Q_{SH}
Free-streaming heat flux	Q_{FS}
Internal energy	R
Gas constant $p/(\rho T)$	R
Ion sphere radius	R_o
Radiation strength parameter	R_r
Poynting flux	$\boldsymbol{S}$
Specific entropy	s
Specific entropy of electrons	s_e
Source of quantity Q	S_Q
Spectral source function	S_ν

(continued)

Table A.3 (continued)

Name	Symbol
Time	t
Temperature	T
Immediate post-shock temperature	T_2 or T_s
Fermi-degenerate temperature	T_d
Electron temperature	T_e
Effective temperature	T_{eff}
Ion temperature	T_i
Temperature corresponding to a radiation flux	T_{min}, T_{eff}
Energetic electron temperature	T_{hot}
Precursor temperature	T_p
Radiation temperature	T_R
Immediate postshock plasma (mainly electron) temperature	T_s
Hohlraum wall temperature	T_w
Fluid velocity	$\boldsymbol{u}$
Zeroth-order fluid velocity	U
Characteristic velocity for scaling arguments	U
First-order components of fluid velocity	$\boldsymbol{u}_1 = (u, v, w)$
Kolmogorov velocity scale	u_k
Particle velocity	$\boldsymbol{v}$
Velocity difference between frames of reference	$\boldsymbol{v}$
Phase velocity	v_p
Oscillating velocity of electron in light wave	$\boldsymbol{v}_{os}$
Electron thermal velocity	v_{th}
Rocket velocity (or capsule velocity)	V
Exhaust velocity	V_{ex}
Vertical component of velocity	w
Vortex rotational velocity	w
Eddy rotational velocity	w_e
Marshak wave scaling variable	W
Magnetic Energy density	W_B
Electric energy density	W_E
Space	x
Marshak wave penetration depth	x_M
Fusion yield	Y
Ionic charge (average)	Z
Albedo	α
Various angles	α
Fraction of incoming photons ionized	α_i
Various angles	β
Relativistic velocity (v/c)	β
Various angles	χ
Coeff of thermal diffusivity	χ
Electron momentum	χ_e

(continued)

Table A.3 (continued)

Name	Symbol
Jet cooling parameter	χ_j
Rosseland-mean opacity	χ_R
Spectral total opacity	χ_ν
Specific internal energy	ϵ
Total specific internal energy density	$\tilde{\epsilon}$
Downstream emissivity	ϵ_d
Specific internal electron energy	ϵ_e
Fermi energy	ϵ_F
Specific internal ion energy	ϵ_{ii}
Specific kinetic ion energy	ϵ_{ik}
Upstream emissivity	ϵ_u
Efficiency of ideal rocket	ϵ_R
Various angles	ϕ
Fusion burn fraction	ϕ
Phase of a wave	ϕ
Phase experienced by an electron	ϕ_e
Scalar electric potential	Φ
Polytropic index	γ
Polytropic index for acoustic waves	γ_s
Polytropic index for heat conduction	γ_h
Relativistic γ	γ_r
Instability growth rate	γ_o
Strong coupling parameter	Γ
Flux of quantity Q	Γ_Q
various angles and fractions	η
X-ray conversion efficiency	η
Kolmogorov length scale	η_k
Spectral emissivity	η_ν
Spectral scattering emissivity	$\eta_{\nu sc}$
Spectral thermal emissivity	$\eta_{\nu th}$
Absorption opacity	κ
Total coefficient of heat conduction	$\tilde{\kappa}$
Opacity of thin layer using cooling function	κ_{astro}
Thermal bremsstrahlung absorption coefficient	κ_b
EM wave absorption coefficient	κ_{EM}
Specific Planck mean opacity	κ_m
Planck mean opacity	κ_P
Radiative coefficient of heat conduction	κ_{rad}
Thermal coefficient of heat conduction	κ_{th}
Spectral absorption opacity	κ_ν
Wavelength of a wave	λ
Vortex characteristic scale	λ
Taylor microscale	λ_T
Debye length	λ_D
Electron Debye length	λ_{De}

(continued)

Table A.3 (continued)

Name	Symbol
Ion Debye length	λ_{Di}
Mean free path	λ_{mfp}
Wavelength in microns	λ_{μ}
Astrophysical cooling function	Λ
Chemical potential	μ
Classical chemical potential	μ_c
Vortex characteristic scale	λ
Atomic mass per charge (A/Z)	μ_e
Electron–ion collision rate	ν_{ei}
Optically thin cooling rate	ν
Kinematic viscosity	ν
Radiation frequency	ν
Extinction rate	ν_e
Cooling rate normalization for thin layer	ν_{rad}^*
Cooling rate more general	ν_1
Cooling rate using cooling function	ν_{astro}
Radiation cooling rate for thin layer	ν_{rad}
Kinematic photon viscosity	ν_{rad}
Scaling variable $n_e/T_e^{3/2}$	θ
Degeneracy parameter	Θ
Mass density	ρ
Charge density	ρ_c
Density of Q	ρ_Q
Scattering opacity	σ_s
Spectral scattering opacity	σ_ν
Viscosity stress tensor	$\underline{\boldsymbol{\sigma}}_\nu$
Kolmogorov time scale	τ_k
Optical depth at frequency ν	τ_ν
Optical depth	τ
Wave frequency	ω
Laser light frequency	ω_o
Normalized frequency	ω_n
Electron plasma frequency	ω_{pe}
Ion plasma frequency	ω_{pi}
Scattered light frequency	ω_s
Irradiance conversion by hohlraum	ξ
General similarity variable	ξ
Gravitational potential	Ψ

Solutions

Problems of Chap. 2

2.1 One approach to deriving the Euler equations is to identify the density, flux, and sources of mass, momentum, and energy and then to use (2.5). Do this for a polytropic gas and then simplify the results to obtain (2.1) through (2.3).

Solution and/or Comments The work of this problem is about the process. The end result is:

$$\frac{\partial \rho}{\partial t} + \nabla \cdot \rho \boldsymbol{u} = 0, \tag{A.1}$$

$$\rho \left(\frac{\partial \boldsymbol{u}}{\partial t} + \boldsymbol{u} \cdot \nabla \boldsymbol{u} \right) = -\nabla p, \text{ and} \tag{A.2}$$

$$\frac{\partial p}{\partial t} + \boldsymbol{u} \cdot \nabla p = -\gamma p \nabla \cdot \boldsymbol{u}, \tag{A.3}$$

where $\boldsymbol{u}$, ρ, and p are the velocity, density, and pressure, respectively.

2.2 Linearize the Euler equations to derive (2.7) and (2.8). Find appropriate divisors to make the physical variables in these equations nondimensional. Then derive the equivalent of (2.9) and an equation for a normalized velocity variable. Comment on the result. (Hint: this is a wave problem not a global-scaling problem, so what you are looking for is not U, L, and etc. as used in that part of the chapter.)

Solution and/or Comments This problem is about the challenge of finding a physically meaningful non-dimensionalization. For normalized density $\tilde{\rho}$ and normalized velocity $\tilde{\boldsymbol{u}} = \boldsymbol{u}/c$, the end result is:

© Springer International Publishing AG 2018
R P. Drake, *High-Energy-Density Physics*, Graduate Texts in Physics,
https://doi.org/10.1007/978-3-319-67711-8

$$\frac{\partial^2}{\partial t^2}\tilde{\rho} - c^2 \nabla^2 \tilde{rho} = 0, \text{ and} \tag{A.4}$$

$$\frac{\partial^2}{\partial t^2}\tilde{u} - c^2 \nabla^2 \tilde{u} = 0. \tag{A.5}$$

2.3 Take the actual, mathematical Fourier transform of (2.9) to find (2.10). Comment on the connection of the result to the substitution used in the text.

Solution and/or Comments This problem is intended to promote thinking about the physical meaning of Fourier transforms. The end result is:

$$\omega^2 - c_s^2 k^2 = 0, \tag{A.6}$$

2.4 Substitute, for the density in (2.9), the actual, mathematical inverse Fourier transform of the spectral density $\tilde{\rho}(\boldsymbol{k}, \omega)$. Show how the result is related to (2.10).

Solution and/or Comments This problem is intended to promote thinking about the physical meaning of Fourier transforms. One finds

$$\tilde{\rho}_a(\boldsymbol{k}, \omega) = \tilde{\rho}_a(\boldsymbol{k})\delta(\omega - c_s k), \tag{A.7}$$

confirming (2.10).

2.5 The Euler equations apply to an ideal gas with $\epsilon = p/[\rho(\gamma - 1)]$, so they should imply (2.14). Demonstrate this by deriving (2.14) from (2.1), (2.2), and (2.4).

Solution and/or Comments This problem is intended to develop facility with the manipulation of fundamental fluid equations. The result is:

$$\frac{Dp}{Dt} - c_s^2 \frac{D\rho}{Dt} = 0$$

2.6 Begin to explore the behavior of longitudinal waves in a charged fluid. Specifically, derive (2.46) from the equations for number and momentum for an electron fluid.

Solution and/or Comments This problem is intended to develop facility with the manipulation of fluid PDEs and experience with the challenge that the electron fluid supports both longitudinal and transverse waves. The result is:

$$\frac{\partial^2 n_{e1}}{\partial t^2} = \frac{n_{eo}e}{m_e}\nabla \cdot \boldsymbol{E_1} + \frac{1}{m_e}\nabla^2 p_{e1}. \tag{A.8}$$

2.7 Collisions do affect electron plasma waves. To see how, derive a replacement for (2.48), keeping an appropriate version of the drag term at the end of (2.43). Comment on the results.

Solution and/or Comments This problem is intended to develop facility with the manipulation of fluid PDEs and a first exposure to equations for damped waves.

Problems of Chap. 3

3.1 Inertial fusion designs typically involve the compression of DT fuel to about 1000 times the liquid density of $0.25\,\mathrm{g\,cm^{-3}}$. Assuming that this compression is isentropic and that the fuel remains at absolute zero, determine the energy per gram required to compress this fuel. Compare this to the energy per gram required to isentropically compress the fuel to this same density, assuming the fuel is an ideal gas whose final temperature is to be the ignition temperature of 5 keV.

Solution and/or Comments Pay careful attention to the need to evaluate the internal energy for each case. One finds $\epsilon = 8.5 \times 10^6$ J/g for Fermi-degenerate compression and $\epsilon = 5.7 \times 10^8$ J/g for ideal-gas compression, which is 68 times larger.

3.2 Generalize the derivation of the Debye length in Sect. 3.2 to a plasma with an arbitrary number of ion species, each of which may have a distinct temperature.

Solution and/or Comments The result, in Gaussian cgs units, is

$$\lambda_D^{-2} = 4\pi e^2 \left(\frac{n_e}{k_B T_e} + \sum_\alpha \frac{n_\alpha Z_\alpha^2}{k_B T_\alpha} \right). \tag{A.9}$$

3.3 Examine the behavior of the integrals for Fermions. Argue conceptually that the contribution of the denominator in (3.29) at large $\mu/(k_B T_e)$ is a step function. Evaluate this integral numerically to determine how rapidly it becomes a step function as $\mu/(k_B T_e)$ increases.

Solution and/or Comments This problem is intended to add some insight into the behavior of the integrals for Fermions. One way to accomplish the numerical demonstration is as follows, using definitions in the Chapter. Transform variables so that $w = \mathcal{E}_e/(k_B T)$ and $a = \mu/(k_B T)$, drop the leading constants, and divide the integral into two pieces to have

$$I_1 = \int_0^a \frac{\sqrt{w}\,dw}{\exp[w-a]+1} \text{ and } I_2 = \int_a^\infty \frac{\sqrt{w}\,dw}{\exp[w-a]+1}. \tag{A.10}$$

As the integral of $[\exp[w-a]+1]^{-1}$ approaches a step function, I_1 will approach

$$I_3 = \int_0^a \sqrt{w}\,dw = 2a^{3/2}/3. \tag{A.11}$$

Fig. A.1 The ratios I_1/I_3 and I_2/I_3 are shown against $a = \mu/(k_B T)$

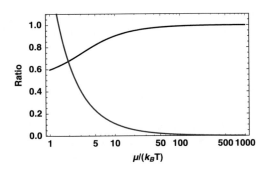

Plotting the ratios I_1/I_3 and I_2/I_3 gives Fig. A.1. One can see that the integral behaves nearly as a step function by $a = 10$ and is not distinguishable from it by $a = 1000$.

3.4 Examine the limiting behavior of the internal energy of Fermi degenerate electrons. Show, in the limit as $T_e \to 0$, that $n_e \epsilon_e = (3/5) n_e \epsilon_F$.

Solution and/or Comments This problem is intended to add some insight into the behavior of the integrals for Fermions. The result is that $n_e \epsilon_e = (3/5) n_e \epsilon_F$, in the limit as $T_e \to 0$.

3.5 What is the relation of heat capacity and entropy? Derive (3.38) and (3.40) and discuss their differences.

Solution and/or Comments The equations to be derived are

$$C_V = \frac{\partial}{\partial T_e}(n_e \epsilon_e)\bigg|_{n_e} = \frac{3}{2} n_e k_B \left[\frac{5}{2} F_{3/2} \Theta^{3/2} + T_e \frac{\partial}{\partial T_e}\left(\frac{\mu}{k_B T_e}\right) \right], \text{ and} \qquad (A.12)$$

$$\frac{S}{V} = \frac{2}{3}\frac{\partial}{\partial T_e}(n_e \epsilon_e)\bigg|_{\mu, V} = \frac{5}{2} n_e k_B \left[\frac{2}{3}\frac{F_{3/2}(\frac{\mu}{k_B T_e})}{F_{1/2}(\frac{\mu}{k_B T_e})} - \frac{2}{5}\frac{\mu}{k_B T_e} \right], \qquad (A.13)$$

but the real point of the problem is to think about the meaning of the chemical potential and how this affects the evaluate of the same partial derivative done while holding different variables constant.

3.6 Make plots comparing Z_{bal} from (3.49) with the estimate $20\sqrt{T_e}$ as a function of T_e, for ion densities of 10^{19}, 10^{21}, and 10^{23} cm^{-3}. Discuss the results.

Solution and/or Comments The point of this problem is to gain experience with the limitations of the simple model that has $Z_{bal} \sim 0.63\sqrt{T_e(eV)}$. This model is fine for doing simple stupid derivations that find the main trends across large variations. However, the author has noticed that students often tend to use it to think about actual data, where it is far too inaccurate. For specific applications, one should at

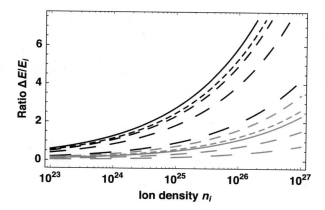

Fig. A.2 The ratios $\Delta E/E_i$ are shown against ion density, with the upper, black curves corresponding to $Z_i = 1$ while the lower, gray curves are for $Z_i = 10$. In each set of curves, in order of increasingly sparse dashing, the models are Stewart-Pyatt, More, Ecker-Kroll, Isolated Ion Sphere, and Griem

minimum solve the Saha equation, and it is really better to apply the Thomas-Fermi model presented in Sect. 3.4.3.

3.7 Carry out the evaluation of the average charge, Z, in (3.53) and compare the result to Z_{bal}, for $T_e = 1\,\text{keV}$, $Z_n = 30$, and $n_i = 10^{21}\,\text{cm}^{-3}$.

Solution and/or Comments The point of this problem is to work through the more accurate evaluation of Z discussed in the text and show that the simple estimate (Z_{bal} is remarkably accurate. One finds $Z_{bal} = 25.182$ for the given parameters and $Z = 24.655$, which is only a difference of 2%.

3.8 Plot the ratio of ΔE to the ionization energy versus ion density for the various models described in Sect. 3.4.4. Discuss the results.

Solution and/or Comments The point of this problem is to gain some experience with the sensitivity of the pressure ionization to the model used. You will generate a plot like that shown in Fig. A.2. Note that the outlier, which is the model given in Griem, is not actually used by anyone.

3.9 The value of R_i used in Sect. 3.4.6 ignores the internal energy in excited states (as well as the energy lost by radiation during ionization, which would properly have to be treated by more general equations). Again assuming hydrogenic ions, estimate what fraction of the internal energy is present in excited states, and how this varies with Z.

Solution and/or Comments One can approach this problem in several ways, and part of the purpose of the problem is to stretch thinking about relative probabilities. There are ways to address this based on material in the text and ways based on either general or specific understanding of statistical physics.

3.10 Complete the derivation of the polytropic index for heat conduction. Derive (3.143) from relations (3.130)–(3.134).

Solution and/or Comments The point of this problem is to provide some experience with the manipulations of thermodynamic variables often needed in the study of fluids, gasses, and plasmas. The result is

$$(\gamma_h - 1)^{-1} = \rho c_V \left(\frac{\partial T}{\partial p} \right)_\rho . \tag{A.14}$$

Problems of Chap. 4

4.1 Add a gravitational force density and gravitational potential energy to (4.2) and (4.3) and derive the modified jump conditions.

Solution and/or Comments The result here may surprise you.

4.2 Suppose that during the shock transition significant energy is lost by radiation. Write down the modified jump conditions.

Solution and/or Comments The point of this problem is to delve more deeply into the physical meaning of the jump conditions. One needs a physical analysis to justify the result that the jump conditions relating to mass and momentum are unchanged, while the one related to energy becomes

$$\rho_1 u_1 \left(\epsilon_1 + \frac{u_1^2}{2} \right) + p_1 u_1 = \rho_2 u_2 \left(\epsilon_2 + \frac{u_2^2}{2} \right) + p_2 u_2 + F_R, \tag{A.15}$$

where $\rho_2 > \rho_1$ and $F_R > 0$ is the radiation energy flux emitted from the region of the shock transition.

4.3 Determine from energy arguments how to generalize (4.20) for a plasma having two ion species.

Solution and/or Comments One must find a physical principle that enables one to relate the immediate heating of each species by the shock transition to the final temperature after all the species have reached some equilibrium. The shock transition is extremely abrupt, so ionization during the transition is not likely to be significant.

Defining f_j as the ratio of the mass density of each species to the total mass density ρ, one will find

$$k_B T_2 = \frac{2 m_p u_s^2}{(1 + Z_2)} \left(\frac{1}{\sum f_j / A_j} \right) \sum f_j \frac{(\gamma_{2j} - 1)}{(\gamma_{2j} + 1)^2} . \tag{A.16}$$

One can see that this has the correct limit for a single species. In the limit that the values of γ_{2j} are the same and $= \gamma$, this becomes

$$k_B T_2 = \frac{2m_p u_s^2}{(1 + Z_2)} \frac{(\gamma - 1)}{(\gamma + 1)^2} \frac{1}{\sum(f_j/A_j)} = \frac{2\langle A \rangle m_p u_s^2}{(1 + Z_2)} \frac{(\gamma - 1)}{(\gamma + 1)^2}, \tag{A.17}$$

defining a specific type of averaged atomic weight, $\langle A \rangle$.

Note that f_j is the fraction of mass, not number density. So for two ion species (and ignoring the negligible contribution of the electrons), one finds

$$f_1 = \frac{f_{a1}A_1}{f_{a1}A_1 + f_{a2}A_2}, \tag{A.18}$$

where the number fraction (the more commonly known quantity) is f_a. For C_1H_1, one finds $\langle A \rangle = 7$, while for Xe_9H_1 (10% H in Xe), one finds $\langle A \rangle = 118$.

4.4 For $\gamma_1 = \gamma_2$, derive the equivalent of (4.18) and (4.20). Express the result in physically clear parameters, so the relation among the terms is evident. Check your result by finding it as a limit of (4.19) and by finding (4.20) as a limit from it. Using a computational mathematics program is suggested.

Solution and/or Comments One will find

$$p_2 == \frac{2}{(\gamma + 1)} \rho_1 u_s^2 \left[1 + \frac{(\gamma - 1)p_1}{2\rho_1 u_s^2} \right], \tag{A.19}$$

which is (4.18), and

$$k_B T_2 = A m_p \frac{p_2}{\rho_2} = \frac{A m_p u_s^2}{(1 + Z)} \frac{2(\gamma - 1)}{(\gamma + 1)^2} \tag{A.20}$$

$$\times \left[1 + \frac{(\gamma^2 - 6\gamma + 1)}{2(\gamma - 1)} \frac{p_1}{\rho_1 u_2^2} + \left(\frac{p_1}{\rho_1 u_2^2} \right)^2 \right],$$

which, in the strong-shock limit, becomes (4.20).

4.5 Find an expression for the entropy production by a shock wave (4.24) as the Mach number approaches 1 from above.

Solution and/or Comments It is fairly straightforward to show that

$$\lim_{M \to 1} \Delta s = c_V \ln(1) = 0$$

4.6 Derive the jump conditions for oblique shocks, (4.28)–(4.31).

Solution and/or Comments One finds, starting with the basic equations in conservative form,

$$\rho_1 u_{n1} = \rho_2 u_{n2}, \tag{A.21}$$

$$\rho_1 u_{n1} \boldsymbol{u}_1 + p_1 = \rho_2 u_{n2} \boldsymbol{u}_2 + p_2, \text{ which implies in part} \tag{A.22}$$

$$u_{\perp 1} = u_{\perp 2}, \text{ and} \tag{A.23}$$

$$\left[\rho_1 u_{n1} \left(\epsilon_1 + \frac{u_{n1}^2}{2} \right) + p_1 u_{n1} \right] = \left[\rho_2 u_{n2} \left(\epsilon_2 + \frac{u_{n2}^2}{2} \right) + p_2 u_{n2} \right]. \tag{A.24}$$

These are the desired jump conditions.

4.7 Derive the relations of the angles for oblique shocks, (4.34) and (4.35).

Solution and/or Comments The point of this problem is to spend time thinking in more detail about oblique-shock behavior. The results are

$$\cos\psi = \cos\phi_2 \left[\cos\phi_1 + \sin\phi_1 \tan\phi_2 \right] \text{ and} \tag{A.25}$$

$$\cos\psi = \frac{\left[\cos\phi_1 + \sin\phi_1 \tan\phi_1 (\rho_2/\rho_1) \right]}{\sqrt{1 + \tan^2\phi_1 (\rho_2/\rho_1)^2}}. \tag{A.26}$$

4.8 Derive (4.42) relating the speeds in different frames of reference. This requires thinking about which frame of reference one is working in, a key element in all such problems.

Solution and/or Comments The result you will derive,

$$u_2 - u_{sR} = u_2' = u_{s12} \rho_1/\rho_2, \tag{A.27}$$

and similar equations, are the key to solving any problem with multiple interfaces moving at different speed. Students often find it challenging to understand this issue, which is key to more than one problem below and occurs often in applications.

4.9 Determine the equations and derive the behavior of the simpler case in which a shock is incident on a stationary wall. Let state 0 be the state of the unshocked fluid, state 1 be that of the once-shocked fluid, and state 2 be the state of the reshocked fluid produced when the shock reflects from the wall.

Solution and/or Comments This problem required careful attention to numbers of equations and numbers of variables, so as to end up with a closed system, and thinking correctly within the relevant inertial frames. The known properties of the incoming shock and the unshocked matter imply the upstream Mach number of the incoming shock, M_1. The flow Mach number of the post-shock flow, $M_{1f} = u_1/c_1$, where c_1 is the sound speed in the once-shocked matter, is related to M_1 by

$$M_{1f} = \frac{u_1}{c_1} = \frac{2(M_1^2 - 1)}{\sqrt{2 + M_1^2(\gamma - 1)}\sqrt{1 + \gamma(2M_1^2 - 1)}}. \qquad (A.28)$$

The upstream Mach number for the second shock is

$$M_2 = (1/4)\left[M_{1f}(\gamma + 1) + \sqrt{16 + M_{1f}^2(\gamma + 1)^2}\right] \qquad (A.29)$$

Figure A.3 shows the results for the fluid parameters.

4.10 Derive the differential equations for self-similar motions of fluids, (4.64)–(4.66). Identify the requirements for quantity in the final curved brackets in the third of these equations to vanish.

Solution and/or Comments For the definitions

$$u = \frac{1}{2}\frac{R}{t}U(\xi), \quad \rho = \rho_0\Omega(\xi), \quad \text{and} \quad p = \left(\frac{1}{2}\frac{R}{t}\right)^2\rho_0 P(\xi), \qquad (A.30)$$

one obtains

$$0 = \frac{\dot{\rho}_0}{\rho_0}\frac{R}{\dot{R}}\Omega(\xi) + \frac{\rho_0'R}{\rho_0}U(\xi)\Omega(\xi) + [U(\xi) - \xi]\Omega'(\xi) \qquad (A.31)$$
$$+ \Omega(\xi)U'(\xi) + \frac{sU(\xi)\Omega(\xi)}{\xi}$$

$$0 = \frac{\rho_0'R}{\rho_0}P(\xi) + \frac{R\ddot{R}}{\dot{R}^2}U(\xi)\Omega(\xi) + [U(\xi) - \xi]U'(\xi)\Omega(\xi) + P'(\xi) \qquad (A.32)$$

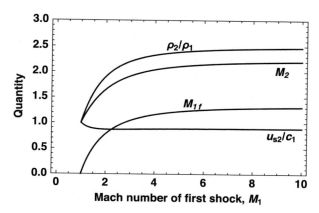

Fig. A.3 $M_2, u_{s2}/c_1, \rho_2/\rho_1$, and M_{1f} are shown vs M_1, for $\gamma = 5/3$

$$0 = \frac{\dot{\rho}_0}{\rho_0}\frac{R}{\dot{R}}(1-\gamma)P(\xi) + \frac{\rho_0'R}{\rho_0}U(\xi)(1-\gamma)P(\xi) + 2\frac{R\ddot{R}}{\dot{R}^2}P(\xi) \tag{A.33}$$

$$+ [U(\xi) - \xi]\left(P'(\xi) - \gamma P(\xi)\frac{\Omega'(\xi)}{\Omega(\xi)}\right).$$

As to the requirement for the quantity in curved brackets to vanish, mathematically this is

$$\frac{P'(\xi)}{P(\xi)} = \gamma\frac{\Omega'(\xi)}{\Omega(\xi)}, \tag{A.34}$$

and one also must explain what that means.

4.11 Show that the conservation of mass in the planar isothermal rarefaction in fact requires $r \geq -c_s t$ in (4.72) and (4.73).

Solution and/or Comments The density profile of the isothermal rarefaction is

$$\rho = \rho_o e^{-(1+\xi)}, \tag{A.35}$$

where $\xi = r/(c_o t)$. By assuming that the minimum value of ξ is some parameter a, one shows that mass conservation requires $a = -1$ and thus that the head of the rarefaction wave moves into the dense matter at c_o.

4.12 Plot the minimum density and pressure in the planar adiabatic rarefaction as a function of U_p. Discuss the meaning of the plots. Reasonable normalizations are recommended.

4.13 Sketch the C_+ and C_- characteristics defined in Sect. 4.4.3 in a fluid flowing uniformly with velocity u.

4.14 Show that the analysis of blast waves that preserves energy conservation produces $\alpha = 1/2$ for cylindrical blast waves and $\alpha = 2/3$ for planar blast waves.

Solution and/or Comments The result is that $\alpha = 1/2$ for cylindrical blast waves and $\alpha = 2/3$ for planar blast waves.

4.15 Find the coefficients α for blast waves treated as cylindrical and planar, momentum-conserving snowplows.

Solution and/or Comments The result is that $\alpha = 1/3$ for both cases.

4.16 Use a computational mathematics program to integrate the relevant equations to find and plot the profiles, and to evaluate Q, of (4.123) for a cylindrical blast wave. Apply this to find the behavior of a lightning channel produced by a deposited energy of 10^{10} ergs/cm.

Solution and/or Comments The result is that $Q = 0.876$ and $R = 1930\sqrt{t}$ cm/s.

4.17 Assuming that a strong shock reaches an interface beyond which the density (ρ_4) is 0.1 times the density of the shocked material behind the interface (ρ_1), solve for the profiles of the fluid parameters in the rarefaction that results.

Solution and/or Comments Defining the incoming flow, the rarefaction, the shocked, lower-density matter, and the unshocked matter as regions 1, 2, 3, and 4, respectively, and using these numbers as subscripts to assign a given variable to a given region, one also defines Mach-like numbers $M_1 = u_1/c_1, M_2 = u_2/c_1$, and $M_3 = u_3/c_1$. For strong shocks and $\gamma = 5/3, M_1 = \sqrt{9/5}$. One also finds $M_3 = 1.64$ and the interface location $r_{23} = 0.736c_1t$. In addition, one finds

$$M_2 = M_1 + \frac{2}{\gamma + 1}[1 + \xi - M_1] \text{ for } \xi \leq 0.736, \tag{A.36}$$

$$p_2 = p_1\left[1 - \frac{\gamma - 1}{2}(M_2 - M_1)\right]^{2\gamma/(\gamma-1)}, \text{ and } p_3 = \left(\frac{\gamma + 1}{2\gamma}\right)\frac{\rho_4}{\rho_1}p_1M_3^2. \tag{A.37}$$

The densities are

$$\rho_2 = \rho_1\left[1 - \frac{\gamma - 1}{2}(M_2 - M_1)\right]^{2/(\gamma-1)}, \text{ and } \rho_3 = \left(\frac{\gamma + 1}{\gamma - 1}\right)\rho_4. \tag{A.38}$$

Figure A.4 shows the resulting profiles for this case ($\gamma = 5/3$).

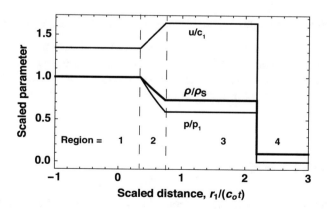

Fig. A.4 Profiles of the indicated parameters for the shock-initiated adiabatic rarefaction against a low-density medium

4.18 Assuming that $\gamma_1 = \gamma_3$ (or not, if you wish), derive (4.128) from (4.44)–(4.52) by letting p_3 approach p_1 as the definition of the transition to a rarefaction. Hint: This one is not easy. Taking a limit will be necessary and the approach to the solution will matter.

4.19 An entertaining aspect of the problem of reshocks in rarefactions is that it is one case where the traditional model in which shocks are driven by moving pistons does not produce correct qualitative behavior. Consider a rarefaction as it approaches a piston that is moving forward at a constant velocity. What will happen?

4.20 To obtain the behavior of oblique shocks at interfaces, one must evaluate the equations in cylindrical polar coordinates. Beginning with the first two Euler equations, carry out this evaluation.

Solution and/or Comments One ends up with

$$0 = \frac{\partial \rho}{\partial t} + \frac{1}{r}\frac{\partial}{\partial r}(r u_r) + \frac{1}{r}\frac{\partial u_\phi}{\partial \phi} + \frac{\partial u_z}{\partial r}, \tag{A.39}$$

$$\rho\left(\frac{\partial u_r}{\partial t} + u_r\frac{\partial u_r}{\partial r} + \frac{u_\phi}{r}\frac{\partial u_r}{\partial \phi} + u_z\frac{\partial u_r}{\partial z} - \frac{u_\phi^2}{r}\right) = -\frac{\partial p}{\partial r}, \tag{A.40}$$

$$\rho\left(\frac{\partial u_\phi}{\partial t} + u_r\frac{\partial u_\phi}{\partial r} + \frac{u_\phi}{r}\frac{\partial u_\phi}{\partial \phi} + u_z\frac{\partial u_\phi}{\partial z} - \frac{u_\phi u_r}{r}\right) = -\frac{\partial p}{\partial \phi}, \text{ and} \tag{A.41}$$

$$\rho\left(\frac{\partial u_z}{\partial t} + u_r\frac{\partial u_z}{\partial r} + \frac{u_\phi}{r}\frac{\partial u_z}{\partial \phi} + u_z\frac{\partial u_z}{\partial z}\right) = -\frac{\partial p}{\partial z}. \tag{A.42}$$

These simplify considerably for the oblique shock problem, in a well-chosen frame.

4.21 Equation (4.133) implies that a property of uniform flow is that $u_r = -\partial u_\phi/\partial \phi$ in any cylindrical polar coordinate system. Landau and Lifshitz use a geometric argument to demonstrate this. Instead, demonstrate this using a vectorial argument. (Hint: Begin by taking dot products of unit vectors along r and ϕ with an arbitrary velocity vector.)

Problems of Chap. 5

5.1 Consider a system with water above oil as described in Sect. 5.1.1. Suppose there is an small, sinusoidal ripple on the surface. Find the vertical profile of the force density between the lower and upper boundaries of the ripple for a region of denser fluid and for a region of less-dense fluid. Discuss the comparison of the two fluids and the shape of the force density profile.

5.2 The final relation in (5.37) is significant for our specific applications, in which one needs to integrate, across an interface, equations that contain discontinuous quantities along with derivatives of discontinuous quantities. By treating the delta function and the step function as limits of appropriate functions (see a mathematical methods book), prove this relation.

5.3 Find the solution for the velocity profiles and the growth rate for the RT instability for two uniform, constant density fluids that are confined by two planar surfaces each a distance d from the interface, which is accelerated at constant g.

Solution and/or Comments Define the situation with the more-dense fluid on top as fluid 1 and the less-dense fluid below as fluid 2. These fluids are bounded at the interface $z = 0$. One finds:

$$w_1 = w_0 \sinh [k(z - d)]$$
$$w_2 = -w_0 \sinh [k(z + d)]$$

and the final growth rate as

$$n = \sqrt{kg} \sqrt{\frac{\rho_1 - \rho_2}{\rho_1 + \rho_2}} \tanh(kd) = \sqrt{A_n kg} \sqrt{\tanh(kd)} \qquad (A.43)$$

5.4 The discussion above (5.55) shows that $\tilde{n} = (n/\sqrt{kg})\sqrt{\tilde{k}}$. This would suggest that it might make more sense to separate the meaning of the axes more cleanly by using $\tilde{\delta} = (n/\sqrt{kg})$ and $\tilde{k} = [(k^2 v)/\sqrt{gk}]^{2/3}$ as the two variables. Recast this equation in terms of these new variables, solve it, and plot the real roots from $\tilde{k} = 0$ to 2. Discuss the results and compare them to $n = \sqrt{A_n gk}$.

Solution and/or Comments The dispersion relation, after also defining $S = s/k = 1 + \tilde{\delta}/\tilde{k}^{3/2}$, becomes

$$0 = A + A^2(S - 1)^3 \tilde{\delta}^3 - (S - 1)S(S + 1)^2 \tilde{\delta}^3, \qquad (A.44)$$

which produces Fig. A.5.

5.5 In the derivation of the dispersion relation for the Rayleigh-Taylor instability with viscosity, some steps were skipped. Derive (5.56) and (5.57) from (5.50). Comment on the nature of the terms that have been dropped.

Solution and/or Comments The results are

$$n = \sqrt{A_n kg + k^4 v^2} - k^2 v \text{ and} \qquad (A.45)$$

$$\tilde{n} = \sqrt{A_n \tilde{k} + \tilde{k}^4} - \tilde{k}^2. \qquad (A.46)$$

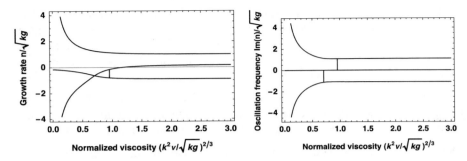

Fig. A.5 The left panel shows the real part of the roots of the dispersion relation. One is always growing and a second one grows weakly at high viscosity. The right panel shows the imaginary part of the roots of the dispersion relation, which correspond to wave propagation. The root with the largest real part has no imaginary part. The comparison to $n = \sqrt{A_n g k}$ is instructive

5.6 Explore the global RT mode in arbitrary directions. Find the plane-wave solutions in x, y and z to (5.60) and discuss their behavior.

Solution and/or Comments One finds

$$
n = \frac{(1 + \cos^2 \theta)}{\left[(1 + \cos^2 \theta)^2 + \frac{1}{L^2} \cos^2 \theta\right]} \sqrt{\frac{g}{L}} - i\frac{1}{2L} \cos \theta \sqrt{\frac{g}{L}}, \tag{A.47}
$$

where $\cos \theta = k_z/k$.

5.7 Consider an exponential density profile that decreases in the direction $\hat{z}$ that opposes that of the acceleration, $\boldsymbol{g}$, as $\rho = \rho_o e^{-z/L}$, and thus is the opposite of the case analyzed in Sect. 5.2.3. Apply the RT instability analysis to find n for this case. Discuss the results.

Solution and/or Comments One finds, for the rightward propagating root,

$$
n = -i\sqrt{\frac{g 4 k^2 L^2}{1 + 4 k^2 L^2}}. \tag{A.48}
$$

5.8 Carry out the calculation described in Sect. 5.3 and find (5.78). Then find the limits when (a) $k_p \to 0$ and $k_x L \gg 1$ and (b) when $A_n = 0$ and $L_p = 0$. Compare these with previous results in the chapter.

5.9 Work out the linear theory of the Rayleigh-Taylor instability to find an expression for the growth rate for the case of an exponential density gradient that extends for a finite distance between two layers of constant density.

Solution and/or Comments One finds

$$n = \sqrt{gk}\sqrt{\frac{4kL}{1 + 4k^2L^2}},\qquad\qquad (A.49)$$

5.10 This problem relates to the derivation of the mode-coupling results. By operating on (5.95) and (5.97), create two scalar differential equations that can be subtracted to eliminate terms involving p. Compare the resulting differential equation to (5.21) and discuss.

5.11 If we take the point of view that the modulations of interest in Kelvin-Helmholtz instabilities are proportional to e^{int}, then we would insist on finding negative imaginary n in order to have growth of the modulations, as opposed to damping, in time. However, this should give us pause because the complex representation is only a mathematical convenience while the physical quantities are real. Considering the real, physical quantities, what is the significance of finding positive or negative imaginary n? (The chapter in Jackson that introduces waves may be of some help regarding the connection of real physical quantities and a complex representation.)

Solution and/or Comments This important problem does not readily lend itself to saying there is a "solution", without short-circuiting the thinking it aspires to stimulate. The author encourages instructors to assign this one.

5.12 Suppose β is small enough that terms involving β in (5.140) can be dropped. Determine whether the two boundaries seen in Fig. 5.11 ever meet, completely eliminating the instability.

Solution and/or Comments One finds that there are four roots describing oscillating waves, except over the range of instability where two of them change to purely growing or damped roots. These growing roots do disappear above some value of η.

5.13 Analyze the shock conditions for a small-amplitude ripple and show that the change in the $\hat{z}$ component of u due to the ripple, relative to the $\hat{z}$ component of u produced by a planar shock, is second order in the ripple amplitude [i.e., generalize (5.143)].

Solution and/or Comments One finds the $\hat{z}$ component of the post-shock velocity to be, for strong shocks,

$$U_{n2} = -\frac{\gamma - 1}{\gamma + 1}U_1 - U_1 a_0^2 k^2 \sin^2(kx).$$

5.14 Delve into the origins of the response to a rippled shock wave. Develop (5.159) and (5.160) from the equations in Chap. 2.

Solution and/or Comments The results are

$$\frac{\partial \boldsymbol{u}}{\partial t} + \boldsymbol{u} \cdot \nabla \boldsymbol{u} = \frac{-1}{\rho}\nabla p + 2\nu\nabla \cdot \underline{\boldsymbol{s}}, \text{ and} \tag{A.50}$$

$$\frac{1}{2}\frac{\partial u^2}{\partial t} + \boldsymbol{u} \cdot \nabla\left(\frac{u^2}{2}\right) = \frac{-1}{\rho}\boldsymbol{u} \cdot \nabla p + 2\nu\nabla \cdot (\boldsymbol{u} \cdot \underline{\boldsymbol{s}}) - 2\nu(\underline{\boldsymbol{s}} \cdot \cdot\underline{\boldsymbol{s}}). \tag{A.51}$$

5.15 Explore further the effects of a rippled shock wave. Solve (5.146) through (5.148) to find the ratio of α, η, and χ to β. Plot the results for various values of γ and comment on what you observe.

5.16 Evaluate the small-angle limit of the equations for a shock at an oblique, rippled interface with a density decrease, and produce a plot similar to Fig. 5.19 for this case.

5.17 Consider the qualitative behavior of the postshock interface when there is a rarefaction but $\chi < 0$. Redraw Fig. 5.20 for this case. Discuss the evolution of the interface.

5.18 Work out the steady state, mean flow equations from the Reynolds decomposition. Derive (5.162) through (5.164). Comment on the meaning of each term.

Solution and/or Comments The results are

$$\rho(\boldsymbol{U} \cdot \nabla)\boldsymbol{U} = \frac{-1}{\rho}\nabla P + 2\nu\nabla \cdot \underline{\boldsymbol{S}} - \nabla \cdot (\rho\overline{\boldsymbol{w}\boldsymbol{w}}) \quad , \tag{A.52}$$

$$\boldsymbol{U} \cdot \nabla\left(\frac{U^2}{2}\right) = \nabla \cdot \left(\frac{-P}{\rho}\boldsymbol{U} + 2\nu\boldsymbol{U} \cdot \underline{\boldsymbol{S}} - \overline{\boldsymbol{w}\boldsymbol{w}} \cdot \boldsymbol{U}\right), \\ - 2\nu(\underline{\boldsymbol{S}} \cdot \cdot\underline{\boldsymbol{S}}) + \overline{\boldsymbol{w}\boldsymbol{w}} \cdot \cdot\boldsymbol{S} \tag{A.53}$$

and

$$\boldsymbol{U} \cdot \nabla\left(\frac{w^2}{2}\right) = -\nabla \cdot \left(\frac{1}{\rho}\overline{w\delta p} - 2\nu\overline{\boldsymbol{w} \cdot \underline{\delta s}} + \frac{1}{2}\overline{w^2\boldsymbol{w}}\right). \\ - \overline{\boldsymbol{w}\boldsymbol{w}} \cdot \cdot\underline{\boldsymbol{S}} - 2\nu\overline{\underline{\delta s} \cdot \cdot\underline{\delta s}} \tag{A.54}$$

5.19 To be more precise about the frozen-in property of vorticity, one should recognize that what moves with the fluid is the vorticity passing through a surface S. Prove this by taking the time derivative of the integral of $\boldsymbol{\omega} \cdot d\boldsymbol{S}$ over a surface S that moves with the fluid and may change its shape in time. Relate the result to (5.169). Hint: The key here is the evaluation of the partial derivative in time of the surface as a contour integral involving the edge of the surface.

5.20 Obtain the various equations describing the behavior of vorticity, (5.168) through (5.170), from the momentum equation. Discuss the point of each one.

Solution and/or Comments You will obtain these equations:

$$\frac{\partial \boldsymbol{u}}{\partial t} = -\nabla \left(\frac{p}{\rho} + \frac{u^2}{2} \right) + \boldsymbol{u} \times \boldsymbol{\omega} - \nu \nabla \times \boldsymbol{\omega}, \tag{A.55}$$

$$\frac{\partial \boldsymbol{\omega}}{\partial t} = \nabla \times (\boldsymbol{u} \times \boldsymbol{\omega}) + \nu \nabla^2 \boldsymbol{\omega}, \text{ and} \tag{A.56}$$

$$\frac{\partial \boldsymbol{\omega}}{\partial t} + \boldsymbol{u} \cdot \nabla \boldsymbol{\omega} = \boldsymbol{\omega} \cdot \underline{s} + \nu \nabla^2 \boldsymbol{\omega} . \tag{A.57}$$

Problems of Chap. 6

6.1 Integrate the thermal intensity B_ν over 2π steradians to find the total radiation power per unit area from a surface at temperature T.

Solution and/or Comments The correct result is σT^4.

6.2 Using the particle treatment of the radiation, derive an expression for the total radiation momentum density, and show that it equals F_R/c^2.

Solution and/or Comments One will show that the spectral momentum density is I_ν/c, so the total radiation momentum density is

$$\int_{4\pi} \int_0^\infty \frac{I_\nu}{c^2} \mathrm{d}\nu \boldsymbol{n} \mathrm{d}\Omega = \frac{F_R}{c^2}. \tag{A.58}$$

6.3 Derive the relation between radiation pressure and energy, (6.14).

Solution and/or Comments The point of this problem is to gain some experience with fundamental quantities describing radiation. By reworking the derivation of p_ν, and applying the definition of E_ν, one will find $p_\nu = E_\nu/3$.

6.4 Graduate students frequently struggle with units, and in particular with the problem posed here. First, integrate B_ν, (6.18), symbolically, over frequency to obtain (6.19). Second, evaluate the coefficients in the integral independently for cgs and mks units, and show that you obtain equivalent results. Third, convert (6.18) to have units of energy per unit area per unit time per unit solid angle per unit photon energy. Integrate this new expression over photon energy and show that you obtain the same result. You would be well-advised to do all of this within a computational mathematics program, either with excellent comments your work file, or with an independent document (in LaTeX at the time of writing this) that describes the calculations and their results.

Solution and/or Comments The author encourages instructors to assign this problem, as problems with such conversions is the source of many headaches for graduate students and their advisors. Scientists need the ability to do this kind of conversion for themselves, and to also do the calculations that enable them to check their work.

6.5 From the uncertainty principle, the spectral width in frequency, $\Delta\nu$, of an emission line is roughly the inverse of the decay time. For a typical decay time of 1 ns, find the normalized spectral width $\Delta\nu/\nu$, for emission lines in the visible and in the soft X-ray with a photon energy of 100 eV. Discuss the significance of this result.

Solution and/or Comments About 10^{-6} for the visible photon and 10^{-8} for the soft X-ray one.

6.6 Derive the radiative transfer equation for a spherically symmetric system, (6.43).

Solution and/or Comments In a standard spherical coordinate system one needs three variables (r, θ, ϕ) to specify the location of a point on the ray and in addition two variables (a polar angle Θ and an azimuthal angle Φ) to specify the direction of the ray with respect to the local radial direction. In a spherically symmetric system, such as a star treated as a symmetric object, the location is fully specified by r. At any specific point, the radiation intensity varies with direction, but it is symmetric about the local radius vector. As a result, one needs a single angle, Θ, to specify the local direction of the ray. Defining $\mu = \cos\Theta$. The result of the derivation is

$$\left[\frac{1}{c}\frac{\partial}{\partial t} + \frac{\partial}{\partial s}\right] I_\nu(\boldsymbol{x}, t, \boldsymbol{n}, \nu) = \left[\frac{1}{c}\frac{\partial}{\partial t} + \mu\frac{\partial}{\partial r} + \frac{(1-\mu^2)}{r}\frac{\partial}{\partial\mu}\right] I_\nu(r, t, \mu, \nu).$$
(A.59)

6.7 Take moments of the radiation transfer equation to derive the equations for radiation energy density and radiation pressure, (6.58) and (6.60).

Solution and/or Comments The results are

$$\frac{\partial E_R}{\partial t} + \nabla \cdot \boldsymbol{F}_R = 4\pi \int_0^\infty \kappa_\nu \left(B_\nu - J_\nu\right) d\nu \equiv 4\pi\kappa \left(B - J_R\right) \text{ and}$$
(A.60)

$$\frac{1}{c^2}\frac{\partial \boldsymbol{F}_\nu}{\partial t} + \nabla \cdot \underline{\boldsymbol{P}}_\nu = \frac{1}{c}\int_{4\pi} \left(\eta_\nu - \chi_\nu I_\nu\right) \boldsymbol{n} d\Omega.$$
(A.61)

6.8 Demonstrate that $\nu d\nu d\Omega$ is Lorentz invariant, i.e. (6.77).

6.9 Given relations (6.78) through (6.80), show that the radiation transfer equation is relativistically invariant.

6.10 Derive the relativistic transformations of opacity and emissivity (6.83) and (6.84), and the implied radiative transfer equation, (6.85). Discuss the limits on v/c for this specific description if the emission and absorption are dominated by (a) continuum emission or (b) line emission.

6.11 Rework the relativistic equation for radiative energy density (6.88) into the form of a conservation equation. Discuss the meaning of the terms that result.

Problems of Chap. 7

7.1 Carry out the calculations of radiation and material energy fluxes and pressures and compare the behavior of pure hydrogen as opposed to C_1H_1 (used in Fig. 7.1).

Solution and/or Comments One obtains Fig. A.6.

7.2 Derive the dispersion relation for isothermal acoustic waves from the Euler equations. That is, demand constant temperature and see what happens.

Solution and/or Comments The result is $\omega^2 = c_i^2 k^2$, where c_i is the isothermal sound speed whose meaning was identified during the derivation (the answer should elucidate this).

7.3 Figure 7.4 shows the wave properties as ω varies for fixed η. Consider how the wave properties vary with η for $\beta = 1$ and fixed $\omega/(\nu_e c_s^2/c^2)$. Plot the normalized phase velocity and damping length for $0.01 \leq \eta \leq 10$ and discuss the results.

7.4 There should be a sensible connection between the present calculation and the optically thick one as the system becomes optically thick. For the limit in which $\kappa \gg k$, seek to reconcile (7.26) and (7.42).

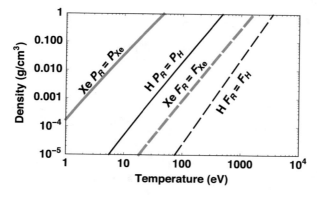

Fig. A.6 Solid curves show where P_R equals the material pressure, and dashed curves show where F_R equals the material energy flux. The gray curves are for Xe and the black curves are for CH

Solution and/or Comments One will have shown that each of this equations has an element that represent a damping rate, and that in each case this damping rate has its origins in $-\nabla \cdot F_R/(\rho c_V)$.

7.5 We did not explore the angular variation in the contributions to (7.37). One might imagine that the largest contributions could come at grazing angles, where μ is very small and the optical depth along a line of sight becomes large. The model used here would be less realistic if most of the emission came at grazing angles, because real systems will have layers that are not truly planar and certainly are not infinite in extent. Use a computational mathematics program to derive (7.37). Then modify the calculation to explore how large the contribution is from such grazing angles. Conclude whether or not the results above might be reasonable estimates for real layers.

Solution and/or Comments Plotting (7.37), we see in the left panel of Fig. A.7 that the intensity is very strongly peaked at highly oblique angles for optically thin cases, as one would intuitively expect. Performing the integral of I_R but excluding angles for which μ is below some minimum value of μ, corresponding to a maximum value of θ, yields the right panel of Fig. A.7.

7.6 It is curious that (7.41) and (7.43) do not depend on β, so that these waves seem not to care whether the system is fully ionized. Beginning with (7.39), derive (7.43) and discuss why there is no β dependence.

Solution and/or Comments The result is

$$-i\omega \left[-\omega^2 + \frac{\gamma_s p_o}{\rho_o}k^2 \right] = -\nu \left[-\omega^2 + \frac{p_o}{\rho_o}k^2 \right]. \tag{A.62}$$

7.7 Beginning with $\rho(\partial \epsilon/\partial t) = \nabla \cdot (\kappa_{\mathrm{rad}}\nabla T)$ derive (7.64).

Solution and/or Comments The result is

$$-\frac{\xi}{2}\frac{df}{d\xi} = \frac{d^2 f^{n+1}}{d\xi^2}. \tag{A.63}$$

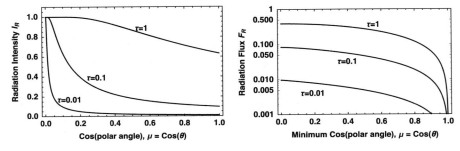

Fig. A.7 Left panel: Radiation intensity vs polar angle produced by layers of varying optical depth. Right panel: F_R corresponding to the integral over all angles above a given value of μ

Fig. A.8 Progress of various waves in gold foam, for the specified parameters

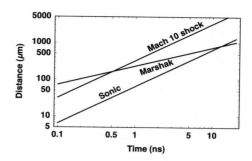

7.8 Work through the constant-flux model for Marshak waves, providing all the missing mathematical steps. Then plot the positions vs. time of the radiation wave and of a disturbance (in the radiation-heated material) moving at Mach 1 or Mach 5, for a wave in Au foam with $T_o = 200$ eV, $\rho = 0.1$ g/cm^{-3}, and $Z = 40$. Discuss the results.

Solution and/or Comments Figure A.8 shows the results.

7.9 For the constant-energy, radiation-diffusion wave, show that (7.87) is a solution to (7.86). Clearly annotated work with a computational mathematics program is preferred.

7.10 Consider a gold container shaped so that a planar approximation is reasonable, having planar walls spaced 1 mm apart in vacuum. Assume $\rho = 20$ g/cm^3 and treat $c_V = 10^{12}$ ergs/(g eV) as constant. Use other parameters from Chap. 6 as appropriate. Suppose 100 kJ/cm^2 is the initial energy content of the vacuum between the walls and that the initial wall temperature is negligible. Assume that the gold material does not move. Apply the self-similar model of the constant-energy radiation diffusion wave to this system, on the assumption that the two walls are touching but contain the specified energy. In doing so, approximate ξ_o from Fig. 7.14. From zero to 10 ns, find the position of the heat front and the temperature of the surface as a function of time. Realizing that the walls are in equilibrium with the temperature of the radiation in the vacuum, plot the ratio of the energy content of the walls to the energy content of the vacuum. Discuss the meaning of this result from the standpoint of the accuracy of an intermediate asymptotic model (see Sect. 4.3).

Solution and/or Comments One obtains the plots shown in Fig. A.9

7.11 Develop the equivalent of (7.90) describing the radius of a spherically symmetric, constant-energy, radiation-diffusion wave.

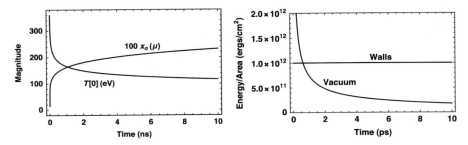

Fig. A.9 Left panel: Central temperature and 100 times the penetration depth in μm for the diffusion wave. Right panel: Areal energy density in the walls and the vacuum, showing the first 10 ps

Problems of Chap. 8

8.1 Demonstrate that the material energy flux coming into a radiative shock does not decrease when lateral losses decrease the upstream energy flux, by considering a system having a planar flow of material within a cylinder of some diameter and of finite length yet losing radiation both radially and axially, and integrating over the cylinder.

8.2 Derive (8.6), relating the radiation flux to the material properties, and discuss the origin and significance of each term.

Solution and/or Comments The result is

$$F_R - F_o = \frac{\rho_o u_s^3}{2} \left[\frac{2\gamma}{\gamma - 1} \frac{\rho_o}{\rho} - \frac{\gamma + 1}{\gamma - 1} \left(\frac{\rho_o}{\rho} \right)^2 - 1 - \frac{p_o}{\rho_o u_s^2} \frac{2\gamma}{\gamma - 1} \left(1 - \frac{\rho_o}{\rho} \right) \right].$$

(A.64)

8.3 Working with the Planck description of blackbody radiation, find and plot the fraction of photons that are ionizing as a function of temperature. You will need a computational mathematics program to generate the plot.

Solution and/or Comments Figure A.10 shows the result.

8.4 Evaluate the net radiation flux $(F_R - F_o)$ for an optically thin precursor using a radiative-transfer calculation similar to that done in (8.20) and (8.24).

8.5 Explore further the behavior of radiative shocks that are optically thin upstream and thick downstream. Beginning with (8.4)–(8.6), derive the final inverse compression (8.30) under the assumptions of the present section.

Solution and/or Comments The result is

$$\frac{\rho_o}{\rho_f} = \sqrt{\frac{\sqrt{1 + 8Q} - 1}{4Q}}.$$

(A.65)

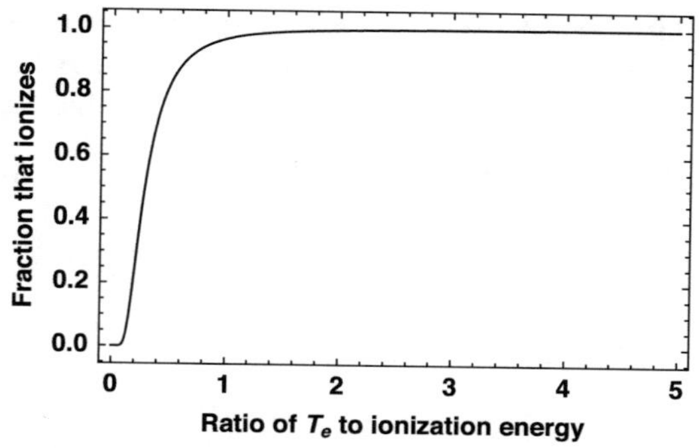

Fig. A.10 Showing the fraction of the photons that are ionizing as a function of the ratio of T_e to the ionization energy, for a Planckian whose temperature equals T_e

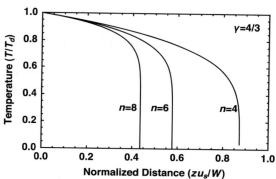

Fig. A.11 Showing profiles of temperature in diffusive radiative precursors for the indicated values of n

8.6 Determine whether the equation for energy flow in a radiative precursor, (8.19), admits a self-similar solution, assuming a diffusive model for F_R.

8.7 Examine the behavior of diffusive precursors. Solve (8.41) numerically, for several relevant values of n. Comment on the results.

Solution and/or Comments Figure A.11 shows the result for the three indicated values of n.

8.8 We saw in Sect. 7.3 that Marshak waves are inherently unsteady, yet in Sect. 8.2.1 we analyzed them using steady jump conditions. Develop a condition for the validity of the use of steady jump conditions for Marshak waves, and determine when this is realistic.

Solution and/or Comments One can see from Fig. 7.10 that the radiation flux does not deviate from a constant value by more than a few percent over the entire Marshak structure, for the realistic value $n = 4.5$, so once any given element of matter has passed through the front of the wave and moved a fraction $\alpha \sim$ few percent of x_o, the matter encounters nearly steady conditions in space. One shows that the flux entering the wave decreases as $1/(2t)$, and that the time required to traverse αx_o is $\alpha 2t$, so that the time taken to traverse the wave front is always smaller than the rate of decay of the radiation flux by the factor α.

8.9 Consider a truly radiation-dominated case, so p can be neglected in (8.50) and (8.51). Solve these equations for p_R and ρ. Find the dependence of the postshock T on the shock velocity, and compare it to the dependence of a non-radiative shock.

Solution and/or Comments One finds

$$p_R = \left(6\rho_o u_s^2 - p_{Ro}\right)/7 \tag{A.66}$$

$$\rho = 7\rho_o \frac{\rho_o u_s^2}{\rho_o u_s^2 + 8 p_{Ro}} \tag{A.67}$$

$$T = \sqrt{\frac{\sqrt{3}c}{2\sqrt{\sigma}}} \left(\frac{6\rho_o u_s^2 - p_{Ro}}{7}\right)^{1/4} \tag{A.68}$$

8.10 Express p and p_R as reasonable functions of T and solve (8.50) and (8.51) to find T and ρ in the postshock state. This may be a numerical solution, for which you should make reasonable choices about the parameters and show a few cases. Provide at least one graph based on these equations as part of the analysis.

Problems of Chap. 9

9.1 Derive the general electromagnetic wave equation (9.3) from Maxwell's equations.

Solution and/or Comments The result, in terms of the vector potential in the Coulomb gauge is

$$\nabla \times \nabla \times A = \frac{-1}{c^2} \frac{\partial^2 A}{\partial t^2} + \frac{4\pi J}{c} - \frac{1}{c} \frac{\partial \nabla \Phi}{\partial t}. \tag{A.69}$$

9.2 Derive an equation for the conservation of charge from (9.3).

Solution and/or Comments The result, for charge density ρ_c, is

$$0 = \frac{\partial \rho_c}{\partial t} + \nabla \cdot J \tag{A.70}$$

9.3 Using the equation of motion for the electron fluid in the fields of an electromagnetic wave in a plasma of constant density, determine the time-averaged distribution of energy among the electric field, the magnetic field, and the kinetic energy of the electrons. Discuss how this varies with density.

Solution and/or Comments The solutions are

$$\text{electron energy } \frac{n_e}{n_c}\frac{k_o^2\hat{A}^2}{16\pi}, \text{ electric-field energy } \frac{k_o^2\hat{A}^2}{16\pi}, \tag{A.71}$$

$$\text{and magnetic-field energy } \frac{k_o^2\hat{A}^2}{16\pi}\left(1-\frac{n_e}{n_c}\right), \tag{A.72}$$

for parameters defined in the standards ways of the chapter and $\hat{A}$ being the amplitude of the vector potential.

9.4 Derive the wave equation for scattering of laser light from density fluctuations (9.20).

Solution and/or Comments The result is

$$\left(\frac{\partial^2}{\partial t^2}+\omega_{pe}^2-c^2\nabla^2\right)A_s = -\omega_{pe}^2\frac{\delta n_p}{n_{eo}}A_L. \tag{A.73}$$

9.5 Derive (9.22) for the longitudinal plasma velocity. Calculate the energy density of the laser light wave and show how this is related to the source term on the right-hand side.

Solution and/or Comments The derived equation and the relation to the energy density of the laser light wave, E_{LW}, is

$$\frac{\partial v_p}{\partial t}-\frac{e}{m_e}\nabla\Phi+\frac{\nabla p_e}{n_e m_e}=-\nabla\frac{v_{os}^2}{2}=-\frac{1}{n_c m_e}\nabla E_{LW}, \tag{A.74}$$

for laser light wave frequency ω_L.

9.6 Develop an energy equation for the electron fluid including a Spitzer–Harm heat flux, and show that it is a diffusion equation.

Solution and/or Comments The gas-energy equation for the electron fluid becomes

$$\frac{3}{2}\frac{D}{Dt}(k_B T_e)-\frac{k_B T_e}{m_e}\frac{Dn_e}{Dt}=\frac{64}{3\pi}\frac{n_e}{m_e \nu_{ei}}\frac{\partial(k_B T_e)^2}{\partial z^2}. \tag{A.75}$$

When n_e is constant and the velocity is zero, this is a nonlinear diffusion equation.

9.7 Determine the range of electron velocities that contribute significantly to the heat flux, by plotting the first-order contribution to the argument of the heat-flux integral (9.28).

Solution and/or Comments Figure A.12 shows the plot.

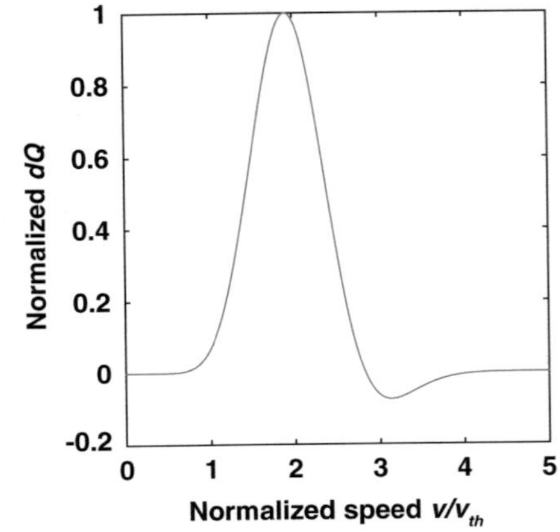

Fig. A.12 Showing the contribution to the heat flux, in the Spitzer-Harm model, as a function of normalized speed

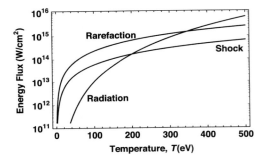

Fig. A.13 Energy fluxes sustaining the indicated power flows

9.8 Find the approximate expression for the ablation efficiency ϵ_R of a rocket, to second order in the quantity m_a/m_o. Plot the corresponding rocket efficiency and the value of (9.52). Discuss the comparison.

9.9 Examine the energy distribution of the dynamics at the surface of a laser-irradiated, mid-Z target, where in the upper rarefaction the plasma temperature is T. Model this region as a material at T and at a density of 1 g/cm^3, growing at the sound speed. Assume the pressure throughout the dense material, up to the shock driven inward, equals that at 1 g/cm^3, and assume the shock is driven into matter whose initial solid density is 10 g/cm^3. Evaluate and plot, as a function of T, the energy fluxes that are radiated outward, that sustain the shock wave driven into the solid material, that sustain the growing region near 1 g/cm^3. Discuss the results.

Solution and/or Comments Figure A.13 shows the resulting plots.

9.10 Assume that a hohlraum of 1 mm radius is heated for 1 ns at a temperature of 200 eV. Estimate the pressure produced at the center of the hohlraum when the plasma expanding from the gold walls reaches the axis. (Note: this is not an application of (9.63). Instead, you will need to think about the rarefaction produced during the heating pulse.)

Problems of Chap. 10

10.1 Find the sizes and directions of the orbits of protons and electrons. Explain from fundamental laws of electromagnetics why their direction is as it is. Show pictorially why the $E \times B$ drift moves particles in the same direction.

10.2 The MHD equations assume charge neutrality, yet MHD plasmas may contain electric fields. Explore this seeming contradiction by, first, evaluating the electric field and relative charge imbalance for a steady electric field in an isothermal plasma having a density gradient (so that $e n_e E = -\nabla p_e$), for reasonable choices of parameters. Compare this to the electric field produced in an electron plasma oscillation for which the amplitude of the electron-density fluctuations is 10%. Express the magnitude of the electric force as $eV/\mu m$.

Solution and/or Comments Expressing the electric field as a force (eE), and choosing $T_e = 1$ keV, $L = 10\mu m$, $n_i = 10^{21}$ cm^{-3}, $k = 2\pi \times 10^4$, one finds $eE = 1.6 \times 10^{-6}$ ergs/cm = 1.1 eV/μm for the field established on a 10 μm ramp, and $eE = 1.8 \times 10^{-4}$ ergs/cm = 100 eV/μm by the density fluctuation.

10.3 Begin with (10.48)–(10.50), keeping the resistivity. Derive the dispersion relation for damped Alfvén waves, starting with the assumptions that k is parallel to B and that u is purely transverse. (You should find this much simpler than the general case just discussed.) For reasonable choices of plasma parameters, plot the ratio of damping rate to real frequency as a function of electron temperature.

Solution and/or Comments The dispersion relation is

$$\frac{\omega}{k v_A} = \frac{1}{\sqrt{1 - i \frac{c\eta}{4\pi} \frac{k}{v_A}}} \sim 1 + i \frac{c^2 \eta}{8\pi} \frac{k}{v_A} = 1 + i \frac{\nu_{ei}}{2\omega_{pe}^2} \frac{c^2 k}{v_A}. \tag{A.76}$$

10.4 Begin with (10.48)–(10.50), assuming small resistivity. Derive the dispersion relation for cross-field sound waves, starting with the assumptions that k is perpendicular to B and parallel to u. (You should find this much simpler than the general case just discussed.)

Solution and/or Comments The dispersion relation is

$$\omega^2 = (c_s^2 + v_A^2) k^2. \tag{A.77}$$

10.5 Show that when the field evolves as (10.73) describes, the quantity $B \cdot \hat{q}$ remains zero to the first-order accuracy of the present model.

10.6 While one can vary the properties of a Z-pinch load from one experiment to the next, one can modify the pulsed-power device itself on a somewhat longer timescale. Such devices are typically characterized by the number of Volt-Seconds they can produce, and operate so that $V\tau = $ constant. First, consider and then explain why

Fig. A.14 Effect of
variations in assumptions on
burn fraction

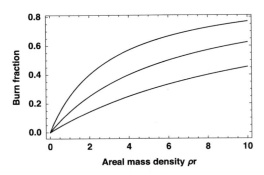

Volt-Seconds is a reasonable way to characterize a pulsed-power device. Second, using the scaling relations developed in Sect. 10.10.1, discuss how to optimize the stagnation power for a device with $V\tau = $ constant.

10.7 Revisit the derivation at the beginning of Sect. 10.10. Consider two infinitely wide, plane parallel conductors carrying opposing currents. Find the force per unit area between them and express it in terms of the magnetic field magnitude. Discuss how the force per unit area compares to the energy density of the magnetic field.

Solution and/or Comments Suppose the currents flow in the $\pm\hat{x}$ direction and that the plates are separated in $\hat{y}$. Describe the current amplitude as I_ℓ amps/m for distance along the plate in $\hat{z}$. One finds the force per unit area, F_A, to be

$$F_A = \frac{\mu_o I_\ell^2}{2}, \tag{A.78}$$

and that it tends to bring the plates together. The total magnetic field is $B = \mu_o I_\ell/2$, so the energy density of the magnetic field is

$$\frac{B^2}{\mu_o} = \frac{\mu_o I_\ell^2}{4}. \tag{A.79}$$

Problems of Chap. 11

11.1 Plot the burn fraction versus ρr. Discuss the impact of the assumptions made in deriving the burn fraction on this curve, and on the size of a system designed to produce a certain quantity of fusion energy.

Solution and/or Comments One can find a plot like that shown in Fig. A.14, in which one will know which assumptions were varied and be able to discuss the meaning and significance for ICF of the variations.

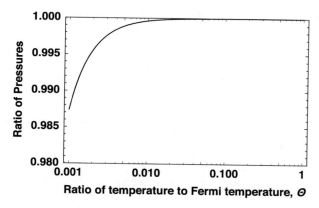

Fig. A.15 Ratio of boson pressure to classical pressure for pure D at 0.1 g/cm^{-3}

11.2 Evaluate the classical pressure and the boson pressure for deuterium as a function of temperature. For pure deuterium at a density of 0.1 g/cm^3, plot the ratio of the pressure for deuterium treated as bosons to that for deuterium treated as a classical gas, as a function of temperature. Discuss the comparison.

Solution and/or Comments Figure A.15 shows the result.

11.3 For DT fuel, derive the classical relation between entropy and pressure (normalized by the Fermi pressure of the electrons).

Solution and/or Comments The result is

$$
s_{tot} = \frac{k_B}{2.5mp} \left[\ln (\Theta)^3 + 18.5 \right] = \frac{k_B}{2.5mp} \left[\ln \left(\frac{p_{tot}}{5p_F} \right)^3 + 18.5 \right] \qquad (A.80)
$$

11.4 Suppose that one could apply a pressure p for a time t, using some energy source. With this source, we could accelerate some amount of mass per unit area, $\rho_o \Delta r$, to $v_{imp} = 300$ km/s. Define a fusion capsule using the reflected pressure due to sunlight for 12 h as the pressure source. Approximate sunlight as light with a wavelength of 580 nm and an energy flux of 1 kW/m^2. How long would such a capsule take to implode?

11.5 Derive the spectrally averaged absorption coefficient for bremsstrahlung in DT. Compare your value and the one below with the value found in the NRL plasma formulary.

Solution and/or Comments The coefficient is

$$
\kappa_{IB} \text{ cm}^{-1} = 1.7 \times 10^{-38} \frac{\ln \Lambda n_i^2 Z^3}{T_e^{3/2} T_s^{3/2}}, \qquad (A.81)
$$

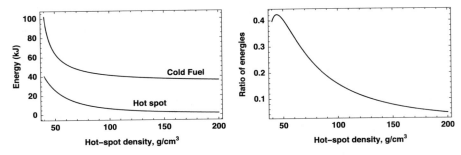

Fig. A.16 Energy comparison for a fuel mass of 3 mg and a hot-spot ρR of 0.3 g/cm². Left: Internal energies in kJ of the cold, compressed fuel and hot spot. Right: Ratio of hot-spot to fuel

in which n_i is cgs, the electron temperature T_e is in eV, and the temperature of the photon spectrum is T_s, also in eV.

11.6 Evaluate the appropriate integral of the radiative transfer equation over solid angle to obtain F_R from a spherical volume of optically thin, spatially uniform DT. Find the value of the characteristic distance. Compare your result to the result in (11.36), which assumes that the integral over solid angle of the distance across the fuel gives πR_h. Extra credit: generalize this calculation to include arbitrary optical depth and discuss the results.

Solution and/or Comments One finds

$$F_R = \frac{4}{3}\sigma T^4 \kappa R_h, \tag{A.82}$$

for spectrally averaged absorption opacity κ, so the integral gives $(4\pi/3)R_h$.

11.7 The Lawson criterion is generally written as $n\tau > 10^{14}$ s/cm³, with density n and energy confinement time τ. Find a way to relate this to (11.38) and comment on the comparison.

11.8 Because the density in a central hot spot is less than that of the cold fuel, a larger fraction of the total energy must be expended to heat it than was estimated above. Revisit the discussion above of the relative energy content of the fuel and the hot spot, and develop an expression for the scaling of the hot-spot energy content with hot-spot density. Find the result as an absolute energy and as a fraction of the energy used to compress the cold fuel.

Solution and/or Comments Figure A.16 shows the energies and their ratio.

11.9 Using the equations of this Sect. 11.3.2 and others as necessary, build yourself a computational model of a fusion target that ignites from a central hot spot. Use it to explore target designs for the National Ignition Facility, using fuel masses of 1 mg or less, a fuel ρR_c of 1.5 or less, and other parameters of your choice. Make some relevant plots and comment on what you find.

11.10 Evaluate the amount of Rayleigh Taylor growth for the sunlight-driven fusion system of the problem11.4.

Problems of Chap. 12

12.1 Show that the Euler equations are in fact invariant under the transformations described in Sect. 12.1.

Solution and/or Comments They are.

12.2 Design a diverging experiment to address the coupling of two structured, unstable interfaces that are affected by a blast wave. Beyond the basic requirements for hydrodynamic scaling, identify other specific parameters that are important to the dynamics. (Hint: review blast-wave propagation and shock stability as part of your work.)

Solution and/or Comments The point of this problem is to have you think through the various elements of such a system, and identify the important parameters for yourself. For instructors, this is a good exercise, integrating material from multiple chapters, but is more involved than many of the homework problems.

12.3 Determine why t_{cc} as defined in Sect. 12.3 is the relevant timescale for the crushing of a cloud.

Solution and/or Comments The point of this problem is to have you think about the details of cloud-crushing dynamics in order to identify the parameters necessary for specific scaling and thereby to draw this conclusion. The purpose is to prepare you to be able to do similar analysis of novel systems.

12.4 Suppose that an astrophysical blast wave of interest is produced by a supernova explosion that is a known distance R from a clump of some radius r_{cl}. Determine the properties of an experimental blast wave and the duration of the experiment that would be required to model the shock–clump interaction in this system.

Solution and/or Comments The point of this problem is to consider both aspects of the specific scaling of this system in designing an experiment. The duration of the experiment will be the sum of the time it takes the blast wave to reach the clump plus the number of cloud-crushing times required, which must be of order 10. On the assumption that the experimental system behaves like a spherical blast wave, the half-width of the pressure impulse from the blast wave, Δ, is a fixed fraction of the distance traveled. To have the specific parameter, Δ/r_{cl}, be the same in both systems, one would require the distance from the initiating location of the experimental blast wave, D and the size of the experimental clump, d, to be related to the astrophysical parameters as $D/d = R/r_{cl}$. The experimental blast wave is not likely to be strictly spherical. Planar systems have similar behavior but may not be self-similar or strictly planar, so this first estimate is a good start but simulations could refine the design.

12.5 An approach that has been used to form hydrodynamic jets is to create an adiabatic rarefaction by allowing a shock wave to emerge from a material into an evacuated tube and then to emerge from this tube into an "ambient medium", at a lower density. Using the simple scaling results from this book, develop a design for a similar experiment to produce a radiative jet.

Solution and/or Comments This problem synthesizes material from multiple chapters, but such a simple model also requires some simplifying assumptions to be tractable. Analysis in terms of key dimensionless parameters is very helpful.

12.6 Magnetized jets must have a ratio of plasma pressure to magnetic field pressure (usually called β in plasma physics) no larger than about 1. For a low-Z plasma with a density of 0.1 g/cm^3 and a temperature of 10 eV, determine how large a magnetic field would be required to satisfy this constraint. How does this compare with the magnetic field of order 1 MGauss that is typically produced in laser–plasma interactions and that might be produced by very clever field-compression experiments?

Solution and/or Comments For $A \sim 2(1 + Z)$, one finds $B \sim 0.98$ MG, so such conditions represent an extreme upper limit on creating $\beta \sim 1$. In fact the density is too high, as the temperature is too low to be good for much.

Problems of Chap. 13

13.1 Design a pulse stretcher. Suppose you have a laser beam with an 800 nm central wavelength and a bandwidth of 20 nm (corresponding to a 50 fs laser pulse). Use two identical gratings, recalling that for the first diffracted order the scattered wavelength λ is given by $\lambda = d(\sin \alpha + \sin \beta)$, where d is the line spacing on the grating and α and β are angles of incidence and reflection relative to the grating normal. Use two identical lenses, recalling that the object distance, o, image distance, i, and focal length f are related by $o^{-1} + i^{-1} = f^{-1}$. Note that the initial grating must be less than one focal length from the lens to obtain stretching.

Solution and/or Comments Hardware design is not so different from experiment design. One must consider a number of interacting factors in order to develop a viable solution.

13.2 Assuming that the electron motion is due to a plane wave with a single frequency and that the electron movement is small compared to the wavelength of the light wave, solve the equations in Sect. 13.2 to find the electron trajectory. Determine how it changes as the electron velocity increases (while remaining $\ll c$).

Fig. A.17 Trajectories of x/λ and $z\lambda$ for $a_o = 0.1$ and 0.5

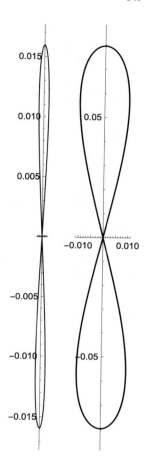

Solution and/or Comments Assuming the $E||\hat{x}$ and $k||\hat{z}$, one finds, for reasonable choices of phase,

$$x(t) = \lambda a_o \frac{\cos(\omega_o t)}{2\pi} \text{ and } z(t) = -\lambda a_o^2 \frac{\sin(\omega t)}{8\pi}. \tag{A.83}$$

Figure A.17 shows the trajectories for two values of a_o. The motion in $\hat{z}$ increases as a_o does. The next order in x motion is cubic in a_o.

13.3 Prove that the definitions (13.15) and (13.16) are equivalent.

Solution and/or Comments The purpose of this problem is to have you reacquaint yourself with some of the basics of relativistic motion, by manipulating some of the common variables.

13.4 Solve (13.25) to see the electron behavior for a range of values of the initial phase (i.e., change π to various other values, for fixed $a_o = 100$ and $\delta(0) = 0.01$. Comment on the variations in the behavior, in comparison to that seen in Fig. 13.3.

13.5 Find the time required to accelerate the electron to $\sim$30 GeV in the example used in discussing (13.43) through (13.45).

13.6 Suppose one has a laser beam that can be focused to 10^{20} W/cm^2 in a 10 μm diameter spot. Would one obtain higher-energy electrons from tunnel ionization (as in Sect. 13.2) or from using the laser for wakefield acceleration, discussed in Sect. 13.7.1?

Solution and/or Comments The energy from tunnel ionization for these conditions is $\sim$4 MeV, based on Fig. 13.4 For wakefield acceleration, obtains a maximum energy near 100 MeV, requiring an electron density near 10^{19} cm^{-3} and a length just over 300 μm.

13.7 Solve for the potential of a spherical cloud of ions having uniform density, and for the energy distribution function of the ions produced by a Coulomb explosion of this cloud.

Solution and/or Comments The derivation will give potential

$$\Phi = 2\pi n_i Ze \left[\left(\frac{8\pi}{3} \right) r_o^2 + (r_o^2 - r^2) \right] \text{ for } r < r_o \text{ and}$$

$$\Phi = 2\pi n_i Ze \left(\frac{8\pi}{3} \right) \frac{r_o^3}{r} \text{ for } r > r_o, \tag{A.84}$$

maximum energy

$$\mathcal{E}_{max} = \frac{8\pi}{3} 2\pi n_i Z^2 e^2 r_o^2, \tag{A.85}$$

and distribution function

$$f(\mathcal{E}_i) = \frac{3}{2} \frac{\sqrt{\mathcal{E}_i}}{\mathcal{E}_{max}^{3/2}} \text{ for } \mathcal{E}_i < \mathcal{E}_{max} \text{ and}$$

$$= 0 \text{ for } \mathcal{E}_i > \mathcal{E}_{max}. \tag{A.86}$$

13.8 Derive the relativistic version of the theory of Sect. 13.8 and find the relativistically correct revision to (13.82).

Index

© Springer International Publishing AG 2018
R P. Drake, *High-Energy-Density Physics*, Graduate Texts in Physics,
https://doi.org/10.1007/978-3-319-67711-8